Plant Pathology

Concepts and Laboratory Exercises

SECOND EDITION

Plant Pathology

Concepts and Laboratory Exercises

SECOND EDITION

Edited by

Robert N. Trigiano
Mark T. Windham
Alan S. Windham

CRC Press
Taylor & Francis Group
Boca Raton   London   New York

CRC Press is an imprint of the
Taylor & Francis Group, an **informa** business

CRC Press
Taylor & Francis Group
6000 Broken Sound Parkway NW, Suite 300
Boca Raton, FL 33487-2742

International Standard Book Number-13: 978-1-4200-4669-4 (Hardcover)

Library of Congress Cataloging-in-Publication Data

Plant pathology concepts and laboratory exercises / editors, Robert N. Trigiano, Mark T. Windham, and Alan S. Windham. -- 2nd ed.
    p. cm.
  Includes bibliographical references and index.
  ISBN 978-1-4200-4669-4 (alk. paper)
  1. Plant diseases--Laboratory manuals. I. Trigiano, R. N. (Robert Nicholas), 1953- II. Windham, Mark Townsend, 1955- III. Windham, Alan S.

SB732.56.P63 2008
632'.3078--dc22   2007019969

Visit the Taylor & Francis Web site at
http://www.taylorandfrancis.com

and the CRC Press Web site at
http://www.crcpress.com

Contents

Part 3
Molecular Tools for Studying Plant Pathogens

Part 4
Plant–Pathogen Interactions

Part 5
Epidemiology and Disease Control

Preface

We thank those instructors who adopted the first edition of *Plant Pathology Concepts and Laboratory Exercises* as a textbook for their classes. We also thank them and their students and colleagues for invaluable feedback and criticism—we have incorporated many of their ideas into this edition. In response to your suggestions, we have included several new chapters, more case studies where appropriate, and a CD containing most of the figures in the book, as well as many supplementary photographs in the form of PowerPoint presentations.

Plant Pathology Concepts and Laboratory Exercises, Second Edition is intended to serve as a primary text for introductory courses and furnishes instructors and students alike with a broad consideration of this important and growing field. It also presents many useful protocols and procedures and advanced laboratories in selected areas. Thus, the book should serve as a valuable reference to researchers and students in plant pathology as well as many allied biological sciences. The textbook is intentionally written to be rather informal; it provides the reader with a minimum number of references, but does not sacrifice essential information or accuracy. Broad topic chapters are authored by specialists with considerable experience in the field and are supported by one or more laboratory exercises illustrating the central concepts of the topic. Each topic begins with a "concept box," highlighting some of the more important ideas contained within chapter and signals students to read carefully for these primary topics. There is an extensive glossary, which collects the **bolded** words and terms found in each of the concept chapters and some of the laboratory chapters. New to many of the concept chapters in this edition are case studies, which emphasize either specific diseases or processes. The case studies stand alone in the chapters and are provided primarily as bulleted lists supplying essential information at a glance.

Collectively, the laboratory exercises are exceptionally diverse in nature, providing something for everyone—from beginning to advanced students to researchers. Importantly, the authors have successfully completed the exercises many times, often with either plant pathology or biology classes or in their own research laboratories. All the laboratory protocols are written in procedure boxes that provide step-by-step, easy-to-follow instructions. A unique feature of this text is that the authors have provided in general terms what results should be expected from each of the experiments. At the end of each exercise, there are a series of questions designed to provoke individual thought and critical examination of the experiment and results. Our intentions are that instructors will not attempt to do all the experiments, but rather select one or two for each concept that serves the needs and interests of their particular class. For an advanced class, different experiments may be assigned among resourceful students. More advanced experiments following the general or beginning class exercises are embedded within some of the laboratory chapters.

We caution instructors and students to obtain the proper documents for transport and use of plant pathogenic organisms and to properly dispose of cultures and plant materials at the conclusion of the laboratory exercises. As always, the mention of products or specific equipment does not constitute product endorsement by either the authors, the various institutions, or the USDA, nor implied criticism of those products not mentioned. There are equally suitable, if not alternative, products and equipment available that may be substituted for those listed herein.

As with the first edition, the textbook is divided into the following six primary parts: Introductory Concepts, Groups of Plant Pathogens, Molecular Tools for Studying Plant Pathogens, Plant–Pathogen Interactions, Epidemiology and Disease Control, and Special Topics. Each section combines related facets of plant pathology and includes one to several concept chapters, usually with accompanying laboratory exercises. Most chapters have been revised to include more up-to-date information as well as additional materials. Four topic chapters have been completely rewritten and we have included five new topic and laboratory exercise chapters on soilborne pathogens, microscopy for students, and plant/fungal interactions.

Part 1 introduces students to the basic concepts of plant pathology, including historical perspectives, fundamental ideas of what is disease, how disease relates to environment, the host, and time, and provides a very broad overview of organisms that cause disease. Part 2 includes chapters that detail the various disease-causing organisms. This section begins with a consideration of viruses, prokaryotic organisms, and plant parasitic nematodes. The next ten chapters are devoted to the various phyla of fungi (classification primarily follows Alexopoulus, Mims, and Blackwell, *Introductory Mycology,* Fourth Edition) followed by chapters that focus on the fungi-like Oomycota, soilborne pathogens, plant parasitic seed plants and other biotic agents, and abiotic diseases. Part 3 introduces students to the basics of molecular tools and is illustrated with laboratory exercises that are adaptable for beginning as well as advanced students. Part 4 explores plant–pathogen interactions including treatments of molecular attack strategies, extracellular enzymes, host defenses, and disruption of plant function. Part 5, Epidemiology and Disease Control, is anchored with an extensive chapter outlining the basic ideas of epidemiology, which is followed in turn by several chapters detailing various strategies for disease control. This section also includes chapters on plant disease diagnosis. Part 6 is devoted to an often neglected topic in plant pathology textbooks—*in vitro* pathology. We have found that students are often fascinated with this topic as it combines several facets of investigation from biological and chemical disciplines. Another unique feature of the textbook is housed in this section as well: how to use compound and

dissecting microscopes. This topic is typically excluded from all plant pathology and biology textbooks, but is essential for most laboratory experiences.

It is our hope that students and instructors will find the format of the book, and level and amount of information it contains, to be appropriate for an introductory course but also for some graduate level work. The informal presentation style has been used very successfully in *Plant Tissue Culture and Laboratory Exercises* and with the inclusion of a glossary, concept boxes, case studies, and supplemental CD, students should find the format stimulating and conducive to learning. As always, we invite and welcome your comments and suggestions for improvements.

R.N. Trigiano
M.T. Windham
A.S. Windham
Knoxville, Tennessee

Acknowledgments

We wish to acknowledge the efforts of all the contributing authors—their creativity, support, and patience throughout the conception and development of the second edition were nothing less than phenomenal; the Institute of Agriculture at the University of Tennessee and especially Dr. Carl Jones, Head of Entomology and Plant Pathology, for providing the time and financial support necessary to complete the book; and Dr. Gary Windham for his photographs and work behind the scenes. We also thank our families for their patience and understanding throughout the project; and special thanks to John Sulzycki, Pat Roberson, and Gail Renard at CRC Press, whose constant encouragement and work were essential for the completion of this textbook. RNT would also like to express his gratitude to Bonnie H. Ownley, who not only authored manuscripts, but unselfishly gave of her time to edit some of the chapters—this book is much better because of your efforts—Thanks! Lastly, RNT thanks RLB, CGT, CAB, and REB for their insights, friendship, and support.

The Editors

Robert N. Trigiano received his B.S. degree with an emphasis in biology and chemistry from Juniata College, Huntingdon, Pennsylvania, in 1975 and an M.S. in biology (mycology) from Pennsylvania State University in 1977. He was an associate research agronomist working with mushroom culture and plant pathology for Green Giant Co., Le Sueur, Minnesota, until 1979 and then a mushroom grower for Rol-Land Farms, Ltd., Blenheim, Ontario, Canada, during 1979 and 1980. He completed a Ph.D. in botany and plant pathology (co-majors) at North Carolina State University at Raleigh in 1983. After concluding postdoctoral work in the Plant and Soil Science Department at the University of Tennessee, he was an assistant professor in the Department of Ornamental Horticulture and Landscape Design at the same university in 1987, promoted to associate professor in 1991 and to professor in 1997. He served as interim head of the department from 1999–2001. He joined the Department of Entomology and Plant Pathology at the University of Tennessee in 2002.

Dr. Trigiano is a member of the American Phytopathological Society (APS), the American Society for Horticultural Science (ASHS), and the Mycological Society of America (MSA), and the honorary societies of Gamma Sigma Delta, Sigma Xi, and Phi Kappa Phi. He received the T.J. Whatley Distinguished Young Scientist Award (University of Tennessee, Institute of Agriculture) and the gamma Sigma Delta Research individual and team Award of Merit (University of Tennessee). He has received an ASHS publication award for the most outstanding educational paper and in the Southern region ASHS, the L.M. Ware distinguished research award. In 2006 he was elected a fellow of the American Society for Horticultural Science. He has been an editor for the ASHS journals, *Plant Cell, Tissue and Organ Culture*, and *Plant Cell Reports* and is currently the co-editor of *Critical Reviews in Plant Sciences* and senior editor of *Plant Disease*. Additionally he has co-edited five books, including *Plant Tissue Culture Concepts and Laboratory Exercises* and *Plant Development and Biotechnology*.

Dr. Trigiano has been the recipient of several research grants from the U.S. Department of Agriculture (USDA), Horticultural Research Institute, and from private industries and foundations. He has published more than 200 research papers, book chapters, and popular press articles and holds six patents for his work with flowering dogwood. He teaches undergraduate/graduate courses in plant tissue culture, mycology, DNA analysis, protein gel electrophoresis, and plant microtechnique. Current research interests include diseases of ornamental plants, somatic embryogenesis and micropropagation of ornamental species, fungal physiology, population analysis, DNA profiling of fungi, and plants, and gene discovery.

Mark T. Windham is a professor of plant pathology and holds the Distinguished Chair in Ornamental Plant Diseases at the Institute of Agriculture, Department of Entomology and Plant Pathology, at the University of Tennessee, Knoxville. He received his B.S. degree and M.S. degree in plant pathology and weed science from Mississippi State University. In 1983, Dr. Windham completed his Ph.D. in plant pathology with a minor in plant breeding from North Carolina State University. After graduation, he accepted a position as a visiting assistant professor at Colorado State University. In 1985, Dr. Windham accepted a position as an assistant professor at the University of Tennessee, Knoxville and was promoted to professor in 1999.

Dr. Windham has taught introductory plant pathology since 1995. He also team-teaches two other courses, Diseases and Insects of Ornamental Plants and Plant Disease Fungi. Dr. Windham's research interests include diseases of ornamental plants, especially flowering dogwood. Dr. Windham has teamed with other scientists to release the first flowering dogwood cultivar resistant to dogwood anthracnose and to patent and release the first white blooming flowering dogwoods resistant to powdery mildew. Dr. Windham has published more than 100 research papers, book chapters, and popular press articles. He has also served as editor of the Plant Pathology Section, Southern Nursery Association Research Conference.

Dr. Windham's research has led to him receiving the Porter Henegar Memorial Award from the Southern Nursery Association and the Research and Team Awards of Merit from Gamma Sigma Delta. He co-authored *Dogwoods for American Gardens*, which was awarded the American Society for Horticultural Science Extension Publication Award.

Alan S. Windham is professor of plant pathology in the Institute of Agriculture, Department of Entomology and Plant Pathology at the University of Tennessee, Knoxville. Dr. Windham is stationed at the Plant and

Pest Diagnostic Center, Ellington Agricultural Center in Nashville, Tennessee.

Dr. Windham received his B.S. degree with an emphasis in plant pathology in 1979 and an M.S. in plant pathology with a minor in botany in 1981 from Mississippi State University, Starkville. He completed his Ph.D. in plant pathology with a minor in soil science at North Carolina State University at Raleigh in 1985. After completing his graduate work, he accepted the position of assistant professor with the University of Tennessee in 1985, and was promoted to associate professor in 1989 and professor in 1995.

Dr. Windham is a member of the American Phytopathological Society and was inducted into the honorary societies Gamma Sigma Delta and Sigma Xi. In 2000, he received the Gamma Sigma Delta Award of Merit for his work with the Dogwood Working Group (University of Tennessee). In 2002, he was awarded the American Society for Horticultural Science Extension Publication Award for *Dogwoods for American Gardens*. He has also served as editor for the Plant Pathology Section, Southern Nursery Association Research Conference.

Dr. Windham has conducted educational programs and consulted with nursery, greenhouse, and turf industries nationally and internationally. He has published over 75 research papers, book chapters, and popular press articles. He has lectured in many undergraduate/graduate courses in plant pathology, mycology, floriculture, and nursery and turfgrass management. His current research interest includes etiology and management of emerging plant diseases of ornamental plants and the development of disease-resistant ornamental plants.

Contributors

Malissa H. Ament
Department of Entomology and Plant Pathology
University of Tennessee
Knoxville, Tennessee

Richard E. Baird
Department of Entomology and Plant Pathology
Mississippi State University
Starkville, Mississippi

Luís M. Batista
Departamento de Botânica e Engenharia Biológica
Instituto Superior de Agronomia
Universidade Técnica de Lisboa
Lisboa, Portugal

D. Michael Benson
Department of Plant Pathology
North Carolina State University
Raleigh, North Carolina

Alexandre Borges
Departamento de Botânica e Engenharia Biológica
Instituto Superior de Agronomia
Universidade Técnica de Lisboa
Lisboa, Portugal

Kira L. Bowen
Department of Entomology and Plant Pathology
Auburn University
Auburn, Alabama

Gustavo Caetano-Anollés
Department of Crop Science
University of Illinois, Urbana
Urbana, Illinois

Martin L. Carson
USDA/ARS Cereal Disease Laboratory
University of Minnesota
St. Paul, Minnesota

Zhenjia Chen
Disease and Stress Biology Laboratory
Instituto de Tecnologia Química e Biológica
Universidade Nova de Lisboa
Oeiras, Portugal

Marc A. Cubeta
Department of Plant Pathology
North Carolina State University
Raleigh, North Carolina

Kenneth J. Curry
Department of Biological Sciences
University of Southern Mississippi
Hattiesburg, Mississippi

Margery L. Daughtrey
Department of Plant Pathology
Cornell University
Ithaca, New York

Renae E. DeVries
Department of Entomology and Plant Pathology
University of Tennessee
Knoxville, Tennessee

João Duarte
Departamento de Botânica e Engenharia Biológica
Instituto Superior de Agronomia
Universidade Técnica de Lisboa
Lisboa, Portugal

Ricardo B. Ferreira
Disease and Stress Biology Laboratory
Instituto de Tecnologia Química e Biológica
Universidade Nova de Lisboa
Oeiras, Portugal

S. Ledare Finley
Department of Entomology and Plant Pathology
University of Tennessee
Knoxville, Tennessee

Regina Freitas
Departamento de Botânica e Engenharia Biológica
Instituto Superior de Agronomia
Universidade Técnica de Lisboa
Lisboa, Portugal

Ann Brooks Gould
Department of Plant Pathology
Rutgers University
New Brunswick, New Jersey

Dennis J. Gray
Mid-Florida Research and Education Center
University of Florida
Apopka, Florida

James F. Green
Department of Chemistry
University of Tennessee
Knoxville, Tennessee

Sharon E. Greene
Department of Entomology and Plant Pathology
University of Tennessee
Knoxville, Tennessee

Kimberly D. Gwinn
Department of Entomology and Plant Pathology
University of Tennessee
Knoxville, Tennessee

Darrell D. Hensley
Department of Entomology and Plant Pathology
University of Tennessee
Knoxville, Tennessee

Donald E. Hershman
Plant Pathology Department
University of Kentucky
Princeton, Kentucky

Kathie T. Hodge
Department of Plant Pathology
Cornell University
Ithaca, New York

Clayton A. Hollier
Department of Plant Pathology and Crop Physiology
Louisiana State University Agricultural Center
Baton Rouge, Louisiana

Subramanian Jayasankar
Department of Plant Agriculture–Vineland Campus
University of Guelph
Vineland Station, Ontario, Canada

Steven N. Jeffers
Department of Plant Pathology and Physiology
Clemson University
Clemson, South Carolina

George H. Lacy
Department of Plant Pathology, Physiology, and Weed Science
Virginia Polytechnic Institute and State University
Blacksburg, Virginia

Kurt H. Lamour
Entomology and Plant Pathology
University of Tennessee
Knoxville, Tennessee

Marie A.C. Langham
Plant Science Department
South Dakota State University
Brookings, South Dakota

Yonghao Li
Department of Entomology and Plant Pathology
University of Tennessee
Knoxville, Tennessee

Larry J. Littlefield
Entomology and Plant Pathology
Oklahoma State University
Stillwater, Oklahoma

Felix L. Lukezic
Department of Plant Pathology
The Pennsylvania State University
University Park, Pennsylvania

Sara Monteiro
Departamento de Botânica e Engenharia Biológica
Instituto Superior de Agronomia
Universidade Técnica de Lisboa
Lisboa, Portugal

Gary Moorman
Department of Plant Pathology
Pennsylvania State University
University Park, Pennsylvania

Sharon E. Mozley-Standridge
Division of Natural Sciences, Mathematics, and Engineering
Middle Georgia College
Cochran, Georgia

Jackie M. Mullen
Department of Entomology and Plant Pathology
Auburn University
Auburn, Alabama

James P. Noe
Department of Plant Pathology
University of Georgia
Athens, Georgia

Bonnie H. Ownley
Department of Entomology and Plant Pathology
University of Tennessee
Knoxville, Tennessee

Jerald K. Pataky
Department of Crop Sciences
University of Illinois
Urbana, Illinois

David Porter
Department of Botany
University of Georgia
Athens, Georgia

Melissa B. Riley
Department of Plant Pathology and Physiology
Clemson University
Clemson, South Carolina

Timothy A. Rinehart
Thad Cochran Southern Horticultural Research
U.S. Department of Agriculture, ARS
Poplarville, Mississippi

Naomi R. Rowland
Biology Department
Western Kentucky University
Bowling Green, Kentucky

Cláudia N. Santos
Disease and Stress Biology Laboratory
Instituto de Tecnologia Química e Biológica
Universidade Nova de Lisboa
Oeiras, Portugal

Nina Shishkoff
Department of Plant Pathology
Cornell University
Ithaca, New York

Artur R. Teixeira
Departamento de Botânica e Engenharia Biológica
Instituto Superior de Agronomia
Universidade Técnica de Lisboa
Lisboa, Portugal

David Trently
Department of Entomology and Plant Pathology
University of Tennessee
Knoxville, Tennessee

Robert N. Trigiano
Department of Entomology and Plant Pathology
University of Tennessee
Knoxville, Tennessee

XinWang Wang
Department of Entomology and Plant Pathology
University of Tennessee
Knoxville, Tennessee

David T. Webb
Department of Biological Sciences
University of Hawaii
Honolulu, Hawaii

Alan S. Windham
Department of Entomology and Plant Pathology
University of Tennessee
Knoxville, Tennessee

Mark T. Windham
Department of Entomology and Plant Pathology
University of Tennessee
Knoxville, Tennessee

Part 1

Introductory Concepts

1 Plant Pathology and Historical Perspectives

Mark T. Windham and Alan S. Windham

CHAPTER 1 CONCEPTS

- Plant pathology is composed of many other disciplines such as botany, microbiology, nematology, virology, bacteriology, mycology, meteorology, biochemistry, genetics, soil science, horticulture, agronomy, and forestry.

- Plant pathology is the study of what causes plant diseases, why they occur, and how to control them.

- Plant pathologists are usually interested in populations of diseased plants and not in individual diseased plants.

- Plant diseases have had a major impact on mankind. Diseases such as ergotism and late blight of potato have led to the deaths of thousands of people.

- Diseases such as coffee rust have changed the way people behave and/or their customs.

- Diseases such as Southern corn leaf spot, chestnut blight, and dogwood anthracnose have appeared suddenly and caused millions of dollars in damage as the pathogen of the diseases spread through the ranges of the hosts.

Plants are the foundation of agriculture and life on this planet. Without plants, there would be nothing to feed livestock or ourselves. Plants are a primary component in building shelter and making clothing. Like humans and animals, plants are plagued with diseases, and these diseases may have devastating consequences on plant populations. Plant pathology is not a pure discipline in the sense of chemistry, mathematics, or physics, but it embodies other disciplines such as botany, epidemiology, molecular biology and genetics, microbiology, nematology, virology, bacteriology, mycology, meteorology, biochemistry, genetics, soil science, horticulture, agronomy, and forestry, among others. Plant pathology encompasses the study of what causes a plant disease, how the pathogen attacks a plant at the molecular, cellular tissue, and whole plant levels of organization, how the host responses to attack, how pathogens are disseminated, how the environment influences the disease process, and how to manage plant pathogens and thereby reduce the effects of the disease on plant populations. Unlike physicians or veterinarians that emphasize treatment of individuals, plant pathologists usually are interested in populations of plants and not individuals. An individual wheat plant has little worth to a farmer. If it dies from a disease, the plants on either side of it will grow into its space and their increased yield will compensate for the loss of the diseased plant.

However, if entire fields become diseased or fields in a region are devastated by disease, economic losses can be staggering. The exception to emphasizing populations of plants to individual plants is specimen plants that include large shade trees or trees planted by a historical figure, such as an oak planted by George Washington at Mount Vernon, or a Southern magnolia planted on the White House lawn by Andrew Jackson. Extraordinary measures may be taken to protect or treat plants of high value or historical significance.

Because of the diversity of questions that plant pathologists are called on to answer, plant pathologists are a heterogeneous group of scientists. Some plant pathologists spend most of their time in the field studying how pathogens move over a large area and what environmental factors play a role in development of epidemics or determining which management tactics are most effective in controlling or reducing the impact of a disease. Other plant pathologists are interested in the processes by which a pathogen induces a disease, or they may be looking for genes that confer resistance in a plant and complete most of their professional activities in a laboratory. Some plant pathologists work in outreach programs, such as the extension service or in private practice, and diagnose disease problems for producers and home gardeners, making recommendations as to how plant diseases may be managed.

Still other plant pathologists work for private companies and are responsible for development of new products (biological control agents, chemicals, and new plant varieties) that reduce the impact of plant diseases on producers and consumers (see Appendix 1).

IMPACT OF PLANT DISEASES ON MANKIND

Diseases have impacted man's ability to grow plants for food, shelter, and clothing since humankind began to cultivate plants. Drawings and carvings of early civilizations in Central America depict corn plants with drooping ears and poor root systems. Crop failures for ancient man and throughout the Middle Ages were common, and plant diseases were often attributed to the displeasure of various deities. The Roman god Robigus was thought to be responsible for a good wheat harvest, and Romans prayed to him to prevent their wheat crop from being blasted with "fire" (rust). In more modern times (since 1800), plant diseases have destroyed the military plans of monarchs, changed cultures, caused mass migrations of people to avoid starvation, resulted in the loss of major components of forest communities, and bankrupted thousands of planters, companies, and banks. In the following paragraphs, some examples of the effects of various plant diseases on the history of mankind and the environment will be illustrated.

ERGOTISM

Ergotism is the result of eating rye bread contaminated with sclerotia (hard survival structures shaped like the spur of a rooster) of *Claviceps purpurea*. Sclerotia are formed in the maturing heads of rye and may contain alkaloids including lysergic acid diethylamide (LSD), a strong hallucinogenic compound. Symptoms in humans eating contaminated bread include tingling of extremities, a high fever, hallucinations, mental derangement, abortions, and loss of hands, feet, and legs due to restricted blood flow and subsequent gangrene. Death often follows consumption of large quantities of contaminated grain. In livestock fed contaminated grain, heifers may abort fetuses, and livestock will lose weight, quit giving milk, and lose hooves, tails, and ears from gangrene. As in humans, death is likely when exposed to high doses of ergot. In the Middle Ages, thousands of people died from this disease in Europe, where the disease was referred to as "the Holy Fire" due to the high fever it produced and the burning and tingling sensations in the hands and feet of victims. An outbreak in France led to the name "St. Anthony's Fire," presumably because monks of the Order of St. Anthony successfully treated inflicted people by feeding them uncontaminated rye bread. The disease continued in Europe for centuries. A number of authors have concluded that the Salem Witch Trials were due to an outbreak of ergotism in the American colonies as rye was the primary grain grown in the New England region. The behavior of the accused "witches" was similar to behavior associated with an outbreak of ergotism in human and livestock populations. In the 1950s, ergotism occurred in several small villages in France and demonstrated that even when the cause of ergotism and how the sclerotia are introduced into grain are known, epidemics of ergotism are still possible.

IRISH POTATO FAMINE

Potatoes were one of the treasures taken from the New World back to Europe and were readily adapted to European farming practices. By the 1840s, potatoes had become the staple food crop in Ireland, and the average Irishman ate approximately seven pounds of potatoes daily. Because so many potatoes could be grown on a relatively small plot of land, the population of Ireland increased dramatically during the first four decades of the 19th century. In the early 1840s an epidemic of a new potato disease was documented in the United States, but little attention was paid to it in Europe. In 1845, an epidemic of potato cholera, later named late blight of potato and attributed to the pathogen *Phytophthora infestans*, swept across Europe. Although starvation was common at this time in continental Europe, it was spared the devastation that was found in Ireland because most of Europe had more diversity in its agricultural production and did not depend on one crop for survival as the Irish did. In Ireland, more than a million people starved to death due to an almost total destruction of the potato crop. Another million people migrated to the United States, taking whatever jobs they could find in the new world. In cities such as Boston and New York, many of the jobs they took were low paying or dangerous, such as firefighting and police work.

COFFEE RUST

In the 1700s and early 1800s coffee was an expensive drink due to the monopoly that Arab traders had on the coffee trade and the careful attention they paid to ensure that viable coffee beans (seeds) did not leave their domain. In the mid-1800s some coffee beans were smuggled to Ceylon (present day Sri Lanka), and the British began growing coffee. Coffee became the preferred drink of British citizens, and coffee houses became as common as pubs. By 1870, more than 400 plantations of coffee, comprising at least 200,000 ha, were found in Ceylon. In the decade of 1870, a new disease, coffee rust, caused by the fungal pathogen *Hemileia vastatrix*, struck Ceylon with terrible consequences and destroyed the island's coffee trade. Planters, banks, and shipping companies went bankrupt, and panic was widespread in British financial markets. By 1880, 140,000 ha of destroyed coffee trees had been replaced with tea plants. Great Britain became a country of tea drinkers, and this custom remains with them through the present.

CHESTNUT BLIGHT

When the first colonists arrived in the New World, they found forests of eastern North America populated with American chestnut. Chestnut wood was resistant to decay, and the bark contained tannins that made the production of leather from animal hides feasible. In many areas, one out of every four trees in the forest was an American chestnut. The crop from these trees was so prolific that the ground could be covered by nearly a foot of nuts. Nuts not only served as a food source for the colonists, but were a major mast crop for wildlife that the colonist depended on for meat. Many of the ships of the American shipping industry in the 19th century were made of rot-resistant chestnut timber. In the early 1900s, a new disease of the chestnut, now named *Cryphonectira parasitica*, was discovered in the northern Atlantic states and named "chestnut blight." The disease spread rapidly south and westward, destroying chestnut stands as it went. The disease finally reached the southern and western extent of the chestnut's range in the 1950s. By this time, millions of trees had been destroyed, which represented billions of dollars in lost timber. The effects of the disease on wildlife populations were also dramatic as wildlife had to adapt to less reliable and nutritional mast crops such as acorns. There have been intensive breeding efforts to incorporate resistance to chestnut blight from Chinese chestnut into American chestnut. Resistant hybrids that have been backcrossed for some generations with American chestnut have resulted in a tree that is resistant to chestnut blight and that strongly resembles the American chestnut. Unfortunately, it will take more than a century before we see forests with the stately giants that Americans marveled at before the onset of chestnut blight.

SOUTHERN CORN BLIGHT

After the advent of hybrid seed corn, corn yields began to skyrocket to unheard of yields and hybrid seed came to dominate the market. To reduce labor costs in producing hybrid seed corn, seed companies began using breeding lines containing a sterility gene that was inherited through the cytoplasm of the female parent. The trait or gene was named the Texas cytoplasmic male sterility (cms) gene. Using this gene in the female parent meant substantial cost savings for the seed companies because they did not have to remove the tassels (detassel) by hand when producing hybrid seed corn. This system worked for several years until an outbreak in 1970 of a new race of the fungus that is currently named *Cochliobolus heterostropus*. This new race caused a disease on corn carrying the male sterility gene (practically all hybrid seed corn at that time) that resulted in tan lesions that covered the leaves. Stalks, ear husks, ears, and cobs were also attacked and destroyed by the pathogen. The disease first appeared in Florida, spread northward and destroyed approximately 15% of the U.S. crop and losses were estimated to be in excess of $1 billion. Experts warned the country that nearly the entire U.S. corn crop would be lost in 1971 if substantial changes could not be quickly made in the way hybrid seed corn was produced. Commercial seed companies leased almost all available space in South America in the winter of 1970 and were able to produce enough hybrid seed corn that did not contain the Texas cms gene and the corn crop of 1971 was saved.

DOGWOOD ANTHRACNOSE

Flowering dogwood, *Cornus florida*, is a popular tree in landscapes throughout much of the United States and is worth more than $100 million in wholesale sales to the U.S. nursery industry. It is also an important natural resource, and its foliage, which is high in calcium, is the preferred browse of lactating deer in early spring in the eastern United States. Its bright red berries are high in fat and are an important mast crop to wildlife, including black bears, squirrels, turkeys, and more than 40 species of neotropical song birds. In 1977, a new fungal disease was reported in Seattle, WA, on flowering dogwood and Pacific dogwood, *C. nuttallii*. The following year the disease was reported on flowering dogwood in Brooklyn Botanical Garden in New York. The origin of this disease organism is unknown; however, genetic data suggests that the disease-causing fungus is exotic to the North American continent. Since first reported, dogwood anthracnose, caused by *Discula destructiva*, has destroyed millions of dogwoods on both coasts. In some areas of the Appalachians, flowering dogwood has nearly disappeared where it was once a common understory tree.

CAUSES OF PLANT DISEASES

Plant diseases are caused by fungi, bacteria, mollicutes, nematodes, viruses, viroids, parasitic seed plants, algae, and protozoa. The largest group of plant pathogens are the fungi. This differs considerably from human pathogens among which the most common pathogen groups include bacteria and viruses. This is not to imply that other groups such as bacteria, mollicutes, nematodes, viruses, viroids, and parasitic seed plants do not cause important and destructive diseases; they do. For example, *Striga* spp. (witchweed) is the limiting factor in sorghum, sugarcane, and rice production in Africa, Asia, and Australia. Dwarf mistletoe, *Arceuthobium* species, severely limits conifer production in some areas of the western United States. Millions of dollars are lost each year to diseases such as root knot, bacterial diseases such as soft rot and crown gall, and virus diseases such as tobacco mosaic virus and impatient necrotic spot virus.

ABIOTIC STRESSES (ABIOTIC DISEASES)

Some abiotic stresses such as air pollution and nutrient deficiencies were once referred to as abiotic diseases. However, this terminology is no longer used in modern plant pathology. Plant stresses such as those listed above—and others, such as extremes in temperature, moisture, pH, and light levels, and exposure to herbicides—are now referred to as *abiotic stresses* or environmental stresses that result in disease-like symptoms. Sometimes the symptoms that these stresses cause in plants—chlorosis, wilting, necrosis, leaf spots, blights, etc.—look like symptoms of diseases caused by plant pathogens.

WHERE TO GO FOR MORE INFORMATION ABOUT PLANT DISEASES

Most plant pathologists belong to professional societies such as the American Phytopathological Society, Nematology Society, the Mycology Society of America, the American Society of Horticultural Science, and so forth. The most prominent society for plant pathology is the internationally recognized American Phytopathological Society. The society's Web page (http://www.apsnet.org) is a clearing house of information concerning new and emerging disease problems, careers in plant pathology, a directory of plant pathology departments at universities in the United States, and featured articles on plant diseases, and is the publisher of several plant pathology journals such as *Phytopathology, Plant Disease, Molecular Plant-Microbe Interactions,* and *Plant Health Progress,* as well as other publications such as books and compendia on specific diseases or diseases affecting specific hosts. It also publishes a monthly online newsletter, *Phytopathology News.* Membership is open to professionals interested in plant pathology and to students at a very reduced rate.

SUGGESTED READING

Agrios, G.N. 2005. *Plant Pathology.* 5th ed. Academic Press. New York. 952 pp.

Campbell, C.L., P.D. Petersen and C.S. Griffith. 1999. *The Formative Years of Plant Pathology in the United States.* APS Press. St. Paul, MN. 427 pp.

Carefoot, G.L and E.R. Sprott. 1967. *Famine on the Wind.* Longmans. Ontario. 231 pp.

Horsfall, J. and E. Cowling. 1978–1980. *Plant Disease: An Advanced Treastise.* Vol. 1–5. Academic Press. New York.

Large, E.C. 1940. *The Advance of the Fungi.* Henry Holt and Co. New York. 488 pp.

Lucas, G.B., C.L. Campbell and L.T. Lucas. 1992. *Introduction to Plant Diseases: Identification and Management.* 2nd ed. Van Nostrand Reinhold. New York. 364 pp.

Schumann, G.L. 1991. *Plant Diseases: Their Biology and Social Impact.* APS Press. St. Paul, MN. 397 pp.

2 What Is a Disease?

Mark T. Windham and Alan S. Windham

CHAPTER 2 CONCEPTS

- A disease is due to the interactions of the pathogen, host, and environment.

- Diseases are dynamic (change over time). Injuries are discrete events.

- Plant stresses are usually due to too much or too little of something.

- The host response to disease is known as symptoms.

- Structures (e.g., mycelia, spores, nematode egg masses) of pathogens on a diseased host are known as signs.

- The interaction between the host, pathogen, and environment is known as the disease cycle.

- A disease cycle is made of a sequence of events including inoculation, penetration, infection, invasion, reproduction, and dissemination.

- Diseases with only a primary disease cycle are known as monocyclic diseases, whereas diseases with secondary disease cycles are know as polycyclic diseases.

- Host plants can be infected by a pathogen, whereas soil or debris is infested by pathogens.

- Koch's postulates (proof of pathogenicity) are used to prove that a pathogen causes a disease.

Before studying plant disease, a framework or concept as to what a plant disease is, and almost as important, what it is not, is useful. There are many definitions of what a plant disease is; however, for this book the following definition will be used: a disease is the result of a dynamic, detrimental relationship between a plant and an organism that parasitizes or interferes with the normal processes of cells and/or tissues of the plant. The organism that incites or causes the disease process with the host is called a **pathogen**.

A pathogen may or may not be a parasite. A **parasite** is an organism that lives on or in another organism and obtains nutrients at the expense of the host. In contrast, pathogens may interfere with plant cell functions by producing toxins that disrupt or destroy cells; by producing growth plant regulators that interfere with the normal growth or multiplication of plant cells; by producing enzymes that interfere with normal cellular functions; or by absorbing water and/or nutrients that were intended for the cellular functions of the host. Pathogens may also incite disease by blocking the vascular system so that water and nutrients cannot be normally moved within the plant. Some pathogens disrupt normal functions of plants by inserting portions of their DNA or RNA into host cells and interfering with replication of nucleic acids.

Pathogenicity is the ability of a pathogen to interfere with one or more functions within a plant. The rate or how well a pathogen is able to interfere with cell functions is referred to as **virulence**. A virulent pathogen is called "very aggressive" and may incite disease over a wide range of environmental conditions. An avirulent pathogen is an organism that rarely is able to interfere with normal cellular functions of the host or does so under very specific environmental conditions. The ability of the pathogen to survive in the environment where the host is grown is a measure of **pathogen fitness**.

Plant stresses or injuries are not diseases because they are not dynamic; that is, they do not change over time. If lightning strikes a tree, the tree may be damaged or killed. However, the lightning does not get hotter or more dangerous to the tree over time. It happens in a discrete instant in time and is therefore an injury and not a disease. The same thing could be said for using a lawn mower on turf. You may severely impede ability of the grass to grow by cutting off 30–50% of the leaf area, but the cut was done in a discrete instant of time. Therefore, it is not a disease,

Disease Triangle

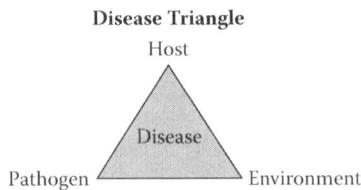

FIGURE 2.1 Disease is dependent on the following three components: host, pathogen and environment. The area within the triangle is the interaction of these components referred to as disease.

but is an injury. A plant stress is usually too much or too little of something. Water stress can be caused by either giving the plant too much (flooding) or too little (drought) water. Other examples of plant stresses can be extremes in temperature, improper pH, and nutrient deficiency or excess. Pollutants, pesticides, and road salt may also cause stresses to plants.

Plant pathogens that cause diseases include organisms such as fungi, prokaryotes (bacteria and mollicutes), viruses, viroids, nematodes, protozoa, algae, and parasitic seed plants. These organisms can detrimentally affect a host in diverse ways. Pathogens can be classified into several groups. **Biotrophs** are pathogens that require living host tissue to complete their life cycle. Examples of biotrophs include fungi such as powdery mildews (Chapter 14) and rusts (Chapter 18); some members of the Oomycota, such as downy mildews and white rusts (Chapters 20 and 21); prokaryotes, such as some species of *Xylella* and mollicutes (Chapters 6 and 7); viruses and viroids (Chapters 4 and 5), phytoparasitic nematodes (Chapters 8 and 9), and protozoa and dwarf mistletoe (Chapter 24). Many pathogens can be parasitic on a host under some conditions and at other times can be saprophytic, living on organic matter. A pathogen that often behaves as a parasite, but under certain conditions behaves as a saprophyte, is a **facultative saprophyte**. A pathogen that often behaves as a saprophyte, but under some conditions becomes a parasite, is known as a **facultative parasite**. Nonbiotrophic organisms kill before feeding on the cells or the cellular contents. These organisms that live on dead tissues are known as **necrotrophs**.

Diseased plants are infected by a pathogen. However, in a few cases, once disease is incited, symptoms may continue to develop even if the pathogen is no longer present (an example is crown gall caused by *Agrobacterium tumefaciens*). Inanimate objects such as soil, pots, or debris are not infected by pathogens, but can be infested by pathogens.

The host plant, pathogen, and environment interact with each other over time and this interaction is referred to as disease (Figure 2.1). The sequence of events that take place during the course of disease, sometimes at set or discrete time intervals, is known as the **disease cycle**. The

Primary Disease Cycle
Monocyclic Disease

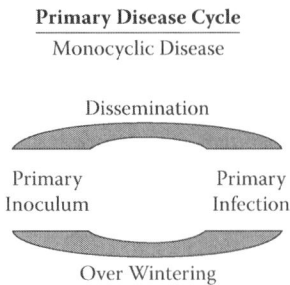

FIGURE 2.2 In a monocyclic disease, the primary disease cycle is composed of discrete events where inoculation and penetration lead to infection. Propagules produced during the disease cycle overwinter and become the primary inoculum (inoculum that begins a new disease cycle) for the next disease cycle. The inoculum is disseminated at the beginning of the next cycle.

disease cycle is not to be confused with the pathogen's life cycle. Sometimes these two cycles follow similar paths, but the cycles are different. The parts of the disease cycle are inoculation, penetration, infection, invasion, reproduction, and dissemination. **Inoculation** is the placement of the pathogen's infectious unit or propagule on or in close proximity to the host cell wall. The propagule will then penetrate the cell wall of the host. In fungi, the propagule may germinate, and the germ tube may penetrate the wall directly or indirectly through a wound or natural opening. Once the pathogen is through the cell wall, a food relationship with the host may develop, and the cell is said to be infected. After infection takes place, the pathogen may grow and invade other parts of the host or reproduce. The pathogen will continue to reproduce and the new propagules will be dispersed or disseminated by a variety of means including in the wind, rain, within or on vectors, by seed, or on contaminated debris or equipment.

Some diseases are monocyclic diseases (Chapter 33), meaning that there is only one disease cycle in a growing cycle. Inoculum that is produced during the disease

Secondary Disease Cycle
Polycyclic Disease

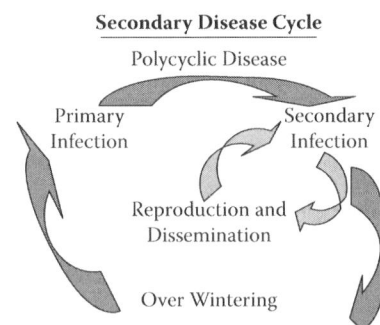

FIGURE 2.3 Polycyclic diseases have primary inoculum that penetrates and infects the plant. This is a part of the primary disease cycle. Inoculum produced after invasion is disseminated and causes more infections during the current growing season. The inoculum contributes to the secondary disease cycle and the secondary cycle may be repeated many times.

TABLE 2.1

Common Symptoms Associated with Plant Host Response to Disease

- **Blight**—extensive area of diseased flowers or leaves.
- **Butt rot**—basal trunk rot.
- **Burl**—swelling of a tree trunk or limb differentiated into vascular tissue; contrast with gall.
- **Canker**—a sunken area in a fruit, stem, or limb caused by disease.
- **Chlorosis**—yellow-green color of foliage due to destruction or lack of production of chlorophyll.
- **Dieback**—generalized shoot death.
- **Flagging**—scattered or isolated dead or dying limbs.
- **Gall**—swollen area of nondifferentiated tissues (tumor) caused by an infection. Galls can arise from hypertrophy (cell enlargement) and/or hyperplasia (increase in cell division). Contrast with burl.
- **Leaf spot**—lesion on leaf.
- **Lesion**—a necrotic (dead) or chlorotic spot that occurs on all plant organs. Anthracnose lesion: a necrotic lesion with a reddish or purplish border. Local lesion: necrotic or chlorotic lesion where infection is limited to a small group of cells and the infection does not spread to other parts of the tissue.
- **Mosaic**—chlorotic pattern, ringspots, and mottles in leaves, petals, or fruit. Mosaics are usually associated with virus infections.
- **Mummy**—shriveled, desiccated fruit.
- **Necrosis**—dead tissue.
- **Rot**—portion of plant destroyed by disease. Root rot: rotted roots.
- **Wilt**—lost of turgor in a plant or plant part.

cycle does not contribute or fuel the disease during the current growing season. A sequence of events of a monocyclic disease cycle is given in Figure 2.2. In many diseases, the inoculum produced during much of the disease cycle contributes to continuing the current disease cycle or epidemic. This inoculum actually fuels the epidemic, and the disease cycle expands to include many more host plants, which in turn contribute more and more inoculum to the disease cycle. In diseases where there are more than one disease cycle, the primary cycle often has a repeating phase known as the secondary disease cycle (Figure 2.3). Such diseases are known as polycyclic diseases.

Plant pathologists study disease cycles to determine where cultural or other types of disease control tactics can be applied to interfere with the disease and thus interrupt the processes. Elimination of infested or infected seed may reduce the primary inoculum used to start a primary disease cycle. Use of resistant cultivars (Chapter 34) that are able to wall off a plant infection and prevent invasion of the host can also stop a disease cycle or prevent the formation of secondary cycles. Elimination of plant debris may reduce the ability of a pathogen to overwinter. By understanding the disease cycle and the series of events that are parts of that cycle, plant pathologists may attack the disease processes and reduce the plant disease's ultimate affects on society.

Host responses to infection are known as **symptoms**. Symptoms include leaf spots, blights, blotches, twig blights, cankers, galls, seed, and root and stem rots. Symptoms of a plant disease may occur on only a small portion of the plant and result in little disruption of the plant's functions, or symptoms may cover the entire plant. Definitions of common plant symptoms are given in Table 2.1.

Structures of the pathogen are referred to as **signs**. Signs can include spores, mycelium, resting structures such as sclerotia, nematodes, bacterial streaming into water, etc. In some cases, symptoms and signs are present together. For example, a plant that is infected with *Sclerotium rolfsii* may exhibit symptoms of wilting, but have signs of the fungus, such as sclerotia.

Once we recognize that a disease is occurring, it is also important to be able to prove that a pathogen is causing a particular disease. To do this, we use a series of rigid rules or postulates known as Koch's postulates or *proof of pathogenicity* (Chapters 10, 39, and 40). Koch's postulates are as follows:

1. The pathogen must be associated with all symptomatic plants.
2. The pathogen must be isolated and grown in pure culture, and its characteristics described. If the pathogen is a biotroph, it should be grown on another host plant and have the symptoms and signs described.
3. The pathogen from pure culture or from the test plant must be inoculated on the same species or variety that was originally described, and it must produce the same symptoms that were seen on the diseased plants originally.
4. The pathogen must be isolated in pure culture again, and its characteristics described exactly like those observed in step 2.

In conclusion, plant pathologists study what causes diseases, how plants are affected by diseases, how plants resist pathogens, and how the host, pathogen, and environment interact with each other. Through their investigations they find what actually is a plant disease, which organisms cause diseases, and how those organisms are classified, observing that how a pathogen, a susceptible host, and the environment interact in a disease relationship is far from static. For example, until a few years ago, diseases such as downy mildews and late blight of potato were thought to be caused by fungi. However, due to molecular studies, the pathogens that cause these diseases are now classified in the kingdom Stramenopila (Chromista) instead of in the kingdom Fungi. Although the interactions of the disease triangle endpoints (host, environment, and pathogen) will continue to be reevaluated and redefined, plant pathologists still agree that without a conducive environment for disease, a susceptible host and a pathogen, no disease will take place.

SUGGESTED READING

Agrios, G.N. 2005. *Plant Pathology.* 5th ed. Academic Press. San Diego, CA. 952 p.

Andrews, J.H. 1984. Life history strategies of plant parasites. *Adv. Plant Pathol.* 2:105–130.

Horsfall, J.G. and E.B. Cowling (Eds.) 1977–1980. *Plant Disease*, Vol. 1–5. Academic Press. New York.

Vanderplank, J.E. 1963. *Plant Diseases: Epidemics and Control.* Academic Press. New York. 349 p.

Zadoks, J.C. and R. Schein. 1979. *Epidemiology and Plant Disease Management.* Oxford University Press, New York. 427 p.

3 Introduction to the Groups of Plant Pathogens

Mark T. Windham

CHAPTER 3 CONCEPTS

- Fungi are acholorophyllous, filamentous, eukaryotic organisms that reproduce by spores and have walls containing chitin.

- Anamorphic spore types include sporangiospores, conidia, and chlamydospores.

- Spores in a sporangium are formed by cleavage of the cytoplasm.

- Sexual spores of fungi are zygospores, ascospores, and basidiospores.

- Members of the Oomycota have oospores and contain cellulose in their cell walls.

- Prokaryotic pathogens include bacteria, which have cell walls, and mollicutes, which have a cell membrane but no cell wall.

- Phytopathogenic nematodes have mouthparts called stylets.

- Viruses are nucleoproteins and are too small to be seen with light microscopy.

Plant pathogens belong to very diverse groups and are found in the kingdoms Animalia, Stramenopila (Chromista), Fungi, Procaryotae, Plantae, and Protozoa. The largest group of plant pathogens is found in the Fungi. This chapter is intended to very superficially acquaint students with the various groups of pathogens and some of the specialized language and terms associated with each of them. More complete descriptions of the different groups follow in subsequent chapters.

THE FUNGI

Fungi are acholorophyllous and eukaryotic. They are generally filamentous, branched organisms that reproduce normally by spores and have walls made of chitin and other polymers. Most of their life cycles are spent in the haploid (N) or dikaryotic (N + N) state. The thread-like filaments of the fungus are known as **hyphae** (sing. **hypha**) (Figure 3.1.).

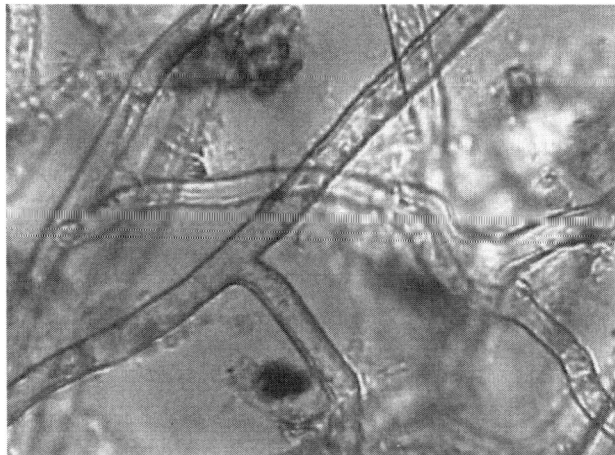

FIGURE 3.1 Hyphae of *Rhizoctonia solani* have crosswalls known as septa.

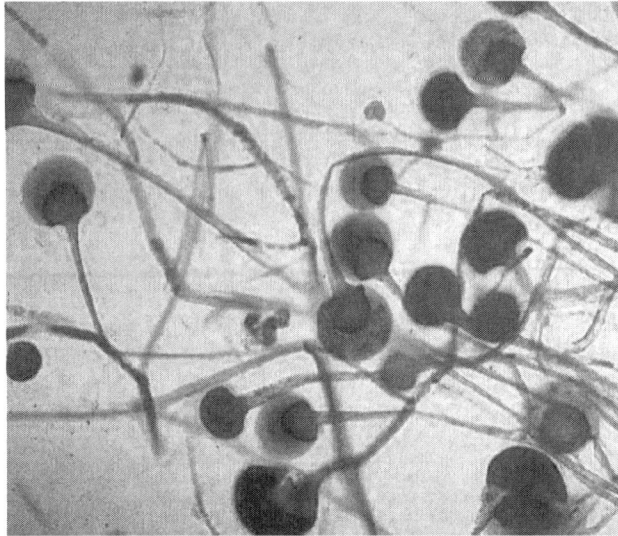

FIGURE 3.2   Sporangia of *Rhizopus niger*. The specialized hyphae or stalk that is attached to each sporangium is a sporangiophore.

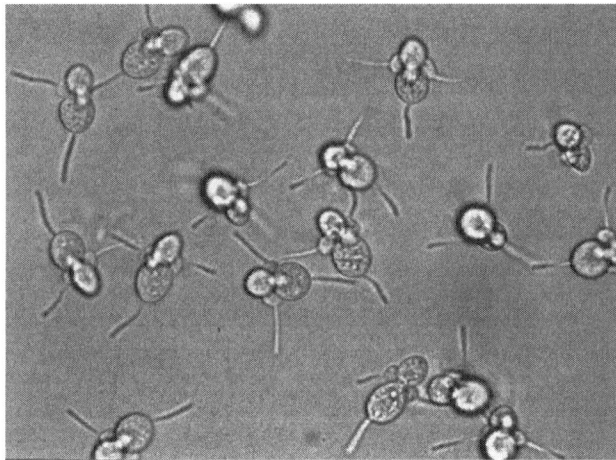

FIGURE 3.3   Conidia of *Entomosporium* species. Conidia can be made of a single cell or be multicellular, as are these conidia.

Hyphae of one body or thallus is known as a **mycelium** (pl. **mycelia**). Spores are the reproductive or propagative bodies of fungi. In some fungi, spores of the **anamorph** (asexual) stage are formed by cleavage of cytoplasm within a sac or **sporangium** (pl. **sporangia**) (Figure 3.2.). Spores produced in the sporangium are referred to as **sporangiospores**. Sporangiospores may be motile (have one or more flagella) or nonmotile. Nonmotile sporangiospores are usually disseminated by air currents. Motile spores within a sporangium are called **zoospores** and dispersed in water. Asexual spores of other fungi are borne on the tips or sides of specialized hyphae called **conidiophores**. Spores borne in this fashion are referred to as **conidia** (sing. **conidium**) (Figure 3.3). Conidia vary in shape, size, color, and number of cells. Some conidia are borne on naked conidiophores, whereas others are borne on conidiophores that are contained in specialized structures. A **pycnidium** (pl. **pycnidia**) is an asexual fruiting body that is flask-shaped and contains conidia and conidiophores. Pycnidia (Figure 3.4) usually have a hole (**ostiole**) from which conidia are pushed out of the structure. An **acervulus** (pl. **acervuli**) is an asexual fruiting body that is found under the cuticle or epidermis of the host. As conidiophores and conidia form, the epidermis and/or cuticle ruptures and spores are released. Spores may be released in a gelatinous matrix referred to as a **cirrhus** (pl. **cirrhi**). Some acervuli have **setae** or sterile hairs (Figure 3.5).

Conidia may be borne singularly or in clusters on branched or unbranched conidiophores. A number of conidiophores may be fused at the base to form a structure known as a **synnema** (pl. **synnemata**) (Figure 3.6A). In other fungi, short conidiophores may be borne on mats of hyphae known as **sporodochium** (pl. **sporodochia**) (Figure 3.6B).

Fungi also form a number of asexual survival structures. **Chlamydospores** are thick-wall resting spores; some may even have a double cell wall. **Sclerotia** (sing.

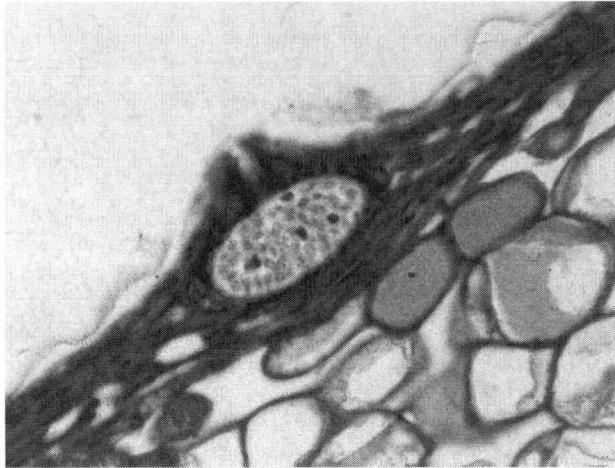

FIGURE 3.4 Pycnidium of *Phoma* species. The hole at the top of the structure is referred to as the ostiole.

FIGURE 3.5 An acervulus of a *Colletotrichum* species with many setae or sterile hairs.

sclerotium) are comprised of hyphae that are so tightly packed that they have lost their individuality. They are very hard and extremely resistant to harsh environmental conditions.

Fungi are usually classified by their **teleomorph** (sexual) stage. In some fungi, **gametes** (sex cells) unite to produce a zygote (Figure 3.7). This is usually how more primitive fungi reproduce. The fusion of gametes that are of equal size and appearance results in a zygote referred to as a **zygospore** and are classified in the Zygomycota (Chapter 10).

In other fungi, there are no definite gametes and instead one mycelium may unite with another compatible mycelium. In the **Ascomycota**, **ascospores**, usually eight in number, are produced within a zygote cell, the **ascus** (pl. **asci**) (Figure 3.8). Asci may be produced naked (not in

any structure) or in specialized structures. **Cleistothecia** (sing. **cleistothecium**) are enclosed structures (Figure 3.9) with asci located at various levels. Cleistothecia usually do

FIGURE 3.6 (A) Fused conidiophores comprise a synnema of *Graphium* species. (B) Sporodochium of an *Epicoccum* species is a mat of densely packed, short conidiophores.

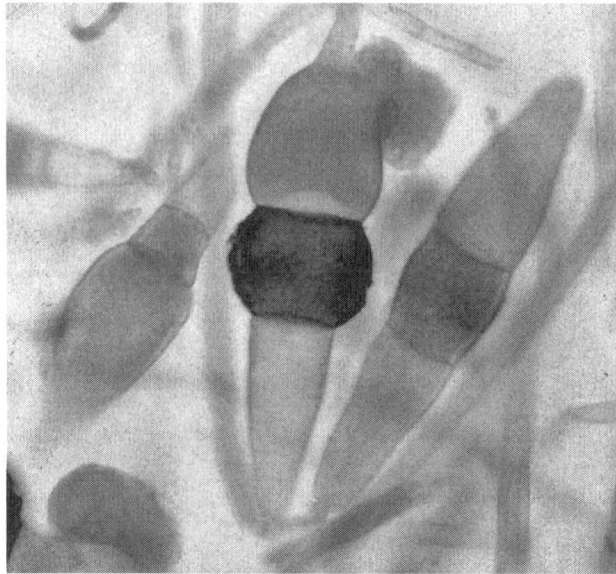

FIGURE 3.7   Zygosprorangium containing a single zygospore (zygote) of *Rhizopus niger*.

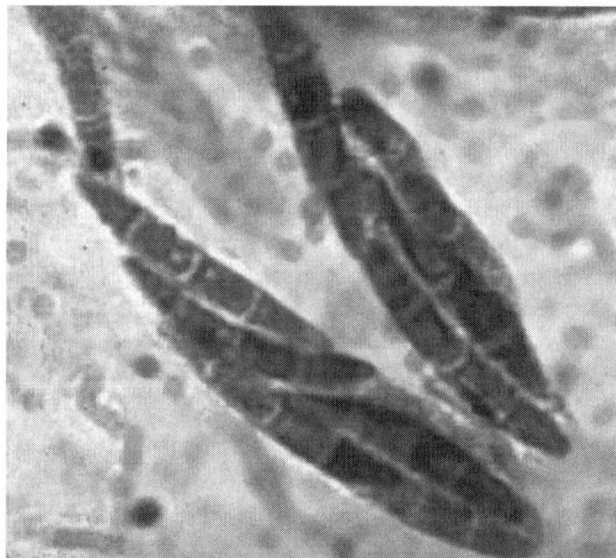

FIGURE 3.8   Asci containing spindle-shaped, multicellular ascospores of *Gibberella* species.

not have openings, and asci and ascospores are usually not released until the cleistothecium ruptures or is eroded by the environment. Ascocarps of powdery mildew have been referred to traditionally as cleistothecia (Figure 3.9), but are now termed perithecia (Chapter 14). **Perithecia** (sing. **perithecium**) are usually flask-shaped structures with an opening (Figure 3.10). Asci and ascospores are formed in a single layer. Ascospores are pushed or forcibly ejected through the opening. They are dispersed via air currents, insects, and water. Some members of the phylum form asci in open, cup-shaped structures known as **apothecia** (sing. **apothecium**) and ascospores are disseminated by modes similar to found fungi that produce perithecia.

In other fungi, sexual spores are produced on the outside of the zygote cell or **basidium** (pl. **basidia**) and are called basidiospores (usually four in number). Fungi that reproduce in this manner are placed in the **Basidiomycota**. In this group, the basidia and basidiospores may be borne naked (rusts and smuts) (Chapter 18) or formed in structures such as mushrooms, puffballs, and conks. Mushrooms are fleshy, sometimes tough, umbrella-like structures, whereas puffballs are white to light tan (dark brown to black, when mature) spongy, spherical bodies formed on the soil surface. Conks are shelf or very hard, bracket-like fruiting bodies and are usually found on stumps, fallen logs, or living trees (Chapter 19).

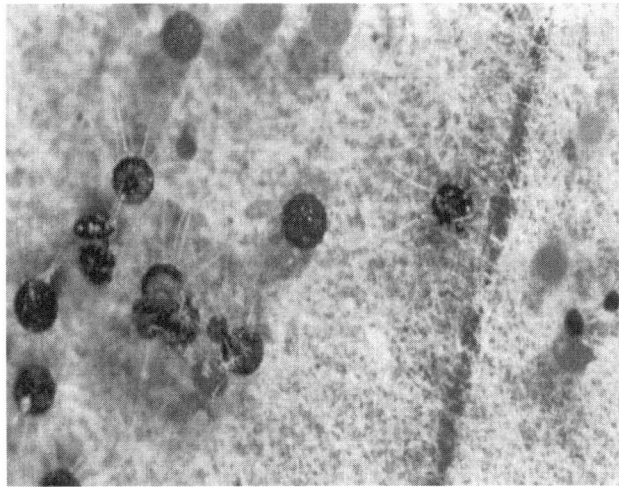

FIGURE 3.9 Cleistothecia (ascomata) of *Erysiphe* species.

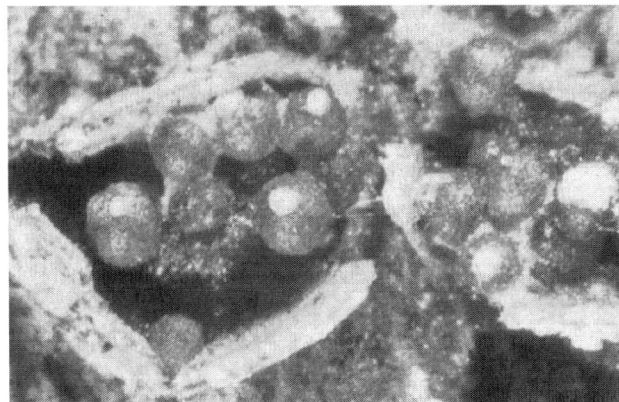

FIGURE 3.10 Perithecia of *Nectria coccinea* var. *faginata*.

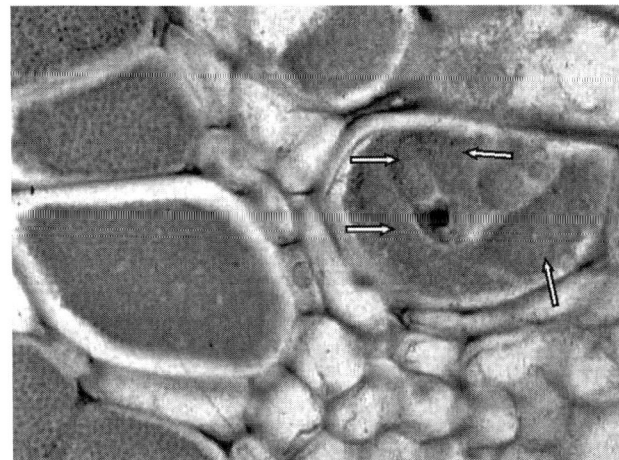

FIGURE 3.11 Plasmodia of *Plasmodiophora brassicae* in a cabbage cell. Each arrow points to a single plasmodium.

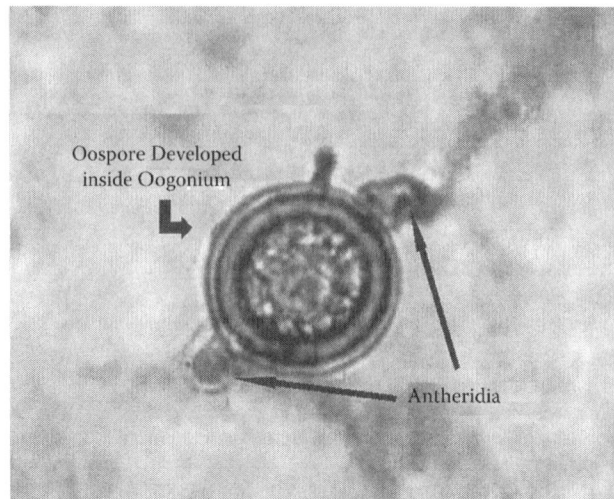

FIGURE 3.12 Antheridia and oospore inside oogonium in a *Pythium* species.

PLASMODIOPHOROMYCOTA—PROTOZOA

Parasitic slime molds are placed in the kingdom Protozoa and the phylum **Plasmodiophoromycota** (Chapter 11). They are unicellular and produce **plasmodia** in root cells (Figure 3.11). Plasmodia are amoeba-like cells without cell walls inhabiting the lumens of host cells. Parasitic slime molds produce zoospores that can function as gametes.

STRAMENOPILA—FUNGI-LIKE ORGANISMS

Other fungal-like organisms are found in the kingdom Stramenopila (Chromista) and include those in the phylum Oomycota (Chapter 20). These organisms were traditionally characterized as fungi because of filamentous growth, lack of chlorophyll, and reproduction by spores. However, with the advent of modern molecular techniques, they are now classified in a different kingdom, which includes the brown algae. These pathogens cause some of the most destructive plant diseases and include the downy mildews and species in the genera *Phytophthora* and *Pythium*. They produce anamorphic spores in sporangia, and the spores may be motile (zoospores). The gametes are of different size and shape—**antheridia** (male) and **oogonia** (female)—as illustrated in Figure 3.12. The sexual spore is an oospore and functions as a survival structure. Members of the Oomycota have cellulose in their cell walls and the majority of their life cycle is diploid (2N).

BACTERIA AND MOLLICUTES

Some of the most important plant diseases are caused by prokaryotic organisms such as bacteria and mollicutes (Chapter 6). Bacteria are prokaryotic (have no nucleus or double membrane-bound organelles) and have a rigid cell wall that is enveloped in a slime layer. Most of the DNA in bacteria is present as a single circular chromosome. Additional DNA is found in many bacteria as independently reproducing plasmids composed of smaller amounts of DNA. Most plant pathogenic bacteria are gram negative with the exception of *Clavibacter* (*Corynebacterium*). Phytopathological bacteria are either rod- or filamentous-shape, may or may not be flagellated, and reproduce by binary fission. Traditionally, bacteria were classified based on Gram stain, cell shape, cultural morphology, and substrate utilization. Today, bacteria are grouped using molecular analysis of genetic material.

Pathogenic bacteria are known as wound pathogens because they usually penetrate the host directly. They may also enter through natural plant openings such as nectaries, hydathodes, and stoma. They are disseminated on air currents, by water and insects, and on plant materials and contaminated equipment.

Mollicutes are smaller than bacteria, do not have a cell wall, and are delimited only by a plasma membrane. Most of the mollicutes are round or elongated and are referred to as phytoplasms. A few members of this group have a helical form and are termed spiroplasms. They are very difficult to culture. Some of the more important diseases caused by this group include aster yellows and X-disease of peach and apple. They are typically disseminated by insects, budding, and grafting.

VIRUSES AND VIROIDS

Viruses and viroids are much smaller than bacteria, cannot be seen with light microscopy, and require the host plant's replication machinery for multiplication. Viruses are nucleoproteins; their nucleic acid (either DNA or RNA) is surrounded by a protein coat. Viruses may or may not be encapsulated with a lipid layer. In some viruses the genome is spread between more than one particle. Viruses may be spherical or shaped like long or short rods that may be rigid or flexible. They may be very sensitive to environmental conditions such as heat and light or may be very stable under most environmental conditions. Viruses are disseminated by budding, grafting, wounding, insects, or infected plant materials. **Viroids** have many attributes of viruses, but differ in being naked strands of RNA that do not have a protein coat.

NEMATODES

Plant parasitic nematodes are small worm-like animals that have a cuticle made of chitin and a piercing mouth-part called a **stylet**, whereas free-living species, which are more common in soil samples, do not have stylets (Chapter 8). Nematodes vary in shape from being very elongated, kidney-shaped, or globose. They reproduce sexually or parthenogenetically as many species do not have males. Plant pathogenic nematodes may be migratory and ecto-parasitic (feed from outside of the root) or endoparasitic (feed inside the root). In some cases they are sedentary. Plant nematodes are disseminated in water, soil, plant materials, by insects, and on contaminated equipment. Although most plant-disease-causing nematodes are parasitic on roots, some nematodes are parasitic on aerial portions of plants.

OTHER PATHOGENS

Other types of pathogens include parasitic seed plants (Chapter 24), flagellate protozoa, algae, and mosses. Some of these pathogens (parasitic seed plants, such as dwarf mistletoe and flagellate protozoa) may cause severe crop losses. Diseases caused by algae and mosses seldom cause damage that lead to economic losses.

This brief chapter has only touched upon the incredibly diverse nature of organisms that cause plant diseases. Plant pathologists, especially Extension pathologists, must be well-versed in many different types of organisms. The succeeding chapters will explore these organisms and diseases more fully and hopefully whet your appetite for more advance study.

SUGGESTED READING

Agrios, G.N. 2005. *Plant Pathology*. 5th ed. Academic Press. San Diego, CA. 952 p.

Alexopoulos, C.J., C.W. Mims, and M. Blackwell. 1996. *Introductory Mycology*. 4th ed. John Wiley & Sons. New York. 868 p.

Horsfall, J.G. and E.B. Cowling. (Eds.) 1977–1980. *Plant Disease*, Vol. 1–5. Academic Press. New York.

Tainter, F.H. and F.A. Baker. 1996. *Principles of Forest Pathology*. John Wiley & Sons. New York.

Part 2

Groups of Plant Pathogens

4 Plant Pathogenic Viruses

Marie A.C. Langham

CHAPTER 4 CONCEPTS

- Viruses are unique, submicroscopic obligate pathogens.

- Viruses are usually composed of RNA or DNA genomes surrounded by a protein coat (capsid).

- Plant viruses replicate through assembly of previously formed components, and replication is not separated from the cellular contents by a membrane.

- Plant virus species are named for the host with which they were originally associated and the major symptom that they cause. Virus species may be grouped into genera and families.

- Plant viruses are vectored by insects, mites, nematodes, parasitic seed plants, fungi, seed, and pollen.

- Plant viruses can be detected and identified by biological, physical, protein, and nucleic acid properties.

PLANT VIRUSES IMPACT CROPS AND THEIR CONTROL FOCUSES ON HOST RESISTANCE

Plant virology is one of the most dynamic research areas in phytopathology. During the last quarter of the 20th century, our understanding of plant viruses and their pathogenic mechanisms has exceeded the imagination of early virologists. Today, new plant viruses are identified rapidly, and our awareness of their pathological impact continues to increase. This impact is most clearly seen in yield and other economic losses. Plant viruses generate economic loss for farmers, producers, and consumers by adversely affecting plant growth and reproduction, causing death of host tissues and plants, sterility, reduction of yield or quality, crop failure, increased susceptibility to other stresses, loss of aesthetic value, quarantine and eradication of infected plants, and the cost of control and detection programs (Waterworth and Hadidi, 1998). Viruses are also unique in the deceptive simplicity of their structure. However, this simplicity leads to a greater dependency on the host, and a highly intricate relationship exists between the two. This complicates strategies for control of plant viruses and the losses caused by them. Control programs depend on our understanding of the virus-host relationship, and control remains one of the greatest challenges for the future of plant virology.

HOW ARE VIRUSES NAMED?

Plant viruses are typically named for the host that they were infecting when originally described, and for the principal symptom that they cause in this host. The word *virus* follows

these two terms. For example, a virus causing a mosaic in tobacco would be *Tobacco mosaic virus* (TMV). This is the species name for the virus. The use of the species concept in plant virology began in recent years following much debate concerning what constitutes a virus species (van Regenmortel et al., 2000). Following the first use of the species name, the virus is referred to by the abbreviation that is given in parenthesis after the first use of the species name. Two levels of taxonomic structure for grouping species are the genus, which is a collection of viruses with similar properties, and the family, which is a collection of related virus genera. *Cowpea mosaic virus* (CPMV) is a member of the genus *Comovirus* and the family *Comoviridae* (van Regenmortel et al., 2000). Table 4.1 lists some of the virus genera and families. Virus species may also be subdivided into **strains** and **isolates**. Strains are named when a virus isolate proves to differ from the type isolate of the species in a definable character but does not differ enough to be a new species (Matthews, 1991). For example, a virus strain may have altered reactions in an important host, such as producing a systemic reaction in a host that previously had a local lesion reaction, or the strain may have an important serological difference. Strains represent mutations or adaptations in the type virus. Isolates are any propagated culture of a virus with a unique origin or history. Typically, they do not differ sufficiently from the type isolate of a virus to be a strain.

WHAT IS A PLANT VIRUS?

Plant viruses are a diverse group infecting hosts from unicellular plants to trees. Despite this diversity, plant

TABLE 4.1
Families and genera of some plant-infecting viruses with their morphology and basic genome characters.
Genera that have been grouped into families are listed under the family*

Family or Genus	Morphology	Nucleic Acid	Number of Strands	Vector
		Single-Strand RNA Viruses		
Benyvirus	Rigid rod	+ Linear	4 or 5	Fungi
Bromoviridae				
Alfamovirus	Bacilliform; icosahedral	+ Linear	3	Aphids
Bromovirus	Icosahedral	+ Linear	3	Beetles
Cucumovirus	Icosahedral	+ Linear	3	Aphids
Ilarvirus	Quasi-isometric	+ Linear	3	Thrips and pollen
Oleavirus	Quasi-isometric to bacilliform	+ Linear	3	Mechanical
Bunyaviridae				
Tospovirus	Spherical[a]	+/- Circular	3	Thrips
Cheravirus	Icosahedral	+Linear	2	Nematodes
Closteroviridae				
Ampelovirus	Filamentous	+ Linear	1	Mealybugs
Closterovirus	Filamentous	+ Linear	1	Aphids
Crinivirus	Filamentous	+ Linear	2	Whiteflies
Comoviridae				
Comovirus	Icosahedral	+ Linear	2	Beetles
Fabavirus	Icosahedral	+ Linear	2	Aphids
Nepovirus	Icosahedral	+ Linear	2	Nematodes, seed, and pollen
Furovirus	Rigid rod	+ Linear	2	Fungi
Flexiviridae				
Allexivirus	Filamentous	+ Linear	1	Eriophyid mites
Capillovirus	Filamentous	+ Linear	1	Seed and grafting
Carlavirus	Filamentous	+ Linear	1	Aphids, whiteflies, mechanical, and seed in some
Foveavirus	Filamentous	+ Linear	1	Graft
Mandarivirus	Filamentous	+ Linear	1	Graft
Potexvirus	Filamentous	+ Linear	1	Mechanical
Trichovirus	Filamentous	+ Linear	1	Grafting and seed
Vitivirus	Filamentous	+ Linear	1	Grafting and mealybugs; Scales and aphids in some
Hordeivirus	Rigid rod	+ Linear	3	Seed and pollen
Idaeovirus	Icosahedral	+ Linear	3	Seed and pollen
Luteoviridae				
Enamovirus	Icosahedral	+ Linear	1	Aphids
Luteovirus	Icosahedral	+ Linear	1	Aphids
Polerovirus	Icosahedral	+ Linear	1	Aphids
Ophiovirus	Filamentous	- Circular	3-4	Vegetative propagation
Ourmiavirus	Bacilliform	+ Linear	3	Mechanical and seed
Pecluvirus	Rigid rod	+ Linear	2	Fungi and seed
Pomovirus	Rigid rod	+ Linear	3	Fungi
Potyviridae				
Bymovirus	Filamentous	+ Linear	2	Fungi
Ipomovirus	Filamentous	+ Linear	1	Whiteflies and grafting
Macluravirus	Filamentous	+ Linear	1	Aphids
Potyvirus	Filamentous	+ Linear	1	Aphids
Rymovirus	Filamentous	+ Linear	1	Eriophyid mites

TABLE 4.1 (Continued)
Families and genera of some plant-infecting viruses with their morphology and basic genome characters. Genera that have been grouped into families are listed under the family

Family or Genus	Morphology	Nucleic Acid	Number of Strands	Vector
Tritimovirus	Filamentous	+ Linear	1	Eriophyid mites
Rhabdoviridae				
Cytorhabdovirus	Bullet	- Linear	1	Leafhoppers, planthoppers, and aphids
Nucleorhabdovirus	Bullet	- Linear	1	Leafhoppers and aphids
Sadwavirus	Icosahedral	+ Linear	2	Nematodes, aphids, and seeds
Sequiviridae				
Sequivirus	Icosahedral	+ Linear	1	Aphids when helper virus is present
Waikavirus	Icosahedral	+ Linear	1	Leafhoppers and aphids
Sobemovirus	Icosahedral	+ Linear	1	Beetles
Tenuivirus	Filamentous	+/- Linear	4 or 5	Planthoppers
Tobamovirus	Rigid rod	+ Linear	1	Mechanical and seed
Tobravirus	Rigid rod	+ Linear	2	Nematodes
Tombusviridae				
Aureuvirus	Icosahedral	+ Linear	1	Mechanical
Avenavirus	Icosahedral	+ Linear	1	Mechanical; Possibly fungi
Carmovirus	Icosahedral	+ Linear	1	Beetles and seed; Fungi and cuttings in some
Dianthovirus	Icosahedral	+ Linear	2	Mechanical
Machlomovirus	Icosahedral	+ Linear	1	Beetles, seed, and thrips
Necrovirus	Icosahedral	+ Linear	1	Fungi
Panicovirus	Icosahedral	+ Linear	1	Mechanical, infected sod, and seed
Tombusvirus	Icosahedral	+ Linear	1	Mechanical and seed
Tymoviridae				
Marafivirus	Icosahedral	+ Linear	1	Leafhoppers and vascular puncture
Maculavirus	Icosahedral	+ Linear	1	Grafting and propagating material
Tymovirus	Icosahedral	+ Linear	1	Beetles
Umbravirus	RNP complex[b]	+ Linear	1	Aphids when helper virus is present
Variscosavirus	Rigid rod	- Linear	2	Fungi
Double-Stranded RNA				
Partiviridae				
Alphacryptovirus	Icosahedral	Linear	2	Seeds and pollen
Betacryptovirus	Icosahedral	Linear	2	Seeds and pollen
Reoviridae				
Fijivirus	Icosahedral	Linear	10	Planthoppers
Phytoreovirus	Icosahedral	Linear	12	Leafhoppers
Oryzavirus	Icosahedral	Linear	10	Planthoppers
Caulimoviridae				
Badnaviruses	Bacilliform	Circular	1	Mealybugs, aphids, and lacebugs
Caulimovirus	Icosahedral	Circular	1	Aphids
Cavemovirus	Icosahedral	Circular	1	Aphids
Petuvirus	Icosahedral	Circular	1	Aphids
Soymovirus	Icosahedral	Circular	1	Aphids
Tungrovirus	Bacilliform	Circular	1	Leafhoppers

Continued

TABLE 4.1 (Continued)
Families and genera of some plant-infecting viruses with their morphology and basic genome characters.
Genera that have been grouped into families are listed under the family

Family or Genus	Morphology	Nucleic Acid	Number of Strands	Vector
		Single-Stranded DNA		
Geminiviridae				
Begomovirus	Geminate	Circular	2	Whiteflies
Curtovirus	Geminate	Circular	1	Leafhoppers
Mastrevirus	Geminate	Circular	1	Leafhoppers and vascular puncture
Topocuvirus	Geminate	Circular	1	Treehoppers
Nanoviridae				
Babuvirus	Icosahedral	Circular	6	Aphids
Nanovirus	Icosahedral	Circular	8	Aphids; not mechanically transmissible

[*] Brunt et al., 1996; Fauquet et al., 2005; Hull, 2002; van Regenmortel et al., 2000.

[a] Spherical particles are more variable in size and morphology than icosahedral particles.

[b] RNP complexes are composed of nucleic acid inside a membrane. These particles lack the traditional protein coat unless they are in mixed infection with a helper virus and are able to utilize the helper's coat protein.

viruses share a number of characteristics. A good definition of plant viruses focuses on characteristics that all plant viruses have in common.

SIZE

Viruses are ultramicroscopic. Their visualization requires an electron microscope, which uses a beam of electrons instead of visible light. The narrower wavelength of the electron beam allows the resolution of smaller objects such as cell organelles and viruses.

GENOMES

Viruses have nucleic acid **genomes**. This nucleic acid may be either ribonucleic acid (RNA) or deoxyribonucleic acid (DNA). In plant viruses, no virus has been discovered that includes both types of nucleic acids. However, there are many variations in the structures of the viral genomes. The nucleic acid may be single-stranded (ss) or double-stranded (ds), and it may be linear or circular. The genome may be on a single piece of nucleic acid or on multiple pieces. These pieces may be encapsidated in a single particle or in multiple particles.

CAPSIDS

Viruses have one or more protein coats or **capsids** surrounding their perimeter. These capsid layers are composed of protein subunits. The subunits may be composed of the same type or different types of protein. For example, *Tobamoviruses* have one type of protein subunit in their capsids, whereas *Comoviruses* have two types of protein

subunits, and *Phytoreoviruses* have six to seven types of structural protein subunits in their capsids (van Regenmortel, 2000). Some viruses may also have a lipoprotein layer associated with them.

OBLIGATE PARASITES

Viruses are obligate parasites. The simplicity of the virus leads to its dependence on the host for many functions. Viruses have no systems for the accumulation of metabolic materials. They have no systems for energy generation (mitochondria), protein synthesis (ribosomes), or capturing light energy (chloroplasts). Thus, viruses are dependent on their host for all of these functions plus the synthesis of nucleic acids and amino acids.

CELLULAR ASSOCIATION

Viruses and their hosts have a more basic relationship than other pathogen systems. Viruses are not separated from their host by a membrane during replication (Matthews, 1991). Viruses infect the cellular structure of the host and control a part of the subcellular systems of the plant.

REPLICATION BY ASSEMBLY

Viral replication is dependent on assembly of new particles from pools of required components (Matthews, 1991). These components are synthesized as separate proteins or nucleic acids using the host enzyme systems and the infecting viral genome. New particles are then assembled using these materials. This type of replication contrasts strongly with binary fission or other methods of replication found in prokaryotic and eukaryotic organisms.

Vector

Plant viruses are not capable of causing an entry wound in order to infect a plant. Thus, they cannot disperse from plant to plant without the assistance of a vector. Plant viruses are dependent on vectors to breach the epidermal layer of the plant and to place them within a living host cell. Virus vectors include insects, mites, nematodes, fungi, seed, and dodder. It also includes man, animals, and other organisms that transfer viruses through mechanical transmission.

WHAT ARE IMPORTANT HISTORICAL DEVELOPMENTS IN PLANT VIROLOGY?

One of the first references to a disease caused by a plant virus occurred during Tulipomania (the craze in Europe for tulips just after their introduction from Turkey) in 17th-century Holland (1600 to 1660) (Matthews, 1991). This is an unusual case of a virus that increased the value of infected plants. It began with the importation of tulips into Holland from Persia. The beautiful flowers became quickly popular and were grown throughout Holland. People began noticing that some flowers were developing streaks and broken color patterns. These tulips were called bizarres, and people quickly learned that they could produce more bizarres by planting a bizarre tulip in a bed or by rubbing the ground bulb onto plain tulip plants. They did not realize that they were transmitting a pathogen. Bizarres became so popular that a single bizarre bulb was worth large sums of money, thousands of pounds of cheese, or acres of land. One case is recorded where a man offered his daughter in marriage in exchange for a single bizarre tulip bulb.

In 1886, Adolf Mayer scientifically confirmed a primary principle of plant virology when he transmitted TMV to healthy tobacco plants by rubbing them with sap from infected plants. The newly rubbed plants displayed the same symptoms as the original infected plants (Scholthof et al., 1999). Mayer's research established the contagious nature of plant viruses and the first procedure for mechanical transmission of a virus. Today, mechanical transmission enables virologists to transmit viruses for experimental purposes and for the evaluation of plants for resistance and tolerance to viral diseases.

The independent experiments of a Russian scientist, Dmitri Ivanowski, in 1892 and a Dutch scientist, M. W. Beijerinck, in 1898 first indicated the unique nature of viral pathogens. Both scientists extracted plant sap from tobacco plants infected with TMV. The sap was then passed through a porcelain bacterial filter that retained the bacteria and larger pathogens. If bacteria or other organisms were the cause of the disease, the filter would have retained the pathogen, and the sap that had been passed through the filter would not transmit the disease. When Ivanowski or Beijerinck filtered sap from TMV infected plants and inoculated the filtered sap onto healthy tobacco plants, the plants became diseased. Beijerinck recognized that this indicated the unique nature of this disease. Beijerinck declared that TMV was a new type of pathogen that he called a *contagium fluidium vivium* (a contagious living fluid) (Scholthof et al., 1999). Later scientists discovered that the ultramicroscopic particle nature of viruses was not truly fluid. However, this was the first indication that viral pathogens represented a new and unique type of pathogen.

Clues to this unique nature would wait to appear until the research of W. M. Stanley in 1935. Working with TMV, Stanley extracted gallons of plant sap and used the newly developed technique of fractionation by precipitation of proteins with salts and other chemicals. Each fraction was tested by inoculation on susceptible host to determine where the infectivity remained. Gradually, Stanley isolated the infective fraction into a pure form in which he crystallized TMV. Stanley was the first person to purify a virus, and in 1946, he was awarded the Nobel Prize for this accomplishment (Scholthof, 2001).

Stanley's first analyses of purified TMV solution only found protein. In 1936, Bawden and Pirie found the presence of phosphorous that indicated that the solution contained nucleic acid (Bawden et al., 1936). Fraenkel-Conrat was able to isolate the RNA from TMV in 1956 and used it to infect healthy tobacco plants, proving that the RNA was the source of infectivity and establishing that RNA contained TMV's genome (Frankel-Conrat, 1956; Creager et al., 1999).

This section has discussed only a few of the most important historical principles in plant virology. However, the brevity of this section should not be used to judge the importance of historical research. The theory and accomplishments of today's research in plant virology is built upon the accomplishments of many researchers who preceded today's researchers, and their contributions are the foundation for tomorrow's research.

WHAT ARE THE SYMPTOMS FOUND IN PLANTS INFECTED WITH VIRUS?

Symptoms, the host's response to infection, are the signal that attracts the attention of pathologist, farmer, producer, or homeowner. However, viruses are sometimes referred to as the great imposters because symptoms are often mistaken for other diseases or conditions. Viral diseases may be misidentified as nutritional deficiencies, genetic abnormalities, mineral toxicities, pesticide damage, environmental stresses, insect feeding, or infection by other classes of plant pathogens. Symptomatology can provide strong first indications of a possible virus infection. However, utilization of symptomatology as the sole basis for diagnosis should be avoided due to the confusion with other conditions.

Classic symptoms of virus infections can be grouped by their similarities. These include the following categories.

SYMPTOMS CAUSED BY LOCALIZED INFECTION

Local Lesions

Local lesions occur when the virus infection fails to spread systemically due to the host response. The virus may overcome this initial reaction to spread systemically, or it may never spread beyond the initial infection site. Local lesions may be either chlorotic or necrotic. Local lesions can often be utilized to quantify the infectivity in a virus solution.

SYMPTOMS BASED ON CHANGES IN CHLOROPHYLL OR OTHER PIGMENTS

Mosaics and Mottles

Patterns of lighter and darker pigmentation are referred to as mosaic or mottles. Areas of lighter pigmentation may be pale green, yellow, or white, and are caused by decreases in chlorophyll, decrease or destruction of the chloroplasts, or other damage to the chlorophyll system. Other theories suggest that mosaics and mottles may also be due to increased pigmentation or stimulation in the areas of darker green. TMV is a classic example of a virus that can produce mosaic in systemically infected hosts.

Stripes and Streaks

When mosaics occur on monocots, the changes in pigmentation are restricted in spread by the parallel venation of the leaves. Thus, the mosaic appears as a stripe or streak either in short segments (broken), or they may extend the entire length of the leaf. Many cereal viruses, such as *Wheat streak mosaic virus* (WSMV) or *Maize dwarf mosaic virus*, produce streaks or stripes.

Ringspots and Line Patterns

Ringspots form as concentric circles of chlorotic or necrotic tissue. *Tomato spotted wilt virus* (TSWV) produces vivid ringspots in many hosts. Line patterns are extensions of this response. They occur as symptomatic patterns near the edge of the leaf and often follow the outline of the leaf. For example, *Rose mosaic virus* (RMV) produces a striking yellow line pattern on rose.

Vein Banding

Vein banding occurs when areas of intense pigmentation form bordering the veins of the leaves. These typically are seen as dark green bands along major leaf veins.

Vein Clearing

Vein clearing is a loss of pigmentation (clearing or translucence of tissue) in the veins and can be best observed by allowing light to shine through the leaf. This clearing is caused by enlargement of the cells near the vein in some viruses (Hull, 2002). Vein clearing sometimes precedes the formation of a mosaic.

SYMPTOMS THAT ARE CAUSED BY GROWTH ABNORMALITIES

Stunting and Dwarfing

Reduction in the size of the infected host plant (stunting or dwarfing) can be one of the most prominent symptoms resulting from a viral infection. Some viruses may stunt the plant only slightly while others affect the host plant dramatically. Stunting may include changes in sizes of all plant parts such as leaves in addition to height. Stunting that includes shortening of the internodes is often referred to as a bushy stunt.

Tumors and Galls

Hyperplasia (enlargement in cell size) and hypertrophy (increase in cell number) of infected cells can result in large overgrowths or tumors. *Wound tumor virus* (WTV) is an example of a virus that produces tumors in its host.

Distortion

Changes in the lamina of the host leaves results in areas that are twisted, deformed, or distorted. The deformations may be described as blisters, bubbles, rumpling, rugosity, or twisting. *Bean pod mottle virus* (BPMV) produces distortion in infected soybean leaves.

Enations

Small overgrowths occurring on the leaf are called enations. *Pea enation mosaic virus* (PEMV) is named for the enations that it produces.

SYMPTOMS AFFECTING REPRODUCTION

Sterility

Host plants infected with plant viruses may lose their ability to produce viable seed. This may be linked with floral abnormalities, decreases in flowering, seed development, or seed set. The production of sterility is also linked with changes in the plant's metabolism and changes in its biochemical signaling.

Yield Loss

Yield loss can take many forms. It may be a reduction in the total reproduction of the plant, or seed or fruit produced

TABLE 4.2
Summary of Mechanical Transmission of Plant Viruses

- Mechanical transmission allows the transmission of plant viruses without a vector.
- An abrasive is used to make wounds that penetrate the epidermal layers, cell wall, and cell membrane. Silica carbide is the most widely used abrasive, but sand, bentonite, and celite have also been utilized.
- These wounds must not cause the cell to die, because the viruses are obligate parasites and require a living cell. This type of wounding is described as nonlethal.
- Mechanical transmission is effective for viruses that infect epidermal cells. Other viruses, which are limited to other tissues such as phloem, are not mechanically transmissible.
- Some viruses that are highly stable, such as TMV, may be mechanically transmitted through accidental contact between healthy and infected plants, wounding the plants and allowing infected plant sap to be transmitted to the healthy plants. Also, contacting healthy plants after handling plants infected with stable viruses may allow you to become the vector by transmitting infected plant sap to the healthy plants.
- Mechanical transmission allows scientists to study the effects of plant viruses with using the vector. It is also used in the evaluation of new plant cultivars for viral disease resistance.

may be shriveled, reduced in size, distorted, or inferior in quality. WSMV is a good example of a virus that causes yield loss due to undersized and shriveled grain in addition to reducing the total number of seeds set. *Plum pox virus* (PPV) is an example of a virus that not only disfigures the fruit, but also reduces the carbohydrate level in the fruit.

WHAT DO VIRUSES LOOK LIKE?

The shape of virons is one of the most fundamental properties of the virus. Plant viruses are based on four types of architecture or morphology. Icosahedral (isometric) viruses seem to be basically spherical in shape. However, on closer examination, they are not simply smooth but are faceted. These viral structures are suggestive of the geodesic domes designed by Buckminster Fuller (Morgan, 2006). Icosahedral viruses have twenty facets or faces. Second, rigid rod viruses are all based on protein coats surrounding a helical nucleic acid strand. Rigid rod viruses are typically shorter and have a greater diameter than flexuous rods. Also, the central canal, an open region in the center of the viral helix, is more apparent in rigid rod viruses. Third, flexuous rods are typically very flexible and may bend into many formations. They are narrow in diameter and are longer than rigid rods. Lastly, bacilliform viruses are short thick particles (rods) that are rounded on both ends. When particles are found with only one rounded end, they are referred to as bullet shaped.

HOW ARE VIRUSES TRANSMITTED?

Plant viruses do not have the ability to penetrate the plant cuticle, epidermis, and cell wall. For experimental transmission, viruses are often transferred using mechanical methods. This is discussed in more detail in Table 4.2 and diagrammatically shown in Figure 4.1. In nature, viruses depend on vectors to breach these defenses and to allow entry to living cells. Vectors may be insects, mites, nematodes, fungi, or parasitic seed plants. Each virus evolves a unique and specific relationship with its vector. Viruses are

dependent on this complex interaction and have developed many methods for capitalizing on the biology of their vectors. Thus, understanding the relationship between virus and vector is vital in developing control programs.

INSECTS AS VECTORS

Aphids

Aphids transmit more viruses than any other vector group. Viruses have developed four types of interactions with aphids. These interactions are dependent on the plant tissue infected, the virus association with the vector, and virus replication (or lack of replication) in the vector (Table 4.3).

NONPERSISTENT TRANSMISSION

Viruses that are transmitted in a nonpersistent manner infect the epidermal cells of the host plant. This type of transmission is dependent on the sampling behavior of the aphid that quickly probes in and out of the epidermal cells in order to determine host suitability. The virus forms a brief association with two sites. The first site is at the tip of the **stylet**, and the second is found just before the **cibereal pump** at the top of the stylet. This association lasts only as long as the next probe of the aphid when it is flushed from the stylet during the process of **egestion**. Nonpersistent aphid transmission requires only seconds for acquisition and transmission, and is increased by preaquisition starvation or by any other condition that increases sampling behavior of the aphid. Evidence suggests that in some nonpersistenly transmitted viruses, such as *Cucumoviruses*, the ability to bind to the aphid is dependent on a property of the coat protein, such as conformation or the binding of metal ions (Ng and Falk, 2006). Nonpersistent transmission in some viruses requires the presence of an additional protein called the helper component, which is theorized to assist in the binding of the virus to the aphid stylet. This binding is referred to as the bridge hypothesis (Ng and Falk, 2006). *Potyviruses* are among the most important viruses transmitted in a nonpersistent manner.

**MECHANICAL
INOCULATION**

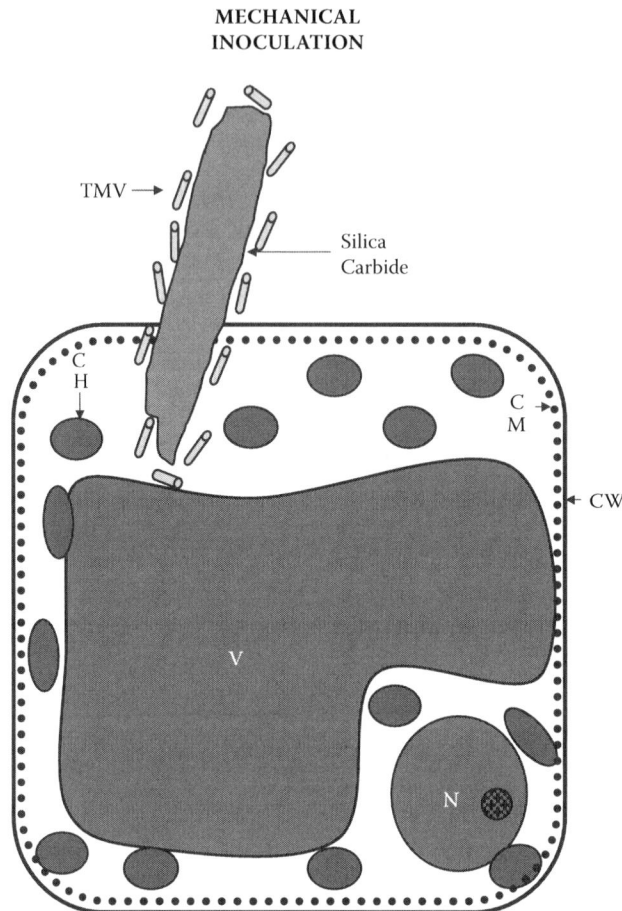

FIGURE 4.1 Mechanical inoculation of plant viruses requires wounding of the plant cell by an abrasive. Silica carbide particles rupture the cell wall (CW) and cell membrane (CM) and allow virus to enter the cell. Other structures shown in the cell include the nucleus (N), chloroplasts (CH), and central vacuole (V).

SEMIPERSISTENT TRANSMISSION

Semipersistent viruses form an association with the lining of the aphid foregut. The acquisition of these viruses requires phloem feeding, leading to longer acquisition times than nonpersistently transmitted viruses; transmission requires minutes, and retention of the virus typically lasts for hours. Virus retention does not last through the aphids' developmental molts as the lining of the foregut is shed with the rest of the cuticular exoskeleton. Semipersistent transmission may require the presence of a helper component or a helper virus (Hull, 2002). *Cauliflower mosaic virus* (CaMV) is semipersistently transmitted by aphids and depends on coat protein and two nonviron proteins in transmission (Ng and Falk, 2006).

PERSISTENT CIRCULATIVE
NONPROPAGATIVE TRANSMISSION

All viruses transmitted in this manner must be taken from the phloem of an infected plant and placed in the phloem of a healthy plant. Thus, they are dependent on the phloem-probing behavior of aphids. Approximately 20 minutes are required for aphids to establish phloem probes. Thus, the minimum acquisition and transmission times for this type of transmission are 20 minutes. Viruses transmitted in a persistent circulative nonpropagative manner form a close association with their aphid vectors. The virus moves through the intestinal tract to the hindgut where it passes into the aphid's **hemolymph** and begins to circulate throughout the hemocoel. The virus moves from the hemoceol into the **salivary glands** by passing through the basal membrane of the salivary membrane. The virus is then injected into the healthy plant with the saliva during egestion. Viruses transmitted in this manner are retained for days to weeks. Retention time is correlated with the amount of virus in the hemoceol and the level of virus in the hemolymph is related to the acquisition time that the vector feeds on the infected host. *Barley yellow dwarf virus* (BYDV) and *Cereal yellow dwarf virus* (CYDV), two of the most widely distributed viruses in the world, are transmitted in this manner.

TABLE 4.3

Comparison of Aphid Transmission Characteristics

	Characteristic Types of Transmission			
	Nonpersistent	Semipersistent	Persistent Circulative Nonpropagative	Persistent Circulative Propagative
Tissue infected by transmitted virus	Epidermis	Mesophyll and phloem	Phloem	Phloem
Virus interaction sites	Stylet tip and preciberium	Foregut	Hindgut and salivary glands	Hindgut and salivary glands; also infects many tissues of the host
Type of feeding behavior associated with transmission	Sampling	Phloem probing	Phloem probing	Phloem probing
Acquisition time	Seconds	Minutes to hours	20 min to hours	20 min to hours
Inoculation time	Seconds	Minutes to hours	20 min to hours	20 min to hours
Retention time	Minutes to hours	Hours to days	Days to life	Days to life
Latent period	No	No	Yes (hours)	Yes (days to weeks)
Retained through molt	No	No	Yes	Yes
Found in the hemolymph	No	No	Yes	Yes
Replicates and infects the host tissues	No	No	No	Yes
Transovarial passage—infects offspring	No	No	No	Yes

PERSISTENT CIRCULATIVE PROPAGATIVE TRANSMISSION

Persistent circulative propagative transmission has many characteristics in common with persistent nonpropagative viruses. Viruses transmitted in this manner are phloem-limited viruses. Transmission is dependent on phloem probes and passage of the virus into the hemolymph. However, after entering the hemolymph, the virus infects and replicates in the aphid. Many tissues of the aphid may be infected. The virus can also pass through the ovaries into the offspring, which are **viruliferous** when they are born. This is termed transovarial passage of the virus. Some viruses in the *Rhabdoviridiae*, such as *Lettuce necrotic yellows virus* (LNYV), are examples of viruses that multiply in their aphid vector (Hull, 2002).

LEAFHOPPERS

Leafhopper transmission of viruses highly parallels aphid transmission with one exception. Nonpersistent transmission does not occur in leafhoppers. Leafhoppers transmit viruses by semipersistent transmission, persistent circulative nonpropagative transmission, or persistent circulative propagative transmission. Important examples of viruses transmitted by leafhoppers include *Beet curly top virus* (BCTV), *Potato yellow dwarf virus* (PYDV), and *Maize stripe virus* (MSV).

BEETLES

Beetle transmission is unique among all the other types of virus transmission in that the specificity of virus transmission is not in the ability of the beetle to acquire the virus, but in the interaction of the virus and host after transmission. Beetles can acquire both transmissible and nontransmissible viruses. Both types of viruses may be found in the hemolymph of some beetles. Other viruses are found only in the gut lumen and mid-gut epithelial cells (Wang et al., 1994). However, the specificity of transmission does not depend on these factors. Beetles spread a layer of predigestive material known as **regurgitant** on the leaves as they feed. This layer contains high concentrations of deoxyribonucleases, ribonucleases, and proteases. When viruliferous beetles spread this layer, they also deposit virus particles in the wound at the feeding site. Beetle transmissible viruses are able to move through the vascular system to an area away from the wound site with its high level of ribonuclease in order to establish infection (Gergerich et al., 1986; Gergerich and Scott, 1988a; Gergerich and Scott, 1988b). Nontransmissble viruses are retained at the wound site where the ribonuclease levels inhibit their ability to infect the plant (Field et al., 1994). *Comovirus* and *Sobemovirus* are among the most important viral genera transmitted by beetles.

WHITEFLIES

The mode of transmission of viruses in whiteflies varies with the genus of virus. *Begomovirus* is transmitted in a persistent circulative manner resembling aphid transmission of *Luteovirus*. However, this relationship may be more complex than it appears due to the extended retention lengths and the transovarial passage of some species of this viral genus (Hull, 2002). In contrast, *Clo-*

steroviruses and *Criniviruses* are transmitted in a fore-gut-borne semipersistent manner (Hull, 2002). Activities of different coat proteins or other proteins may also be necessary for transmission. *Lettuce infectious yellows virus* (LIYV), a *Closterovirus*, has a minor coat protein (CPm) that is necessary for transmission (Ng and Falk, 2006). Regardless of which virus is transmitted or what components are required for transmission, whiteflies present constant challenges as virus vectors due to their dynamic population increases, resistance to control, and changes in their biotype.

THRIPS

Virus transmission by thrips has been a dynamic area of research in recent years. The rapid increase in the importance of *Tospovirus* and the diseases that they cause in both greenhouses and fields has stimulated much of this research. *Tospovirus* is persistently and propagatively transmitted by thrips. Thrip **larvae** acquire the virus while feeding on virus-infected tissue, and the virus crosses through the midgut barrier and enters the salivary glands. The virus must be acquired by immature thrips because adult thrips cannot acquire the virus (Moyer et al., 1999); the virus passes from larvae to adult thrip as it undergoes pupation and the changes associated with maturity. This is known as transstadial passage (Whitfield et al., 2005). There is no evidence for virus in thrip hemolymph. Thus, viruses cannot travel the same path through the thrip that persistent circulative viruses use in aphids and leafhoppers (Hull, 2002). Alternate routes that have been proposed for viral movement in thrips include the tubular salivary glands and ligament-like structures that connect the midgut to the primary salivary glands or the changes in organ proximity during the thrip's development may cause direct contact between organs, which are widely separated at maturity (Whitfield et al., 2005). Thrips retain infectivity for their lifetime, and the virus titer has been shown to increase as the virus replicates in the thrip. No evidence of virus passage through the egg (transovarial passage) has been found (Moyer et al., 1999), but question of whether or not *Tospoviruses* are pathogenic to thrips is an active research area (Whitfield et al., 2005).

Thrips have also been shown to transmit three viral genera, *Ilarvirus, Sobemovirus*, and *Carmovirus*, by movement of virus-infected pollen. The virus from the infected pollen is then transmitted to the host plant through wounds caused by thrip feeding (Hull, 2002).

OTHER VECTORS

Mites

Eriophyid mites are tiny **arthropods** (0.2 mm length) known to transmit several plant viruses, including WSMV. Mites acquire virus during their larval stages. As in thrips, adult mites cannot acquire the virus, but both the larvae and adults transmit the virus (Slykhuis, 1955). Mites have been shown to remain infective for over two months (Hull, 2002). WSMV particles have been found in the midgut, body cavity, and salivary glands of the mite (Paliwal, 1980). However, there has been no evidence to prove replication of virus in the mite.

Nematodes

Nematodes that transmit plant viruses are all migratory ectoparasites. Three genera of nematodes, *Longidorus, Xiphinema,* and *Trichodorus*, are primarily associated with transmission of viruses. Nematodes feeding on virus-infected plants retain virus on the stylet, buccal cavity, or esophagus. When the nematodes are feeding on healthy host plants, the retained virus is released into the feeding site to infect the new host. Viruses in the *Tobravirus* and the *Nepovirus* genera are transmitted by nematodes.

Fungi and Fungi-Like Organisms

The Chytridiomycete, *Olpidium*, and Plasmodiophoro-myetes, fungal-like *Polymyxa,* and *Spongospora* transmit viruses as they infect the root systems of their hosts. Zoospores released from infected plants may carry virus either externally or internally. Viruses absorbed to the external surface of the zoospore, such as the *Tombusviridae*, are released to infect the new plant. Virus absorbed to the zoospore flagellum can enter the zoospore when its flagellum is retracted to encyst (Hull, 2002). The process through which viruses are carried internally is undefined and remains a topic for future research. *Bymovirus* and *Furovirus* are examples of viral genera that are transmitted in this manner (Hull, 2002). Rhizomania of sugarbeets, caused by *Beet necrotic yellow vein virus* (BNYVV) transmitted by Polymyxa, is a good example of a viral disease transmitted in this manner that is a major economic problem.

Seed and Pollen

Viruses may be transmitted in seed by two methods. In the first method, the virus infects the embryo within the seed, and it is already infected when it emerges. This is often referred to as true seed transmission. The second transmission method is through contamination of the seed, especially the seed coat. As the germinating seedling emerges from the seed, the virus infects the plant through wounds or through microfissures caused through cell maturation. TMV is transmitted to tomato seedlings by contamination of the seed coat.

Pollen may also transmit viruses. In addition to infecting the ovule during pollination, infected pollen may be moved to uninfected plants and infected during pollination (Hull, 2002). It may also carry virus into wounds.

Barley stripe mosaic virus (BSMV) is an example of a virus transmitted by pollen.

Dodder

The parasitic seed plant dodder (*Cuscuta* species) sinks haustoria into the phloem of the plants that it parasitizes. This connection allows the carbohydrates and other compounds to move into the dodder's phloem. When a dodder plant connects a healthy and virus-infected host plant, viruses can be transmitted from the infected plant through the dodder to the phloem of the noninfected plant.

Vegetative Propagation and Grafting

Viruses that systemically infect plants can be transmitted by the vegetative propagation of a portion of the infected plant. This portion can range from leaves, stems, branches, and roots to bulbs, corms, and tubers. Grafting is a form of vegetative transmission, and transmission occurs through the newly established vascular system linking the graft and the scion (Hull, 2002).

HOW ARE PLANT VIRUSES DETECTED AND IDENTIFIED?

Detection and identification of plant viruses are two of the most important procedures in plant virology. Correct identification of viruses is critical to establish control tactics for the disease. Procedures for the identification of plant viruses can be divided into the following categories.

BIOLOGICAL ACTIVITY

Infectivity assays, indicator hosts, and host range studies are all types of bioassays that are based on defining the interaction of the viral pathogen and its hosts. Infectivity assays measure the range of viral infectivity by determining the number of host plants infected at different dilutions. Indicator hosts are certain species of plants that have known reactions to a wide range of viruses. *Chenopodium quinoa* and different species and cultivars of *Nicotiana* are widely used as indicator hosts. Host range studies test the ability of the virus to infect different plant species. Some types of viruses have a very narrow host range and infect only closely related plants. MDMV infects only monocots. TMV infects a wide range of host plants, including plants from the Solanaceae, Chenopodiaceae, and Asteraceae, to name a few. Although indicator plants and bioassays were once used as the principle method of virus identification, they are currently employed only for the primary characterization of new viruses. Another type of biological activity that is an important identification characteristic is the mode of transmission. Identification of the vector association helps indicate the relationship of the unknown virus to characterized groups.

PHYSICAL PROPERTIES

The most important physical property used in detection and identification is morphology, which is usually determined by electron microscopy. Viruses can be visualized by negative stains of sap extracts or purified virus solutions, thin sections of infected tissue to localize the virus within the cellular structure, and immunospecific electron microscopy, which combines electron microscopy and **serology** to capture virus particles on coated electron microscope grids. Visualization of the virus provides the viron shape and size. It can also determine the presence of features such as spikes or other features of the capsid surface.

Other physical properties that have been classically utilized for identification of viruses measure the stability of the particle outside the host. These properties include the following: thermal inactivation point—the temperature at which a virus loses all infectivity; longevity *in vivo*—the length of time a virus can be held in sap before it loses its infectivity; and dilution endpoint—the greatest dilution of sap at which the titer of virus is capable of causing infection in a susceptible host. Many of these characteristics take several weeks to obtain and are no longer used as routine diagnostic techniques. However, they continue to be utilized as part of the official virus description.

Protein

Serology

Serology is based on the ability of the capsid protein to elicit an antigenic response. Areas of the capsid with unique shape and amino acid composition are capable of inducing immune responses in avians and mammals. These uniquely shaped areas are referred to as epitopes and are only a few amino acids in length. Animals injected with plant viruses produce antibodies (**immunoglobulins**) that can recognize and attach to the epitopes. These antibodies can be isolated from the serum or utilized as the serum fraction (**antiserum**). These antibodies are termed polyclonal antibodies due to the presence of many **antibody** types in the serum. Antisera can be exchanged between researchers around the world in order to compare diseases from many countries. Antisera can also be frozen and utilized to compare viruses for many years. This allows researchers to compare viruses or virus strains across time to follow the evolution and epidemiology of the disease.

Monoclonal antibodies are produced by the fusion of an isolated spleen cell from a mouse immunized to the plant virus and a murine myeloma cell. The resulting hybridoma cell line produces only one type of antibody. Advantages to monoclonal antibodies include a single antibody that can be well characterized, identification of the eliciting epitope, production of large amounts of

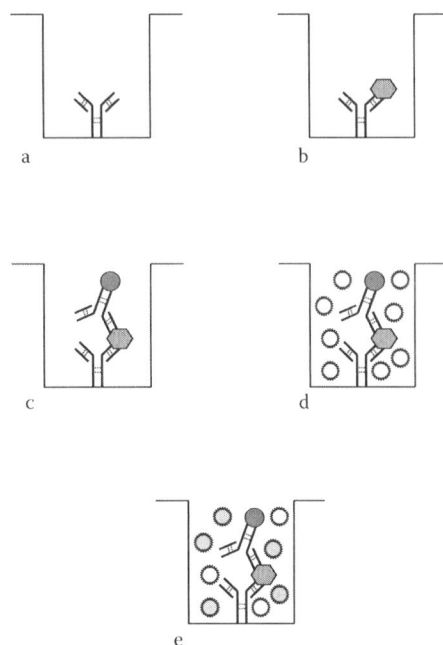

FIGURE 4.2 The principal steps in a double-antibody sandwich (DAS). (a) Antibodies (isolated immunoglobin-G) to a virus are attached to a polystyrene micotiter plate well by incubation with an alkaline pH carbonate buffer. Unattached antibodies are removed from the well by washing with phosphate buffered saline with Tween 20 added. (b) Sap extracted from plant samples is diluted and added to the well. Virus contained in infected samples binds to the matching antibody. Washing removes unbound materials from the well. (c) A second layer of antibodies that have been conjugated to an enzyme (typically alkaline phosphatase) is added to the well and attaches to the virus. If the virus has not been trapped by the primary layer of antibody, these detecting antibodies are removed by the wash and thus are not present to react in the remaining steps. (d) A substrate solution is added to the wells. (e) The enzyme attached to the detecting antibody causes a color change in the substrate. The intensity of the color is proportional to the amount of enzyme present and can be quantified by spectrophotometry.

antibody, and the ability of the producing cell line to be multiplied and frozen.

Antisera have allowed the development of many rapid and widely utilized detection assays. The most commonly utilized serological assay is the enzyme-linked immunosorbent assay (ELISA). Many variations exist in the ELISA procedure, but the most widely adopted protocol is the double antibody sandwich (Figure 4.2). This procedure starts by trapping a layer of antibodies on the well surfaces of a polystyrene microtiter plate. Attaching the antibodies to a solid surface is important as it allows all reactants that do not attach to the antibody and well to be washed away between steps. The attached antibodies are used to trap virus particles from sap solutions. A second layer of antibodies conjugated to an enzyme is then used to label the virus. Alkaline phosphatase is the most widely utilized enzyme, but other enzymes, such as horseradish peroxidase,

may also be used. After the final washing, a substrate solution is added to the wells. Substrates are chosen because they change color after being acted on by the conjugated enzyme. The color change can be quantified by reading the absorbance of a known wavelength of light when passed through each well. The amount of virus in the test solution is proportional to the light absorbance. ELISA is the basis for the development of many of the rapid diagnostic tests used by growers, producers, and agricultural consultants.

MOLECULAR PROPERTIES

The molecular weight and number of the proteins forming the capsid layer are also important characteristics of plant viruses. These characters can be determined by standard gel electrophoresis procedures. Determination of the amino acid content and sequence are also important in understanding particle structure and function.

NUCLEIC ACID

The ability to rapidly compare nucleic acid characteristics opened a new dimension in the identification of plant viruses. Identification based on the nucleic acid genome has enhanced the ability to identify strains and new viruses with similar characteristics. Comparisons of viral genomes have allowed the organization of plant viruses into the current taxonomic system and have had important implications in concepts of viral evolution. Nucleic acid tests provide some of the best tools for viral identification. The most powerful of these new tools is polymerase chain reaction (PCR).

POLYMERASE CHAIN REACTION

Polymerase chain reaction amplifies a portion of a nucleic acid genome. This area is defined by the use of short sequences of matching nucleotides called *primers*. Initially, the use of PCR was limited to DNA viruses; however, the use of reverse transcriptase to generate DNA strands (cDNA) from RNA viral genomes has expanded our ability to utilize this technology with RNA viruses.

HOW ARE PLANT VIRUSES CONTROLLED?

Control of plant viruses has a different primary focus than many other pathogens because there are no practical therapeutic or curative treatments for plant viruses. Thus, plant virus control focuses on preventative measures.

HOST PLANT RESISTANCE

Host plant resistance is the major approach to control of viral diseases but has not been identified in many crop species. It is typically the most economical control measure because it requires low input from the producer. Also, resistant plants eliminate the need for controlling the vector and are selective against the primary pathogen

(Khetarpal et al., 1998). Resistance to plant viruses can be due to the inability to establish infection, inhibited or delayed viral multiplication, blockage of movement, resistance to the vector, and viral transmission from it (Jones, 1998), and resistance to symptom development, also known as tolerance (Khetarpal et al., 1998). Development of plant cultivars with viral resistance is the primary goal of plant virology and breeding projects.

In addition to traditional breeding, two other methods are available for increasing the resistance of the host plant. In **cross protection**, a viral strain that produces only mild or no symptoms is inoculated as a protecting strain. When a second more severe strain is used as a challenge inoculation, the presence or effects of the first virus block its infection. This effect was first observed by McKinney (1929), and it has been used successfully in the control of some viruses, such as *Citrus tristeza virus* (CTV) in Brazil and South Africa (Lecoq, 1998). Limitations to the usefulness of cross protection include the inability of viruses to cross protect, yield reductions caused by mild strains, and the possible mutation of the mild strain to a more virulent form.

Genetic engineering provides a second method for enhancing host plant resistance. It is particularly valuable in situations where no natural source of resistance has been identified. Genes coding for coat protein (Kaniewski and Lawson, 1998), replicase (Kaniewski and Lawson, 1998), **antisense** RNA (Tabler et al., 1998), and **ribozymes** (Tabler et al., 1998) have been used to confer resistance to different example viruses. This technology holds much promise for the future. However, problems with stability of the inserts, expression of the inserted genes, unanticipated effect on the host plant, and public acceptance of genetic engineering remain as challenges to this technique.

VECTOR CONTROL

Controlling pests to minimize viral transmission differs from simply controlling pests. To control pests, the principal goal is to reduce the levels of pests below the levels that cause economic damage. However, low levels of pests may still be significant in the transmission of viral diseases (Satapathy, 1998). In order to determine the suitability of vector control, it is necessary to determine the vector and virus affecting the host (Matthews, 1991). This information establishes the type of relationship involved in transmission. For example, aphids transmitting MDMV require only seconds to accomplish this because the virus is transmitted in a nonpersistent manner. *Oat blue dwarf virus* (OBDV), in contrast, is transmitted by leafhoppers in a persistent circulative propagative manner that requires the leafhopper to feed on the plant for over 20 min. Thus, chemical treatments may be more effective on the leafhoppers transmitting OBDV than on the aphids transmitting MDMV. Also, effects of the controls on the feeding behaviors of the hosts should be considered.

Vectors moving into the field from surrounding hosts or due to seasonal migration may affect control measures. Each of these affects the probability for successful control of the virus. In the future, understanding the biological interactions between viruses and vectors may allow development of unique controls based on these intricate molecular mechanisms (Andret-Link and Fuchs, 2005).

QUARANTINE AND ERADICATION

Control of viral movement between fields, states, countries, and continents is often difficult and time consuming. **Quarantine** acts as the first line of defense against the introduction of many foreign viruses. However, the ability to detect viruses and other submicroscopic pathogens for exclusion is limited in comparison to other pathogens (Foster and Hadidi, 1998). Viruses may enter into new areas through importation of seeds, nursery stock, viruliferous vectors, plants, or experimental material (Foster and Hadidi, 1998). When exclusion of these pathogens fails, the cost of control and **eradication** of the virus (if possible) is often overwhelming. A current example of this can be examined in the PPV outbreak in the northeastern United States. Costs for this outbreak have already included loss of hundreds of nursery trees, cost of the quarantine (detection, identification, and tree removal), and loss of income to the growers.

CULTURAL CONTROLS

Cultural controls for viruses can be referred to as "wise production practices." These include any production practice or method that eliminates or significantly reduces the threat of viruses. Virus control is based on breaking the bridge of living hosts that is necessary to complete the seasonal movement of a virus. Any cultural practice that modifies the time or spatial interaction of virus and host can be an effective control. One common cultural control is modifying the date of planting. In South Dakota, delaying the planting of winter wheat until the seasonal end of wheat curl mite movement is one of the most effective controls available for WSMV. However, other concerns such as early planting of winter wheat to increase the amount of residue cover for the winter conflict with virus control. Thus, the producers must decide the wise balance point between these goals for their farm. Other common cultural controls include production of virus-free seeds and plants through indexing and **seed certification** programs, elimination of volunteer plants, elimination of alternate host plants, elimination of overlap between the planting of two crops, and sanitation of equipment.

Plant virology is a diverse area of plant pathology. Research with plant viruses includes determining the basic molecular structure and mechanisms of viruses, studying the cellular interaction, understanding vector relationships, or working with farmers and producers in the field. However, the ultimate goal of assisting farmers and producers across the world in the control of plant viruses remains the same.

REFERENCES

Andret-Link, P. and M. Fucha. 2005. Transmission specificity of plant viruses by vectors. *J. Plant Pathol.* 87: 153–165.

Bawden, F.C., N. W. Pirie, J.D. Bernal, and I. Fankuchen. I. 1936. Liquid crystalline substances from virus-infected plants. *Nature* 138: 1051–1055.

Brunt, A.A., K. Crabtree, M.J. Dallwitz, A.J. Gibbs, L. Watson, and E.J. Zurcher. (Eds.) 1996 onwards. Plant Viruses Online: Descriptions and Lists from the VIDE Database. Version: 20th August 1996. URL http://biology.anu.edu. au/Groups/MES/vide/.

Creager, A.N.H., K.B.G. Scholthof, V. Citovsky, and H.B. Scholthof. 1999. Tobacco mosaic virus: pioneering research for a century. *Plant Cell* 11: 301–308.

Faquet, C.M., M.A. Mayo, J. Maniloff, U. Desselberger, and L.A. Ball. 2005. *Virus Taxonomy—Classification and Nomenclature of Viruses.* Eighth Report of the International Committee on the Taxonomy of Viruses. Elsevier Academic Press. San Diego, CA.

Field, T.K., C.A. Patterson, R.C. Gergerich, and K.S. Kim. 1994. Fate of virus in bean leaves after deposition by *Epilachana varivestis*, a beetle vector of viruses. *Phytopathology* 84: 1346–1350.

Foster, J. A., and A. Hadidi. 1998. Exclusion of plant viruses. In *Plant Virus Disease Control*, Hadidi, A., Khetarpal, R.K., and Koganezawa, H. (Eds.). APS Press. St. Paul, MN, pp. 208–229.

Fraenkel-Conrat, H. 1956. The role of the nucleic acid in the reconstitution of active tobacco mosaic virus. *J. Am. Chem. Soc.* 78: 882–883.

Gergerich, R.C., H.A. Scott, and J.P. Fulton. 1986. Evidence that ribonuclease in beetle regurgitant determines the transmission of plant viruses. *J. Gen. Virol.* 67: 367–370.

Gergerich, R.C., and H.A. Scott. 1988a. The enzymatic function of ribonuclease determines plant virus transmission by leaf-feeding beetles. *Phytopathology* 78: 270–272.

Gergerich, R.C. and H.A. Scott. 1988b. Evidence that virus translocation and virus infection of non-wounded cells are associated with transmissibility by leaf-feeding beetles. *J. Gen Virol.* 69: 2935–2938.

Hull, R. 2002. *Matthews' Plant Virology* (4th edition). Academic Press. New York.

Jones, A.T. 1998. Control of virus infection in crops through breeding plants for vector resistance. In *Plant Virus Disease Control,* Hadidi, A., Khetarpal, R.K., and Koganezawa, H. (Eds.). APS Press. St. Paul, MN, pp. 41–55.

Kaniewski, W., and C. Lawson. 1998. Coat protein and replicase-mediated resistance to plant viruses. In *Plant Virus Disease Control*, Hadidi, A., Khetarpal, R.K., and Koganezawa, H. (Eds.). APS Press. St. Paul, MN, pp. 65–78.

Khetarpal, R.K., B. Maisonneuve, Y. Maury, B. Chalhoub, S. Dinant, H. Lecoq, and A. Varma. 1998. Breeding for resistance to plant viruses. In *Plant Virus Disease Control*, Hadidi, A., Khetarpal, R.K., and Koganezawa, H. (Eds.). APS Press. St. Paul, MN, pp. 14–32.

Lecoq, H. 1998. Control of plant virus diseases by cross protection. In *Plant Virus Disease Control*, Hadidi, A., Khetarpal, R.K., and Koganezawa, H. (Eds.). APS Press. St. Paul, MN, pp. 33–40.

Matthews, R.E.F. 1991. *Plant Virology* (3rd edition). Academic Press. San Diego, CA.

McKinney, H.H. 1929. Mosaic diseases in the Canary Islands, West Africa and Gibraltar. *J. Agric. Res.* 39: 557–578.

Morgan, G.J. 2006. Virus design, 1955–1962: Science meets art. *Phytopathology* 96: 1287–1291.

Moyer, J.W., T. German, J.L. Sherwood, and D. Ullman. 1999. An update on tomato spotted wilt virus and related tospoviruses. APSnet Feature (April 1999). http://www.apsnet. org/online/feature/tospovirus/.

Ng, J.C.K., and B.W. Falk 2006. Virus-vector interactions mediating nonpersistent and semipersistent transmission of plant viruses. *Annu. Rev. Phytopahol.* 44: 183–212.

Paliwal, Y.C. 1980. Relationship of wheat streak mosaic and barley stripe mosaic viruses to vector and nonvector eriophyid mites. *Arch. Virol.* 63: 123–132.

Satapathy, M.K. 1998. Chemical control of insect and nematode vectors of plant viruses. In *Plant Virus Disease Control*, Hadidi, A., Khetarpal, R.K., and Koganezawa, H. (Eds.). APS Press. St. Paul, MN, pp. 188–195.

Scholthof, K.B. 2001. 1898—The beginning of Virology … time marches on. *The Plant Health Instructor.* DOI: 10.1094/PHI-I-2001-0129–01.

Scholthof, K.B., J.G. Shaw, and M. Zaitlin (Eds.). 1999. *Tobacco Mosaic Virus—One Hundred Years of Contributions to Virology.* APS Press. St. Paul, MN.

Slykhuis, J.T. 1955. *Aceria tulipae* Keifer (Acarina: Eriophyidae) in relation to the spread of wheat streak mosaic virus. *Phytopathology* 45: 116–128.

Tabler, M., M. Tsagris, and J. Hammond. 1998. Antisense RNA and ribozyme-mediated resistance to plant viruses. In *Plant Virus Disease Control*, Hadidi, A., Khetarpal, R.K., and Koganezawa, H. (Eds.). APS Press. St. Paul, MN, pp. 79–93.

van Regenmortel, M.H.V., C.M. Fauquet, D.H.L. Bishop, E.B. Carstens, M.K. Estes, S.M. Lemon, J. Maniloff, M.A. Mayo, D.J. McGeoch, C.R. Pringle, and R.B. Wickner. 2000. *Virus Taxonomy—Classification and Nomenclature of Viruses.* Seventh Report of the International Committee on Taxonomy of Viruses. Academic Press. San Diego, CA.

Wang, R.Y., R.C. Gergerich, and K.S. Kim. 1994. The relationship between feeding and virus retention time in beetle transmission of plant viruses. *Phytopathology* 84: 995–998.

Waterworth, H.E., and A. Hadidi. 1998. Economic losses due to plant viruses. In *Plant Virus Disease Control*, Hadidi, A., Khetarpal, R.K., and Koganezawa, H. (Eds.). APS Press. St. Paul, MN, pp. 1–13.

Whitfield, A.E., Ullman, D.E., and German, T.L. 2005. Tospovirus-thrips interactions. *Annu. Rev. Phytopathol.* 43: 459–489.

5 Mechanical Inoculation of Plant Viruses

Marie A.C. Langham

Mechanical transmission or inoculation is a widely used technique in experimentation with plant viruses. It can be utilized for propagation of plant viruses, determination of infectivity, testing of host range, and evaluation of plants for host plant resistance. However, successful mechanical inoculation is dependent on the condition of the virus-infected tissue source, condition of the host plants for inoculation, inoculum preparation, and the use of abrasives.

The ideal propagation host for the preparation of inoculum is a young, systemically infected plant with a high concentration (titer) of virus in the tissue. The host should not develop necrotic symptoms. It should contain neither a high concentration of inhibitors that interfere with the establishment of infection nor tannins or other phenolic compounds that can precipitate virus particles. Unfortunately, it is not always possible to utilize an ideal host. Propagation hosts with less desirable characteristics may require modification of inoculation protocols. For example, utilizing a host with a low virus concentration or one that produces local lesions may require modification of buffer to tissue ratios. Additionally, poor growing conditions of an ideal propagation host may cause changes in the virus concentration or in compounds present in the tissue.

Release of virus from the host cellular structure necessitates maceration of infected tissue (Figure 5.1). Small amounts of tissue may be ground in a mortar and pestle. Large amounts of tissue are macerated in blenders. Disrupting plant cells not only releases the virus, but also breaks the central vacuole as well as other smaller vacuoles releasing their contents into the extracted sap. The first effect of the vacuole contents is to increase the acidity (lowering the pH) of the plant sap extract. As the pH of the sap decreases, proteins (including viruses with their protein capsid) precipitate out of solution. Thus, buffers (Figure 5.2) are added to the tissue during grinding to stabilize the extract at a pH favorable to virus infectivity, integrity, and suspension. Although the majority of viruses can be transmitted in the pH 7 (neutral) range, the desired pH varies with the virus being inoculated. This is not the only role that buffers serve in preparing inoculum. The ionic composition of the buffer contributes to the success of the inoculation. Addition of phosphate to an inoculation buffer increases the infection rate during inoculation, and phosphate is one of the most commonly utilized buffer (Hull, 2002). Other ions may be added to the inoculation buffer if required for virus stability. For

FIGURE 5.1 Infected plant material is ground in a mortar and pestle to macerate the tissue as the beginning step in the preparation of sap extracts for inoculation.

FIGURE 5.2 Sap extract is prepared by adding buffer to macerated leaf tissue and grinding.

example, some viruses require the presence of calcium or magnesium to maintain their structure, and these ions must be added to the inoculation buffer.

Reducing and chelating agents, and competitors are other categories of chemicals that may be added to inoculation buffers for viruses with specific requirements. Sodium sulfite, 2-mercaptoethanol, sodium thioglycollate, and cysteine hydrochloride are examples of frequently utilized reducing agents. Sodium sulfite is often used in inoculation buffers for *Tomato spotted wilt virus* (TSWV). Quinones, tannins, and other phenolic compounds in plant sap can cause the precipitation of plant viruses. This activity is countered by chelating agents and competitors. Chelating agents, such as ethylene diamine tetraacetic acid (disodium salt), are added to bind the copper ions found in polyphenoloxidase, the enzyme that oxidizes phenolic compounds in the plant. Competitors, such as hide powder, have been used to dilute the effects of phenolic compounds by serving as alternate sites for phenolic activity (Matthews, 1991).

Finally, mechanical inoculation is based on our ability to create nonlethal wounds in plant cells for virus entry (Figures 5.3 and 5.4). Wounds in the cell wall and membrane are created by the use of an abrasive. The most commonly used abrasive is silica carbide; however, celite, bentonite, and acid-washed sand have also been utilized.

EXERCISES

EXPERIMENT 1. TRANSMISSION OF TOBACCO MOSAIC VIRUS (TMV)

The objective of this exercise is to demonstrate the mechanical inoculation of a plant virus.

Materials

Each team of students will require the following items:

- TMV-infected host plants (*Nicotiana tabacum* is a good systemic host for providing inoculum.)
- Healthy host plants (*N. tabacum*)
- Mortars and pestles
- Inoculation pads (sterile cheesecloth rectangles approximately 2–3 cm × 10–15 cm)
- Silica carbide (600 mesh)
- 0.02 M phosphate buffer, pH 6.8–7.2 (to make add 3.48 g of K_2HPO_4 and 2.72 g of KH_2PO_4 to 1 L of distilled water.
- *Nicotiana tabacum* plants with 3–4 well expanded leaves (4 plants per student or group)
- Plants of *Chenopodium quinoa*, *N. glutinosa*, and *N. tabacum* cv. Samsun NN and/or *Phaseolus vulgaris* cv. Pinto (reaction is dependent on TMV strain)
- 10-mL pipettes
- Top loading balance
- Paper tags and string

Follow the protocol outlined in Procedure 5.1 to complete this part of the experiment.

Anticipated Results

Plants should be observed closely for 2–3 weeks. Mechanical damage due to excessive pressure during inoculation appears first and can be observed as necrotic damage to the epidermis. It appears on both the plants inoculated with virus and those inoculated with healthy tissue. Local lesion hosts begin to display lesions in 3–5 days. Develop-

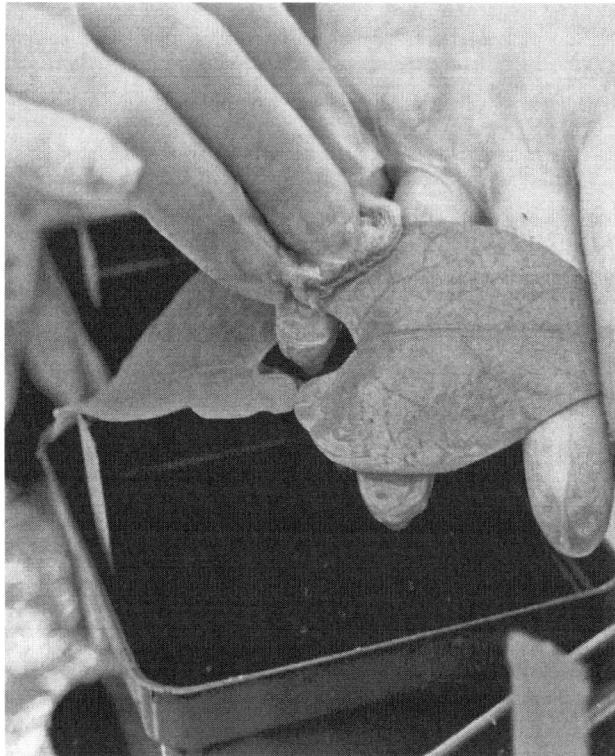

FIGURE 5.3   Mechanical inoculation (BPMV) of *Phaseolus vulgaris* cv. Provider is accomplished by gently rubbing the surface of plant leaves with a mixture of sap extract and silica carbide.

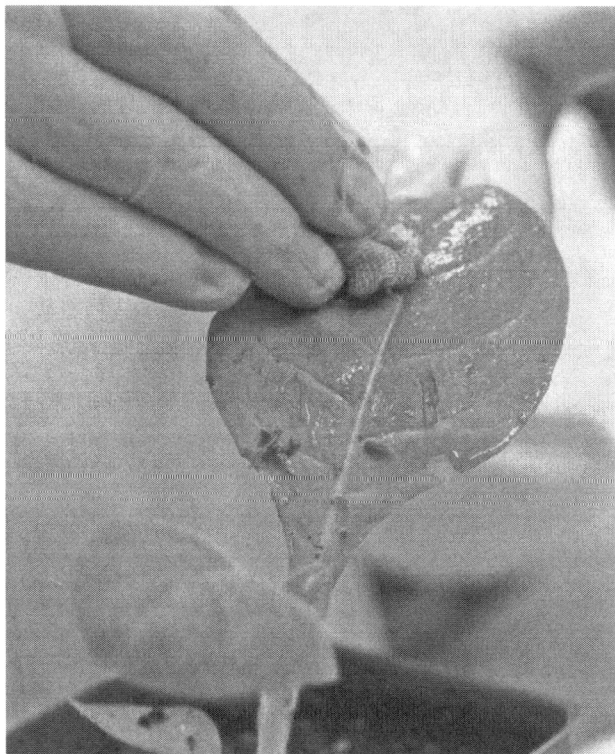

FIGURE 5.4   Mechanical inoculation (TMV) of tobacco demonstrating the technique for supporting the leaf and coating it with inoculum during inoculation.

Procedure 5.1

Mechanical Transmission of Tobacco Mosaic Virus (TMV)

Step	Instructions and Comments
1	Control plants. Collect 0.1 g of healthy plant tissue. Grind the tissue thoroughly in a mortar and pestle. Add 9.9 mL of phosphate buffer. Grind the mixture again. Remove any plant material that has not been thoroughly macerated. Add 1% silica carbide to the sap extract (silica carbide can also be dusted on the leaves with a sprayer; however, adding it to the sap mixture decreases the amount that may be accidentally inhaled). Stir the mixture well and thoroughly wet the inoculation pad. Silica carbide settles to the bottom of the sap extract and must be stirred each time the inoculation pad is soaked. With the soaked inoculation pad, firmly but gently rub the upper surface of leaves on two tobacco plants. Excessive pressure will result in lethal damage to the epidermal cells from the silica carbide; however, too little pressure will not wound the cells for viral entry.
2	Experimental plants. Repeat step 1 utilizing infected plant tissue.
3	Observe plants for the next 2–4 weeks for symptom development.

---

ment of the local lesions from pinpoint size to larger size can be observed. Systemic symptoms begin from 10 days to 2 weeks depending on growth conditions. Symptoms first appear in the newly expanding leaves.

Dilution of Inoculum

Experiments to compare the effects of inoculum dilution can be accomplished by utilizing the above procedure. Follow the protocols outlined in Procedure 5.2 to complete this part of the experiment.

Anticipated Results

This experiment demonstrates the effects of inoculum dilution on the infectivity of the virus. As the dilution factor of the inoculum increases, the number of local lesions produced by inoculation decreases. The infectivity curve from plotting these results demonstrates this relationship.

Host Range

Experiments to study host range can be done by utilizing the inoculation procedure in the transmission exercise to inoculate a number of different host plants, including

---

Procedure 5.2

Dilution of Plant Extract and Calculation of Infectivity Curve

Step	Instructions and Comments
1	Grind the infected tissue and buffer at the lowest dilution ratio to be utilized for the experiment as previously described. The remaining dilutions for the series can be prepared by diluting the initial sap extract. For TMV, the dilution series might be 1:50, 1:100, 1:500, 1:1,000, and 1:10,000. An alternative for this experiment is to user *Bean pod mottle virus* (BPMV) or *Tobacco ring spot virus* (TRSV). Dilutions of 1:10, 1: 20, 1: 50, 1:100, and 1:500 would be good choices for these viruses.
2	Using the materials from Procedure 5.1, the initial sap extract was prepared by macerating tissue in buffer at a 1:10 ratio. To make a 1:50 ratio, 1 mL of the sap extract is added to 4 mL of buffer. Make the remaining dilutions.
3	A host that produces local lesions should be chosen as the host for this experiment. For TMV, *Chenopodium quinoa*, *Nicotiana glutinosa*, and *N. tabacum* cv. *samsun* NN are good local lesion hosts. For BPMV, *Phaseolus vulgaris* cv. *pinto* is a good choice for local lesion assays. If a local lesion producing tobacco with at least four expanded leaves is used, half leaves can be used for comparison in a Latin square design. When using half leaves, a paper tag on a string loop can be used to designate the solution used on each side.
4	Inoculate plants as described in Procedure 5.1. An inoculation of a half leaf with buffer only should be included as a control.
5	Plants should be examined for developing local lesions in 3–4 days. Plot number of local lesions against the dilution of sap extract to obtain the infectivity curve of the virus.

TABLE 5.1

Examples of Plant Species Utilized in a Host Range Experiments

Virus	Plant Species	Reaction[a]
TMV		
	Chenopodium amaranticolor	Susceptible
	C. quinoa	Susceptible
	Cucumis sativus	Susceptible
	Cucurbita pepo	Susceptible
	Lactuca sativa	Susceptible
	Nicotiana glutinosa	Susceptible
	N. tabacum	Susceptible
	Avena sativa	Not susceptible
	Pisum sativum	Not susceptible
	Spinacia oleracea	Not susceptible
	Triticum aestivum	Not susceptible
	Zea mays	Not susceptible
	Zinnia elegans	Not susceptible
BPMV		
	Glycine max	Susceptible
	Lens culinaris	Susceptible
	Phaseolus vulgaris cvs. Black Valentine, Tendergreen, or Pinto	Susceptible
	P. sativum	Susceptible
	Vigna unguiculata	Susceptible
	C. sativus	Not susceptible
	N. glutinosa	Not susceptible
	N. tabacum	Not susceptible
	Petunia × hybrida	Not susceptible
TRSV		
	Beta vulgaris	Susceptible
	C. amaranticolor	Susceptible
	C. sativus	Susceptible
	Cucurbita pepo	Susceptible
	N. tabacum	Susceptible
	Petunia × hybrida	Susceptible
	P. vulgaris	Susceptible
	Brassica campestris ssp. Rapa	Not susceptible
	Helianthus annuus	Not susceptible
	Secale cereale	Not susceptible
	Trifolium hybridum	Not susceptible
	T. aestivum	Not susceptible

[a] Variation in reaction is possible due to virus isolate and inoculation conditions.

susceptible and nonsusceptible. Table 5.1 provides some examples of plants for some sample viruses, and the VIDE provides a listing of useful host plants for many viruses (Brunt et al., 1996). Utilize five plants per species (for each student or group) for the experimental and five for the control group inoculated with extract from healthy sap or with buffer. Follow the protocols listed in Procedure 5.3 to complete this part of the experiment.

Anticipated Results

The inoculated plants should be observed closely over the next 2–3 weeks noting the type of symptom, the date of appearance, and which plant species display symptoms. Compare host ranges among students for variation in these factors. Virus or virus strain variation and variation in inoculation efficiency produce differences in host range. The list of infected plants should be correlated with

Procedure 5.3
Inoculation and Assessment of Host Range

Step	Instructions and Comments
1	Prepare the sap extract as described in the first exercise and inoculate in the same manner. Changing inoculation pads between species is a good procedure in case one species is contaminated with a virus. Plants with viscous sap, such as *Chenopodium*, should be inoculated last as the sap may contaminate the extract and cause inhibition of the virus.
2	Plants should be observed daily. Record the date of symptom appearance, symptom type, and changes in symptoms as plants mature.

the plant families represented to note other possible susceptible or nonsusceptible species.

Questions

- Why is nonlethal wounding required for plant virus infection?
- Why do local lesion infections differ from systemic infections?
- Can a single virus particle infect a plant?
- How can host ranges assist in predicting the susceptibility of an untested plant species?
- What is needed to prevent people from becoming plant virus vectors by mechanical inoculation in field or greenhouse situations?
- How can mechanical inoculation be utilized in developing host plant resistance in crops?
- Would dilution curves be the same for viruses with multiple particles as for those with single particles?
- Why should plants inoculated with healthy plant sap be included in all mechanical inoculation studies?
- Can a local lesion reaction ever become a systemic reaction?

EXPERIMENT 2. OUCHTERLONY DOUBLE DIFFUSION TECHNIQUE FOR DETECTING AND IDENTIFYING VIRUSES

The objective of this exercise is to confirm the presence of stable icosahedral viruses, such as BPMV or TRSV. This protocol was originally developed from Ball (1990) and from personal communication with Drs. Howard Scott and Rose Gergerich, University of Arkansas.

Materials

The following items will be needed by each team of students:

- Agarose (low EEO)
- 1% solution of sodium azide. (Sodium azide is a toxic chemical. Avoid contact with this solution or the agarose gel containing it.)

- Two 100 mm × 15-mm petri dishes. (Allow at least two plates per student team as it often takes more than one attempt to make a double diffusion plate for students.)
- Antisera (diluted to experimental concentrations with phosphate buffered saline)
- Distilled water (or phosphate buffered saline)
- Well punch, cork bore, etc., for making wells
- Humidity box or ziplock bags with damp paper towels
- Pipettes (borosilicate glass) and bulbs
- Light source for observing plates. (Although light boxes or specially constructed double diffusion light boxes are excellent, a flashlight will work well and provides a low cost alternative.)

Follow the protocol listed in Procedure 5.4 to prepare agarose Ouchterlony dishes.

Preparation of Sap Extracts

Plant sap extracts can be obtained by grinding the tissue in a mortar and pestle in combination with a buffer. They may also be prepared by expressing the sap with a hand squeezer or with a sap extractor. Another alternative is to seal the tissue and buffer in a plastic bag and use a heavy smooth object to macerate the tissue by rubbing the surface. With this method, avoid sealing excessive amounts of air in the bag as this will make maceration more difficult. Follow the protocols in Procedure 5.5 to complete the experiment.

Anticipated Results

After filling the plate, diffusion of the extracts and antiserum begin. When compatible antibodies and antigens come together, the antibody attaches to the antigen or antigens. The typical antibody has two attachment sites, making it possible for it to bind to more than one antigen. The virus particle repeats antigenic sites on each identical protein subunit, which allows it to bind to multiple antibodies. As more and more antibody and antigen bind, the mass begins to precipitate. This precipitant forms the

Procedure 5.4	
Preparation of Agarose Plates	
Step	Instructions and Comments
1	Mix 1 g of agarose with 98 mL water in a flask (more than twice the volume of the liquid). Record the weight of the flask with solution in it. Heat the mixture slowly in a microwave with an additional beaker of water, stopping to swirl the mixture occasionally. Agarose crystals become clear before they dissolve. Thus, observe the solution carefully while swirling to determine if the crystals have dissolved. The flask should be observed during the entire heating period to avoid boiling over.
2	When all crystals have dissolved, allow the flask to sit until the steaming has slowed. Weigh the flask a second time and use the heated water in the second beaker to return the agarose solution to its original weight. Water has been lost as steam and its replacement maintains the original concentration of agarose.
3	Add 2 mL of sodium azide to the mixture and swirl gently to avoid bubbles. Use a level surface to pour the dishes. Measure 15 mL of the agarose solution into each petri dish. The agarose should solidify without being moved or replacing the lids to keep condensation from accumulating. When cool, cover and store them in a humid container.

Procedure 5.5	
Preparation of Experimental Ouchterlony Plates	
Step	Instructions and Comments
1	Using a pattern of six wells surrounding a central well, punch and carefully remove (by suction or other means) the wells in the agarose plates prepared in Procedure 5.4. Damage to the walls of the wells can result in leaking wells or confused reactions.
2	The well where numbering originates should be marked with a small mark on the plate bottom. Mark the alignment of the top and bottom of the dish if using the top lid for marking the wells. In each pattern of six wells, leave one well empty and fill one well with healthy sap as checks for false and healthy plant reactions. Fill wells using a small amount of fluid in the pipette until the surface of the fluid is level with the surface of the agarose. The level can be determined by watching the light reflectance off the fluid surface. When the fluid is level, the light reflectance will dull rather than shine as it does when the surface is convex or concave. The outer wells should be filled first. Antiserum is added to the central well last. Optimum concentrations for sap extracts and antisera should be determined prior to the exercise. After filling, dishes should not be moved until the extracts and antisera have begun to absorb into the agarose. Gently place the dish in a humid container and allow it to incubate undisturbed.
3	Bands of precipitation begin formation within 24 h and continue to grow stronger over the next week (Figure 5.5). The observation of bands is improved by shining a light through the bottom of the dish. Record the reactions and any spur formation that is observed.

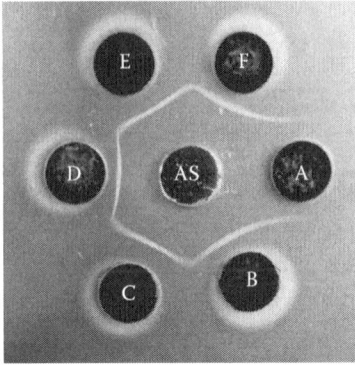

FIGURE 5.5 Double diffusion dish with six wells surrounding the center antiserum (AS) well demonstrates the white precipitin bands formed in the plate by a positive reaction. The antiserum being utilized was made to detect *Bean pod mottle virus* (BPMV). Sample A is sap from a healthy bean plant. Samples B-F are samples from beans infected with BPMV.

typical white band in positive reactions. An example of typical results in an Ouchterlony can be found in Figure 5.5. Bands begin to form within 24 h and grow stronger with more time. An extension of the precipitin line beyond where it meets the line formed by a neighboring well is sometimes seen. This is caused when one of the test samples contains some antigens recognized by the antiserum that are not in the neighboring sample. This extension is called a spur, and their formation can be an indication of a strain difference in the virus contained in the two samples.

Questions

- Why should a healthy sample be included in each test?
- What advantages does serological identification of the virus provide?
- Why would a precipitin line form closer to the antiserum well?
- Why would a precipitin line form closer to the antigen well?
- How can this technique be utilized in addition to an inoculation technique to provide more information about an unknown virus sample?
- What would happen if a test sample was infected with more than one virus?
- What is a spur in the Ouchtolony test and what does it tell us about the virus?

Notes

These exercises were written as examples of exercises that can be utilized in a laboratory and are based on exercises that have been utilized in my classes. They have been written with the hope that you will modify them to meet the requirements of your classes. One of the most important

modifications could be the choice of virus. Examples of common viruses have been provided. However, these viruses may not be available to you. The VIDE provides information on host range that can be used in modifying the given exercises. If you do not have any plant viruses available, TMV can often be isolated by grinding the tobacco from a cigarette that has not been treated to add flavor or reduce the tar or nicotine, and inoculating the extract on a susceptible tobacco. Dilution may also require adjustment, depending on your virus and infected plant tissue.

If you desire to also utilize a viral detection exercise, the choice of virus for use in this exercise should be based on which antisera and virus combination is available to your class since this is often a limiting factor in this exercise. Double diffusion is applicable for many icosahedral viruses. However, some viruses, such as in the *Cucumovirus*, require additional ions to form precipitin bands. Flexuous and rigid rod viruses diffuse through the agarose gel very poorly and require the addition of sodium dodecyl sulfate (SDS) to remove the capsid from the particles. SDS also solubilizes the precipitin band that forms in plates after approximately 24 h and makes these plates less useful in classes with laboratories meeting once a week. For those who would like to demonstrate how Ouchterlony double diffusion works, but do not have the antisera available, the following protocol that demonstrates double diffusion by forming a chemical precipitate in the agarose was provided by Dr. Robert N. Trigiano, University of Tennessee.

SIMULATION OF THE OUCHTERLONY TECHNIQUE WITHOUT VIRUSES AND ANTISERA

Antisera to specific viruses can be expensive and cost prohibitive to use in introductory plant pathology laboratories with many students. Also, maintaining virus-infected plants may not be possible nor desirable for any number of reasons. The simple technique in this experiment demonstrates the basic principles of diffusion and precipitation found in the Ouchterlony test. It requires a minimum amount of materials that are inexpensive and readily available.

Materials

Each student or team of students will require the following materials to complete the experiment:

- Silver nitrate solution—0.1 M (1.70 g in 100 mL of distilled water)
- Sodium chloride solution—0.1 M (0.58 g in 100 mL of distilled water)
- No. 3 cork borer or plastic drinking straws
- Two eyedroppers or bulbs with glass Pasteur pipettes

Procedure 5.6
Demonstration of the Ouchterlony Double Diffusion Technique with Silver Nitrate and Sodium Chloride

Step	Instructions and Comments
1	Using a No. 3 cork borer or plastic straw, remove agar plugs from two petri dishes as depicted in Figure 5.5.
2	For petri dish 1, fill the center well with silver nitrate solution. Next, fill three of the outside reservoirs with sodium chloride solution and the remaining three outside wells with distilled water. Different pipettes should be used for each of the fluids.
3	For petri dish 2, fill the center well with sodium chloride solution. Next fill three of the outside reservoirs with silver nitrate solution and the remaining three wells with distilled water. Different pipettes should be used for each of the fluids.
4	Label drawings of each of the dishes and record the results.

- 60 mm-diameter plastic petri dishes containing 0.6% water agar (6 g agar per liter of distilled water; autoclave at 121 C for 15 min and dispense about 15 mL per dish)
- Distilled water

Follow the protocol in Procedure 5.6 to complete this experiment.

Anticipated Results

White precipitate should form between the center well and the surrounding wells within 15–30 min, depending on the temperature of the room. The precipitate is analogous to the interaction of antisera to the antigen on the coat protein of the virus.

ELISA

A basic ELISA protocol was not included in these exercises. For laboratories that have antisera and supplies, ELISA protocols are available from many sources, such as Converse and Martin (1990). Any ELISA protocol requires modification based on the activity of available antisera. Another option for laboratories with antisera is dot immunobinding assays (Hammond and Jordan, 1990). These assays require shorter incubation periods and are often more adaptable to laboratory schedules. For laboratories that are not doing ELISA routinely, the equipment, antisera, supplies, and time investment for a basic protocol is considerable. To minimize the investment needed with ELISA for laboratories that must buy all components or for ELISA with a virus for which the antiserum is not available, AGDIA produces a large variety of kits utilizing ELISA with several enzyme detection systems or immunostrips that adapt easily to classrooms. AGDIA can be contacted at the following address:

AGDIA, Inc.
30380 County Road 6
Elkhart, IN 46514
phone: 1-574-264-2014 or 1-800-62-AGDIA
fax: 1-574-264-2153
email: info@agdia.com

LITERATURE CITED

Ball, E.M. 1990. Agar double diffusion plates (Ouchterlony): Viruses. In Hampton, R., Ball, E., and DeBoer, S. (Eds.). *Serological Methods for Detection and Identification of Viral and Bacterial Plant Pathogens*. APS Press, St. Paul, MN, pp. 111–120.

Brunt, A.A., K. Crabtree, M.J. Dallwitz, A.J. Gibbs, L. Watson and E. J. Zurcher. (Eds.). 1996 onwards. Plant Viruses Online: Descriptions and Lists from the VIDE Database. Version: 20th August 1996. URL http://biology.anu.edu.au/Groups/MES/vide/.

Converse, R.H. and R.R. Martin. 1990. ELISA methods for plant viruses. In Hampton, R., Ball, E., and DeBoer, S. (Eds.). *Serological Methods for Detection and Identification of Viral and Bacterial Plant Pathogens*. APS Press, St. Paul, MN, pp. 179–196.

Hammond, J. and R.L. Jordan. 1990. Dot blots (viruses) and colony screening. In Hampton, R., Ball, E., and DeBoer, S. (Eds.) *Serological Methods for Detection and Identification of Viral and Bacterial Plant Pathogens*. APS Press, St. Paul, MN, pp. 237–248.

Hull, R. 2002. *Matthews' Plant Virology* (4th edition). Academic Press, New York.

Matthews, R.E.F. 1991. *Plant Virology* (3rd edition). Academic Press, San Diego, CA.

6 Pathogenic Prokaryotes

George H. Lacy and Felix L. Lukezic

CHAPTER 6 CONCEPTS

- Phytopathogenic prokaryotes are a diverse group including over 30 genera.

- Fastidious phytopathogenic prokaryotes are recognized as more economically important as methods improve for their detection and identification.

- Epiphytic, nonpathogenic resident phase of plant colonization is important for pathogen survival and disease control.

- Biofilms composed of bacteria and extracellular polysaccharides, such as zoogloea, are important in pathogen colonization, survival, and dissemination.

- Chemical control of diseases caused by phytopathogenic prokaryotes is made difficult because of short generation times.

- Cultural control of diseases caused by phytopathogenic prokaryotes is critical because chemical controls may not be adequate.

CHARACTERISTICS OF PHYTOPATHOGENIC PROKARYOTES

All **prokaryotes** have cell membranes, cytoplasmic 70S ribosomes, and a nonmembrane-limited nuclear region (Figure 6.1). Phytopathogenic prokaryotes have a unique character — the ability to colonize plant tissues. Pathogens increase from a low to a high number in a remarkably short time within a diseased host plant. The essence of pathogenicity is this ability to live and multiply within plant tissues.

TAXONOMY

Taxonomy is a classification devised by humans of living organisms according to their current understanding of phylogenetic groupings. Current understanding of phylogenetics has been determined by a variety of methods, each with its own strengths and weaknesses. Therefore, taxonomies have strengths and weaknesses and change in response to development of the new technologies used to explore natural relationships among prokaryotes. Listed next are some groups of phytopathological prokaryotes, a brief description of the groups, and some of the diseases they cause.

GENERA OF PLANT PATHOGENIC PROKARYOTES

Gram-Negative Aerobic/Microaerophilic Rods and Cocci

- *Acetobacter* — Ellipsoid- to rod-shaped organisms motile by peritrichous or lateral flagella that oxidize ethanol to acetic acid. *Acetobacter aceti* and *A. diazotrophicus* cause pink disease of pineapple.
- *Acidovorax* — Rods motile by a polar flagellum. *Acidovorax avenae* subspecies are pathogenic on oats (subsp. *avenae*), watermelon (subsp. *citrulli*), and orchids (subsp. *cattleyae*).
- *Agrobacterium* — Rods motile by 1 to 6 peritrichous flagella. *Agrobacterium* species cause proliferations on many plants: smooth and rough galls (*A. tumefaciens*), root proliferations (*A. rhizogenes*), and galls on berry roots (*A. rubi*). Tumorigenic *Agrobacterium* species carry parasitic plasmids vectoring tumor DNA (T-DNA). *Note:* The International Society of Plant Pathology has indicated that members of this genus are indistinguishable from species of *Rhizobium,* except that they induce rhizogenic or tumorigenic plant growths rather than symbiotic nodules.

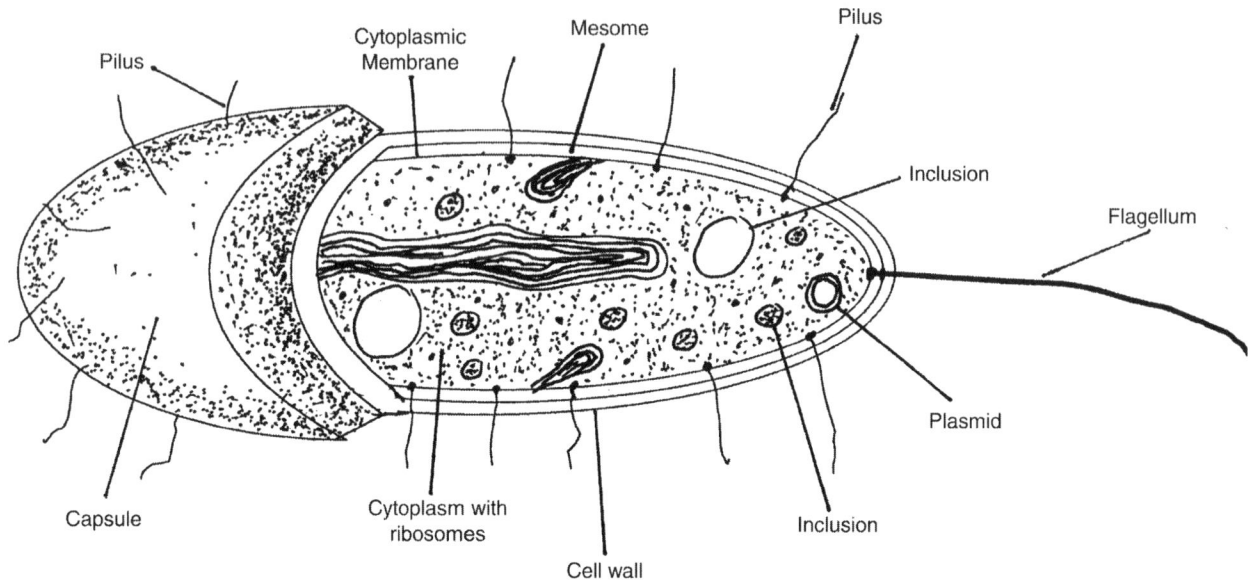

FIGURE 6.1 Three-dimensional view of a bacterium.

- *Bradyrhizobium* — Rods motile by one polar or subpolar flagellum. *Bradyrhizobium japonicum* is a phytosymbiotic, nitrogen-fixing bacterium on soybean. Under high soil nitrogen conditions, it produces phytotoxic rhizobitoxin and damages its host.
- *Burkholderia* — Straight rods motile by polar flagella. *Burkholderia cepacia* is an onion pathogen and *B. gladioli* is pathogenic on gladiolus and iris. *Burkholderia cepacia* may cause disease in persons with cystic fibrosis. It is not known whether strains causing plant disease also cause human disease.
- *Gluconobacter* — Ellipsoid- to rod-shaped, occurring singly and in pairs, either nonmotile or motile by several polar flagella, and oxidizing ethanol to acetic acid. *Gluconobacter* species causes discolorations in apples, pears, and pineapple fruit.
- *Herbaspirillum* — Usually vibroid, occasionally helical, motile by 1 to 3 flagella at one or both poles. Strict aerobes, associated with roots. *Herbaspirillum rubrisubalbicans* is a pathogen of cereal roots, and was previously known as *Pseudomonas rubrisubalbicans.*
- *Pseudomonas* — Rods motile by one or more polar flagella. Many plant pathogens produce water-soluble pigments that fluoresce light blue to greenish-yellow with ultraviolet light. *Pseudomonas marginalis* is a soft-rotting pathogen of many plants. *Pseudomonas syringae*, subdivided into many pathovars (pvs.),

includes the leaf spot pathogens on tobacco (pv. *tabaci*) and field beans (pv. *phaseolicola*). *Pseudomonas syringae* pv. *syringae* causes warm temperature frost damage (WTFD). *Pseudomonas savastanoi* has several pathovars including *P. savastanoi* pv. *glycinea,* which causes leaf spot of soybean.
- *Ralstonia* — Rods motile by one or more polar flagella. *Ralstonia solanacearum* causes wilts on solanaceous crops (potato, tobacco, and tomato), peanuts, and banana.
- *Rhizobacter* — Rods nonmotile or motile by polar or lateral flagella. *Rhizobacter dauci* causes bacterial gall of carrot.
- *Sphingomonas* — Rods motile by one lateral, subpolar, or polar flagellum and accumulates polyhydroxybutyrate granules. *Sphinogomonas suberifaciens* causes corky root of lettuce.
- *Xanthomonas* — Rods motile by one polar flagellum. Colonies are yellow due to xanthomonadin pigments. *Xanthomonas campestris* has over 70 pathovars, including pathogens of cole crops such as cabbage (pv. *campestris*). *Xanthomonas axonopodis* has more than 40 pathovars, including pathogens of field beans (pv. *phaseolus*) and cotton (pv. *malvacearum*). *Xanthomonas oryzae* pv. *oryzae* is an important pathogen of rice.
- *Xylella* — Long flexuous rods, nonmotile, nutritionally fastidious inhabitants of insects and plants. *Xylella fastidiosa* is a leafhopper-vectored, xylem-limited bacterium causing

scorches of almond, elm, and oak as well as citrus leaf blight.

- *Xylophilus* — Rods motile by one polar flagellum. *Xylophilus ampelinus* causes bacterial necrosis and canker of grapes.

Facultative Anaerobic Gram-Negative Rods

- *Brennaria* — Rods occur singly, in pairs, and occasionally in short chains. Various species cause diseases of alder, walnut, banana, oak, and cricket-bat willow.
- *Enterobacter* — Rods motile by peritrichous flagella. *Enterobacter cloacae* causes brown discoloration of papaya fruits. *Enterobacter cloacae* is also an opportunistic pathogen of humans, causing problems during burns and wounds and in urinary tracts. It is not known whether strains causing plant disease also cause human disease.
- *Erwinia* — Rods motile by peritrichous flagella. This group is divided into the "true erwinias," including nonpectolytic species such as *E. amylovora*, which causes fire blight of pear and apple, and the soft-rotting "pectobacteria" (described later).
- *Klebsiella* — Nonmotile, straight rods. These pectolytic plant-associated bacteria (*K. planticola* and *K. pneumoniae*) are rarely associated with soft rots. *Klebsiella pneumoniae* might also cause human disease, but it is unknown whether strains that cause plant disease also cause human disease.
- *Pantoea* — Rods motile by peritrichous flagella. *Pantoea stewartii* causes Stewart's wilt of sweet corn. Many strains of epiphytic *P. agglomerans* (e.g., *E. herbicola*) cause warm temperature frost damage to many plants.
- *Pectobacterium* — Rods motile by peritrichous flagella. This group is separated from the "true" erwinias. *Pectobacterium* species are strongly pectolytic. *Pectobacterium* (*Erwinia*) *carotovorum* and *P. chrysanthemi* cause soft rot in numerous plant hosts.
- *Serratia* — Straight rods usually motile by peritrichous flagella. *Serratia marcesens* causes yellow vine of cucurbits. *Serratia marcesens* is also a pathogen of insect larvae and an opportunistic pathogen of humans, causing septicemia and urinary tract problems. It is not known whether strains causing plant disease also cause human disease.

Irregular, Nonsporing Gram-Positive Rods

- *Arthrobacter* — Young cultures contain irregular rods often V-shaped with clubbed ends. In older cultures, rods segment into small cocci. Cells are nonmotile, Gram-positive, but easily destained. *Arthrobacter ilicis* causes bacterial blight of American holly.
- *Clavibacter* — Irregular, wedge-, or club-shaped, nonmotile rods arranged in pairs or in V shapes or palisades. *Clavibacter michiganensis* and subspecies (*michiganensis*) cause tomato canker [ring rot of potato (*C. sepedonicus*)].
- *Curtobacterium* — Young cultures have small, short, irregular, nonmotile rods in pairs arranged in V shapes. In older cultures, rods segment into cocci are Gram-positive, but easily destain. *Curtobacterium flaccumfaciens* pv. *poinsettiae* causes poinsettia canker.

Nocardioform Actinomyces (Gram-Positive, Acid Fast, and Nonmotile)

- *Nocardia* — Partially acid-fast, branching, mycelium-like hyphae with coccoid spore-like units. Colonies have sparse aerial mycelia. *Nocardia vaccinii* cause galls on blueberry.
- *Rhodococcus* — Rods to extensively branched substrate mycelium. Weakly acid-fast. *Rhodococcus fascians* is the causes fasciation of sweet pea, geranium, and other hosts.

Streptomyces (Gram-Positive, Fungus-Like Bacteria)

- *Streptomyces* — Vegetative, extensively branched hyphae. Aerial mycelium matures to form three or more spores in chains. *Streptomyces scabiei* and *S. acidiscabies* cause potato scab.

Endospore-Forming, Motile Gram-Positive Rods

- *Bacillus* — Large rods motile by peritrichous flagella with oval central endospores. Strongly Gram-positive, aerobic to facultative anaerobes. *Bacillus* species may cause rots of tobacco leaves, tomato seedlings, and soybean and white stripe of wheat.
- *Paenibacillus* — Resemble *Bacillus* and cause rots of terminal buds in date palms.
- *Clostridium* — **Obligate anaerobes**, rods motile by peritrichous flagella with subpolar to polar endospores. Plant pathogenic, *Clostridium* species are usually Gram-negative despite typical Gram-positive cell walls. Pectolytic *Clostridium*

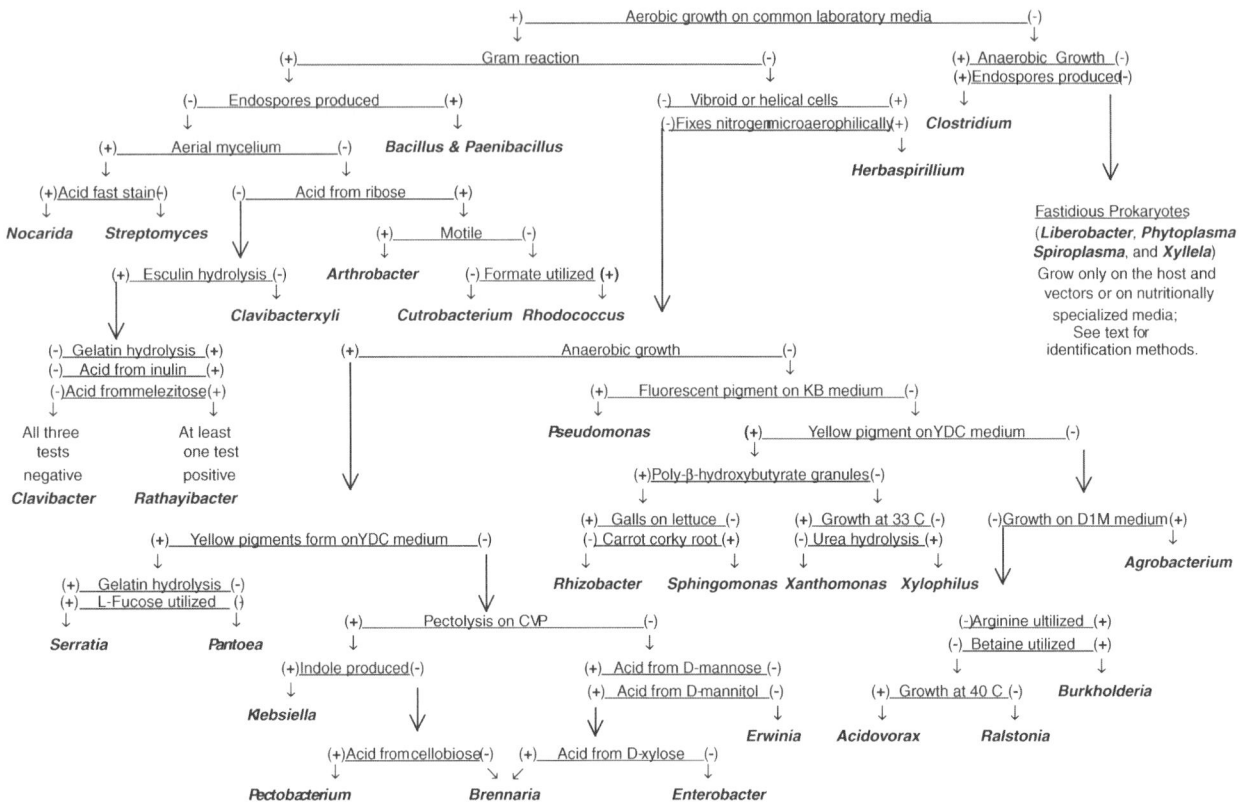

FIGURE 6.2 Flow chart to identify genera of phytopthogenic bacteria that grow readily on common laboratory media [based on Schaad et al. (2001) and Holt et al. (1994)]. Fastidious bacteria are not dealt with here. The genera *Acetobacter* and *Gluconobacter* contain opportunistic bacteria that affect fruit only and are not treated here. This scheme assumes that the bacteria tested are pathogens. In other words, Koch's postulates have been satisfied or approximated.

species are associated with wetwood of poplar and elm and soft rots of potato and carrot.

Mollicutes

Mollicutes lack cell walls, phloem-limited, fastidious mollicutes vectored usually by leafhoppers.

- Phytoplasma — Polymorphic. The aster yellows phytoplasma causes proliferation or yellows in over 120 host plants. These organisms are identified using PCR-amplified elements of ribosomal RNA (rRNA).
- Spiroplasma — Spiral shape supported around a central protein "rod" embedded in the membrane. Plant pathogenic spiroplasmas cause stubborn disease of *Citrus* species (*S. citri*) and corn stunt *(S. kunkelii)*.

Organisms of Uncertain Affiliation

- Phloem-limited bacteria — A group of insect-transmitted, phloem-limited bacteria have been

poorly studied because they cannot be cultured in the laboratory. Apparently, they represent a wide selection of bacteria. *Liberobacter* species include the citrus-greening pathogens.

RAPID IDENTIFICATION OF PHYTOPATHOGENIC PROKARYOTES

Rapid identification of prokaryotes for disease control differs from taxonomic description. Speed rather than precision is the goal. The clinician identifies pathogens only to the level required to understand what is needed for disease control. Dr. Norman W. Schaad has compiled the *Laboratory Guide for Identification of Plant Pathogenic Bacteria*. This authoritative reference presents a combination of selective media, diagnostic tests, and serological and PCR methods to guide the clinician to successful identification of prokaryotes from diseased plants. Figure 6.2 shows an abbreviated flow chart for identifying plant pathogenic bacteria.

ANATOMY AND CYTOCHEMISTRY OF PROKARYOTES

FLAGELLA (SING. FLAGELLUM)

Flagella are narrow, sinuous structures 12 to 30 nm wide and 3.7 mm long composed of flagellin (protein) subunits arranged helically. Flagella propel bacteria through water at speeds approaching 50 μm/sec. This is accomplished by the counterclockwise motion of flagella, which pulls (rather than pushes) the bacterium. A flagellar "motor" called the basal body is located in the cell wall and requires biological energy equivalent to 64 NADH oxidations or the formation of 128 ATP molecules per flagellar rotation. A system that requires such high energy must be significant in the disease process. Flagella of a *Pseudomonas fluorescens* stimulating plant growth enhanced colonization of potato roots. Motile strains of ice-nucleation-active (INA) *P. syringae* pv. *syringae* more efficiently colonized expanding bean leaves. Motile strains of *Erwinia amlyovora* caused a greater incidence of disease in apple.

PILI (SING. PILUS)

Pili are rod-like protein structures (length 0.2 to 2 mm and diameter 2 to 5 nm) composed of helically arranged pilin units. Pili are often correlated with parasitic plasmids in bacteria, and the genes controlling pili production are often located on these plasmids. For the plasmid, pili function in the process of bacterium-to-bacterium transfer of plasmid DNA (type IV secretion of proteins and nucleic acids). The transfer mechanism for T-DNA from tumorigenic *Agrobacterium* species and strains to plants is also accomplished by type IV secretion. Other pili are encoded within the hypersensitive response/pathogenicity (hrp/pth) region of the bacterial genome and involved with secretion of proteins into or close to plant cells.

EXTRACELLULAR POLYSACCHARIDES (EPSS)

EPSs are high-molecular-weight carbohydrates either loosely attached (slime polysaccharides are easily removed by washing) or more tightly attached (capsular polysaccharides are less easily removed) *Erwinia amylovora* EPS is made up of 98% galactose. Bacterial cells embedded in EPS are extruded from lenticels and dried into very fine filaments, which are dispersed by air currents over long distances. *Pseudomonas savastanoi* pv. *glycinea* produces acetylated alginate EPSs in diseased leaves. The best-studied EPS is xanthan gum (MW 10^6 Da) produced by *Xanthomonas campestris* pv. *campestris* and composed of D-glucose, D-mannose, and D-glucuronic acid with some pyruvate. Xanthan gum, important for black rot pathogen of crucifers, is also used

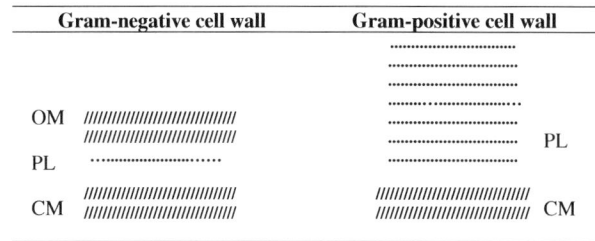

Gram-negative cell wall	Gram-positive cell wall
OM ////////////////////////// //////////////////////////	························ ························ ························ ························ ························ PL
PL ··················	························
CM ////////////////////////// //////////////////////////	////////////////////////// ////////////////////////// CM

FIGURE 6.3 Schematic comparison of Gram-negative and Gram-positive cell walls. Gram-negative cell walls are 15 to 20 nm thick and composed of a 1.4-nm-thick peptidoglycan layer sandwiched in the periplasm, a space between the 7.5 nm-thick cellular lipid bi-layer (////) outer (OM) and inner cellular membranes (CM). Gram-positive cell walls are thicker (22 to 25 nm) with a dense peptidoglycan layer (15 nm) enclosing a single 7.5–nm-thick cellular membrane.

commercially as an emulsifer in foods and is found in ice cream, salad dressings, and sauces. EPS is a primary pathogenic determinant in wilting plants. *Clavibacter michiganensis* subsp. *insidiosus* (alfalfa wilt), *Ralstonia solanacearum* (Granville wilt of banana, peanut, and tobacco), *Erwinia amylovora* (fire blight pathogen of pear and apple), and *Pantoea stewartii* (Stewart's wilt of maize) block xylem vessels in wilting plants with high-molecular-weight EPSs.

CELL WALL

Cell walls are Gram-negative, Gram-positive, or absent in prokaryotes. Cell walls protect bacteria against osmotic extremes. Low osmotic events occur during water dispersal or colonization of xylem sap (<1 mg/mL dissolve nutrients). Phytoplasmas and spiroplasmas lack cell walls but live in protected high osmotic phloem sap (30 mg/mL dissolved nutrients) and insect hemolymph. Most plant pathogenic bacteria are Gram-negative. The cell wall of Gram-negative bacteria consists of an outer lipid membrane sandwiching a peptidoglycan layer over an inner lipid membrane. The peptidoglycans provide cell rigidity. Gram-positive bacteria have much thicker peptidoglycan layers and lack the outer membrane. Figure 6.3 schematically compares Gram-negative and Gram-positive cell walls.

PROTEINS

Proteins play many important roles in plant–pathogen interactions. For instance, acetosyringone from wounded plants diffuses to and interacts with receptor proteins to trigger a cascade of T-DNA-interacting proteins that configure *Agrobacterium tumefaciens* to transfer T-DNA into host plants. Porins and permeases allow bacteria to move nutrients and wastes across cell membranes. Binding proteins concentrate substrates in proximity to permeases. Enzymes associated with the cell wall, such as ATPase, provide energy for secretion. Pectate lyases and proteases from soft rot pathogens [*Pectobacterium* (*Erwinia*) *carotovorum*] digest pectic materials and plant proteins, respectively, releasing nutritional components to the pathogen. Other proteins are structural in nature. Lipoproteins bind cell membranes to the peptidoglycan layer, and actins in *Spiroplasma citri* form the central spindle around which the spiral-shaped cell is supported. Finally, in *Pseudomonas syringae* pv. *syringae* and *Pantoea agglomerans*, causal agents of WTFD, ice-nucleation proteins and cell-surface proteins catalyze ice formation. Plant cells broken by ice crystals leak nutrients to the pathogens.

LIPOPOLYSACCHARIDES (LPSs)

LPSs are anchored into the outer membrane of bacteria. Specificity is imparted depending on which sugars are incorporated and linkage of the sugars one to another. LPS bind bacteria to their plant hosts if the charges and shape of bacterial LPSs complement sugar moieties on plant glycoproteins. The interaction between plant glycoproteins (also known as lectins) and bacterial lipopolysaccharides gave rise to the lectin hypothesis for binding bacteria to host plants. For instance, soybean binds most *Bradyrhizobium* strains at high frequencies. However, soybean lacking lectin is also nodulated at low frequencies by the bacterium, suggesting that LPSs are not wholly responsible for bacterial binding.

MEMBRANE-BOUND PIGMENTS

Bacteria contain lipid-soluble carotenoid pigments. In xanthomonads, pantoeas, and curtobacteria, carotenoid pigments have been shown to protect cells from light damage. The major protection mechanisms are quenching of triplet sensitizer, quenching of singlet oxygen, and inhibition of free radical reactions. Other roles suggested for pigments include modifying membrane permeability, altering antibiotic sensitivity, electron transport, and enhancing enzyme activity.

CYTOPLASM

Bacterial cytoplasm seems homogeneous; however, on careful examination, several structures are apparent. Mesosomes, invaginations of the cytoplasmic membrane, occur with septa formation; bacterial nuclear region, skeins of double-stranded DNA (dsDNA) fibrils are visible by transmission electron microscopy; storage granules may be observed and include glycogen, poly-hydroxbutyric acid (HBA or lipid), and metachromatic granules (polymerized inorganic metaphosphate); and endospores are present. Endospores are special survival structures composed of a thin exosporium; layers of protein; a cortex

of dipicolinic acid, peptidoglycan, and calcium; a cytoplasmic membrane; and desiccated cytoplasm containing ribosomes, mRNA, and DNA. Endospores of *Bacillus*, *Paenibacillus*, and *Clostridium* survive at 80°C for 15 min.

HOW PROKARYOTES DAMAGE PLANTS

Plant pathogenic prokaryotes cause soft rots with enzymes; tissue proliferations with phytohormones; wilts with extracellular polysaccharides; and leaf spots, blights, and necroses with toxins and enzymes.

ENZYMES

Enzymes are catalytic proteins. Cell degrading enzymes (CDEs) produced by plant pathogenic bacteria reduce plant components to compounds useful for pathogen nutrition (Table 6.1). Enzymes may be located extracellularly and secreted into the environment, intracellularly in the cytoplasm, inserted into membranes or periplasmically (between the cytoplasmic and the outer membranes of Gram-negative bacteria or within the cell wall of Gram-positive bacteria). Coordinated batteries of enzymes are required for plant pathogenesis.

Pectic Enzymes

Pectate lyase is the major pectolytic enzyme involved in bacterial soft rot pathogenesis by species of *Pectobacterium* (*Erwinia*), *Pseudomonas*, and *Xanthomonas*. *Pectobacteria* have a battery of several pectate lyases that have different locations in the cell (extracellular or periplasmic), pH optima, endo- or exoabilities, and digestion products (pentamers, dimers, or monomers of galacturonic acid). Pectate lyases are major determinants of disease. X-ray diffraction studies suggest that efficient folding makes pectate lyase molecules small enough to diffuse into spaces (about 4 nm) between cellulose fibrils in plant cell walls. Pectin-degrading enzymes also have a role in pathogenesis. Pectate lyases cannot degrade pectin (methylated pectic acid). Pectin methylesterase demethylates pectin to pectic acid. Some bacteria, including *Ralstonia solanacearum*, have polygalacturonases that degrade pectin directly.

Cellulases

Degradation of crystalline cellulose requires C_1 cellulase to cleave cross-linkages among cellulose fibrils, C_2 endocellulase to break primary cellulose polymers, C_x cellulase [endo-(β1,4 glucanase] to cleave soluble cellulose into cellibiose, and cellobiase [β-glucosidase (β-glucanase)] to degrades cellobiose to D-glucose. The Granville wilt pathogen, *Ralstonia solanacearum*, has several cellulases (endoglucanases).

TABLE 6.1
Plant Cell Structures, Putative Nutrients or Macromolecular Building Blocks Provided to the Pathogen, and Corresponding Cell-Degrading Enzyme Found in the Pathogen

Affected Plant Cell Structure	Nutrient for Pathogen	Degradative Enzyme
Cuticle		
Cutin	Fatty acid peroxides	Cutinase
Suberin		
Suberin	Fatty acid polyesters	Suberin esterase
Cell wall		
Pectic substances	Galacturonans	
	Nonmethylated	Pectate lyase
		Oligogalacturonase
	Methylated	Pectin methylesterase
		Pectin lyase
		Polygalacturonase
Cellulose	Glucose monomer	Cellulases
Native cellulose cross links		C_1 cellulase
Native cellulose main strand		C_2 cellulase
Soluble cellulose to cellibiose		C_x cellulase
Cellibiose to glucose		β-glucanase
Hemicellulose	β-1.4-linked xylans	Xylanases
Proteins		
Cytoplasmic membrane (and other organelle membranes)	Complex structure	
Proteins	Polypeptides	Proteases, proteinases
Phospholipids	Phospholipids	Phospholipase
Phosphatidyl compounds	Phospholipids	Phosphatidase
DNA	3-deoxy polynucleotides	Deoxyribonucleases (DNases)
RNA	Ribopolynucleotides	Ribonucleases (RNases)

Proteases (Proteinases)

These enzymes cleave proteins into peptones and peptides. Protein degradation products such as di- and tripeptides are assimilated by bacteria. Proteins in plants occur in cell walls, cell membranes, and cytoplasm. Proteases apparently coordinate with pectate lyases to produce a synergistic increase in soft rot damage by digesting extensins, hydroxyproline-rich cell wall structural proteins, and loosening plant cell wall so that endopectate lyases penetrate more effectively into its cellulose–pectin matrix.

Cutinases and Suberin Esterases

Cutinases and suberin esterases digest cuticular waxes and suberin, respectively. These enzymes are produced by bacteria as diverse as *Pseudomonas syringae* pv. *tomato* and *Streptomyces scabiei* and aid in bacterial penetration of the host and nutrition.

TOXINS

Bacteria produce many metabolites that may harm plants. However, toxins are those metabolites present during pathogenesis in concentrations great enough to cause plant damage. Although toxins extend the amount of damage a pathogenic bacterium may cause, the presence of a toxin alone is not enough to make a bacterium a pathogen. Toxins enter plants by several methods. Syringotoxin, produced by *Pseudomonas syringae* pv. *syringae*, is a cyclic peptide with a hydrophobic lipid tail. The lipid tail dissolves into the plant cell membrane and the cyclic peptide causes an aberration in the membrane, causing it to leak. Phaseolotoxin, produced by *P. syringae* pv. *phaselolicola*, and tabtoxin, produced by *P. syringae* pv. *tabaci*, resemble di- and tripeptides and are taken up by the plant as nutrients. Tagetitoxin, produced by *P. syringae* pv. *tagetis* affects chloroplast thylakoid membranes and inhibits chloroplast RNA polymerase. Rhizobitoxin, produced by *Bradyrhizobium japonicum*, inhibits production of homocysteine and blocks ethylene production. The toxic degradative product of phaseolotoxin, octicidin or psorn, inhibits production of citrulline, causing ornithine to accumulate. Ornithine blocks arginine and citrulline biosynthesis, resulting in disruption of chloroplast membranes. Tabtoxinine, the toxic degradative product of tabtoxin, inhibits glutamine synthetase, causing ammonia accumulation and chlorosis.

WILTS

Xylem sap, with a constant flow of dilute nutrients (<1 mg/mL aqueous sugars, organic acids and amino acids), is an excellent habitat for a plant pathogenic bacterium. Bacteria that colonize xylem vessels and cause wilts are a heterogeneous group. These organisms include *Pantoea stewartii*, causing Stewart's wilt of maize; *Clavibacter michiganensis* subsp. *michiganensis*, causing tomato canker; *C. michiganensis* subsp. *insidiosus*, causing bacterial wilt of alfalfa; *Ralstonia solanacearum*, causing Granville wilt of tobacco, other solanaceous plants, bananas, and peanut; and *Xanthomonas campestris* pv. *campestris*, causing blackrot of crucifers such as cabbage. Wilts are caused by high-molecular-weight polysaccharides restricting the flow of xylem sap by collecting on endplates and lateral pits of xylem vessel cells.

TUMORIGENESIS

This group of pathogens includes those that produce tumors (galls) as well as overproduce (proliferating) roots or shoots. Common to production of galls and proliferations of roots or shoots are plant growth regulators, especially auxins and cytokinins. Tumor-inducing bacteria include *Agrobacterium tumefaciens,* causing crown galls on many plants; *Rhizobium* and *Bradyrhizobium* species, causing nitrogen-fixing nodules on legume roots; *Pseudomonas savastanoi*, causing olive and oleander knot; and *Nocardia vaccini*, causing galls on blueberries. Bacteria that cause proliferations of plant organs include *A. rhizogenes* (root proliferation on many plants), *Rhodococcus fascians* (proliferation of lateral shoots on many plants including pea), and, most likely, phytoplasmas that cause "witches' broom."

ECOLOGY OF PLANT PATHOGENS

EPIPHYTES

Epiphytic bacteria obtain their nutrition on the surfaces of plants. Because most plant pathogenic bacteria lack specialized resistant structures to survive environmental conditions not conducive to pathogenesis, plant pathogens must coexist as epiphytes either on their host or on other plants without causing disease. Thus, **endophytic** growth probably represents a survival stage (Figure 6.4). Examples of bacteria with endophytic residence phases include *Pectobacterium* (*Erwinia*) *carotovorum*, which colonizes potato lenticels and causes bacterial soft rot, and *Pseudomonas syringae* and *Pantoea agglomerans*, agents of WTFD.

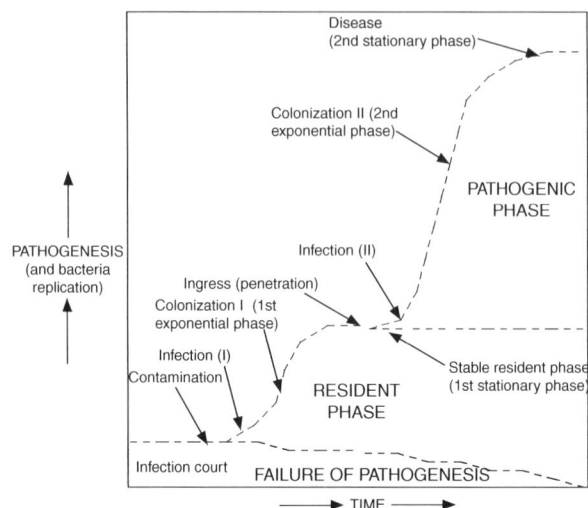

FIGURE 6.4 Pathogenesis and the resident phase of bacteria. This figure indicates steps in pathogenesis based on the growth curves of bacteria (dashed lines) and consisting of lag, exponential, and stationary phases. One sequence, including contamination, infection (I), and colonization (I; first exponential phase), leads to establishment of a stable resident phase population on or in plants (first stationary phase). This resident phase may extend for days or months before the pathogenic phase is indicated. A second sequence, ingress, infection (II) colonization (II; second exponential phase) and disease (indicated by the second stationary phase), occurs when conditions are correct for pathogenesis. In this figure, infection and the establishment of a nutritional association with the host is indicated by the end of the lag phases of growth and the points where bacterial growth curves begin to increase.

ENDOPHYTES

Endophytic plant pathogens colonize epidermal cells (e.g., *Streptomyces scabiei*, causal agent of scab of potato), apoplast or the space outside living protoplasts, including cell walls and free space (e.g., *Pseudomonas savastanoi* pv. *phaseolicola*, causal agent of haloblight of bean), xylem vessels (e.g., *Ralstonia solanacearum*, causal agent of Granville wilt), and phloem (e.g., *Spiroplasma citri*, causing citrus stubborn disease). The pathogens causing WTFD (*Pseudomonas syringae* and *Pantoea agglomerans*) do not fit the four ecological niches indicated. Instead, in WTFD, pathogenesis may be a joint effort of epiphytic and apoplastic bacteria. Ice nucleation most likely occurs on the phylloplane, but the cells of the pathogen that benefit from nutrients leaking from ice-damaged cells are those colonizing the apoplast.

Nutritional Classes of Endophytes

Phytopathogenic prokaryotes are nutritionally nonfastidious (may be cultured on general laboratory media), nutritionally fastidious (require special, chemically complex ingredients), or obligate endophytes (cannot be cultured on media at this time). Nonfastidious endophytic prokaryotes may be cultured on simple agar media containing chemically defined inorganic or simple organic compounds. They have some portion of their disease cycle separate from the host plant and insect vector — as plant epiphytes or soil saprophytes. They include toxin-producing pathogens (e.g., *Pseudomonas savastanoi* pv. *phaseolicola*), enzyme-producing pathogens [e.g., *Pectobacterium* (*Erwinia*) *carotovorum*], tumor-inducing pathogens (e.g., *P. savastanoi*), and wilt-inducing pathogens (e.g., *Clavibacter michiganensis* subsp. *michiganensis*). Fastidious phloem- or xylem-limited prokaryotes vectored by insects do not have any part of their life cycle outside of their plant and insect hosts. These fastidious organisms include *Xyllela fastidiosa* (leaf scorch of many hosts) and *Spiroplasma citri* (citrus stubborn disease) as well as obligate endophyte *Candidatus Liberobacter asiaticus* (citrus greening) and *Phytoplasma* species (yellows and proliferations).

PHYLLOSPHERE

The **phyllosphere** is the environment created by leaf surfaces. For this discussion, the phyllosphere encompasses all open plant surfaces, including stems and flowers, in contact with the atmosphere.

NATURE OF PROKARYOTES IN THE PHYLLOSPHERES

Bacteria found in the phyllosphere originate from soil, water, other plants, seeds, pollen, or dust, or they may be vectored by insects. Population numbers are reduced compared to the soil and vary from $10^{2-6}/cm^2$. Casual inhabitants are bacteria derived from airbiota deposited from the soil and water. Populations of these organisms, although occasionally very high, do not establish nutritional relationships with plants and decrease with time. Provided a nutritional relationship is established and populations are maintained over a lengthy period of time, organisms are classified as resident. Because populations of residents may fluctuate greatly over time, this relationship may be difficult to recognize.

ENVIRONMENT

The phyllosphere is a harsh environment. Temperature fluctuates up to 35°C because the atmosphere does not provide as much insulation as soil. Ultraviolet and infrared radiation may be lethal to bacteria. On the upper (adaxial) leaf surfaces, infrared radiation causes heating, and ultraviolet radiation, at 256 nm, is bactericidal and mutagenic. Bacteria escape some radiation by persisting on the lower (abaxial) leaf surfaces. Bacteria require free water for replication, but on the leaf surface, the relative humidity (RH) also varies greatly each day. Environmental rigors cause wide fluctuations in bacterial populations on plant surfaces. Surprisingly, the epiphytes *Pseudomonas syringae* pv. *syringae* and *Pantoea agglomerans* multiply even during times of low humidity. Hydrophilic and absorptive bacterial EPSs may be important in maintaining holding water. The nutrient supply is also limited. Water of guttation from hydathodes supplies less than 1.0 mg/mL combined amino acids, carbohydrates, and organic acids. Additional nutritional input from the plant is derived from leakage via ectodesmata directly through the waxy cuticle and deterioration of the cuticle itself.

Leaf attachment is important for survival and successful pathogenesis. For example, *Pseudomonas syringae* pv. *lachrymans*, causal agent of angular leafspot of cucumbers (*Cucumus sativus*), adheres more tightly to its host than to nonhost plants. On host plant leaves inoculated by dipping into a suspension of a pathogen, up to 90% of the colony forming units (cfu) are retained after gently washing. However, 60 to 89% cfu are removed from leaves of nonhost cultivars.

Biofilms are important to attachment as well as survival. In the phyllosphere, biofilms are limited to aggregations of cells embedded in EPS. EPSs are important for attachment to phylloplane surfaces because they retain absorbed water when plant surfaces are dry As bacteria require free water for growth and replication, EPSs are an extremely important part of survival. For dissemination, bacterial ooze (EPSs mixed with bacterial cells) is exuded from stoma, lenticels, nectaries, and hydathodes, where it is available for water splash dispersal.

RHIZOSPHERE AND RHIZOPLANE

The **rhizosphere** is the volume of soil affected biologically, chemically, or physically by the presence of roots. The **rhizoplane** is the two-dimensional surface of the root. Soil is a rich environment. Agricultural soils contain more than 10^9 bacteria/g. Bacteria colonize anticlinal intercellular depressions on the surfaces of elongating roots. Only about 10^2 bacteria/cm^2 root are on the rhizoplane because of the geometry of the surface and bacterial microcolonies. Rhizosphere soil is enriched for Gram-negative bacteria and contains fewer actinomyces than soil away from roots. Competition is intense, with predatory bacteria, protozoa, nematodes, and mites harvesting bacteria. Plant products affect the metabolism of root-associated phytopathogenic bacteria. Exudates, including most commonly low-molecular-weight metabolites and inorganic ions, leak passively from living cells. Pectins and hemicelluloses are secreted into the root cap and the cells in the elongation zone. Lysates are compounds released by the autolysis of the older epidermal cells and include sugars, amino acids, nucleotides, enzymes, growth factors, and terpines. Diffusion of substances from roots stimulates bacterial motility by nutrient concentration gradients. Motility is restricted to water films clinging to soil particles and is limited especially at the low water potentials encountered near the root.

PENETRATION OF PLANT SURFACES

PASSIVE PENETRATION

Bacteria enter plants mostly through breaches in physical barriers. Natural openings such as stoma, hydathodes, and lenticels also provide avenues of entrance as well as areas of higher moisture for bacterial colonization. During plant development, wounds develop by the growth of lateral or adventitious roots, abscission of leaves, or the breaking of root hairs during root elongation among abrasive mineral particles and provide sites for bacterial entry. Wounding, whether by developmental processes or other processes, bypasses the integrity of the host surfaces.

ACTIVE PENETRATION

Zoogloea are masses of phytopathogenic bacteria embedded in EPSs living in the apoplastic spaces in the roots, stems, leaves, and vascular systems of plants. These zoogloea represent biofilms trapped among parenchymous cells. Zoogloea expand with bacterial growth and replication and EPS production. Therefore, zoogloea exert hydrostatic pressure that forces bacteria into adjacent spaces among cells. Very likely, enzymes that degrade the cell wall aid zoogloeal separation of cells.

Other enzymes may aid direct penetration. Cutinases degrade cutin protecting epidermal plant cells. The potato scab pathogen *Streptomyces scabiei* produces cutinase that allows direct penetration of tuber epidermal cells.

DISEASE CONTROL

HOST RESISTANCE

Plants are not universally compatible with disease development. A pathogen on its susceptible host cultivar comprises a compatible interaction and disease will develop provided environmental conditions permit. Conversely, a pathogen on a resistant or nonsusceptible host cultivar yields an incompatible interaction. Breeding or engineering host plants for disease resistance is the most important method to control plant diseases.

PREFORMED RESISTANCE

Plants remain attractive ecological niches and are underexploited nutrient sources practically free of competition from microorganisms. The barriers to basic compatibility

are strong because only about 200 of 2400 named species of bacteria cause plant disease. Preformed resistance, or passive plant defense to entry and colonization, resembles ancient defenses of plants to prokaryotes. Physical barriers include the cuticle, epidermis, plant cell walls, turgor and hydrostatic pressure within vascular elements, special anatomic features of natural plant openings, and inhibitory chemicals. These defense mechanisms have broad activity and protect plants against entry and colonization by most prokaryotes. Plants have developed methods to prevent damage and possible entry of plant pathogens by making plant surfaces less appealing for colonization. Modified trichomes (leaf hairs) form gland cells containing antimicrobial terpenoids. The root tip is protected by rapidly dividing cells in the root cap. Their lubricating ability proceeds from the ease with which they are sloughed off and the softness of the pectic gel containing them. These cells are sacrificed to protect the actual root tip from abrasion damage by mineral particles or attachment or penetration by pathogens. Plants may also develop protective symbioses. Epiphytic microbial populations in the rhizosphere and phyllosphere often have protective properties, which resist through microbial competition or antagonism colonization by the resident phases of phytopathogenic bacteria or fungi. Ecto- and endomycorrhizal associations provide additional protection from pathogens.

Preformed chemical barriers protecting plants from pathogens must occur in the plant in a form not damaging to the plant itself, occur at a concentration effective against the target pathogen, and be either in its active form or be able to be readily released in an active antimicrobial form. In resistant pea cultivars, antimicrobial homeoserine occurs in concentrations inhibitory to *Pseudomonas syringae* pv. *pisi*. Glycosylation, the addition of sugar moieties to toxic compounds, renders these compounds nontoxic within the cytoplasm of the plant. Glucosidases, stored in plant vacuoles, deglycosylate or remove the sugars. Thus, in pathogen-damaged cells, the contents of vacuoles mix with the cytoplasm resulting in the formation of the aglycone (deglycosylated) toxin. The aglycone suppresses pathogen growth.

INDUCED NONSPECIFIC RESISTANCE SYSTEMS

In the coevolution of plants with their pathogens, some pathogens have developed the ability to evade passive or preformed defenses. In response, plants had to develop resistance in a different way. In induced resistance, challenge of the plant by living or dead pathogens or certain chemicals induces a general resistant response against several types of pathogens. It is now clear that at least three types of nonspecific induced resistance occur: systemic acquired resistance (SAR) signaled by salicylic acid, induced systemic resistance (ISR) signaled by jasmonic acid, and riboflavin-induced systemic resistance (RIR). These systems often induce pathogenesis-related proteins, including chitinases and endoglucanases in plants (Chapter 28).

SPECIFIC INDUCED RESISTANCE

Specific induced resistance, or the hypersensitive reaction (HR), to pathogens requires that the plant chemically recognize the pathogen to trigger its defenses (see also Chapter 28). Systems in which HR occur typically involve specialization of both pathogens and hosts, including the development of races of the pathogen and resistant plant lines or cultivars within the susceptible plant species. Incompatible pathogens are recognized by the plant and induced HR prevents disease development. Compatible pathogens are not recognized by the plant and may cause disease because no HR is triggered. HR is apoptosis (programmed plant cell death) triggered by the interaction between a plant and a pathogen incompatible with disease development in that plant. Plant cells in the immediate vicinity of the site of bacterial ingress die rapidly, preventing further spread of the pathogen into the plant.

Recognition of an incompatible pathogen by a plant results from the identification of a chemical signal (effector) from the pathogen by the host. Basic to the concept is the existence of specific receptor proteins on the plant surface that detect the effector and trigger a multiplying cascade of signals inducing many genes in the resistant response. Genes for phytoalexins (antibiotic compounds) as well as pathogenesis proteins are induced. Bacterial effectors are encoded by and secreted through a type III secretion system of the hypersensitive response and pathogenicity (hrp/path) region of xanthomonads, pseudomonads, erwinias, and pectobacteria. Effectors are secreted through pili either very close to or within plant cells. Effectors such as harpin, encoded by hrpN in *Erwinia amylovora*, may have separate roles in pathogenesis.

ENGINEERING RESISTANCE INTO PLANTS

The importance of receptor proteins (derived from R genes; see Chapter 31) is clear. The Pto locus (an R gene) was engineered into a cultivar susceptible to tomato spot. Transfer of Pto caused the susceptible line to become resistant to *Pseudomonas syringae* pv. *tomato* strains containing the effector protein locus avrPto. Similarly, a rice R gene, Xa21, was introduced to rice cultivars grown on over 9 million hectares (22 million acres) in Asia and Africa and provides resistance to rice blight caused by *Xanthomonas oryzae* pv. *oryzae*.

CULTURAL CONTROL

Diseases caused by bacteria occur very rapidly due to short generation times of the pathogen and tend to follow wet weather closely, when fields and orchards are difficult or impossible to work with heavy machinery and are usually sequestered in tissue and protected from treatment with bactericides. Consequently, they are difficult to manage by chemicals alone. Fastidious prokaryotes are sequestered in phloem and xylem and are often difficult to eradicate as chemical delivery to these locations is irregular, resulting in pockets of surviving pathogens. Therefore, prophylactic management by cultural methods is very important for bacterial diseases of plants. Cultural controls include using pathogen-free propagating materials, controlling weed hosts, removing diseased plants (roguing or pruning), doing crop rotation, avoiding overhead irrigation, and planting wind breaks.

CHEMICAL CONTROL

A few chemical controls, including copper- and zinc-containing fungicides, have utility against bacteria. Antibiotics are low-molecular-weight compounds usually produced by microbes as antagonists of other microbes. Streptomycin, which interferes with the translation of bacterial proteins, has been useful as a control for fireblight, caused by *Erwinia amylovora* and pepper spot, caused by *Xanthomonas vesicatoria*. High-level resistance arises in bacteria in a single mutation of ribosomal subunit protein gene. In many apple- and pear-growing regions, streptomycin is now useless for fire blight control. Oxytetracycline (OTC) is less effective than streptomycin in fire blight control. OTC also inhibits protein synthesis, but low-level resistance develops slowly by stepwise chromosomal mutation. Currently, gentamycin has the best potential for controlling fire blight; however, environmental and health reviews and registration are incomplete.

SUGGESTED READING

Andrews, J.H. and R.F. Harris. 2000. The ecology and biogeography of microorganisms on plant surfaces. *Annu. Rev. Phytopathol.* 38: 145–180.

Barras, F., G. van Gijsegem and A.K. Chaterjee. 1994. Extracellular enzymes and pathogenesis of soft-rot erwinia. *Annu. Rev. Phytopathol.* 32: 201–234.

Beattie, G.A. and S.E. Lindow. 1995. The secret life of foliar bacterial pathogens on leaves. *Annu. Rev. Phytopathol.* 33: 145–172.

Burr, T.J. and L. Otten. 1999. Crown gall of grape: Biology and disease management. *Annu. Rev. Phytopathol.* 37: 53–80.

Denny, T.P. 1995. Involvement of bacterial polysaccharides in plant pathogenesis. *Annu. Rev. Phytopathol.* 33: 173–197.

Dow, M., M.-A. Newman and E. von Roepenak. 2000. The induction and modulation of plant defense responses by bacterial lipopolysaccharides. *Annu. Rev. Phytopathol.* 38: 241–261.

Hahn, M.G. 1996. Microbial elicitors and their receptors in plants. *Annu. Rev. Phytopathol.* 34: 387–412.

Hammerschmidt, R. 1999. Phytoalexins: What have we learned after 60 years? *Annu. Rev. Phytopathol.* 37: 285–306.

Holt, J.G., N.R. Krieg, P.H.A. Sneath, J.T. Staley and S.T. Williams. 1994. *Bergey's Manual of Determinative Bacteriology,* 9th ed. Williams & Wilkins, Baltimore.

Kinkel, L.L. 1997. Microbial population dynamics on leaves. *Annu. Rev. Phytopathol.* 35: 327–347.

Madigan, M.T., J.M. Martinko and J. Parker. 2000. *Brock: Biology of Microorganisms,* 9th ed. Prentice Hall, Upper Saddle River, NJ, p. 991.

Mount, M.S. and G.H. Lacy (Eds.). 1982. *Phytopathogenic Prokaryotes,* Vols. 1–2, Academic Press, New York.

Romantschuk, M. 1992. Attachment of plant pathogenic bacteria to plant surfaces. *Annu. Rev. Phytopathol.* 30: 225–243.

Salmond, G.P.C. 1994. Secretion of extracellular virulence factors by plant pathogenic bacteria. *Annu. Rev. Phytopathol.* 32: 181–200.

Schaad, N.W., J.B. Jones and W. Chun. 2001. *Laboratory Guide for Identification of Plant Pathogenic Bacteria,* 3rd ed. American Phytopathological Society Press, St. Paul, MN.

Young, J.M., Y. Takikawa, L. Gardan and D.E. Stead. 1992. Changing concepts in the taxonomy of plant pathogenic bacteria. *Intl. J. System Bact.* 30: 67–105.

7 Laboratory Exercises for Plant Pathogenic Bacteria

George H. Lacy and Felix L. Lukezic

Phytopathogenic prokaryotes are very diverse. Further, several different phytopathogenic prokaryotes and fungi cause similar-appearing diseases that confuse identification of the specific pathogens. Finally, phytopathogenic prokaryotes may easily be overgrown by saprophytic bacteria and fungi. Therefore, they are surprisingly difficult to isolate from plant materials.

Identification of the actual pathogen is essential for selecting the best method(s) for disease control. Methods for isolation, purification, and characterization, including pathogenicity assays, of phytopathogenic prokaryotes are very important. This chapter illustrates some of these basic methods. See Schaad et al. (2001) for more details.

EXERCISES

EXPERIMENT 1. OBSERVATIONS OF BACTERIA IN PLANT TISSUE

Direct microscopic observation for the pathogen is important to determine whether bacteria are involved in plant disease. Selection of the diseased tissues is important because pathogenic bacteria may occupy different locations in the plant. For instance, leaf spotting bacteria are found in the apoplastic spaces in leaves, wilting pathogens are found in vascular tissues, and soft rotting pathogens are found among macerated and dead cells. Some prokaryotes, especially the mollicutes (including phytoplasmas and spiroplasmas), which lack a cell wall, and phloem-limited bacteria are very small and difficult to observe microscopically unless their DNA is stained with 4′, 6′-diamidino-3-phenylindole (DAPI; Lee et al., 2001). Mature phloem sieve tubes, the plant host habitat for these pathogens, lack nuclei, allowing detection of the pathogen DNA. DAPI staining is beyond the scope of this exercise. Simple methods for visualizing bacteria from plant tissues are the subject of this protocol.

Materials

Each student or team of students will require the following materials:

- Compound light microscope with 40× or 60× objective lens
- Microscope slides and coverslips
- Razor blade
- Eyedropper bottle with distilled water
- Infected plant material — We suggest the following bacteria and diseases for this exercise, although other materials will also be appropriate: *Ralstonia solanacearum* (Granville wilt of tobacco), *Pseudomonas syringae* pv. *tabaci* (angular leaf spot of tobacco), and/or *Pectobacterium* (*Erwinia*) *carotovorum* subsp. *carotovorum* (hand rot of tobacco/soft rot of potato)

Follow the protocols outlined in Procedure 7.1 to complete the experiment.

Questions

- What is Brownian motion and how does it differ from cell motility?
- Why would microscopic evaluation of tissue for bacterial streaming not be used for crown gall disease caused by *Agrobacterium tumefaciens*? *Note:* The International Society for Plant Pathology has indicated that this species name is a synonym for tumorogenic *Rhizobium radiobacter.*

EXPERIMENT 2. ISOLATION OF BACTERIA FROM PLANT TISSUE

A critical step for working with plant pathogenic bacteria is obtaining a pure culture of the organism. Pure cultures are absolutely essential for pathogenicity assays and characterizing the pathogen for identification. General methods for isolation and purification are the subject of this protocol.

Materials

Each student or team of students will require the following items:

| Procedure 7.1
Observing Bacteria in Diseased Tissue ||
Step	Instructions and Comments
1	Tissue selection — Lesions are often colonized by secondary organisms that may be either saprophytes living on dead material or secondary pathogens. To detect the primary pathogen, select tissue that has recently been colonized. Select tissue at or near the active margin of lesions to avoid observing saprophytic bacteria.
2	Cut tissues — With a single blade razor, slice strips of tissue thin enough to observe microscopically. (*Hint*: Make oblique sections of stems, fruits, tubers, and flowers.) Place the strips on a microscope slide in a drop of water.
3	Microscopic evaluation — Observe the tissue at 400× to 600× magnifications. Locate and focus sharply on the interface of the cut edge of the tissue and the water. Bacteria are not resolved at this magnification; they are only visible by refracted light and the Brownian movement of cells. Reduce the amount of light by either closing the iris beneath the stage or lowering the condenser. Bacteria streaming out of the plant tissue will be visible as bright points of light by refraction. Use apparently healthy tissue (lacking bacteria) as a control.

- Infected plant material — We suggest the following bacteria and diseases for this exercise, although other materials will also be appropriate: *Ralstonia solanacearum* (Granville wilt of tobacco), *Pseudomonas syringae* pv. *tabaci* (angular leaf spot of tobacco), and/or *Pectobacterium* (*Erwinia*) *carotovorum* subsp. *carotovorum* (hand rot of tobacco/soft rot of potato)
- 10% commercial bleach solution (90 mL of water and 10 mL of bleach; equal to 0.5 to 0.6% sodium hypochlorite)
- Distilled or sterile distilled water
- Several 12- × 75-mm snap cap sterile test tubes
- Glass rod
- Several sterile transfer pipettes
- 1.5-mL Eppendorf tubes and micro centrifuge
- Nutrient agar [8 g powdered medium (Difco, Detroit, MI), 20 g agar and 1 l of water]
- Bacterial loop
- Incubator set for 25°C (optional)

Follow the protocol outlined in Procedure 7.2 to complete this experiment.

Anticipated Results

Growth of bacteria on the initial dish of nutrient agar may be heavy and individual colonies may be difficult to select. Restreaking a small portion of an individual colony onto the second dish should result in growth of a number of individual colonies (Figure 7.1). Pure cultures should be obtained with as few as one or two restreakings.

Questions

- Ideally during isolation, how many bacteria form an individual colony on a nutrient agar dish?

- Why is it important to have pure cultures of plant pathogenic bacteria?

EXPERIMENT 3. PATHOGENICITY ASSAYS

Bacteria require free water for growth and they are surprisingly fragile when subjected to extremes of temperature, drying, or light intensity. Therefore, pathogenicity assays are most successful when these conditions for growing host plants are stabilized. Temperature is usually maintained between 16°C and 25°C (61°F and 75°F) and lighting is subdued. In glasshouses, shade inoculated plants with one to two layers of cheesecloth and inoculated plant organs, such as rhizomes, tubers and roots, should be maintained in the dark. Moisture should be maintained such that the surfaces of the inoculated plants do not dry out for 48 to 96 h after inoculation. A mist chamber may be inexpensively constructed of plastic sheeting (painting drop cloths) stretched over a frame constructed of snap-fit polyvinylchloride (PVC) pipes and moisturized by a bedside humidifier.

Susceptible Plant

Generally, pathogenesis is encouraged by inoculating pathogenic bacteria onto succulent, young tissue of compatible plants. New, fully expanded leaves and new stems are usually ideal. Matching plant species and cultivar with the diseased host from which the pathogen was isolated avoids genetic resistance. The same organ that was affected in the host of isolation should be inoculated (e.g., inoculate flowers if flowers were blighted; inoculate leaves if leaf spots developed; inoculate tubers if tubers were rotted; or inoculate stems if the plant wilted).

Procedure 7.2

Isolation of Pathogenic Bacteria

Step	Instructions and Comments
1	Select tissue to be sampled. Tissue selected for sampling can be any plant part with symptoms. Remove any unnecessary material. For example, do not use the whole leaf, but a portion of the leaf with the suspect lesion surrounded by some apparently healthy tissue. Select the portion to fit the container used in Step 3.
2	Surface-disinfest plant organs with 0.5 to 0.6% sodium hypochlorite for 30 to 60 sec, and then rinse three times with freshly deionized, distilled, or autoclaved water.
3	Immerse the tissue in 1 to 2 mL of distilled water in a clean or sterile 12- × 75-mm snap cap tube. Crush the tissue with a clean glass rod or other convenient implement. Let the mixture stand for 5 to 10 min to allow any bacteria to diffuse out of the tissue and into the water.
4	Agitate the mixture briefly to suspend any bacteria evenly in the water. Allow the mixture to stand until the larger portions of remaining tissue settle to the bottom of the tube. Using a transfer pipette, remove as much of the supernatant as possible to one or more sterile, 1.5-mL Eppendorf tubes.
5	Briefly centrifuge at 14 to 17 K for 30 to 60 sec in a microcentrifuge to pellet bacteria. Gently pour or pipette the supernatant off allowing only the pellet and about 50 to 100 μl to remain.
6	Streak directly from the remaining volume onto either a general bacteriological medium such as nutrient agar. Consult Schaad et al. (2001) for the medium and temperature of incubation recommended for the pathogen you expect to isolate. Incubate the medium for 72 to 96 h at 25°C.
7	Select colonies that are well isolated and represent the most numerous colony morphology (Figure 7.1). Generally, pathogens are in high populations in lesions.
8	Restreak the selected colony at least once onto fresh isolation medium. Each time, select isolated colonies representing the most numerous colony morphology. Bacteria isolated from nature may be contaminated with saprophytic species; restreaking for isolation ensures a pure culture.
9	Bacterial cells from isolated colonies may be suspended in 15% sterile aqueous glycerol (w/v) and stored at −20 to −80°C in snap-cap, 1.5-mL conical tubes. It is wise to make several glycerol tubes so that you have a fresh tube for each experiment. The bacterial strain isolated from plant tissue is ready for pathological studies and identification.

Materials

Each student or team of students will require the following items:

- Pure culture of pathogenic bacteria isolated in Procedure 7.2
- Nutrient agar or nutrient broth (8 g powder medium and 1 l of water) or both
- Rotary shaker
- Spectrophotometer set at 550 nm and disposable cuvettes
- Graduated cylinder and 1- and 5-mL pipettes
- Host plants — See introduction for host plants and diseases
- Microliter pipette and tips
- Hypodermic syringe (1-mL tuberculin) and 23- to 26-gauge needle
- Sewing needles (#7 sharp)
- Potting soil

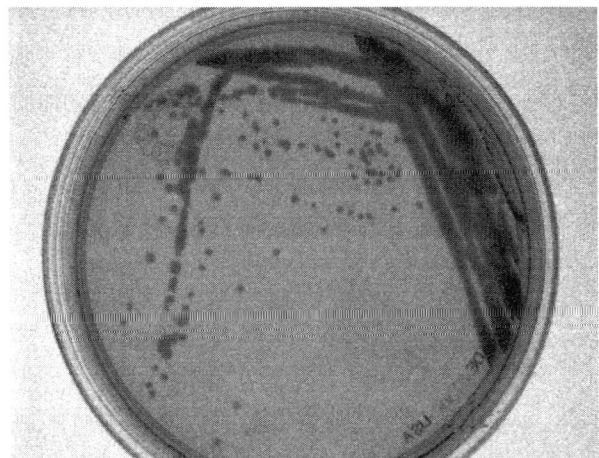

FIGURE 7.1 Isolation (streaking) of bacteria. Note the individual colonies used to make final isolations and cultures. Photo supplied by R.N. Trigiano, University of Tennessee.

Procedure 7.3
Pathogenicity Assays

Step	Instructions and Comments
1	Inoculum preparation — From a freshly restreaked, isolated colony from Procedure 7.2, inoculate either an agar dish or broth. Incubate the dishes statically or the broth cultures with shaking (150 to 200 rpm/min) at a temperature similar to the temperature at which the inoculated plants will be incubated. Grow the bacteria until colonies form on the agar (48 to 96 h) or the broth culture reaches the stationary phase of growth (turbidity reaches an optical density [OD] of 0.7 to 1.0 at 550 to 600 nm as read on a colorimeter with an optical path of 1.0 cm). Suspend bacteria from colonies in water such that OD at 550 to 600 nm = 0.7 to 1.0. Dilute broth-grown or agar-grown bacterial suspensions 100-fold with water. The final concentration of bacterial cells will be approximately 10^6 to 10^7 colony forming units (CFU)/mL. Use this suspension for inoculation immediately.
2	Inoculation — Inoculation methods vary with the organs affected by the pathogen. • Leaves — Using a sprayer (window cleaner sprayer from grocery store disinfested with 70% isopropanol and rinsed carefully with freshly distilled water), adjust the spray to a fine mist and spray the underside (abaxial) surface of the leaf until water-soaked areas are obvious. Water-soaked areas are where the bacterial suspension has been forced through stomata into the leaf, which appear darker and more translucent to light than surrounding leaf tissue. • Flowers — With a microliter pipettor, deliver 10:1 of bacterial suspension to the area of the nectartodes. • Stems — Mildly stress the plants for water before inoculation. Stress plants until leaves are slightly flaccid without being so wilted that they do not revive with watering. With a hypodermic syringe, slide the needle (23 to 26 gauge) about 2 to 5 mm into an axil where a leaf petiole joins the stem. With the needle in the plant, apply gentle pressure to the plunger until a drop of bacterial suspension forms in the axil. Remove the needle and water the plant. The bacterial suspension will be drawn into the plant through the needle wound by transpiration. • Tubers, rhizomes, and fruits — Fill the eye of a sewing needle (#7 sharp) with either bacteria from a colony on an agar dish or a turbid broth culture (OD at 550 to 600 nm = 0.7 to 1.0) and press the eye of the needle 0.5 to 1.0 cm into the organ, rotate the needle and remove. • Roots — Uproot young, mildly water-stressed seedlings and gently rinse their roots free of potting medium. Immerse their roots in a bacterial suspension (10^6 to 10^7 CFU/mL) for 1 to 5 min. Replant in fresh potting medium and water.
3	Symptoms — Incubate inoculated plants in a moist chamber for 48 to 72 h. Remove plants to a shaded area in the glasshouse and observe for symptom development.

Follow the protocol listed in Procedure 7.3 to complete the experiment.

Anticipated Results

Generally, blighting and rotting occur in 2 to 6 days, leaf spots appear in 5 to 14 days, wilting occurs in 10 to 21 days, and proliferation of plant tissue happens in 14 to 28 days. Well-watered plants often do not wilt. Mildly water-stress the plants by increasing the rate of transpiration with a small electric fan.

Questions

• Why is it necessary to reinoculate a host plant with the isolated bacterium?

• What would be the next step in this experiment to complete Koch's postulates?

EXPERIMENT 4. BACTERIAL CELL WALL

Two methods are available for determining bacterial cell wall structure: the KOH lysis technique and Gram's stain. Both techniques are easy to perform and take little time. Students may need some instruction on how to use oil immersion. Cellular morphology (rods, helical, vibroid, mycelial, etc.) is important to identify prokaryotic plant pathogens (see Chapter 6). The Gram stain is usually the standard for visualizing cellular morphology. Although the KOH method damages or lyses cells and is not suitable for microscopic examination, it is a rapid and reliable method for determining cell wall structure. In Gram-pos-

	Procedure 7.4
	Determination of Bacterial Cell Wall Structure

Step	Instructions and Comments
1	KOH procedure — In this test, mix a loopful of bacteria with 3% (w/v) potassium hydroxide. Gram-negative bacteria lyse, releasing DNA, and causing the mixture to become very viscous. Strands of viscous DNA are visualized by lifting the inoculating loop out of the mixture. Although the Gram-positive cells also lyse, their thick peptidoglycan cell walls trap DNA and strands do not form.
2	Gram's stain procedure — On a clean slide, dry a very thinly spread bacterial film in air without heat. (*Hint*: Use controls. On the same microscope slide with the unknown bacterium, prepare flanking smears of Gram-positive *Bacillus subtilis* and Gram-negative *Escherichia coli*.) Then lightly flame the underside of the slide twice to fix the bacteria to the slide. Flood the smear with crystal violet solution for 1 min. Wash in tap water a few seconds. Drain off excess water and lightly blot dry on a paper towel. Flood the smear with iodine solution for 1 min. Wash in tap water a few seconds blot dry. Decolorize with decolorizer solution until the solvent flows colorlessly from the slide (10–15 sec). Blot dry. (If decolorizer is used longer the Gram-positive bacteria may lose color.) Rinse in tap water for about 2 sec. Counterstain for 30 sec with safranin solution. Wash briefly in tap water. Blot dry and examine with an oil immersion lens. Gram-positive bacteria retain color after decolorizing whereas Gram-negative bacteria rapidly decolorize and are counterstained with the red dye safranin.
3	Endospores — Endospores of clostridia and bacilli may be detected by steaming slides flooded with malachite green. Wash the smear under running water until the green color no longer runs off the slide. Counterstain 30 to 60 sec with safranin. Examine preparation by using oil immersion. Endospores appear green and the cells surrounding them appear red. Use as controls endospore-forming *Bacillus subtilis* and non-spore-forming *Escherichia coli*.

itive reactions, the higher amounts of peptidoglycans in the cell wall may retain the crystal violet–iodine complex more completely than in Gram-negative cells. In Gram-negative cells, the crystal violet–iodine complex is more easily removed by alcohol treatment. Excessive destaining of Gram-positive cells can cause them to appear to be Gram-negative. Likewise, inadequate destaining can make Gram-negative cells appear to be Gram-positive. Use of controls (mixed Gram-negative and Gram-positive bacteria) aid pathologists in determining the optimum amount of destaining.

The malachite green method for staining endospores also depends on differential retention of dye. Steaming fixes the malachite green dye in the tough endospores of clostridia and bacilli. Washing removes the malachite green dye from the surrounding cytoplasm. Counterstaining with safranin red allows the pathologist to determine whether the endospores are central, subterminal, or terminal in the cells — these are important criteria for bacterial identification.

Materials

Each student or team of students will require the following items:

- KOH (3 g in 100 mL water)
- Microscope slides
- Alcohol lamp and bacterial loop
- Cultures of *Pectobacterium* (*Erwinia*) *carotovorum* subsp. *carotovorum*, *Bacillus subtilis*, and *Escherichia coli*
- Crystal violet solution (Fisher)
- Gram's iodine solution (Fisher)
- Decolorizer solution (Fisher)
- Safranin solution (Fisher)
- Compound microscope with oil immersion lens and lens paper
- Paper towels
- Malachite green (5 g malachite green dye dissolved in 100 mL water)
- Safranin (0.25 g safranin dissolved in 100 mL of 10% ethanol)

Follow the protocols in Procedure 7.4 to complete this exercise. Anticipated results are provided as comments in the protocols.

Question

- What property of the Gram-positive bacterial cell wall allows the stain to be retained?

EXPERIMENT 5. SOME PHYSIOLOGICAL PROPERTIES OF PLANT PATHOGENIC BACTERIA

Physical and morphological characteristics that can be used to distinguish bacterial genera are limited. Therefore, phytobacteriologists have relied on some simple physiological tests to characterize different groups. Although with molecular techniques, including the PCR methods outlined in Schaad et al. (2001), reflecting phylogenetic groupings of bacteria are being developed, currently relatively few phytopathogens can be identified by these procedures. No doubt this will change in the near future. Therefore, the information gained from physiological tests is still critical for pathogen identification. This protocol demonstrates several basic methods for physiological testing of bacteria that are also important for taxonomical identification.

Materials

- King's medium B (Difco, Detroit, MI)
- Small disposable test tubes and bacterial loop
- Cultures of *Pseudomonas fluorescens, Escherichia coli, Pectobacterium carotovorum* subsp. *carotovorum, Pseudomonas syringae, Burkholderia cepacia, Ralstonia solanacearum,* and bacterium isolated in Procedure 7.2
- YS broth (800 mL water; 0.5 g HNH_4PO_4, 0.5 g K_2HPO_4, 0.2 g $MgSO_4 \cdot 2H_2O$, 5 g NaCl, 1 g yeast extract, and 16 mg cresol red) to which add two parts urea stock solution (20 g urea in 200 mL water and filter sterilized)
- CVP medium — Using a blender at low speed, rapidly mix 500 mL of boiling water, 4.5 mL of 1 N NaOH, 3 mL of 10% aqueous $CaCl_2 \cdot 2H_2O$, 1 g $NaNO_3$, 2 g agar, and 1.0 mL 0.075% aqueous crystal violet (w/v). While blending at high speed, add 10 g sodium polypectate. Pour into a 2-l flask containing 0.5 mL 10% aqueous sodium dodecyl sulfate (SDS), mix, and autoclave. Pour into 10-cm-diameter petri dishes.
- DM1 medium — 5 g cellibiose, 1 g NH_4Cl, 1 g NaH_2PO_4, 1 g K_2HPO_4, 3 g $MgSO_4 \cdot 7H_2O$, 10 mg malachite green, 15 g agar, and 1 l water
- Carbohydrate testing medium — Add 0.1% (w/v) filter-sterilized carbohydrate (e.g., galactose, glucose, or lactose) to a minimal medium consisting of 0.5 g HNH_4PO_4, 0.5 g K_2HPO_4, 0.2 g $MgCl_2 \cdot 7H_2O$, 5 g NaCl, and 1 l water. Adjust to pH 7.2.
- Dye's medium C — Add 0.5% (w/v) filter-sterilized carbohydrate (e.g., glucose, galactose, or lactose) to basal Dye's medium C consisting of 0.5 g HNH_4PO_4, 0.5 g K_2HPO_4, 0.2 g

$MgCl_2 \cdot 7H_2O$, 5 g NaCl, 1 g yeast extract, and 0.7 mL 1.5% bromocresol purple (w/v) in 95% ethanol. Adjust pH to 6.8.
- Kovac's reagent — Add 75 mL amyl or isoamyl alcohol, 5 g *p*-dimethylaminobenzaldehyde, and 25 mL concentrated hydrochloric acid — make under fumehood.
- PHB — 0.2 g $(NH_4)_2SO_4$, 0.2 g KCl, 0.2 g $MgCl_2 \cdot 7H_2O$, and 5 g DL-hydroxybutyrate per liter of water
- Sudan black B (0.3% dye in 70% ethanol)
- 0.5% aqueous safranin

Follow the protocols in Procedure 7.5 to complete this experiment. Anticipated results are provided as comments in the protocols.

Questions

- What product is important in the identification of phytopathogenic bacteria produced on King's medium B?
- D1M semiselective medium is important for the identification of what important group of phytopathogenic prokaryotes?
- Why do the carbohydrate sources in Dye's medium C vary?
- What bacterial staining agent collects in granules of polyβ–hydroxybutyrate?

EXPERIMENT 6. THE HYPERSENSITIVE TEST FOR PATHOGENS

Hypersensitive response (HR) in tobacco is useful to determine whether a bacterium may be a plant pathogen. Briefly, pathogens of plants (other than tobacco) cause the HR in tobacco whereas pathogens of tobacco cause symptoms of disease. Saprophytes or nonpathogens of plants cause no symptoms of disease and do not elicit a response. The HR test takes advantage of the specific induced resistance mechanism determined genetically by plants. The HR test is especially useful to quickly screen possible pathogens before the more time- and labor-intensive pathogenicity assays are performed. Although HR tests are extremely useful for a large number of leaf spotting and blighting pathogens (e.g., erwinias, pseudomonads, ralstonias, and xanthomonads) other pathogens may not be identified readily using this test, such as soft-rotting bacteria (e.g., pectobacteria) and agrobacteria. For those pathogens, pathogenicity assays are necessary.

Materials

Each student or team of students will require the following items:

Procedure 7.5
Physiological Characteristics of Bacteria

Steps	Instructions and Comments
1	Fluorescent pigments — Fluorescent pyoverdine siderophores may be detected by growing bacteria in test tubes on King's medium B, which is high in glycine-rich peptones that chelate iron. Creating an iron-depleted environment induces siderophore synthesis in the bacteria. Under long-wave ultraviolet light, water-soluble pyoverdine pigments fluoresce green to blue. (*Hint:* Use fluorescent *Pseudomonas fluorescens* and nonfluorescent *Escherichia coli* as controls.)
2	Urease hydrolysis — Urease activity may be detected by growing bacteria on modified YS broth. Use a tube without urease as a color control. An increase to about pH 9.0 accompanied by the development of a magenta color indicates urease activity.
3	Pectinase — Streak bacteria onto CVP medium and incubate at 25°C for 1 week. Use the following controls: *Pectobacterium carotovorum* subsp. *carotovorum* and *Pseudomonas fluorescens* create conical and flat pits, respectively, and nonpectolytic *Escherichia coli* does not pit CVP medium.
4	Selective growth medium — Streak bacteria onto D1M medium and incubate at 25°C. Use the following controls: *Agrobacterium* species grow on D1M whereas *Pseudomonas, Xanthomonas,* and *Pectobacterium* species do not.
5	Utilization of carbon sources — Dispense 5 mL of sterile carbohydrate testing medium into test tubes, inoculate with a loop of bacteria, and incubate at 25°C. Make three serial transfers on this medium. Turbidity will develop if the substrate is utilized and growth occurs. Use a control that has no added carbohydrate.
6	Acid production from carbohydrates — Dispense 5 mL of Dye's medium into test tubes. Inoculate the medium with a loop of bacteria and incubate at 25°C. The medium turns yellow if acid is produced from the carbohydrate.
7	Indole production — Culture organism(s) in broth (5 g yeast extract, 10 g tryptone, and 1 l water) at 25°C. Add 1 mL of turbid broth to 1 mL Kovac's reagent. Indole is produced from tryptophan and results in a cherry-red reaction with Kovac's reagent.
8 ·	Poly β-hydroxybutyrate (PHB) granules — Culture bacteria in 10 mL of PHB broth. Smear, air dry, and fix the cells to a microscope slide. Flood with Sudan black B for 10 to 15 min. Drain and blot dry and decolorize with water. Counter stain with safranin. Examine cells with oil immersion. Cells appear red or pink and PHB inclusions are black. *Burkholderia cepacia* or *Ralstonia solanacearum* produce PHB granules and *Pseudomonas syringae* does not.

- Nutrient agar and nutrient broth (Difco, Detroit, MI)
- Rotary shaker
- Spectrophotometer and cuvette
- Centrifuge and tubes
- Tobacco plants
- Syringe with 23- to 26-gauge needle
- Spray bottle
- Cultures of *Ralstonia solanacearum, Pectobacterium carotovorum* subsp. *carotovorum, P. syringae* pvs. *syringapisi,* or *tabaci,* and *Escherichia coli*

Follow the protocols in Procedure 7.6 to complete the experiment. Anticipated results are provided as comments in the protocols.

Question

- In the hypersensitivity test, what reactions would you expect tobacco leaves to exhibit when treated separately with each of the following organisms: a pathogen of tobacco, *Escherichia coli*, and a pathogen of lilac?

Procedure 7.6

Hypersensitive Test for Bacterial Pathogens

Step	Instructions and Comments
1	Grow bacterial cells in NA broth overnight or on NA agar for 48 h. Suspend cells grown on agar in sterile distilled water. Transfer centrifuge tubes and centrifuge at 14,000 rpm. Pour off water and resuspend the cells in sterile water to a density of 10^6 cfu/mL. (See Procedure 7.3 for determining density by spectrophotometry.)
2	Grow tobacco plants in styrofoam drinking cups with a hole in the bottom for drainage. Tobacco seedlings (one per cup) may be grown in peat potting media under fluorescent lights or in window boxes. Plants may be reused, using a different leaf for each test.
3	Soak a leaf panel (the space between two major veins perpendicular to the mid-vein) with a bacterial suspension. Water-soaked areas (with water forced through stomata into the intercellular region of the leaf) appear dark and are more translucent than normal leaf tissue. If the leaf is large enough, use four leaf panels on the same leaf to accommodate and control organisms. Inoculate three panels, one each with (a) an incompatible (hypersensitive) pathogen such as *Pseudomonas syringae* pv. *pisi*, a pathogen of pea and a nonpathogen of tobacco; (b) a compatible (disease) pathogen such as *Pseudomonas syringae* pv. *tabaci*, a pathogen of tobacco; and (c) a saprophytic (nonplant pathogen) control, *Escherichia coli*. Inoculate the fourth panel with an unknown bacterium.
4	Inoculation methods — Three methods are available: (1) Gently press the flat surface of a 25-gauge needle against the abaxial (bottom) side of a leaf stretched over your finger. With much practice a whole panel may be flooded without tearing the leaf or injuring your finger. (2) With a 1-mL tuberculin syringe with the needle removed, gently press the tip against the abaxial (bottom) side of a leaf stretched over your finger. With some practice a panel may be flooded without crushing the leaf. (3) By a hand-pumped sprayer (window cleaner pump), adjust the nozzle for a fine mist. With practice, an area of leaf panel may be water soaked without tearing the leaf.
5	Incubate plants in the window box or under fluorescent lights for 24 to 96 h. Complete collapse of leaf tissue with drying to a paper-like consistency is typical of an incompatible or hypersensitive reaction. In 72 to 120 h, lesions with water-soaked edges with or without chlorotic halos indicate a compatible or disease interaction. After incubation, a normal appearing leaf without water soaking or dry, dead tissue indicates no reaction with a saprophytic bacterium.

REFERENCES

Lee, I.-M., R.E. Davis and J. Fletcher. 2001. Cell-wall free bacteria, pp. 283–320 in *Laboratory Guide for Identification of Plant Pathogenic Bacteria*, 3rd ed. Schaad, N.W., J.B. Jones and W. Chun (Eds.), American Phytopathological Society Press, St. Paul, MN.

International Society for Plant Pathology. 2007. http://www.isp-pweb.org/names_bacterial_revised.asp.

Schaad, N.W., J.B. Jones and W. Chun. 2001. *Laboratory Guide for Identification of Plant Pathogenic Bacteria*, 3rd ed. American Phytopathological Society Press, St. Paul, MN.

8 Plant-Parasitic Nematodes

James P. Noe

CHAPTER 8 CONCEPTS

- Plant-parasitic nematodes are important pests in most plant production systems. There are many different species of plant-parasitic nematodes. Almost every plant is attacked by some species of nematode, and many species have very wide host ranges.

- Plant-parasitic nematodes are all obligate parasites, meaning that they can only feed on living plant cells.

- Plant-parasitic nematodes cannot feed on dead or decaying plants, nor on bacteria or fungi. Without a living host, plant-parasitic nematodes will die of starvation. As a result, they almost never kill their hosts.

- Plant-parasitic nematodes always have a hardened spear-like stylet for feeding on living plant cells. However, not all stylet-bearing nematodes are plant-parasites.

- Root-knot nematodes (*Meloidogyne* species) are the most common and widespread plant-parasitic nematodes. They cause diagnostic galls or knots on roots.

- Most nematode species do not produce any obvious diagnostic symptoms on plants.

- Plant damage resulting from nematodes is dependent on the number of nematodes or eggs present at planting. Soil and root assays can provide information on nematode populations prior to planting that may be used to predict crop loss and formulate management strategies.

Plant-parasitic nematodes are members of a primitive group of animals called nonsegmented roundworms, and are all **obligate parasites**, meaning that they can only feed on living plants. They are usually found in the soil and in plant roots, but a few species may attack above-ground parts of the plant. Most species are microscopic, with lengths ranging from 300–4000 μm and diameters of 15–35 μm, which are within the range of large fungal hyphae. Nematodes undergo four molts during their life cycles. Most nematodes have a worm-like shape, with the body much longer than it is wide. The word *nematode* is derived from Greek words meaning "thread like." The bodies of these simple animals are arranged in a tube-shape contained within the tough, flexible cuticle. Within the outer layer of cuticle, the front part of the nematode contains the esophagus, and the rear part contains the intestines and reproductive organs. Each of these organ systems is also arranged as a simple tube-shaped structure. Nematode morphology is often referred to as a tube-within-a-tube. There is no circulatory or respiratory system, but nematodes contain most of the organ systems of higher animals, including digestive, reproductive, excretory, nervous systems, and several types of muscles. The adults are larger than **juveniles**, and in some species may be shaped differently. Most nematodes may be observed easily under a dissecting microscope at 40–60×. Detailed observations needed to identify nematode species, however, require much higher magnifications (600–1000×). The amount of time it takes for plant-parasitic nematodes to complete their life cycles ranges from a few weeks to more than a year depending on the nematode species, plant host status and ambient temperatures.

All plant-parasitic nematodes have a hardened spear-like feeding structure, called a **stylet**, in the anterior portion of their head region (Figure 8.1). For most plant-parasitic nematodes, this stylet is hollow and operates much like a hypodermic needle, allowing the removal of nutrients from within plant cells. The stylet also is used to penetrate plant tissues directly, allowing the nematode to burrow through the plant tissues and move toward preferred feeding sites. Glands located in the posterior portion of the esophagus produce and secrete compounds through the stylet into the plant cells, sometimes causing extreme structural and physiological modifications of the targeted cells. Plant cells fed on by nematodes usually are not killed. Nematodes feeding on roots from outside are referred to as **ectoparasites**, and those feeding from

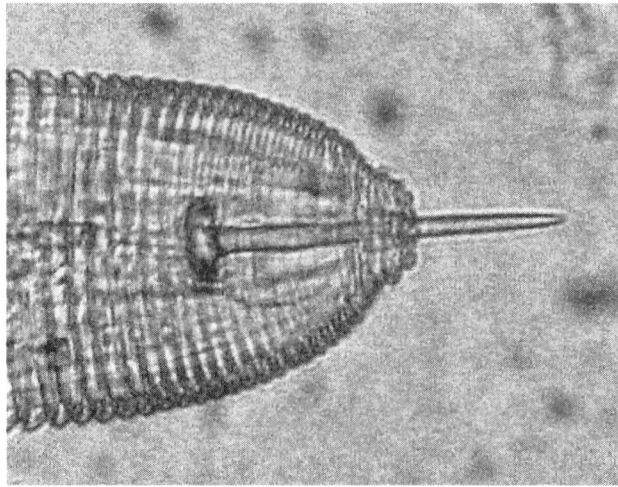

FIGURE 8.1   Head of a ring nematode. Note the stylet (or spear) used to penetrate plant tissues when feeding.

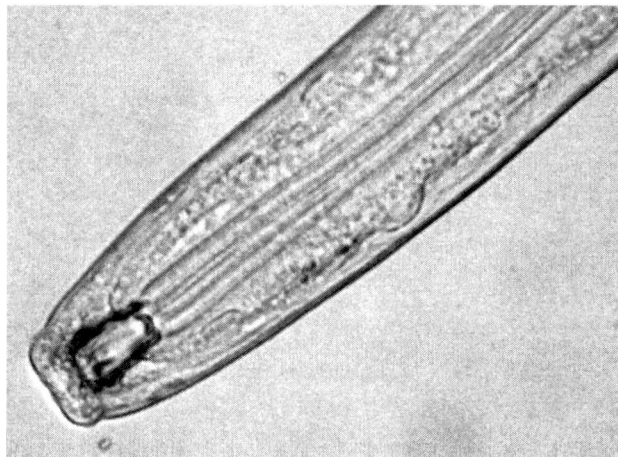

FIGURE 8.2   Free-living nematode. These nematodes are commonly found in soil and play an important role in nutrient recycling.

inside the root are **endoparasites**. Within both feeding types, some species may be migratory, moving around throughout their life cycles, whereas others are sedentary and settle in one location to establish a permanent feeding site.

Free-living nematodes (non-plant-parasitic nematodes) are often found in soil samples in large numbers along with plant-parasitic nematodes (Figure 8.2). These nematodes can be distinguished from plant-parasitic nematodes by their lack of stylets or by the absence of knobs or flanges at the base of their stylets. When observing fresh specimens collected from soil, free-living nematodes are usually moving much more rapidly than plant-parasitic nematodes, which tend to lie on the bottom of the container and move very slowly in a serpentine manner. Free-living nematodes feed primarily on microorganisms associated with the decomposition of organic matter. These nematodes recycle nutrients in soil and are an important component of the ecosystem. Some species of free-living nematodes are predators, feeding primar-

ily on other nematodes. Types and numbers of free-living nematodes in soil samples are often used by ecologists as an indicator of the health of an ecosystem. Greater numbers of free-living nematodes, and more important, more diversity in the number of species present, usually indicates better soil health.

Plant-parasitic nematodes spend large portions of their life cycles in the soil environment. Soil texture, structure, and chemical composition are of primary importance in determining the number and types of plant-parasitic nematodes found in a given geographical area on a suitable host plant. On a given host, such as cotton, root-knot nematodes (*Meloidogyne incognita*) are usually a problem in sandier soils having relatively large spaces between the soil particles and good drainage. In contrast, reniform nematodes (*Rotylenchulus reniformis*) are found more often on cotton grown in clay soils with finer textures, relatively small spaces between the soil particles, and high bulk densities. Specific soil relationships have been demonstrated for many nematodes species. This type of information is valuable to crop managers and advisors when determining the potential risk for a specific nematode problem.

Within the soil, plant-parasitic nematodes can move only a short distance, whereas long-distance dissemination is usually by movement of soil, water, or plant-propagative parts. Movement of soil across fields, within regions, countries, and even globally is a major concern for the dispersal of plant-parasitic nematodes. Increased emphasis on biosecurity and preventing the importation of new pests has led to numerous quarantines on the movement of plants or any other product that may contain agricultural soil.

Plant-parasitic nematodes attack most economically important plants in agriculture, horticulture, ornamentals, and turf. Nematode problems are usually more prevalent in warmer, humid climates, but some species attack plants even in the coldest and most arid climates. The prevalence in warmer climates is largely due to the fact that nematodes cannot regulate their internal body temperatures. Rates of development and maturation are dependent on ambient temperatures. Within an optimum range, nematodes are usually more active and reproduce more rapidly at warmer temperatures. Also, host plants are usually available longer and are growing more rapidly in warmer climates, which is a critical factor to these obligate parasites.

Above-ground symptoms attributed to plant-parasitic nematodes include generally vague growth problems related to root impairment, such as stunting, yellowing, and wilting. Root symptoms may include galls, lesions, stunting, stubby appearance, excessive branching, and a generalized darkening or rotting of the root tissues (with very high numbers of nematodes). Crop losses from nematodes range from 10–60% of yield, depending on species, host status, and environmental conditions. Crop losses due to plant-parasitic nematodes are not evenly distributed among farming systems. Losses in more developed countries range from 5–10% each year, whereas losses in less developed countries may range from 30–60%, particularly in tropical and subtropical countries. Similarly, some crops experience much less nematode damage than other crops, even without nematode management. Many plant-parasitic nematodes enhance crop losses by forming **disease complexes** with other soilborne pathogens (Chapter 22), such as fungi and bacteria, making it difficult to decide exactly how much damage the nematodes are causing.

Most plant-parasitic nematodes do not produce any obvious or easily distinguishable symptoms on their host plants. Because most nematodes are microscopic and feed below ground, many nematode problems may remain undiagnosed. A key indication of a nematode problem, however, is the manifestation of one or more of the typical root-impairment symptoms of nematode damage in an irregular or patchy distribution within a growing area.

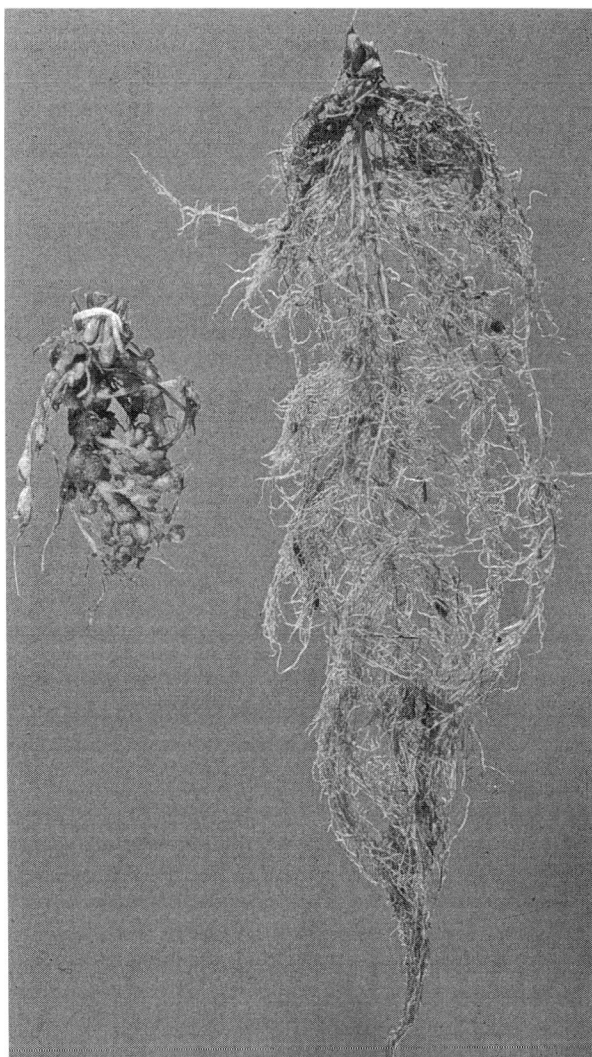

FIGURE 8.3 White clover root systems infected with the southern root-knot nematode. The root system on the left is severely galled and stunted. (Photo provided by Dr. Gary Windham, USDA-ARS Crop Science Research Laboratory at Mississippi State University.)

Plant-parasitic nematodes are almost never evenly distributed across a field, and the damage they cause will reflect that uneven distribution. An assay of soil samples completed by technicians trained in **nematology** is usually the only way to confirm a nematode diagnosis. Proper collection of soil samples is essential for reliable and accurate diagnosis of nematode problems. Plant-parasitic nematodes are found typically in the root zones of suitable host plants with most of the population occupying the top 15 cm of soil. Samples must be collected from within this root zone for effective diagnosis. Soil samples for nematode diagnosis also are more reflective of the targeted growing area if numerous small samples are collected systematically across the entire area and then composited (bulked) for analysis.

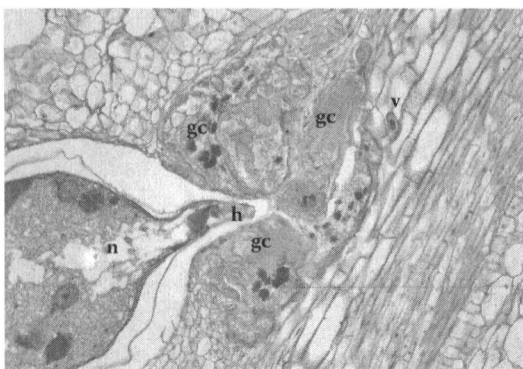

FIGURE 8.4 Giant cells (gc) in tomato roots caused by the feeding activities of root-knot nematode (n). Giants cells are specialized feeding cells that form around the head of the nematode (h) near the vascular tissues of the root (v). They are similar in appearance to the syncyia of cyst nematodes (see Figure 8.7) but differ in the way they are formed. (Photo courtesy of R. Hussey, University of Georgia.)

ROOT-KNOT NEMATODES

Root-knot nematodes (*Meloidogyne* species) are a widespread and diverse group of sedentary endoparasitic nematodes. Although more than 100 species have been described, four root-knot species (*M. arenaria, M. hapla, M. incognita,* and *M. javanica*) cause most of the damage reported on agricultural crops. Of these four, *M. incognita* is the most common globally, perhaps because it has a very wide host range among commonly grown crops. Root-knot nematodes are found worldwide but are more common in warmer climates and in sandier soils. Plants infected with these nematodes may be stunted, chlorotic, and may show moderate wilting during the hottest part of the day. Galls or "knots" form on host roots due to nematode infection and are often used to diagnose the disease (Figure 8.3). This is one of the few nematode genera that can be positively diagnosed from examination of the roots, which may be another reason why it appears to be the most common plant-parasitic nematode.

Root-knot nematode females lay eggs in a gelatinous matrix, and first-stage **juveniles** undergo one molt while still in the egg. Worm-shaped, or **vermiform** (15-μm diameter; 400-μm length), second-stage juveniles hatch from the eggs using their stylets to break through the tough egg casing. Second-stage juveniles are the only infective stage of root-knot nematodes. After penetrating suitable host roots at the root tips, the juveniles migrate to the developing vascular cylinder and begin feeding on several cells near the endodermis. Secretions from the esophageal glands of the nematode are injected through the stylet into nearby cells. These cells enlarge, become multinucleate and serve as feeding cells for the rest of the nematode life cycle. The enlarged feeding cells for root-knot nematodes are called **giant cells** (Figure 8.4). These specialized feeding cells have extensive cell wall in-growths, which increase the surface area of the cellular membrane. This increased surface area allows the feeding cells to transfer large amounts of nutrients into the cell, to feed the nematode. The giant cells function as a large siphon, diverting the downward flow of nutrients from the phloem into feeding the nematodes. A gall rapidly begins to develop around the feeding juvenile as a result of cell division and enlargement, which is also stimulated by the nematode feeding activities.

Once juveniles begin feeding, they undergo a series of three additional molts. The third- and fourth-stage juveniles are short-lived stages and are slightly swollen to sausage-shaped in appearance. Root-knot nematodes exhibit **sexual dimorphism** at maturity. Females (Figure 8.5) become enlarged and spherical (400-μm diameter; 700-μm length), having a flask-like shape called **pyriform,**

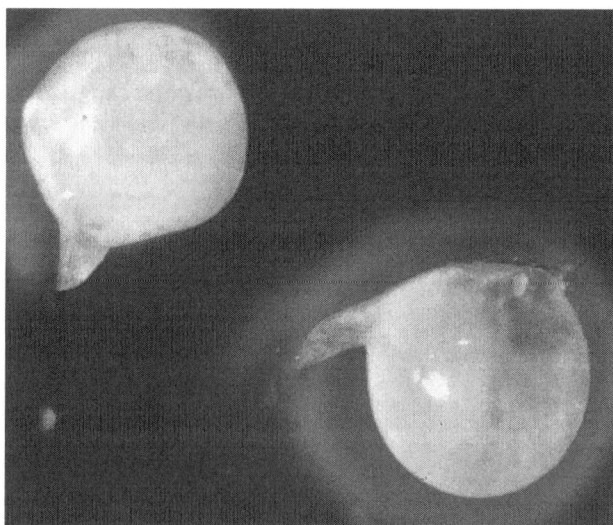

FIGURE 8.5 Mature root-knot nematode females that have been excised from galled roots. (Photo courtesy of Dr. Gary Windham, USDA-ARS Crop Science Research Laboratory at Mississippi State University.)

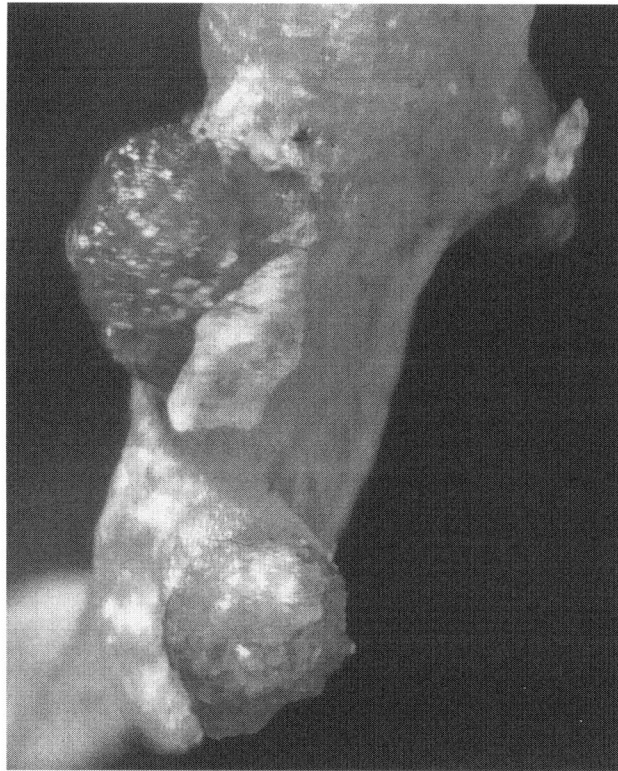

FIGURE 8.6 Root-knot nematode egg masses on root galls. Egg masses may contain as many as 500 eggs. (Photo courtesy of Dr. Gary Windham, USDA-ARS Crop Science Research Laboratory at Mississippi State University.)

whereas males molt to a relatively large vermiform shape (30-μm diameter; 1400-μm length) and migrate from the root. The posterior end of the adult female usually protrudes from the surface of the root gall, where an egg mass containing 300–500 eggs is produced (Figure 8.6). The life cycle typically requires 21 to 50 days, depending on root-knot species, plant host, and environment. Root-knot nematodes survive intercrop periods primarily as eggs in the soil. Survival rates are enhanced by protection from the egg-mass matrix and host plant debris.

CYST NEMATODES

Cyst nematodes (*Heterodera*, *Globodera*, and *Punctodera* species) are sedentary endoparasites that infect vegetables, small grains, soybean, corn and legumes. *Heterodera glycines*, the soybean cyst nematode, is one of the most studied plant-parasitic nematodes. This nematode is found in most soybean production areas and is among the most important diseases of soybeans. Soybean plants infected with *H. glycines* may be stunted, exhibit foliar chlorosis, have necrotic roots, and have suppressed root **nodulation**. Dead seedlings or plants may be found in fields with high numbers of the soybean cyst nematode. Other important cyst nematodes are the potato cyst nematodes (*G. rostochiensis* and *G. pallida*) and the sugar-beet cyst nematode (*H. schachtii*).

Cyst nematodes are unusual among plant-parasites in that their geographic distributions typically extend into cooler climates, including the northern United States and European areas, and higher elevations in the tropics.

The life cycle of soybean cyst nematodes is similar to that of root-knot nematodes in several ways. These nematodes are another example of sexual dimorphism; juveniles undergo four molts before becoming adults, and eggs are deposited in a gelatinous matrix. Eggs of cyst nematodes may respond to specific hatching factors, such as host-root exudates, that stimulate the eggs to hatch in greater numbers. Second-stage juveniles may penetrate roots anywhere although there is some preference for the region behind the root tip. Cyst juveniles migrate by penetrating through cells, causing more destruction than do root-knot juveniles. The second-stage juveniles migrate to vascular tissues, begin feeding, and become sedentary. Nematode feeding results in the production of very large specialized feeding cell called a **syncytium** (Figure 8.7). A syncytium differs from the giant cells formed by root-knot nematodes in that syncytia are formed by dissolving the cell walls of adjacent cells (up to 250 cells) to create large feeding cells. In contrast, root-knot nematodes cause a single cell to enlarge without dividing.

The bodies of adult cyst females eventually erupt through the root surface and can be seen easily with the naked eye or with the aid of a hand lens. Adult males are

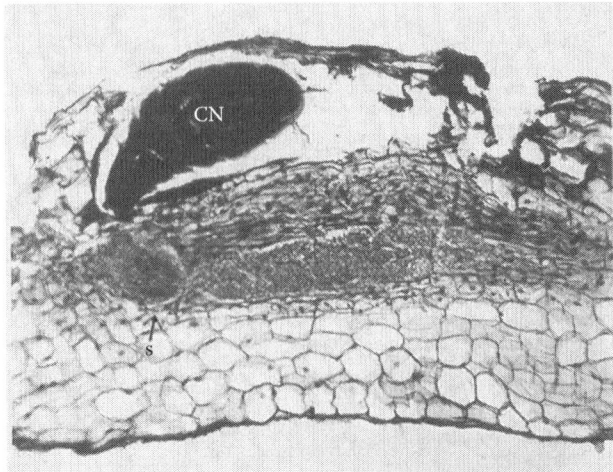

FIGURE 8.7 A syncytium (S), or specialized food cell, in a soybean root system initiated by feeding of soybean cyst nematode (CN). (Photo courtesy of R. Hussey, University of Georgia.)

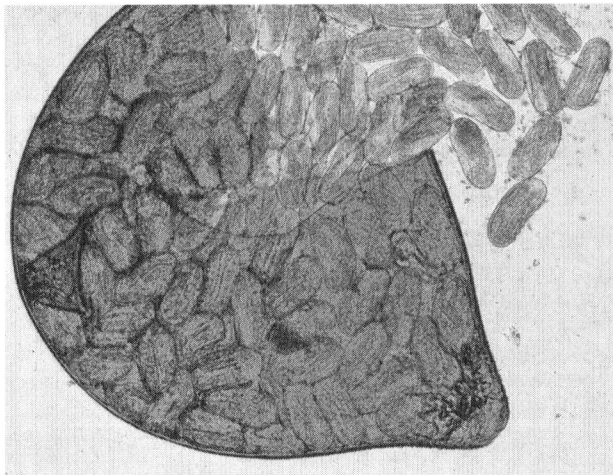

FIGURE 8.8 The body of a soybean cyst nematode female becomes brown at death and is filled with 200–600 viable eggs. (Photo courtesy of Dr. Gary Windham, USDA-ARS Crop Science Research Laboratory at Mississippi State University.)

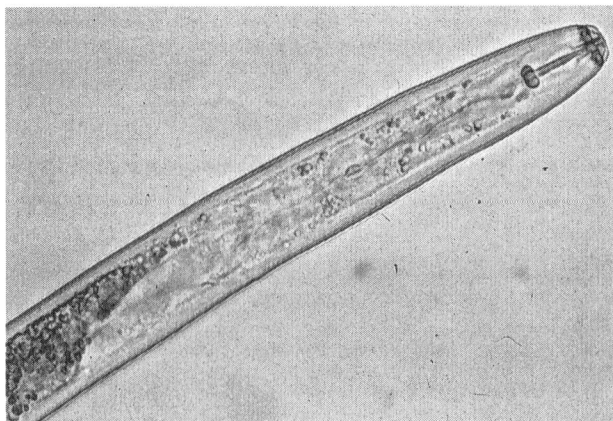

FIGURE 8.9 Anterior end of a lesion nematode. Note the well-developed stylet of this migratory endoparasite. (Photo courtesy of Dr. Gary Windham, USDA-ARS Crop Science Research Laboratory at Mississippi State University.)

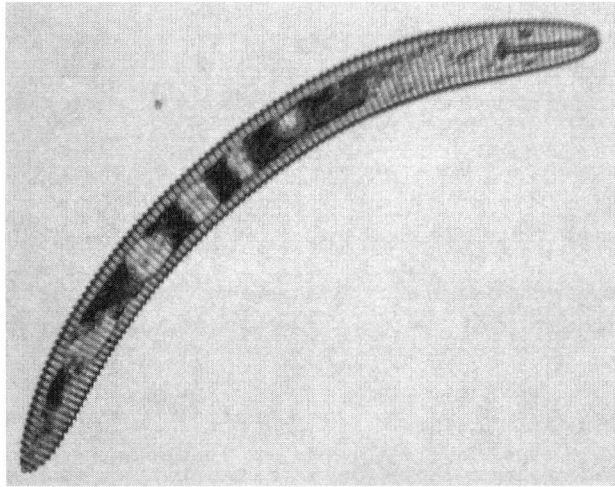

FIGURE 8.10 A ring nematode. The cuticles of these nematodes give them a segmented or ring-like appearance (courtesy of D. Cook, University of Tennessee.)

necessary for reproduction. Many nematodes, such as root-knot nematodes, do not require the presence of males for reproduction. The color of the adult female body or cyst changes from white to yellow, becoming brown at death and is filled with 200–600 eggs (Figure 8.8). Cysts protect the eggs from unfavorable environmental conditions, and eggs within the cysts may survive for more than seven years. The cysts are readily spread by contaminated soil, wind, and water.

LESION NEMATODES

Lesion nematodes (*Pratylenchus* species) attack many crops including vegetables, row crops, legumes, grasses, and ornamentals (Figure 8.9). Important species include *P. brachyurus, P. penetrans, P. scribneri,* and *P. vulnus.* These nematodes can be found in host roots in large numbers—up to 3000 per gram of root—and cause large, spreading necrotic lesions on fibrous or coarse roots. Lesions may cover the entire root system, and root pruning may occur in heavily infested fields. Lesion nematodes are among the relatively few nematode species that cause economic damage on woody ornamentals and trees. Lesion nematodes also have been associated with **disease complexes** involving other plant pathogens. *Pratylenchus penetrans* and the fungus *Verticillium dahliae* cause potato early death, which severely reduces potato yields in many parts of the world.

The life cycle of lesion nematodes is typical for many plant-parasitic nematodes. Adult females lay eggs singly in root tissue or in the soil. The first-stage juvenile molts to the second-stage within the egg. The second-stage juvenile hatches from the egg and then molts three more times to become an adult. The presence of host roots has been shown to stimulate egg hatch. All juvenile stages outside the egg and adults can infect host roots. Lesion nematodes

are classified as migratory endoparasites because they enter and exit roots numerous times to feed during a growing season. Invasion of roots may occur at the root tip, the root hair region, and occasionally in young lateral root junctions. As these nematodes migrate through the cortex of host roots, using their stout, well-developed stylets to destroy cells in their path, they briefly feed on nearby cells and then continue moving. Death of nearby cells is caused by the nematode movement and feeding activities, resulting in elongated, spreading lesions just below the root surfaces. With large numbers of lesion nematodes, the lesions may completely encircle the roots, causing death of the distal portion of the root segment.

RING NEMATODES

Ring nematodes are commonly found on perennial hosts, such as turf, woody ornamentals, and trees (Figure 8.10). There are about 80 species of ring nematodes, but many of the species have been grouped into a single genus, *Mesocriconema.* Ring nematodes have very deep **annulations,** or grooves, around the outside of their cuticles that appear to be segments, or rings across the body. They are very easy nematodes to identify under a dissecting microscope. These nematodes were often overlooked because their recovery rates from earlier nematode assay methods was very low. Ring nematodes are sedentary ectoparasites with relatively short, wide bodies (30–60 µm diameter; 400–700 µm length), giving them a cigar-like shape referred to as **fusiform**. This wide shape and their very slow rate of movement prevented detection in assays that relied on movement of the nematode for collection of the specimens. With improved assay methods, ring nematodes were found to be quite common, and were detected at very high numbers in the soil on some hosts.

CASE STUDY 8.1

SLOW DECLINE OF CITRUS—A WORLD OF TROUBLE

- Citrus slow decline, caused by the citrus nematode, *Tylenchulus semipenetrans*, is the most significant nematode problem on citrus and is found throughout the world, wherever citrus is grown.
- In the United States, the number of groves infested varies from state to state, but ranges from 50–90% of commercial groves.
- Resistant citrus root stocks are available, but the citrus nematode occurs in numerous physiological host races that can attack resistant rootstocks.
- The citrus nematode has a life cycle and feeding habit very similar to the reniform nematode, *Rotylenchulus reniformis*, although the two species are not closely related.
- The citrus nematode causes a very slow, generalized decline in the health and vigor of citrus.
  - The citrus nematode interacts strongly with water, nutrients, and temperature stress, and with root rot caused by the fungus *Phytophthora parasitica*
  - It is very difficult to assess the damage caused by the citrus nematode because of interacting factors, and because the damage observed in the present growing season may be due to nematode attack in the previous season, or even several years prior
- Although most citrus-producing areas are infested with the citrus nematode, all groves within these areas are not infested.
  - Sanitation, preventing the movement of soil and plants from one grove to another, is essential to maintaining healthy production areas
  - Planting stock from nurseries must be certified as disease-free before use
- Control of the citrus nematode relies primarily on the use of chemical nematicides.
  - Nematicides that can be used after citrus trees are planted are usually systemic in the plant and water soluble
  - Use of these postplant nematicides has been problematic in terms of human health risks and contamination of groundwater
- Groves infested with the citrus nematode should not be replanted in citrus because this nematode can live in the soil for several years without a host.

Life cycle completion for ring nematodes takes 25–35 days, with second-stage juveniles through adult stages feeding on host roots. Most populations do not produce males. Each ring nematode female may lay only 3–5 eggs per day in the soil, but on their typically perennial hosts, these nematodes have the capacity to eventually increase to extremely high population densities. Ring nematodes have very long, stout stylets (50–60 μm length) and feed from the outside of plant roots on cells near the root surface. The nematode stops moving and may feed for an extended period of time on a single feeding cell. There is very little cell death at ring nematode feeding sites, and it usually takes very high numbers of nematodes to visibly damage a host plant. Ring nematodes have been implicated in a number of disease complexes with pathogenic bacteria and fungi. *Mesocriconema xenoplax* is a common and widely distributed ring nematode that has a large host range, including most plant species of *Prunus* (peach, plum, apricot, cherry, and almond), as well as walnut and apple. This species predisposes trees to bacterial canker caused by the above-ground pathogenic bacterium *Pseudomonas syringae,* and is a critical component of the peach tree short life disease complex, interacting with bacterial canker and several fungal plant pathogens.

RENIFORM NEMATODE

The reniform nematode, *Rotylenchulus reniformis*, has the largest and most diverse host range of any single, economically-important, plant-parasitic nematode species (Figure 8.11). In terms of global economics, it is well worth discussing this single species at the same level of importance as the previously described major groups of plant-parasitic nematodes. The reniform nematode is found throughout the world in tropical and subtropical areas, and has more than 140 economically important hosts, spread across 30 plant families. It causes severe damage on most of its hosts. Where reniform nematodes occur on any of these susceptible crops, growers have little choice except to treat their production areas with chemical nematicides. Crops on which reniform nematodes are a major limiting factor in production include cotton in the southern United States and pineapples in Hawaii. Other hosts include banana, cassava, citrus, coffee, kale, lettuce, mango, papaya, soybean, sweet potato, and tea. Above-ground

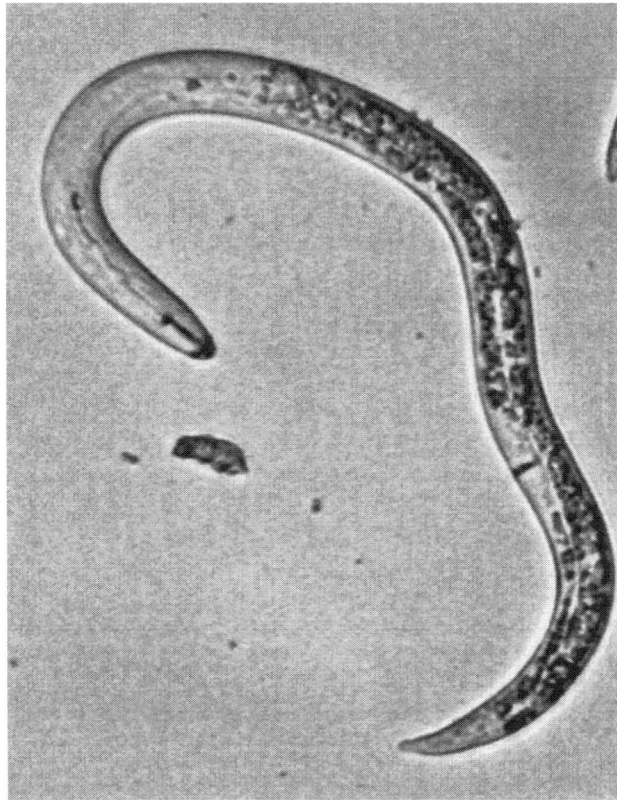

FIGURE 8.11 A preadult reniform nematode. After the nematode enters the root and begins feeding, the posterior portion of the body becomes kidney shaped. (Photo courtesy of T. Stebbins, University of Tennessee.)

symptoms of reniform nematodes include the typical root-impairment symptoms of stunting, nutrient deficiencies, and wilting. Below ground, the roots are stunted, coarse, and may be discolored and decayed with high population densities of nematodes. Reniform nematodes also cause symptoms from fungal wilt diseases in cotton, caused by either *Verticillium albo-atrum* or *Fusarium oxysporum*

f. sp. vasinfectum, to be more severe. Infection by reniform nematodes has been reported to cause *Fusarium* wilt resistant cultivars of cotton to become susceptible.

The life cycle of reniform nematodes takes 24–30 days to complete. The first molt occurs within the egg. Second-stage juveniles hatch and undergo three molts in the soil without feeding. Young adult females are the only infec-

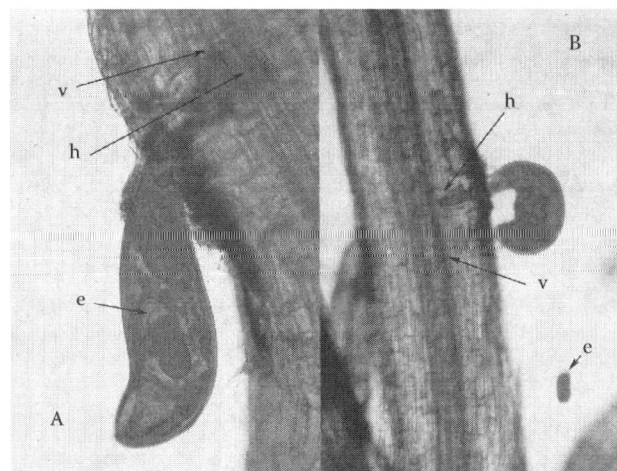

FIGURE 8.12 Swollen adult reniform nematode females outstretched (A) and in the typical kidney-shape (B), with an egg inside the female body (A-e) and outside, nearby (B-e). The nematode head (h) extends into the root system near the vascular tissues (v). (Photo courtesy of R. Hussey, University of Georgia.)

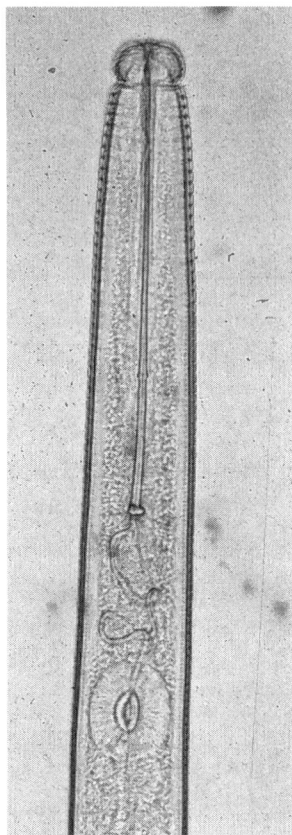

FIGURE 8.13   Head of a sting nematode. Note the very long stylet that the nematode uses to feed on root tips. (Photo courtesy of Dr. Gary Windham, USDA-ARS Crop Science Research Laboratory at Mississippi State University.)

(15 μm diameter; 500 μm length) penetrate host roots with only the anterior portion of their bodies, leaving the posterior portion in the soil outside the root. This type of feeding habit is called a sedentary semi-endoparasite. The nematode begins to feed on a single cell of the root endodermis, which is then transformed into a large syncitial feeding cell, incorporating up to 200 plant cells by causing the adjacent cell walls to dissolve. After the feeding site is formed, the posterior part of the female swells up to a large, 100-μm-diameter kidney shape (reniform means kidney-shaped) (Figure 8.12), and up to 200 eggs are laid by each female into a sticky, gelatinous mass in the soil.

STING NEMATODES

The sting nematode (*Belonolaimus* species) can have a devastating effect on the growth and yields of vegetables, row crops, ornamentals, and turf. The most important species of sting nematode is *B. longicaudatus*. Sting nematodes are nearly always found in very sandy soils that have a sand content of greater than 80%. Plants parasitized by these nematodes may be severely stunted and appear to be nutrient deficient. Seedlings in heavily infested areas may be severely stunted and devoid of a root system, which can lead to plant death. Below-ground symptoms may include roots with coarse, stubby branches and necrotic lesions may develop on the root surface. These nematodes are such severe pathogens that the presence of one sting nematode in a soil assay sample may necessitate the use of **nematicides**.

Sting nematodes are large, slender worms that reach a length of nearly 2.5 mm and have very long stylets, which may be 60 to 150 μm (Figure 8.13). The life cycle of this nematode is similar to other plant-parasitic nematodes. Reproduction is sexual with males comprising about 40% of the population. Adult females deposit eggs in soil, and

tive stage of reniform nematodes. Males do not feed, but the species reproduces sexually. The young females retain previous layers of cuticle after molting, and this sheath of old cuticles is thought to aid the nematode in surviving in the soil for up to two years. Vermiform young adult females

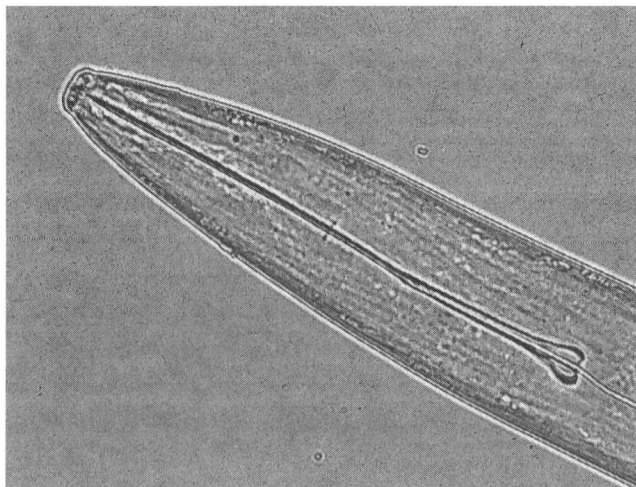

FIGURE 8.14   Head of a dagger nematode. Dagger nematode stylets have flanges and are called odontostylets. (Photo courtesy of Dr. Gary Windham, USDA-ARS Crop Science Research Laboratory at Mississippi State University.)

CASE STUDY 8.2

PINE WILT NEMATODE—A TRIPLE THREAT

- Pine wilt nematode, *Bursaphelenchus xylophilis*, causes a rapid wilt and death in as little as 3 weeks on susceptible pine trees.
  - The symptoms are caused by nematodes feeding in resin canals in the tree, in association with the blue stain fungus, *Ophiostoma piceae*, and possibly other microorganisms.
  - Pine wilt nematodes reproduce rapidly from egg to adult in 4–5 days, and spread rapidly throughout the tree, inhibiting the functioning of the xylem through their feeding activities.
- The pine wilt nematode is not a typical plant parasite, in that it can also feed on fungi, as do many members of its nematode family. Also, its feeding activities cause rapid death of the affected cells.
- The pine wilt nematode is believed to be native to the United States. American pine species are extremely tolerant to the nematode and typically show no symptoms.
- Pine wilt nematodes are carried from tree to tree by pine sawyer beetles (*Mononchamus* spp.) in a very specific biological association.
  - The pine sawyer beetles are attracted to dead or declining trees to lay their eggs.
  - In the presence of the pine sawyer pupae, pine wilt nematodes will form a fourth stage juvenile specialized for dispersal that migrates into the insect tracheids.
  - The newly emerged pine sawyer adults bore out of the tree and fly to healthy trees to feed, thus transporting the pine wilt nematodes to their next host.
- The considerable economic damage this nematode has caused comes from three sources:
  - The nematode has been introduced from America to Asia, where it has devastated natural stands of native Asian pine species. Up to 50% of the trees have died in infested areas.
  - Pine wood nematode is also killing imported exotic pines in the United States, particularly valuable landscape specimens in the Midwest and Northeast.
  - The threat posed by this nematode has led to a complete ban on the import of untreated pine products from the United States and Canada into the European Union and China.
- Researchers have developed a heat treatment (56°C for 30 min) that will free wood products from pine wilt nematodes, but this treatment adds cost for the producers.

---

the first-stage juvenile molts once before emerging from the egg. Sting nematodes are considered ectoparasites because they feed from the root surface and rarely enter the root tissue. These nematodes feed on root tips and along the sides of succulent roots by inserting their long slender stylets into epidermal cells.

DAGGER NEMATODES

Dagger nematodes (*Xiphinema* species) are ectoparasites and are commonly found associated with fruit or nut trees and in vineyards. Agriculturally important dagger species include *X. americanum*, *X. californicum*, *X. index*, and *X. rivesi*. Root systems of plants parasitized by dagger nematodes may be stunted, discolored, and have a limited number of feeder roots. In 1958, *X. index* was documented as the first soil nematode to **vector** a plant virus (grapevine fanleaf virus). Nematodes found in the *X. americanum* group transmit the following North American nepoviruses: cherry rasp leaf virus, tobacco ringspot virus, tomato ringspot virus, and peach rosette mosaic virus. There are very few reports of direct crop damage from

dagger nematodes, and most of their economic importance derives from their roles as plant-virus vectors.

Dagger nematodes are in a different taxonomic class within the nematode phylum, along with the stubby-root nematodes, and are only distantly related to all the other plant-parasitic nematodes. Dagger nematodes are slender, up to 2 mm long, and have a flanged **odontostyle** (instead of having knobs on the stylet, like the other class of plant-parasitic nematodes) that may be 130 μm long (Figure 8.14). Females lay eggs singly in the soil near host plants, and the nematodes hatch as first-stage juveniles. These nematodes may have three or four juvenile stages that can be recognized by the lengths of their developing stylets. Male dagger nematodes are very rare. Dagger nematodes can live up to 3 years under favorable environmental conditions.

STUBBY-ROOT NEMATODES

Stubby-root nematodes, *Paratrichodorus* and *Trichodorus* species, are important parasites of corn, vegetable crops, and turf in the coastal plain region of the southeastern

United States. The most important species is *P. minor*. These nematodes are most widely distributed in sandy soils, but have also been found in organic-based soils. The most characteristic symptom of damage by stubby-root nematodes is a stunted, stubby root system. Above-ground symptoms are similar to plants deprived of a root system. Plants are stunted and chlorotic, wilt easily, and have very little ability to withstand drought. *Paratrichodorus* species are vectors of the following tobraviruses: tobacco rattle virus, pea early browning virus, and pepper ringspot virus.

Stubby-root nematodes are migratory ectoparasites and feed almost exclusively at the root tips. As these nematodes feed in the root apical meristem area, root elongation and growth is stopped, leading to the characteristic stubby-root symptoms. Feeding by stubby-root nematodes is different from other plant-parasitic nematodes in that food is not ingested through a hollow tube in the stylet. These nematodes have a dorsally curved stylet referred to as an **onchiostyle**, with a simple groove on one side. The life cycle of this nematode consists of an egg stage, four juvenile stages, and the adult stage, and can be completed in 16 to 17 days. Soil populations of stubby-root nematodes can increase rapidly and then decline with equal abruptness. These nematodes have commonly been found at deep soil levels, which may create sampling problems and limit detection of these parasites.

ADDITIONAL PLANT-PARASITIC NEMATODES

The awl nematode, *Dolichodorus heterocephalus*, is an ectoparasite of vegetables and is similar to the sting nematode in general appearance. This nematode is generally found in fields with high soil moisture and causes symptoms on roots similar to the sting nematode. The lance nematode (*Hoplolaimus columbus*) is a serious pest of cotton and soybean in the coastal plain regions of Georgia, South Carolina, and North Carolina. Lance nematodes have a heavily sclerotized stylets with large, tulip-shaped knobs. These nematodes are classified as semi-endoparasites or as migratory endoparasites.

A number of plant-parasitic nematodes feed within above-ground plant tissues. The seed gall nematode, *Anguina tritici*, invades florets of rye and wheat and can survive for long periods in seed-like galls. This nematode has been all but eliminated from developed countries by maintaining supplies of certified nematode-free seed for growers. There are still active quarantine measures to limit its spread and reintroduction. *Aphelenchoides* species, the bud and leaf nematodes, attack the foliar parts of ornamentals and floral crops, potato, strawberry, and onions. These nematodes require a constantly moist, humid environment, as they move up the outside of shoots and leaves,

entering the plant through stomata. Bud and leaf nematodes often cause problems in greenhouse ornamental and floral production areas. *Bursaphelenchus* species, including *B. xylophilis*, the pine wilt nematode, typically have insect vectors and attack trees above-ground. The stem and bulb nematode, *Ditylenchus dipsaci*, is an important pest of alfalfa, onion, strawberry, and nursery crops. This nematode has a fairly low temperature optimum and is a problem in cool, moist temperate regions of the world.

NEMATODE MANAGEMENT

Plant-parasitic nematodes typically cause crop damage only at high population densities. Control methods should be used to keep nematode numbers below damaging levels instead of trying to **eradicate** these organisms. Host plants are often able to compensate for damage caused by moderately high numbers of nematodes, especially where the plants are grown under optimum environmental conditions. Although nematodes such as root-knot nematodes may establish hundreds of galls on a root system, very large numbers of these nematodes are usually required to cause significant crop losses. This density-dependent damage relationship for plant-parasitic nematodes is the foundation of nematode advisory programs. In these programs, data from soil assays are used to predict the nematode hazard to anticipated crops and to recommend control practices if necessary. In order to accurately advise farmers of nematode hazards, a relationship between nematode numbers in the soil and crop performance must first be established. This information is best obtained from infested fields, but controlled studies in greenhouse pots are also useful.

In agricultural fields, nematode management usually relies on the application of chemical pesticides to the soil. These pesticides, called nematicides, are expensive, exceedingly dangerous to apply, and are cause for serious environmental concerns. Nematicides do not eliminate all the nematodes from a field. Fields treated with nematicides may have higher numbers of nematodes at the end of the growing season compared to fields not treated with chemicals. Nematicides include both fumigant and nonfumigant compounds. The fumigant types are active in a gaseous phase in the soil, and typically are very toxic to other nontarget organisms in the soil. These compounds usually must be applied well in advance of planting because they may damage the crop. Nonfumigant nematicides nearly all work by attacking the nervous system of the nematodes. These compounds are usually systemic within the plant and can be applied at or after planting time, but are extremely toxic to animals and people. In greenhouses and limited acreage high-value crops, soil sterilization with heat or application of a broad-spectrum biocidal fumigant is commonly practiced.

For most field crops, resistance is the only other practical means for controlling plant-parasitic nematodes. Resistant crops are economical in that little or no additional costs are assessed to the grower, and no special equipment is needed to utilize these resistant plants. Also, from the standpoint of impact on the environment, resistant cultivars are considered the best choice of control practices. Cultivars resistant to the sedentary endoparasitic nematodes such as root-knot and cyst nematodes are available, but only for a relatively few crops. Use of resistant plants is complicated by the existence of physiological host races among both root-knot nematodes and cyst nematodes. The existence of these physiological races means that some populations of the nematodes can overcome the host resistance. A continual battle must be waged to derive new sources of resistance to newly discovered nematode races. Very little progress has been made in identifying cultivars resistant to ectoparasitic plant-parasitic nematodes.

Other methods used to control plant-parasitic nematodes include **sanitation**, **cultural practices** (Chapter 35), **biological control** (Chapter 37), quarantines, and the use of organic soil amendments. Sanitation methods prevent the spread of nematodes from infested fields or plant materials to uninfested fields by cleaning or sanitizing tools, implements, farm and nursery equipment, and any other source of nematode-infested soil and plants. Cultural control practices include crop rotation with nonhosts and using tillage practices to destroy plant roots at the end of the growing season to reduce intercrop nematode survival rates. Soil-inhabiting plant-parasitic nematodes cannot move very far under their own power. Rotation to a nonhost crop will starve the nematodes present in a field, allowing planting of a susceptible crop in subsequent growing seasons. Rotation crops available to growers, however, are severely limited by the wide host ranges of some plant-parasitic nematodes, and by production economics that preclude use of a good rotation crop with no economic market.

Biological control methods, which use bacterial or fungal pathogens to control nematodes, are still under development but may be used in combination with other control methods in the future to enhance nematode management. Difficulties have been encountered with deployment of biological control agents due to failure of the biological agents to establish successfully in the soil environment. Incorporation of organic amendments into nematode-infested soil stimulates population increases of soil organisms that attack plant-parasitic nematodes. Decomposition products from the organic amendments may further reduce nematode population densities through direct toxic effects. Safer, more effective options for the control of nematodes are badly needed. Nematode problems continue to worsen globally as more agricultural production is concentrated on smaller areas of arable land, and urbanization of agricultural production areas reduces the availability of toxic nematicides.

ACKNOWLEDGMENT

The author and editors gratefully acknowledge the contributions of Dr. Gary Windham to this chapter.

SUGGESTED READING

Barker, K.R, C.C. Carter and J.N. Sasser, Eds. 1985. *An Advanced Treatise on Meloidogyne*. Volume 2: Methodology. North Carolina State Graphics. Raleigh, NC. 223 pp.

Barker, K.R., G.A. Pederson, G.L. Windham, Eds. 1998. *Plant and Nematode Interactions*. Monograph 36. American Society of Agronomy. Madison, WI. 771 pp.

Dropkin, V.H. 1989. *Introduction to Plant Nematology*. John Wiley. New York. 304 pp.

Mai, W.F., and H.H. Lyon. 1975. *Pictorial Key to Genera of Plant-Parasitic Nematodes*. Cornell University Press. Ithaca, NY. 219 pp.

Southey, J.F., Ed. 1986. *Laboratory Methods for Work with Plant and Soil Nematodes*. Her Majesty's Stationery Office. London. 202 pp.

Zuckerman, B.M., W.F. Mai, and M.B. Harrison. 1985. *Plant Nematology Laboratory Manual*. Univ. Mass. Agric. Exp. Stat., Amherst, MA. 212 pp.

9 Pathogenicity and Isolation of Plant-Parasitic Nematodes

James P. Noe

Plant-parasitic nematodes are found in the root zone of most plants grown in urban and rural areas. A diverse mixture of plant-parasitic and free-living nematodes (nonparasitic) are present in most soil samples. These nematodes can be readily isolated from both soil and root samples for viewing and identification. Damage to plants caused by plant-parasitic nematodes in the field can be reproduced easily in the greenhouse. The amount of damage caused by nematodes can be related to the number of nematodes (or eggs) present at planting or at the start of an experiment. The laboratory exercises in Experiment 1 will allow students to observe symptoms caused by root-knot nematodes and to observe various life stages. In Experiment 2, students will isolate nematodes from soil samples and observe the diverse nematode populations present in several settings.

EXERCISES

EXPERIMENT 1. THE EFFECTS OF ROOT KNOT NEMATODES ON PLANT GROWTH

This experiment will require two laboratory periods about 6–8 weeks apart. In the first part of the exercise, tomato plants will be inoculated with eggs of root-knot nematodes and in the second part the effects of the nematodes on the plants will be observed. This experiment will familiarize students with the root-knot nematode and demonstrate the effect of various concentrations of inoculum (eggs) on plant growth and nematode reproduction. Tomato plants inoculated in the first laboratory will have sufficient time to show the effects of the nematode for the second lab exercise in 6–8 weeks. Additional objectives are to observe the various life stages of this plant-parasitic nematode and to learn basic methods for inoculation and evaluation of host plants.

Materials

Each student or group of students will need the following materials to complete the experiment:

- One each of root-knot nematode resistant and susceptible tomato plants. Check with your local cooperative Extension office for variety recommendations. Tomato cultivar names change frequently and this step will ensure that you have cultivars that are locally available and appropriately classified. Start plants in seedling trays 6 weeks before the laboratory exercise. Transplant to 15-cm diam. pots in nematode soil mix 3–4 weeks after germination
- Two 15-cm diameter greenhouse pots
- Sterile greenhouse soil mix suitable for nematode experiments (>85% sand, low organic content) in sufficient amount to fill required number of 15-cm diam. pots. Avoid common potting mixtures as the nematodes will not do well, and it is very difficult to clean the roots for observation
- Metal sieves for preparation of nematode inoculum. Two sieves are required, 75 µm-pore opening (200-mesh) and 26 µm-pore opening (500-mesh) each with a 15-cm diameter frame
- Several cultures of root-knot nematodes (*Meloidogyne incognita*). These nematodes must be maintained on living plants, preferably on a nematode-susceptible tomato cultivar, such as Marglobe or Rutgers. Start at least three cultures for the laboratory about 60 d before they are needed in experiment 1. At least one million eggs can be expected from each tomato plant if they are originally inoculated with 10,000 or more eggs. Do not keep cultures too long (>75 days) since they may deteriorate rapidly causing recovered egg numbers to drop dramatically
- Household bleach and 2–3-L flask or large plastic bottle for collection of nematode eggs
- Greenhouse bench space adequate for the number of pots anticipated. Usually, 130-ft² is sufficient for 30–35 students.
- Phloxine B stain solution (150 mg/L tap water)
- Stereo dissecting microscope and light
- Large beaker and plastic wash bottles
- Rubber aprons
- Cork borer
- Permanent marker and pot labels
- 10-mL pipettes and small beakers

TABLE 9.1
Data Sheet for Experiment 1 (Procedure 9.2)
Group no:_____

		Measurement				
		Shoot (top)		Root		
Plant #	Number of Eggs Inoculated	Length	Weight	Length	Weight	Number of Galls
1						
2						
3						
4						
5						

General Considerations—Inoculum Preparation

To prepare nematode inoculum, cut the shoots off from stock tomato plants at the soil line and gently remove the roots from the pot. Wash the roots well in water and cut galled sections into 5–10-cm lengths. Immerse and shake sections in 0.5% NaOCl solution (100 mL household bleach in 900 mL tap water) in a 2–3-L flask or plastic bottle. Caution: wear rubber apron—even diluted bleach will damage clothing. This solution will dissolve the egg masses and free the eggs. Remove the eggs from the solution immediately as leaving the eggs in the solution for more than 4 min will reduce their viability. Pour the solution and roots over two stacked sieves with the 200-mesh above the 500-mesh. Most of the roots will be caught on 200-mesh sieve, and the nematode eggs will pass though and be retained on the 500-mesh sieve. Wash the sieves gently for several minutes with tap water to remove all traces of the bleach and wash most of the eggs onto the 500-mesh sieve. Discard the roots and gently rinse the eggs from the 500-mesh sieve into a beaker. A plastic wash bottle works well for this step and allows back washing of the sieve to remove all the eggs. Wash eggs to one side of the 500-mesh sieve and gently tip the sieve to pour eggs into a 600-mL beaker. Use the wash-bottle to spray a stream of water on the backside of the sieve to wash eggs into the beaker. Bring the level of water in the beaker up to 500 mL.

The resulting egg suspension then must be calibrated. Shake or stir the egg suspension and remove several 1 mL aliquots. Place the aliquots in a small counting dish or plastic tray, wait for the eggs to settle to the bottom (1–2 min), and count them with a dissecting stereomicroscope. When observing nematode eggs or juveniles, the light source must be transmitted from below the specimen using a mirrored stage. From the sample counts determine the average number of eggs per mL in the stock solution and prepare dilutions to yield 2000, 1000, 500, and 250 eggs per mL. Remember always to shake or stir the egg solutions before pouring or sampling as they sink very rapidly to the bottom. The prepared inoculum can be stored overnight at 5–15°C. Susceptible and resistant tomato plants should have been transplanted to 15-cm pots several weeks before and will be ready for inoculation.

Follow the protocols outlined in Procedure 9.1 to complete this part of the experiment, then go to the protocols outlined in Procedure 9.2 to finish this experiment.

Anticipated Results

The purpose of this exercise was to demonstrate the effects of root-knot nematodes on susceptible tomato plants at different inoculum levels. More galls should have been observed on the roots at higher inoculum levels than at lower inoculum levels. This difference may be difficult to see between the two highest inoculum levels since competition for infection sites would be quite high. The shoot weights and lengths should be similar between the controls and the lower two inoculum levels. At the lowest inoculum level the shoots may actually be longer and weigh more. At the highest inoculum levels the shoot weight and length should be lower. The root lengths should be shorter at the higher inoculum levels compared to the control since galling generally tends to reduce root growth. However, root weights may actually increase with increasing nematode inoculum levels up to the highest inoculum rates used. The galls generally make the roots heavier as the root-knot nematodes are also causing more photosynthate to be directed toward the roots than shoots, as compared to uninfected plants. This effect is, in part, how the shoots become stunted as a result of nematode infection.

Procedure 9.1	
Inoculating Tomato Plants with Root Knot Nematodes	
Step	Instructions and Comments
1	Inoculate young susceptible tomato plants with calibrated suspensions of fresh root knot nematode eggs prepared from galled roots as described under inoculum preparation. Working in groups of five students, each group should inoculate five pots. Each pot will be inoculated with a different inoculum level including 0, 250, 500, 1000, and 2000 eggs/mL. The number of eggs added per pot equals 0 (10 mL of water only; this is the control), 2500 (10 × 250), 5000 (10 × 500), 10000 (10 × 1000) and 20,000 (10 × 2000), respectively. Make sure the soil in the pots is evenly moist (not soaked). Nematode eggs will die immediately if exposed to dry soil.
2	Make three equally spaced holes, 3–5-cm deep and about 3 cm from the stem of each plant. A marker pen or large cork-borer makes a good diameter hole for the inoculation.
3	While swirling the suspension constantly, carefully pour equal amounts (a little more than 3 mL) of the 10 mL of inoculum into each of the three holes. Squeeze the holes shut with your fingers and water lightly immediately. Do not over water the plants. Label each pot with your group number, the date, and number of eggs per pot.
4	Repeat steps 1–3 using a nematode-resistant tomato variety, but only use the 10,000 eggs (10 mL of 1000 eggs/mL) inoculum rate. Label this pot as "resistant" and use it for comparison only in the second exercise.

Procedure 9.2	
Assessing the Effects of Nematodes on Tomato Plants	
Step	Instructions and Comments
1	Carefully wash the soil, causing as little damage as possible, from the roots of plants inoculated in Procedure 9.1. Turn the pot upside-down and gently tap the rim of the pot on a bench until the root ball separates from the pot. Slide the root mass out of the pot and immerse in a trashcan filled with water. Gently shake the root mass up and down until most of the soil has fallen out of the root ball. Wash the remaining soil from the roots under a stream of water in the sink. Keep the correct pot label with each plant so that treatments do not get confused.
2	Remove the excess water from the plants by blotting with paper towels and bring them to the laboratory.
3	Separate the roots from the top by cutting the plant at the soil line.
4	Make the following measurements and record in Table 9.1. Overall length of shoot (top) in cm. Overall length of root system in cm. Weight of shoot in grams. Weight of root system in grams. Estimated number of galls. This estimate can be based on counts made on a part of the root system. The number found on a total of 30 linear cm of individual root segments is an acceptable estimate. A total of 30 cm of short segments of root should be collected from different parts of the root mass.
5	Visually compare the resistant tomato plant to the susceptible plant inoculated with the same number of eggs. The resistant plant should be larger, more vigorous, and have no galls on the roots. However, there may be dark and decaying lesions on the roots (necrosis). The mechanism of resistance to root knot nematodes in tomato is called hypersensitivity. As juveniles migrate into the roots and attempt to establish a feeding site, the plant cells die, thus preventing further feeding by the nematode. In response to a large inoculum level, this hypersensitive response may cause considerable root necrosis and damage to the plant, even though the nematodes do not complete their life cycle and reproduce.
6	Place several 2–3 cm segments of galled roots in a small beaker containing phloxine B stain. Remove segments after 15 min and observe the stained egg masses on the outer surface of the roots under the dissecting microscope. It is possible to estimate the total number of eggs on the root system by staining and counting egg masses on several 1-g aliquots of root and multiplying the number of egg masses by 300. This is a conservative estimate of the number of eggs per gram of root. Multiply again by the total weight of the root system to get the total number of eggs. This value can be compared to the initial inoculum levels to estimate the reproduction of the nematode.

| | **Procedure 9.3** |
| | Isolation of Parasitic and Free-Living Nematodes from Soil |

Step	Instructions and Comments
1	Collect 20 soil cores (15–20-cm deep) in the root zone of an annual crop from the selected field, place in a plastic bucket, and mix thoroughly. Soil (500–1000 cm^3) should be sealed in a plastic bag, labeled, and placed in an insulated ice chest. Soil samples around the shrubs should be collected and handled in the same manner.
2	Add 250 cm^3 of soil to 800–1200 mL of water in a 2-L beaker and mix vigorously for 30 sec. Allow soil to settle for 30 sec. Decant the suspension slowly onto a 60-mesh sieve nested on top of a 400-mesh sieve. Use a wash bottle to backwash nematodes and soil into a 150-mL beaker. Nematode samples must be allowed to settle for 1 h before the centrifugation procedure.
3	Gently pour off most of the water from beakers containing soil and nematodes. Swirl contents and pour into 50-mL centrifugation tubes. Make sure tubes are balanced using a top-loading balance; the weight of the tubes and contents should be within 1 g of each other. Weight may be adjusted using a squeeze bottle of water. Centrifuge at 420 g for 5 min (do not use brake). Gently pour water from tubes. Nematodes are in the soil pellet in the bottom of the tubes. Refill tubes with sucrose solution and mix with vortex mixer to resuspend nematodes. Centrifuge for 15 sec at 420 g (do not use brake). Decant sucrose solution/nematode suspension on to the 400-mesh sieve (pour very slowly). Gently rinse with water to remove sugar solution and backwash nematodes into 150-mL beaker. Beaker should contain no more than 25 mL of water.
4	Pour samples into petri dishes or Syracuse watchglasses and place on a dissecting scope. Nematodes are best observed using a light source from underneath. Use a pictorial key to identify nematode genera.

Questions

- Sketch the life cycle of the root-knot nematode beginning with the egg stage. What are vulnerable stages of the life cycle that could be targeted to keep nematode numbers below damaging levels?
- Are there other pathogens that cause root galling that might be confused with symptoms caused by root-knot nematodes?

EXPERIMENT 2. ISOLATION OF PLANT-PARASITIC AND FREE-LIVING NEMATODES FROM SOIL

The objectives of this experiment are to isolate plant-parasitic nematodes from soil and to demonstrate the diversity of nematode populations in row crop fields and ornamental plantings.

Materials

Each student or group of students will require the following to complete the procedure:

- *Pictorial Key to Genera of Plant-Parasitic Nematodes* (Mai and Lyon, 1975)
- Soil sampling tubes, 1-qt plastic bags, plastic buckets, and ice chest
- Metal sieves for separating the nematodes from the soil; two sieves are required, 250-µm pore

openings (60 mesh) and 38-µm pore openings (400 mesh), each with a 20-cm diameter and 5-cm depth
- Centrifuge equipped with a hanging bucket rotor and 50-mL, non-sterile centrifugation tubes
- Vortex mixer
- 2-L beakers, 150-mL beakers, and 1-L wash bottles
- Sucrose solution (454 g sugar and water to make 1 L)
- Dissecting microscopes with light source
- Top-loading balance
- Plastic buckets and bags
- 150-mL beakers
- Petri dishes or Syracuse watchglasses
- Plastic squeeze bottles for water

General Considerations

Locate a row crop or vegetable field that has been in continuous production for 5–10 growing seasons that can be sampled during the laboratory period. Also, find some 10–20 year old foundation plantings (shrubs) that can be sampled on the same day. Make the sucrose solution mentioned previously in the materials section and refrigerate (4°C) until needed.

Follow the protocols outlined in Procedure 9.3 to complete this experiment.

Anticipated Results

Students should be able to distinguish between free-living nematodes (nonparasitic types) and plant-parasitic nematodes. Free-living nematodes do not have stylets and are usually more active than plant-parasitic nematodes. Plant-parasitic nematodes have stylets that have knobs or flanges on the posterior end and are usually quite sluggish (see Figure 8.1). Predatory nematodes, which have rasp-like teeth or a stylet without knobs, may be present in samples. Because of the build-up of organic matter, samples from the shrubs should contain more free-living nematodes than the samples from the field. Plant-parasitic nematodes should be isolated from both samples. Some of the larger plant-parasitic nematodes such as dagger, ring, or lance (Chapter 8) may be found in the samples and students should be able to identify them using the pictorial key.

Questions

- What characteristics and/or behaviors distinguish free-living nematodes (nonparasitic types) from plant-parasitic nematodes?
- Are any predatory nematodes present in the samples?
- Which sample has more free-living nematodes? Which sample has more plant-parasitic nematodes?
- Can any of the plant-parasitic nematodes be identified using the pictorial key?

ACKNOWLEDGMENT

The author and editors gratefully acknowledge the contributions of Dr. Gary Windham to this chapter.

LITERATURE CITED AND SUGGESTED READING

Mai, W.F. and H.H. Lyon. 1975. *Pictorial Key to Genera of Plant-Parasitic Nematodes*. Cornell University Press. Ithaca, NY. 219 p.

Southey, J.F., Ed. 1986. *Laboratory Methods for Work with Plant and Soil Nematodes*. Her Majesty's Stationery Office. London. 202 p.

Zuckerman, B.M., W.F. Mai and M.B. Harrison. 1985. *Plant Nematology Laboratory Manual*. Univ. Mass. Agric. Exp. Stat. Amherst Massachusetts. 212 p.

10 Plant Pathogenic Fungi and Fungal-like Organisms

Ann Brooks Gould

CHAPTER 10 CONCEPTS

- Fungi are eukaryotic, heterotrophic, absorptive organisms with cell walls.

- Organisms that were collectively grouped as fungi have been placed into the following three different kingdoms: Stramenopila (Oomycota), Protozoa (Plasmodiophoromycota), and Fungi (Chytridiomycota, Zygomycota, Ascomycota, and Basidiomycota)

- The Deuteromycota (Fungi Imperfecti) is an artificial group of true fungi that is placed in a form or artificial phylum because its teleomorphic stage is unknown or very rarely produced. Some members of this artificial group are referred to as *Mycelia sterilia* because they produce no spores at all.

- Chitin is a primary component of cell walls of true fungi and is not found in the cell walls of fungal-like organisms in the Oomycota. The Oomycota contain cellulose in their cell walls, which is not found in the cell walls of true fungi.

- Most plant pathogenic fungi can live as saprophytes or parasites (facultative saprophytes). However, some plant pathogenic fungi require living plant cells for their nutrition (biotrophs).

- The sexual state of a fungus is known as the teleomorph and the asexual state is known as the anamorph.

- The anamorphic state of most fungi is the conidium (Ascomycota, Basidiomycota, and Deuteromycota) and amotile sporangiospores (Zygomycota). In Plasmodiophoromycota (kingdom Protozoa), the Chytridiomycota (kingdom Fungi), and most species of the Oomycota (kingdom Stramenopila), the anamorphic state is motile sporangiospores referred to as zoospores.

- Examples of teleomorphic spores include oospores (Oomycota), zygospores (Zygomycota), ascospores (Ascomycota), and basidiospores (Basidiomycota).

- Most fungi are dispersed as spores through air currents, water, and animals (primarily insects). Fungi may also be spread in or on infected plant parts, movement of soil, and on agricultural equipment.

Of the biotic plant disease agents known to humankind, fungi are the most prevalent, and the study of these organisms has some very rich history. Fungi were once considered to be plants and thus were in the domain of botanists. Fungi are actually microorganisms, however, and differ from plants in that they are now placed in their own kingdom (Table 10.1). With the development of techniques needed for studying microorganisms, the discipline of mycology (which is Greek for *mycos,* meaning fungus, and *logy,* meaning study) developed, also, and now the fungi encompass organisms from many different groups. Indeed, not all fungi that cause plant disease are true fungi.

Although the formal, systematic study of fungi is a mere 250 years old, fungi have played important roles in the history of mankind for thousands of years. In their most important role, fungi are agents of decay, breaking down complex organic compounds into simpler ones for use by other organisms. Fungi are important in the production of food (wine, leavened bread, and cheese) and can also be a source of food. Fungi may be poisonous, may produce hallucinogens, and have played a role in religious rites in many cultures. Fungi can attack wood products, leather goods, fabrics, petroleum products, and foods at any stage of their production, processing, or storage. They produce toxins (called **mycotoxins**), are a source of antibiotics, cyclosporin (an immunosuppressant), vaccines, hormones, and enzymes, and can also cause diseases in both animals and plants. Those fungi found consistently

TABLE 10.1

Some Differentiating Characteristics of Plants, Animals, and Fungi

	Animals	Plants	Fungi
Mode of obtaining food	Heterotrophy	Autotrophy	Heterotrophy
Chloroplasts	No	Yes	No
Chief components of cell wall	Cell wall lacking	Cellulose and lignin	Chitin, glucans, or cellulose (in some fungal-like organisms)
Chief sterols in cell membrane	Cholesterol	Phytosterols (ß-sistosterol, campesterol, stigmasterol)	Ergosterol
Food storage	Glycogen	Starch	Glycogen
Chromosome number in thallus	Diploid	Haploid, diploid, or polyploid	Haploid, diploid, or dikaryotic
Alternation of generations	No	Yes	Yes
Specialized vascular tissues for transport	Yes	Yes	No

in association with a particular plant disease are called fungal **pathogens**.

Fungi can destroy crops, and the economic consequences of this have been enormous throughout human history. Fungi reduce yield, ruin crops in the field and in storage, and produce toxins poisonous to humans and animals. Blights, blasts, mildews, rusts, and smuts of grains are mentioned in the Bible. The Greek philosopher Theophrastus (370–286 B.C.) recorded his speculative (but not experimental) studies of grain rusts and other plant diseases. Indeed, wheat rusts have been important from ancient times until the present wherever wheat is grown. A Roman religious ceremony, the Robigalia, appealed to the rust gods to protect grain crops from disease.

When societies depend on a single crop for a major degree of sustenance, plant diseases can have a devastating impact. The potato was the major food of Irish peasants in the 1800s; the disease late blight destroyed the potato crop in 1845–1846, resulting in massive starvation and in emigration for years to come. The discipline of plant pathology was born over the scientific and political controversy caused by this disease. Until that time, fungi seen in plants were thought to be the result of plant disease, not the cause of plant disease. The famous German botanist Anton deBary worked with the late blight pathogen and proved experimentally that this was not the case. The causal agent of light blight of potato is now known as *Phytophthora infestans*. Other fungal diseases that played major roles in human history include chestnut blight, coffee rust, downy mildew of grape, and white pine blister rust.

So, what are these fascinating organisms called fungi, and what role do they play in plant disease?

FUNGAL CHARACTERISTICS

A **fungus** (pl. fungi) is a **eukaryotic**, **heterotrophic**, absorptive organism that develops a microscopic, diffuse, branched, tubular thread called a **hypha** (pl. hyphae). A group of hyphae is known collectively as a **mycelium** (pl. mycelia) and is often visible to the unaided eye. The mycelium makes up the vegetative (nonreproductive) body or thallus of the fungus.

Fungi have definite cell walls, have no complex vascular system, and except for a few groups, are not motile. The thallus of some fungi is single celled (as in the yeasts) or appears as a **plasmodium** (as in the slime molds) (Chapter 11). Many fungi can be grown in pure culture, facilitating their study, and they reproduce sexually or asexually by means of spores. As a group, the fungi encompass a way of life shared by organisms of different evolutionary backgrounds.

The hyphae of most fungi are microscopic and differ in diameter among species (3 to 4 µm to 30 or more µm wide). **Septa** (crosswalls) may or may not be present, and usually contain small pores to maintain continuity with other cells. Those hyphae without crosswalls are called aseptate or **coenocytic** (Chapter 20). Hyphae may branch to spread over a growing surface, but branching usually occurs only when nutrients near the tip become scarce. In culture, fungi form **colonies**, which can be discrete or diffuse, circular collections of hyphae or spores, or both, that arise from one cell or one grouping of cells.

Hyphal cells are bound by a cell envelope called the plasma membrane or **plasmalemma**. The plasmalemma of fungi differs from plants and animals in that it contains the sterol ergosterol, not cholesterol, as in animals, or a phytosterol, as in plants. Outside the plasmalemma is the **glycocalyx**, which is manifest as a slimy sheath (as in slime molds) or as a firm cell wall (as in most other fungi). This cell wall is composed chiefly of polysaccharides. In the fungal phyla Ascomycota, Basidiomycota, and Deuteromycota (mitosporic fungi), **chitin** [ß-(1→4) linkages of *n*-acetylglucosamine] and **glucans** (long chains of glucosyl residues) are major cell wall components. Zygomycota contain chitosan, chitin, and polyglucuronic acid; and Oomycota contain **cellulose** [ß-(1→4) linkages of glucose] and glucans. Chitin, glucans, and cellulose form strong fibers called microfibrils which, embedded in a matrix of glycoprotein and polysaccharide, lend support to the hyphal wall. Cell walls also contain proteins and, in some fungi, dark pigments called **melanins**.

Fungal cells may contain one or many nuclei. A hyphal cell with genetically identical haploid nuclei is **monokaryotic**; cells with two genetically different but compatible haploid nuclei are **dikaryotic** (a characteristic of fungi in the Basidiomycota). Compared with animals and plants, the nuclei of fungi are small with fewer chromosomes or number of DNA base pairs. **Plasmids** (extrachromosomal pieces of DNA that are capable of independent replication) are found in fungi; the plasmid found in *Saccharomyces cerevisiae*, the common yeast fungus used to make wine and leavened bread, has been intensively studied. **Vacuoles** in hyphal cells act as storage vessels for water, nutrients, or wastes, or may contain enzymes such as nucleases, phosphatases, proteases, and trehalase. Cells accumulate carbon reserve materials in the form of lipids, **glycogen**, or low molecular weight carbohydrates such as **trehalose**. **Mitochondria** (energy producing organelles) in fungi vary in size, form, and number.

Fungal cells from different hyphal strands may often fuse in a process called **anastomosis**. Anastomosis, which is very common for some fungi in the Ascomycota, Basidiomycota, or Deuteromycota, results in the formation of a three-dimensional network of hyphae and permits the organization of some specialized structures such as **rhizomorphs** and **sclerotia**, as well as fruiting structures, also known as **sporocarps**.

HOW FUNGI INFECT PLANTS

NUTRITION

Fungi have the advantage over organisms in the plant kingdom in that they do not have chlorophyll and are not dependent on light to manufacture food. Thus, fungi can grow in the dark and in any direction as long as there is an external food source and water.

Fungal hyphae elongate by apical growth (from the tip). Unlike animals, fungi do not ingest their food and then digest it; they instead obtain their food through an absorptive mechanism. As the fungus grows through its food, hyphae secrete digestive **exoenzymes** into the external environment (Chapters 28 and 30). Nutrients are carried back through the fungal wall and stored in the cell as glycogen. Free water must be present to carry the nutrients back into fungal cells.

Fungi can grow rapidly over a surface stratum and may penetrate it or may produce an aerial mycelium. Fungi have the potential to utilize almost any carbon source as a food. This is restricted only by what exoenzymes the

fungus produces and releases into the environment. Some of these enzymes are listed in Table 10.2.

Fungal growth may continue as long as the appropriate nutrients and environmental conditions are present. Growth ceases when the nutrient supply is exhausted or the environment is no longer favorable for development. Nutrients required by fungi include a source of carbon in the form of sugars, polysaccharides, lipids, amino acids, and proteins; nitrogen in the form of nitrate, ammonia, amino acids, polypeptides, and proteins; sulfur, phosphorus, magnesium, and potassium in the form of salts; and trace elements such as iron, copper, calcium, manganese, zinc, and molybdenum.

TYPES OF FUNGAL PATHOGENS: HOW DO FUNGI GET THEIR FOOD?

Most fungi are **saprophytes** in that they use nonliving organic material as a source of food. These organisms are important scavengers and decay organisms and, along with bacteria, recycle carbon, nitrogen, and essential mineral nutrients. Most plant pathogenic fungi are versatile organisms that can live as saprophytes or parasites. These fungi are called **facultative saprophytes**. They attack their hosts, grow and reproduce as parasites, and then act as saprophytic member of the normal soil microflora between growing seasons (as saprophytes). This makes them very difficult to control.

Many plant pathogenic fungi are classified as **necrotrophs**. These fungi are usually saprophytes and survive well as sclerotia, spores, or as mycelia in dead host material in the absence of a living host. Given the opportunity, however, these fungi become parasitic, kill, and then feed upon dead plant tissues. Necrotrophs produce **secondary metabolites** that are toxic to susceptible host cells. Fungal enzymes degrade tissues killed by the toxins, and the cell constituents are used as food. Disease symptoms caused by necrotrophs are manifested as small to very large patches of dead, blackened, or sunken tissue. A classic example of a necrotroph is the fungal pathogen *Monilinia fructicola*, which causes brown rot of peaches.

Biotrophic pathogens (sometimes called obligate parasites) grow or reproduce only on or within a suitable host. The biotrophic relationship is highly host specific. Biotrophic pathogens, especially fungi that attack foliar plant parts, derive all their nutrients from the host. They invade host tissues, but do not kill them (or may kill them gradually), thus ensuring a steady supply of nutrients. These fungi may produce special penetration and absorption structures called **haustoria**. Haustoria penetrate the host cell wall, but not the plasmalemma, essentially remaining outside the host cell while nutrients are transferred across the host plasmalemma into the fungus. Classic examples of pathogenic biotrophs include

TABLE 10.2

Exoenzymes Produced by Some Fungi

Exoenzyme	Substrate	Utility
Cutinase	Cutin: A long chain polymer of C_{16} and C_{18} hydroxy fatty acids, which, with waxes, forms the plant cuticle	Facilitates direct penetration of host cuticle by pathogenic fungi
Pectinase (pectin methyl esterase, polygalacturonase)	Pectin: Chains of galacturonan molecules [α-(1→4)-D-galacturonic acid] and other sugars; main components of the middle lamella and primary cell wall	Facilitates penetration and spread of pathogen in host; causes tissue maceration and cell death
Cellulase (cellulase C_1, C_2, C_x, and ß-glucosidase)	Cellulose: Insoluble, linear polymer of ß-(1→4) linkages of glucose; skeletal component of plant cell wall	Facilitates spread of pathogen in host by softening and disintegrating cell walls
Hemicellulase (e.g., xylanase, arabinase)	Hemicellulose: Mixture of amorphous polysaccharides such as xyloglucan and arabinoglucan that vary with plant species and tissues; a major component of primary cell wall	Role of these enzymes in pathogenesis is unclear
Ligninase	Lignin: Complex, high molecular-weight polymer (made of phenylpropanoid subunits); major component of secondary cell wall and middle lamella of xylem tissue	Few organisms (mostly basidiomycetes) degrade lignin in nature; brown rot fungi degrade lignin but cannot use it as food; white rot fungi can do both
Lipolytic enzymes (lipase, phospholipase)	Fats and oils (fatty acid molecules): Major component of plant cell membranes; also stored for energy in cells and seeds and found as wax lipids on epidermal cells	Fatty acid molecules used as a source of food by pathogen
Amylase	Starch: Main storage polysaccharide in plants; enzymes hydrolyze starch to glucose	Glucose readily used by pathogen as a source of food
Proteinase or protease	Protein: Major components of enzymes, cell walls, and cell membrane; enzymes hydrolyze protein to smaller peptide fractions and amino acids	Role of these enzymes in pathogenesis is unclear

the powdery mildew fungi (Chapter 14), and smuts and rusts (Chapter 18).

Some biotrophs are part of a mutually beneficial symbiosis with other organisms. They are called **mutualists** and include the **mycorrhizal fungi** (fungi that grow in association with plant roots) and **endophytes**. Another group of organisms, the **hemibiotrophs**, function as both biotrophs and necrotrophs during their life cycle. An example of a hemibiotroph is the soybean anthracnose pathogen, *Colletotrichum lindemuthianum*. This fungus first lives as a biotroph by growing between the plasma membrane and cell wall of living cells. The fungus then suddenly switches to a necrotrophic phase and kills all the cells it has colonized.

DISEASE SYMPTOMS CAUSED BY FUNGI

Details of the infection process are described in Chapters 28 and 31. The visual manifestation of this process (or **symptoms**) varies, depending on the infected plant part, the type of host, and the environment. Symptoms caused by fungal pathogens can be similar to those caused by other biotic and abiotic disease agents. Generally described, these symptoms are the following:

- **Necrosis**—Cell death; affected tissue may be brown or blackened, sunken, dry or slimy; in leaves, often preceded by yellowing (chlorosis, or a breakdown in chlorophyll).
- **Permanent wilting**—Blockage or destruction of vascular tissues by fungal growth, toxins, or host defense responses; affected tissue wilts and dies, associated leaves may scorch and prematurely abscise.
- Abnormal growth—**Hypertrophy** (excessive cell enlargement), **hyperplasia** (excessive cell division), **etiolation** (excessive elongation); affected tissue appears gall- or club-like, misshapen, or curled.
- **Leaf and fruit abscission**—Premature defoliation and fruit drop.
- **Replacement of host tissue**—Plant reproductive structures replaced by fungal hyphae and spores.

- **Mildew**—Surfaces of aerial plant parts (leaves, fruit, stems, and flowers) are covered with a white to gray mycelium and spores.

Common manifestations of these symptoms are listed in Table 10.3.

FUNGAL REPRODUCTION

The reproductive unit of most fungi is the **spore**, which is a small, microscopic unit consisting of one or more cells that provides a dispersal or survival function for the fungus. Spores are produced in asexual or sexual processes.

ASEXUAL REPRODUCTION

Asexual reproduction is the result of mitosis, and progeny are genetically identical to the parent. Most true fungi (Ascomycota, Basidiomycota, and Deuteromycota—the mitosporic fungi) produce nonmotile, asexual spores called **conidia** (Figure 10.1). Conidia are produced at the tip or side of a stalk or supporting structure known as a **conidiophore**. Conidiophores can be arranged singly or in fruiting structures. A **synnema** consists of a group of conidiophores that are fused together to form a stalk. An **acervulus** is a flat, saucer-shaped bed of short conidiophores that grow side by side within host tissue and beneath the epidermis or cuticle. A **pycnidium** is a globose or flask-shaped structure lined on the inside with conidiophores. A pycnidium has an opening called an **ostiole** through which spores are released.

Another type of asexual spore is the **chlamydospore**. Chlamydospores are thick-walled conidia that form when hyphal cells round up and separate. These spores function as resting spores and are found in many groups of fungi. Some members of the Oomycota (Chapter 20), or water mold fungal-like organisms, produce a motile spore called a **zoospore** in a sac-like structure called a **sporangium**. The stalk that supports the sporangium is called a **sporangiophore**. Zygomycota, best known for *Rhizopus stolonifer*, the fungus that causes bread mold and soft rot of fruit and vegetables, produce asexual spores called **sporangiospores**, also borne in a sporangium.

SEXUAL REPRODUCTION

Genetic recombination is the result of sexual reproduction, and offspring are genetically different from either parent. The heart of the sexual reproductive process is fertilization. In fertilization, sex organs called **gametangia** produce special sex cells (gametes or gamete nuclei), which fuse to form a zygote. The fertilization process occurs in two steps: (1) **plasmogamy**, when the two nuclei, one from each parent, join together in the same cell, and (2) **karyogamy**, when these nuclei fuse together to form a zygote.

In most fungi, the vegetative body, or thallus, has one set of chromosomes (haploid). The diploid (having two sets of chromosomes) zygote that results from fertilization therefore must undergo meiosis or a chromosome reduction division before it can develop into a haploid, multicellular organism. In contrast, most plants and animals, and some fungal-like species in the kingdom Stramenopila, possess a diploid thallus. Meiosis, therefore, occurs in the gametangia as the gametes are produced.

Homothallic fungi produce both male and female gametangia on a single mycelium that are capable of reproducing sexually (i.e., they are self-fertile). This is analogous to a monoecious plant, where both male and female flowers are found on the same plant. Homothallic gametangia may be obviously differentiated into male and female structures (antheridia and oogonia, respectively), or may be morphologically indistinguishable (sexually undifferentiated). Homothallic species occur in all phyla of fungi.

Conversely, the sexes (or **mating types**) in **heterothallic** fungi are separated in two different individuals, which are self-sterile. This is analogous to dioecious plants, where male and female flowers are produced on different plants. The different mating types in heterothallic fungi are usually differentiated as plus (+) or minus (−), or by using letters (e.g., A and A'). Heterothallic fungi are not as common as homothallic ones.

Examples of sexual spores include **oospores**, **zygospores**, **ascospores**, and **basidiospores**. The name of a particular spore type usually reflects the parent structure that produced it. For example, ascospores are produced in a parent structure called an ascus, and basidiospores are produced from a parent structure called a basidium.

ANAMORPH–TELEOMORPH RELATIONSHIPS

Most fungi are capable of reproducing both asexually as well as sexually. The asexual state of a fungus is called the **anamorph**, and the sexual state is the **teleomorph**. The **holomorph** encompasses the whole fungus in all its facets and forms (both anamorph and teleomorph). Some fungi have no known teleomorph, whereas others have no known anamorph.

Intriguingly, many fungi, particularly Ascomycota and some Basidiomycota, produce anamorphs and teleomorphs at different points in their life cycle. For this reason, many fungi have been given two different names, one for the anamorph when it was described, and one for the teleomorph when it was described. When the anamorph is finally associated with its corresponding teleomorph, the whole fungus (holomorph) is more properly referred to by its teleomorph name. For example, the sexual state or teleomorph of *Claviceps purpurea*, which causes ergot of rye and wheat, has an asexual state called *Sphacelia segetum*, a mitosporic fun-

TABLE 10.3
Symptoms of Plant Diseases Caused by Fungi

Symptom	Manifestation	Description	Example
Necrosis	Leaf spot	Discrete lesions of dead cells on leaf tissue between or on leaf veins, often with a light-colored center and a distinct dark-colored border and sometimes accompanied by a yellow halo; fruiting structures of fungus often evident in dead tissue	Strawberry leaf spot (*Mycosphaerella musicola*)
	Leaf blotch	Larger, more diffuse regions of deaf or discolored leaf tissue	Horse chestnut leaf blotch (*Guignardia aesculi*)
	Needle cast	Needles of conifers develop spots, turn brown at the tips, die, and prematurely fall (cast) to the ground	Rhabdocline needle cast of Douglas-fir (*Rhabdocline pseudotsugae*)
	Anthracnose	Sunken lesions on leaves, stems, and fruit; on leaves, regions of dead tissue often follow leaf veins and/or margins	Anthracnose (or cane spot) of brambles (*Elsinoë veneta*)
	Scab	Discrete lesions, sunken or raised, on leaves, fruit, and tubers	Apple scab (*Venturia inaequalis*)
	Dieback	Necrosis of twigs that begins at the tip and progresses toward the twig base	Sudden oak death, Ramorum leaf blight (*Phytophthora ramorum*)
	Canker	Discrete, often elliptical lesions on branches and stems that destroy vascular tissue; can appear sunken, raised, or cracked; causes wilt, dieback, and death of the branch; fruiting structures of fungus often evident in dead tissue	Cytospora canker of spruce (*Cytospora kunzei* or *Leucostoma kunzei*)
	Root and crown rot	Necrosis of feeder roots to death of the entire root system; often extends into the crown and can girdle the base of the stem; causes wilt, dieback, and death of the canopy	Rhododendron wilt (or Phytophthora root rot) (*Phytophthora cinnamomi*; *P. parasitica*)
	Cutting rot	Cuttings in propagation beds are affected by a blackened rot that begins at the cut end and travels up the stem, rapidly killing the cuttings	Blackleg of geranium (*Pythium* species)
	Damping-off	Seeds and seedlings are killed before (preemergent damping-off) or after (postemergent damping-off) they emerge from the ground	Damping-off caused by species of *Pythium*, *Rhizoctonia*, and *Fusarium*
	Soft rot	Fleshy plant organs such as bulbs, corms, rhizomes, tubers, and fruit are macerated and become water soaked and soft; tissues may eventually lose moisture, harden, and shrivel into a mummy; common post-harvest problem	Rhizopus soft rot of papaya (*Rhizopus stolonifer*)
	Dry rot	Dry, crumbly decay of fleshy plant organs	Fusarium dry rot of potato (*Fusarium sambucinum*)
Permanent wilting	Vascular wilt	Vascular tissue is attacked by fungi, resulting in discolored, non-functioning vessels and wilt of canopy	Dutch elm disease (*Ophiostoma ulmi*)
Abnormal growth	Leaf curl	Leaves become discolored, distorted, and curled, often at the leaf edge	Peach leaf curl (*Taphrina deformans*)
	Club root	Swollen, spindle- or club-shaped roots	Clubroot of crucifers (*Plasmodiophora brassicae*)
	Galls	Enlarged growths, round or spindle-shaped, on leaves, stems, roots, or flowers	Cedar-apple rust (*Gymnosporangium juniperi-virginianae*)
	Warts	Wart-like outgrowths on stems and tubers	Potato wart disease (*Synchytrium endobioticum*)
	Witches' broom	Profuse branching of twigs that resembles a spindly broom	Witches' broom of cacao (*Crinipellis perniciosa*)
	Etiolation	Excessive shoot elongation and chlorosis, induced in poor light or by growth hormones	Foolish seedling disease of rice (*Gibberella fujikuroi*)
Leaf and fruit abscission		Infection of petioles, leaves, and fruit cause tissues to drop prematurely	Anthracnose of shade trees (*Apiognomonia* species)

TABLE 10.3 (Continued)

Symptoms of Plant Diseases Caused by Fungi

Symptom	Manifestation	Description	Example
Replacement of host tissue		Plant reproductive structures are replaced by fungal hyphae or spores, or both	Common smut of corn (*Ustilago zeae*)
Mildew		Surfaces of aerial plant parts (leaves, fruit, stems, and flowers) are covered with white to gray mycelium and spores	Powdery mildew of rose (*Sphaerotheca pannosa* var. *rosa*)

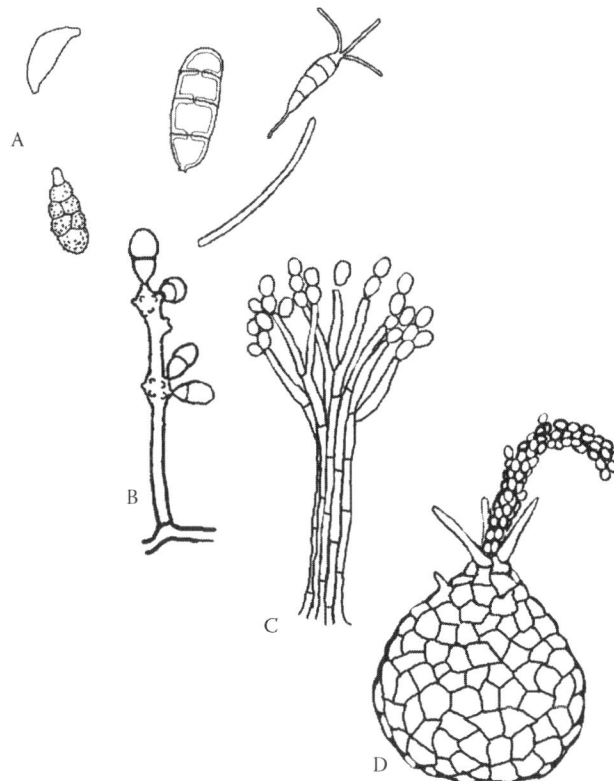

FIGURE 10.1 Asexual reproduction in mitosporic fungi. (A) Single conidia varying in shape, color (hyaline or brown), and number of cells. (B–D) Asexual fruiting structures. (B) Conidia produced on a distinct conidiophore. (C) Synnema. (D) Pycnidium. (Courtesy of N. Shishkoff.)

gus. The holomorph of this fungus is properly referred to as *C. purpurea*. Molecular biology has been useful for assessing the relationships between anamorphs and teleomorphs. Although a fungus may produce different taxonomic structures at different points in its life cycle, its genetics are the same. Examples of holomorphs include *Thanatephorus cucumeris* ana. *Rhizoctonia solani* Kuhn (brown patch of turf), *Venturia inaequalis* ana. *Spilocaea pomi* (apple scab), and *Magnaporthe poae* ana. *Pyricularia grisea* (summer patch of turfgrass).

FUNGI AND THE ENVIRONMENT

Fungi are ubiquitous and grow in many different habitats and environmental conditions. Of these, a source of moisture is most critical for growth and reproduction, and most fungi grow best in a damp environment. As stated previously, moisture is needed to move nutrients into hyphal cells. Indeed, some thin-walled species require a continuous flow of water to prevent desiccation. For foliar pathogenic fungi, relative humidity is important for germination and penetration of leaf tissue. For soil fungi, free moisture in the soil is important for dispersal as well.

Some fungi can adapt to very low moisture availability by regulating the concentration of solutes (or osmotic potential) in cells. A cellular osmotic potential higher than that of the environment will cause water to enter cells. When the environment becomes drier, concentration of solutes outside the cell becomes higher than that within the cells, and water leaves the cells. The net result is that the cell desiccates and the plasmalemma shrinks. Active release or uptake of solutes in hyphal cells helps certain fungi maintain adequate hydration under fluctuating environmental conditions. Other fungi produce resting structures with impermeable walls that withstand drying. These structures germinate when conditions become favorable for growth.

Because oxygen is required for respiration (generation of energy), most fungi do not grow well when submerged. Fungi that require oxygen for respiration are called **aerobic** or obligately oxidative. Those that do not need oxygen for respiration are **anaerobic** or fermentive. Fungi that can derive energy by oxidation or fermentation are **facultative fermentives**. Most plant pathogenic fungi are aerobic or facultatively fermentive.

Fungi must adapt to, or tolerate, considerable temperature fluctuations that occur daily or seasonally. Most fungi are **mesophilic** and grow well between 10°C and 40°C with the optimum temperature for most species between 25°C and 30°C. Some species are **thermophilic** (grow at 40°C or higher) or **psychrophilic** (grow well at less than 10°C). Fungi function best at a range of pH 4 to 7 and some produce melanins in cell walls to protect against damage from sunlight.

SURVIVAL AND DISPERSAL

With the exceptions of biotrophs, most plant pathogenic fungi spend part of their life cycle as parasites and the remainder as saprophytes in the soil or on plant debris. When a fungus encounters adverse environmental conditions, drains its environment of nutrients, or kills its host, growth usually ceases. Inactive hyphae are subject to desiccation or attack by insects or other microbes, so pathogenic fungi must survive in a reduced metabolic or dormant state in plant debris or soil until conditions improve, or disperse to find other hosts.

Fungi produce spores both to survive adverse environmental conditions (e.g., chlamydospores) and for dispersal (e.g., zoospores). Indeed, the different reproductive strategies exhibited by many fungi have different ecological advantages. Generally speaking, fungi reproduce asexually when food sources are abundant and dispersal and spread is of prime importance; they reproduce sexually at the end of the season when food is limited or in cases where dispersal is not as critical. For example, the biotrophic powdery mildew pathogens produce copious quantities of asexual spores (conidia) during the growing season when environmental conditions are favorable and there is abundant host material. These spores are easily dispersed in air and serve to spread disease rapidly over the summer months. At the end of the growing season, environmental and host factors trigger sexual reproduction of the fungus. The ascocarp produced (called a **cleistothecium** or **perithecium**) (see Chapter 14) overwinters

in plant parts or debris, releasing ascospores (spores that result from meiosis) to initiate new infections the following spring.

Spore dispersal can be critical for fungal survival; spores released quickly from fruiting structures have a better chance of finding new hosts and initiating new infections. Spores can be passively or actively liberated from the parent mycelium. Passive spore release mechanisms include rain splash, mechanical disturbance (wind, animal activity, and cultivating equipment), and electrostatic repulsion between a spore and its **sporophore** (supporting stalk). Spores can also be actively released; ascospores are forcibly discharged from the ascus in many members of the Ascomycota (e.g., *C. purpurea*), and sudden changes in cell shape may launch spores into the air (e.g., the aeciospore stage of the rust fungus *Puccinia*).

Most spores, once liberated, are passively dispersed (or **vectored**) through air or soil short or long distances. The most common modes of dispersal for plant pathogenic fungi include air currents, water (via rain splash and dispersal through flowing water), and animals (notably insects). Insects have an intimate role in the life cycle and spread of fungi that cause many important diseases. For example, beetles that vector the Dutch elm disease pathogen *Ophiostoma ulmi* pick up fungal spores on their bodies and inoculate new hosts during feeding. Some plant pathogenic fungi such *Tilletia caries* (stinking smut of wheat) are dispersed on seed, which conveniently places them with their host when the time is appropriate for infection to commence. Soil pathogens that produce motile zoospores (some Oomycota and plasmodial slime molds) can more actively move through soil. Soil zoospores, attracted to roots by root exudates (**chemotaxis**), encyst and penetrate the root cortex to initiate new infections.

Some fungi, such as the basidiomycete *Armillaria mellea*, disperse through potentially inhospitable environments by producing rhizomorphs, root-like structures composed of thick strands of somatic hyphae. Rhizomorphs have an active meristem that grows through soil from diseased plants to healthy roots, thus facilitating the dispersal of the fungus to new substrates. In cross section, a rhizomorph has a dark outer rind and an inner cortex.

More commonly, however, fungi produce resting structures called sclerotia. Sclerotia are spherical structures 1 mm to 1 cm in diameter with a thick-walled rind and a central core of thin-walled cells that have abundant lipid and glycogen reserves. Sclerotia may remain viable in adverse conditions for months or years. Many plant pathogenic fungi in the Ascomycota, Basidiomycota, and Deuteromycota produce sclerotia, especially those that infect herbaceous plants as a means of surviving between crops. Sclerotia may germinate to form sexual structures as in *C. purpurea* or asexual conidia, as in *Botrytis cinerea*.

FUNGAL CLASSIFICATION

The classification of organisms (taxonomy) including fungi is based on criteria that changes as more information becomes known about them. The taxonomy of fungi has been based classically on sexual and asexual spore morphology and how they are formed, followed by secondary considerations such as hyphal and colony (the fungus in culture) characteristics. Molecular techniques that examine and compare the genetics of organisms are increasing in importance to taxonomists.

As previously mentioned, the "fungi" as a group used to be classified with the plants in the kingdom Planta, but are now placed in three different kingdoms (Table 10.4). The endoparasitic slime molds belong to the kingdom Protozoa. The water mold fungi (also called straminopiles) belong to the phylum Oomycota within the kingdom Stramenopila. These two groups are more closely related to protozoans (endoparasitic slime molds) or brown algae (water molds) than to true fungi and are often referred to in the literature as fungal-like, lower fungi, or pseudofungi and are discussed further in Chapters 11 and 20. The other fungal groups (e.g., Chytridiomycota, Zygomycota, Ascomycota, and Basidiomycota) are members of the kingdom Fungi (or Mycota). These fungi are also often referred to as the true fungi or higher fungi. Interestingly, the Chytridiomycota, once placed in the kingdom Protozoa, may be regarded as ancestors of the other true fungi.

Fungi are further divided into subcategories. The most basic classification of a fungal organism is the **genus** and **species** (or specific epithet). The generic name and specific epithet is constructed in Latin according to internationally recognized rules. Fungal names are also associated with an authority, which is the name of the person who discovered or named the species. For example, the species *Claviceps purpurea* (Fr.:Fr.) Tul. was described by Elias Fries and modified by Tulanse. There are more than 70,000 species of fungi described.

In the classification scheme, related genera are grouped into families, families into orders, orders into classes, classes into phyla, and phyla into kingdoms. Each grouping has a standard suffix (underlined). For example, the Dutch elm disease pathogen *Ophiostoma ulmi* is classified as follows:

Kingdom **Fungi** (or **Mycota**)

Phylum Asc**omycota**
Class Asco**mycetes**
Order Ophiostomat**ales**
Family Ophiostomat**aceae**
Genus *Ophiostoma*
Species (and authority) ***Ophiostoma ulmi* (Buisman) Nannf**.

Fungi are probably best identified by the phylum to which they belong. Fungi are placed in phyla based on the sexual phase of the life cycle. Following are brief descriptions of the different phyla, the kingdoms to which they belong, and examples of each.

KINGDOM PROTOZOA

Plasmodiophoromycota (Chapter 11) is one of four groups of slime molds, including the endoparasitic slime molds that produce zoospores with two flagella and are obligate parasites. Sexual reproduction in this group includes fusion of zoospores to form a zygote. Some of these fungi vector plant viruses (Chapter 4).

Plasmodiophora brassicae (clubroot of crucifers) has a multinucleate, amoeboid thallus (plasmodium) that lacks a cell wall. The plasmodium invades root cells, inducing hypertrophy and hyperplasia (thus increasing its food supply). The "clubs" that result interfere with normal root function (absorption and translocation of water and nutrients). Plants affected by this disease wilt and stunt. *P. brassicae* survives in soil between suitable hosts as resting spores (product of meiosis) for many years (Chapter 11).

KINGDOM STRAMENOPILA

Many members of the **Oomycota** (Chapter 20) produce motile, biflagellate zoospores that have a tinsel flagellum and a whiplash flagellum. The thallus is diploid and their cell walls are coenocytic, containing cellulose and glucans. Oomycetes reproduce asexually by producing motile zoospores in a zoosporangium (Figure 10.2), and a few species produce nonmotile sporangiospores. Chlamydospores occur in some species. Sexual reproduction results in an oospore (zygote) produced from contact between male (antheridia) and female (oogonia) gametangia (gametangial contact). Most members of the Oomycota attack roots, but some species of *Phytophthora*, the downy mildews, and white rusts also attack aerial plant parts. The Oomycota are often called "water molds" because most species have a spore stage that swims (biflagellate zoospores) and thus require free water for dispersal.

The ubiquitous soil water mold *Pythium* causes seed rots, seedling damping-off, and root rots of all types of plants, and is especially troublesome in greenhouse plant production and in turfgrasses. *Pythium* species are also responsible for many soft rots of fleshy vegetable fruit and other organs in contact with soil particles in the field, in storage, in transit, and at market.

Phytophthora species cause a variety of diseases on many hosts ranging from seedlings and annual plants to fully developed fruit and forest trees. They cause root rots; damping off; rots of lower stems, tubers, and corms; bud or fruit rots; and blights of foliage, twigs, and fruit. Some species of *Phytophthora* are host specific, whereas oth-

TABLE 10.4

Characteristics of the Different Phyla of Plant Pathogenic Fungi

	Protozoa	Stramenopila	Kingdom	Fungi (Mycota)		
	Plasmodiophoromycota	Oomycota	Chytridiomycota	Zygomycota	Ascomycota	Basidiomycota
Habitat	Aquatic	Aquatic	Aquatic/terrestrial	Terrestrial	Terrestrial	Terrestrial
Form of thallus	Plasmodium (multinucleate mass of protoplasm without a cell wall)	Coenocytic hyphae	Globose or ovoid thallus (lacks a true mycelium)	Well-developed coenocytic hyphae; rhizoids, stolons	Well-developed septate hyphae; some are single celled (yeasts)	Well-developed septate hyphae
Chromosome number in thallus	Haploid (cruciform nuclear division)	Diploid	Diploid	Haploid	Haploid	Dikaryotic, haploid
Chief component of cell wall	Cell wall of thallus lacking	Cellulose/glucans	Chitin	Chitin/chitosan	Chitin/glucans	Chitin/glucans
Motile stage	Zoospores with two anterior, unequal, whiplash flagella	Zoospores with two flagella: one anterior tinsel and one posterior whiplash	Zoospores with one posterior whiplash flagellum	None	None	None
Pathogenic relationship	Obligate parasites	Facultative or obligate parasites	Obligate parasites	Facultative parasites; obligate symbionts (endomycorrhizal fungi)	Facultative or obligate parasites; endophytes	Facultative or obligate parasites
Asexual reproduction	Secondary zoospores in zoosporangia	Zoospores in sporangia; chlamydospores	Holocarpic (entire thallus matures to form thick-walled resting spores)	Sporangiospores in sporangia; chlamydospores (endomycorrhizal fungi)	Conidia on conidiophores produced on single hyphae or in fruiting bodies; budding (yeasts)	Budding, fragmentation, arthrospores, oidia, or formation of conidia
Sexual reproduction	Resting spores with chitin in cell walls, the result of fusion of zoospores to form a zygote	Oospores, the result of fusion between male antheridia and female oogonia	Not confirmed	Fusion of gametangia to produce a zygospore	Formation of ascospores within an ascus	External formation of four basidiospores on a basidium; fusion of spermatia and receptive hyphae (rusts); fusion of compatible mycelia (smuts)
Major plant diseases (or groups of diseases)	Clubroot of crucifers, powdery scab of potato	Root, stem, crown, seed, and root rots, white rust, downy mildew	Potato wart, crown rot of alfalfa	Soft rot, fruit rot, bread mold, postharvest disease, endomycorrhizae	Powdery mildew, root, foot, crown, and corm rot, vascular wilt, canker, wood decay, anthracnose and other foliar diseases, needle cast, apple scab, postharvest disease	Root rot, heart rot, white and brown wood decay, fruit rot, smuts, rusts, snow mold of turfgrass, fairy ring

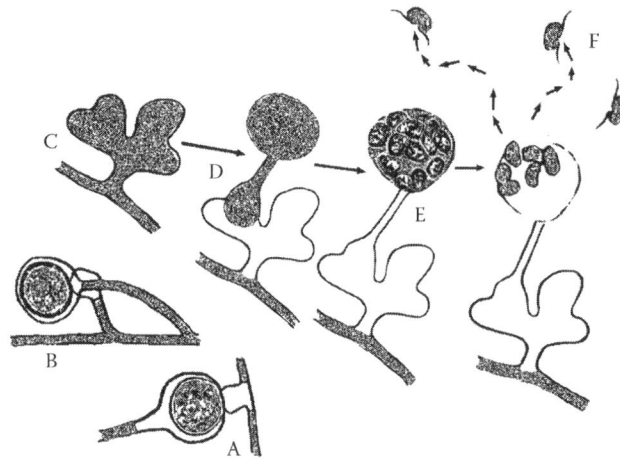

FIGURE 10.2 Reproduction in *Pythium aphanadermatum* (Oomycota), a root, stem, seed, and fruit rot pathogen with a very wide host range. (A–B) Sexual stage. Oogonium (female gamete) with an (A) intercalary or (B) terminal antheridium (male gamete). (C–F) Asexual stage. (C) inflated, lobate sporangium. (D) Sporangium with vesicle. (E) Sporangium with vesicle containing zoospores. (F) Biflagellate zoospores. (Courtesy of N. Shishkoff.)

ers have a broad host range. The genus is best known for its root and crown rots (most *Phytophthora* species), late blight of potato and tomato (*P. infestans*), rhododendron wilt (*P. cinnamomi*), and sudden oak death or Ramorum blight (*P. ramorum*; see Case Study 20.1).

The downy mildews (e.g., *Plasmopara* spp.) are obligate parasites that occur on many cultivated crops. They primarily cause foliage blights that attack and spread rapidly in young and tender green leaf, twig, and fruit tissues. These fungi produce sporangia (asexually) on distinctively dichotomously branched sporangiophores, which emerge through stomates on the lower surfaces of leaves, forming a visible mat with necrotic lesions on leaf surfaces. The downy mildews require a film of water on the plant and high relative humidity during cool or warm, but not hot, weather. Downy mildew of grape (*P. viticola*) destroyed the grape and wine industry in Europe after the disease was imported from the United States in 1875. It led to the discovery of one of the first fungicides, **Bordeaux mixture** (copper sulfate and lime) (see Chapter 36).

KINGDOM FUNGI

Chytridiomycota

Chytridiomycota (Chapter 11), the chytrids, inhabit water or soil. They are coenocytic and lack a true mycelium. The thallus is irregular, globose, or ovoid; diploid; and its cell walls contain chitin and glucans. They have motile cells (zoospores) that possess a single, posterior, whiplash flagellum. The mature vegetative body transforms into thick-walled resting spores or a sporangium.

Olpidium brassicae is an **endobiotic** parasite of the roots of cabbage and other monocots and dicots. The entire thallus converts to an asexual reproductive struc-

ture (**holocarpic**); the sexual cycle in this organism, however, has not been confirmed. *Olpidium* is a known vector of plant viruses such as *Tobacco necrosis virus* and *Lettuce big vein virus*.

Zygomycota

The **Zygomycota** are true fungi with a well-developed, coenocytic, haploid mycelium with chitin in their cell walls. Zygomycetes produce nonmotile spores called sporangiospores. Zygospores (sexual spores) are produced by fusion of two morphologically similar gametangia of opposite mating type (heterothallic) (Figure 10.3). These fungi are terrestrial and are saprophytes or weak pathogens, causing soft rots or molds. Members of this group also include parasites of insects (Entomophthoralean fungi), nematodes, mushrooms, and humans, as well as dung fungi and the endomycorrhizal fungi (see discussion on mutualists).

Rhizopus stolonifer is a ubiquitous saprophyte, notably recognized as a common bread mold, which also causes a soft rot of many fleshy fruit, vegetables, flowers, bulbs, corms, and seeds. Black, spherical sporangia that produce asexual sporangiospores (also known as mitospores) appear on the swollen tip (**columella**) of a long, aerial sporangiophore. The fungus produces **rhizoids** (short branches of thallus that resemble roots) within its food substrate and hyphal **stolons** that skip over the surface of the substratum. *Rhizopus* does not penetrate host cells; it produces cellulases and pectinases that degrade cell walls, causing infected areas to become soft and water soaked, and it absorbs the nutrients released by the enzymes. The fungus emerges through the broken epidermal tissue and asexually reproduces, forming a fluffy covering of fungal growth. When the food supply diminishes, gametangia

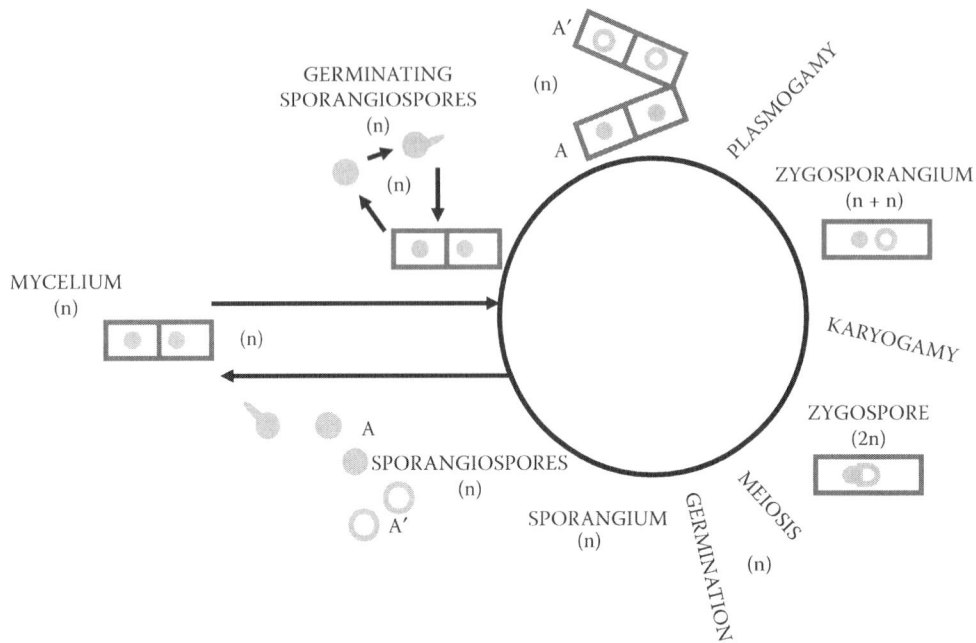

FIGURE 10.3 Generalized life cycle of the Zygomycota. This life cycle is for heterothallic members of the Zygomycota.

combine to form a zygosporangium with a single zygospore inside.

Like *Rhizopus*, species of *Mucor* are common soil inhabitants and are often associated with decaying fruits and vegetables. They differ from *Rhizopus* fungi, however, in that they lack rhizoids. *Mucor amphibiorum* causes a dermal and systemic infection of toads, frogs, and platypus; other species in this genus are human parasites and are implicated in dermal and respiratory allergies. Post-harvest rots of apple, peach, and pear are caused by *M. piriformis*. This species is a soil inhabitant that contaminates fruit that falls to the orchard floor. The fungus penetrates at the calyx or through wounds and within two months in cold storage causes complete decay of the fruit. Fungal mycelia can often be seen emerging in tufts from the surface of affected fruit. The disease is managed by destroying fallen fruit and using sanitation practices where contaminated soil is not allowed to contact harvested fruit.

Ascomycota

The **Ascomycota** (Chapters 13 to 15) are true fungi with a well-developed, septate mycelium containing chitin in their cell walls. The sexual structure of the ascomycetes is an **ascus**, which is shaped like a sac and contains the products of meiosis, called **ascospores**, which normally number eight per ascus. Asci are either **unitunicate** (with a single wall) or **bitunicate** (with a double wall).

Ascospores are forcibly liberated from asci, which are produced in fruiting structures called **ascocarps**.

Ascocarps produced by various ascomycetes include the following:

- **Cleistothecia**—completely closed ascocarps lined with one or more asci
- **Perithecia**—flask-shaped ascocarps with an ostiole at the tip (Chapter 15)
- **Apothecia**—open, cup-shaped ascocarps where asci are exposed (Chapter 15)
- **Ascostroma** or **pseudothecia**—asci produced in a cavity or **locule** buried within a matrix (stroma) of fungal mycelium (Chapter 15)

Asci not produced in an ascocarp are called **naked asci** (Chapter 13).

Asexual reproduction in the Ascomycota includes fission, budding, fragmentation, or formation of conidia or chlamydospores. Conidia are produced on conidiophores. The species of the Ascomycota are a diverse group of fungi that can use many different substrates and occupy a variety of niches.

One of the most infamous diseases caused by an already introduced ascomycete is ergot of rye and wheat (Chapter 1). The pathogen *C. purpurea* occurs worldwide. Grain in the seed head is replaced with fungal sclerotia (called **ergots**) that are poisonous to humans and animals. When ingested, fungal alkaloids restrict blood vessels and affect the central nervous system, causing gangrene in the extremities, hallucinations, and miscarriage. The resulting disorder in humans, known as ergotism or St. Anthony's fire, was common in the Middle Ages, especially in

---

CASE STUDY 10.1

DUTCH ELM DISEASE

- Dutch elm disease is an example of a vascular wilt disease. The pathogens that cause vascular wilts live in the xylem and disrupt the flow of water and nutrients.
- **Significance**: In two 20th century epidemics, Dutch elm disease destroyed over 40 million American elm trees planted throughout the United States as street tree monocultures.
- **Pathogens**: *Ophiostoma ulmi*, *O. novo-ulmi*
- **Hosts**: American elm (*Ulmus americana*) is most susceptible; winged elm, slippery elm, rock elm, and cedar elm vary from susceptible to tolerant; Siberian and Chinese elms are the most resistant.
- **Symptoms and signs**:
  - In summer, leaves flag (or wilt) on individual branches, followed by branch death shortly thereafter.
  - Affected xylem is discolored in a characteristic streaking pattern just beneath the bark.
- **Transmission**:
  - The native elm bark beetle (*Hylurgopinus rufipes*) and the smaller European bark beetle (*Scolytus multistriatus*) serve as vectors. Female insects lay eggs in dead or dying elm wood. Eggs hatch into larvae that feed in wood to form characteristic tunnel patterns called galleries. If the fungus is present, conidia stick to the adult beetles as they emerge in spring to feed on healthy twigs. Spores deposited in xylem during feeding germinate and form a mycelium that penetrates xylem pit openings. Conidia are produced that disperse through the xylem sap.
  - The pathogens also spread from infected to healthy trees through root grafts.
- **Management**: Dutch elm disease is best managed by combining cultural and chemical controls.
  - Sanitation measures include prompt removal of dead or dying elms as well as diseased limbs in mildly affected trees. Wood must be debarked or burned prior to beetle emergence the following spring.
  - Insecticide is used to control the beetle vector, and valuable trees may be injected on a preventive basis with fungicide.
  - Trenching midway between diseased and healthy trees is necessary to prevent root graft transmission.
  - From breeding programs, hybrids of American elm and elms of Asian origin, which are resistant to Dutch elm disease, have been released. These trees are disease resistant yet possess the desired horticultural qualities of the American elm. In addition, a number of American elms with a moderate level of tolerance to Dutch elm disease have been recently introduced to the horticultural trade.

---

peasants who subsisted on bread made with contaminated rye flour.

Claviceps purpurea is homothallic. In the spring, sclerotia germinate on fallen seed heads to form perithecia at the periphery. Each perithecium produces many asci, which forcibly discharge eight ascospores, which are wind-disseminated and infect the ovaries of young flowers. Within a week, droplets of conidia exude in a sticky liquid (honeydew) from the young florets of infected heads. Insects are drawn to the honeydew and carry the conidia on body parts to new flowers. Conidia are also spread by splashing rain. The ergots mature at the same time as seed and are harvested or fall to the ground where they overwinter. Grain containing as little as 0.3% of sclerotia by weight can cause ergotism.

Basidiomycota

The **Basidiomycota** (Chapters 18 and 19) are true fungi that often have large, macroscopic fruiting bodies. They have a well-developed, septate mycelium with chitin in their cell wall. The mycelium of basidiomycetes can be dikaryotic, each cell containing two haploid nuclei. The sexual structure is called a **basidium** (or club), on which four haploid basidiospores (the result of karyogamy and meiosis) are produced on stalks called sterigmata. Basidia are produced on **basidiocarps**. Basidiomycetes reproduce asexually by budding and fragmentation, or by formation of conidia, **arthrospores** (formed when hyphal fragments break into unicellular sections), or **oidia** (produced by short branches, or oidiophores, that cut off oidia in succession from the tip). Like ascomycetes, basidiomycetes utilize a variety of substrates and occupy many different niches. Some of the most important plant diseases, the rusts and smuts (Chapter 18), are basidiomycetes.

Some basidiomycetes are best known for the production of the classic mushroom, which is a basidiocarp. A mushroom basidiocarp consists of a **pileus** (cap) lined with **gills** and a **stipe** (stalk). A **volva** may remain at the base as a remnant of the universal veil that once enveloped the entire developing mushroom (button). The **annulus** is the ring around the stipe as a remnant of a partial or inner veil that once enveloped only the gills or pores on the underside of the basidiocarp. Basidia line the outer surface of gills.

Armillaria mellea affects hundreds of species of trees, shrubs, fruits, and vegetables worldwide. The pathogen is also an important decay fungus in forest ecosystems. *Armillaria* forms rhizomorphs and **mycelial fans** (fan-shaped hyphae) under the bark near the crown of infected trees. Rhizomorphs consist of cord-like "shoestring" threads of mycelium 1 to 3 mm in diameter. They have a compact outer layer (or rind) of black mycelium and a core of white mycelium, and form a branched network that develops over wood and into surrounding soil. Rhizomorphs initiate new infections when in contact with fresh substrate. Plants affected by the fungus show symptoms similar to those caused by other root diseases (reduced growth, small and yellow leaves, dieback, and gradual decline and death). *Armillaria* produces honey-colored mushrooms at the base of dead or dying trees in early fall.

Deuteromycota (Mitosporic Fungi)

Some fungi are classified in an artificial group (not a true phylum) called the Deuteromycota or the *Fungi imperfecti* (Chapter 16). These fungi have no known sexual state in their life cycle and thus have not been classified in one of the other groups. Mitosporic fungi usually reproduce asexually (mitotically) via conidia and are considered to be anamorphic partners of associated teleomorphs in the Ascomycota or Basidiomycota, whose identity may or may not be known. They have hyphae that possess the characteristics of the teleomorph. Some mitosporic fungi have no known reproductive process and are called the sterile fungi or **Mycelia Sterilia**, although some of these fungi do produce sclerotia. Molecular techniques are very useful for placing mitosporic fungi with their associated teleomorphs, and, indeed, may one day eliminate the need for artificial names such as Mycelia Sterilia.

Mitosporic fungi occupy the same habitats as their teleomorph relatives, thus they are largely terrestrial. Most are saprophytes, some are symbionts of lichens, others are grass endophytes or mycorrhizal fungi, and many are plant pathogens. These fungi are divided into three informal classes: **Hyphomycetes** (spores are produced on separate conidiophores), **Agonomycetes** (Mycelia Sterilia), and **Coelomycetes** (spores are produced in fruiting structures called **conidiomata**). Classification within these groups is based on conidiophore, conidia, and hyphal characteristics.

MUTUALISTS

Certain biotrophs, such as mycorrhizal fungi and endophytes live all or most of their life cycles in close association with their plant host. In many cases, this symbiosis is a mutualistic one because both the fungus and the plant host benefit from the association.

Turfgrass Endophytes

Simply stated, endophytes are organisms that grow within plants. When searching the literature, one finds that the term is often used to describe a specific relationship between certain fungi and turfgrass hosts. Unlike turfgrass pathogens such as *Magnaporthe poae*, however, endophytic fungi do not harm the host; indeed, they appear to enhance stress tolerance and resistance to feeding by certain insects.

Endophytic fungi associated with turfgrass are ascomycetes in the family Claviciptaceae and include species of *Epichloë* [ana. *Neotyphodium* (= *Acremonium*)] and *Balansia*. Although these fungi colonize all above-ground plant parts, they are found most readily within the leaf sheath of the turfgrass host and do not colonize the roots. Turfgrass hosts include perennial ryegrass and several fescues (e.g., tall fescue, hard fescue, chewings fescue, and creeping red fescue).

The ability to reproduce sexually varies among the different types of endophytic fungi. In the heterothallic endophyte *Epichloë*, a stroma, composed of fungal mycelium embedded together with the flowers and seeds of the grass itself, forms on the grass stem just below the leaf blade. The stroma is situated in such a way that nutrients flow from living host tissues into the fungus. Clusters of spores called **spermatia** are produced on the surface of the stroma. Flies (*Phorbia* species) visit the stroma and eat the spermatia, which remain undigested in the fly intestinal tract. When a female fly then visits another stroma of opposite mating type to lay eggs, she defecates and deposits spermatia all over the stroma. Fertilization follows, and then perithecia form and ascospores are released to infect nearby florets. Interestingly, the fly larvae that hatch from the eggs then eat the stroma. Some endophytes such as *Neotyphodium* do not produce stroma and cannot reproduce sexually. These fungi grow into the seed at flowering time and are thus seed transmitted. Other endophytes exhibit both sexual and asexual reproductive characteristics.

Benefits conferred to the host by this endophytic association include enhanced nutrition, drought tolerance, and increased hardiness and resistance to disease. Notably, surface feeding by insect pests is deterred. Toxic alkaloids such as peramine are produced in infected hosts, and these compounds deter feeding by insects such as chinchbug, sod webworm, and billbug. *Neotyphodium* endophytes are used in turfgrass breeding programs to enhance these beneficial characteristics in turf intended for sports or landscape (not pasture) use.

Unfortunately, the endophyte association in pastures grasses can have disastrous impact on grazing livestock. Host plants containing *Neotyphodium* endophytes produce ergot alkaloids that reduce blood flow to extremities, causing tails and hooves to rot and fall off. In other associations, ingested lolitrems cause animals to spasm uncontrollably in a syndrome known as ryegrass staggers. Indeed, sleepy grass (*Achnatherum robustrum*) infected by a *Neotyphodium* endophyte produces lysergic acid, a relative of LSD; horses that ingest relatively small amounts of such compounds go to sleep for several days.

Mycorrhizal Fungi

A mycorrhiza (or "fungus-root," coined by A.B. Frank in 1885) is a type of mutualism where both partners benefit in terms of evolution and fitness (ability to survive and reproduce). This is primarily a nutritional relationship where the fungal partner increases the efficiency of nutrient uptake (notably phosphorus) by the host and in turn receives carbon made by the host during photosynthesis. The mycorrhizal association is intimate, diverse, and the fungi involved vary in taxonomy, physiology, and ecology. Mycorrhizal fungi infect more than 90% of vascular plants worldwide, including many important crop plants, and thus the association is economically, ecologically, and agriculturally significant.

Based on the morphology of the association, mycorrhizae are classified into the following three major groups: ectomycorrhizal (outside of root), endomycorrhizal (within root), and ectendomycorrhizal (both inside and outside the root).

Ectomycorrhizae

In this group, a fungal **mantle** of septate hyphae forms over the entire root surface, replacing root hairs and the root cap. Hyphae proliferate between root cortex cells, forming an intercellular network, or **Hartig net**, and it is within this net that nutrients are exchanged between partners. The fungus does not penetrate cells of the cortex or stele (vascular bundle). The ectomycorrhizal relationship is very long lasting; lasting up to 3 years. The mycelium can extend and retrieve nutrients up to 4 m from the surface of roots.

Ectomycorrhizal fungi are mostly members of the Basidiomycota and the Ascomycota and include tooth fungi, chanterelles, mushrooms, puff balls, truffles, and club fungi. Indeed, many fungal sporocarps seen beneath the tree canopy on the forest floor are ectomycorrhizal fungi. Approximately 5000 fungal species form associations with more than 2000 hosts, including plants in the families Betulaceae (birch), Fagaceae (beech and oak), Pinaceae (pine), and also with some Pteridophytes (ferns and horsetail) and eucalyptus. The ectomycorrhizal association comprises about 20% of all mycorrhizal associations. The taxonomy of this group is based on identification of the fungal sporocarp.

Endomycorrhizae

The endomycorrhizae, in contrast, do not change the gross morphology of the root, and hyphae proliferate both between and within the cells of the root cortex, but do not penetrate the stele. There are three different types of endomycorrhizae: arbuscular, orchidaceous, and ericaceous.

Arbuscular Endomycorrhizae

The arbuscular mycorrhizal (AM) relationship, also known as glomeralean or vesicular-arbuscular (VAM), is worldwide in distribution and occurs in many different habitats. AM fungi have a very broad host range; a relatively small number of species (approximately 150) in the Glomeromycota form endomycorrhiza.

In the infection process, chlamydospores resting in soil germinate in the vicinity of plant roots. They penetrate the epidermis, and an aseptate, irregular mycelium grows between the cells of the root cortex. Curious hyphal coils called **pelotons** may form between and within the cells of the outer cortex. **Arbuscles** (special, dichotomously branched haustoria) then develop intracellularly and serve as the site of nutrient and carbon exchange. These structures last only 4 to 15 days before the hyphal content is withdrawn and the arbuscules disintegrate. **Vesicles** are produced intra- or intercellularly. These ovate to spherical structures contain storage lipids and may also serve as propagules. The fungus reproduces asexually via chlamydospores that form internally or on external hyphae that may extend as far as 1 cm from the surface of the root. As the sexual state of these fungi is unknown, the taxonomy of this group has been traditionally based on characters such as chlamydospore morphology and content. Newer molecular methods are helping to redefine phylogenetic relationships among these species of fungi.

Although there are relatively few glomalean species, many host plants can accept more than one mycorrhizal endophyte. The ability of an endophyte to infect a certain host often depends on environmental conditions rather than host specificity. For example, a strain of a single species indigenous to a certain area may successfully compete with a non-indigenous strain of the same species introduced into the soil, suggesting that certain strains may adapt to local environmental conditions.

Orchidaceous Endomycorrhizae

This group includes those fungi that form associations with members of the orchid family. In nature, fungal hyphae penetrate the protocorms of orchids during the saprophytic stage, enabling the seedlings to continue development. Orchidaceous mycorrhizal fungi are primarily basidiomycetes, most of which form imperfect stages in the genus *Rhizoctonia*.

Ericaceous Endomycorrhizae

In this group, fungi in the Ascomycota and Basidiomycota form associations with plants in the Ericales (e.g., rhododendron, blueberry, heather, Pacific madrone, bearberry, Indian pipe). In most associations, penetration of cortical cells by the fungus occurs, but arbuscules do not form. A mantle or sheath may form on the root surface and in some cases a Hartig-net may be present. An interesting example of an ericaceous mycorrhizal relationship is the one between several *Boletus* species (Basidiomycota) and *Monotropa*, or Indian pipe. This plant is a white, unifloral (produces a single flower), achlorophyllous (without chlorophyll), parasitic plant (Chapter 24) that receives all of its carbon and nutrients from a neighboring trees via a connecting mycorrhizal fungus.

Ectomycorrhizae

In this last major group, typical ectomycorrhizae form on forest trees. Under certain conditions, however, fungal hyphae penetrate the cortical cells close to the stele. The perfect stages of these fungi are Basidiomycetes.

Significance of the Mycorrhizal Association

Mycorrhizal fungi, in essence, act as root hairs. Fungal mycelia explore large volumes of soil and retrieve and translocate nutrients that may be otherwise unavailable to the host. The success of the mycorrhizal relationship is probably due to its long evolutionary history, its economy (fungal mycelium is less "expensive" for a host to maintain than an extensive system of root hairs), and its efficiency (fungi produce phosphatase enzymes that readily solubilize phosphorus from soil particles).

Hosts associated with a mycorrhizal fungus often exhibit an increase in growth, which is attributed to an increase in phosphorus nutrition of the plant. External mycelium effectively increases the phosphorus depletion zone around each root, and the phosphorus absorbed by the fungal mycelium is translocated to the site of nutrient exchange (such as arbuscules or the Hartig net) and is released to the host. Fertilizer high in phosphorus and nitrogen reduces mycorrhizal infectivity and sporulation in host plants.

The mycorrhizal relationship benefits plants in other ways, as well. Ectomycorrhizae, for example, play a role in protecting plants from plant pathogenic organisms; the mantle acts as a physical barrier. Some ericoid mycorrhizae produce antimicrobial compounds that inhibit other competing organisms. In a broader sense, mycorrhizae serve to interconnect individuals within a plant community, thus mature plants can "nurture" seedlings in a community buffering effect.

Of course, the fungi benefit from the association in that a high percentage of photosynthate manufactured by the host is translocated to the fungal symbiont. Host sugars (glucose and sucrose) obtained by the fungus are used to produce new hyphae or are converted to glycogen and stored in vesicles or other structures.

SUGGESTED READING

Agrios, G.N. 2005. *Plant Pathology*, 5th ed. Elsevier Academic Press, Burlington, MA.

Barr, D.J.S. 2001. Chytridiomycota, in *The Mycota VII. Part A: Systematics and Evolution*, D.J. McLaughlin, E.G. McLaughlin and P.A. Lemke (Eds.). Springer-Verlag, Berlin, pp. 93–112.

Braselton, J. 2001. Plasmodiophoromycota, in *The Mycota VII. Part A: Systematics and Evolution*, D.J. McLaughlin, E.G. McLaughlin and P.A. Lemke (Eds.). Springer-Verlag, Berlin, pp. 81–91.

Carlile, M.J., S.C. Watkinson and G.W. Gooday. 2001. *The Fungi*, 2nd ed. Academic Press, San Diego, CA.

Deacon, J. 2006. *Fungal Biology*, 4th ed. Blackwell Publishing, Malden, MA.

Hawksworth, D.L., B.C. Sutton and G.C. Ainsworth (Eds.). 1983. *Ainsworth & Bisby's Dictionary of the Fungi*. Commonwealth Mycological Institute, Kew Surrey.

Kyde, M.M. and A.B. Gould. 2000. Mycorrhizal endosymbiosis, in *Microbial Endophytes*, C.W. Bacon and J.F. White (Eds.). Marcel Dekker, New York, pp. 161–198.

Schumann, G.L. 1991. *Plant Diseases: Their Biology and Social Impact*. APS Press, St. Paul, MN.

Schumann, G.L. and C.J. D'Arcy. 2006. Essential Plant Pathology. APS Press, St. Paul, MN.

Sinclair, W.A. and H.H. Lyon. 2005. *Diseases of Trees and Shrubs*, 2nd ed. Cornell University Press, Ithaca, NY.

Taylor, J.W., J. Spatafora, and M. Berbee, M. 2006. Ascomycota. Tree of Life Web Project. http://tolweb.org/tree?group=Ascomycota&contgroup=Fungi.

Ulloa, M. and R.T. Hanlin. 2000. *Illustrated Dictionary of Mycology*. APS Press, St. Paul, MN.

Volk, T.J. 2001. Fungi, in *Encyclopedia of Biodiversity*, Vol. 3. Academic Press, New York, pp. 141–163.

Volk, T.J. 2007. *Tom Volk's Fungi*. Department of Biology, University of Wisconsin-LaCrosse. http://botit.botany.wisc.edu/toms_fungi.

White, J.F., P.V. Reddy and C.W. Bacon. 2000. Biotrophic endophytes of grasses: a systemic appraisal, in *Microbial Endophytes*, C.W. Bacon and J.F. White (Eds.). Marcel Dekker, New York, pp. 49–62.

11 Slime Molds and Zoosporic Fungi

Sharon E. Mozley-Standridge, David Porter, and Marc A. Cubeta

CHAPTER 11 CONCEPTS

- Slime molds and zoosporic fungi represent several phylogenetically distinct groups of microorganisms.

- *Labyrinthula zosterae*, *Plasmodiophora brassicae*, and *Synchytrium macrosporum* have been classified at various times with protists, slime molds, and zoosporic fungi.

- *Synchytrium macrosporum* is a zoosporic fungus; *L. zosterae* is a member of the newly named kingdom Straminipila, whereas the taxonomic placement and phylogenetic relationship of *P. brassicae* to other microorganisms is less certain.

- Knowledge and understanding of the phylogeny and taxonomy of genetically diverse assemblages of plant pathogenic microorganisms can contribute to improved diagnosis and management of plant disease.

Three lower eukaryotic microorganisms (*Plasmodiophora brassicae*, *Labyrinthula zosterae,* and *Synchytrium macrosporum*) represent important pathogens of agricultural, aquatic, grassland, wetland, and woodland species of plants. The phylogeny (evolution of organisms over time) and taxonomy (classification of living organisms) of *L. zosterae*, *P. brassicae*, and *S. macrosporum* will be discussed in relation to their importance in plant pathology.

The scientific discipline of plant pathology is focused largely on the study of organisms that cause disease (i.e., pathogens) on economically important species of plants. Plant pathologists must be familiar with a wide variety of pathogens to ensure accurate identification of the causal agent for successful deployment of economical and environmentally sound disease management strategies. In addition, plant pathologists must also be able to recognize a diverse range of plant-associated organisms typically studied in related disciplines outside of plant pathology (e.g., biology, botany, mycology, and zoology). Slime molds and certain groups of zoosporic fungi represent good examples of organisms that cause a relatively small number of diseases of economically important species of plants.

Zoosporic fungi are found in several phylogenetically unrelated groups of microorganisms that produce motile, flagellated **spores** (**zoospores**), usually as a result of **asexual reproduction** during some stage of their life (Fuller and Jaworski, 1987).

Thus, the term zoosporic fungi is descriptive and not of evolutionary or phylogenetic significance. Zoosporic fungi are found in two different kingdoms of organisms: the kingdom Fungi and the recently named kingdom Straminipila (e.g., Stramenopila) (Dick, 2001). The oldest branch of the kingdom Fungi is the **phylum** Chytrid-iomycota, all of which are zoosporic fungi. Also in the kingdom Fungi are the more familiar (but not zoosporic) Ascomycota, Basidiomycota, Glomeromycota, and Zygomycota. The zoosporic fungi in the kingdom Straminipila include the hyphochytrids, labyrinthulids, and oomycetes. The kingdom Straminipila represents an extremely diverse group of organisms that also include photosynthetic organisms such as brown algae, chrysophyte algae, and diatoms. Organisms classified in the phylum Chytridiomycota produce zoospores with usually a single, posteriorly-directed and "hair-less" (smooth) **flagellum**, whereas organisms classified in the kingdom Straminipila (straminipiles) are characterized by the production of zoospores with usually two flagella (biflagellate) but always with tripartite tubular hairs decorating one of them (Dick, 2001) (Figure 11.1).

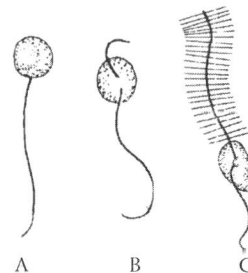

FIGURE 11.1 Zoospores of (A) *Synchytrium macrosporum,* (B) *Plasmodiophora brassicae,* and (C) *Labyrinthula zosterae.* The tripartite tubular hairs on the posterior flagellum of *L. zosterae* are not visible with a light microscope. (Parts [A] and [B] from Dick, M.W., 2001. *Straminipilous Fungi.* Kluwer, Dordrecht, The Netherlands, and Part [C] from Sparrow, F.K. 1960, *Aquatic Phycomycetes*, 2nd ed., University of Michigan Press, Ann Arbor.)

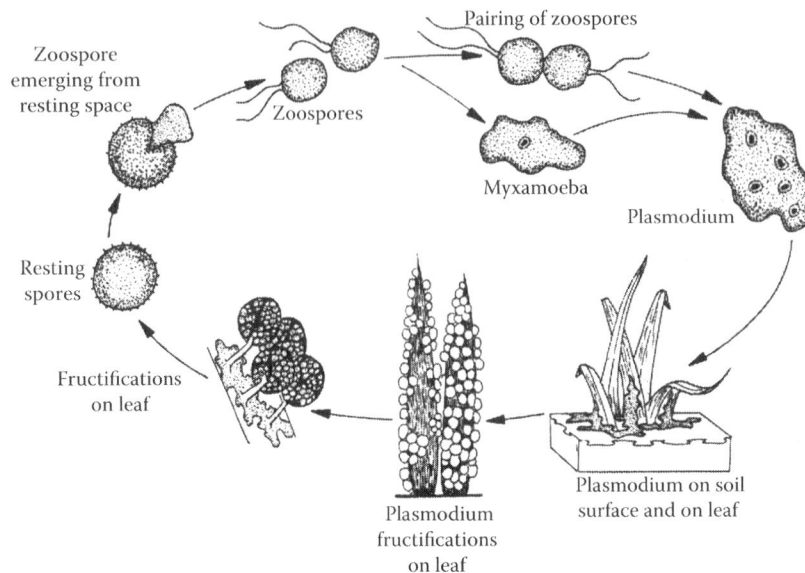

FIGURE 11.2 Life cycle of slime mold. (From Agrios, G.N. 2005. *Plant Pathology*, 5th ed., Academic Press, New York. With permission.)

A flagellum that possesses two rows of tripartite tubular hairs is also referred to as a heterokont or tinsel flagellum (Kirk et al., 2001).

The slime molds or Mycetozoans include three related groups, the protostelids, dictyostelids, and myxomycetes, and a fourth unrelated group, the acrasids, a polyphyletic (i.e., of multiple evolutionary origin) assemblage of organisms (Blanton, 1990; Spiegel et al., 2004; Schaap et al., 2006). The monophyly of the protostelids, dictyostelids, and myxomycetes is supported by phylogenetic analyses of protein gene data (Baldauf and Doolittle, 1997). Slime molds are heterotrophic organisms that form an amoeboid stage during their life and can be isolated from bark, decaying organic matter, dung, plants, and soil. Slime molds commonly ingest bacteria and fungal spores as a food source and can often be observed on grass or wood during periods of wet weather. In general, slime molds are not usually damaging to plants. However, slime molds may grow on the surfaces of plants or mulch and cause concerns to the public. Although different species of slime molds may occur in a limited geographic area, as a group they are found in most temperate and tropical regions of the world (Figure 11.2).

Two groups of organisms often classified with both slime molds and zoosporic fungi include the plasmodiophorids and labyrinthulids (Braselton, 2001). The connection between slime molds and plasmodiophorids is tenuous and based only on the following two common characteristics: (1) the presence of a naked multinucleate protoplast similar to a myxomycete **plasmodium** and (2) the production of anteriorly directed, biflagellate zoospores with two hairless flagella. The plasmodiophorids have a distinct phylogenetic origin separate from the mycetozoans and

the acrasids (Olive, 1975; Castlebury and Domier, 1998). The plasmodiophorids are usually considered as members of the zoosporic fungi because of their osmotrophic mode of nutrition and production of zoospores, but are distantly related to both the straminipiles and zoosporic fungi (Dick, 2001). As a group, the plasmodiophorids are obligate endoparasites (**biotrophs**) of algae, fungi, protists, and plants. Some plasmodiophorids are important pathogens of agricultural crops and include *Plasmodiophora brassicae* (clubroot of crucifers), *Polymxya graminis* (root diseases of cereals), and *Spongospora subterranea* (powdery scab of potato). The latter two pathogens can also transmit plant viruses (Chapter 4).

From a taxonomic perspective, labyrinthulids have also been associated with slime molds, fungi, and various protists. Labyrinthulids produce a network of fine hyaline filaments through which the characteristic spindle-shaped cells move (called an ectoplasmic net), thus prompting some researchers to classify them with the slime molds. However, morphological, ultrastructural, and recent molecular data place the labyrinthulids into the kingdom Straminipila as a sister group to the oomycetes (Chapter 20) and clearly separating them from both the plasmodiophorids and slime molds (Porter, 1990; Honda et al., 1999; Leander and Porter, 2001; Dick, 2001). The labyrinthulids include organisms that are important decomposers and parasites of algae and plants in coastal marine habitats. *Labyrinthula zosterae* is a pathogen of eelgrass (*Zostera marina*), an ecologically important seagrass that forms vast subtidal meadows in estuarine communities in temperate regions of the world. Between 1934 and 1935, most populations of eelgrass present in the North Atlantic were killed due to a

---

CASE STUDY 11.1

WART DISEASE OF POTATO

- Potato wart is caused by the chytrid fungus *Synchytrium endobioticum* and is distributed worldwide. Wart disease is of economic significance to the potato industry on an international scale.
- *Synchytrium endobioticum* is related genetically to *S. macrosporum*, but has a narrower host range and infects plants in the Solanaceae, such as potato and tomato.
- Potato wart disease is challenging to manage with traditional approaches such as breeding plants for resistance and crop rotation.
- The fungus can survive for many years in soil in absence of a plant host by producing resting spores. These spores germinate to produce swimming spores (zoospores) that penetrate host cells directly to initiate disease.
- *Synchytrium endobioticum* is a biotroph and obtains nutrients inside of plant cells, which results in abnormal cell enlargement (**hypertrophy**) and division (**hyperplasia**), and in the formation of galls (warts) on roots, stems and tubers of potato.
- The fungus is dispersed mainly by the activities of humans by moving potatoes, animal manure, soil, and equipment that harbor the organism.
- Potato wart is managed primarily by plant inspection, legislation to establish quarantine laws, and eradication to prevent potential spread of the fungus.
- Plant and soil bioassays coupled with DNA-based molecular methods are providing useful tools for detecting and identifying *S. endobioticum*.

---

pandemic "wasting disease" caused by *L. zosterae*. Since that time, Atlantic eelgrass beds have recovered, but are still not as extensive as before the pandemic. However, *L. zosterae* is still associated with localized dieback of eelgrass. Another pathogenic species of labyrinthulid, *L. terrestris*, is responsible for rapid blight of turfgrass (Bigelow et al., 2005). The unexpected occurrence of a marine organism as the cause of an emerging disease of terrestrial grass may be the result of changes in our cultural practices for turfgrass management, such as the increased use of saline or reclaimed water for irrigation.

Chytrids are members of the phylum Chytridiomycota, the most basal clade within the kingdom Fungi (James et al., 2006). Because of their early evolutionary divergence, most chytrids exhibit certain characteristics not shared by fungi in the phyla Ascomycota and Basidiomycota, such as determinate growth and production of zoospores via asexual reproduction. Chytrid sporangia are microscopic in size and much smaller than the fruiting bodies of most Ascomycota and Basidiomycota. Although a large number of chytrid species are parasitic on a wide variety of other organisms, including fungi, only a few plant hosts are of economic importance. Chytrids often appear simple in form and structure on initial examination, but their morphological characteristics can exhibit considerable variation in shape and size. Chytrids are of great ecological importance as decomposers of a wide variety of biological substrates. For example, a large number of chytrids are known to degrade cellulose in the leaves and stems of plants, whereas others can degrade chitin, keratin, and sporopollenin, a biopolymer associated with pollen grains highly resistant to biological degradation (Sparrow, 1960). Chytrids present within the rumen of certain herbivores (e.g., cows and sheep) are also associated with the break down of cellulose substrates and provide much needed energy and nutrients for the growth and development of animals.

Similar to other true fungi, chytrids have chitin in their cell walls, flattened plate-like mitochondrial cristae, and exhibit an absorptive mode of nutrition. However, unlike all other members of the kingdom Fungi, chytrids reproduce by the formation of zoospores and lack true mycelium. Chytrid zoospores differ from straminipile zoospores by having only one posteriorly directed, smooth flagellum (Figure 11.1). Because chytrids are microscopic and usually determinant in their growth, they are often overlooked in the environment but can be readily found and observed using baiting techniques and a simple dissecting microscope. Although chytrids are sometimes referred to as aquatic fungi or water molds, chytrids are present and can be found nearly anywhere that water is available, including in soil. A large number of chytrids are parasites of algae, fungi, plants, and even frogs. Many chytrids can also exist as **saprotrophs**.

The genus *Synchytrium* has over 200 species that are known to be parasites of a wide range of algae and plants in freshwater and terrestrial habitats. *Synchytrium endobioticum* is the best-known species, and is the causal agent of potato wart disease. The fungus can also attack tomato and noncultivated species of *Solanum*. Since the initial discovery of potato wart in Hungary in 1896, this disease has been identified on most continents. Because of the serious nature of potato wart, *S. endobioticum* was placed on the

---

CASE STUDY 11.2

RAPID BLIGHT OF COOL SEASON TURFGRASSES—AN UNEXPECTED PATHOGEN

- Rapid blight is a disease of cool season turfgrasses and a major concern to golf course managers in the southeast and western United States and in the United Kingdom.
- The symptoms include a patchy water-soaked appearance and browning followed by chlorosis and death of annual grasses such as rough bluegrass (*Poa trivialis*), perennial rye (*Lolium perenne*), annual bluegrass (*Poa annua*), and creeping bentgrass (*Agrostis stolonifera*) on golf courses during autumn and spring.
- Rapid blight is an emerging disease that was unknown before the 1990s, and this disease is costly and difficult to manage.
- An endobiotic chytrid fungus was originally thought to cause rapid blight, but in 2005 *Labyrinthula terrestris* was identified as the causal agent of this disease.
- Other species of *Labyrinthula* grow in estuarine and marine habitats where some are known to cause diseases of seagrasses. Prior to the description of *L. terrestris*, on turfgrasses none were known to be pathogens of terrestrial plants.
- An obvious question: how has a pathogen, whose known relatives are inhabitants of marine habitats, become a serious and emerging disease problem on turfgrasses?
- The answer is not known, but turfgrass management practices may be important in selecting for this emergent pathogen.
  - Some fungicides provide protection against *L. terrestris*, whereas other fungicides reduce competition for the pathogen.
  - Irrigation of golf courses with saline water encourages growth of *L. terrestris*, which can grow in saline conditions.
  - Frequent and close mowing allows for entry of the pathogen into grass leaves.
- Rapid blight is an excellent reason for plant pathologists to expect the unexpected, and to be broadly trained.

---

list of quarantined organisms established in 1912 by the United States Department of Agriculture (USDA) under the Plant Protection Act. Canada and other countries have also established quarantine laws to prevent movement of this fungus on potatoes and in soil. A recent outbreak of potato wart in a single field in Prince Edward Island (PEI) Canada in 2000 prompted the USDA to impose a quarantine that banned importation of seed potatoes from PEI into the United States for one year. Potato wart was also recently discovered in two additional fields on PEI in 2002. The **resting spores** of *S. endobioticum* can survive for 20 to 30 years in soil (Agrios, 2005).

Other chytrid pathogens of economically important species of agricultural plants include; *Olpidium* spp. (infects roots of many plants and serve as a vector for at least six plant viruses) (see Chapter 4), *Physoderma alfalfae* (causal agent of crown wart of alfalfa), and *P. maydis* (causal agent of brown spot of corn).

PLASMODIOPHORA BRASSICAE

Plasmodiophora brassicae is an important pathogen of cultivated agricultural crops that belong to the Brassicaceae (mustard family). Members of this plant family are often referred to as crucifers (because of their cross-shaped flowers) or cole (which is German for "stem") crops, and include the following: broccoli, Brussels sprouts, cabbage, Chinese cabbage, canola, cauliflower, collards, kale, kohlrabi, mustard, radish, rape, rutabaga, and turnip. Several additional genera of cultivated and weed species of plants in the genera *Alyssum, Amoracia, Brassica, Camelina, Capsella, Erysimum, Iberis, Lepidium, Lobularia, Lunaria, Matthiola, Nasturtium, Raphanus, Rorippa, Sinapis, Sisymbrium*, and *Thlaspi* are also hosts for *P. brassicae* (Farr et al., 1995). *Arabidopsis thaliana*, a mustard species known as mouse-ear cress and widely employed as a model system in the genetic research of plant development and plant-microbe interactions, has been reported recently as a host for *P. brassicae* (Siemens et al., 2002).

Plasmodiophora brassicae is a biotrophic (obligate) parasite that causes a devastating disease of crucifers known as clubroot. The disease occurs throughout the world in commercial crucifer production fields, but is also a serious problem in home gardens. Clubroot has been known since the 13th century in Western Europe and was first studied in detail by Woronin in Russia in the late 1870s. Woronin originally described *P. brassicae* as a slime mold. Despite the tremendous amount of research on clubroot, it still remains as one of the most serious diseases of crucifers and is largely responsible for the dis-

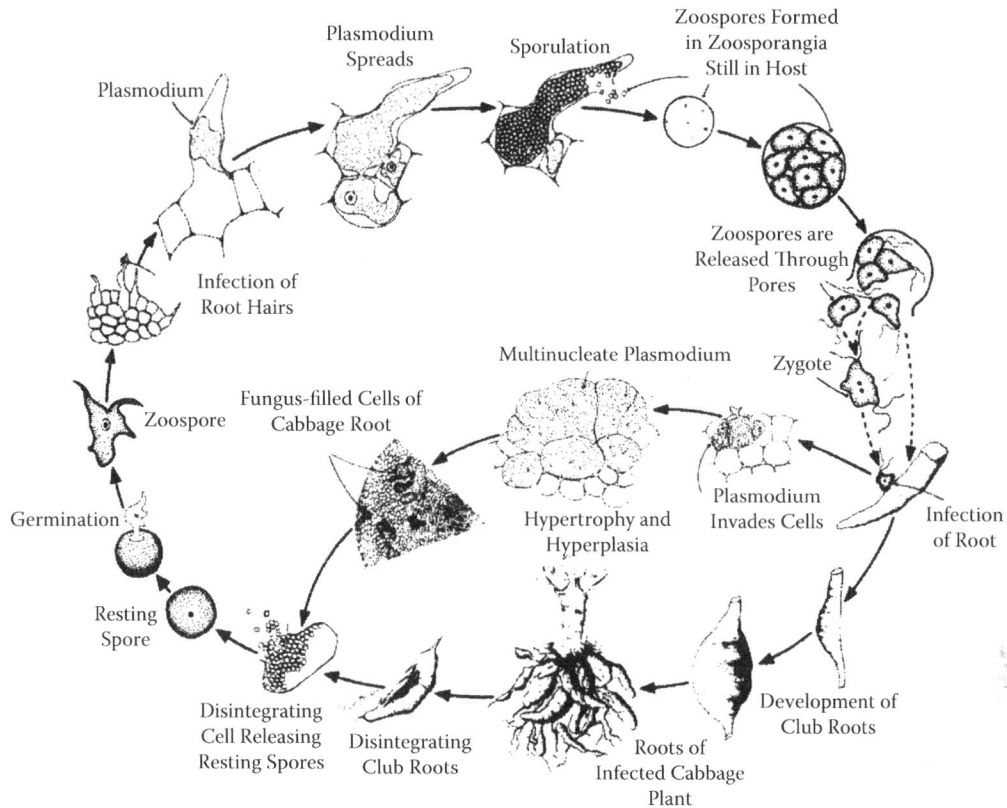

FIGURE 11.3 Life cycle of *Plasmodiophora brassicae*. (From Agrios, G.N. 2005. *Plant Pathology*, 5th ed., Academic Press, New York, NY. With permission.)

appearance of commercial cabbage production in many regions throughout the world.

Because *P. brassicae* produces resting spores that can persist in soil for many years, traditional approaches such as crop rotation are of limited value in managing clubroot. Some success has been achieved in breeding plants for resistance to *P. brassicae*. However, field populations of *P. brassicae* are genetically diverse and composed of many races of the pathogen. Races are represented by genetically distinct individuals and typically identified by artificially inoculating a series of well-defined species of crucifers and observing them for disease symptoms (Williams, 1966; Cubeta et al., 1998). Although some resistant varieties of crucifers have been developed and are commercially available, none of them are resistant to all known races of *P. brassicae*. One approach used for centuries for managing clubroot involves the modification of the soil environment by adding lime (either calcium carbonate or calcium oxide) to increase soil pH to at least 7.2. This approach provides an economical and effective means of reducing the damaging effects of clubroot, and it is hypothesized that the increased soil pH interferes with zoospore motility and the initial root infection process. Unfortunately, a soil pH of 7.2 or higher is not a favorable growth environment for many cultivated agricultural crops.

LIFE HISTORY

Plasmodiophora brassicae can survive for at least 10 years in soil by forming resting spores (Agrios, 2005) (Figure 11.3). During periods of cool, wet weather when the soil becomes saturated with water, resting spores (also referred to as cysts) germinate to produce usually one primary zoospore with two hairless flagella (Figure 11.1). These zoospores swim to the root and penetrate hairs of young roots or enter the plant through wounds in secondary roots (Williams, 1966). Once inside the root, *P. brassicae* produces an amoebae-like structure called a **plasmodium** (pl. **plasmodia**) that passes through the cells and becomes established in them. In plant cells, the nucleus of each plasmodium divides and becomes transformed into a multinucleate structure called a zoosporangium (pl. zoosporangia) that contains 4 to 8 secondary zoospores. Secondary zoospores are discharged through exit pores in the plant cell wall and usually fuse with each other to cause more infection of roots to form additional plasmodia. These diploid plasmodia divide by meiosis and produce clusters (**sorus**, pl. **sori**) of haploid resting spores with a single nucleus (uninucleate).

As plasmodia continue to grow and develop they ingest proteins and sugars in the plant cells as a source of nutrients while stimulating them to divide (**hyper-**

plasia) and enlarge (**hypertrophy**). This abnormal plant growth results in the production of small, spindle shaped swellings on roots that later develop into larger-sized galls or clubs. Root galls interfere with nutrient and water movement in the plant and initial symptoms on infected plants often appear as a yellowing and wilting of the lower leaves, particularly on warm, sunny days. Severely infected plants are often stunted and smaller than healthy plants. Eventually the galls become a food source for other soil dwelling microorganisms.

SYNCHYTRIUM MACROSPORUM

Synchytrium macrosporum is a biotrophic pathogen of more than 1400 different species of plants representing 185 families and 933 genera and ranging from liverworts (hepatophytes) to flowering plants (angiosperms), especially those in the families Asteraceae, Brassicaceae, Cucurbitaceae, Fabiaceae, and Solanaceae (Karling, 1964). *Synchytrium macrosporum* has the largest and widest host range of any biotrophic fungus in the kingdom Fungi (Karling, 1964). *Synchytrium macrosporum* is a weak pathogen that primarily attacks young seedlings.

Although most plants survive early infection and grow to maturity, some seedling death may occur in rare cases of severe infection, particularly if environmental conditions are favorable. As with *P. brassicae*, galls caused by the hypertrophy and hyperplasia of infected host epidermal cells are the most recognizable symptom produced by *S. macrosporum* on plants. In general, galls form on leaves and stems of developing plants and ranging in size from 350 μm to 1.3 mm. However, galls can also form on the roots and underground fruits of certain legumes. The characteristic galls are composed of a central infected cell with a single resting spore and a sheath of surrounding cells of increased size. Occasionally, some portion of the host cell cytoplasm is retained around the resting spore. Galls associated with infection by *S. macrosporum* only containing resting spores are referred to as monogallic, whereas galls produced by other species of *Synchytrium* containing either a resting spore or a zoosporangium are referred to as digallic.

Life History

Synchytrium macrosporum survives primarily as resting spores in soil and infected plant debris (Figure 11.4). Depending on the host and geographic location, dormant resting spores usually germinate in the presence of moisture in late winter or early spring. When resting spores germinate, they function as prosori (sing. prosorus), or containers for cellular contents that will later become sori. During germination, contents of the resting spore and prosorus exit the thick-walled casing through an opening or exit pore in the wall that is eventually

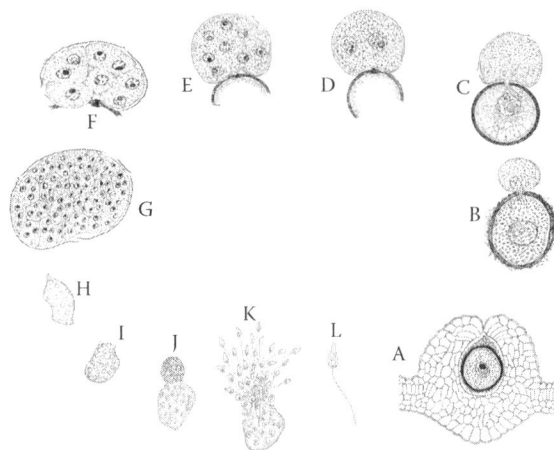

FIGURE 11.4 Life cycle of *Synchytrium macrosporum*. (A) composite gall on leaf of ragweed (*Ambrosia trifida*); (B) early germination stage of resting spore/prosorus; (C) later stage of resting spore germination; (D) binucleate incipient sorus (notice the plug of dark material between the sorus and resting spore case); (E) multinucleate incipient sorus; (F) cleavage of soral contents into multiple uninucleate protospores, the precursor to sporangia; (G) sorus of multinucleate sporangia; (H) individual sporangium; (I) individual sporangium with a ruptured sporangial wall showing an inner membrane prior to spore discharge; (J) dehiscence of the sporangium; (K) dispersal of the zoospores; (L) an individual zoospore. (From Karling, J.S., 1964. *Synchytrium*, Academic Press, New York. With permission).

filled by a plug of dark pigmented material. The cytoplasm, surrounded by a plasma membrane, undergoes a number of mitotic divisions before partitioning into numerous sporangia. The number of sporangia within a single sorus can range from 120 to 800, whereas the diameter of an individual **sporangium** can range from 18 to 60 μm. Sporangia within a sorus can remain dormant for 1 to 2 months; infected plant tissue needs to dry out completely and then be rehydrated to induce resting spores to germinate (Karling, 1960). These conditions may simulate events that occur in nature where infected material senesce, dry out, and then become rehydrated with water from dew or rain.

The cytoplasm within each sporangium cleaves up into individual zoospores, each with its own nucleus. In order for zoospores to be released, the sorus has to open up (dehisces) and releases individual sporangia from the soral membrane. The sporangia, in turn, release zoospores by a splitting of the sporangial inner membrane. Zoospores swim to new hosts, through the thin film of water present either on the surface of a plant or in soil, to infect young seedlings. Zoospores can also be dispersed in water splashed from one plant to another. The zoospores are ovoid to slightly elongate (3 to 3.8 × 4 to 4.5 μm, with a single yellowish-orange refractive lipid globule) and possess a single, posteriorly directed, hairless flagellum (12 to 14 μm in length). When zoospores are released from the sporangium in the spring, they alternate between swimming and moving in an amoeboid fashion for as long as 24 h before settling down on the surface of a host. Once settled, encystment occurs, and the flagellum is either retracted into the zoospore, later referred to as a zoospore cyst, or cleaved off completely, and a membrane is produced on the outside of the zoospore. A narrow germ tube develops from the encysted zoospore and penetrates the host cell wall. Cytoplasm flows into the host cell from the zoospore cyst and both the zoospore cyst and germ tube disintegrate. The cytoplasm can assume a variety of different shapes from round to amoeboid once inside the host and moves within the cell, positioning itself near the host nucleus. The cytoplasm, now called an initial cell or uninucleate thallus, increases in size and develops a thick wall as it matures. After the thick-walled initial cell goes into a state of dormancy, it becomes a resting spore.

Resting spores can either be spherical or ovoid in shape and range in diameter from 80 to 270 μm. The color of the resting spore wall varies from dark amber to reddish-brown and is usually 4 to 6 μm thick. The walls of resting spores and sporangia are yellow-orange colored, as are the characteristic galls that form as a result of infection. Although Karling (1960) observed the production of zygotes from the fusion of two zoospores and their nuclei, no one has substantiated this observation or determined the role that sexual reproduction plays in resting spore formation.

LABYRINTHULA ZOSTERAE

The devastating eelgrass wasting disease of 1934 and 1935 brought attention to the obscure protist *Labyrinthula,* which at the time was known only from a few 19th-century German publications. Although *Labyrinthula* was thought to be associated with the disease in the early 1930s (Porter, 1990), it was not until 1988 that, by satisfying Koch's postulates, Muehlstein et al. (1988) demonstrated that *L. zosterae* was the causal agent responsible for the necrotic lesions on eelgrass and dieback symptoms observed in seagrass meadows.

Ten species of *Labyrinthula* are recognized (Dick, 2001), but there are many yet to be described, particularly those associated with different species of seagrass. For example, a different species of *Labyrinthula* is associated with turtle grass (*Thalassia testudinum*) and has been implicated in a devastating dieoff of that seagrass in Florida Bay during the late 1980s (Porter, 1990). In addition, the turfgrass pathogen *L. terrestris* represents a different species than either the eelgrass or turtlegrass pathogens. While some labyrinthulas are agents of plant disease, other species are not pathogens of seagrasses or turfgrasses and exist primarily as saprotrophs.

In contrast to *P. brassicae* and *S. macrosporum,* most species of *Labyrinthula* can be easily grown on nutrient

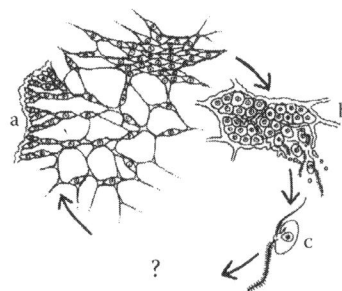

FIGURE 11.5 Life cycle of *Labyrinthula vitellina* (the sexual phase of the life cycle of *L. zosterae* has not been observed). (a) colony of trophic cells within an ectoplasmic network; (b) sorus within which occurs meiosis and the release of haploid zoospores (meiospores); (c) biflagellate zoospore (which possibly develops into a new spindle-shaped trophic cell, although the developmental process is not known). (From Porter, 1990, Phylum Labyrinthulomycota, pp. 388–398, in *Handbook of Protoctista*, Margulis, L. et al. Eds., Jones and Bartlett, Boston, MA. With permission.)

medium in culture, and none appear to be biotrophs. In culture, labyrinthulids are most easily grown with a yeast or bacterium coinoculated on the nutrient medium as a food organism (Porter, 1990).

LIFE HISTORY

Labyrinthula species are known primarily from their trophic (or feeding) stage. This stage, which can be observed on nutrient medium in a petri dish, is composed of a colony of cells within a network of interconnecting filaments (called the "ectoplasmic network") (Figure 11.5). The cells are spindle shaped and move with a gliding motion within the ectoplasmic network with speeds as rapid as 150 μm/min. The network plays an important role in the biology of the *Labyrinthula* colony by: (1) providing a structure through which the cells move, (2) aiding in the attachment of the colony to the substrate, (3) housing digestive enzymes necessary for feeding by the cells, and (4) serving as a conduit for transmitting signals within the colony to coordinate communal motility (Porter, 1990).

In *L. zosterae,* the trophic colony is the only stage of the life history that has been observed. In other closely related species such as *L. vitellina,* a sexual life cycle is observed and characterized by biflagellate, heterokont zoospores produced by meiosis (Porter, 1990). However the fusion of gametes (syngamy) has never been observed. The life cycle of *L. vitellina* is presented in Figure 11.5 (Porter, 1990).

Slime molds and zoosporic fungi represent phylogentically and taxonomically distinct assemblages of microorganisms commonly associated with plants. The interdisciplinary scientific study of slime molds and zoosporic fungi has provided a foundation of knowledge for understanding their biology, ecology, genetics, and interactions with plants. In the following laboratory exercises (Chapter 12), students will be provided with an opportunity to observe characteristic structures of three zoosporic plant pathogens; *P. brassicae, S. macrosporum,* and *L. zosterae* and their parasitic interactions with different species of host plants. The biotrophic feeding behavior of *P. brassicae* and *S. macrosporum* that prevents their culturing on a nutrient medium also provides a unique opportunity for students to understand how Koch's postulates are modified to examine the disease-causing activities of biotrophic plant pathogens.

REFERENCES

Agrios, G.N. 2005. *Plant Pathology,* 5th ed., Academic Press, New York.

Baldauf, S.L. and W.F. Doolittle. 1997. Origin and evolution of the slime molds (Mycetozoa). *Proc. Natl. Acad. Sci. (USA)* 94: 12007–12012.

Bigelow, M., D. Olsen and R.L. Gilbertson. 2005. *Labyrinthula terrestris* sp. nov., a new pathogen of turf grass. *Mycologia* 97: 165–190.

Blanton, R.L., 1990. Phylum Acrasea in *Handbook of Protoctista,* L. Margulis, J.O. Corliss, M. Melkonian and D.J. Chapman (Eds.), Jones and Bartlett, Boston, MA, pp. 75–87.

Braselton, J.P. 2001. Plasmodiophoromycota, in *The Mycota VII, Part A, Systematics and Evolution,* D.J. McLaughlin, E.G. McLaughlin, and P.A. Lemke (Eds.), Springer-Verlag, Berlin-Heidelberg, Germany, pp. 81–91.

Castlebury L.A. and L.L. Domier. 1998. Small subunit ribosomal RNA gene phylogeny of *Plasmodiophora brassicae. Mycologia* 90: 102–107.

Cubeta, M.A., B.R. Cody and P.H. Williams. 1998. First report of *Plasmodiophora brassicae* on cabbage in Eastern North Carolina, *Plant Dis.* 81: 129.

Dick, M.W. 2001. *Straminipilous Fungi,* Kluwer Dordrecht, The Netherlands.

Farr, D.F., G.F. Bills, G.P. Chamuris and A.Y. Rossman. 1995. *Fungi on Plants and Plant Products in the United States.* APS Press, St. Paul, MN.

Fuller, M.S. and A. Jaworski. 1987. *Zoosporic Fungi in Teaching and Research.* Southeastern Publishing Corporation, Athens, GA.

Honda, D., T. Yokochi, T. Nakahara, S. Raghukumar, A. Nakagiri, K. Schaumann and T. Higashihara. 1999. Molecular phylogeny of labyrinthulids and thraustochytrids based on the sequencing of 18S ribosomal gene. *J. Eukaryot. Microbiol.* 46: 637–647.

James, T.Y., P.M. Letcher, J.E. Longcore, S.E. Mozley-Standridge, D. Porter, M.J. Powell, G.W. Griffith and R. Vilgalys. 2006. A molecular phylogeny of the flagellated fungi (Chytridiomycota) and description of a new phylum (Blastocladiomycota). *Mycologia* 98: 860–871.

Karling, J.S. 1960. Inoculation experiments with *Synchytrium macrosporum. Sydowia* 14: 138–169.

Karling, J.S. 1964. *Synchytrium.* Academic Press, New York.

Kirk, P.M., P.F. Cannon, J.C. David and J.A. Staplers 2001. *Ainsworth & Bisby's Dictionary of the Fungi.* 9th ed., CAB International, Wallingford, U.K.

Leander, C.A. and D. Porter. 2001. The Labyrinthulomycota is comprised of three distinct lineages. *Mycologia* 93: 459–464.

Muehlstein, L.K., D. Porter and F.T. Short. 1988. A marine slime mold, *Labyrinthula,* produces the symptoms of wasting disease in eelgrass, *Zostera marina. Mar. Biol.* 99: 465–472.

Olive, L.S. 1975. *The Mycetozoans,* Academic Press, New York.

Porter, D. 1990. Phylum Labyrinthulomycota, in *Handbook of Protoctista,* L. Margulis, J.O. Corliss, M. Melkonian and D.J. Chapman (Eds.), Jones and Bartlett, Boston, MA, pp. 388–398.

Schaap, P., T. Winckler, M. Nelson, E. Alvarez-Curto, B. Elgie, H. Hagiwara, J. Cavender, A. Milano-Curto, D.E. Rozen, T. Dingermann, R. Mutzel and S.L. Baldauf. 2006. Molecular phylogeny and evolution of social amoebas. *Science* 314: 661–663.

Siemens, J., M. Nagel, J. Ludwig-Mueller and M.D. Sacristan. 2002. The interaction of *Plasmodiophora brassicae* and *Arabidopsis thaliana*: parameters for disease quantification and screening of mutant lines. *J. Phytopath.* 150: 592–605.

Sparrow, F.K. 1960. *Aquatic Phycomycetes,* 2nd ed., University of Michigan Press, Ann Arbor, MI.

Spiegel, F.W., S.L. Stephenson, H.W. Keller, D. Moore and J.C. Cavender. 2004. Mycetozoans in *Biodiversity of Fungi, Inventory and Monitoring Methods,* G.M. Mueller, G.F. Bills and M.S. Foster (Eds.), Elsevier, Amsterdam, The Netherlands, pp. 547–576.

Williams, P.H. 1966. A system for the determination of races of *Plasmodiophora brassicae* that infect cabbage and rutabaga. *Phytopathology* 56: 624–626.

12 Laboratory Exercises with Zoosporic Plant Pathogens

Marc A. Cubeta, David Porter, and Sharon E. Mozley-Standridge

Plant diseases have been observed and recorded by humans for more than 2000 years (Agrios, 2005). Many plant diseases were described initially based on the observation of visible signs (vegetative reproductive structures of the pathogen) and symptoms (reactions of the plant to infection) on fruits, leaves, roots, and stems. Since then, scientists in the discipline of plant pathology have continued to investigate the causal role that microorganisms play in plant disease and how their biology, ecology, and genetics influence pathogenesis (i.e., the disease-causing ability of an organism). The majority of scientific studies conducted by plant pathologists have focused primarily on plant species of economic importance to agriculture and fostered the development of experimental methods to examine plant pathogens and their associated diseases. In general, most agricultural crops are subject to many diseases caused by a wide array of plant pathogens. Some of these pathogens have a narrow host range and can infect only a single species or variety of plant, whereas other pathogens have the ability to infect a wider range of hosts, often in genetically different plant families. The incidence and severity of disease can also vary depending on environmental conditions and genetic composition of the pathogen and plant.

In the following laboratory exercises, the biology of *Plasmodiophora brassicae*, a well-studied pathogen of crucifers in agricultural production systems (Sherf and MacNab, 1986) and *Labyrinthula zosterae* and *Synchytrium macrosporum*, two important pathogens of plants in natural ecosystems, will be examined. These organisms produce motile spores (zoospores) that are an important component of their ecology and disease biology (Agrios, 2005; Fuller and Jaworski, 1987; Karling, 1960; Porter, 1990). Differences in the life cycle, feeding (trophic) behavior, and symptom expression of each organism will provide the basis for determining differences in susceptibility and how these organisms cause plant disease. In each laboratory exercise, students will examine infected plant material to familiarize themselves with signs and symptoms of each organism. Various extraction and artificial inoculation methods (depending on the organism) will be employed to monitor and record disease development. Because *L. zosterae* can be readily isolated in pure culture on nutrient medium, the students will initially isolate this organism from infected seagrass plants and then reisolate the organism from plants they have artificially inoculated to fulfill Koch's postulates and offer "proof of pathogenicity" (Chapter 39).

EXERCISES

EXPERIMENT 1. SUSCEPTIBILITY OF CRUCIFERS TO *PLASMODIOPHORA BRASSICAE*

The selection of species or varieties of crucifers with reduced susceptibility to *P. brassicae* can often be used to manage clubroot disease. However, because of the inherent genetic diversity that exists in field populations of *P. brassicae*, no variety of plant is likely to be resistant to all genetic individuals of the pathogen. In this laboratory experiment, a modification of the procedures developed by Williams (1966) and Castlebury et al. (1994) for isolation of resting spores and plant inoculation will be employed to examine the susceptibility of different crucifers to infection by *P. brassicae*. Each type of crucifer host will be critically examined for the incidence and severity of disease symptoms and compared with the noninoculated control. Students will also have an opportunity to examine the characteristic microscopic structures of *P. brassicae* in infected plant tissue.

Materials

Each student or team of students will require the following laboratory items:

- 50- and 100-mL beakers
- Blender
- Centrifuge (table top or swinging bucket)
- Centrifuge tubes
- Cheesecloth
- Compound light microscope
- Dissecting microscope
- Distilled water
- 500- and 1000-mL Erlenmeyer flasks
- Forceps
- Funnel
- Glass slide and cover slips
- 1000-mL graduated cylinder

	Procedure 12.1
	Germination of Seeds
Step	Instructions and Comments
1	Prepare five petri dishes of the following seeds: broccoli (*Brassica oleracea* var. *italica*), cabbage (*B. oleracea* var. *capitata*), canola (*B. napus*), cauliflower (*B. oleracea* var. *botrytis*), Chinese cabbage (*B. pekinensis*), collard (*B. oleracea* var. *acephala*), kale (*B. oleracea* var. *acephala*), mustard (*B. nigra*), radish (*Raphanus sativus*), rutabaga (*B. napus*), and turnip (*B. rapa*).
2	Place a piece of Whatman #1 filter paper into a plastic petri dish (10-cm diameter) and moisten with distilled water. Arrange seeds (30 to 50 per dish) on filter paper 1- to 2-cm apart and gently press each seed into the filter paper with a pair of forceps. Offset seeds in each row to allow roots of each seedling to grow straight and not become entangled.
3	Incubate seeds at 20°C to 25°C with 12 h of supplemental lighting from two cool watt fluorescent bulbs placed 10-cm above plates (250 μmol s^{-1} m^{-2}). After seeds have germinated (usually 24 to 48 h), carefully place petri dishes on wooden holder at a 45° angle (Figure 12.1).
4	Check dishes daily to ensure filter paper remains moist. If filter paper begins to dry out, remove lid and add 1 to 3 mL of distilled water to base of the plate.

- Hemacytometer
- Mortar and pestle
- 6- to 8-cm deep paper cups
- Pasteur pipettes with rubber bulb
- Plastic petri dishes (10-cm diameter)
- Plastic stakes and trays
- Potting soil (peat moss and vermiculite, 1:1 v/v)
- Seeds
- Staining solution (0.005% cotton blue in 50% acetic acid)
- Stir plate and stir bars
- 9-cm diameter Whatman #1 filter paper
- Wooden petri dish holder

Small packages of seed can usually be purchased at a local garden center, Asian market, or seed company. Try to include as many crucifers as possible in the laboratory exercise. Have each student or group of students select one or two species of plants for the inoculation experiments.

Follow the protocol in Procedure 12.1 to germinate seeds for the experiment.

Contact a local county agricultural extension agent or plant pathologist to obtain a sample (100 to 500 g fresh weight) of infected roots. Students could also collect infected roots from crucifers grown in a home garden or commercial field with a history of clubroot disease. Plants infected with *P. brassicae* are usually smaller and stunted compared to healthy plants and have characteristic spindle-shaped galls (Figure 12.2) on their roots. Infected plants should be placed in a paper bag and transported to the laboratory in a cooler for processing. Infected roots not used for the laboratory exercise can be stored for 2 to 3 years in a nondefrosting freezer at −20°C.

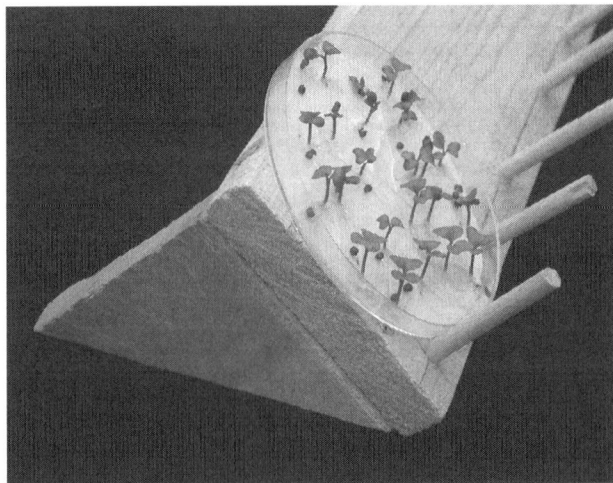

FIGURE 12.1 Wooden apparatus for holding petri dishes at a 45° angle. (Courtesy of B.R. Cody, North Carolina State University.)

FIGURE 12.2 Infected (left) and healthy (right) and Chinese cabbage roots. (Courtesy of L.A. Castlebury.)

Follow the protocol described in Procedure 12.2 to isolate and quantify resting spores of *P. brassicae* and Procedure 12.3 to inoculate seedlings and collect data.

Anticipated Results

Crucifers will vary in their susceptibility to infection by *P. brassicae*, and galls of various sizes will be produced on infected roots. No galls should be observed on roots of crucifer plants in the control treatment. Resting spores should be readily observed inside of root cells with the microscope after sectioning and staining.

Questions

- Why is it important to include a noninoculated control and replicates of each crucifer crop in the experiment?
- What was the response of each crucifer species and variety to infection? Were there differences in disease incidence and severity?
- What are some differences in morphology of infected and healthy roots based on macroscopic and microscopic examination?

EXPERIMENT 2. CHARACTERIZATION AND COMPARISON OF PLANT INFECTION BY THE GALL-INDUCING CHYTRID FUNGUS SYNCHYTRIUM MACROSPORUM

The plant pathogen *Synchytrium macrosporum* has the widest host range of any known biotrophic fungus (Karling, 1960). The organism can infect more than 1400 different species of plants, most of which grow in natural rather than agricultural ecosystems. Plants infected with *S. macrosporum* can also exhibit considerable variation in symptom expression. In this laboratory exercise, students will examine diseased chickweed plants for resting spores, sporangia, and sori of *S. macrosporum* and observe the unique swimming pattern of zoospores associated with this organism. Subsequent experiments will be conducted to compare and contrast disease symptoms of asparagus, bean, corn, and turnip artificially inoculated with *S. macrosporum*. The inability to culture *S. macrosporum* on a nutrient medium also requires the use of a modified method for isolation of resting spores and inoculation of plants.

Materials

Each student or team of students will require the following laboratory items:

- 50-mL beakers
- Commercial peat-based soil mix
- Commercial soluble fertilizer
- Compound light microscope
- Dissecting microscope
- Dissecting needles
- Glass slides and coverslips
- Kimwipes
- Pasteur pipettes with rubber bulb
- 10-cm plastic petri dishes
- Plastic bags with twist ties
- Plastic pots (10- or 15-cm diameter)
- Plastic or wooden stakes
- Seeds (asparagus, bean, corn, and turnip)
- Tape (double-sided)
- Tooth picks (wooden)
- Tween 80 (0.1% solution)
- 9-cm Whatman #1 filter paper

Follow the protocol in Procedure 12.4 to prepare squash mounts.

Procedure 12.2	
Isolation and Quantification of *Plasmodiophora brassicae* Resting Spores	
Step	Instructions and Comments
1	To release resting spores of *Plasmodiophora brassicae* from roots, place 100 g of diseased roots (either fresh or frozen) in 400 mL of sterile distilled water, and macerate in a blender for 2 min at high speed.
2	Place a glass funnel with five layers of cheesecloth on top of a 1000-mL Erlenmeyer flask and collect filtrate from blended solution. If cheesecloth becomes clogged with plant material and does not filter properly, remove the plant material and liquid from the cheesecloth, and repeat blending and filtering procedure.
3	Remove plant debris from cheesecloth and place in a mortar with 5 to 10 mL of distilled water. Grind debris with a pestle for 1 to 2 min, filter the solution as described above, and combine with previously collected filtrate.
4	Place equal volumes of filtrate into centrifuge tubes (10- or 50-mL tubes, depending on size of rotor) and adjust their weight by adding appropriate amounts of distilled water to each tube. Once centrifuge tubes have a similar weight and are balanced, place them in the rotor and centrifuge for 10 to 15 min at 2000 g at room temperature.
5	Gently remove centrifuge tubes and place in a rack. Carefully remove the top, gray-colored fraction with a Pasteur pipette (this fraction contains the resting spores and often will appear above a whitish-colored layer in the middle of the tube) and place into another centrifuge tube. Repeat the centrifuge process as needed, particularly if filtrate is cloudy or contaminated with excessive plant material. Place the collected filtrate containing the resting spores in a glass beaker with a stir bar for quantification.
6	Gently mix collected filtrate using a stir plate to evenly distribute resting spores. Pipette a small drop of the collected filtrate on a clean glass slide, add a cover slip and examine with a microscope at 400× to 1000× magnification. Resting spores are round and approximately 4 μm in diameter (Figure 12.3). For better resolution of resting spores, add one drop of staining solution (0.005% cotton blue in 50% acetic acid) to the slide preparation prior to examination (Figure 12.3). If resting spores are observed, determine their concentration in collected filtrate with the following procedure. Spores without cytoplasm are not viable and should not be counted.
7	Measure the volume of collected filtrate in a graduated cylinder; pour collected filtrate into a glass beaker and gently swirl the mixture to evenly distribute resting spores. Place one drop of spore solution at the edge of a coverslip on a hemacytometer (counting chamber with a ruler) and allow solution to be drawn into area between coverslip and glass slide. Let slide sit for 2 to 3 min to allow spores to settle on the slide.
8	Examine center of slide and locate the area of the ruler with 25 cells (5 rows and 5 columns) and consisting of 16 smaller squares each. Count the number of resting spores in each corner and center squares (5 squares total). Calculate mean number of spores per square, apply correction factor for area (2.5×10^5) and determine concentration of resting spores per mL of collected filtrate. Example: upper left square = 74; upper right square = 56; lower left square = 45; lower right square = 60; and center square = 80 (74 + 56 + 45 + 60 + 80)/5 = 63; $63 \times (2.5 \times 10^5) = 1.575 \times 10^7$ resting spores per mL of collected filtrate.
9	After determining concentration of resting spores, adjust spore concentration to 1×10^7 spores per mL and pour equal amount of stock solution into 50-mL beakers. The number of beakers required is determined by the total number of plant species and varieties to be inoculated. Freshly extracted resting spores are preferred for inoculation. However, resting spore solutions may be stored for 3 to 5 days at 4°C or 3 to 4 months at –20°C prior to use.

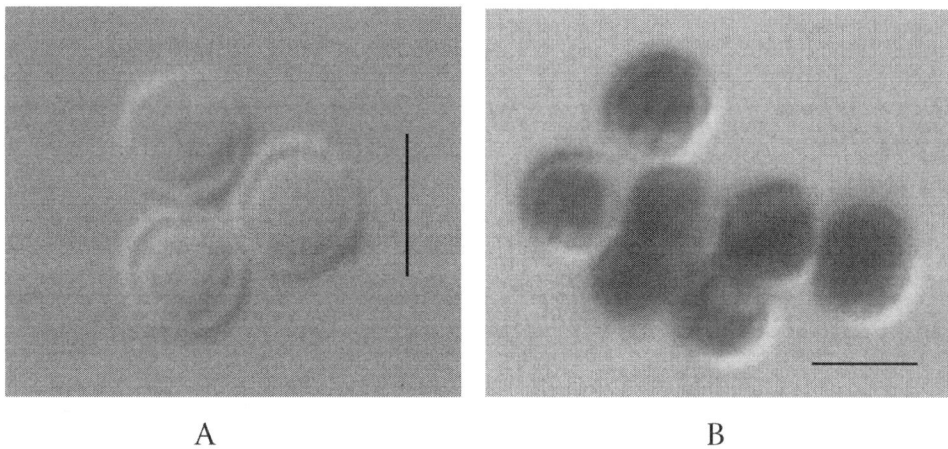

FIGURE 12.3 Unstained (A) and stained (B) resting spores of *Plasmodiophora brassicae*. Bar represents 4 μm (Courtesy of L.A. Castlebury.)

Procedure 12.3
Inoculation of Seedlings and Collection of Data

Step	Instructions and Comments
1	For each crucifer examined, fill 20 paper cups (6- to 8-cm deep) with a potting mixture consisting of 1 part peat moss and 1 part vermiculite. Poke several small holes in the bottom of cup to allow for adequate drainage and water potting mixture until moist. Place one set of 10 cups into a plastic tray and repeat this process for a second set of cups (two replicates).
2	Remove 20 seedlings from petri dishes and dip their roots into quantified resting spore solution of *Plasmodiophora brassicae* for 10 sec. Create a 5-cm deep hole with a pencil and transplant one seedling into each individual paper cup. In a separate set of 20 cups, also include seedlings dipped in sterile distilled water as a control. Incubate seedlings at 18°C to 28°C for 6 to 8 weeks. During the first week of incubation, keep potting mixture saturated with water. Thereafter, water seedlings as needed and fertilize after 3 weeks (1 g of 20-20-20 fertilizer in 3.78 L of water).
3	After incubation, remove inoculated seedlings and gently wash soil from roots in running tap water. Examine roots of each seedling for galls. Determine the percentage of infected seedlings based on gall symptoms (disease incidence) and severity of infection with the 0–4 rating scale (Figure 12.4). Follow the example presented in Table 12.1. Repeat the procedure above for each of two replicates of each plant species examined and then calculate a separate average for disease incidence and severity. Also examine seedlings in each control treatment for the incidence and severity of clubroot disease.
4	With a sharp scalpel or razor blade cut a thin section of healthy and diseased roots. Place section in distilled water on a glass slide and observe cells for the presence of resting spores (Figure 12.5).
5	Cut infected roots from the remaining seedlings, place in a plastic freezer bag, and store at –20°C for future laboratory exercises.
6	At the completion of the experiment, place all paper cups, plant material, and soil in an autoclave for 1 h at 121°C.

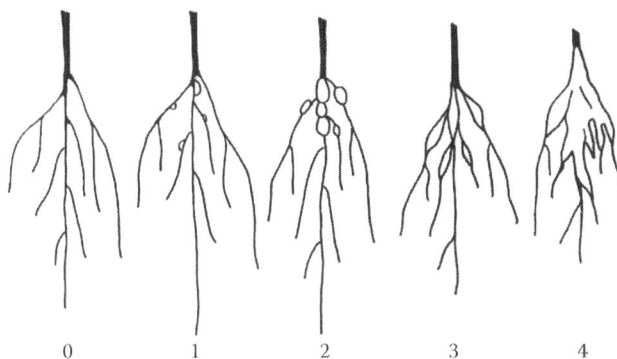

FIGURE 12.4 Clubroot disease rating scale (0–4) for crucifer seedlings (Modified by L.J. Gray from Williams, P.H., 1966. Phytopathology 56: 624-626.)

TABLE 12.1

Example Data Sheet to Calculate Disease Incidence and Severity

Crucifer Tested	# of Healthy Plants with Rating = 0	# of Diseased Plants with Rating = 1	# of Diseased Plants with Rating = 2	# of Diseased Plants with Rating = 3	# of Diseased Plants with Rating = 4
Cabbage	2	1	3	2	2

Note: Disease incidence = (number of diseased plants/total number of plants) × 100 (1 + 3 + 2 + 2)/(2+1 + 3 + 2 + 2) × 100 = 80%.

Disease severity = (number of diseased plants in each rating category × correction factor associated with each disease rating category)/total number of plants [(0 × 2) + (1 × 1) + (3 × 2) + (2 × 3) + (2 × 4)]/ (2 + 1 + 3 + 2 + 2) = 2.1.

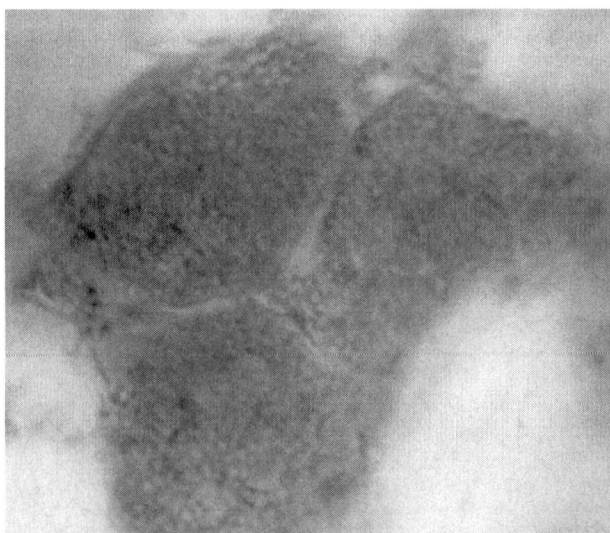

FIGURE 12.5 Resting spores of *Plasmodiophora brassicae* in root cells of Chinese cabbage. (Courtesy of L.A. Castlebury).

	Procedure 12.4
	Squash Mount Procedure
Step	Instructions and Comments
1	Collect several chickweed (*Stellaria media*) plants infected with *Synchytrium macrosporum*. Diseased plants will have galls with dark amber to reddish brown resting spores or bright orange-yellow sori from germinated resting spores on aboveground stems (Figure 12.6A and 12.6B). Infected chickweed plants can be collected from February to March in the southeastern United States (e.g., Alabama and Georgia). This may translate into a later date for areas farther north depending on when chickweed seeds germinate.
2	Place a piece of infected chickweed tissue in a clean petri dish, secure with double-sided tape and then affix petri dish on the stage of a dissecting microscope with double-sided tape.
3	Add six separated drops of sterile double distilled water on an inverted petri dish lid and place lid next to dissecting scope. Flame sterilize tips of two dissecting needles, cool for 10 sec; and carefully remove orange sori and resting spores from tissue.
4	Rinse sori and resting spores by serially running them through the six drops of sterile water before placing on a clean glass slide in a small drop of sterile water.
5	Carefully place a clean cover slip over the drop of water and gently tap it with the eraser end of a pencil to break open sori and release the individual sporangia. Examine the prepared slide with a compound light microscope.
6	Place slide in a petri dish with a moist Kimwipe folded into the bottom. Take a small wooden toothpick, break in half, and use the two halves to hold slide above moistened Kimwipe.
7	Examine slide after 1 h and again after every 2 h. Sporangia from mature sori are usually ready to release zoospores after 1 to 2 h. Periodically moisten Kimwipe by adding a small amount of water to ensure that the slide does not dry out.

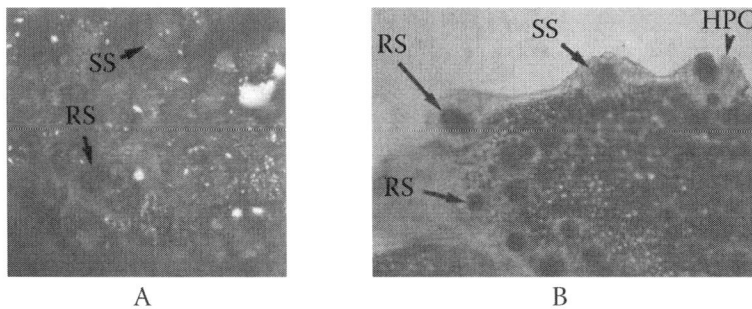

A   B

FIGURE 12.6 Galls with sporangia, sori and resting spores of *Synchytrium macrosporum* on infected chickweed leaves. (A) Galls filled with either dark orange resting spores (RS) or yellow sori containing sporangia (SS); (B) Microscopic view of galls showing hypertrophied plant cells (HPC) containing resting spores (RS) and sori with sporangia (SS). (Courtesy of M.J. Powell and P. Letcher).

Anticipated Results

The students should observe the multicellular sheath produced from host epidermal cells surrounding both sporangial and resting spore galls. The students should also observe zoospore discharge from individual sporangia, be able to distinguish different parts of the life cycle, and observe the "jerky" swimming pattern of the zoospores which is typical for chytrid fungi.

Questions/Activities

- Draw an infected chickweed plant.

- Draw resting spores, sori, and sporangia observed with the microscope. Indicate magnification of each drawing.
- What kind of swimming pattern do the zoospores have, and how are they released from the sporangia?
- Using the life cycle diagram in Chapter 11 (Figure 11.4), label the structures observed.

To prepare host plants for inoculation, the following list of germination times for each species of plant will be useful: bean (*Phaseolus vulgaris*) 6 to 10 days; corn (*Zea mays*) 5

	Procedure 12.5
	Preparation of Inoculum of *Synchytrium macrosporum*
Step	Instructions and Comments
1	Soak fresh or dried leaves of chickweed infected with *Synchytrium macrosporum* in sterile distilled water for 5 to 7 days to soften tissues around prosori with resting spores of the fungus.
2	With dissecting needles and a dissecting microscope, tease prosori from tissue and rinse gently in distilled water.
3	Transfer prosori to a petri plate lined with two layers of moistened, Whatman #1 filter paper and incubate in the laboratory for 2 to 3 weeks or until prosori germinate and sori mature.
4	Transfer 25 to 50 mature sori individually with a fine dissecting needle into a drop of distilled water on a clean glass slide.
5	Cover the suspension of sori with a cover slip and press gently with a pencil eraser to break open the sori to release zoosporangia.
6	Zoospores are released from zoosporangia within 1 to 2 h and can be collected by gently washing slide with distilled water into a 10-mL beaker 2 to 2.5 h after zoospores have formed.

to 7 days; turnip (*Brassica rapa*) 7 to 10 days, and asparagus (*Asparagus officinalis*) 14 to 21 days. Asparagus seeds should be soaked in warm water for 48 h prior to planting. Be sure to replenish warm water when it has cooled down at least two or three times during the 48 h soaking period. Asparagus seed germinate slowly and should be planted 2 to 3 weeks before planting bean, corn, and turnip seeds. This will ensure that all species of plants can be inoculated with *S. macrosporum* at the appropriate growth stage.

Follow the protocol in Procedure 12.5 to prepare inoculum of *S. macrosporum*.

A well-drained, commercial peat based soil mix should be used to propagate each host plant. The soil mix should be kept moist but not saturated with water. Host plants can be grown in a greenhouse or in the laboratory at room temperature with supplemental fluorescent lighting (14-h photoperiod and 0.3 m from the light source). When plants are 21 days old, fertilize once a week with half the recommended rate of a complete soluble fertilizer. Plant six seeds of each host plant in a 10 or 15-cm diameter pot and thin to three plants after seedlings have emerged from the soil.

Because *S. macrosporum* is not available commercially, cultures for use in the laboratory exercise will be prepared from infected chickweed plants collected from the field. Infected plant tissue can be dried and stored at room temperature in a low humidity environment to provide a source of viable resting spores for future experiments.

Follow the protocol in Procedure 12.6 to inoculate plants with *S. macrosporum*. The inoculation procedure is a modification of the method developed by J.S. Karling (1960).

Anticipated Results

Synchytrium macrosporum should not infect asparagus, but will usually cause moderate to severe infection of bean, corn, and turnip. The type, size, and structure of galls produced by *S. macrosporum* will vary on chickweed, corn, bean, and turnip.

Questions

- What differences can you note in appearance of the galls on asparagus, bean, corn, and turnip? Are these galls similar in appearance to the galls observed on chickweed?
- Do you think these differences are attributable to the plant, the fungus, or both? Explain your answer.
- Complete Table 12.2. Answer yes or no in the "Infected" column. Put "YES" in the appropriate box corresponding to each species of plant if galls, resting spores, and/or prosori are present on the plant tissue and "NO" if no symptoms (i.e., galls) and/or signs of the fungus are evident. For the "Degree of Infection" column, indicate the severity of disease with the following scale; sparse = 2–10 galls per plant, moderate = 10–20 galls per plant, heavy = 20–100 galls per plant, and severe = >100 galls per plant. For the "Type of Gall" column, indicate whether galls were single-celled (e.g., simple) or multicelled (e.g., composite). If the degree of infection was severe, then place the term "Confluent" in the type of gall column. Confluent galls occur when several epidermal cells adjacent to one another become infected and the individual sheaths of enlarged host cells that form around the developing gall merge together (Karling, 1960).
- Summarize the results of your observations in Table 12.2. What can you conclude from these observations?

Procedure 12.6	
Inoculation of Plants with *Synchytrium macrosporum*	
Step	Instructions and Comments
1	Swab the emerging leaves of each seedling with Tween 80 (0.1% v/v) and rinse with sterile distilled water. This procedure provides a wet surface for zoospores to swim on the leaf. The Tween solution should be applied to leaves of similar age 1 to 2 weeks prior to conducting the experiment to determine whether it is toxic to each species of plant. If phytotoxicity is observed, dilute the Tween solution with water to a concentration that does not damage the leaves.
2	Dilute the zoospore mount from steps 5 and 6 of Procedure 12.5 with 5 mL of sterile distilled water and place a drop of the zoospore mount onto the treated emerging leaf. Inoculate nine plants (three plants in three separate pots). Also, include another pot of three seedlings on which a few drops of sterile distilled water are placed on emerging leaves (control treatment).
3	Place a bell jar or plastic container over the seedling to maintain high humidity. If seedlings are too large to cover, then place a wet pad of absorbent cotton around the inoculated leaf. For corn and asparagus, several pots can be covered with a plastic dome or a large plastic bag with a twist tie.
4	Each seedling should be inoculated once a day for six consecutive days.
5	Once the inoculations are completed, remove the bell jar or plastic container, bag or dome, and maintain plants in a greenhouse or laboratory for 3 to 4 weeks after inoculation. Galls should be apparent 2 weeks after infection and will mature in 2 to 4 months.

TABLE 12.2

Susceptibility of Asparagus, Bean, Corn, and Turnip to *Synchytrium macrosporum*

	Infected	Degree of Infection	Type of Gall
Asparagus officinialis			
Brassica rapa			
Phaseolus vulgaris			
Zea mays			

- Karling advocated the use of host range for identification of species of *Synchytrium*. What modern techniques could be used to aid in the identification of a fungus rather than conducting a host range study? Do you think host range studies are still important? Explain your answer.

EXPERIMENT 3. OBSERVATIONS OF EELGRASS WASTING DISEASE

Labyrinthula zosterae is an important pathogen of sea grass in estuarine environments. The organism produces zoospores that are involved in the disease cycle and can be grown in pure culture by using yeast cells to supplement their nutrition. In this laboratory exercise, students will observe eelgrass (*Zostera marina*) wasting disease, and isolate *L. zosterae* from diseased eelgrass exhibiting typical symptoms. To demonstrate that *L. zosterae* is the causal agent of eelgrass wasting disease, students will follow Koch's postulates to establish "proof of pathogenicity."

Materials

Each student or team of students will require the following laboratory items:

- Agar (Difco)
- Air pump and tubing
- Alcohol lamp
- Antibiotics (ampicillin, penicillin G, and streptomycin sulfate)
- Artificial or natural seawater
- Buchner funnel with rubber cork
- 25-L Carboy
- Commercial bleach (sodium hypochlorite; NaOCl)
- Compound light microscope
- Culture of yeast (any nonfilamentous, nonmucoid yeast—preferably one cultured from nonsurface disinfested and decaying eelgrass leaves)
- Dissecting microscope

Procedure 12.7

Preparation and Filtering of Seawater from Natural Sources

Step	Instructions and Comments
1	Place a Buchner funnel containing two pieces of Whatman #1 filter paper into a 2-L side-arm flask connected to a vacuum source with flexible rubber tubing.
2	Once the vacuum has been established, slowly pour the seawater into the Buchner funnel and proceed until the flask is full.
3	Filtered seawater should be stored at 4°C to 10°C in clean plastic carboys and covered with black plastic. For the production of sterile seawater use a filtration system (0.4 μm pore diameter) or autoclave for 20 min at 121°C.

- Dissecting needles and scissors
- 2-L Erlenmeyer flasks with cotton stoppers and glass tube for aeration
- Ethanol (80% EtOH)
- Filters (≤0.4 μm pore diameter)
- 2-L sidearm flasks
- Forceps
- Germanium dioxide (GeO$_2$)
- 5-mL glass pipettes
- Glass slides and coverslips
- Paper towels (sterile)
- 4.5 and 10-cm diameter plastic petri dishes
- 50-mL plastic screw top tubes
- 3.78-L plastic Ziploc bags
- Scalpel
- Serum seawater plus medium (see recipe below)
- Sterile distilled water
- Tygon tubing
- Vacuum pump with rubber tubing
- 9-cm Whatman #1 filter paper

Approximately 20 to 30 L of artificial or natural seawater (28 to 32 parts per thousand salinity) is required for this laboratory experiment. The seawater will be used to prepare serum seawater plus medium to isolate *L. zosterae* and for seagrass inoculation experiments. Various prepackaged mixtures that approximate seawater chemistry are com-

mercially available and can be easily prepared by mixing these ingredients with distilled water (e.g., Instant Ocean).

Follow the protocol outlined in Procedure 12.7 to prepare and filter seawater from natural sources.

Follow the protocol outlined in Procedure 12.8 to prepare serum seawater plus (SSA+) medium for isolation of *L. zosterae*.

Eelgrass (*Zostera marina*) is found in coastal estuaries throughout the Northern Hemisphere. In North America, eelgrass is distributed in seagrass meadows along the Atlantic coast from Labrador to North Carolina and along the Pacific Coast from Alaska to Baja California. Because most seagrass meadows are protected by law, check with local authorities to identify appropriate eelgrass beds where plants can be collected. Eelgrass shoots can be collected at the lowest tides by wading or swimming a short distance into the water.

Follow the protocol outlined in Procedure 12.9 to collect healthy and diseased samples of eelgrass.

One week before beginning the isolation procedure, streak several plates of SSA+ medium with a culture of yeast (see list of materials provided previously). The yeast will serve as a food source for *L. zosterae*. Monthly transfer on SSA without antibiotics can maintain dual axenic cultures of *L. zosterae* and yeast.

Follow the protocol outlined in Procedure 12.10 to isolate *L. zosterae* from eelgrass.

Procedure 12.8

Preparation of Serum Seawater Plus (SSA+) Isolation Medium

Step	Instructions and Comments
1	Add 12 g of agar and 3 mg of germanium dioxide (a diatom inhibitor) to 1 L of filtered seawater (Procedure 12.7). Autoclave medium for 20 min at 121°C .
2	After medium has cooled to 50°C, add 4 to 10 mL of sterile horse serum (1% v/v, BBL or Gibco) and 250 mg each of the antibiotics ampicillin, penicillin G, and streptomycin sulfate. Gently swirl to mix the medium and pour into 10-cm plastic petri dishes. Caution: All of the SSA+ medium should poured immediately into petri dishes because reheating (remelting) of this medium after it has solidified will cause the horse serum protein to coagulate and render SSA+ useless as an isolation medium.

	Procedure 12.9
	Collecting Healthy and Diseased Samples of *Zostera marina*
Step	Instructions and Comments
1	Collect approximately 20 healthy plants (without disease symptoms manifested by the appearance of blackened or dead necrotic areas present on leaves) with at least three inner (youngest) green leaves from the eelgrass bed.
2	While collecting, also include a portion of the rhizome and roots with each collected shoot. Place shoots in a 3.78-L plastic Ziploc bag with a seawater dampened paper towel.
3	Do not submerge collected plants directly in seawater. For transport, keep plants cool, but not directly on ice.
4	After collecting healthy plants, carefully examine plants in the eelgrass beds forwasting disease symptoms. Plants infected with *Labyrinthula zosterae* will have black streaks and patches of necrotic tissue (lesions) on older leaves (Figure 12.7).
5	Collect approximately 30 leaves with disease symptoms by selecting leaves with some areas healthy green tissue adjacent to the necrotic lesions. Place diseased leaves into a new 3.78-L plastic Ziploc bag and keep moist and cool until ready to isolate the pathogen as described above.
6	If you are unable to collect healthy and diseased eelgrass plants, request them from a colleague in a coastal area. Eelgrass plants can be shipped overnight in an insulated container and successfully used for this laboratory experiment.

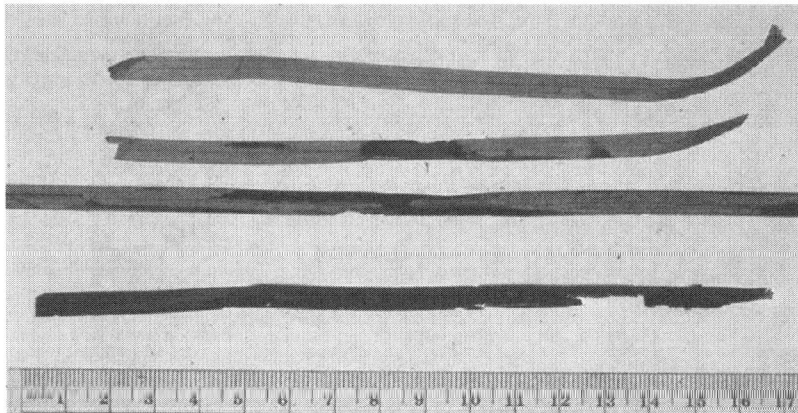

FIGURE 12.7   Eelgrass leaves with necrotic lesions symptomatic of wasting disease.

	Procedure 12.10 Isolation of *Labyrinthula zosterae* from Eelgrass
Step	Instructions and Comments
1	Dip forceps and scissors in 50-mL tube containing 80% EtOH and flame to disinfest.
2	Cut eelgrass leaves into small pieces (5 to 10 mm²) using the disinfested forceps and scissors. Use leaf pieces from the edge of the blackened, necrotic lesions where *Labyrinthula zosterae* is likely to be most active.
3	Place cut leaf pieces to a sterile 4.5-cm petri dish, add 0.5% sodium hypochlorite to cover leaf pieces and gently swirl them for 1 min. Aseptically transfer each disinfested leaf piece to new 4.5-cm petri dish, add sterile distilled water and gently swirl for 2 min.
4	Repeat the rinsing process with sterile, filtered seawater. Transfer leaf pieces to a sterile paper towel or filter paper to remove excess water, and place 4 to 5 disinfested pieces of eelgrass leaf tissue on a 10-cm diameter petri dish containing SSA+ medium
5	Observe each SSA+ plate daily for growth of *L. zosterae* from each piece of eelgrass tissue (Figure 12.8).
6	Continue to observe plates for at least 1 week or until they become overgrown by bacteria and/or fungi making observation and isolation of *Labyrinthula* difficult.
7	When an appropriate *Labyrinthula* colony is located and ready for transfer, first streak a clean dish of SSA+ with a small amount of yeast from an actively growing culture.
8	Then transfer a portion of the actively growing *Labyrinthula* colony to the yeast streak on the new petri dish of SSA+.

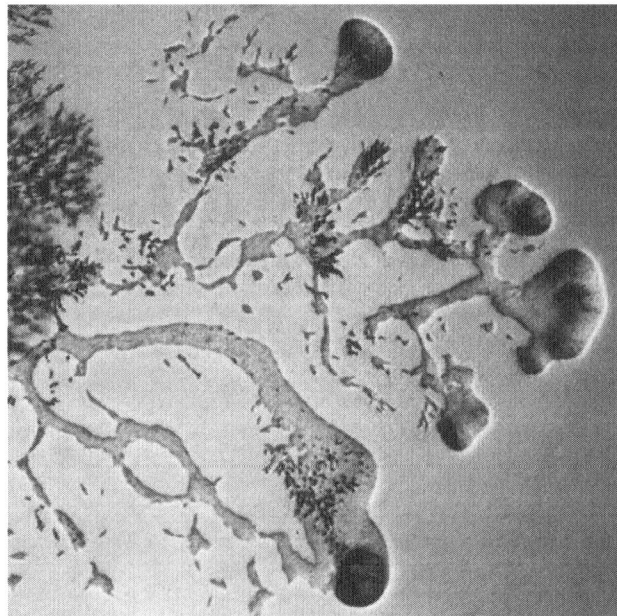

FIGURE 12.8 *Labyrinthula zosterae* growing on SSA+ medium from diseased eelgrass leaf (bar = 1.0 mm).

Procedure 12.11
Demonstrating that *Labyrinthula zosterae* is the Causal Agent of Eelgrass Wasting Disease Using Koch's Postulates

Step	Instructions and Comments
1	Fill 2-L Erlenmeyer flasks to the neck with artificial or natural sterilized seawater. Place a single healthy eelgrass shoot in each flask and weigh down with a short piece of heavy rubber tubing slipped over the rhizome. Stopper flasks with cotton through which a glass tube extends to near the bottom of the flask (a sterile 5-mL pipette is a good substitute). Attach an air pump or air line with a cotton plug filter to the glass tube and adjust the rate to deliver about one bubble per second.
2	Place several 1-cm pieces of green eelgrass leaf in distilled water and autoclave for 20 min at 121°C. Place some of these sterilized leaf pieces in cultures of *Labyrinthula* isolated for Procedure 12.10. Place them on the agar surface adjacent to the spreading colonies of *Labyrinthula* but not on top of the yeast cells. Allow the *Labyrinthula* cells to grow into the leaf piece for 24 to 48 h.
3	Cut 0.5-cm pieces of thin Tygon tubing and slit the tube wall along one radius to create a small clip to attach the inoculated leaf pieces to healthy eelgrass plants. Clips should be sterilized before use.
4	Remove a green shoot of eelgrass from a flask and place on a sterile paper towel. Using sterile forceps pick up a piece of inoculated carrier leaf and clip it to a green leaf of the eelgrass shoot with a Tygon clip. Replace the inoculated shoot into the flask.
5	As an experimental control, clip a sterilized piece of leaf tissue to a green leaf of an eelgrass shoot in a separate flask.
6	Repeat the experimental and control inoculations for at least five flasks each.
7	Place the flasks on a lighted bench or greenhouse where the plants will receive at least 20% full sunlight.
8	Observe plants for wasting disease symptoms 1 to 7 days after inoculation (Figure 12.9).
9	Complete Koch's postulates by reisolating *Labyrinthula zosterae* on SSA+ medium from diseased leaves with necrotic lesions in the flasks.

Follow the protocol outlined in Procedure 12.11 to demonstrate that *L. zosterae* is the causal agent of eelgrass wasting disease, using Koch's postulates (Muehlstein et al., 1988).

Anticipated Results

Wasting disease should be readily observed on eelgrass, and *L. zosterae* should be easily cultured from infected eelgrass leaves on SSA+ isolation medium. Eelgrass plants inoculated with *L. zosterae* will produce typical wasting disease symptoms of leaves, followed by isolation and observation of the microorganism from diseased leaves on SSA+ isolation medium. No symptoms should be observed, and *L. zosterae* should not be isolated from eelgrass plants in the control treatment.

Questions

- Why is it necessary to clip a noninoculated, disinfested leaf piece to a green eelgrass plant as an experimental control?
- Low salinity has been reported to inhibit wasting disease. How could you test this hypothesis with

FIGURE 12.9 Eelgrass shoot with necrotic lesion 60 h after inoculation with *Labyrinthula zosterae*.

the inoculation apparatus set up for this laboratory experiment?

- How could you determine whether necrotic lesions on leaves of other seagrasses are caused by *L. zosterae* or other plant pathogenic microorganism(s)?

ACKNOWLEDGMENTS

We would like to thank the following persons: Paul H. Williams, emeritus professor, University of Wisconsin, for his many useful suggestions and clubroot resistance screening protocol that served as a template for developing Laboratory Exercise 1; Timothy James (Duke University) and Lisa Castlebury (USDA Systematic Botany and Mycology Laboratory) for presubmission review, and to LC for providing photographs of resting spores and diseased cabbage; Celeste A. Leander for providing comments and suggestion for Laboratory Exercise 3; Andy Tull (University of Georgia) for advice on growing plants for the Laboratory Exercises; Lynnette J. Gray for illustrations; Bryan R. Cody for digital images; and Pamela E. Puryear (North Carolina State University) for assistance in literature searches.

REFERENCES

Agrios, G.N. 2005. *Plant Pathology*, 5th ed., Academic Press, New York.

Castlebury, L.A., J.V. Maddox and D.A. Glawe, 1994. A technique for the extraction and purification of viable *Plasmodiophora brassicae* resting spores from host root tissue, *Mycologia* 86: 458–460.

Fuller, M.S. and A. Jaworski. 1987. *Zoosporic Fungi in Teaching and Research*. Southeastern Publishing Corporation, Athens, GA.

Karling, J.S. 1960. Inoculation experiments with *Synchytrium macrosporum*. *Sydowia* 14: 138–169.

Muehlstein, L.K., D. Porter and F.T. Short. 1988. A marine slime mold, *Labyrinthula*, produces the symptoms of wasting disease in eelgrass, *Zostera marina*. *Mar. Biol.* 99: 465–472.

Porter, D. 1990. Phylum Labyrinthulomycota, in *Handbook of Protoctista*, L. Margulis, J.O. Corliss, M. Melkonian and D.J. Chapman (Eds.), Jones and Bartlett, Boston, MA, pp. 388–398.

Sherf, A. F. and A.A. MacNab, *Vegetable Diseases and Their Control*, 2nd ed., John Wiley & Sons, New York, 1986.

Williams, P.H. 1966. A system for the determination of races of *Plasmodiophora brassicae* that infect cabbage and rutabaga. *Phytopathology* 56: 624–626.

13 Archiascomycete and Hemiascomycete Pathogens

Margery L. Daughtrey, Kathie T. Hodge, and Nina Shishkoff

CHAPTER 13 CONCEPTS

- *Taphrina* and *Protomyces* are two genera of Archiascomycetes, an early diverging clade of Ascomycota.

- Both *Taphrina* and *Protomyces* have a saprobic yeast stage and a parasitic mycelial stage during which asci are formed.

- The asci of *Taphrina* and *Protomyces* are naked (not contained in an ascoma).

- Typical symptoms of diseases caused by *Taphrina* or *Protomyces* are galls, leaf curls, and leaf spots.

- Peach leaf curl is a common disease caused by *T. deformans*.

- *Eremothecium* is unusual in that it is a plant pathogen within the Hemiascomycetes (ascomycetous yeasts).

Archiascomycete and hemiascomycete **pathogens** cause a number of fairly obscure diseases as well as some commonly recognized problems such as peach leaf curl and oak leaf blister. Many of the diseases discussed in this chapter affect weeds or native plants rather than cultivated species, which tend to receive the most attention from plant pathologists. The plant–pathogen interactions in these groups are unique and quite fascinating.

Phylogenetic studies by Nishida and Sugiyama (1994) and others reveal the following three major lineages within the Ascomycota: Archiascomycetes, Hemiascomycetes, and Euascomycetes (Sugiyama, 1998). The Archiascomycetes appear to have diverged before the Hemiascomycetes (ascomycetous yeasts) and Euascomycetes (filamentous ascomycetes).

HEMIASCOMYCETE PATHOGENS

Most ascomycetous yeasts (Hemiascomycetes) are not plant pathogens, but the genus *Eremothecium* (syn. *Ashbya, Holleya, Nematospora*) is an exception (Kurtzman and Sugiyama, 2001). *Eremothecium* is a filamentous fungus that also produces yeast cells within the host (Batra, 1973). **Asci**, which are formed directly from the **mycelium**, contain needlelike **ascospores**.

Sucking insects, especially the true bugs, often vector this genus. *Eremothecium coryli* fruit rot associated with stink bug feeding was responsible for losses of more than 30% in field tomatoes in California in 1998 (Miyao et al., 2000). In addition to tomatoes, *Eremothecium coryli*

infects cotton, citrus, hazelnuts, and soybeans, and a related species, *E. sinecaudii*, infects the seeds of mustards. The three species *E. cymbalariae*, *E. ashbyi,* and *E. gossypii* all infect *Hibiscus* species and coffee, and also cause surface lesions on citrus fruits and cotton bolls. *Eremothecium ashbyi* is also used for the industrial production of riboflavin (vitamin B_2) via a fermentation process.

ARCHIASCOMYCETE PATHOGENS

The Archiascomycetes undergo sexual reproduction that is **ascogenous**, but lack **ascogenous hyphae** and **ascomata**. The following four orders are currently grouped within the newly-proposed class Archiascomycetes: the Taphrinales, the Protomycetales, the Pneumocystidales, and the Schizosaccharomycetales (Kurtzman and Sugiyama, 2001). The last two are animal pathogens and fission yeasts, respectively. The Protomycetales and Taphrinales are exclusively plant pathogens and each contains a single family, the Protomycetaceae and Taphrinaceae, respectively. Members of these families produce both yeast-like and mycelial states; **asexual reproduction** occurs by budding.

THE PROTOMYCETACEAE

The Protomycetaceae includes five genera (*Burenia, Protomyces, Protomycopsis, Taphridium,* and *Volkartia*) comprising 20 species (Alexopoulos et al., 1996). The genus *Protomyces* causes galls on leaves, stems, flowers, and

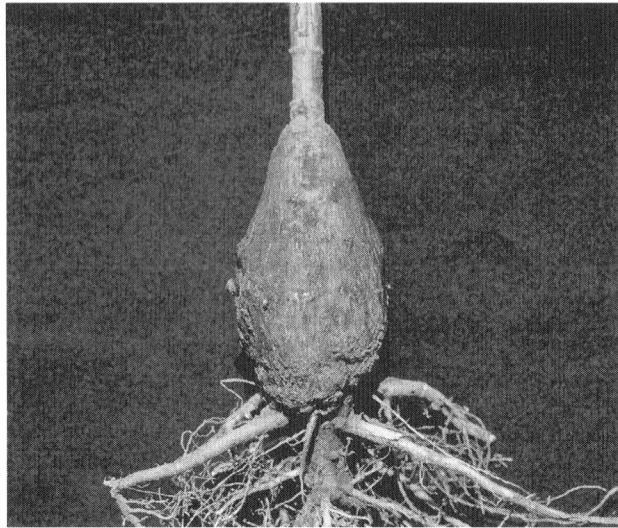

FIGURE 13.1   Stem gall of giant ragweed (*Ambrosia trifida*) caused by *Protomyces gravidus*. (From Holcomb, Gordon E., 1995, *Plant Disease* Vol. 79[8] [cover].)

fruit of plants in the Apiaceae (dill family) and Asteraceae (composite family). *Protomyces macrosporus*, for example, causes a leaf gall of wild carrot (*Daucus carota*); it is reported to occur on 26 other genera, including 14 genera in the Asteraceae. *Protomyces gravidus* causes a stem gall on giant ragweed (*Ambrosia trifida*) in the U.S. (Figure 13.1) and also affects tickseed (*Bidens*). Of the other members of the family Protomycetaceae, *Burenia* and *Taphridium* occur only on Apiaceae, whereas *Protomycopsis* and *Volkartia* occur only on Asteraceae. These three genera cause color changes or galls and spots on their hosts.

Protomyces produces an intercellular diploid mycelium and thick-walled intercalary ascogenous cells that form in the host tissue. These cells are sometimes mistaken for the spores of smuts such as the white smut, *Entyloma* (Preece and Hick, 2001). The ascogenous cells of *Protomyces* over-winter and germinate to form asci in the spring. Inside each multinucleate ascus, four spore mother cells formed by mitosis continue to divide to form many ascospores. Hundreds of these spores are forcibly released en masse, and they continue budding after release. Fusion of two of these ascospores forms a diploid cell that can reinfect leaves.

THE TAPHRINACEAE

The genus *Taphrina* was created in 1832 by Fries; it is the only genus of the Taphrinaceae, and includes some 95 species (Mix, 1949). *Taphrina* species are **parasites** of members of many plant families. Those on ferns generally differ from species that attack **angiosperms** in having thin, **clavate** asci. The fern hosts of *Taphrina* show some very elaborate symptoms: *T. cornu-cervi* causes antler-shaped galls

---

CASE STUDY 13.1

OAK LEAF BLISTER

- Oak leaf blister is caused by the fungus *Taphrina caerulescens.*
- Leaf infections result in chlorotic, sometimes bright yellow, patches that may bulge and cause crinkling and buckling of the leaf.
- Often the symptoms are not noticed until the chlorotic areas turn necrotic in midsummer.
- Red oaks are much more susceptible to oak leaf blister than white oaks.
- The disease is most likely to be noticed in a year with a cool, rainy spring, when infections will be more numerous.
- Although a curiosity and responsible for minor cosmetic injury, oak leaf blister is not especially damaging to overall tree health.
- Because the disease does not usually result in leaf drop or reduced vigor, oak trees need not be sprayed for protection from oak leaf blister.
- Were fungicide sprays to be used, the effective timing would be early spring, just before buds swell—before ascospores germinate and the fungus gains access to the immature leaf tissues.

TABLE 13.1

Taphrinas and Their Woody Plant Hosts Reported in the United States and Canada[a]

Host	Species
Acer spp.	*Taphrina aceris,* UT
	T. carveri, c, s US, ONT
	T. darkeri, nw US, w Can
	T. dearnessii, c, e NA
	T. letifera, e, nc US, NS
	T. sacchari, c, e US, QUE
Aesculus californicus	*T. aesculi,* CA, TX
Alnus spp.	*T. alni* (catkin hypertrophy), AK, GA, c Can, Eur, Japan
	T. japonica, nw NA, Japan
	T. occidentalis (catkin hypertrophy), nw NA, QUE
	T. robinsoniana (catkin hypertrophy), c, e, NA
	T. tosquinetii (catkin hypertrophy), NH, NS, Eur
Amelanchier alnifolia	*T. amelanchieri* (witches' broom), CA
Amelanchier sp.	*T. japonica,* CA
Betula spp.	*T. americana* (witches' broom), NA
	T. bacteriosperma, n hemis
	T. boycei, NV
	T. carnea, ne US, e, w Can, Eur
	T. flava, ne US, e Can
	T. nana, WY, e, w Can, Eur
	T. robinsoniana, NF
Carpinus caroliniana	*T. australis,* e US, ONT
Castanopsis spp.	*T. castanopsidis,* CA, OR
Corylus	*T. coryli, US,* Japan
Malus	*T. bullata,* WA, Eur
Ostrya virginiana	*T. virginica,* c, e US, e Can
Populus spp.	*T. johansonii* (catkin hypertrophy), US, e Can, Eur, Japan
	T. populi-salicis, nw NA
	T. populina, n hemis
	T. rhizophora, NY, WI, Eur
Potentilla spp.	*T. potentillae,* US, Eur
Prunus spp.	*T. armeniacae,* US
	T. communis (plum fruit hypertrophy), NA
	T. confusa, (chokecherry fruit hypertrophy, witches' broom), US, e Can
	T. farlowii, (cherry leaf curl, fruit and shoot hypertrophy) c, e US
	T. deformans cosmopolitan
	T. flavorubra (cherry and plum fruit and shoot hypertrophy)
	T. flectans (cherry witches' broom), w US, BC
	T. jenkinsoniana, NV
	T. pruni (plum pockets), n hemis
	T. pruni-subcordatae (plum witches' broom), w USA
	T. thomasii (witches' broom on holly-leaf cherry), CA
	T. wiesneri (cherry witches' broom) cosmopolitan
Pyrus spp.	*T. bullata,* WA, BC, Eur

Continued

TABLE 13.1 (Continued)
Taphrinas and Their Woody Plant Hosts Reported in the United States and Canada[a]

Host	Species
Quercus spp.	*T. caerulescens*, NA, Eur, n Africa
Rhus spp.	*T. purpurescens*, c, e US, Eur
Salix laevigata	*T. populi-salicis*, CA
Sorbopyrus auricularia	*T. bullata*, BC
Ulmus americana	*T. ulmi*, c, e US, QUE, Eur

[a] Diseases caused by the fungi listed for each host genus are leaf blisters or curls unless otherwise noted. Geographic distributions are identified with postal abbreviations of states and provinces or more broadly as follows: c, central; e, east; n, north; s, south; w, west; Can, Canada; Eur, Europe; hemis, hemisphere; NA, North America.

Source: Table courtesy W. A. Sinclair, Cornell University.

on the fern *Polystichum aristatum* Presl. and *T. laurencia* causes highly branched bushy protuberances on the fronds of *Pteris quadriaurita* Retz. in Sri Lanka. Although Taphrinas affect many herbaceous and woody plants, the symptoms they cause are often overlooked. The woody plant hosts occurring in the United States and Canada are shown in Table 13.1. Some of the most frequently encountered *Taphrina* diseases are those affecting the genus *Prunus*; a number of species occur on plums, apricots, and cherries. *Taphrina pruni* causes deformed fruits called "plum pockets" or "bladder plums" on plums and related species. The **witches' brooms** caused by *T. wiesneri* on Japanese flowering cherries caused some concern in Washington, DC, in the 1920s, but these perennial infections were successfully eradicated by pruning (Sinclair et al., 1987).

Taphrina species often cause small, yellow leaf spots that may or may not be thickened or blistered into con-cave or convex areas; these often turn brown with age (Figure 13.2). Spots on maples (*Acer* species) caused by a number of different *Taphrina* species may be brown to black. Asci generally appear as a white bloom on one or both leaf surfaces within the areas colonized by *Taphrina*. Twig deformation, galls, witches' brooms, and distorted **inflorescences** or fruit are also possible. Alder cones infected by *T. robinsoniana* show antler-shaped outgrowths (Figure 13.3).

Within the host, *Taphrina* forms a mycelium that is intercellular, subcuticular, or stays within the epidermal wall. There are no ascomata. The naked asci develop from ascogenous cells either in a subcuticular layer or within a wall locule (Figure 13.4). Ascospores in many species bud within the ascus, packing the ascus with blastospores. This budding may continue after spores are released from the ascus. Much of the *Taphrina* life cycle is spent in the sap-

FIGURE 13.2 *Taphrina purpurescens* leaf blister of dwarf sumac (*Rhus copallina*). Affected leaves display red-brown blisters.

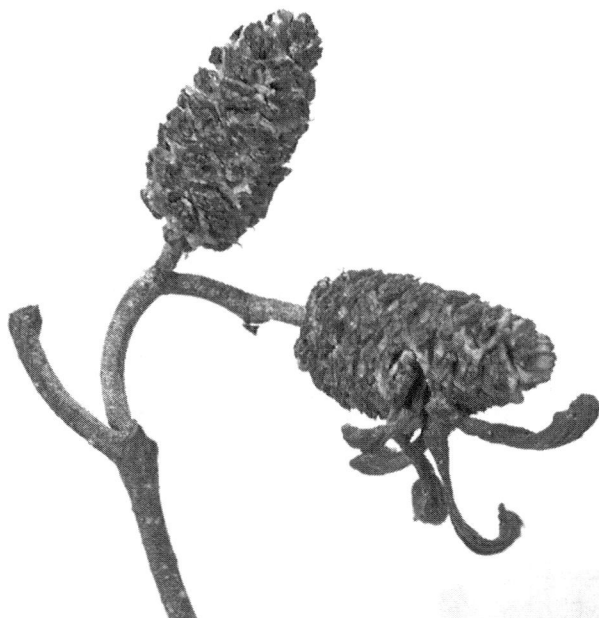

FIGURE 13.3 Catkin hypertrophy on hazel alder, *Alnus serrulata,* caused by *Taphrina robinsoniana.*

rotrophic yeast state, during which the fungus is haploid and uninucleate. The yeast-like anamorph has been named *Lalaria*, after a pebbled beach on a Greek island (Moore, 1990). A dikaryotic mycelium develops only within the parasitized host tissue, and asci are produced. *Taphrina* species generally overwinter as blastospores in buds or on bark, although in a few cases overwintering is accomplished by means of perennial mycelium within the host.

Cultures made from ascospores or blastospores of *Taphrina* grow slowly on artificial media as pale pink yeast colonies. Without knowledge of the organism's parasitic phase, the fungus in culture might be mistaken for one of the pink basidiomycetous yeasts that are common on plant tissue.

Taxonomic understanding of *Taphrina* is being revolutionized by modern molecular research. Assignment of *Taphrina* to the early-diverging Archiascomycete **clade** based on molecular traits fits with earlier knowledge of the unique morphological and physiological traits of this genus, some of which are more typical of **Basidiomycota** than of **Ascomycota**. Molecular methods have also been employed for detection of *T. deformans* in washings from symptomless peach buds (Tavares et al., 2004). PCR fingerprints have been used to distinguish a number of species of *Taphrina* occurring on alder (*Alnus* spp.) that would otherwise be inseparable on the basis of yeast phase characteristics (Bacigalova et al., 2003).

SOME IMPORTANT DISEASES CAUSED BY *TAPHRINA*

Most diseases caused by *Taphrina* attract the attention of only the most curious observer. The following two dis-

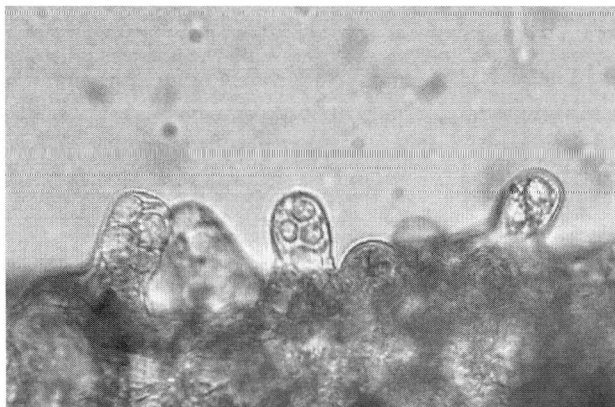

FIGURE 13.4 Asci of *Taphrina purpurescens.* The asci are "naked," that is, they are produced directly from the host tissues and not inside an ascoma. Several ascospores can be seen in the ascus at center.

FIGURE 13.5   Leaf blister of red oak (*Quercus* sp.) caused by *Taphrina caerulescens*. (Photographed by H. H. Whetzel and H. S. Jackson, [CUP 1938a]. Courtesy Cornell Plant Pathology Herbarium.)

eases stand out as more obvious and damaging to cultivated plants: oak leaf blister caused by *T. caerulescens* and peach leaf curl caused by *T. deformans*.

Oak leaf blister is a minor disease in most climates, but is relatively damaging in the southern United States on shade trees in the red oak group. About 50 oak (*Quercus*) species are susceptible, including members of both the white and red oak groups. Ascospores that have overwintered on bud scales germinate and infect young leaves during rainy periods in the spring. Even in areas where the disease is not especially harmful to the oaks, the peculiar leaf symptoms attract comment. The symptoms may include yellowish convex blisters on the leaves, leaf curling, and leaf drop (Figure 13.5). Sporulation usually occurs on the concave side of the blisters. A layer of asci forms between the cuticle and epidermis; the asci swell, break the cuticle, and forcibly discharge their ascospores. The leaf blisters typically turn brown after ascospore production. Control measures are not usually necessary for oak leaf blister.

Taphrina deformans causes the best-known disease in this group: peach leaf curl. This disease is notorious because it causes striking symptoms and significant damage on economically important hosts. Peaches, nectarines, and almonds are affected; the disease occurs all over the world where peaches are grown, but has a more limited geographic occurrence on its other hosts. Peach leaves that are infected when they are still undifferentiated tissue later show yellow, pink or reddish areas that are buckled and curl into weird shapes; entire leaves may be distorted (Figure 13.6). Cynthia Westcott, noted ornamentals pathologist, described peach leaves as sometimes looking "as if a gathering string had been run along the midrib and pulled tight" (Westcott, 1953). The leaf distortion results from plant growth regulators (Chapter 32) secreted by the fungus; **cytokinins** and **auxins** are known to be produced by some *Taphrina* species. Leaf drop may lead to poor fruit quality and weaken trees. The young peach fruits may be distorted or show reddish warty spots lacking the usual fuzz; infected twigs are swollen. The blistered portions of leaves or fruit develop a powdery gray coating of naked asci on the upper surface. After the ascospores are released, the diseased leaves sometimes turn brown, wither, and drop.

Ascospores that land on the surface of the peach tree will bud to produce a saprobic, epiphytic yeast phase. During the summer the fungus lives thus invisibly and harmoniously on the peach tree. This haploid inoculum can persist on the host plant for several years. In the spring, the expanding leaf buds are subject to fungal invasion when

FIGURE 13.6 Peach leaf curl, caused by *Taphrina deformans*. (Photographed by W. R. Fisher [CUP 36520]. Courtesy Cornell Plant Pathology Herbarium.)

plant surfaces are wet for a minimum of 12.5 h (Rossi et al., 2006). Cool temperature (lower than 16°C) during the wet period allows infection and temperature below 19°C is required during incubation. Disease susceptibility is negatively correlated with the rate of host shoot development. The fusion of two yeast cells is thought to form the mycelium that infects the plant. Asci form beneath the cuticle of the host and push to the outside of the plant to release ascospores once again.

There is a single infection period for *T. deformans* in the spring, with no secondary cycles later in the growing season. Thus, by the time the symptoms are noticed, it is too late to achieve any disease control in that same year.

The timing of spring rains and the severity of the preceding winter will influence how extensive the disease symptoms are from one year to the next. Infection of immature peach leaves is optimum at temperatures from 10°C to 21°C and requires rain. Because the fungus overwinters on the surface of the host, this is one of the few diseases that can be controlled with spray treatments at the close of the previous growing season: dormant sprays in late fall or treatments before bud swell in the spring can curb infection. Lime-sulfur and bordeaux mixture have traditionally been used in this fashion. Other effective fungicides include chlorothalonil, fixed coppers, ferbam, or ziram. Chemical control is used for only a few other dis-

---

CASE STUDY 13.2

PEACH LEAF CURL

- The cause of peach leaf curl is *Taphrina deformans.*
- The infection of developing leaves results in blistered, puckered, discolored areas, and may lead to premature leaf drop that damages tree vigor.
- Rarely, young fruit will become infected; they develop warty reddish spots and drop from the tree.
- Years with mild winters followed by cool rainy springs foster this disease and increase the chance of serious effects on peach yield.
- In a year in which it is unusually cool while peach leaves are emerging from buds, damage from *T. deformans* is increased because leaf tissue matures more slowly; once they reach a certain stage of maturity, the leaves are no longer vulnerable to the fungus.
- Often in home landscapes the disease is not noticed until it is too late in the season to take any action that will reduce the problem for that year.
- **Pesticide** applications made after the onset of symptoms are wasted because mature leaf tissue is not susceptible.
- Because of this pathogen's unusual life cycle, it is vulnerable to control actions taken during the tree's **dormant** season.
- A single preventive spray of fungicide during dormancy eliminates the overwintering stage of the pathogen, and thus controls the disease. Trees can be sprayed either after 90% of the leaves have fallen in the autumn or before bud swell in the spring.

eases caused by *Taphrina* species, primarily those affecting plums and cherries.

So, although most of the curious swellings, discolorations, and spots due to fungal pathogens in the Hemiascomycetes and Archiascomycetes remain little studied and largely ignored, the interaction of *T. deformans* with peach trees is considered a major disease because it can cause significant economic loss. Societal value placed on the host plant is often what determines whether a disease is of minor or major import; it is not just a matter of how many plants are killed or disfigured, but which plant species are affected that determines whether control efforts and research dollars will be focused on the problem.

LITERATURE CITED

Alexopoulos, C.J., C.W. Mims and M. Blackwell. 1996. *Introductory Mycology.* Wiley & Sons, New York.

Bacigalova, K., K. Lopandic, M.G. Rodrigues, A. Fonseca, M. Herzberg, W. Pinsker and H. Prillinger. 2003. Phenotypic and genotypic identification and phylogenetic characterisation of *Taphrina* fungi on alder. *Mycological Progress* 2: 179–196.

Batra, L.B. 1973. Nematosporaceae (Hemiascomycetidae): taxonomy, pathogenicity, distribution, and vector relations. U.S. Dept. of Agric., *Agric. Res. Ser. Tech. Bull.* 1469: 1–71.

Kurtzman, C.P. and J. Sugiyama. 2001. Ascomycetous yeasts and yeastlike taxa, pp. 179–200. In: D.J. McLaughlin, E.G. McLaughlin, and P.A. Lemke (Eds.). The Mycota VII Part A, Systematics and Evolution. Springer-Verlag, Berlin.

Mix, A.J. 1949. A monograph of the genus *Taphrina. Univ. Kan. Sci. Bull.* 33: 1–167.

Miyao, G.M., R.M. Davis and H.J. Phaff. 2000. Outbreak of *Eremothecium coryli* fruit rot of tomato in California. *Plant Dis.* 84: 594.

Moore, R.T. 1990. The genus *Lalaria* gen. nov.: Taphrinales Anamorphosum. *Mycotaxon* 38: 315–330.

Nishida, H. and J. Sugiyama. 1994. Archiascomycetes: detection of a major new lineage within the Ascomycota. *Mycoscience* 35: 361–366.

Preece, T.E. and A.J. Hick. 2001. An introduction to the Protomycetales: *Burenia inundata* on *Apium nodiflorum* and *Protomyces macrosporus* on *Anthriscus sylvestris. The Mycologist* 15: 119–125.

Rossi, V., M. Bolognesi, L. Languasco and S. Giosue. 2006. Influence of environmental conditions on infection of peach shoots by *Taphrina deformans. Phytopathology* 96: 155–163.

Sinclair, W.A., H.H. Lyon and W. A. Johnson. 1987. *Diseases of Trees and Shrubs.* Cornell University Press, Ithaca, New York.

Sjamsuridzal, W., Y. Tajiri, H. Nishida, T.B. Thuan, H. Kawasaki, A. Hirata, A. Yokota and J. Sugiyama. 1987. Evolutionary relationships of members of the genera *Taphrina, Protomyces, Schizosaccharomyces*, and related taxa within the archiascomycetes: Integrated analysis of genotypic and phenotypic characters. *Mycoscience* 38: 267–280.

Sugiyama, J. 1998. Relatedness, phylogeny, and evolution of the fungi. *Mycoscience* 39: 487–511.

Tavares, S., J. Inacio, A. Fonseca and C. Oliveira. 2004. Direct detection of *Taphrina deformans* on peach trees using molecular methods. *European Journal of Plant Pathology* 110: 973–982.

Westcott, C. 1953. *Garden Enemies.* D. Van Nostrand Co., New York.

14 The Powdery Mildews

Margery L. Daughtrey, Kathie T. Hodge, and Nina Shishkoff

Chapter 14 Concepts

- Powdery mildews are obligate parasites that show interesting morphological adaptations to herbaceous vs. woody hosts.

- Powdery mildews have specialized feeding cells called haustoria that absorb nutrients from their hosts.

- Molecular genetic studies and scanning electron microscope studies have led to a recent reversal in taxonomic thought and a new paradigm: the anamorphs reflect phylogeny much better than the teleomorphs.

- Many powdery mildew fungi recently have been renamed to reflect the newly apparent relationships as revealed by molecular biology.

- Powdery mildews injure many ornamental crops and garden plants because of the highly conspicuous colonies that cause aesthetic injury; vegetable, field, and fruit crops suffer from yield and quality reduction.

The powdery mildew fungi in the phylum Ascomycota cause easy-to-recognize diseases. The fungus grows across the surface of the host in conspicuous colonies, creating whitish circular patches that sometimes coalesce until the entire leaf surface is white. The colonies may form on either the upper or lower surface of leaves, as well as on stems, flower parts, and fruits. Because energy from photosynthesis is diverted into growth of the pathogen, infected plants may be stunted and produce fewer or smaller leaves, fruits, or grain. The impacts of powdery mildews on their hosts may be mainly aesthetic, or may reduce yield or quality. Many floral and nursery crops (Figure 14.1) and also fruit, vegetable, and field crops are affected.

Powdery mildews are spread over great distances by wind and can also be moved around the world on plants with inconspicuous or latent infections. Within the powdery mildew family, Erysiphaceae, some members have very narrow host ranges, whereas others have broad host ranges and affect plants in multiple plant families. Powdery mildews differ from the vast majority of other fungal pathogens in that, with few exceptions, the mycelium grows superficially over host tissues. Only specialized feeding cells called **haustoria** (Chapter 10) penetrate the host epidermis (Figure 14.2). Powdery mildew life cycles (Figure 14.3) are entirely **biotrophic**. No species has been grown in **axenic** culture apart from its host for any significant duration and none can grow on dead plant material.

Conidiophores develop directly from the mycelium on the host surface throughout the growing season; they produce infective **conidia** either one at a time (Figure 14.4)

FIGURE 14.1 Powdery mildew of poinsettia caused by *Oidium* sp. White colonies on the red bracts of this Christmas favorite make the plants unsaleable. The sexual state of this pathogen is as yet unknown.

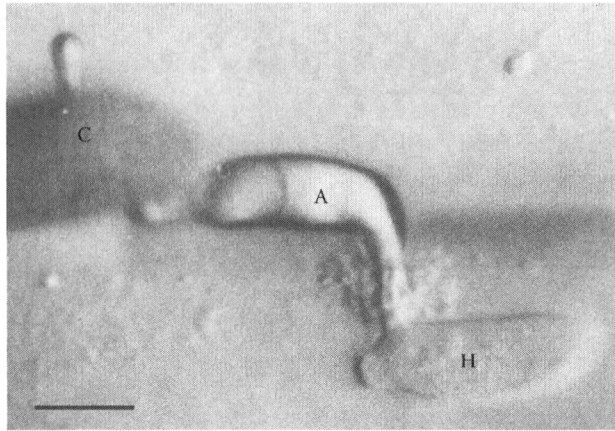

FIGURE 14.2 Penetration structures of *Blumeria graminis* f. sp. *hordei*, powdery mildew of barley. A conidium (C, upper left) that is slightly out of focus on the leaf surface has germinated to produce a primary (top) and a secondary germ tube. The secondary germ tube has produced a large appressorium (A, center) and penetrated the epidermal cell wall. A haustorium (H, bottom right) has been formed inside the epidermal cell. This basic unit of infection may be replicated in hundreds of epidermal cells underlying a single mildew colony. (Photo courtesy J. R. Aist, Cornell University.)

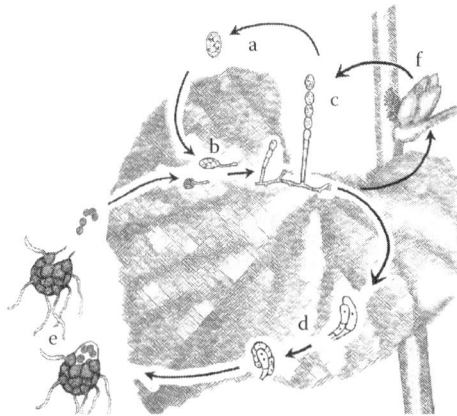

FIGURE 14.3 Life cycle of a powdery mildew. (Drawing by N. Shishkoff.)

or in short chains. In tropical climates or greenhouses, this may be the only spore stage in the life cycle. Conidia are wind borne or splash dispersed, and serve as secondary inocula during the growing season. Crops grown in greenhouses may suffer repeated cycles of infection via conidia year-round. Spore release typically follows a diurnal pattern, with the greatest number of conidia being released around midday in response to a decrease in relative humidity. Outdoors, rain may serve as a trigger for the release of conidia. Powdery mildews are unusual among fungi in that their conidia do not require large amounts of free water to germinate. Significant disease spread can thus occur even during dry weather.

Most species of the Erysiphales are **heterothallic**; two compatible individuals must mate before asci can be formed. Compatibility is genetically determined by the mating type locus. Two different **alleles** confer two possible

FIGURE 14.4 Conidiophores and conidia of *Oidium* sp., cause of powdery mildew of poinsettia. (Photo courtesy of Gail Celio, University of Minnesota.)

mating types. Some homothallic species are known, but the genetic basis of homothallism has not been determined.

After mating, ascomata are produced superficially on the host, typically toward the end of summer (Figure 14.5). Ascomata can be an important overwintering stage in temperate climates as they provide the primary inoculum at the beginning of each growing season for a number of powdery mildews. In arid climates, they aid in survival during hot, dry periods. In direct contrast to conidia, free water is required to stimulate ejection of the ascospores and allow their germination. In some species that attack perennial hosts, such as *Podosphaera pannosa* (sect. Sphaerotheca) f. sp. *rosae* on rose, the fungus can also survive as mycelium inside infected buds. In powdery mildews of tropical climates, the ascomata are of little importance and have apparently been lost altogether in some species.

The appendages on the ascocarps function in dispersal by helping the ascocarps to adhere to plant surfaces and **trichomes**. For powdery mildews on woody plants, the elaborate appendages allow the ascomata to cling to rough bark surfaces. In *Phyllactinia*, which causes the powdery mildew of alders and numerous other hardwoods, the moisture-sensitive appendages play a unique role in dispersal (Figure 14.5). On maturity, the appendages press down on the substrate, breaking the ascocarp away and releasing it into the wind. Sticky secretions produced from a second set of appendages on top of the ascocarp promote its adherence when it lands. The ascocarp passes the winter upside-down. In spring, the ascocarp splits around the circumference and flips open completely, like a hinged jewel box, so that the formerly upside-down asci in the "lid" now point upward to discharge their infective spores. In other powdery mildews, the appendages are less complex, and the ascocarps typically open from the pressure of asci against the upper surface.

HOST RELATIONSHIPS

Powdery mildews engage in fascinating interactions with their host. Unlike other plant pathogens, a single individual penetrates its host at many different sites and does not proliferate within the plant tissues. A few powdery mildews are atypical in that they do penetrate the stomatal space and form a limited **hemiendophytic** mycelium. The complex organs of infection formed by powdery mildews are among the best-studied host–pathogen interfaces in plant pathology (Figure 14.2).

THE INFECTION PROCESS

Powdery mildew conidia deposited on a hydrophobic leaf surface excrete an adhesive matrix within minutes of contact (Nicholson and Kunoh, 1995). This matrix is believed to mediate host recognition. In a compatible interaction, conidia rapidly germinate by producing one or more germ tubes that follow the contours of the host surface (Figure 14.6). At the germ tube's apex, an **appressorium** (Figures 14.2, 14.6, and 14.7) is produced at the site where penetration will occur. In different species, the appressoria can be undifferentiated, simple, forked, or lobed, and these morphological variations can be useful

---

CASE STUDY 14.1

POWDERY MILDEW OF VERBENA

- Powdery mildew, caused by the fungus *Podosphaera xanthii*, causes a disease that ruins the appearance of flowering verbena crops.
- Sales of bedding plants in packs, specimens in hanging baskets, and trailing plants in patio containers are harmed, although symptoms usually disappear after plants are moved outdoors.
- This same fungus can also infect members of the squash family (Cucurbitaceae).
- Sometimes the disease spreads by airborne **conidia** from verbenas to cucumbers or squash within the same greenhouse.
- Sometimes disease can spread from field-grown plants to those in greenhouses, or vice versa.
- Cultivars of verbena vary a great deal in susceptibility.
- Chemical and biological controls vary in their effectiveness.
- **Integrated pest management (IPM)** is recommended:
  - Choose to grow less-susceptible verbenas.
  - Space plants in the greenhouse to allow air penetration.
  - Use fans to circulate the air.
  - Use heating and ventilation to keep greenhouse humidity below 85%.
  - Scout for the first signs of powdery mildew in the crop.
  - Use contact and systemic fungicides in rotation for resistance management.

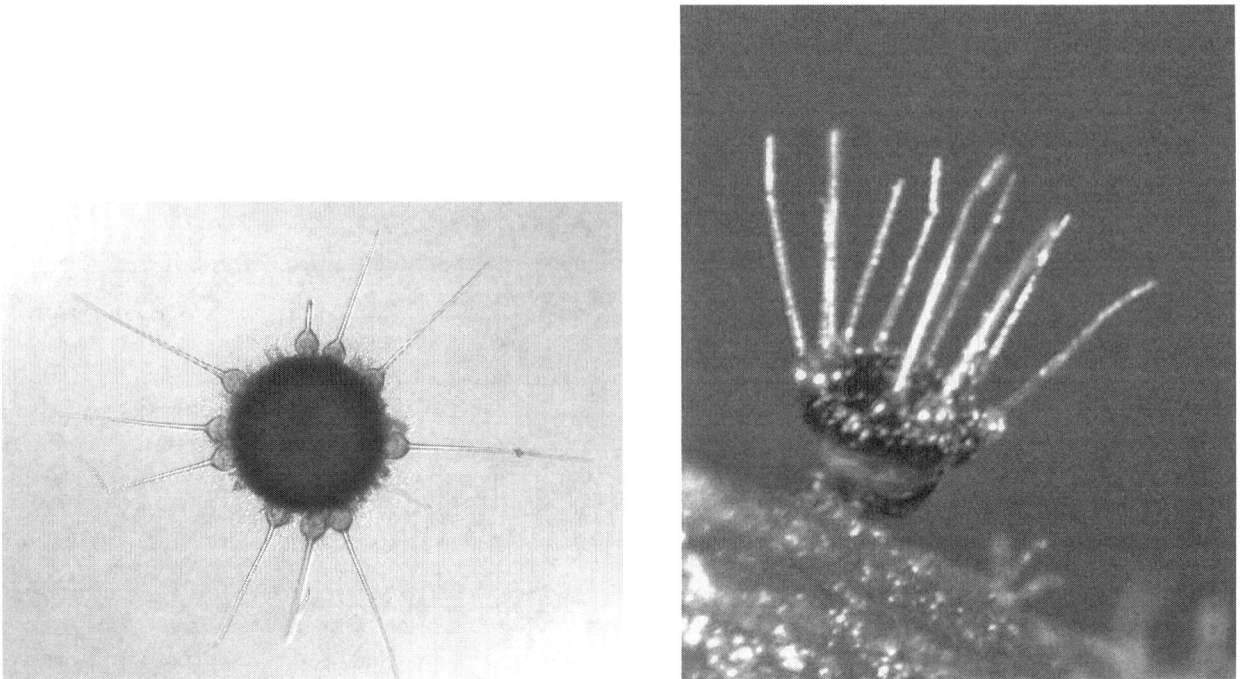

FIGURE 14.5 The ascoma of *Phyllactinia guttata*, the fungus causing one of the powdery mildew diseases of oak, (left) as it appears when first formed, with the bulbous-based appendages lying flat against the oak leaf surface; (right) after liberation. The appendages have successfully flexed to lift the ascoma from the leaf undersurface. After a brief tumble, the top of the liberated fruiting body has adhered by its sticky cushion to nearby plant material. It will overwinter in this upside-down position. In spring, the ascoma will split open around its equator, revealing asci in its "lid" in the perfect position to discharge infective spores into the air. (Photo courtesy of Kathie T. Hodge, Cornell University.)

CASE STUDY 14.2

SNOW-ON-THE-POINSETTIAS? POWDERY MILDEW FOR THE HOLIDAYS?

- Powdery mildew on poinsettias is caused by an *Oidium* sp.
- This pathogen first appeared in the United States in the 1980s and had not been described in the world before.
- It has now been reported from Europe and North America as well as Central America, and appears in some U.S. greenhouses each year.
- Conspicuous white powdery mildew colonies form on the leaves as well as on the bracts of poinsettias, which are the brightly-colored petal-like leaves surrounding the inconspicuous flowers.
- Because **fungicide** labels specify use on particular crops for particular problems, few were available for the first few years to use on poinsettias to control this unheard of disease.
- Growers are reluctant to use fungicide sprays on poinsettias once they have begun to develop bracts, which sometimes show **phytotoxicity** in response to pesticides.
- The disease can be **latent** during hot summer months in the greenhouse, but reappear in the autumn and begin to spread, once temperatures fall below 85°F.
- **Scouting** for this disease requires checking the reverse side of any leaf that shows a yellow spot.
- Control of the disease requires the following:
  - Scouting, because there is no need for foliar fungicides if the pathogen is not present
  - Roguing out infected plants to reduce **inoculum**
  - Use of nonphytotoxic fungicides in rotation if the disease is detected
  - Reducing humidity and increasing air circulation to slow progress of the disease

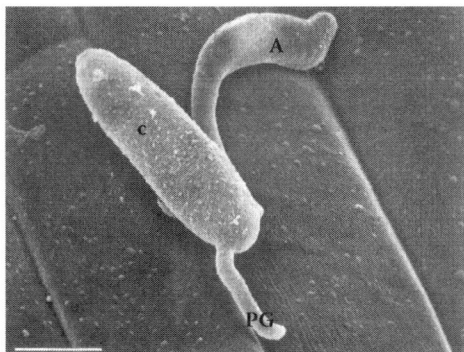

FIGURE 14.6 Scanning electron micrograph of a conidium (C) of *Blumeria graminis* f.sp. *hordei* (the cause of powdery mildew of barley) on a leaf surface. A primary (PG) and a secondary germ tube are visible; the secondary germ tube (top, right) has started to differentiate into an appressorium (A). In this species, the primary germ tube serves a sensing function, and the secondary germ tube leads to penetration of the host. (Photo courtesy of J.R. Aist, Cornell University.)

in identification (Figure 14.8). The fungus also continues to grow across the host's surface, forming the distinctive superficial mycelium.

The difficult traverse of the plant cell wall is the next step in infection. At this stage, a hypha with narrow diameter, known as a **penetration peg**, grows from the bottom surface of the appressorium. Through a combination of enzymatic and mechanical actions, it penetrates the cell wall of an epidermal cell. Having breached the cell wall, the fungus produces a **haustorium** (Figure 14.2), a special feeding cell that assimilates nutrients from the plant. Haustoria can be simple, lobed, or digitate. The haustorium is not in direct contact with the host cytoplasm; rather, it is enfolded by the cell membrane of its host cell. The fungus induces changes in the plant membrane surrounding the haustorium that result in leakage of nutrients that are

then taken up by the fungus. The invaded cell remains alive and functioning for some time and receives nutrients from surrounding cells that are then leaked to the fungus. This basic pathogenic interaction is repeated in many different cells, and the superficial mycelium spreads across the leaf surface and occasionally develops appressoria, each leading to penetration and parasitism of a single cell. Although the host is seldom killed outright, the continual drain of nutrients often depresses growth and sometimes causes puckering or chlorotic or necrotic spotting of the leaf surface.

Host Responses

Host plants resist powdery mildew attack through passive and active defense systems (Chapters 28 and 31). Leaf surface features, such as the thick waxes that protect the undersurfaces of some grass leaves, may inhibit germination. The cuticle and cell wall present formidable barriers to penetration because their thickness and durability affect the penetration ability of fungal germ tubes. Many plants actively respond to penetration attempts by forming **papillae** (sing. papilla)—tiny, thick deposits of callose and other materials that are laid down on the inner surface of the cell wall. Papillae are a physical barrier to invasion and also serve as foci for the release of antimicrobial compounds. The **hypersensitive response** may be induced in incompatible interactions, resulting in the death of an invaded host cell and starvation of the pathogen. All of these mechanisms are genetically controlled and provide important sources of resistance for plant breeders.

GENERAL TAXONOMY

Powdery mildews are classified in the class Leotiomycetes (with primitive cup-fungi), the order Erysiphales

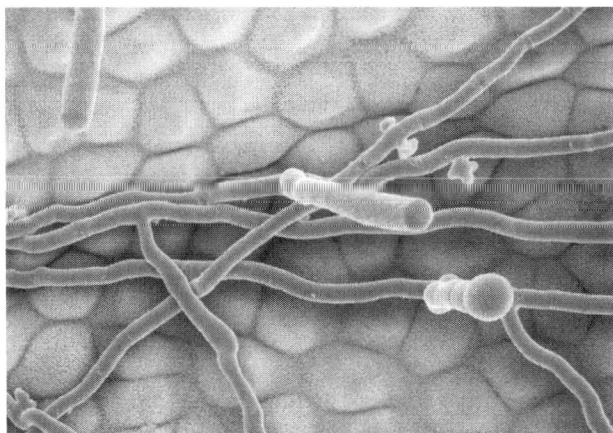

FIGURE 14.7 Scanning electron micrograph of a powdery mildew colony on a leaf surface. The small, lobed appressoria mark points of penetration into the leaf. Several conidiophores can be seen projecting upwards off the leaf surface. (Photo courtesy of Gail Celio, University of Minnesota.)

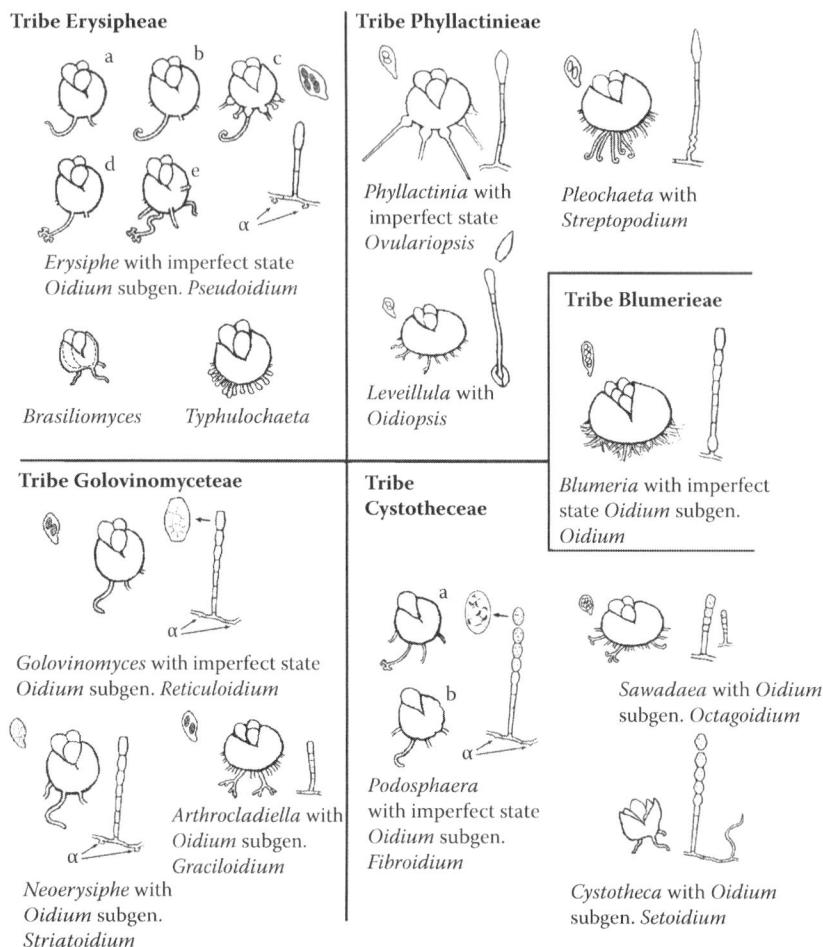

FIGURE 14.8 The genera of powdery mildews. **Tribe Erysipheae:** *Erysiphe* is shown with its anamorph, *Oidium* subgenus *Pseudoidium,* and its characteristic lobed appressoria (α). (This genus now embraces several former genera: (a) *Erysiphe*, (b) *Uncinula*, (c) *Bulbouncinula*, (d) *Microsphaera*, and (e) *Medusosphaera*). *Brasiliomyces* has a one-layered transparent outer ascomatal wall and no known anamorph. *Typhulochaeta* has club-shaped appendages and no known anamorph. **Tribe Golovinomycetinae:** *Golovinomyces* has two-spored asci, the anamorph *Oidium* subgenus *Reticuloidium* and mycelium with simple appressoria (α). *Neoerysiphe* has asci that only mature after overwintering and anamorph *Oidium* subgenus *Striatoidium* with lobed apressoria (α). *Arthrocladiella* has dichotomously branched appendages and anamorph *Oidium* subgenus *Graciloidium*. **Tribe Blumerieae:** *Blumeria*, sole genus in this tribe, is found on grasses; its anamorph is *Oidium* subgenus *Oidium*. **Tribe Phyllactinieae:** *Phyllactinia* has bulbous-based appendages and anamorph *Ovulariopsis*. *Pleochaeta* has curled appendages and the anamorph *Streptopodium*. *Leveillula*, which is unusual in having internal mycelium and conidiophores that grow out of stomatal pores, has undifferentiated appendages and the anamorph *Oidiopsis*. **Tribe Cystotheceae:** *Podosphaera* has simple appressoria (α) and the anamorph *Oidium* subgenus *fibroidium*. Conidia contain fibrosin bodies. (This genus now embraces the former genera (a) *Podosphaera* and *(b) Sphaerotheca*). *Sawadaea* has curled appendages (that can be uni, bi, or trifurcate) and anamorph *Oidium* subgenus *Octagoidium*. *Cystotheca*, which is unusual in having thick-walled aerial hyphae, has an ascomatal outer wall that easily splits in two, and anamorph *Oidium* subgenus *Setoidium*. (Drawing by N. Shishkoff.)

and the family Erysiphaceae. For years the taxonomy of the powdery mildews relied largely on the morphology of the ascomata and their appendages, because they showed more obvious differences (Braun, 1987; 1995). Recent advances in molecular phylogeny have shown the morphology of the anamorph to be a better indicator of relationships (Saenz and Taylor, 1999; Takamatsu et al., 1999). The characteristics of the conidial chains and the ornamentation of the conidial surface are particularly useful (Cook et al., 1997). Now it is thought that differences in the ascomata merely reflect adaptation to particular hosts.

Braun and Takamatsu (2000) and Braun et al. (2002) have renamed many powdery mildews based on new genetic and morphological data.

THE ANAMORPH

Conidia (meristemic arthrospores) are produced on unbranched conidiophores, singly or in chains. The name *Oidium* is often used informally to refer to any **anamorph** of a powdery mildew. The genus *Oidium* in the more formal sense refers to powdery mildews producing ovoid conidia singly or in chains. *Oidium* species are divided into sub-genera based on morphological features, and these neatly correlate with particular **teleomorphs**, so that *Oidium* subgen. *Oidium* is associated with the teleomorph *Blumeria*, the subgen. *Pseudoidium* with the genus *Erysiphe*, and so on. Other powdery mildew anamorphs have conidia that are not ovoid or are borne singly (Figure 14.8).

THE TELEOMORPH

The ascomata of powdery mildews are initially fully closed, and they have often been referred to as **cleistothecia**. They differ from other cleistothecial fungi, however, in that the ascomata do eventually rupture and the ascospores are forcibly discharged. Developmental features also suggest that the ascomata are more similar to perithecia than cleistothecia.

Mycologists once assumed that primitive powdery mildews were *Erysiphe*-like, with undifferentiated ascomatal appendages (Figure 14.8). New genetic evidence suggests that the primitive powdery mildews lived on woody hosts and possessed curled appendages (Takamatsu et al., 2000). Species on herbaceous hosts typically lack elaborate appendages. This suggests that appendages may be "costly" to maintain, valuable in anchoring ascomata to persistent plant parts, but lost quickly when the plant host has no persistent parts on which to anchor. Related species thus display conserved anamorph morphology, whereas teleomorph morphology may be plastic and strongly influenced by host and environment.

IDENTIFICATION OF POWDERY MILDEWS

The new nomenclature (Braun et al., 2002) is sure to be confusing for a time because many familiar names have been changed in order to reflect our new understanding of relationships among the powdery mildew fungi. In this system, the Erysiphaceae is divided into five tribes including, in all, 13 current genera. A simple overview of modern powdery mildew genera is presented in Figure 14.8. It is easy enough to recognize that a disease is caused by a powdery mildew fungus by noting the white powdery colonies on infected plant parts. Morphological differences separating the species can be subtle, however, so the best route to diagnose powdery mildew diseases is through the use of a thorough host index, such as Farr et al. (1989) (also available on-line).

SOME IMPORTANT DISEASES CAUSED BY POWDERY MILDEWS

POWDERY MILDEW OF GRAPE

Powdery mildew of grape is caused by *Erysiphe* (sect. *Uncinula*) *necator*. The disease can reduce yield and fruit quality, as well as stunt vines and decrease their winter

FIGURE 14.9 Powdery mildew of grape, caused by *Erysiphe* (sect. *Uncinula*) *necator*. The powdery white bloom on the fruit makes these grapes unfit for consumption or processing. Photo by W.R. Fisher, courtesy of the Cornell University Herbarium (CUP 11853a).

hardiness. *Vitis vinifera* and hybrids such as Chardonnay and Cabernet Sauvignon are more susceptible to powdery mildew than are varieties derived from American species. Wine quality can be affected when as few as 3% of berries are infected. Leaves, green shoots, and fruits are susceptible to infection. White mildew colonies appear on upper and lower surfaces of leaves, and may coalesce. Immature leaves may be distorted and stunted. Shoot infections appear as dark patches. Infected berries can display white colonies (Figure 14.9), abnormal shape, or rust-colored spots, and may split open, rendering them vulnerable to fruit rot by *Botrytis cinerea*. Infected leaves may be cupped-up or scorched and may fall prematurely.

Erysiphe necator overwinters as ascomata in crevices on the surface of the bark of vines or as mycelium in buds (Gadoury and Pearson, 1988). Ascospore release is triggered by even minuscule amounts of rain in the spring when the temperature is warm enough. The ascospores are spread by wind to the green tissues of the grapevine from bud break through bloom. The short time required from infection until the production of new inoculum gives this powdery mildew powerful epidemic potential. Because little free moisture is needed for conidial germination, powdery mildew can be a serious problem even in years when black rot and downy mildew are hampered by dry conditions. Relative humidity of 40 to 100% is all that is required for the formation of microdroplets of water used by conidia for germination and infection. Temperature is optimum for disease development from 20°C to 25°C, but infection is possible over a wide temperature range of 15°C to 32°C.

All cultural practices that promote the drying of plant surfaces will reduce powdery mildew. Full sun exposure is desirable; alignment of rows with the direction of the prevailing wind is recommended. Vines should be pruned and trained to reduce shading and allow air circulation within the canopy. Excessive nitrogen should be avoided. Irrigation should be done with trickle systems rather than over-the-vine impact sprinklers. Control of infections during the immediate prebloom through early postbloom period is stressed for grapes because this is when fruit are most susceptible to the disease. The more susceptible *V. vinifera* and hybrid cultivars require continued suppression of powdery mildew on the foliage until at least **veraison** in order to maintain a functional leaf canopy to ripen the crop and avoid premature defoliation. Controlling later-season infection also helps to reduce the overwintering **inoculum** and may improve winter hardiness in colder regions.

Elemental sulfur has long been used internationally for powdery mildew suppression, although it is phytotoxic to some native American (e.g., Concord) and hybrid cultivars. Modern fungicides such as the sterol inhibitors (both the demethylation inhibitor [DMI] and, more recently, morpholine groups) and strobilurins are also widely used in disease management programs (Chapter 36). Certain nontraditional products (e.g., oils, potassium bicarbonate, and monopotassium phosphate salts and dilute hydrogen peroxide) also provide suppression and are used to variable extents where they are labeled for this purpose. It appears that much of the activity of such materials is eradicative, presumably due to the susceptibility of the exposed fungal colony to topical treatments of these substances.

POWDERY MILDEW OF CEREALS

Powdery mildew diseases affect **monocots** as well as **dicots**. The pathogen *Blumeria graminis* has several host-specialized **formae speciales** that infect different cereal crops. An example is *Blumeria graminis* f. sp. *tritici*, a pathogen of wheat (*Triticum aestivum*) (Wiese, 1977).

White colonies of powdery mildew are most common on the upper surface of older wheat leaves. The haustoria of *Blumeria* are unusual because they have long finger-like projections. The conidiophores generate long chains of conidia that can germinate over a wide temperature range

(1°C to 30°C), with an optimum temperature of 15°C to 22°C and an optimum relative humidity of 85% to 100%. A complete disease cycle takes place in 7 to 10 days, so epidemics can develop quickly. Wheat plants are most susceptible when they are heavily fertilized and rapidly growing.

As wheat plants mature, ascomata begin to appear on the leaf surface as long as both mating types are present. The ascomata are important, because sexual recombination allows new races of powdery mildew to be produced. These new races may be able to grow on wheat cultivars that were not susceptible to the original powdery mildew population. Ascomata on wheat stubble also function as overwintering structures, and in mild climates the mycelium itself can survive. Infections by ascospores require rain and may occur in midsummer or fall. Powdery mildew of wheat prospers in both humid and semiarid climates. Yield is decreased by powdery mildew, because photosynthesis is reduced whereas transpiration and respiration are increased. Wheat yields may be reduced by as much as 40%.

In spite of the strong impact of this disease, it is not always cost-effective to use fungicides to control powdery mildew on wheat. Systemic fungicides are used for powdery mildew control in Europe, primarily. Host-plant resistance is used worldwide for powdery mildew management. Plant breeders work to supply wheat lines that are not susceptible to the powdery mildew **races** in a given geographic area; new lines are needed frequently to keep up with adaptations of the pathogen. Rotating crops and eliminating "volunteer" wheat plants and crop debris aid in disease management.

POWDERY MILDEW OF ROSE

Powdery mildew diseases are especially important on ornamental plants, on which any visible infection may lower the aesthetic appeal. Powdery mildew of rose (*Rosa × hybrida*) affects roses in gardens as well as in field or greenhouse production. It is the most common disease of roses grown in greenhouses for cut flowers and also on potted miniature roses. Young foliage and pedicels are especially susceptible and may be completely covered with mildew (Figure 14.10). Young shoots and flower petals may also be infected. The rose powdery mildew is *Podosphaera* (sect. *Sphaerotheca*) *pannosa* f. sp. *rosae*.

Infections may develop quickly and conidia may start to germinate within just a few hours of landing on the host (Horst and Cloyd, 2007). Chains of conidia are produced only 3 to 7 days after the initial infection. Disease development is optimum at 22°C. The fungus overwinters in buds unless the climate is so mild that conidia are continually produced year-round, as in greenhouses. Control of rose powdery mildew in greenhouses is achieved by air circulation with fans plus heating and ventilation to reduce humidity, coupled with the use of fungicides. Strobilurins, DMIs, and morpholine fungicides are commonly

FIGURE 14.10 Powdery mildew of rose. The arrow indicates a severe infection of the flower pedicel that may affect the longevity and saleability of the flowers.

utilized for disease control in greenhouses. Products featuring potassium bicarbonate, botanical extracts, and biocontrols, including the fungus *Pseudozyma flocculosa,* have recently been developed. Gardeners wishing to circumvent the problem of powdery mildew on outdoor roses can seek disease resistant cultivars and species.

POWDERY MILDEW ON CUCURBITS

Numerous vegetable crops are susceptible to powdery mildew, but cucurbits are arguably the group most severely affected. The powdery mildew on cucurbit crops (*Cucumis* species, including squash, pumpkin, and cucumber) is *Podosphaera* (sect. *Sphaerotheca*) *xanthii* (referred to in earlier literature as *Sphaerotheca fuliginea* or *S. fusca*). Infections on both upper and lower leaf surfaces occur, and these reduce yield by lowering plant vigor and increasing the number of sun-scorched fruits (Figure 14.11). Resistant cultivars of cucumbers and melons are available, but most squash varieties are susceptible. Susceptible melons may have significantly poorer fruit quality due to low sugar content. Even the flavor of winter squash can be harmed because the fruits of mildew-infected plants have fewer stored soluble solids that affect taste. The color and handle quality of pumpkins can be ruined by powdery mildew. The disease develops on cucurbits during warm summer weather (a mean temperature of 68°F to 80°F is most favorable). Conidia are thought to be airborne over long distances to initiate infections.

The fungicides used for disease management in cucurbits often are systemic materials with specific, single-site, modes of action that make them vulnerable to the development of resistance in the pathogen (McGrath, 2001). The fungicide benomyl is no longer effective against cucurbit powdery mildew because of resistance development. Due to the rapid epidemic development of powdery mildews, they are more capable of developing resistance to fungicides than many other fungi. The pathogen population can shift quickly to resistant individuals even when these are initially at an undetectable level in the population. Growers of any crop with high susceptibility to a powdery mildew are encouraged to rotate among chemical classes with different modes of action to delay the development of resistance. The use of materials with multisite modes of

FIGURE 14.11 Powdery mildew of pumpkin, showing colonies on the upper and lower leaf surfaces. Photo by Margaret Tuttle McGrath (Reprinted, with permission, from M.T. McGrath, 2001. Fungicide resistance in cucurbit powdery mildew: experiences and challenges, *Plant Disease* 85:236–245.)

action, such as chlorothalonil, copper, sulfur, horticultural oil, and potassium bicarbonate, is also a good strategy to avoid resistance.

An **integrated pest management** (IPM) program for cucurbit powdery mildew can be very effective (McGrath and Staniszewska, 1996; McGrath, 2001; Chapter 38). Elements of such a program include using resistant varieties when available, scouting for the first colonies beginning at the time of fruit initiation, using air-assist sprayers to maximize spray coverage on difficult-to-cover lower leaves, and using fungicides according to a resistance-management strategy. For greenhouse-grown cucumbers, silicon amendment of the nutrient solution has reduced disease. A number of organisms including the fungi *Pseudozyma flocculosa*, *Sporothrix rugulosa*, *Tilletiopsis* spp., *Ampelomyces quisqualis* and *Verticillium lecanii* have been used for biocontrol (Chapter 37) of cucumber powdery mildew in greenhouses (Dik et al., 1998). Mechanisms of antagonism used by these fungi include **hyperparasitism** or **antibiosis**. As powdery mildews may thrive at lower relative humidity than the antagonistic fungi that might be deployed against them, environmental conditions in a given situation may determine whether biocontrol is effective.

POWDERY MILDEW ON FLOWERING DOGWOOD

Many woody plant species are subject to one or more powdery mildews, but these diseases rarely have a significant economic impact. Lilac foliage, for example, frequently shows conspicuous powdery mildew in late summer, but the shrubs are attractive again during bloom the following spring. Powdery mildew has been noted on dogwoods (*Cornus* species) since the 1800s, but a serious powdery mildew disease on *C. florida*, the flowering dogwood, became apparent in the eastern United States only in the mid-1990s (Figure 14.12). *Erysiphe* (sect. *Microsphaera*) *pulchra* (Cooke and Peck) is the powdery mildew that now affects flowering dogwood. Aesthetic injury is caused on landscape trees by whitening and twisting of the leaves at the tips of branches in mid-summer, whereas economic injury is seen during nursery production when young trees with powdery mildew are severely stunted. In many cases, infection of landscape trees is subtle, such that the thin coating of mycelium on the leaves is often overlooked. Symptoms of reddening or leaf scorch are often seen on infected leaves in dry summers, and it is thought that the mildew may contribute to the drought stress on the host tree.

Ascomata on leaf debris are the only documented overwintering mechanism for this powdery mildew. The abundance of ascomata varies from year to year, perhaps due to variation in weather conditions during the fall when they are maturing. The nonnative Kousa dogwood, *C. kousa*, is not appreciably affected by powdery mildew. Several selections of powdery mildew resistant

FIGURE 14.12 Dogwood powdery mildew causes curling and deformation of leaves at the tips of branches; the fungus may entirely coat the leaves. The eastern flowering dogwood (*Cornus florida*) is quite susceptible, but most cultivars of the Kousa dogwood (*C. kousa*) are quite resistant. (Photo by Kerry Britton, USDA-FS.)

cultivars of *C. florida* have been released (Windham et al., 2003). Fungicides such as propiconazole and chlorothalonil are used for disease suppression during nursery crop production.

CONCLUSION

In some ways, powdery mildews seem to be mild diseases, because the fungi only rarely kill the plants that they parasitize. Powdery mildew fungi and their hosts come closely together in a carefully balanced parasitic relationship that results in nutrient flow to the powdery mildew without extensive death of plant host cells. Perennial plants may be infected by powdery mildew summer after summer, and still leaf out vigorously every spring. Annual plants usually bear fruit and set seed in spite of powdery mildew. Yet to the horticultural producers aiming to have high yield, high-quality crops, or to gardeners wanting attractive and vigorous plants, powdery mildews are formidable diseases. Curbing powdery mildew epidemics on economic hosts is essential to maximize the growth potential of the plants and maintain the marketability of the products harvested from them.

REFERENCES

Braun, U. 1987. *A Monograph of the Erysiphales (powdery mildews)*. J. Cramer, Berlin-Stuttgart.

Braun, U. 1995. *The Powdery Mildews (Erysiphales) of Europe*. G. Fischer Verlag Jena.

Braun, U., R.T.A. Cook, A.J. Inman and H.-D. Shin. 2002. In *The Powdery Mildews: A Comprehensive Treatise*. Eds. R.R. Belanger, A.J. Dik and W.R. Bushnell. APS Press, St. Paul, MN.

Braun, U. and S. Takamatsu. 2000. Phylogeny of Erysiphe, Microsphaera, Uncinula (Erysipheae) and Cystotheca, Podosphaera, Sphaerotheca (Cystotheceae) inferred from rDNA ITS sequences—some taxonomic consequences. *Schlechtendalia* 4: 1–33.

Cook, R.T.A., A.J. Inman and C. Billings. 1997. Identification and classification of powdery mildew anamorphs using light and scanning electron microscopy and host range data. *Mycol. Research* 101: 975–1002.

Dik, A.J., M.A. Verhaar and R.R. Belanger. 1998. Comparison of three biocontrol agents against cucumber powdery mildew (*Sphaerotheca fuliginea*) in semi-commercial scale glasshouse trials. *Eur. J. Plant Path.* 104: 413–423.

Farr, D.F., G.F. Bills, G.P. Chamuris and A.Y. 1989. Rossman. *Fungi on plants and plant products in the United States*. APS Press, St. Paul, MN.

Gadoury, D.M. and R.C. Pearson. 1988. Initiation, development, dispersal and survival of cleistothecia of *Uncinula necator* in New York vineyards. *Phytopathology* 78: 1413–1421.

Horst, R.K. and R.A. Cloyd. 2007. *Compendium of Rose Diseases and Pests*, APS Press, St. Paul, MN.

McGrath, M.T. 2001. Fungicide resistance in cucurbit powdery mildew: experiences and challenges. *Plant Dis.* 85: 236–245.

McGrath, M.T. and H. Staniszewska. 1996. Management of powdery mildew in summer squash with host resistance, disease threshold-based fungicide programs, or an integrated program. *Plant Dis.* 80: 1044–1052.

Nicholson, R.L. and H. Kunoh. 1995. Early interactions, adhesion, and establishment of the infection court by *Erysiphe graminis*. *Can. J. Bot.* 73(Suppl.): S609–S615.

Saenz, G.S. and J.W. Taylor. 1999. Phylogeny of the Erysiphales (powdery mildews) inferred from internal transcribed spacer (ITS) ribosomal DNA sequences. *Can. J. Bot.* 77: 150–169.

Takamatsu, S., T. Hirata, Y. Sato and Y. Nomura. 1999. Phylogenetic relationship of *Microsphaera* and *Erysiphe* sect. *Erysiphe* (powdery mildews) inferred from the rDNA ITS sequences. *Mycoscience* 40: 259–268.

Takamatsu, S., T. Hirata and Y. Sato. 2000: A parasitic transition from trees to herbs occurred at least two times in tribe Cystotheceae (Erysiphaceae): evidence from nuclear ribosomal DNA. *Mycol. Research* 104: 1304–1311.

Wiese, M.V. 1977. *Compendium of Wheat Diseases*. APS Press, St. Paul, MN.

Windham, M.T., W.T. Witte and R.N. Trigiano. 2003. Three white-bracted cultivars of *Cornus florida* resistant to powdery mildew. *HortScience* 38: 1253–1255.

15 Ascomycota
Pyrenomycetes, Discomycetes, and Loculoascomycetes

Kenneth J. Curry and Richard E. Baird

CHAPTER 15 CONCEPTS

- Recognition of ascomycetes is based on sexual reproduction culminating in a specialized sporangium, the ascus.

- Asci develop within a fruiting body or ascoma (pl. ascomata).

- An ascoma may be a perithecium, an apothecium, or an ascostroma, each of which corresponds with a major fungal evolutionary group.

- Some ascomycetes have asexual states that are classified as deuteromycetes.

The fungi considered in this chapter are commonly called ascomycetes. This is the most diverse group of fungi. They are characterized by a specialized unicellular sporangium called the **ascus** (pl. asci) in which sexually derived spores, called ascospores, are formed. The ascomycetes considered here are further characterized by the development of a protective structure called the **ascoma** (pl. ascomata) in which asci develop in a single layer or plane, the **hymenium**. The basic construction of the ascomycetes (and most other fungi) is filamentous, which gives them tremendous surface area for the absorption of nutrients from their substrate.

The ascomycetes may be divided into two major lifestyles. About 40% of the ascomycetes are lichenized (Hawksworth et al., 1995). These lichenized ascomycetes have an obligate symbiotic relationship with a photosynthetic symbiont (symbiotic partner) that can be a unicellular green alga or one of the cyanobacteria. The ascomycete member is never found under natural conditions without its photosynthetic partner. The photosynthetic partner, however, is sometimes found free-living in nature.

The remaining 60% of the ascomycetes are either saprophytes or are engaged in symbiotic relationships that range from **parasitism** through **commensalism** to **mutualism**. Most of these ascomycetes, in sharp contrast to lichenized ascomycetes, grow with their hyphae immersed in the substrate from which they obtain nutrients. Their reproductive structures are usually the only parts exposed. Hyphae grow through a substrate exploiting fresh nutrient sources. Older hyphae gather nutrients and transport materials to growing tips. The oldest hyphae die as the fungus grows into new areas. Thus, the fungus is constantly replacing its senescent parts.

The majority of the ascomycetes are saprophytes. Their carbon and their energy come from the organic remains of dead organisms. They grow in dead plant material where they degrade, among other things, cellulose and lignin (Chapter 30). Fungi excel at degrading these two tremendously abundant molecules. They grow in soil associated with organic remains primarily of plants.

A minority of the ascomycetes are parasites. Their carbon and their energy come from living organisms. A small number of these ascomycetes are animal parasites. A few grow in or on vertebrates including humans. Some produce structures to trap nematodes and rotifers. Most ascomycete parasites have plant hosts with which they exhibit a wide range of relationships. Some ascomycetes grow within plant tissues as endophytes. The relationship is parasitic in that the endophyte takes nutrients from its host, but no disease symptoms are manifested. A parasite that causes disease symptoms is a pathogen (Bos and Parlevliet, 1995). Some ascomycete pathogens are obligate **biotrophs**. They invade living plant cells, keeping the host cells alive while withdrawing nutrients. Some ascomycete pathogens are **necrotrophs**, killing host cells to obtain nutrients. Between the extremes of biotroph and necrotroph is the **hemibiotroph** (Luttrell, 1974). Hemibiotrophs invade and maintain living host cells for a period of hours or days, and then revert to necrotrophy. Hemibiotrophs and necrotrophs can also exist in nature as saprophytes. Their ability to function as parasites allows them to create their own saprophytic substrates.

REPRODUCTION

We recognize the following three reproductive categories for the ascomycetes: (1) ascomycetes that only reproduce sexually, (2) ascomycetes known to reproduce sexually, but also observed in asexual reproduction, and (3) fungi (deuteromycetes) thought to be ascomycetes, but only observed reproducing asexually.

Sexual reproductive structures all function with one end—to bring together two compatible nuclei, where compatible means the nuclei can fuse. Typically for ascomycetes this results in two haploid nuclei fusing to form a diploid nucleus. The act of nuclear fusion, called karyogamy, usually does not follow immediately upon bringing the two nuclei into a single cell. Since bringing the nuclei together typically involves bringing the cytoplasm of two cells together, this act is called plasmogamy (literally the marriage of cytoplasm). The nuclear pair resulting from plasmogamy is called a dikaryon. The two haploid nuclei, though physically separate, function in some ways as if they were diploid. Typically the nuclear pair undergo a synchronized division with the resulting new nuclear pair migrating into a growing hyphal tip to be delimited from the first pair by a septum. The successive growth of hyphae and synchronized division of the dikaryotic nuclei can continue for some time. We refer to the hyphae thus formed as dikaryotic hyphae and contrast this with monokaryotic hyphae. Ultimately, particular dikaryotic cells differentiate into asci, karyogamy or nuclear fusion occurs, meiosis occurs, and the sexual process culminates in the production of ascospores. Thus, sexual reproduction is defined as comprising two steps, nuclear fusion and meiosis. Other events might be closely associated with the sexual process, such as the occurrence of gametangia to effect nuclear transfer, but they are not necessary to the definition of sexual reproduction.

The transfer of nuclei in many ascomycetes can only occur between two compatible organisms. Ascomycetes that require a partner for sexual reproduction are referred to as heterothallic. Both members of the pair may produce both male and female gametangia, so gender is not ascribed to one organism, but rather ascomycetes, with a few exceptions, are hermaphroditic. A number of ascomycetes do not require a partner to complete sexual reproduction. Formation of ascogonia and antheridia, dikaryotic hyphae, nuclear fusion, and meiosis occur within a single organism. We refer to these ascomycetes as homothallic and note that, although they have met our definition for sexual reproduction, they will not participate at the same level of genetic variation through gene exchange as heterothallic ascomycetes.

The principal propagule of dissemination of ascomycetes is a spore. Asexual reproduction, in contrast with sexual reproduction, involves the production of propagules of dissemination via mitosis. Nuclear fusion and meiosis are both notably absent. The most common asexual propagule of dissemination for ascomycetes is a spore called a **conidium**. Think of a spore as any propagule, whether derived sexually through meiosis or asexually through mitosis, which from its inception is designed for dissemination. A spore may comprise one cell or two or more cells. Typically the number of cells is regulated genetically and precisely.

NOMENCLATURAL CONVENTIONS

Identifying ascomycetes is sometimes complicated when we find only the asexual state. The connection between sexual and asexual states within a single species is often unknown. We have developed three pragmatic terms that allow us to discuss combinations of sexual and asexual states (Reynolds and Taylor, 1993). When we refer to a fungus that we are discussing as the **holomorph**, we mean to include reference to both the sexual and asexual states of the fungus, even if one or both states are not present at the time of observation (Figure 15.1). When we refer to the **teleomorph**, we mean to indicate only a sexual state of the fungus even if the asexual state is present. When we refer to the **anamorph**, we mean to indicate only the asexual aspect of the fungus even if the sexual state is present. The anamorph of ascomycetes is usually given a separate binomial name from the teleomorph out of pragmatic convenience. We frequently see the asexual state without the sexual state. The practice of assigning two names to a holomorphic fungal species is sanctioned in the International Code of Botanical Nomenclature and by long tradition. The anamorphic name is assigned to the deuteromycetes or fungi imperfecti (described in Chapter 16). The deuteromycetes are considered to be a form-group with members referred to form-species, form-genera, etc. The designation of form is applied to indicate a less formal taxonomic status to the asexual state compared to the sexual state, while recognizing the ubiquity of the asexual state in nature. Our formal taxonomy is tied to the sexual state, to the teleomorph, indicating our belief that sexual reproduction is somehow more important in our evolutionary model then asexual reproduction. Thus, for example, we apply the binomial name *Glomerella cingulatum* to teleomorphic fungi meeting certain morphological criteria. If members of this fungus are seen expressing only the asexual state, we acknowledge this to be the anamorph and apply the name *Colletotrichum gloeosporioides*. The anamorph and teleomorph expression of the fungus are not commonly seen together, but if we wish to discuss the fungus in both its sexual and asexual states—that is, to discuss the holomorph—we use the teleomorph name because of its formal taxonomic priority over the anamorph. Thus, the holomorph and teleomorph share one name—in this example, *Glomerella cingulatum*—and the anamorph of

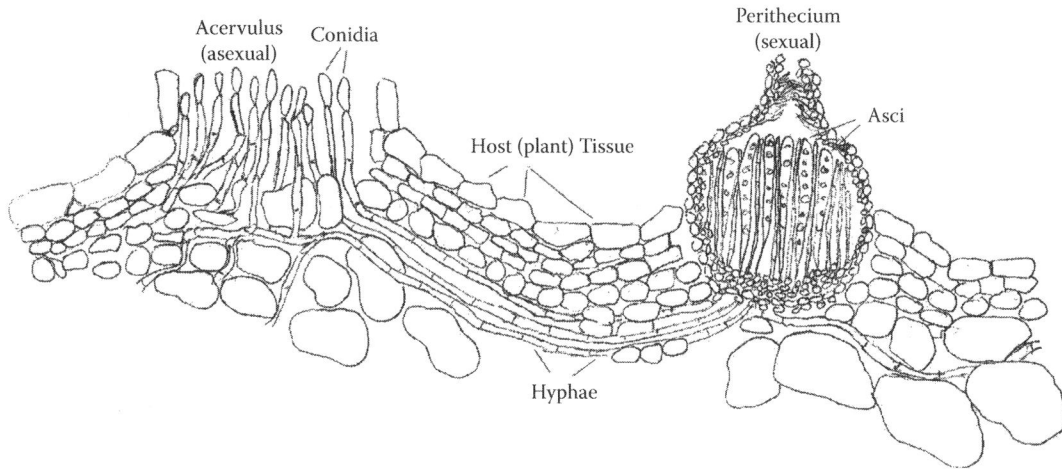

FIGURE 15.1 Nomenclatural conventions. The plant pathogen, *Glomerella* sp., is shown here developing in host plant tissue. The anamorph or asexual aspect of the fungus *Cercosporella* sp. is shown on the left developing as an acervulus. The teleomorph or sexual aspect of the fungus *Glomerella* sp. is shown on the right. The holomorph name is also *Glomerella* sp.

the same fungus has a separate name, in this example *Colletotrichum gloeosporioides* (Sutton, 1992).

GENERALIZED ASCOMYCETE LIFE CYCLE

Considerable value will be gained in understanding a generalized ascomycete life cycle, which may serve as a starting point for understanding specific details of development of individual organisms among the species of ascomycetes. For this generalized life cycle we shall assume the presence of male and female gametangia, the development of dikaryotic hyphae, the formation of asci in a single layer or plane called the hymenium, and the formation of ascospores (Figure 15.2). A modern account is found in Alexopoulos et al. (1996).

We shall begin the life cycle with the differentiation of hyphae to form an **ascogonium** (female gametangium) with a trichogyne and to form an **antheridium** (male

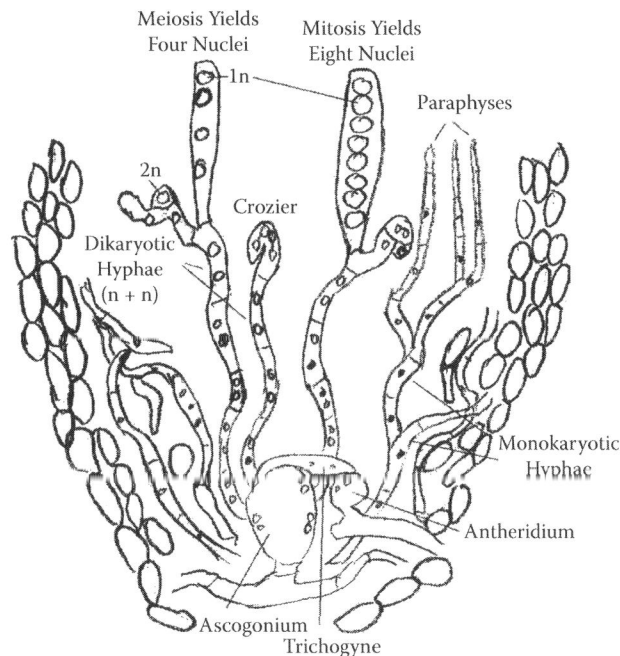

FIGURE 15.2 A diagrammatic representation of an apothecium is shown with the gametangia (ascogonium ♀ and antheridium ♂) producing dikaryotic (n + n) hyphae from which croziers develop. The dikaryon fuses to form a diploid nucleus in a cell called the ascus mother cell. The ascus mother cell develops into the mature ascus while the diploid nucleus undergoes meiosis to yield four nuclei, each of which typically divide by mitosis to give eight nuclei in the ascus. A spore wall develops around each nucleus, resulting in eight ascospores.

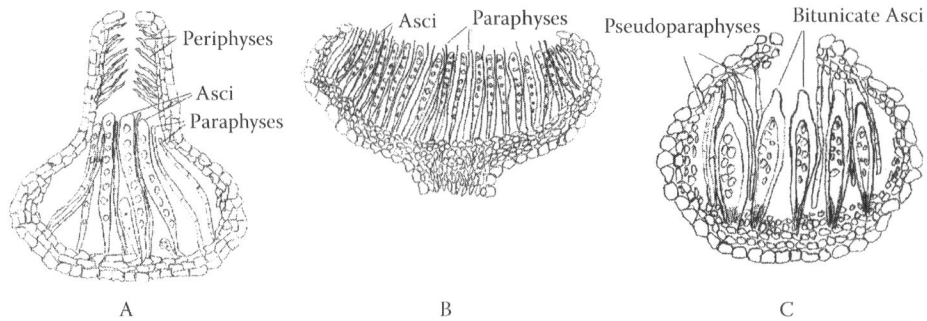

FIGURE 15.3 Diagrammatic representations of ascomata. (A) A perithecium with periphyses and paraphyses and unitunicate asci; (B) an apothecium with paraphyses and unitunicate; and (C) a pseudothecium with pseudoparaphyses and bitunicate asci.

gametangium). Nuclei from the antheridium migrate through the trichogyne into the ascogonium where they are thought to pair with ascogonial nuclei. Hyphae develop from the walls of the ascogonium, and the paired nuclei, dikaryons, migrate into these dikaryotic hyphae. The dikaryotic hyphae grow along with monokaryotic hyphae to form a protective fruiting body or ascoma in which asci will develop. The dikaryons undergo synchronized divisions, distributing themselves in pairs through the growing dikaryotic hyphae. The dikaryotic hyphae form a layer called the hymenium within the developing ascoma. The tip of each dikaryotic hypha destined to become an ascus bends over 180° to form a structure called a crozier (resembles a crochet hook). The nuclear pair (dikaryon) in the tip of the crozier undergoes nuclear fusion to form a diploid nucleus, whereas the crozier grows into the structure we call an ascus.

The diploid nucleus undergoes meiosis to yield four haploid nuclei. These four nuclei typically undergo a synchronized mitotic division to yield eight haploid nuclei. Double membranes are elaborated within the ascus and surround each of the eight nuclei. A spore wall is laid down between these two membranes, and thus, eight ascospores are formed. Ascospores can be disseminated passively by disintegration of the ascus, or more commonly, disseminated actively by forcible discharge from the ascus. In the latter case, the ascus develops internal turgor pressure and forcibly ejects the ascospores one at a time through a preformed opening at the tip of the ascus. Ascospores may be disseminated by wind, water, or vectors, usually insects. Spores landing on suitable substrates germinate and grow into the substrate to exploit nutrient sources. Environmental conditions that stimulate reproduction are not well understood, but exhaustion of the nutrient source is commonly associated with reproduction.

ASCOMATA

The taxa addressed in this chapter produce asci in structures that, in the most general terminology, are fruiting bodies or ascomata (sing. ascoma) (Figure 15.3). These can

be further distinguished developmentally as ascocarps, that include perithecia and apothecia, and ascostromata (sing. ascostroma), that include pseudothecia (Alexopoulos et al., 1996). The distinction between ascocarp and ascostroma hinges on developmental studies that indicate the ascocarp as developing from sexual stimulation of the activity of an ascogonium and antheridium. In contrast, the ascostroma develops as a stroma first with gametangia and asci developing later in locules dissolved in the stromatic tissue.

Two major architectural types of ascocarp are represented in this chapter. The **perithecium** is typically flask-shaped enclosing its asci in a single locule with a single pore or ostiole through which ascospores are discharged at maturity. A few of the species represented here produce ascocarps without ostioles. These ascocarps might be called cleistothecia (introduced in Chapter 14), but differ from the typical cleistothecium in that their asci are organized in a hymenium within the ascocarp instead of the scattered arrangement that characterizes cleistothecia. The morphological situation may be complicated if the perithecium develops embedded in a stroma. The stroma develops first, just as an ascostroma, but the sexual structures stimulate the development of an ascocarp wall around the developing asci so the final structure is a perithecium encompassing asci and surrounded by stroma.

The basic plan for the **apothecium** is an open cup with or without a stipe (stalk) and with asci organized in a hymenium across the surface of the cup. Convolutions of the basic cup lead to a variety of morphologies and corresponding taxa.

The ascostroma may express a predictable or an irregular morphology. Asci may develop in multiple, adjacent locules formed in the stroma or they may develop in one locule. Some taxa produce a flask-shaped stroma resembling a perithecium and called a **pseudothecium**.

TAXONOMY

The ascomycetes addressed in this chapter are organized into three major groups characterized principally by type

---

CASE STUDY 15.1

TAR SPOT OF STRIPED MAPLE

- The tar-spot pathogen, *Rhytisma punctatum*, is an apothecial-forming ascomycete. Within a leaf lesion, numerous flat and circular tar-like stroma are seen containing sexual structures called apothecia (Figure 15.4). The apothecia or ridge-like swellings on the stromata break open exposing asci containing needle-shaped ascospores.
- The apothecia embedded in the stroma are superficially similar to hysterothecia, specialized ascocarps of Loculoascomycetes.
- In the areas where the stromata are observed on the leaves, chlorophyll continues to be produced by the plant making the lesions appear as dark green islands of leaf tissue. These green islands containing stromata become even more evident during senescence.
- Following senescence, the infected leaves fall onto the ground, and *R. punctatum* over-winters in the litter. In the spring, as the new leaves open, the stromata-containing apothecia on the fallen leaves split open, and needle-like ascospores are released and windblown to infect the new foliage.
- For maples planted near homes, sanitation or removal of the leaves is the best method for control if no other infections are within the neighborhood. However, some people do not consider the stromata of the pathogen unsightly, and the disease is rarely known to seriously affect the health of the plants.

FIGURE 15.4 Tar spot of striped maple caused by *Rhytisma punctatum*, an apothecial-forming ascomycete. Insert is an enlargement of a "tar spot." Notice the numerous apothecia (dots). Photographs courtesy of Alan Windham, University of Tennessee.

of ascoma: the pyrenomycetes, the discomycetes, and the loculoascomycetes. These groups have had both formal taxonomic names with positions in the Linnaean hierarchy and trivial names of convenience. Trivial names will be used here in deference to the state of flux in higher ascomycete taxonomy. Taxonomy within the three major groups is organized around major centrum types (i.e., the ascoma and its internal organization). Various types of molecular information are currently being used to support, refine, and refute morphological and developmental morphological data on which past taxonomic schemes have been constructed (Taylor et al., 1994; Samuels and Seifert, 1995). The morphological groups described here are generally supported by molecular data as being monophyletic, i.e., each group represents a single phylogenetic lineage and all of its descendants.

PYRENOMYCETES

Fungi producing ovoid to cylindrical asci that develop in a hymenium within a perithecium are included in this group. Three plant pathogens representing this group are described here.

Glomerella spp. are responsible for anthracnose diseases, commonly of annual plants, and fruit rots including bitter rot of apple and ripe rot of grape. The asexual state

---

CASE STUDY 15.2

CHESTNUT BLIGHT CAUSED BY *CRYPHONECTRIA PARASITICUS*

- American chestnut trees comprised nearly 25% of the forest canopy in the Appalachian Mountains in the early 1900s. Following introduction of the pathogen from Asia on exotic chestnut wood, the trees were virtually eliminated throughout their native range within 50 years.
- Fortunately, the fungus cannot kill the roots of the trees; they continue to produce shoots and some stems develop viable nuts before being killed.
- The chestnut blight fungus is a perithecial forming ascomycete. The perithecia or flask-shaped structures containing ascospores form long necks that protrude from clusters of fungal tissue called stromata.
- Ascospores of *C. parasitica* are disseminated by wind or rain but will only infect trees following injury or wounding. Once established the fungus encircles the boles growing throughout the bark and killing the vascular tissue. The infected areas or **cankers** are observed as sunken areas of bark containing orange stroma-perithecia. In age, the bark begins to peel away exposing the inner tissues. Once the cankers can be seen around the circumference of the boles, the tree dies above the area of the damage.
- Without living host tissues, the pathogen can survive either saprophytically or as a facultative parasite on other trees such as oaks. When new American chestnut sprouts form and host tissue is again available, the infection process begins.
- Potential control methods are available but were not effective or economical in the United States. Resistant hybrids have been developed which contain 99% American chestnut tree genes and are starting to be planted throughout the eastern mountains.

(anamorph), *Colletotrichum* sp., is commonly encountered in these diseases. A conidium landing on a host surface produces a short germ tube terminating in a specialized swelling, the appressorium. A penetration peg from the appressorium grows through the host cuticle, thus allowing the fungus to gain entry into the plant. The fungus grows usually as a necrotroph, destroying host tissue. Eventually the fungus reemerges at the host surface, developing a tissue called stroma just below the host cuticle. Expansion of the stroma and the developing conidiophores and conidia rupture the host cuticle, allowing conidia to disperse. This expanded stroma rupturing the host cuticle is called an acervulus. The conidia can reinfect other hosts. The fungus overwinters at the end of the temperate growing season in various ways, not all known to us. Taking bitter rot of apple as an example, the infected fruit dries and hardens, a process called mummification. The mummy may remain on the tree or fall to the ground. Perithecia form on the surface of the mummy at the beginning of the following spring where they discharge their ascospores to initiate a new season of infection. Controls for *Glomerella* spp. are generally through sanitation or removal of mummified fruits and dead or diseased tissues that were invaded by the pathogen. Fungicides are also effective, but require full season applications when environmental conditions are suitable for fungal growth and sporulation.

Claviceps purpurea causes ergot of rye and wheat. It is occasionally found on oats, barley, and corn and occurs on some wild grasses. *Claviceps purpurea* overwinters as compact hyphal masses called sclerotia. Stipitate stromata develop in the spring from these sclerotia. Perithecia form within the enlarged heads of the stromata. The ascus produces eight thread-like, multicellular ascospores that are forcibly discharged. Spores that happen to reach host flowers germinate with hyphae invading and destroying the ovary. Conidia are produced in a sticky matrix from the mycelium that develops in the ovary. The conidia are dispersed to other host flowers by insects attracted to a sticky exudate or honey dew. Honey dew production gradually diminishes toward the end of the growing season, and the mycelium develops into a hard mass, the sclerotium. This sclerotium may fall to the ground to overwinter or be harvested with healthy seed to be sown the following year. The sclerotium is the source of ergot, a group of alkaloids including secoergolenes, ergolines, and lysergic acid derivatives that protect it from herbivores. Control practices include use of sclerotium-free seed, rotation with resistant crops, and deep tillage practices for burial of sclerotia.

DISCOMYCETES

This group includes fungi that produce asci in a hymenium within an apothecium. Two major phylogenetic groups of the discomycetes are recognized based on asci that are operculate or inoperculate. Operculate asci have an apical apparatus that is a hinged lid opening under turgor pressure during spore discharge. Inoperculate asci have a variety of pore types that direct ascospores through the ascus tip during forcible, turgor-driven spore discharge. The basic cupulate architecture of the ascocarp is expressed in a broad variety of modifications including a closed cup that develops underground (truffles). One representative plant pathogen is described here.

Species of *Monilinia* and *Sclerotinia* are responsible for brown rot of stone fruits (e.g., peach). We shall consider *M. fruticola* and its anamorph *Monilia* sp. on peach. Young peaches are generally resistant to brown rot, but as they approach maturity, they become increasingly susceptible. Fungal invasion occurs through hair sockets, insect punctures, and other wounds. The mycelium spreads through the host tissue rapidly using pectinases and other enzymes to destroy cell walls. Mycelium reemerges on the surface of the fruit to produce conidia that are spread variously by wind, water, and insects to other fruit. Some fruit falls to the ground where it quickly disintegrates under the action of saprophytic fungi. Heavily infected fruit remaining on the tree loses water content and becomes a shriveled mass of plant tissue and mycelium called a mummy. The mummy may persist on the tree through the winter or it may fall to the ground and persist in the soil for up to three years. In either case the fungus overwinters in the mummy. The following spring, mummies on trees express the anamorph and begin a new infection cycle with conidial production. Mummies on the ground express the teleomorph and begin the infection cycle by producing stipitate brown apothecia with inoperculate asci. The ascospores are forcibly discharged and wind dispersed. Those ascospores landing on susceptible host tissue are thought to infect in a manner similar to conidia. Control of the brown rot pathogen is primarily by the use of select fungicides on a time schedule. Cultural practices such as sanitation of mummies and pruning and removal of infected or dead material will reduce disease levels. Also, insect control can significantly reduce disease incidence.

LOCULOASCOMYCETES

Fungi producing asci in an ascostroma are included in this group. The asci are constructed with two functionally distinct walls. The inner wall is thick and expands rapidly during ascospore discharge, rupturing the thin, outer wall. This is called a bitunicate ascus in contrast with the unitunicate ascus characteristic of the pyronomycetes and discomycetes. One representative plant pathogen is described here.

Venturia spp. includes several species of plant pathogens of which we shall consider *V. inaequalis* that causes apple scab and attacks some related plants. Both leaves and fruit can be infected. Infection begins as a velvety, olive-green lesion that gets darker and dries as it matures. *Venturia inaequalis* overwinters in leaves on the ground. A stroma, which in this species is a pseudothecium, develops slowly within the leaf tissue. An ascogonium develops within the pseudothecium and projects its trichogyne into the surrounding plant tissue. The fungus requires an opposite mating strain to proceed (i.e., it is heterothallic). If that strain exists in the leaf, an antheridium will develop near the trichogyne and transfer nuclei to it. Dikaryotic hyphae will develop from the ascogonium and form asci through the crozier mechanism described earlier in this chapter. This very slow development culminates in spring with asci forcibly discharging ascospores each comprising two cells of unequal size, hence the name *inaequalis*. Spores are dispersed by air currents. Those landing on suitable host surfaces produce short germ tubes with appressoria by which they penetrate the host cuticle and invade the tissue. Hyphae reemerge at the surface to form acervuli and conidia that continue spreading infection throughout the growing season. This anamorphic state is *Spilocaea pomi*. Conidia infect via appressoria and penetration of host cuticle just as with ascospores. Late in the season, hyphae grow deep into leaf tissue to start the overwintering phase of the life cycle. Control of the apple scab pathogen is primarily with fungicides on a timed schedule. Removal of infected or dead plant tissues levels and use of resistant varieties are other recommended practices.

CONCLUSION

The ascomycetes described in this chapter include a significant number of plant pathogens (although this is still a small percent of ascomycete diversity) and some of the most agriculturally devastating pathogens are represented here. These pathogens are not confined to a few families within the larger group, but are scattered throughout many families of the group. They represent the widest range of relationships from the necrotrophs that quickly kill their host cells for nutrients, through the hemibiotrophs to the genetically intimate biotrophs, carefully balanced to keep their hosts alive while obtaining nutrients from them. Our understanding of the taxonomy of the group has been confounded by the plethora of life cycles and reproductive strategies and the extraordinary adaptability of fungi. We know so much about these fungi and yet so little, that students of this group can expect a lifetime of challenges, frustrations, and satisfactions. The key to controlling these fungi with respect to agricultural lies in understanding much more about them.

LITERATURE CITED

Alexopoulos, C.J., C.W. Mims, and M. Blackwell. 1996. *Introductory Mycology*, 4th ed. John Wiley & Sons, New York.

Bos, L., and J.E. Parlevliet. 1995. Concepts and terminology on plant/pest relationships: toward a consensus in plant pathology and crop protection. *Annual Review of Phytopathology* 33: 69–102.

Hawksworth, D.L., P.M. Kirk, B.C. Sutton, and D.N. Pegler. 1995. *Ainsworth & Bisby's Dictionary of the Fungi*, 8th ed., CAB International, University Press, Cambridge, U.K.

Luttrell, E.S. 1974. The parasitism of vascular plants. *Mycologia* 66: 1–15.

Reynolds, D.R., and J.W. Taylor (Eds.). 1993. *The Fungal Holomorph: Mitotic, Meiotic and Pleomorphic Speciation in Fungal Systematics*. CAB International, Wallingford, U.K.

Samuels, G.J., and K.A. Seifert. 1995. The impact of molecular characters on systematics of filamentous ascomycetes. *Annual Review of Phytopathology* 33: 37–67.

Sutton, B.C. 1992. The genus *Glomerella* and its anamorph *Collectotrichum*. In J.A. Bailey and M.J. Jeger (Eds.). *Colletotrichum: Biology, Pathology and Control*. CAB International, Redwood Press, Melksham, U.K., pp. 1–26.

Taylor, J.W., E. Swann, and M.L. Berbee. 1994. Molecular evolution of ascomycete fungi: phylogeny and conflict. In D.L. Hawksworth (Ed.). *Ascomycete Systematics: Problems and Perspectives in the Nineties*. Plenum Press, New York, pp. 201–211.

16 Deuteromycota
An Artificial Assemblage of Asexually Reproducing Fungi

Richard E. Baird

CHAPTER 16 CONCEPTS

- Taxonomy of the Deuteromycota is based on asexual spore formation, or no spores produced.

- Sexual stages of these fungi are mostly in the Ascomycetes, but a few are in the Basidiomycetes.

- Species can be parasitic or saprophytic.

- Asexual spores, called conidia, are nonmotile and are produced on conidiophores.

- Conidia are formed on conidiophores either singly or grouped in sporodochia, pycnidia, acervuli, or synemmata.

Species of the Deuteromycota (deuteromycetes), also known as the "imperfect fungi" or **mitosporic**, are among the most economically destructive group of fungi, causing leaf, stem, root, fruit, and seed rots, and blights and other diseases. The Southern corn leaf blight epidemic in the 1970s, which resulted in a loss of one billion dollars, was incited by *Helminthosporium maydis*, the anamorph or the asexual form of the ascomycete, *Cochliobolus heterostrophus*. Other deuteromycetes, such as *Aspergillus flavus*, produce mycotoxins (aflatoxins) in infected corn kernels. Mycotoxins, when ingested by humans or animals, can cause cancer of the digestive tract, other serious illnesses, or death.

The deuteromycetes were called imperfect fungi in the early literature because they were thought not to produce sexual spores like species of the Ascomycota (Chapters 13 and 15) and Basidiomycota (Chapter 19). Descriptions and classifications of these fungi were based solely on production of **conidia** and/or on mycelial characteristics. Deuteromycetes are now known to be the anamorphic stage of members of the Ascomycota and Basidiomycota. For example, *Fusarium graminearum* is the imperfect (asexual) stage of *Gibberella zeae*. The ubiquitous pathogen *Rhizoctonia solani*, which does not produce asexual spores, is the anamorph of the basidiomycete *Thanatephorus cucumeris*.

DEUTEROMYCOTA OR FUNGI IMPERFECTI

A brief history of this group may be helpful in understanding why a fungus may be known by two scientific names. During the 1800s, fungal identification was based strictly on morphological characters. The object of these studies was to identify pathogenic fungi when very little was known about their anamorph-teleomorph (sexual spore) relationship. The asexual fruiting bodies were often the only structures present on infected host tissues and scientists were unaware that sexual reproductive states existed. Early mycologists, such as Persoon (1801), Link (1809), and Fries (1821), described the genera and species of imperfect fungi, which seemingly lacked a sexual stage; they were later classified as *Fungi imperfecti*. These studies initiated the description of many deuteromycetes and other fungi, and are considered the starting point for fungal classification. Saccardo (1899) compiled descriptions of the known fungi into one unified source in his *Sylloge Fungorum* series. Taxonomic keys and descriptions to the genera and species of the deuteromycota using spore shape, size, presence of cross-walls (septa) in the hyphae, and fruiting body type were provided by numerous scientists over the next half a century.

Conidiospore development (ontogeny) was used as a basis for the identification of genera and species after the 1950s. Asexual spore development on conidiophores and within fruiting bodies was considered a more natural basis for classification of these fungi. Though this is considered a more natural classification scheme than previous systems based on spore and conidiophore morphology, the earlier systems continue to be used by plant pathologists and disease diagnosticians because of the ease in identifying the genera and species. Since these early works, hundreds of

CASE STUDY 16.1

GREY LEAF SPOT OF MAIZE

- The disease caused by the fungus *Cercospora zeae-maydis* is one of the most destructive foliar diseases of maize. Conidiospores of the pathogen are formed in acervuli and are disseminated primarily by wind and rain, and can infect new host tissues.
- The lesions are tan, then become brown and are rectangular. As the fungus grows and expands, the lesions can become 3 to 4 in. long and 1/16 to 1/8 in. wide.
- Under favorable weather conditions, conidia are produced, and the brown lesions appear silvery-gray. The spores are released and can cause secondary infections throughout the plant's foliage. Yield reduction is dependent upon when during a growing season the initial infection occurs.
- The pathogen overwinters in plant debris and can infect the host in subsequent years if the debris is not managed.
- Crop rotation and clean plowing reduce severity of the disease. The use of resistant cultivars is the most effective and economical way to reduce crop losses. These cultivars show symptoms in the latter part of the season, but losses are greatly reduced. Other controls such as fungicides can be used but are often not economical.

papers and monographs have been developed that include additional genera and species with information on their associated sexual stages.

Many of the imperfect fungi do not readily form a sexual stage in culture or on host tissue. Therefore, artificial systems for identification were developed and are currently being used. Recognizing that the majority of deuteromycetous fungi with a teleomorphic stage also have a second name for the sexual stage may be critical in understanding how to identify these fungi.

LIFE HISTORY

The deuteromycetes are primarily terrestrial in distribution, but can occur in salt- or freshwater. They survive by deriving nutrients as saprophytes on plant debris or as parasites on living hosts. The degree of parasitism or pathogenicity varies depending upon the fungus, the isolate, and host they invade. Many species of deuteromycetes are not only parasitic on plants, but also infect animal cells.

Deuteromycetes may cause damage directly by infecting a host or indirectly by producing toxins. Direct infections by deuteromycetes to living hosts other than plants are termed mycoses, and approximately 15 types are known. The more common mycoses include candidiosis (*Candida albicans*) and superficial infections called dermatophytosis, which include athlete's foot or dandruff caused by several *Tinea* species. A common species of deuteromycete associated with animal, avian, and human disease is *Aspergillus,* which causes aspergillosis. Infections may induce lesions on the surface of the skin or may damage internal organs such as the lungs and liver. Infection frequently occurs when hosts are under stress, and immunosuppressive situations result from poor health (e.g., AIDS).

Indirect damage from some of these fungi results from ingestion of infected food (feed) or through inhalation of particles containing mycotoxins that are either carcinogenic or cause other health problems. Mycotoxins are produced by fungi during the growth of a crop or during transportation, processing, or storage. For example, aflatoxin levels in corn increase during drought and when high levels of nitrogen fertilizers are applied. Different types of mycotoxins are produced by species of *Aspergillus,* *Fusarium,* and *Penicillium.* One of the most important groups of toxins are the aflatoxins produced by *Aspergillus* species and occur primarily on crops such as corn, cotton seed, and peanuts.

Another group of mycotoxins are fumonisins that are produced by *F. verticillioides* (= *moniliforme*) and *F. proliferatum* (Gelderblom et al., 1988). Fumonisin is primarily associated with *F. verticillioides*, which can routinely be cultured or identified in corn tissues (Bacon and Nelson, 1994). Several forms of the toxin exist, but FB_1, FB_2, and FB_3 are the most common and important that are associated typically with food and feed (Gelderblom et al., 1988). If ingested, fumonisin can cause a neurological disorder in horses called leukoencephomalacia, pulmonary edema in pigs, and esophageal cancer in humans.

TAXONOMY OF THE DEUTEROMYCETES

The size, shape, and septation pattern of conidia are used as the primary characters for the practical and working identification of deuteromycetes genera and species. Conidia are defined as asexual, nonmobile spores that belong to the anamorphic stage of a fungus life cycle. The Saccardian system, which used spore type to identify the deuteromycetes, was the first major tool employed by mycologists to identify genera (Saccardo, 1899). The Saccardian sys-

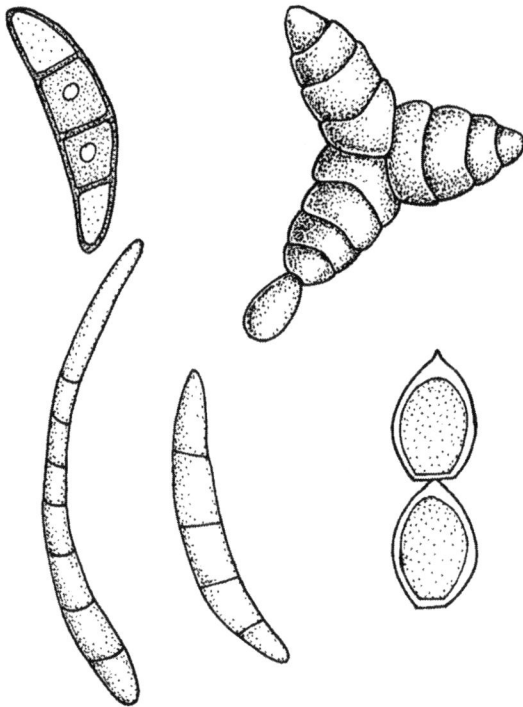

FIGURE 16.1 Conidia or asexual spores can be simple or complex and can be single or multicellular. (Drawing courtesy of Joe McGowen, Mississippi State University.)

tem was incorporated into other systems that included more natural classification schemes based on conidial ontogeny or development. This more advanced system is usually referred to as the Hughes–Tubaki–Barron system of classification. The work by Barnett and Hunter (1986) includes keys and descriptions for identification of genera. Finally, Hennebert and Sutton (1994) identified subtle differences in spore development on conidiophores for identifying genera and species.

MORPHOLOGICAL STRUCTURES

Conidia (Figure 16.1) are produced on specialized hyphae called **conidiophores**. Because there are a large number of deuteromycete fungi, much variation can occur in their reproductive structures. Conidia vary in shape and size and can be one- or two-celled or multicellular depending on the number of septa present. Septa within the spores vary from transverse (across) to longitudinally oblique. Shapes range from filiform (thread-like), ovoid (egg-shaped), and clavate (club-shaped) to cylindrical (cylinder-shaped), stellate (star-like), or branched. Conidia can be ornamented with appendages, and appear hyaline (colorless) to colored.

Conidiophores are specialized hyphae, branched or unbranched, bearing specialized conidiogenous cells at the points where conidia are produced. Conidiophores may occur singly (separate) or in organized groups or clusters.

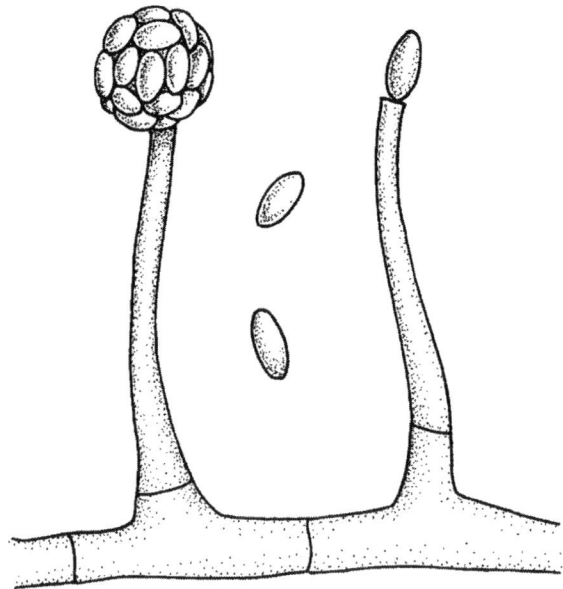

FIGURE 16.2 Conidia that are borne on loosely spaced conidiophores are called Hyphomycetes. (Drawing courtesy of Joe McGowen, Mississippi State University.)

If conidiophores are formed individually (Figures 16.2 and 16.3) and not enclosed in specialized structures, then the fungi that produce these forms are called Hyphomycetes. The Hyphomycetes are subdivided into two groups based on color of hyphae and spores. The dematiaceous group has dark hyphae and spores, whereas the moniliaceous species possess light or pale-colored hyphae and spores. Dematiaceous genera of deuteromycetes include plant pathogenic species such as *Alternaria, Aspergillus, Bipolaris*, and *Penicillium*.

An example of an economically important dematiaceous hyphomycete is *A. solani*, the causal organism of early blight of tomato. The disease caused by *A. solani* is considered by many to be the most economically damaging to tomatoes in the United States. The disease occurs every year because the pathogen overwinters in plant debris in soil as **chlamydospores** (thick-walled survival spores). As the temperature warms up in spring, the chlamydospores germinate within the plant debris or soil, and hyphae continue to grow saprophytically forming conidia on individual conidiophores. Conidia produced during the saprophytic stage are either disseminated by wind, rain, or insects, or are transported in soil on farm machinery. For long-distance dissemination, infected seed is the primary source of inoculum. Under wet and warm conditions, the conidia present on the tomato plant tissue germinate, and hyphae can either penetrate the host indirectly through stoma or directly through the cuticle. Infections usually occur first on the mature foliage. Lesions develop, and conidia are produced within the necrotic areas of the tomato foliage or stems. Secondary infections can occur from local dissemination of new conidia formed on the

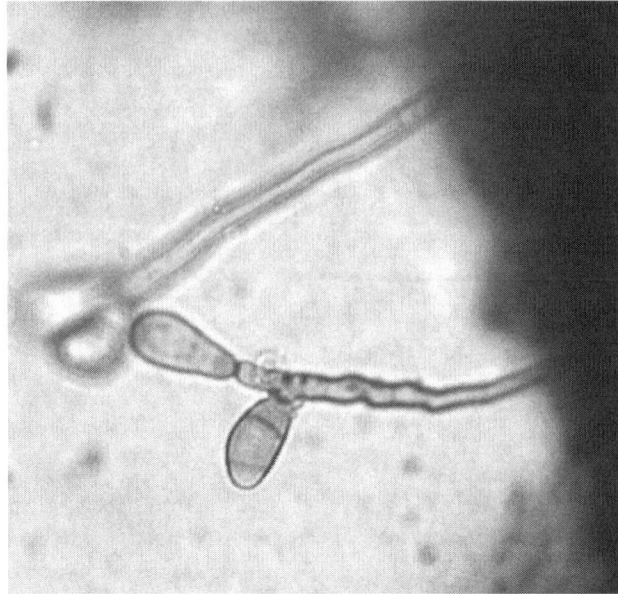

FIGURE 16.3 *Curvularia* sp. with two conidiospores attached to a conidiophore. (Photograph by A. Windham, University of Tennessee.)

host, and plants can become completely defoliated. The fungus can develop chlamydospores that remain dormant in the dead plant tissues and the cycle starts over the following season. Symptoms of early blight can be observed on all tomato parts above the ground. Following germination, pre- or post-emergence damping-off of plants can occur. Because lesions are generally first observed on mature leaves, the disease appears to progress from the lower portion, moving upward to the top of the plant. Infections increase, forming circular lesions of up to 4 to 5 mm in diameter, which become brown with

concentric rings, giving the necrotic area a target-shaped appearance. Leaves that are infected often are observed with yellowing areas. As multiple lesions occur from secondary infections, leaves turns brown and die. The entire plant can become defoliated and die at this stage. Controls for the disease include avoiding purchase of infected seed and soil for transplants that harbor the fungus. Other cultural control measures include crop rotation with other than solanaceous plants, such as potatoes, eggplant, and peppers; removal or burial of crop residue; and the use of disease-free plants, fungicides, and resistant cultivars.

CASE STUDY 16.2

CHARCOAL ROT DISEASE CAUSED BY *MACROPHOMINA PHASEOLINA*

- *Macrophomina phaseolina*, the cause of charcoal rot disease, is a soilborne pathogen that belongs to the Deuteromycetes. Isolates, which are gray to black in culture, contain hundreds of microsclerotia. Pycnidia are often present, but their occurrence varies per isolate of the fungus. Microsclerotia can be observed on living hosts along the lower stems and roots of susceptible plants. Microsclerotia are the primary propagule for survival.
- Charcoal rot pathogen occurs worldwide and infects over 500 plant species including many economic species in the United States.
- During hot and dry weather conditions, the pathogen is more destructive to crops such as soybean. The adverse environmental conditions cause plant stress, making the plant more susceptible to increased colonization. The pathogen can survive on the roots of crops causing little problem when growth conditions are optimum for the plant.
- Under poor growing conditions, *M. phaseolina* causes stalk rot of maize, and early maturation and incomplete pod yields of soybean.
- Control of the pathogen would include using nonhosts, reducing plant stress through increased fertility, avoiding high plant population to reduce competition for water, and maintaining good soil moisture conditions.

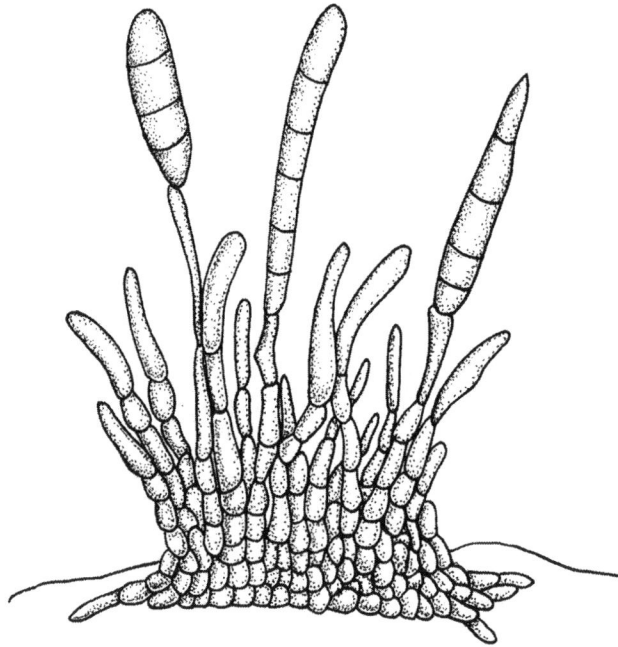

FIGURE 16.4 Sporodochium form on the surface of the host plant containing clusters or groups of conidiophores. (Drawing courtesy of Joe McGowen, Mississippi State University.)

DEUTEROMYCETE CONIDIOMATA

If conidiophores are grouped into organized clusters, then they are formed within specialized structures called conidiomata. The different types of conidiomata include **acervuli**, **pycnidia**, **sporodochia,** and **synnemata**. Modern references assign acervuli and pycnidia to the group Coelomycetes. Sporodochia forming species are considered under the dermatiaceous or moniliaceous subgroups of Hyphomycetes. Synnemata have been placed under the Hyphomycetes in the subgroup stilbaceous fungi (Alexopoulus et al., 1996).

Conidiophores and conidiomata on hosts or in culture are used for identification. The ability of the deuteromycetes to form these structures in culture versus on a host varies per genus and species. A structure that routinely is formed on a host may not often be observed when grown in a culture media. An example is setae (sterile hyphae or hairs) that are associated with acervuli of *Colletotrichum* species. Setae (hair-like appendages) routinely form in the acervuli produced on living hosts, but may be absent when grown on selective medium. Because most identification keys are based partially on the morphology of the pathogen on host tissue, a proper identification may require the direct observation of conidiomata development on plant tissue rather than on artificial media.

Sporodochia are similar to acervuli, except that the cluster or rosettes of conidiophores form on a layer or cushion hyphae on the host surface (Figure 16.4), whereas acervuli are imbedded in epidermal tissue or on the plant cuticle. Sporodochia also appear as mat-like cottony struc-

tures due to the clustering of conidiophores. However, in culture, sporodochia that resemble those on host materials are rarely observed, making species identification difficult. Examples of specific genera of fungi forming sporodochia include *Epicoccum, Fusarium,* and *Strumella.*

Fusarium oxysporum f. sp. *vasinfectum,* causal agent of Fusarium wilt of cotton is a good example of a sporodochia-forming fungus. The conidia of this pathogen overwinter in plant debris or can be introduced into fields by infected seed or in soil transported by farm equipment. The fungus forms chlamydospores in the soil or in plant debris. Under optimum weather conditions, conidia or chlamydospores germinate, and the fungus grows saprophytically, producing conidia on conidiophore rosettes. Germinating conidia or hyphae that come in contact with host root tissue can invade by direct penetration or indirectly through wounded areas. Root wounding is often increased by the root-knot nematode, *Meloidogyne incognita* (Chapter 8), and the occurrence of this pest has been associated directly with fields having increased levels of fusarium wilt. Following invasion into the root, the hyphae then grows inter- and intracellularly through the cortex and endodermis. The fungus penetrates the vascular system, and conidia are produced rapidly and distributed systemically into the transpiration stream of the cotton plants. The fungus physically obstructs the lumens of the vessel elements in the xylem tissue preventing water movement, which eventually results in wilting and death of the plants. Chlamydospores form in dead host tissues and overwinter in the plant residue. Control practices include using clean seed from uninfested fields,

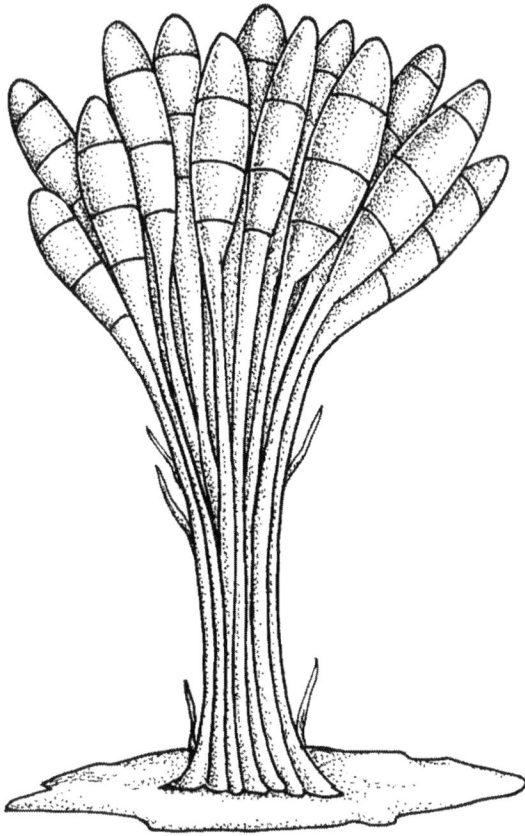

FIGURE 16.5 Synnemata consisted of fused conidiophores at the base forming conidia at the apex or on the sides of the structure. (Drawing courtesy of Joe McGowen, Mississippi State University.)

are usually elongate and are easy to identify from cultures because of their upright, whiskerlike appearance. Synnemata are generally formed in culture, unlike the sporodochial-forming species. Common genera that form synnemata are *Graphium, Arthrosporium, Isaria,* and *Harpographium.*

Graphium ulmi, the anamorph for the causal agent of Dutch elm disease (see Case Study 10.1), produces synnemata during the asexual or conidial stage of the life cycle. In spring, the pathogen, which overwinters in plant debris, grows saprophytically and forms mycelia where the synnemata are produced. The spores of the pathogen can then infect healthy trees by an insect vector, and if roots of healthy plants are grafted to an infected tree, the pathogen can be transmitted to the healthy tree. The conidia are primarily disseminated by the elm bark beetle (*Scolytidae*) carried on their body parts. As the insects feed on the tree tissues, they deposit conidia into the wound sites. The fungus germinates, becomes established, and grows into the xylem tissue. Blockage of the vascular (water conducting) system occurs from the fungus and by the defense responses of the host tree. Symptoms of Dutch elm disease include yellowing and wilting on just one or many branches early in the season, depending on when the infection occurs. Leaves of infected trees turn brown and die in portions of the tree, or the entire tree may be affected. If the tree survives the first year following invasion, death will occur sometime during the second year. If stems of the tree are sectioned, a brown discoloration is observed in the outer xylem of twigs, branches, and sometimes roots. Also, in dead and dying trees, insect larval galleries from the elm bark beetle can be observed under the bark of the tree trunk. Controls for Dutch elm disease include methods to eliminate the vector and pathogen by removing dying and dead wood (for example, sanitation) (Chapter 32). Fungicides injected into the tree will stop the spread of the pathogen, but treatments must be repeated continuously from one year to the next.

using resistant varieties, and reducing root-knot nematode levels through chemical control, rotation, or by selecting nematode-resistant varieties.

Synnemata conidiomata form conidiophores that are fused, and the conidia often form at or near the apex (Figures 16.5 and 16.6). Species that produce conidia in this fashion belong to the Stilbellaceae group of deuteromycetes. Conidiophores of the synnemata forming species

FIGURE 16.6 *Graphium ulmi* synnemata. (Photograph by A. Windham, University of Tennessee.)

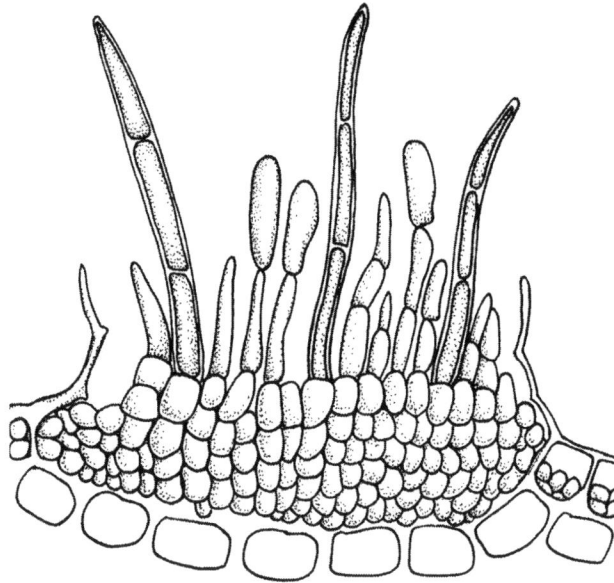

FIGURE 16.7 Acervulus embedded in host tissue containing clusters or groups of conidiophores. (Drawing courtesy of Joe McGowen, Mississippi State University.)

FIGURE 16.8 Acervulus with setae (large hairs) of *Colletotrichum* sp. (Photograph by A. Windham, University of Tennessee.)

COELOMYCETES CONIDIOMATA

Acervuli (sing. acervulus) contain a defined layer of conidiophores and conidia formed just below the epidermal or cuticle layer of plant tissues (Figures 16.7 and 16.8). Conidiophores and conidia erupt through the host epidermis or cuticle exposing the acervulus. The conidiophores within these structures are generally short and simple compared to the hyphomycetes, such as *Aspergillus* and *Penicillium*. Once exposed, acervuli usually assume a saucer shape (Barnett and Hunter, 1986). In cul-

ture, fungi that typically form acervuli on host tissue can often appear to produce sporodochia. The eruption of host cuticle or epidermis, which defines an acervulus, cannot be observed in culture.

An example of an acervulus-forming pathogen is *Colletotrichum lagenarium*, the causal agent of anthracnose of cucurbits. *Colletotrichum lagenarium* overwinters on plant debris and in seeds obtained from infected fruits. The fungus will grow saprophytically in the soil and produce conidia on dead tissue. The conidia are disseminated locally by rainfall and by soil transported on

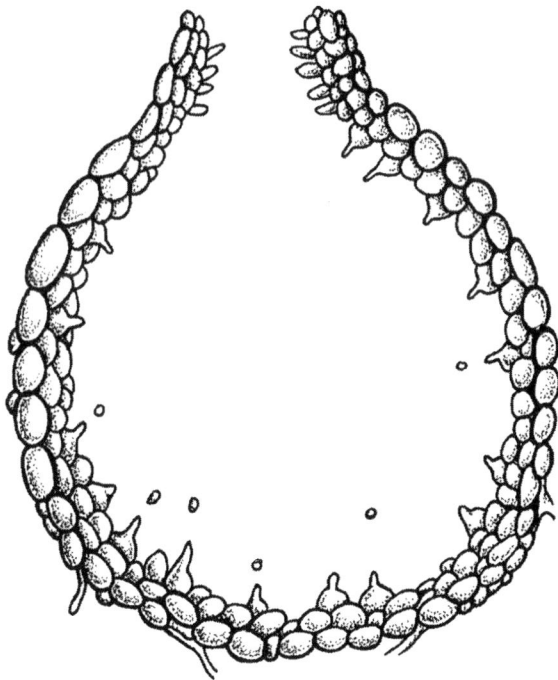

FIGURE 16.9 Pycnidium, a flask-shaped structure composed of fungal tissue, can be either imbedded in or superficial on host tissue. Large numbers of conidiophores and conidia formed within the structure. (Drawing courtesy of Joe McGowen, Mississippi State University.)

farm equipment and, over longer distances, by infested seed. When weather conditions are suitable for infection, conidia germinate on the host. The germ tubes (hyphae) form appressoria at the point of contact with the host cell, quickly followed by formation of infection pegs, which allow direct entry into the host. Hyphae then grow intracellularly, killing host cells. As the fungus continues to grow, angular light to dark brown or black lesions are formed between the leaf veins. The lesions are elongate, narrow, and water-soaked in appearance, and become sunken and yellowish-to-brown. When conditions are favorable, acervuli form on stromal tissue, and conidiophores containing conidia erupt through the cuticle of the host. Spores are exposed to the environment and disseminated. Girdling of the stems or petioles can occur, and defoliation results. Control practices include using disease-free seed, crop rotation with resistant varieties, cultural practices that remove or bury plant debris, fungicide sprays, and the use of resistant cucurbit varieties, when available.

Pycnidia (sing. pycnidium) differ from acervuli by the formation of the flask-shaped structures composed of fungal tissue that enclose the conidia and conidiophores (Figures 16.9 and 16.10). Pycnidial shapes described by Alexopoulus et al. (1996) include the following: papillate, beaked, setose, uniloculate, and labyrinthiform. Conidiophores that form within the pycnidium can be extremely short, as in *Phoma*, or longer, as in *Septoria* or *Macrophoma*. Pycnidia resemble perithecia, which are sexual reproductive structures of some species of the Ascomycota (Chapter 15). If observed microscopically, the spores of the Ascomycete are borne *in asci* and not on conidiophores. Another important consideration when identifying pycnidia-producing deuteromycetes is the presence or absence of an ostiole, which is located at the apex where conidia are exuded in a thick layer or cirrhus. Keys to the identification of deuteromycetes often refer to imbedded pycnidia within the host material, but in pure culture, pyc-

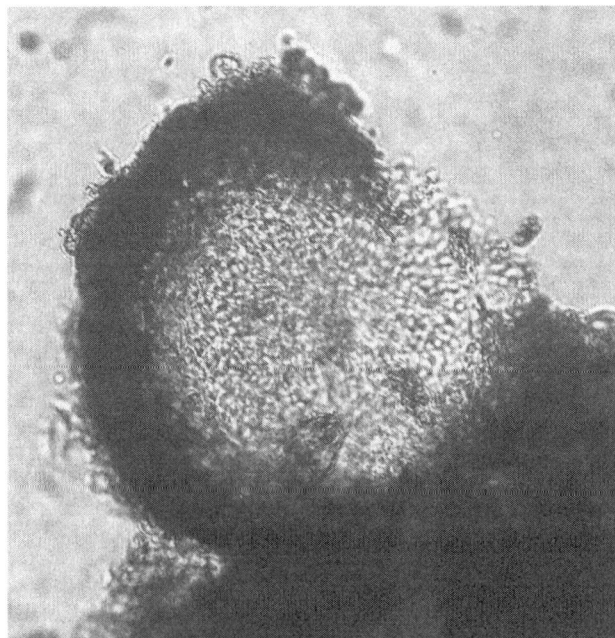

FIGURE 16.10 Cross section of pycnidium of *Phomopsis* sp. (Photograph by A. Windham, University of Tennessee.)

FIGURE 16.11 Sclerotia of *Sclerotium rolfsii.* (Photograph by A. Windham, University of Tennessee.)

nidia of the same fungal species will occur superficially on the media.

Septoria glycines, the causal agent of brown spot of soybean, produces conidia within pycnidia. The fungus can reinfect plants within the same field during subsequent years because the conidia or mycelium overwinter on debris of host stem and leaf tissues. During warm and moist weather, sporulation occurs as the fungus is growing saprophytically and the conidia are disseminated by wind or rain. The spores germinate, and hyphae invade by passing through stomata. The pathogen grows intercellularly, killing adjacent host cells. Lesions and the new conidia form, which can serve as a source for secondary infections on the same host plants. The pathogen primarily invades foliage causing a flecking appearance on mature leaves, but infections can also occur on stems and seed. If environmental conditions are optimum, second-

ary infections can occur, resulting in defoliation that generally moves from the lower to the upper leaves. Lesions are usually irregular, become dark brown, and can be up to 4 mm in diameter. During initial development of the pathogen, lesions often coalesce to form irregular-shaped spots. On young plants, leaves turn yellow and abscise, but late in the growing season, infected foliage can turn rusty brown before falling off. Control methods are limited to using resistant varieties, rotating to nonhost crops, and spraying with fungicides. The latter methods are generally not considered to be economically feasible.

Species in the **Mycelia Sterilia** are traditionally included in the Deuteromycota as a group that does not form asexual spores. Identification of these fungi is based upon hyphal characteristics, absence or presence of sclerotia (survival structures), and number of nuclei per hyphal cell (Figure 16.11). *Rhizoctonia solani* and *Sclerotium*

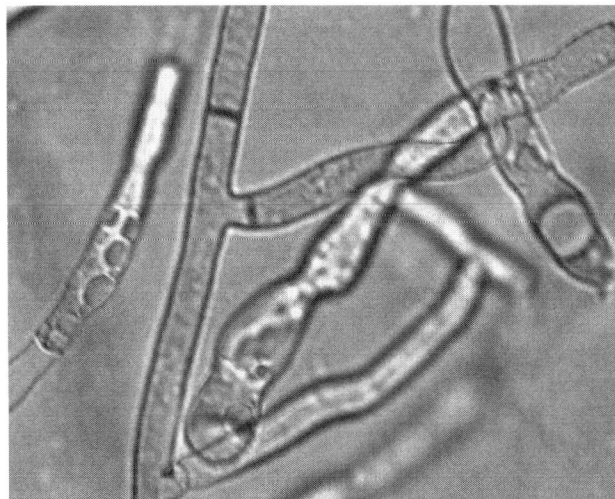

FIGURE 16.12 Right angle branching of hyphae typical of *Rhizoctonia solani.* (Photograph by A. Windham, University of Tennessee.)

rolfsii have teleomorphs that place them into the Basidiomycetes (Chapter 19). Both of these fungi are important plant pathogens that occur worldwide and attack agronomic vegetable and ornamental crops. *Rhizoctonia solani* has many morphological and pathogenic forms, called anastomosis groups (Figure 16.12).

Rhizoctonia species, which are responsible for brown patch of turfgrass, survives as sclerotia in plant debris in the soil. The sclerotia germinate over a wide range of temperatures, and the fungus grows saprophytically until a suitable host becomes available. Once the fungus becomes established in the host, circular lesions develop. The leaves and sheaths lose their integrity and appear water soaked. The damaged tissue at first has a purplish-green cast, then becomes various shades of brown depending on weather conditions and the type of grass. Often, dark purplish or grayish brown borders can be observed around the infected areas. Fungicide applications are effective in preventing or reducing brown-patch severity. The disease may also be controlled by the following cultural practices: avoiding excessive nitrogen applications that enhance fungal growth, increasing surface and subsurface drainage, removing any source of shade that reduces direct sunlight and increased drying of the leaf surface, and reducing thatch build-up when possible.

In summary, the deuteromycetes are a very diverse group based on the presence of conidiophores and conidia, except for the fungi that do not reproduce asexually. The reader should keep in mind that when mycologists first created this artificial group, the sexual reproductive stages were unknown. Although the teleomorphs have now been determined for many of the deuteromycetes, their sexual stages are often rare or almost never seen on a host or in culture. Because many deuteromycetes are plant pathogens, the group has been maintained for the purpose of identification, based on their asexual reproductive structures and cultural characteristics when identifying members of the Mycelia Sterilia.

REFERENCES

Alexopoulus, C.J., C.W. Mims, and M. Blackwell. 1996. *Introductory Mycology* (4th ed.). John Wiley, New York, p. 869.

Bacon, C.W. and P.E. Nelson. 1994. Fumonisin production in corn by toxigenic strains of *Fusarium moniliforme* and *Fusarium proliferatum. J. Food Protect.* 57: 514–521.

Barnett, H.L. and B.B. Hunter. 1986. *Illustrated Genera of Imperfect Fungi* (4th ed.). Burgess, Minneapolis, MN, p. 218.

Fries, E.M. 1821. *Systema Mycologicum 1.* Gryphiswaldiae, Lund, Sweden. p. 520.

Gelderblom, W.C.A., K. Jaskiewicz, W.F.O. Marasas, and P.G. Thiel. 1988. Novel mycotoxins with cancer-promoting activity produced by *Fusarium moniliforme. Appl. Environ. Microbiol.* 54: 1806–1811.

Hennebert, G.L. and B.C. Sutton. 1994. Unitary parameters in conidiogenesis. In *Ascomycetes Systematics: Problems and Perspectives in the Nineties.* Plenum, New York, pp. 65–76.

Link, J.H.F. 1809. Observationes in ordines plantarum naturales. *Mag. Ges. Naturf. Freunde*, Berlin, 3: 3–42.

Persoon, D.C.H. 1801. Synopsis Methodica Fungorum. H. Dieterich. Göttingen. p. 706.

Saccardo, P.A. 1899. Sylloge Fungorum Omnium Hueusque Cognitorum, Vol. 14. (Self published) Pavia, Italy.

17 Laboratory Exercises with Selected Asexually Reproducing Fungi

Richard E. Baird

The Deuteromycota (deuteromycetes) is a heterogeneous and artificial (does not reflect phylogeny) phylum erected on the production of asexual spores called conidia that are formed on conidiophores. The conidiophores develop either free on mycelium or are enclosed in structures called conidiomata (fruiting structures). Laboratory exercises for the deuteromycetes are designed to teach students how to recognize the different asexual reproductive structures and the role of environmental conditions or nutrition on the shapes and forms of the conidiomata. The Mycelia Sterilia include genera of fungi within the deuteromycetes that do not produce asexual spores at all. Many of the fungi within the Mycelia Sterilia are economically important pathogens (e.g., *Rhizoctonia* and *Sclerotium* species) and are incorporated into this chapter to teach students the important characteristics used in their identification.

EXERCISES

EXPERIMENT 1: IDENTIFICATION OF DEUTEROMYCETOUS FUNGI BY OBSERVING ASEXUAL REPRODUCTION STRUCTURES

When attempting to identify deuteromycetes, it is important to note the morphological features of the fungus under the dissection microscope. Fungi placed on microscope slides often fragment, or spores do not remain attached to conidiophores, making identification difficult. Morphological features to observe using the dissection microscope include the following: conidia borne singularly, in chains, clusters, or clumps; conidia borne in slime droplets; conidia borne at the apex or laterally on conidiophores; and the presence of pycnidia, sporodochia, synnemata, or acervuli (Chapter 16). The purpose of this experiment is to demonstrate variations in conidiomata that are used in identification of the deuteromycetes in host material and on artificial media. This exercise will train participants how to recognize the different asexual structures involved in spore production that are important in identifying the deuteromycetes. Table 17.1 suggests the use of fungi representative of each conidiomata type that can be compared during this exercise. Depending on availability, other fungi can be used.

Materials

Each student or team of students will require the following materials:

- Cultures of at least two species of fungi showing each of the four types of asexual development in Table 17.1
- Fresh materials with diseased tissue—Table 17.2
- Dissection and light microscopes
- Dissecting needles or probes
- Slides, coverslips, distilled water for mounting hyphae, and spores
- A stain such as aqueous aniline blue (0.05% or 5 mg/L)
- Eye dropper with bottle
- Cellophane tape
- Single-edge razor blade or scalpel with #11 blade
- Pencil eraser
- Alcohol lamp for surface-sterilizing the instruments

TABLE 17.1

Genera of Fungi Used to Demonstrate Deuteromycete Conidiophores and Conidiomata

Conidiophore Location				
Naked Hyphae	**Sporodochia**	**Pycnidia**	**Synnemata**	**Acervuli**
Alternaria	*Epicoccum*	*Diplodia*	*Briosia*	*Colletotrichum*
Bipolaris	*Fusarium*	*Macrophoma*	*Graphium*	*Discula*
Cladosporium	*Tubercularia*	*Phomopsis*	*Harpographium*	*Pestalotia*
Penicillium	*Volutella*	*Septoria*	*Trichurus*	*Melanconium*

TABLE 17.2

Fungi for Cultures and Host Material
for Comparison

Fungus	Host
Colletotrichum lindemuthianum	Green beans
Colletotrichum gloeosporiodes	Apple fruit
Colletotrichum circinans	Onions
Cytospora species	Spruce
Entomosporium mespili	Red tip photinia
Melanconium species	Grapefruit
Phomopsis species	Soybeans

Follow the protocols listed in Procedure 17.1 to complete this exercise.

Anticipated Results

Students should be able to differentiate the various ways conidia are produced, either naked as individual conidiophores or clustered together into synnemata, pycnidia, acervuli, and sporodochia.

Questions

- Describe and draw the following structures: individual conidiophores (Hyphomycetes), single or

---

Procedure 17.1

Observation of Asexual Structures of Deuteromycetous Fungi

Step	Instructions and Comments
1	Observe the fungal culture with a dissection microscope at low power. Note the shape and form of the fruiting structures. Fruiting structures can be observed more easily on the hyphae nearest the margin of active growth in culture. Conidia in mass may obscure fruiting structures. The mycelium (hyphae) becomes much thicker and often obscures the asexual structures near the center of the dishes.
2	Viewing asexual fruiting structures of deuteromycetes in plant material is often essential to identify the pathogen. Transition areas in the zone between healthy and necrotic plant tissue are usually good areas to look for structures. If fruiting structures are not evident in this area, move the tissue piece so that areas with greater tissue damage are viewed. If fruiting structures are not observed within the lesions, place the plant tissue overnight in a moist chamber. To obtain fruiting bodies, place a moistened tissue paper in the bottom of a plastic food storage container (a plastic bag will also work), set the plant tissue onto tissues, and close the box. Store at room temperature and check in 12 to 24 h for sporulation. Adding a drop of water directly to the lesion may also stimulate sporulation.
3	After viewing the diseased plant material with a dissection microscope, closer examination with a compound or light microscope is necessary to identify the fungus. Slice thin sections (1 mm) of a leaf lesion with a single-edge razor blade or scalpel and mount on a microscope slide with a drop of analine blue or water. Look for conidia and fruiting bodies, or either, on the plant tissue and floating in the mounting medium.
4	To observe sporulation of a fungal colony growing in a petri dish, place a drop of water or analine blue to the area of interest, and add a coverslip directly onto the agar medium. Fruiting structures can be observed with minimal damage to them. For closer observation, mount fruiting structures on a slide in a drop of distilled water and a coverslip. If fruiting bodies cannot be seen on the agar plate, simply remove a tiny portion of the mycelium (≤1 cm) and place it into the water on the slide. Flame-sterilize the dissection needles and allow to cool before removing the mycelium from the petri dish. Sometimes the coverslip does not appress directly to the microscope slide when a mass of mycelium and agar are mounted. If this happens, use a pencil eraser to gently press on the coverslip and flattened the preparation. First use the lowest power (generally a 4× or 10× objective lens) on a specific area containing the fruiting structures, and then proceed to high, dry objective lens (40×). As with the dissection microscope, observe the hyphae nearest the growing margin and move inward until light can no longer penetrate through the mycelium. Observe and draw the sporulating structures in detail.
5	A third method of viewing the deuteromycetes is with the use of cellophane tape. This method is particularly useful for obtaining fruiting structures and spores from fresh material. Place a piece (1 to 2 cm) of cellophane tape with the adhesive side downward onto the plant material. Press slightly and remove from the tissue. Take the piece of cellophane and place the side that was in contact with plant tissue downward onto a microscope slide containing water. Observe spores and/or fruiting structures with the 10× and 40× objectives lens and record your observations.

multiple branched, spore or conidia attachments; sporodochia, conidiophore arrangements, and locations on host tissues (living materials).

- How do fruiting structures vary among Hyphomycetes?
- How do pycnidia differ from sporodochia and acervuli?
- Where are the conidiophores located in a pycnidium?
- Does the pycnidium have a pore (ostiole) in the neck? What is the purpose of an ostiole (when present)?
- Are the pycnidia embedded in or on the surface of host tissue?
- What is a synemma?
- How do the conidiophores in synemmata differ from the conidiophores in sporodochia?
- Where are the spores borne on synemmata, and how are they attached?
- Discuss in detail the strategies for observing fungi on host material and in culture.
- Why is a dissection microscope important for fungal identification?

EXPERIMENT 2: THE EFFECTS OF LIGHT AND TEMPERATURE ON GROWTH AND SPORULATION OF FUNGI

The purposes of this experiment are to show how temperatures affect growth of fungi and to compare sporulation and growth potential of fungi when placed under different lighting regimens. Through direct observation, students will learn that growth conditions will affect spore production, cultural characters, and growth rates of fungi.

Common fungi that can be used in this exercise include the *Alternaria* species (single conidiophores), *Epicoccum nigrum* or *Fusarium solani* (sporodochia types), *Colletotrichum graminicola* or *Pestalotia* spp. (acervuli types), and *Phoma* or *Phomopsis* (pycnidia types). Many other fungi may be substituted in this experiment. Experiments typically require up to 14 days to complete but will obviously vary depending upon the fungi included in the study.

The fungi should be grown on potato dextrose agar (PDA: Difco, Lansing, MI) for several days, perhaps a week, at room temperature to provide sufficient inoculum. The number of conditions (treatments) may vary depending upon the available space and number of replicate plates (V-8 medium) used. We suggest that the class be divided into four groups, each with its own fungal species.

Materials

Each team of students will need the following items to complete the experiment:

- Four isolates of common deuteromycetes (one for each group). Inoculate each fungus onto five 10-cm-diameter petri dishes containing PDA.
- A minimum of twenty-one 10-cm-diameter petri dishes containing V-8 agar medium (Diener, 1955), 200 mL of V-8 medium (Campbell Soup Co.), 3 g $CaCO_3$, 15 g Difco agar, and 800 mL of water
- Aluminum foil
- Parafilm®
- Sterile cork borer, dissection needle, or spatula for subculturing of fungi
- Alcohol burner, matches, and container of 95% ethanol for sterilization of subculturing tools
- Incubators set for different temperatures and lighting conditions
- Millimeter ruler for measuring growth of colonies
- Light source for observing cultures

Follow the protocol outlined in Procedure 17.2 to complete the experiment.

Anticipated Results

Students should observe that changes in temperatures affect growth and sporulation of fungi. However, different fungal species used in the experiment may not respond with the same growth and sporulation patterns as the other species when compared under similar growth conditions.

Questions

- How did growth vary between species? Hint: graph the data, and calculate a growth rate.
- Did the fungi exhibit the same growth rate throughout the experiment? If not, what may have caused the changes in growth rates?
- When was sporulation first observed for each fungus? How did it vary per light regime?
- How did the effects of different light regimes compare between species? (Consult other groups.)
- Compare different ways you might be able to collect data on sporulation (e.g., number of conidiophores per unit area, spore counts, etc.).

EXPERIMENT 3: IDENTIFICATION AND VARIABILITY OF FUNGI GROWING ON HOST MATERIAL AND IN CULTURE

The purpose of this experiment is to compare and contrast the morphological variability of "imperfect fungi" on plant material with isolates grown in nutrient cultures. Asexually reproducing fungi, like other fungi, are affected by environmental conditions such as light, tem-

Procedure 17.2

Effects of Temperature and Light on Growth and Sporulation of Fungi

Step	Instructions and Comments
1	Use a flame-sterilized cork borer to cut agar pieces from cultures growing on PDA medium. If possible, remove plugs from actively growing margins (see Figure 17.2). Mycelium closest to the original inoculation plugs can differ physiologically from mycelium nearest to the margin, resulting in different growth rates.
2	Each group should transfer a single plug (mycelium side down) of their fungus onto the center of a petri dish containing V-8 agar. Inoculate a minimum of 21 dishes and wrap with parafilm®. Label the dishes with the name of the fungus and number them from 1 to 21.
3	Each group should place at least three of the inoculated petri dishes in incubators set at 10, 20, 30, or 40°C. Record the number of the petri dishes at each temperature. If sufficient incubator space is not available, we suggest that the 20°C and either the 10 or 30°C treatments be used. These cultures should be exposed to light for a minimum of 8 h per day; otherwise sporulation could be adversely affected.
4	Each group should place at least three of the inoculated petri dishes in incubators set for the following light treatments: continuous artificial light, alternating light and darkness (12 h each), and continuous darkness. The continuous dark treatment may be achieved by wrapping the cultures in aluminum foil. All incubators should be maintained at the same temperature (e.g., 20 to 25°C).
5	Record growth data every 2 days for 14 days (maximum) or until the mycelium reaches the edge of the dishes for both temperature and light treatments. Using a millimeter ruler, measure the radial growth from the inoculation plug to the perimeter of the colony. The growing edge of the mycelium can be viewed easily by placing the dishes toward a light source. Repeat in three locations for each dish and find the average to the nearest millimeter for each dish. Now average the radial growth for all three replicates for each fungus at each temperature. Plot this data on a graph with the y-axis as growth in millimeter and the x-axis as days in culture. Calculate the standard deviation (SD) for each point using the following formula: $$SD = \sqrt{\sum (x-\bar{x})^2/(n-1)} = \sqrt{x^2/(n-1)}$$ where x = sample measurement, $\bar{x}$ = calculate mean of data, $X = x - \bar{x}$, and n = number of samples used to calculate the mean (3).
6	Sporulation should also be recorded when growth is measured. A rating scale for sporulation can be developed such as none = 1, slight = 2, moderate = 3, and heavy = 4. Find the average rating for each fungus at each treatment. Plot sporulation on the y-axis and days on the x-axis.
7	Prepare a final report from the data. Include graphs showing the growth and sporulation of species under the various temperatures and lighting conditions.

perature, and moisture or nutrition found in the different growth substrates. Different nutrient agars may affect the morphology of the fruiting bodies and can complicate identification of the organism. Fruiting bodies often look different in culture compared to host tissue (saprophytic or parasitic). Keys to the deuteromycetes are based on their appearance on plant tissue; therefore, correct identifications made strictly from growth on culture media can be difficult if not impossible at times.

Materials

Each group of students will need the following materials:

- Dissection and light microscopes
- Slides, coverslips, and distilled water or 10% KOH (10 g KOH in 90 mL water)
- Several cultures of each of six fungi growing in 10-cm-diameter petri dishes with representative samples of the same six species on plant material (see Table 17.2 for suggestions); herbarium material may be substituted for fresh
- Dissection needle, sectioning knife for woody tissue, and forceps

Follow the protocol outlined in Procedure 17.3 to complete the experiment.

	Procedure 17.3
	Comparison of Fungal Structures from Host Plant and Culture
Step	Instructions and Comments
1	Observe with a dissecting microscope the corresponding cultures and plant material for a specific fungus. Cut very thin sections of diseased host tissue with a razor blade or scalpel, and mount on a microscope slide in water or in 10% KOH for dried specimens. Make a corresponding slide of the fungus growing on nutrient agar. Note: The dissection microscope may be the best method for examining any variation of the fruiting bodies from nutrient agar medium and on plant tissue.
2	Make drawings and record similarities and differences in structures.

Anticipated Results

Fungi that grow in culture often appear morphologically different on nutrient agar than in host materials. The differences may be slight or extreme depending upon the species used. The differences, however, may be enough to affect correct identification of the pathogen based on growth and sporulation on artificial media.

Questions

- Detail the fruiting structure shapes (morphology) for each fungus observed in this exercise. How are they similar when compared on the plant tissue and the agar medium? How do they differ when compared on the plant tissue and the agar medium?
- How might the variability of the fruiting bodies on the different growth sources affect their identification?

Experiment 4: Identification of *Rhizoctonia* species (Mycelia Sterilia) using hyphal characteristics

Rhizoctonia species do not form conidia in culture (or in nature), but are identified based on cultural characters such as mycelial shapes and color, sclerotia formation, and hyphal morphology. The hyphae are brown, elongate, inflated, and branch at right angles. This exercise will familiarize students with the general cultural characteristics of this important genus.

Materials

The following materials will be needed by each student or group of students:

- Several 10-cm-diameter petri dishes of *R. solani* growing on PDA

	Procedure 17.4
	Identification of *Rhizoctonia solani*
Step	Instructions and Comments
1	Compare *Rhizoctonia solani* and BNRLF cultures for the following characteristics (see step 2): • Hyphae are generally darker for *R. solani*. • Cells of the hyphae are elongate and often slightly inflated. • Hyphae branch at right angles and constrict at the area of branching, followed by a septum (cross wall) formation. • Sclerotia common in culture, are light at first, becoming dark brown to black, and are variable in size. • Asexual fruiting bodies are absent.
2	Cut a small plug (5 mm^2) of agar containing fungal mycelium and place on microscope slide in a drop of water. Cover with a plastic coverslip and flatten.
3	Observe the preparation with a compound microscope at 100× and 400×.

	Procedure 17.5
	Determining the Number of Nuclei in Cells of *Rhizoctonia* spp.
Step	Instructions and Comments
1	Inoculate known isolates of *Rhizoctonia* species onto thin layers of water agar in 10-cm-diameter petri dishes and incubate at room temperature for 48 to 72 h. Label dishes with the species name.
2	After the fungi have grown at least 5 cm from the inoculation plug, place a drop of 0.5% analine blue in lactophenol onto the hyphae at the edge of the colony. Place a coverslip over the colored area, lightly press the coverslip down and onto the mycelium, and observe the hyphal cells at 400× using a compound microscope. Be careful not to move the coverslip so that the agar comes in contact with the objectives lens. Count the number of nuclei in each hyphal cell and for each isolate.
3	Another method for nuclear staining is to remove a section (1 cm^2) of the fungal mycelium from the thin-layer water agar plates and place onto a microscope slide. Then add the dye as in step 3. Place a coverslip over the material and heat slightly with a small burner until the agar begins to melt. Place the slide onto the microscope stage, view at 400×, and count the nuclei as before.

- Several 10-cm-diameter petri dishes of a binucleate *Rhizoctonia*-like fungus (BNRLF) growing on PDA
- Dissection needle or microspatula for removing the fungal tissue to be observed
- Dissection and light microscopes, slides, coverslips, and distilled water for mounting the hyphae

Follow the protocol outlined in Procedure 17.4 to complete the experiment.

Anticipated Results

Students should observe the morphological differences between the two groups of *Rhizoctonia*, including thickness and color of the hyphae (mycelium), presence of sclerotia, and patterns of sclerotial occurrence in the petri dishes. Hyphae observed microscopically should appear as described earlier.

Questions

- What do the various *R. solani* isolates look like in culture?
- Discuss how the sclerotia varies in shape and size, and the patterns formed in the petri dishes.
- What distinguishes this group of fungi from others?

Experiment 5: Determine the number of nuclei in hyphal cells of *Rhizoctonia* species

The genus *Rhizoctonia* is divided into groups based on number of nuclei (binucleate or multinucleate) per hyphal cell. The multinucleate isolates are *R. solani* (**anastomosis groups**), and the binucleate isolates are either *R. cerealis* or BNRLF and belong to the **Ceratobasidium**

anastomosis group (CAG). Nuclear stains can be used to differentiate the groups.

Materials

Each student or team of students will need the following materials:

- 10-cm-diameter petri dishes containing 4 to 6 different anastomosis groups (AG and CAG) of *Rhizoctonia* species growing on PDA
- 10-cm-diameter petri dishes containing 5 mL of water agar
- Dissection needle or spatula for removing the fungal tissue to be observed
- Dissection and light microscopes
- Slides and coverslips
- Aniline blue, 0.5%, in lactophenol, or 0.05% trypan blue in lactophenol in dropper bottles
- Alcohol lamps

Follow the protocol outlined in Procedure 17.5 to complete the experiment.

Anticipated Results

The nuclear condition of the hyphal cells should be determined easily for the different isolates. From these observations, CAG or AG isolates can be determined for the unknowns.

Questions

- How many nuclei did you observe in the hyphal cells of *R. solani*?
- How many nuclei did you observe in the hyphal cells of CAG types?

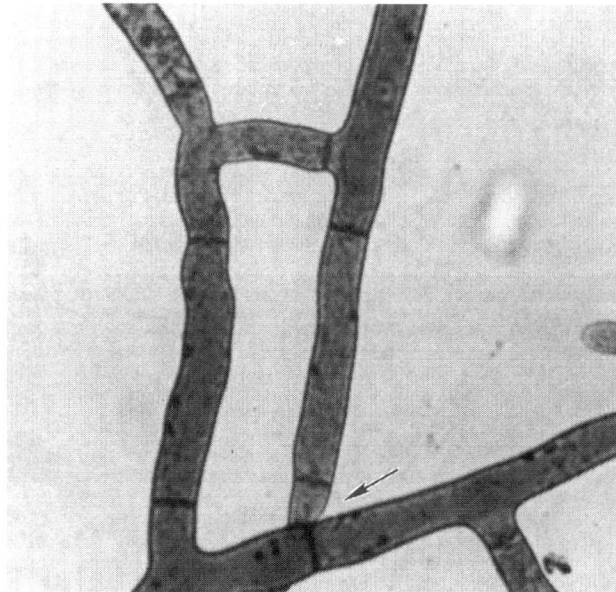

FIGURE 17.1 "Perfect fusion" during the anastomosis pairing experiment for *Rhizoctonia* species isolates. This reaction occurs with closely related isolates from the same genetic source or population. (Photo courtesy of Donald Carling, University of Alaska.)

EXPERIMENT 6: DETERMINATION OF ANASTOMOSIS GROUPING FOR ISOLATES OF *RHIZOCTONIA* SPP.

The genus *Rhizoctonia* is unusual because the asexual stage does not produce spores, and the sexual stage is rarely observed. As identification of these fungi is very difficult using cultural characters alone, alternative procedures were developed to identify the various AG using a method known as **anastomosis** pairing reactions.

Anastomosis reactions that may occur when *Rhizoctonia* isolates are paired include the following:

1. "Perfect fusion" occurs when the cell wall membranes combine and a mixture of continuous living cytoplasm occurs (Figure 17.1). This reaction only happens when pairing the same isolates or ones from the same anastomosis and **vegetative compatibility** groups.
2. "Imperfect fusion" occurs when the connection of cell walls and membrane contact is not clearly seen (Figure 17.2). Hyphal anastomosis occurs, but isolates are from different vegetative compatibility groups. Often the hyphae at the point of contact are smaller in diameter than the rest of the hyphae (Carling, 1996). Those isolates showing imperfect fusion are related, such as two AG-4 isolates identified from different sources.
3. "Contact fusion" occurs when hyphal cell walls of *Rhizoctonia* isolates come into contact, but no fusion, lysis, or wall penetration or membrane contact occurs (Figure 17.3). Hyphal cells at the contact point and adjacent cells may occasionally die but not anastomose.
4. "No fusion" occurs among totally unrelated isolates. The unknown and tester isolates show no reaction to the presence of the others.

The following experiment is intended for graduate students or advanced undergraduate special topics.

Materials

The following supplies are needed for each student or team of students:

- Four known *Rhizoctonia* AG (CAG) cultures with four AG (CAG) grown on PDA (Selection should include isolates capable of showing the different types or degrees of anastomosis that occur during pairing.)
- Cultures of isolates from unknown anastomosis groups
- Compound microscope and coverslips
- Twelve 10-cm-diameter petri dishes containing 1.5% water agar
- Twelve microscope slides
- Forceps, dissection needles or microspatula
- Alcohol burner and jar of 95% ethanol for flame sterilization of tools

Follow the protocol outlined in Procedure 17.6 to complete the experiment.

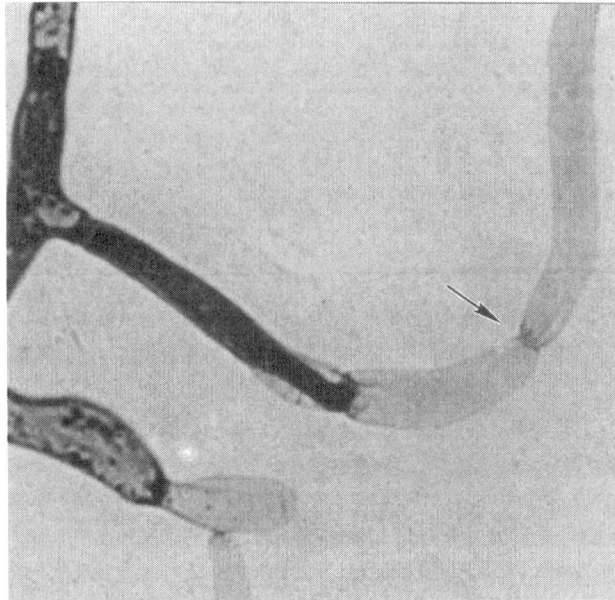

FIGURE 17.2 "Imperfect fusion" during the anastomosis pairing experiment for isolates of *Rhizoctonia* species. Isolates causing this reaction are related, but from different genetic sources or populations. (Photo courtesy of Donald Carling, University of Alaska.)

Anticipated Results

The four types of fusion ranging from perfect, imperfect, and contact, to no fusion should be observed. Students will see perfect fusion reactions and cellular death associated with true anastomosis pairing versus the contact types. From these observations, they will be able to determine which unknown anastomosis groups they have been screening.

Questions

- How are tester isolates chosen to use for comparison with an unknown *Rhizoctonia*?
- What are the major types of anastomosis pairing reactions that can occur?

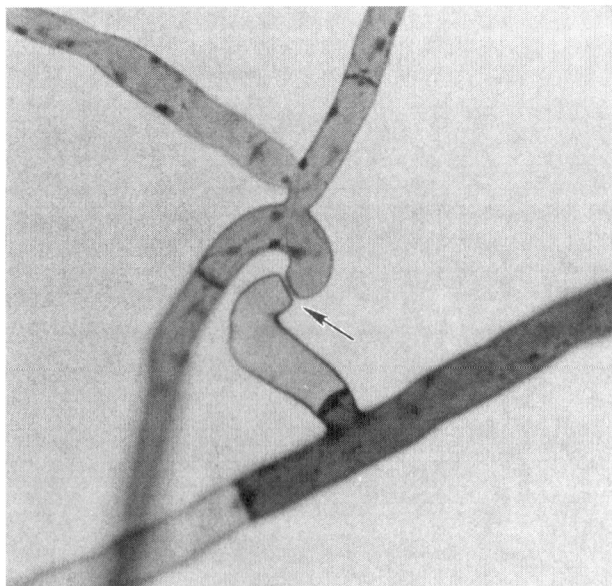

FIGURE 17.3 Photograph showing "contact fusion" during the anastomosis pairing experiment for isolates of *Rhizoctonia* species. Isolates come into contact, but cell wall fusion does not occur. Isolates exhibiting this pairing reaction are distantly related. (Photo courtesy of Donald Carling, University of Alaska.)

Procedure 17.6	
Determination of Anastomosis Grouping of Iolates of *Rhizoctonia* spp.	
Step	Instructions and Comments

1   "Tester isolates" will be used to compare morphological or visible (growth form, color, and shape in culture) differences between the isolates. In this exercise, unknown AG (*Rhizoctonia solani*) isolates will be compared to known (tester) AG isolates. Selection of tester isolates are based on similar growth forms, shapes, and colors of the unknown *Rhizoctonia*. All isolates should be inoculated onto Difco PDA dishes at the same time and grown for a minimum of 10 to 14 days at room temperature before the comparisons can be made.

2   *Rhizoctonia* isolates may be separated by anastomosis reactions using the water agar plate method or water agar slide method.

3   Water agar dish method: Place a mycelial plug of the unknown *Rhizoctonia* isolate off center in the 10-cm-diameter petri dish containing 1.5% water agar. Transfer a plug of the unknown tester to the dish, opposite and off center so that the two plugs will be approximately 3 to 5 cm from each other (Figure 17.4).

4   Anastomosis pairing reaction is made immediately after the hyphae from the two isolates meet (slightly touch) as seen visually with a dissection microscope. Place a drop of 0.5% analine blue in lactophenol at the point where the two meet, add a coverslip, and view at 400×. The following four types of hyphal interactions are possible: perfect fusion (C3), imperfect fusion (C2), contact fusion (C1), and no fusion (C0) (Matsumoto et al., 1932; Carling, 1996).

5   Water slide method: Using forceps, surface-sterilize a microscope slide by dipping in 95% ethanol and flaming. Alternatively, slides may be placed in a beaker covered with aluminum foil or in a glass petri dish and autoclaved for 20 min. Holding the slide with the forceps, dip the sterile slide into 1.5% PDA and place onto petri plates containing 1.5% water agar. The water agar in the plate prevents the agar on the slide from drying out during the pairing studies.

6   After the PDA coated slides have hardened, place the mycelial plugs 3 to 5 cm apart on the slides. One side of the slide will have the unknown isolate and the other side the tester isolate, of *R. solani*.

7   Incubate a petri dish containing the inoculated slides at room temperature for 1 to 2 days and observe regularly for hyphal growth of the two isolates. View the hyphae of the two isolates microscopically when the hyphae come into contact. Avoid overlapping hyphal growth. Too much overlap makes it extremely difficult to locate the hyphae of the two isolates.

8   Remove the slide from the petri dish and stain the zone where the hyphae of the two isolates intersect with 0.5% analine blue and lactophenol. Place a coverslip over the top of the stained area and determine the anastomosis pairing reaction using the compound microscope.

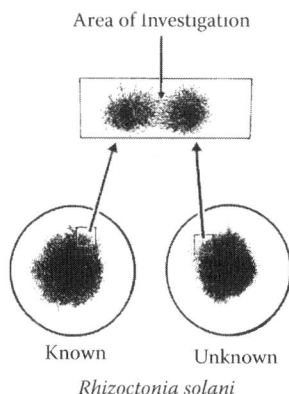

FIGURE 17.4 Diagrammatic representation of pairing of *Rhizoctonia* isolates on an agar slide. (Drawing by Joe McGowen, Mississippi State University.)

REFERENCES

Carling, D.E. 1996. Grouping in *Rhizoctonia solani* by hyphal anastomosis reaction. In Sneh, B., Jabaji-Hare, S., Neate, S., and Dijst, G. (Eds.). *Rhizoctonia Species: Taxonomy, Molecular Biology, Ecology, Pathology and Disease Control*. (pp. 37–47). Kluwer Academic, Boston, MA.

Diener, U.L. 1955. Sporulation in pure culture by *Stemphylium solani*. *Phytopathology* 45: 141–145.

Matsumoto, T., Yumamoto, W. and Hirane, S. 1932. Physiology and parasitology of the fungi generally referred to as *Hypochnus sasakii* Shirai. I. Differentiation of the strains by means of hyphal fusion and culture in differential media. *J. Soc. Trop. Agric.* 4: 370–388.

Yokoyama, K. and Ogoshi, A. 1986. Studies on hyphal anastomosis of Rhizoctonia solani IV. Observation of imperfect fusion by light and electron microscopy. *Trans. Mycol. Soc. Japan* 27: 399–413.

18 Smut and Rust Diseases

Larry J. Littlefield, Yonghao Li, and Darrell D. Hensley

CHAPTER 18 CONCEPTS

- Smut and rust diseases are important; rusts are far more important worldwide and attack a wider range of host plants than smuts.

- Smut diseases are important primarily on cereals; many can be managed effectively by chemical seed treatment and/or resistant cultivars. The three major types of smut infection, each requiring different management strategies, are: (1) seedling infection, (2) floral infection, and (3) meristem infection.

- Rust diseases cause millions of dollars damage annually to numerous agronomic, horticultural, and timber species.

- Rust fungi are highly specific, obligate parasites, often attacking limited numbers of host species or cultivars. The degree of host specificity is expressed as formae speciales and/or physiological races in many rust fungi.

- The life cycle of rusts may contain up to five different spore stages; some rust species (autoecious forms) complete their life cycle on one host species; others (heteroecious forms) require two.

- Many rusts are disseminated by wind for long distances, which contributes greatly to their widespread occurrence.

- Several management practices are used to reduce severity and economic impact of rust diseases depending on the particular host and rust species. They include genetic resistance, foliar fungicide sprays, elimination of alternate host species, and other cultural practices.

The pathogens responsible for smut and rust diseases are members of the kingdom Fungi and phylum Basidiomycota, and are placed in the orders Ustilaginales and Uredinales, respectively. These orders differ from most other Basidiomycota by not having their sexual reproductive spores (**basidiospores**) borne on or in fruiting bodies, as in mushrooms and puffballs. Rather, their basidiospores are borne on specialized germ tubes (basidia) produced upon the germination of thick-walled, typically dormant, resting spores (**teliospores**). Smut fungi can be cultured on artificial media much more easily than the rusts. However, as pathogens in nature, both the smuts and rusts function as obligate parasites, i.e., they require living host tissue as a nutrient source. Both types of pathogens cause important plant diseases, but rust fungi attack a much broader host range and are far more important economically worldwide than are the smuts.

SMUT DISEASES

Smuts are most important on cereal crops, e.g., barley, wheat, oats, grain sorghum, and corn, although violets, onions, sugarcane, forage grasses, date palms, and several other crops can also be infected. The economic impact of smut infection is most commonly evident in individual kernels or entire heads of wheat (Figure 18.1a and b) and barley (Figure 18.1c) (Figure 18.1a,b,c), or ears of corn (Figure 18.1d,e) being largely replaced by a dark brown, dry, powdery mass of teliospores. The most common smut diseases in North America include corn smut (Figure 18.1d–g), stinking smut or common bunt of wheat (Figure 18.1a), and loose smut of wheat (Figure 18.1b), barley, and oats. With the development of highly effective seed-treatment chemicals for small grains and many smut-resistant corn varieties, smuts are now far less damaging economically than as recently as 50 years ago. However, smuts, as well as many other plant pathogens, can spread easily over great distances by wind and modern transportation, or either, and such pathogens, once controlled, have a history of reappearing sometimes in yet more aggressive forms. Thus, it is important to understand the biology of smut diseases and the fundamentals of their management. As recently as 1996, Karnal bunt first appeared in the United States, coming from Mexico, where it had previously been introduced inadvertently from India on seed utilized in international wheat breeding programs. Although Karnal

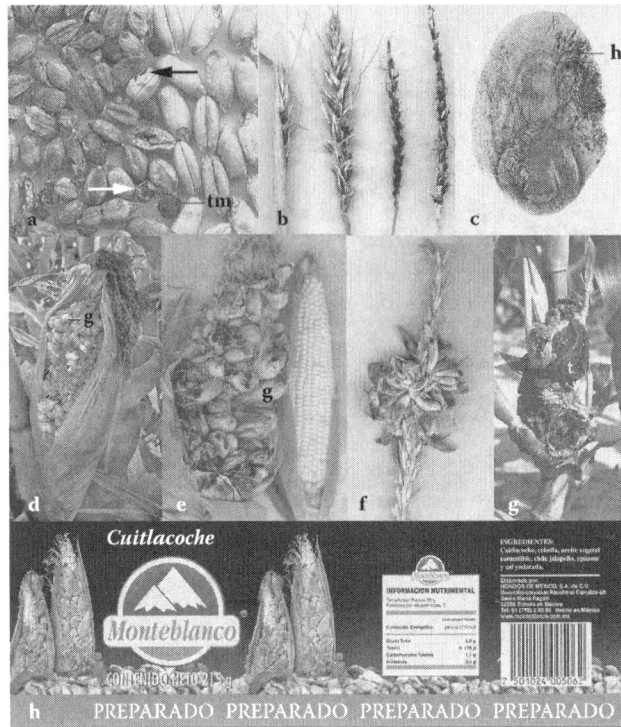

FIGURE 18.1 (**a**) Common bunt of wheat (*Tilletia tritici*); diseased kernels (left), with partially to completely ruptured pericarp (arrows), revealing dense teliospore masses (tm) that eventually disintegrate into individual teliospores; healthy kernels (right). (**b**) Loose smut of wheat (*Ustilago tritici*); progressive emergence (left to right) of the diseased head and transformation of kernels to masses of dry teliospores, providing wind-borne inoculum for immature ovaries in healthy heads. (**c**) Excised, hydrated, loose smut-infected embryo of barley, containing *U. nuda* hyphae (h) that can later infect apical meristem of the young seedling. ×25. (With permission from North Dakota State University.) (**d**) Corn smut (*U. maydis*) converts kernels into large gall-like masses of teliospores (g) initially surrounded by an outer, white membrane that eventually disintegrates. (With permission from University of Illinois.) (**e**) Corn-smut-infected ear (left), healthy ear (right). Smut galls (g) are mature, with a partially ruptured outer membrane exposing masses of black teliospores within. (With permission from University of Illinois.) (**f**) Corn smut galls on infected tassels. (**g**) Mature corn smut gall on an infected stalk. Millions of teliospores (t) are exposed upon disintegration of the gall's outer membrane. (**h**) Label from can of corn smut galls ("*Cuitlacoche*") sold in Mexico. (With permission from Hongos de Mexico, S.A., Mexico City.)

bunt has little effect on yield, it has had major economic impacts on export sales of American wheat.

Smut fungi, as do the rusts, reproduce **sexually**, but can do so only once per growing season. However, smuts lack the highly prolific **asexual** reproductive stage characteristic of many rust fungi. That latter character precludes secondary infection during a growing season, thus limiting their geographic distribution as compared to many rusts.

Although **life cycles** of most smut fungi are similar, their **disease cycles** can be very different, resulting in different approaches to disease management for various smut diseases. The most important differences among smut disease cycles, and hence, the major differences in their management, are related to the stage of host development at which primary infection occurs. Although there are three major stages of host plant growth at which different smut fungi can initiate infection, they all require the presence of meristematic host tissue for infection to actually occur. Understanding those differences is essential to implementing an appropriate disease management

strategy. Different smuts can infect plants in the seedling stage (where infection typically occurs in below-ground cotyledon tissue), in the flowering stage (where infection of ovaries occurs early in their development), or in juvenile to mature stages (where infection can occur in meristems of any above-ground parts of the plant). Different types of smut diseases include the following examples.

SEEDLING INFECTION

Examples of such smut diseases in wheat include common bunt or stinking smut, *Tilletia tritici* and *T. laevis*; dwarf bunt, *T. controversa*; Karnal bunt, *T. indica;* and flag smut, *Urocystis agropyri*. Symptoms of common bunt or stinking smut (Figure 18.1a) appear in mature kernels, resulting from infection in the seedling stage earlier that same growing season (Figure 18.2a). That infection resulted from teliospores present in soil of the previous years' crops, particularly in drier soils, or from teliospores deposited onto the surface of seed during previous seasons' harvest, or subsequent shipping or storage.

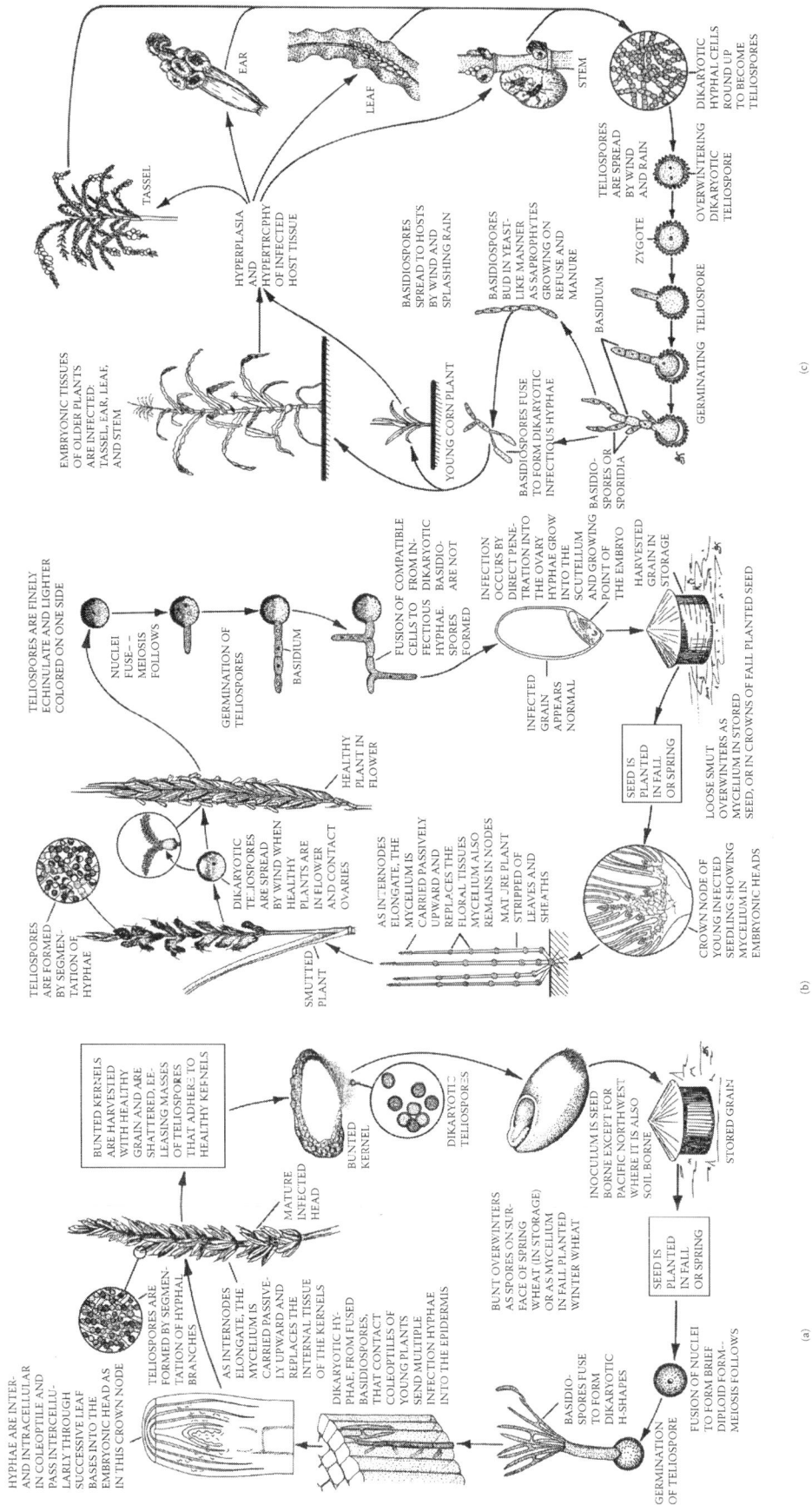

FIGURE 18.2 (a) Disease cycle of wheat common bunt disease (*Tilletia tritici*). (b) Disease cycle of wheat loose smut disease (*Ustilago tritici*). (c) Disease cycle of corn smut disease (*U. maydis*). (All with permission from Kenaga, C.B., E.B. Williams, and R.J. Green. 1971. *Plant Disease Syllabus*, Balt Publishers.)

Such teliospores can remain viable in soil or on the seed surface up to 10 years or more, until moisture and temperature conditions are satisfactory for germination, and then proceed to infect the cotyledon of germinating seedlings. **Mycelium** within a newly infected seedling invades the apical meristem of the plant; as the stem differentiates and elongates, the mycelium is carried upward, eventually residing in the head and the developing kernels. Smut mycelium that differentiates to form teliospores eventually replaces all internal tissues of the kernel and is surrounded by the **pericarp**. The pericarp breaks down during harvesting or handling, releasing teliospores that can be wind borne and can infest otherwise healthy seed, thus completing the disease cycle. The most economical, widely used management procedure for common bunt is fungicide seed treatment to kill surface-contaminating teliospores and to protect seedlings from soilborne teliospores. Other management strategies include using only pathogen-free seed or smut-resistant cultivars, and planting when the soil temperature is greater than 26°C (68°F), the optimum for teliospore germination.

FLORAL INFECTION

Examples in wheat and barley include loose smut of wheat, *Ustilago tritici,* and loose smut of barley, *U. nuda.* The major symptom of loose smut (Figure 18.1b) is the often naked rachis bearing only the remnants of kernels that had been totally replaced by brown to black, dry, powdery masses of teliospores. That condition results from infection that occurred one or more years earlier during the flowering process that led to the production of the current year's planted seed (Figure 18.2b). Infected heads emerge slightly before the uninfected heads; the former are transformed into masses of teliospores that provide **inoculum** for the current season's crop as the new, uninfected heads begin to emerge. Infection occurs in the young florets; there, mycelium grows into the young embryos (Figure 18.1c), where it can remain viable for several years. When that infected seed is planted and it germinates, the mycelium in the ovary invades the apical meristem and is eventually carried passively into the developing head. There it converts the entire kernels, including the pericarp, into the mass of dry, black teliospores, thus completing the disease cycle. Those spores can then function as primary inoculum for that year's crop as new heads emerge and begin to form kernels. Visual evidence of that infection will not be seen until the heading time of the crop produced from those infected kernels. Successful chemical control of loose smut requires seed treatment with one of the several systemic fungicide seed treatments available. Such materials must enter into the seed where they can kill smut mycelium within the embryo. Surface, contact fungicide treatments are ineffective for controlling floral infecting smuts. Other recommendations include planting

pathogen-free seed, as determined by laboratory analysis, or resistant cultivars, if available.

MERISTEM INFECTION

Corn smut, *U. maydis,* is the most common example of this type of smut infection. It is also more easily visible to the untrained eye than any of the small grain smuts due to large galls (up to 2 to 3 in. diameter) that often occur in clusters, sometimes occupying nearly the entire volume of an ear of corn while still attached to the stalk (Figure 18.1d,e). Galls of various sizes are also common on the tassels (Figure 18.1f), stalks (Figure 18.1g), brace roots, and leaves. Before maturity, galls appear as somewhat white, translucent, rather brittle masses (Figure 18.1d). As galls mature, their internal portions become progressively dark, with the white exterior membrane initially remaining intact. Later, the outer membrane disintegrates, and the entire gall becomes a dark brown, dry, powdery mass consisting of millions of teliospores formed from hyphae that originally comprised the gall. Teliospores become wind borne and contaminate soil, or the galls fall to the ground where they overwinter in soil as crop debris (Figure 18.2c). The following season, soilborne teliospores from galls germinate, producing basidiospores that may land on susceptible host tissue (i.e., *any* meristematic, above-ground corn tissue). Following the fusion of germ tubes from sexually compatible basidiospores, the newly fused infection hyphae can then penetrate the corn plant, e.g., into individual silks, where they grow toward and into the developing kernels; or they may penetrate and infect leaf, stem, or tassel meristems. Once established in the host, smut mycelium can continue to invade host tissue, eventually resulting in gall formation.

Corn smut is no longer a serious disease in hybrid dent corn production due to the many resistant varieties available. However, smut remains a significant problem in sweet corn as resistance in those cultivars is limited. Management of smut in sweet corn is facilitated by maintaining well-balanced fertility, without excess nitrogen. Fungicide sprays or seed treatments provide no benefit in control of corn smut.

A major difference in the epidemiology of smut diseases compared to rusts is the absence of asexually produced **repeating spores** in the life cycle of smuts. Consequently, smuts, lacking the ability to form billions of asexual reproductive spores easily disseminated over hundreds of miles by wind, cannot build up major epidemics compared to many rust diseases.

Although considered only a plant disease in the United States, corn smut is regarded as a choice food in Mexico, where young smut galls are harvested prior to turning black and powdery. Commercially, the product is sold under its Spanish name "*Cuitlacoche*" (often spelled "*Huitlacoche*"; pronounced "wheat-la-coach-e") (Figure 18.1h). The 19th century Hidatsa Indians of present southwestern North Dakota harvested, boiled, dried, and later mixed corn smut galls ("*Mapëdi*") into boiled corn as a flavor additive.

RUST DISEASES

Currently and historically, rusts are among the most damaging of all plant diseases worldwide. The devastation they caused was justification for annual sacrifices to ancient Roman gods and for 17th and 18th century legislation in France and the American colonies. Rusts continue to cost untold millions of dollars annually to producers of cereals, vegetables, ornamentals, timber, sugarcane, coffee, and other crops, with those costs being passed on inevitably to processors and consumers. In this highly abbreviated coverage of rust diseases and their pathogens, it is essential to choose only those most important examples that illustrate particular concepts of pathogen biology, host–pathogen relationships, and disease management. Suggested readings at the end of the chapter provide greater breadth and depth of information.

Symptoms and signs of cereal and other herbaceous rusts are commonly seen as shriveled seed (Figure 18.4a) and erumpent lesions (pustules) on leaves and stems (Figure 18.4b–d), respectively. Rust fungi have a highly complex life cycle that often contains up to five different spore stages; some rusts require two unrelated, alternate hosts for completion of their life cycle (Figure 18.3a–c). The five sequential stages of the complete rust life cycle, beginning and ending with the teliospores, are shown in microscopic view in Figures 18.4e and 18.5e. The overwintering stage (or oversummering stage, depending on the rust and the climate in which it exist) is that of teliospores (Figures 18.4e and 18.5e) borne in telia (Figure 18.3c; 18.4d). Teliospores germinate to form basidiospores (Figures 18.3a and 18.4e); if basidiospores land on a susceptible host, they initiate infection, resulting in production of **pycniospores** (Figure 18.4g) in pycnia (Figure 18.4f). Pycniospores function as spermatia (Figure 18.3a) and accomplish fertilization of the female flexuous hyphae, thus establishing the N+N, i.e., dikaryotic, nuclear condition (Figure 18.3b). The next stage is that of **aeciospores** (Figure 18.4i), borne in aecia (Figure 18.4h,i); that is followed by production of **urediniospores** (Figure 18.5a–c) in uredinia (Figures 18.4b and 18.5a) and, eventually, by teliospores (Figure 18.5d,e) in telia, to complete the cycle. The life cycle, showing the nuclear condition of different hyphal and spore stages, is summarized in Figure 18.3b. Rusts that complete their entire life cycle on one host species (not necessarily on one host plant) are termed *autoecious* rusts, whereas those requiring two alternate host species for that process are *heteroecious* rusts. The terminology used to identify primary versus alternate hosts of autoecious rusts lacks consistency and can be rather confusing. Some define

"primary host" as the host that produces the telial stage, and "**alternate host**" as the one that produces the pycnial/aecial stage. Others define primary host as the host of greater economic importance, with alternate host being less important economically. Figure 18.3a illustrates the classic hetereocious disease cycle of *Puccinia graminis* (wheat stem rust), which requires wheat or some related species and common barberry for completion of its life cycle. In wheat-stem rust disease, wheat is most commonly referred to as the primary host, with barberry being the alternate host, though it is not always so.

Similar cycles requiring two unrelated hosts occur in many other heteroecious cereal rusts; e.g., wheat leaf rust (*P. triticina* = *P. recondita* f. sp. *tritici*) and oat crown rust (*P. coronata*). Autoecious rusts, such as those on bean, asparagus, flax, or rose complete their entire life cycle on only their one respective host. Many rusts, known as short-cycle rusts, lack one or more spore stages in their life cycle. Cedar-apple rust (Figure 18.3c) lacks the uredinial stage; consequently, it has no repeating urediniospores, the most important spores epidemiologically in rust diseases of most agronomic or horticultural crops. As a result, no major "explosions" of cedar-apple rust occur as is possible in cereal, bean, or coffee rusts. However, significant damage can occur in apple orchards due to rust resulting from massive primary infection by untold numbers of basidiospores originating from the cedar (botanically, a juniper) alternate host, not from repeating generations of urediniospores as in cereal rusts. Hollyhock rust (Figure 18.5d,e), which produces only teliospores that overwinter and germinate to form basidiospores capable of initiating new infections the following year, exemplifies of the greatest degree of shortened life cycle possible, i.e., the microcyclic form, and yet it contains the essential steps for sexual reproduction. Other abbreviated forms of rusts lack the sexual stage completely, or it is yet to be discovered. Many such rusts, e.g., coffee rust (*Hemileia vastatrix*), survive very successfully in the uredinial stage only.

The economically important stage of a rust pathogen's life cycle can be different, depending on the rust. A few heterocieous, herbaceous rusts (e.g., hops rust) produce pycnia and aecia rather than uredia and telia on their more economically important hosts. More commonly, e.g., in rusts of cereals, bean, sunflower, flax, alfalfa, asparagus, and many other herbaceous Angiosperm hosts, the most damaging effect on yield results from the uredinial stage (Figures 18.4b and 18.5a) irrespective of their being autoecious or heteroecious. The subsequent telial stage on the same host species is essential for completion of the life cycle but, typically, it develops too late in the growing season to seriously reduce yield. The uredinial stage is not only the major cause of severe yield reduction in herbaceous hosts, it is also the spore stage most commonly and widely spread by wind, potentially resulting in

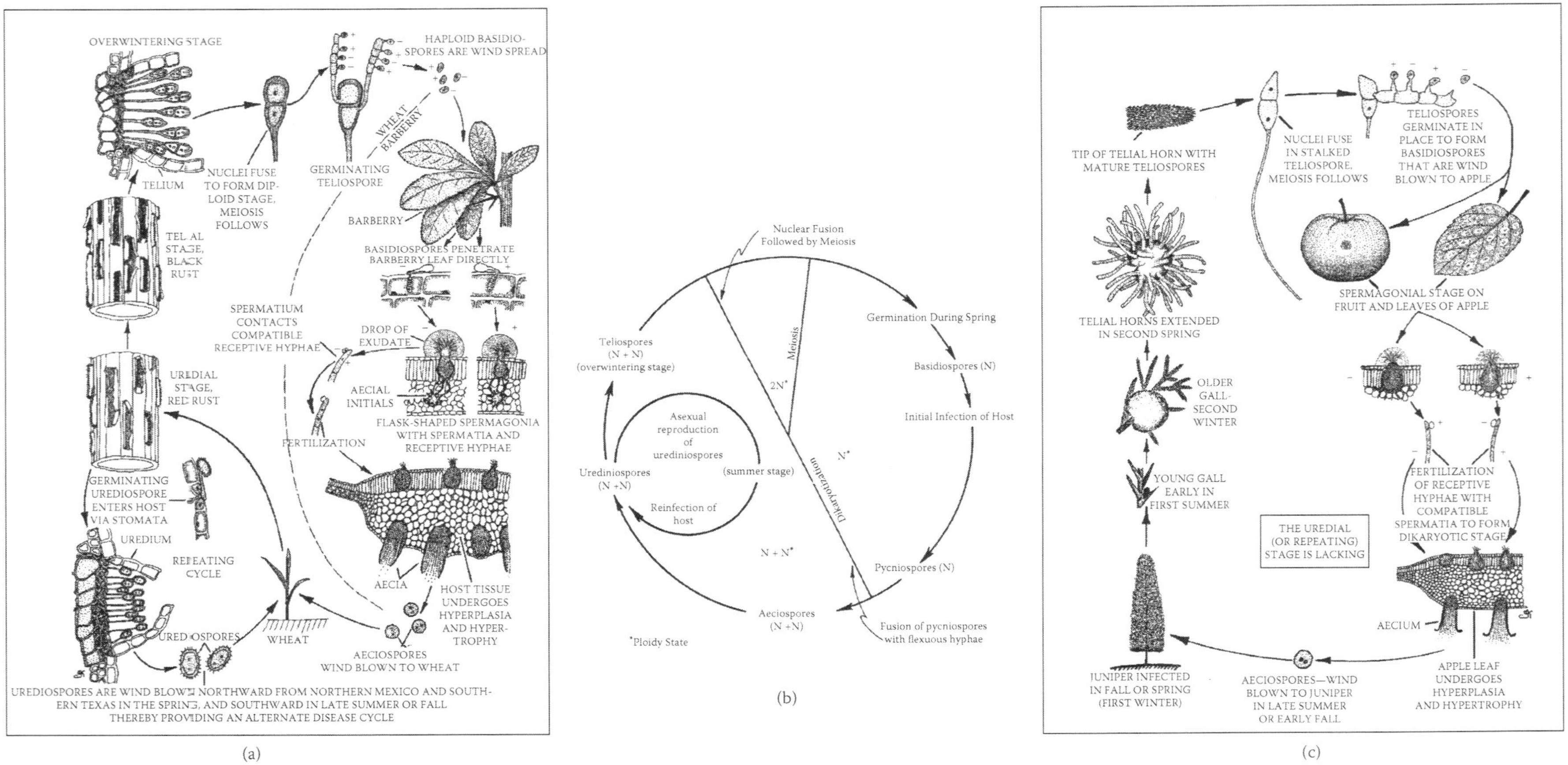

FIGURE 18.3 (a) Disease cycle of wheat-stem rust disease (*Puccinia graminis* f. sp. *tritici*). (With permission from Kenaga, C.B., E.B. Williams, and R.J. Green. 1971. *Plant Disease Syllabus*. Balt Publishers.) (b) Generalized life cycle of rust fungi showing the sexual and asexual reproductive stages and ploidy number of those stages. (With permission from Littlefield, L.J. 1981. *Biology of the Plant Rusts: An Introduction*. Iowa State University Press, Ames, IA.) (c) Disease cycle of cedar-apple rust (*Gymnosporangium juniperi-virginianae*). (With permission from Kenaga, C.B., E.B. Williams, and R.J. Green. 1971. *Plant Disease Syllabus*. Balt Publishers.)

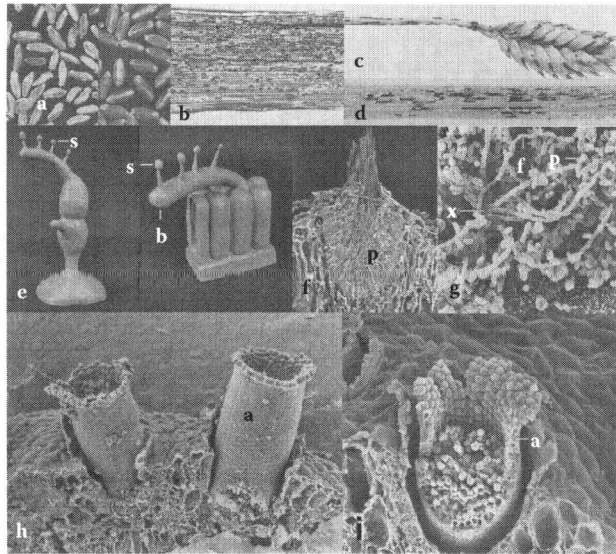

FIGURE 18.4 (**a**) Stem rust (*Puccinia graminis* f. sp. *tritici*) infection induces shriveling and reduced kernel size (left) compared to kernels from healthy plants (right). (With permission from United States Department of Agriculture, Agricultural Research Service). (**b**) Uredinial pustules of *P. graminis* f. sp. *tritici* on wheat leaf. (**c** and **d**) Telial pustules of *P. graminis* f. sp. *tritici* on stem and glumes (c) and stem (d). (With permission from United States Department of Agriculture, Agricultural Research Service.) (**e**) Models of germinating teliospores of the stalked, two-celled teliospore (t) of *Puccinia* spp. (left) and the sessile, single-celled teliospores (t) of *Melampsora* spp. (right). Each teliospore cell can potentially germinate to form an elongated, septate basidium (b) that produces four basidiospores (s) borne on thin, peg like sterigmata. (With permission from Littlefield, L.J. 1981. *Biology of the Plant Rusts: An Introduction*. Iowa State University Press, Ames, IA.) (**f**) Longitudinal fracture through a pycnium (p) (= spermagonium) of *P. triticina*. Pycniospores (= spermatia), not present in this micrograph, are produced inside pycnia and belong to one of two possible mating types. They exude from the pycnial opening and are carried passively by leaf contact, splashing water, or insects to the receptive ("female") hyphae (see Figure 18.4g) extending from pycnia of a second, compatible mating type, thus accomplishing spermatization. ×200. (With permission from Brown, M.F. and H.G. Brotzman. 1979. *Phytopathogenic Fungi: A Scanning Electron Stereoscopic Survey*. University of Missouri–Columbia Extension Division, Columbia, MS.) (**g**) Leaf surface of pycnia/aecial host of *P. triticina*, showing elongated, flexuous hyphae (f) (the receptive "female" hyphae) that extend from the apical opening of an underlying pycnium (x). Pycniospores (p) (= spermatia) have exuded from that same underlying pycnium, but pycniospores and flexuous hyphae produced from the same pycnium are sexually incompatible. ×750. (**h**) Two aecia (a) of *P. graminis* f. sp. *tritici* extending beyond the surface of the barberry host leaf. ×100. (With permission from Brown, M.F. and H.G. Brotzman. 1979. *Phytopathogenic Fungi: A Scanning Electron Stereoscopic Survey*. University of Missouri–Columbia Extension Division, Columbia, MS). (**i**) Diagonal fracture through an aecium (a) of *P. graminis* f. sp. *tritici*, exposing aeciospores (s) contained within the tubular aecium. ×200.

epidemics spanning hundreds of miles during a growing season. The annual south-to-north dispersal of both wheat leaf and stem rust urediniospores occurs along the well-documented "**Puccinia Path**." Both those pathogens can overwinter as uredinial stage infections on winter wheat in southern Texas and Mexico; urediniospores produced there in early spring are blown northward and have been trapped at elevations up to 14,000 ft (about 4300 m). Progressive infections of susceptible winter wheats occur from northern Texas through South Dakota and, eventually, into susceptible spring wheats in Minnesota, North Dakota, and central Canada by midsummer. Similarly, the rapid spread of coffee rust in Asia and Africa resulted largely from wind-borne urediniospores, as in much of South America, following the rust's introduction into the western hemisphere, possibly carried there on plants from Africa in 1970 (see Figure 18.5k).

In contrast to rusts of most agronomic crops, rusts of conifers cause the greatest damage in the pycnial and aecial stages of the life cycle. Trunk and branch cankers (Figure 18.5f), and galls (Figure 18.5g), resulting from infections two or more years old, initiated by basidiospores from an alternate host, can eventually girdle infected regions, resulting in death of all tissue beyond the canker. Several species of pine rusts cause millions of dollars loss annually in both northern and southern forests of the United States, as well as in pine, spruce, fir, larch, and other timber species worldwide. In contrast, an important example of a uredinal stage rust in commercial trees is leaf rust of poplars, which causes severe defoliation and reduced pulp-wood production.

Rust pathogens are highly host specific, with most species being limited to one or only a few closely related hosts. Within some rust species there are also "specialized forms" (Latin plural, *formae speciales*; abbreviated "f. sp.") that are limited to certain host species, e.g., *P. graminis* f.sp. *tritici* on wheat and barley, *P. graminis* f. sp. *avenae* on oats, *P. graminis* f. sp. *secalis* on rye, and

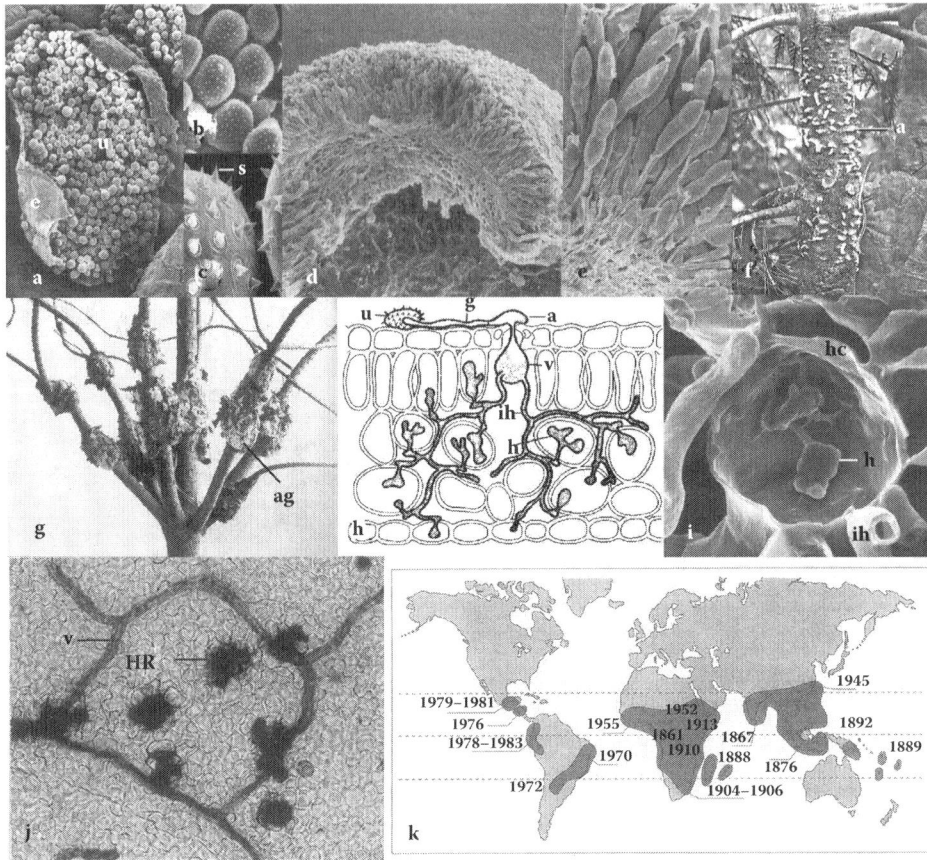

FIGURE 18.5 (**a**) Surface view of a uredinium of flax rust (*Melampsora lini*) containing several hundred urediniospores. Host epidermis (e) around the periphery of the uredinium is ruptured upon expansion of that structure. ×100. (With permission from Littlefield, L.J. 1981. *Biology of the Plant Rusts: An Introduction*. Iowa State University Press, Ames, IA.) (**b**) Higher magnification of urediniospores in Figure 18.5a. ×750. (**c**) Spines (s) cover the surface of those urediniospores. ×2200. (With permission from Brown, M.F. and H.G. Brotzman. 1979. *Phytopathogenic Fungi: A Scanning Electron Stereoscopic Survey*. University of Missouri–Columbia Extension Division, Columbia, MS.) (**d**) Longitudinal fracture of a telium of hollyhock rust (*Puccinia malvacearum*) on surface of host leaf. ×50. (**e**) Higher magnification of a portion of Figure 18.5d reveals two-celled teliospores (t) characteristic of the genus *Puccinia*. ×130 (d and e with permission from Littlefield, L.J. 1981. *Biology of the Plant Rusts: An Introduction*. Iowa State University Press, Ames, IA.) (**f**) Aecial canker of white pine blister rust (*Cronartium ribicola*) on trunk of a young tree. Aecia (a), containing aeciospores, are the white, blister-like lesions on the canker surface. All portions of the trunk and branches beyond the canker will likely die. (**g**) Multiple aecial galls (ag) of western gall rust (*Endocronartium harknessii*) on lodgepole pine (f and g with permission from Ziller, W.G. 1974. *The Tree Rusts of Western Canada*. Canadian Forestry Service Publication No. 1329, Department of the Environment, Victoria, B.C.). (**h**) Diagrammatic illustration of a typical uredinial-stage rust infection in a host leaf. The uredinio-spore (u) germinates to form a germ tube (g) that bears at its apex an appressorium (a) over a host stoma. From that, infection occurs through the stoma, forming a substomatal vesicle (v), intercellular hyphae (ih), and ultimately numerous haustoria (h) that absorb host nutrients and transport them to other portions of the expanding and/or spore-producing regions of the colony. (**i**) Internal view of a rust-infected coffee leaf, showing intercellular hyphae (ih) and several haustoria (h) of *Hemileia vastatrix* that extend into the host cell (hc). ×1400 (h and i with permission from Littlefield, L.J. 1981. *Biology of the Plant Rusts: An Introduction*. Iowa State University Press. Ames, IA). (**j**) Cleared, stained leaf of flax expressing immune, necrotic, hypersensitive reactions (HR), each resulting from an unsuccessful infection by an incompatible race of *Melampsora lini*. v = leaf vein. ×100. (With permission from Littlefield, L.J. 1973. Histological evidence for diverse mechanisms of resistance to *Melampsora lini*. *Physiol. Plant Pathol.* 3: 241–247. Academic Press, London.) (**k**) Current distribution of coffee rust (*Hemileia vastatrix*) in outlined, dark shaded areas and the years in which infections were first noted in those areas. Mode of its introduction into the western hemisphere is uncertain, i.e., wind and/or inadvertent human activity. Its widespread distribution within much of Asia, Africa, and South America is thought to have resulted primarily from wind dispersal of urediniospore inoculum. (With permission from Lucas, J.A., 1997. *Plant Pathology and Plant Pathogens*. 3rd ed., Blackwell Science, Oxford, U.K.)

CASE STUDY 18.1

DAYLILY RUST

- Daylily rust, caused by *Puccinia hemerocallidis*, has become one of the most important diseases since it was first reported in the United States (Williams-Woodward et al., 2001).
- The disease cycle of daylily rust can be accomplished in the urediniospore stage in daylily.
- Differences in resistance to rust exist in daylily cultivars.
  - On the susceptible cultivars, symptoms are chlorotic speckles with abundant uredia and urediniospores (Figure 18.6a).
  - On resistant cultivars, symptoms show as necrotic lesions (hypersensitive reaction) with no urediniospore or a small number of urediniospores (Figure 18.6b).
- Host resistance is a major strategy for daylily rust management.
- Fungicide application is one of the components in the integrated disease management (Mueller et al., 2004).
- Carefully remove and destroy infected leaves from plants and beds, which can reduce the amount of inoculum.

---

several others, as demonstrated by Ericksson in Sweden in 1894. There are also "physiologic races" within many species of several rusts, e.g., *Melampsora lini* on flax and in many *formae speciales* of rusts (e.g., *P. graminis* f. sp. *tritici*) that can attack only certain cultivars of host species. The identity and, sometimes, the actual genotypes of the physiologic races are determined by inoculating defined sets of "differential varieties" with urediniospores, and approximately 10 to 14 days later, observing the various degrees or types of immune, resistant, or susceptible responses of those varieties. The distribution of reaction types across the set of differential varieties tested determines the identity of the rust race or races so tested.

Parasitically, rust fungi obtain nutrition through specialized branches (**haustoria**) (Figure 18.5h,i) that extend from intercellular hyphae into living host cells. Nutrients absorbed by haustoria are transported through the intercellular hyphae, supporting further expansion of the fungus colony and the production of reproductive spores. Nutrients diverted from the host into the pathogen contribute to reduced yield or quality of the plant's economic product (e.g., see Figure 18.4a).

Resistance to rusts is often expressed as "hypersensitivity" or the "hypersensitive reaction" (HR) (Chapters 28 and 31), a very localized reaction in which only a few host cells are killed (Figure 18.5j), thus depriving the obligate parasite of a source of nutrition. Other mechanisms, e.g., production of toxic compounds (phytoalexins) associated with HR, can contribute to death of the pathogen in infected tissue. Other types of resistance include "tolerance," where the host is capable of producing an acceptable yield in spite of rust infection, and "slow rusting," when rust development in host tissues is significantly retarded rather than halted (killed), as in hypersensitivity. These latter types of resistance are thought to result from more widely based, multigenic responses than hypersensitivity,

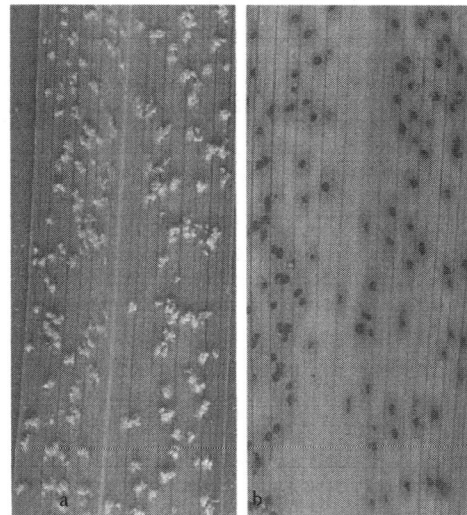

FIGURE 18.6 The symptoms of daylily rust caused by *Puccinia hemerocallidis* on the susceptible and resistant cultivars. (**a**) Chlorotic specks with abundant uredia and urediospores on the susceptible cultivar. (**b**) Necrotic lesions (hypersensitive reaction) with a few uredia and scant urediospores on the resistant cultivar.

which often results from the action of a single resistance gene in the host. Rapid development and spread of new physiologic races of rusts results in part from the selection pressure placed on rusts by the planting of resistant cultivars over large regions of crop production. The genetic control of resistance and susceptibility of many hosts, and the pathogenicity of many rust pathogens, respectively, are governed by the "gene-for-gene" relationship, first demonstrated by Flor in the 1950s with flax rust.

A wide array of management techniques for rust diseases is dictated by the many variations in life cycles of rust fungi, the disease cycles of the different diseases, the diversity of host plants and differences in their cultural methods, the variably important significance of wind dispersal of spores, different economic impacts of rust diseases on different hosts, and differences in breeding techniques for herbaceous versus timber species. Examples of some more commonly used methods are discussed in the following text.

Genetic resistance (Chapter 34). Wherever genetically resistant cultivars are available, they are generally the most effective and least expensive means of managing rust diseases. Many cultivars of rust-resistant agronomic, orchard, and vegetable crops are available to commercial growers and homeowners. If they are not available, or when new races of rusts arise, different means of disease management must be utilized.

Foliar fungicidal sprays (Chapter 36). Both protective and eradicant sprays are widely recommended for control of many rust diseases of herbaceous and orchard crops and ornamentals; many are quite effective. Unfortunately, the widespread, continued use of such fungicides sometimes provides selection pressure for evolution of new, fungicide-resistant or fungicide-tolerant strains of rust pathogens. Judicious use of foliar sprays is thus essential.

Cultural practices (Chapter 35). Practices such as planting susceptible conifers in low-risk environments, based on environmental evaluations, is recommended for some conifer rusts and is suggested even for some cereal rusts. Sanitation procedures, e.g., inspecting and destroying infected seedlings, and pruning out infected branches well below cankers or galls, another type of cultural practice, are recommended for some pine rusts. Other examples include providing adequate greenhouse ventilation to reduce high humidity favorable for geranium rust; similarly, for producers of organic asparagus, planting crop rows parallel to the prevailing wind is recommended to facilitate air movement and to reduce humidity in asparagus fields for management of rust disease. Removing or plowing under crop debris, including overwintering teliospores, is recommended for some vegetable diseases. However, such efforts may be futile if other growers in the region fail to do the same, thus providing the region an ample supply of wind-borne inoculum, similar to the near impossibility of controlling corn smut by sanitation.

Elimination of alternate hosts. This has long been a recommended practice for many heteroecious rusts, dating back to the 1660 French legislation and similar laws in the American colonies, spanning several decades in the 1700s that prohibited growing barberry bushes in close proximity to wheat fields. Those laws, based on empirical observation, far predated scientific proof for the linkage of wheat and barberry by DeBary in Germany in 1865. Influenced by a severe rust epidemic in 1916 and crucial demands for North American wheat during World War I, massive barberry eradication campaigns were implemented in 18 northerly, wheat-producing states of the United States and three adjacent Canadian provinces. That campaign lasted

for some 50 years, although less in some states and provinces than others, and resulted in over 100 million barberry bushes being destroyed. The practice greatly reduced localized infections of wheat rust early in the growing season, which definitely increased yields in the participating states and provinces and reduced the potential for development of new races by genetic recombination on the barberry host. However, barberry eradication did little to reduce the annual northward advance of urediniospores and the consequent wheat stem rust infections along the Puccinia Path, as the latter was more a result of meteorological than biological factors. Removal of junipers near apple orchards was formerly recommended as a control for cedar-apple rust, but the impracticality of that, given the widespread distribution of junipers and the wind-borne distribution of basidiospores from that host, was later realized. Similarly, in the United States, only partial success was achieved by eradicating wild currant bushes in western states for control of white pine blister rust, although the practice was more successful in the eastern states.

EXERCISES

Unfortunately, rusts do not lend themselves easily to classroom laboratories situated in departments or universities in which rust research is not being conducted. For the necessary supply of urediniospore inoculum, it is best if one can provide that personally or obtain it from colleagues working with rust fungi. Purchasing an initial supply of spores and increasing that sufficiently for class use is difficult to accomplish if one is not already familiar with the basic methods of rust culture. Also, because rusts are plant pathogens, they cannot be legally shipped across state lines without obtaining proper permission from USDA/APHIS, which can take several weeks. Two alternative rust disease laboratory exercises are suggested, however, depending on local availability of host plant seeds and rust urediniospores. Both exercises have similar objectives and demonstrate similar methods. The following are the educational objectives of the experiments:

  To learn methods of maintaining rust cultures on living host plants;
  To demonstrate different manifestations of susceptible and resistant responses, depending on the availability of host cultivars having different levels of resistance;
  To demonstrate effects of different temperatures on rate of rust development and production of urediniospores, depending on the availability of growth chambers or greenhouses of different temperatures;
  To observe anatomical properties of urediniospores, e.g., surface ornamentation, germ pores (numbers and distribution), and "normal" anatomy,

i.e., size and shape of partially dehydrated spores under normal, ambient air conditions versus "textbook" anatomy of hydrated, liquid-mounted spores typically used for microscopic examination (Littlefield and Schimming, 1989).

EXPERIMENT 1: WHEAT LEAF RUST

Depending on the local availability of differentially resistant and/or susceptible cultivars of wheat and differentially virulent races of *P. triticina*, one can demonstrate different manifestations of resistance and susceptibility to leaf rust infection. Total time required to complete the experiment is about 3 to 4 weeks.

Materials

The following materials are needed per student or group of two or three students, and are based on supplies needed for two pots per student group (the minimum recommended numbers of items needed may increase depending on availability of growth chambers in which to observe effect of temperature on disease development):

- Two 3 to 4 in. diameter plastic pots

- Sufficient commercial potting mix for two pots
- 20 to 30 seeds of a rust-susceptible wheat cultivar
- 4 to 5 mL dry talc
- Microspatula
- 0.5 to 1.0 mL fresh urediniospores of leaf rust fungus (or one pot of young, leaf-rust infected wheat plants to use for "brush-on" inoculation alternative described)
- A test tube or other small vial
- A piece of cheesecloth, approximately 6 × 6 in.
- A household water-mist applicator bottle (one bottle sufficient for many students)
- A 1-gallon zip-lock or similar food storage bag
- Sufficient pot labels, one per pot, depending on total number of pots used

Follow the protocol outlined in Procedure 18.1 to complete this experiment.

Anticipated Results

If favorable infection conditions are created, urediniospores should germinate and infect cells within 12 to 24

Procedure 18.1
Production of Wheat Leaf Rust in the Laboratory

Step	Instructions and Comments
1	Grow wheat seedlings in 3 to 4 in. pots containing commercial potting mix (10 to 15 plants per pot) until 2 to 3 weeks old.
2	Inoculate plants with a mixture of urediniospores and dry talc (mix thoroughly about 4 to 5 mL dry talc and less than 1 mL fresh or stored urediniospores—about the volume of dry spores that can be picked up on the terminal centimeter of a microspatula). Place that inoculum mixture into a test tube, a screw cap vial, or similar vessel and affix 4 to 5 layers of cheesecloth over the open top of that vessel with a rubber band. The inoculum mixture can then be applied to the wheat leaves by gently shaking the inverted test tube above the leaves, allowing the talc and spores to settle onto the leaves. Only a faint layer of talc–spore deposit on the leaves is sufficient for the experiment to succeed. Alternatively, healthy wheat plants can be inoculated by simply "brushing" them with an inverted pot of rust-infected wheat inoculated about 14 days earlier. (Be sure that pustules are evident.) This method will transfer sufficient urediniospore inoculum to the healthy plants and result in infection.
3	Very lightly atomize water onto the plants, being careful not to wash off the inoculum–talc mixture. Place pots into a 1-gallon plastic bag containing a water-soaked paper towel, seal the bags, and allow inoculated plants to incubate at room temperature for 18 to 24 h. Partially open the bags to allow plants to dry gradually over the next 3 to 4 h and then remove the pots to a greenhouse or growth chamber.
4	By maintaining inoculated plants at different temperatures ranging from about 15 to 30°C, the effect of temperature on rust development can be observed over the next 7 to 14 days.
5	Spores can be examined microscopically in aqueous mounts to observe and record spore size, ornamentation, number, and distribution of germ pores. Examine dry urediniospores on microscope slides, not mounted in liquid, to observe their normal size and partially collapsed shape under typical ambient atmospheric conditions compared to those mounted in liquid.

CASE STUDY 18.2

ASIAN SOYBEAN RUST—A POSSIBLE ECONOMIC THREAT

- Asian soybean rust (ASBR) is caused by the fungus *Phakopsora pachyrhizi*. ASBR is an obligate fungal parasite that produces both uredospores and teliospores. *Phakopsora meibomiae*, a closely related rust species, is less aggressive and does not cause severe yield losses in soybeans. This species was present in Puerto Rico in 1976 and may also occur in the southern United States.
- ASBR was reported from South America in 2001 and was discovered in North America near Baton Rouge, LA, in November 2004.
- It is believed that spores of ASBR were transported to North America via the strong winds of Hurricane Ivan, and by November 2004, nine southern states reported soybean rust. Morphological and molecular testing (polymerase chain reaction) confirmed the presence of ASBR in the United States in 2004.
- Fungal survival throughout the year is dependent on the continued production of urediospores on a suitable host. ASBR has been reported to overwinter on alternate hosts, such as kudzu (*Pueraria lobata*), in southern coastal states.
- Development of this disease is greatest during excessive leaf moisture caused by heavy, prolonged dews or frequent rains. Optimum temperature for development of the fungus is between 15 and 28°C, but may take longer during periods of temperature extremes (Yeh et al., 1982).
- In addition to soybeans (*Glycine max*), the host range for soybean rust includes more than 90 species of legumes, including wild soybean (*G. soja*), yam bean (*Pachyrhizus erosus*), snap bean (*Phaseolus vulgaris*), yellow lupine (*Lupinus arboreus*), cowpea (*Vigna unguiculata*), and kudzu in the southern United States.
- Current control measures include early planting, planting of sentinel plots to determine presence of the pathogen, scouting, and proper timing of foliar fungicide applications. It is recommended that growers increase row spacing to allow greater aeration of the canopy.

---

h. Symptoms and signs should develop over 7 to 14 days after inoculation. Initially small chlorotic spots ("flecks") develop on leaves, which, in immune reactions, will remain as such; in intermediate and susceptible reactions, they will continue to develop into larger infections, eventually producing uredinial pustules that rupture the leaf epidermis and bear urediniospores.

Questions

- What results would you expect if you had used the urediniospores to inoculate the pycnia/aecial (alternate) host (*Thallictrum* sp.) rather than wheat?
- What anatomical stages of rust fungus developed that could have been observed microscopically during the 18- to 24-h period in the moist plastic bags?
- If you inoculated host varieties having different levels of rust resistance, describe and compare development of symptoms and visible signs of the pathogen in the different host varieties. How was hypersensitivity expressed, if it occurred?

EXPERIMENT 2: BEAN RUST

The basic objectives and procedures are similar to those for the wheat leaf rust exercise. This alternative is included for those who might have better access to the bean rust pathogen, *Uromyces appendiculatus*, or for classes comprised of students more interested in horticultural than cereal crops. Total time required to complete the experiment is about 3 to 5 weeks.

Materials

The following materials are needed per student or group of two or three students:

- Basically, the same number and kinds of supplies needed for wheat leaf rust study (Procedure 18.1), except plant three bean seeds per pot
- Edible dry pinto beans available in grocery stores provide inexpensive, susceptible seeds for this experiment
- Tween 20 solution if mist inoculation method is preferred
- Bean rust fungus, *U. appendiculatus*

Follow the instructions provided in Procedure 18.2 to complete this experiment.

Questions

- With this rust, what spore stage would have been produced as a result of your inoculation had you

---

Procedure 18.2

Production of Bean Rust in the Teaching Laboratory

Step	Instructions and Comments
1	Grow seedlings in commercial potting mix in 4-in. pots until 2 to 3 weeks old (three plants per pot, thin to two plants before inoculation with rust). Soaking seeds in wet paper towels or disposable diapers 24 h prior to planting will hasten germination. For best disease development, inoculate plants when leaves are about 50 to 70% fully expanded. (If leaves are expanding too rapidly before the scheduled class time, place pots into refrigerator. Cold temperatures will reduce the rate of leaf expansion without hurting the plant).
2	Inoculate plants with a mixture of *Uromyces appendiculatus* urediniospores and dry talc; subsequently, incubate plants and examine for development of symptoms, signs, and urediniospore anatomy as described for wheat leaf rust in Procedure 18.1. Alternatively, mix dry spore inoculum in water containing Tween 20 (two drops/liter) using about 20 to 30 mL of the water/Tween solution. Use atomizer to moist-inoculate plants with urediniospores suspended in the water/Tween solution. Lightly mist plants with water, moist-incubate them 24 h at room temperature, gradually dry plants, and place into greenhouse or growth chamber (see Procedure 18.1). Observe infection development and urediniospore morphology as described for wheat leaf rust in Procedure 18.1.

---

used aeciospores rather than urediniospores as inoculum?

- Describe the anatomical properties of urediniospores produced as a result of the inoculation, including pigmentation, spore ornamentation and number, and distribution of germ pores.
- What effect did incubation temperature have on development of the rust? What appeared to be the optimum temperature for this rust's development?

The following are some variations in the foregoing experiments, depending on availability of supplies, growth chambers, or greenhouse space:

- Include, incubate, and maintain controls in each exercise. Mock-inoculate other plants with water mist only to show that rust will not develop on the plants without having been inoculated with the pathogen.
- Cross-inoculate plants, i.e., bean plants with *P. triticina* and wheat plants with *U. appendiculatus*, to show restricted host range of each pathogen. Nonhost plants will not get rust infection.

SUGGESTED READING AND LITERATURE CITED

Agrios, G.N. 2005. *Plant Pathology* (5th ed.). Academic Press. San Diego, CA.

Alexopoulos, C.J., C.W. Mims, and M. Blackwell. 1996. *Introductory Mycology* (4th ed.). John Wiley & Sons. New York.

Brown, M.F. and H.G. Brotzman. 1979. *Phytopathogenic Fungi: A Scanning Electron Stereoscopic Survey.* University of Missouri–Columbia Extension Division, Columbia, MS.

Buller, A.H.R. 1950. *Researches on Fungi*, VII; The Sexual Process in the Uredinales. University of Toronto Press. Toronto.

Bushnell, W.R. and A.P. Roelfs. 1984. *The Cereal Rusts*, Vol. 1. Academic Press. New York.

Kenaga, C.B., E.B. Williams, and R.J. Green. 1971. *Plant Disease Syllabus*, Balt Publishers.

Littlefield, L.J. 1973. Histological evidence for diverse mechanisms of resistance to *Melampsora lini. Physiol. Plant Pathol.* 3: 241–247. Academic Press, London.

Littlefield, L.J., 1981. *Biology of the Plant Rusts: An Introduction.* Iowa State University Press. Ames, IA.

Littlefield, L.J. and W.K. Schimming. 1989. Size and shape of urediniospores as influenced by ambient relative humidity. *Mycotaxon* 36: 187–204.

Lucas, J.A., 1997. *Plant Pathology and Plant Pathogens.* 3rd ed., Blackwell Science, Oxford, U.K.

Mueller, D.S., S.N. Jeffers, and J.W. Buck. 2004. Effects of timing of fungicide applications on development of rusts on daylily, geranium, and sunflower. *Plant Dis.* 88: 657–661.

Roelfs, A.P. and W.R. Bushnell. 1985. *The Cereal Rusts*, Vol. 2. Academic Press. New York.

Schieber, E. and G.A. Zentmyer. 1984. Coffee rust in the Western Hemisphere. *Plant Dis.* 68: 89–93.

Schumann, G.L. 1991. *Plant Diseases: Their Biology and Social Impact.* APS Press. St. Paul, MN.

Williams-Woodward, J.L., J.F. Hennen, K.W. Pardar, and J.M. Fowler. 2001. First report of daylily rust in the United States. *Plant Dis.* 85: 1121.

Yeh, C.C., J.B. Sinclair, and A.T. Tschanz. 1982. *Phakospora pachyrhizi*: Uredial development urediospore production and factors affecting teliospore formation on soybeans. *Aust. J. Agric. Res.* 33: 25–32.

Ziller, W.G. 1974. *The Tree Rusts of Western Canada*. Canadian
    Forestry Service Publication No. 1329, Department of the
    Environment, Victoria, B.C.

WEBSITE

United States Department of Agriculture, Pest Information Plat-
    form for Extension and Education: http://www.sbrusa.
    net/.

19 Basidiomycota
Fleshy Mushrooms and Other Important and Symbiotic Associations

Richard E. Baird

CHAPTER 19 CONCEPTS

- Fungi in this group all produce haploid **basidiospores** on **basidia**. The basidia line the surface of gills (typical mushroom), tubes (boletes), pores and teeth (shelf and conks), and internally (puffballs).

- Some fleshy basidiomycetes cause heart, butt, and root rots of trees.

- The hyphae of many basidiomycetes have **clamp connections**.

- Hyphal cells are separated by a very complex **dolipore septum**.

- Many fleshy basidiomycetes are associated with **wood decay**; some of them form **ectomycorrhizae** with plant partners, and relatively few mushrooms are **parasitic**.

Approximately 30% of all known fungi are classified into the phylum Basidiomycota (basidiomycetes; Alexopoulus et al., 1996). A large percentage of this total belongs to the group known as the fleshy basidiomycetous fungi that produce soft to flexible fruiting bodies that decay or decompose rapidly. Commonly called mushrooms (Figure 19.1), fleshy basidiomycetes include gilled and pored fungi, the puff balls, bird's nest fungi, jelly fungi, and the stinkhorns (Christensen, 1950). Species that form structures associated with wood decay are generally shelf- to conk-like (lacking a stalk; Figures 19.2 and 19.3) or resupinate (appressed to the plant tissue; Figure 19.4).

In nature, basidiomycetes exist as somatic mycelium (hyphae grown in a mass or layer) on wood, litter, and in

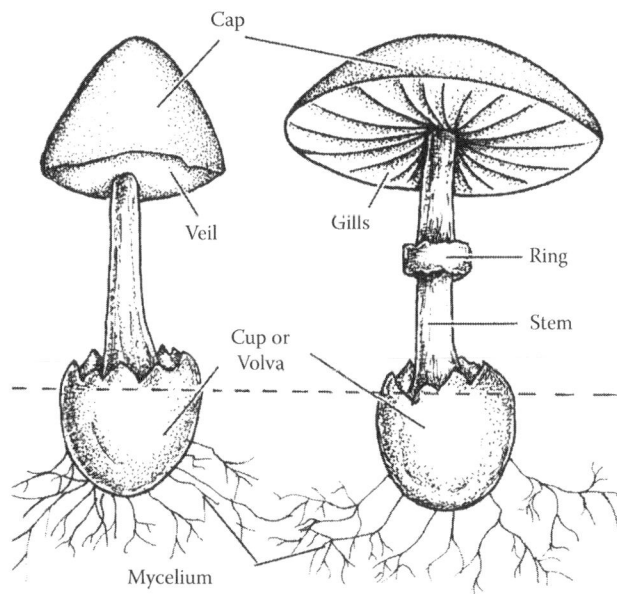

FIGURE 19.1 Diagrammatic representation of a mushroom. (Illustration provided by J. McGowen, Mississippi State University.)

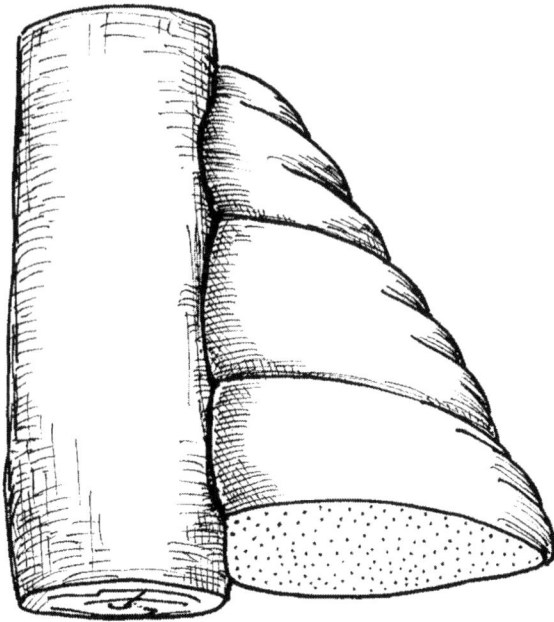

FIGURE 19.2 Woody mushrooms called conks are shelf-like basidiocarps where basidiospores are formed. Conks grow on living or dead woody plants or trees. (Illustration provided by J. McGowen, Mississippi State University.)

soil. When environmental conditions are favorable, they form fruiting bodies of various sizes and shapes. Fleshy basidiomycetes occur worldwide in diverse habitats in the cold, temperate, and tropical regions. Individual species, however, are restricted in distribution based on habitats or flora constituents.

Although the majority of basidiomycetous fungi are saprophytic, certain mushroom-forming species are necro-

trophic, and still others, the rusts and smuts (Chapter 18), are obligate biotrophic plant pathogens. Saprophytic fungi, such as bird's nest fungi and stinkhorns, may be brought to the attention of plant pathologists by the public when these fungi are found associated with landscape plantings. Many species form mycorrhizal associations with woody plants and forest trees, whereas some species of fleshy basidiomycetes are associated with aquatic or wetland ecosystems such as bogs (Alexopoulus et al., 1996). In addition to the types of basidiomycetous fungi already mentioned, the genera *Ceratobasidium* and *Thanatephorus*, the teleomorphs of *Rhizoctonia* species (Chapters 16 and 17), produce basidia on extremely small mycelial mats on soil or plant surfaces. A discussion of *Ceratobasidium* and *Thanatephorus/Rhizoctonia* is included later in this chapter because of the economic importance of these genera to agriculture.

MORPHOLOGY

Species of fleshy basidiomycetes range from saprophytic to parasitic, and the morphological structures of the fruiting bodies within this group are classified by several important characteristics. One of the most important is the presence of **basidia**, of which several forms or types occur. The basidium (singular) is the site where karyogamy and meiosis occur. The resulting nuclei that form following meiosis migrate through sterigmata, small stalk-like outgrowths, on which spores are formed externally. The **basidiospores** (haploid or N) are formed from the products of meiosis and are also called meiospores.

Hyphae of the basidiomycetes appear to be similar to other phyla such as the Ascomycota, except that the

FIGURE 19.3 Conk of *Fomes fomentarius.* (Photograph by A. Windham, University of Tennessee.)

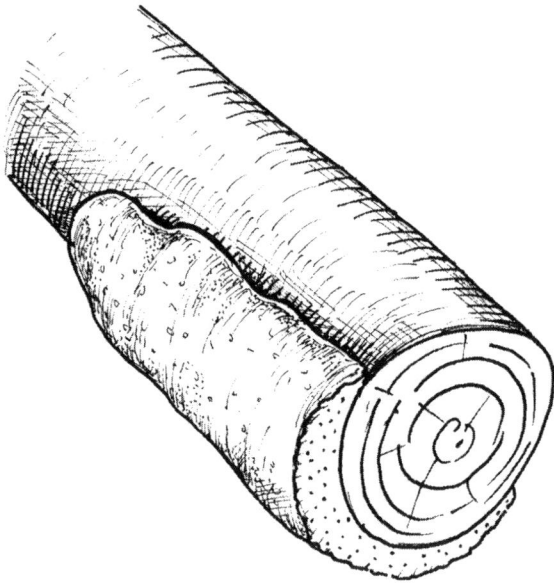

FIGURE 19.4 Resupinate basidiocarp growing on a dead branch. (Illustration provided by J. McGowen, Mississippi State University.)

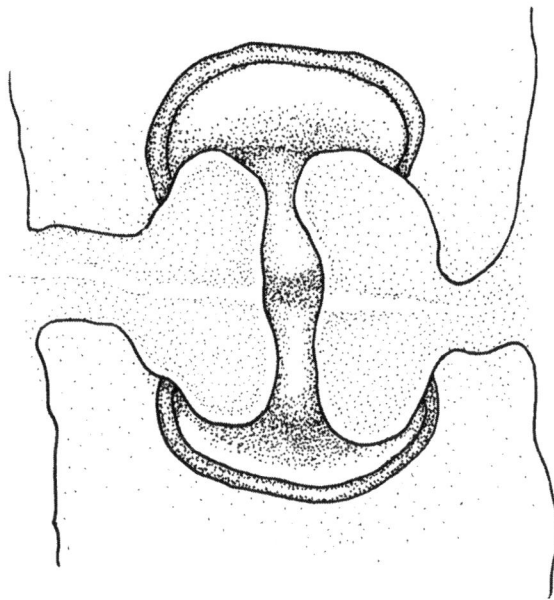

FIGURE 19.5 Hyphae containing a dolipore septum that occurs only in certain genera and species of basidiomycetes. The structure of the septum restricts movement of nuclei between hyphal cells and maintains nuclear condition. (Illustration provided by J. McGowen, Mississippi State University.)

hyphae contain a specialized septum called the **dolipore septum**. The dolipore septum (Figure 19.5) is unique to most members of the Basidiomycota and restricts movement of nuclei and other organelles between the hyphal cells. The exception is the rust fungi (Chapter 18); these basidiomycetes have cross walls similar to species of the Ascomycota. Another important characteristic that dis-

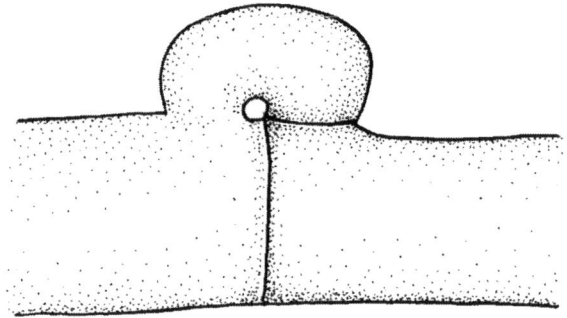

FIGURE 19.6 Hyphae with clamp connections that occur only in some basidiomycetes. (Illustration provided by J. McGowen, Mississippi State University.)

tinguishes the basidiomycetous fungi from other fungal phyla is the presence of **clamp connections** (Figure 19.6) on the hyphae where septa have formed. This structure occurs in less than 50% of all species of basidiomycetes. Clamp connections help ensure the dikaryotic (N + N) condition of new cells during compartmentalization (septa formation) of the mycelium. This enables the basidiomycete fungi to reproduce sexually.

Although most fleshy basidiomycetes are associated primarily with decay of organic material, other species form unions with plants and forest trees in what is called a mycorrhizal or symbiotic association. However, several mushroom-forming genera can be pathogenic on economically important crops such as corn, vegetables, grasses, and forage crops. Members included in the genus *Marasmius, Xerompholina,* and *Armillaria* have also been reported to be pathogenic on agricultural and horticultural crops.

BASIDIOMYCETE TAXONOMY

The Basidiomycota is a diverse group of fungi. Molecular studies have caused mycologists to subdivide the group into the Hymenomycetes, Ustilaginomycetes, and Urediniomycetes (Swann and Taylor, 1993; McLaughlin et al., 1995; Berres et al., 1995). The hymenomycetes, as indicated by the name, produce aggregates or clusters of basidia on a layer of specialized hyphae called a **hymenium**. All are members of the fleshy basidiomycetes. This group consists of the Agaricales (mushrooms), Gastromycetes (e.g., puff-balls and bird's nest fungi), Aphyllophorales (e.g., wood decay fungi), the groups of jelly fungi, and the Ceratobasidiales. Ceratobasidiales is an order in which the economically important genus *Rhizoctonia* (asexual state) belongs. Members of the deuteromycete genus *Rhizoctonia* may belong to either of the two sexual or basidia-forming states. These include the genera *Ceratobasidium* that has two nuclei per cell, and *Thanatephorus,* which has more than two nuclei per cell.

The Agaricales or gilled fungi belong to an order that are referred to as toadstools or mushrooms, which are the

---

CASE STUDY 19.1

Honey Mushroom Cause Armillaria Root Disease

- Honey mushrooms or *Armillaria* spp. causes root disease throughout temperate and tropical regions, attacking conifers and declining hardwood tree species such as oaks. These fungi belong to the order Agaricales, forming basidiocarps that typically have caps, gills, and stems.
- This mushroom-forming species is found in clusters, are honey or yellow-brown in color, with small scales on the caps, and with white gills and spore prints. A complex of different species of *Armillaria* can be found throughout the United States.
- The honey mushrooms, which are common in the forests of the United States, are also known as the "shoestring fungus." They become established on the roots of trees and attack sapwood, colonizing the tissue under the bark, moving along the boles as black rhizomorphic strands or "shoestrings."
- Long-range basidiospore dispersal is by wind. Short-range infections occurs by mycelial or rhizomorphic contact with roots. After establishment, the fungus can survive saprophytically up to 50 years in stumps, waiting for the next generation of trees to infect.
- Control of *Armillaria* root disease is by stump removal following tree harvest. Foresters can plant resistant tree species if management practices allow for those species. Also, increased plant vigor and planting practices, which increase root growth, reduce damage by the pathogen.

---

fruiting bodies. Although there is extreme variation in morphology, mushrooms generally consist of a stalk (stipe) and a cap (pileus). The cap and sometimes the stalk support gills (lamellae) or pores where the basidia and basidiospores are produced. When a stalk is absent, the fruiting bodies are called woody conks, resupinate or bracket fungi. The gills or pores have a hymenial layer where basidia and basidiospores are formed (Figure 19.1).

HABITAT

Members of the Agaricales are found on a broad range of plants and are commonly associated with decay of dead plants and woody tissues (lignicolous). Others types are more specific or occur in select ecosystems such as the coprophilous (dung inhabiting) fungi or the hallucinogenic mushroom species. *Psilocybe cubensis* is a very common coprophilous species found in pastures in the southern United States, and is often collected for its hallucinogenic properties.

Certain mushroom-forming species are extremely host specific. Examples of this specific host–fungi relationship is the saprophytic mushroom-forming species *Strobilurus esculentus*, which forms mushrooms only on fallen spruce cones, or *Auriscalpium* species (tooth-forming fungus) observed only on cones of pine and firs. The mycorrhizal-associated mushroom species *Suillus pictus* and *S. americanus* are found primarily where white pine trees grow (Weber et al., 1988).

MYCORRHIZAE

Almost all vegetation and trees require a beneficial symbiotic relationship with certain genera and species of fungi.

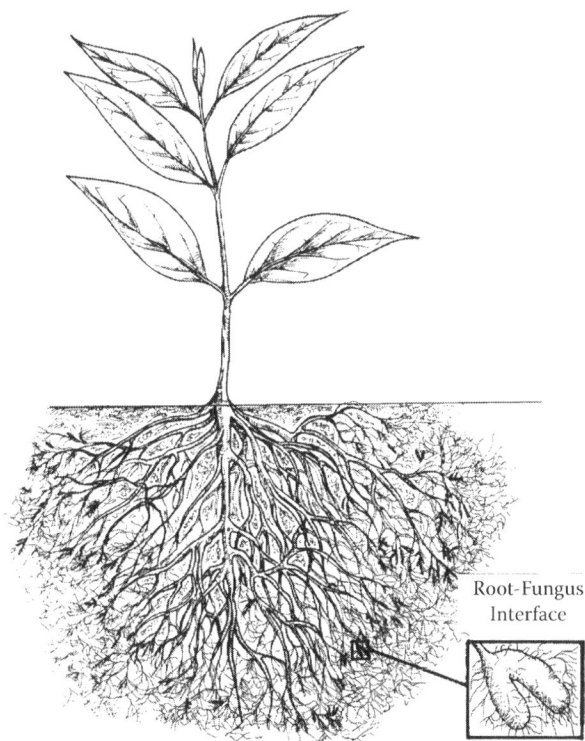

Root-Fungus
Interface

FIGURE 19.7 Ectomycorrhizal associations between a fungus and a tree. Fungus acts as a "secondary root system" to provide nutrients and water for the host tree species. (Illustration provided by J. McGowen, Mississippi State University.)

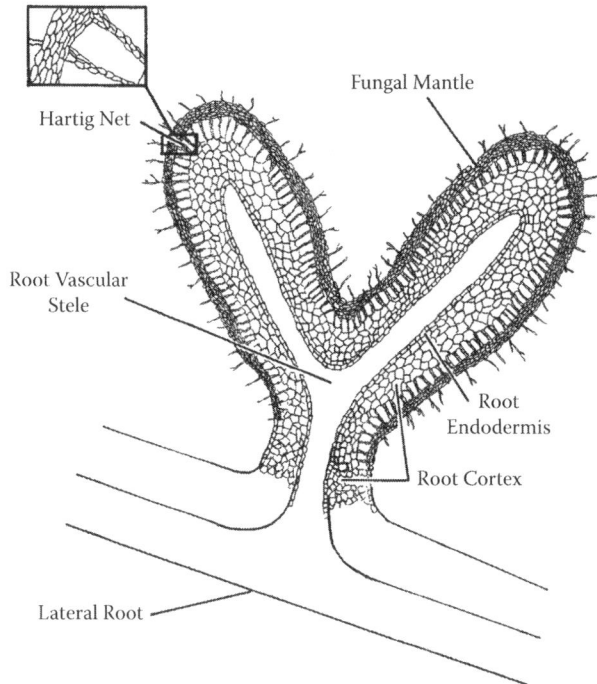

FIGURE 19.8 Endomycorrhizal fungus growing into the cortical cells of the tree roots. "Hartig Net" (box) from an ectomycorrhizal association. Hyphae from the fungus form an outer layer around the root and grow between the cortical cells. (Illustration provided by J. McGowen, Mississippi State University.)

In a mycorrhizal relationship (Chapter 10), fungal hyphae envelope the thin rootlets of trees but do not harm the roots. The symbiotic relationship occurs when carbohydrates or sugars from photosynthesis are passed from the plant to the fungal associate, whereas the fungus acts as a secondary root system providing water and nutrient uptake to the plant or tree host (Figure 19.7). The transport or movement of the materials to either partner in the symbiotic association involves complex chemical pathways not discussed here. There are generally considered to be two types of mycorrhizal fungal associations—endomycorrhizae and ectomycorrhizae. Hyphae of endomycorrhizal fungi grow into the cortical cells of the root and do not form a thick layer of tissue around the root hairs (Figure 19.8). These fungi do not belong to the fleshy basidiomycetes but are primarily members of the Zygomycota (Chapter 10). Alexopolous et al. (1996) estimated that there were over 2000 species of ectomycorrhizal fungi, and the majority were fleshy basidiomycete mushroom-forming species. The hyphae of these fungi attach to the feeder rootlets of plants, forming a fungal mantle or thick layer of mycelium around the surface of the roots. In a true ectomycorrhizal association, the fungus initially forms a mantle (Figures 19.8, 19.9, and 19.10) around the roots before growing between the outer cortical cell layer to form what is called the "Hartig Net" (Wilcox, 1982). The color of the mantle varies depending upon the fungal species. This network of fungi is the main distin-

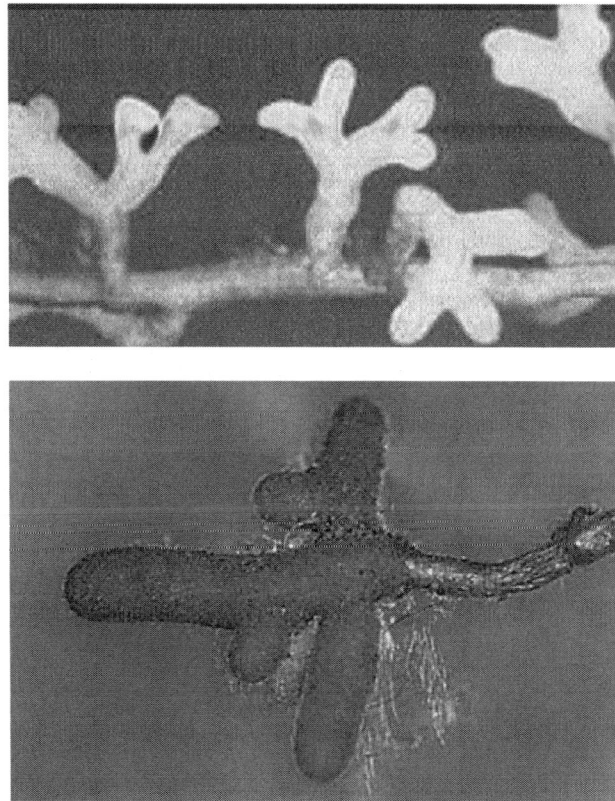

FIGURE 19.9 Feed roots with "Hartig Net" of ectomycorrhizal fungus. (Photographs by A. Windham, University of Tennessee.)

FIGURE 19.10 Ectomycorrhizal mushroom forming *Russula* sp. (Photograph by A. Windham, University of Tennessee.)

guishing feature that defines ectomycorrhizae because a number of fungi might be able to form what could be interpreted as a mantle but do not have intercellular growth. The root hairs in an ectomycorrhizal association become forked or multiforked, or are irregularly shaped. Other fungi that form ectomycorrhizal associations are members of the phylum Ascomycota, including the cup fungi and truffles, but the number of species and genera are fewer than for the Basidiomycota.

Ectomycorrhizal fungi are associated with many plant and tree species. Important tree partners include members of the Pinaceae (pines, true firs, hemlock, spruce, and others), Fagaceae (oaks, beach, and others), Betulaceae (birch, alders, and others), Salicaceae (poplars, willows, and others), and Tiliaceae (basswood). On a given tree root, many ectomycorrhizal-forming species can occur. A specific tree species can partner with many different fungal species at one time. For example, Douglas fir in the northwestern United States has more than 200 ectomycorrhizal associates. In turn, a single fungus can associate with a broad range of plants or be specific to select tree species. An example is the mushroom-forming species *S. americanus* that primarily occurs under white pine. In contrast, *Amanita vaginata* occurs under many species of pines and hardwoods.

Pathogenic genera and species of fleshy basidiomycetous fungi can infect agricultural crops and forest tree species. For agricultural crops, a complex of fungi called the "sterile white basidiomycetes," containing genera that produce white mycelium with clamp connections in culture, have been identified as mushroom-forming species, including *Marasmius* and *Xeromphalina*. Cultures of these fungi have been reported to infect roots and lower stems of corn, sorghum, and millet, and vegetable crops such as snap beans.

Mushroom, conk, or bracket-forming basidiomycetous fungi have been shown to be pathogenic on forest trees. A classic example of pathogenicity by a mushroom-forming species is *Armillaria mellea* or *A. tabescens* (shoestring fungus–honey mushrooms). The fungus becomes established on trees or woody plants by forming rhizomorphs, black shoestring-like structures that transport nutrition and water, extending out from the mycelial mat. Rhizomorphs are a clustering of hyphae into strands forming a complex mycelial structure (Figure 19.11). *Armillaria mellea* is one of the most common, best known, and most damaging fungus that cause diseases of forest, shade, and ornamental trees and shrubs. It flourishes on an extraordinary broad host range and under diverse environmental conditions. *Armillaria mellea* becomes established when host plants are under stress (e.g., nutrients, water), whereas *A. tabescens* can be highly virulent. The latter fungus invades bark and cambium tissues of the roots and the root collar of coniferous tree species. The extent and speed of invasion depends on host species and degree of stress to the plants. Plant stress increases by plant competition and by mechanical injury, resulting in declining root systems. After the host dies, the fungus continues to decay the host tissue and may survive in the soil for decades.

Another fleshy basidiomycete species pathogenic to forest trees is *Heterobasidium annosum* (the causal agent of annosum root rot). This basidiomycetous fungus occurs in coniferous forest ecosystems and can invade tree roots when under stress from environmental conditions such as drought or when management practices are minimal. The fungus occurs throughout the United States and Canada, but particularly causes severe economic damage in the southeastern United States. Young seedlings can be killed by *H. annosum*, but in older trees the heartwood is invaded, and chronic infections can continue for years. When roots of different trees graft, the fungus can invade the uninfected trees. The conks or shelf-like woody mushrooms can develop on infected living trees or on dead hosts. The

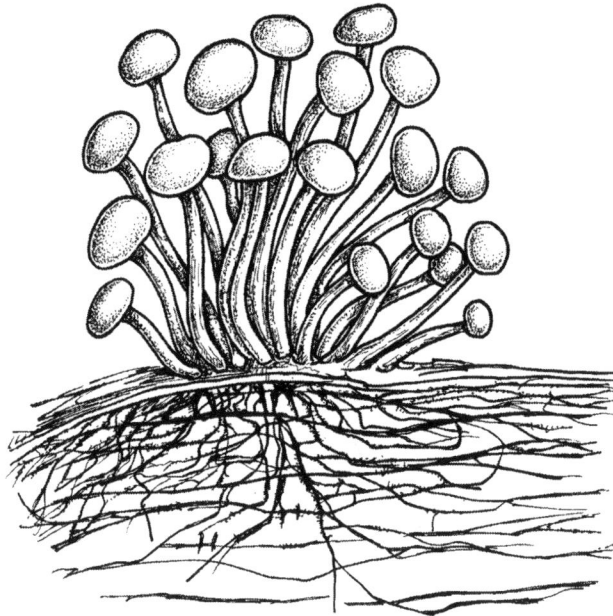

FIGURE 19.11   Rhizomorphs are black, shoestring-like structures deriving nutrients and water for fungal growth and sporulation. (Illustration provided by J. McGowen, Mississippi State University.)

conks generally are found slightly buried in the humus or can be seen just above the ground surface on the butt of the tree (Figure 19.2). Recognition of the basidiocarps or conks of *H. annosum* is useful for diagnosis.

WOOD ROTS CAUSED BY FLESHY BASIDIOMYCETES

The majority of fleshy basidiomycetes are saprophytic and often have the ability to decay wood. Fungi in this group derive nutrients from dead plant or woody tissues by enzymatically and chemically degrading cellulose, hemicellulose, and/or lignin (Fergus, 1966) (see Chapters 10, 29, and 30). Some of these fungi occur on living trees, but they generally decay only the dead tissues of the heartwood. Two groups of fleshy basidiomycetous fungi are most important in deriving nutrition in their role as wood rotters. The groups include the resupinate fungi (form flat basidiomata on host) (Figure 19.3) placed into the Thelephoraceae or Corticiaceae (Alexopoulus et al., 1996), and the remainder were traditionally placed into the group known as the Polypores. Within these groups the fungi can either be referred to as white rot or brown rot decay species.

WHITE ROTS

White rot fungi can generally utilize all components of the wood including lignin, cellulose, and hemicellulose. The wood degraded by these fungi is light in color due to the loss of lignin (brown color). Because lignin gives strength to the wood, the tissue becomes spongy, stringy, or mottled. Genera in the Agaricales, such as *Agaricus, Coprinus,* and

Pleuteus, are common plant- and wood-decay fungi. Many genera in the Aphyllophorales, such as polypores (pored fungi), chanterelles, tooth and coral fungi, and the corticiod-forming fungal species, also decay wood. An example of a white rot fungus is *Oxyporus populinus* that attacks numerous hardwood tree species and causes trunk and limb rots. *Trametes versicolor* attacks the sapwood of hardwood trees such as sweet gum, *Liquidambar styraciflua*.

BROWN ROTS

Brown rot fungi, also known as dry rot, generally degrade cellulose and hemicellulose and either lack or have limited ability to utilize lignin. Lignin is more or less brown and remains after other wood polymers have been decayed, so wood decomposed by these fungi appears brown, cubicle, and crumbly. There is a defined margin between healthy and decayed wood. This form of rot is more common in coniferous trees than hardwood tree species. *Poria placenta* and *Phaeolus schweinitzii* are often associated with brown rot of timber and lumber.

FAIRY RINGS

Fairy rings are fleshy mushroom-forming basidiomycetes that grow in circular zones. Although there are many species that form fairy rings, some of the more common genera include *Coprinus, Chlorophyllum, Lepiota,* and *Marasmius*. The rings are readily observed in lawns and golf courses, but can also be found in forests and other ecosystems. The rings occur from the expansion or outward circular growth of the mycelium that initially originated from one central point. The mushrooms, however,

---

CASE STUDY 19.2

RHIZOCTONIA SOLANI

- *Rhizoctonia solani* is one of the most important soilborne pathogens (Chapter 22) and can cause severe economic loss to agricultural crops. The fungus causes damping-off of seedlings for crops such as cotton, soybeans, and vegetables. The pathogen is also important in ornamental and forest nurseries.
- The pathogen is particularly damaging when growth conditions of seedlings are poor. If the germinating seed is not destroyed below ground, seedlings being attached can be seen with sunken dark brown cankers at the ground surface. Often, they will appear to be wind-blown and broken at the canker (damping-off).
- Different anastomosis groups of the *R. solani* occur and attack either a wide range of host or are host specific.
- The fungus does not produce asexual spores, but basidiospores are known to be produced on exposed basidia formed directly on soil or plant debris. *Rhizoctonia solani* form sclerotia as a means for surviving in the soil and in plant debris. The *R. solani* anastomosis group that attacks rice has adapted to survive in water by forming a thick outer layer around the sclerotia.
- During subsequent growing season, sclerotia germinate and can reinfect the host crop planted in the field. Because this fungus attacks many hosts, rotation to economically important nonhost crops is not a very effective means of control.
- Control is primarily by use of fungicides as seed treatments, and when appropriate, applying treatment in-furrow or drenching with fungicides. Growing conditions that increase plant health reduce post-emergence damage. Biological controls using bacterial species (*Bacillus* sp.) and other fungi (*Trichoderma* sp.) have also been successful as seed treatments.

---

almost always form near the outer edge of the active growth zones. Some fairy rings have been found to be hundreds of years old and extend over very long distances. A ring can be disrupted due to physical barriers or loss of the nutrient base, forming arcs or partial zones. In golf course greens or home lawns, fairy rings may be considered unsightly or may affect growth of the grass, causing obvious discolored patches and dead tissues within the zones. The mycelium from the fungus can act as physical barrier preventing adequate moisture and reduced vigor of the invaded lawn. Control or prevention of rings includes reducing thatch and maintaining adequate moisture to the thatch layers. Labeled fungicides are effective for controlling these fungi.

OTHER BASIDIOMYCETES (NONFLESHY)

An unusual plant pathogenic basidiomycete genus, *Exobasidium*, forms basidia directly from dikaryotic mycelium. A layer of hyphae forms between the epidermal cells of the plant tissue, causing tissue distortion, including swelling, blistering, and galling. Infections can occur on host species, including azalea, rhododendron, camellia, blueberry, ledum, and leucothoe, and other ornamental and forest plants. Hyphae of *Exobasidium* grow intercellularly and invade cells by the formation of haustoria. Controls included removal or sanitation of galls in landscape situations or small greenhouses. Fungicides can be used, but success depends on applications prior to bud break and follow-up sprays. Resistant host plants are generally not available.

The genus *Rhizoctonia* (Chapters 16 and 17) is one of the most economically important groups of pathogenic species in the world. Billions of dollars are lost annually from seed decay, root rots, seedling disease, and aerial or foliar blights. Although *Rhizoctonia* is known primarily for disease of agronomic, vegetable, and ornamental crops, several species were reported to form mycorrhizal associations with plants (e.g., orchids). *Rhizoctonia* species do not produce conidia in culture and can be difficult to identify. Isolates of *Rhizoctonia* species may vary greatly in the number and size of sclerotia (survival structures), hyphal growth (appressed or aerial) and color in culture, and presence of the dolipore septum (Chapter 17). Without conidia, these cultural characters alone make it difficult to identify *Rhizoctonia* species.

Identification is further complicated by different mating types, called anastomosis groups (AG) (Carling, 1996). These types will only mate by hyphal fusion to compatible anastomosis types. During complete anastomosis, the hyphal cell walls of the two compatible types fuse, lysis occurs, and the nuclei are exchanged between cells. Until recently, mating studies were the only techniques available for isolate identification. Isolates of *Rhizoctonia* that do not mate are called vegetatively incompatible.

The hyphae of the different groups have two levels of nuclei per cell. The binucleate (two nuclei per cell) group is known as the *Ceratobasidium* anastomosis group (CAG) and is named after the teleomorphic state for this group of *Rhizoctonia* species. Approximately 20 different mating types have been identified in this group. The

---

CASE STUDY 19.3

BLACK LOCUST DECAY BY *PHELLINUS RIMOSUS*

- *Phellinus rimosus* causes a yellow heart rot of living black locust throughout their native range. The fungus is specific to black locust trees.
- The fungus forms a conk or shelf-like structure that is perennial. The conks can be up to 1 ft wide, brown, becoming black on the upper surfaces, later with cracked and furrowed tops and with reddish brown pore surfaces on their underside with age.
- Basidiospores are wind blown to trees. The fungus becomes established in old branches and tunnels made by the locust borer insect.
- From the initial infections, decay progresses into the heartwood. The heartwood then becomes spongy-yellow or brown in mass over time.
- No real controls are noted for this disease, especially because black locust is not an economically important forest tree.

---

second group, which includes *R. solani*, has three or more nuclei per cell (multinucleate), and isolates are divided among 13 AG.

The number of nuclei per cell and anastomosis pairing studies have been the most useful and effective methods for identification of *Rhizoctonia* species. Recently developed molecular techniques are more accurate in identification of CAG and AG forms of *Rhizoctonia*. The anastomosis pairing groups have been correlated with groups defined by the newer molecular techniques (Vilgalys and Cubeta, 1994).

The teleomorph for the AG forms of *Rhizoctonia* has been identified as *Thanatephorus cucumeris*. At least two other teleomorphic genera have been identified for *Rhizoctonia* but will not be mentioned here. The sexual stage for the CAG and AG types of *Rhizoctonia* species are generally nondescript. A hymenium layer of hyphae may form on the soil surface or occasionally in culture, with the basidia and basidiospores being produced directly from that area (Sneh et al., 1991). Sexual structures are either rare or extremely difficult to find in nature.

Target spot of tobacco, also called Rhizoctonia leaf spot, is caused by basidiopore infections from *T. cucumeris*. The fungus overwinters in soil or plant debris as sclerotia. The sclerotia germinate, and hyphae grow saprophytically in the soil. Under favorable environmental conditions, basidiospores are formed on tiny resupinate mycellal tissues that form a hymenial surface containing basidia and basidiospores. The spores are wind-borne and disseminated to host tobacco plant foliage. The spores germinate, appressoria form from the growing hyphae, allowing direct entry of the fungus and subsequent colonization of the tissue. Once established in the host, target-shaped lesions develop from these initial infections. New basidiospores are produced within the lesions, causing secondary infections and ultimately resulting in defoliation of the tobacco plant. New sclerotia form and overwinter in the dead host tissue or soil, completing the disease cycle.

The fleshy basidiomycetes include a very large and diverse group of fungi. Species within the group occur in different ecosystems including agricultural fields, forests, and ornamental landscapes. These fungi survive and reproduce either as saprophytes or parasites, and are involved in mycorrhizal associations. Fruiting bodies of basidiomycetes can vary in shape and size, ranging from mushroom-like to those that form woody conks on dead wood, or species can be very tiny and wispy with only a hymenial layer present in soil or on plant tissues. The key morphological structures used by taxonomists to group basidiomycetes together include the basidium, dolipore septum and, when present, clamp connections. The basidium is where meiosis and subsequent basidiospore formation occur.

LITERATURE CITED

Alexopoulos, C.J., C.W. Mims, and M. Blackwell. 1996. *Introductory Mycology* (4th ed.). John Wiley & Sons, New York. 869 pp.

Berres, M.E., L.J. Szabo, and D.J. McLaughlin. 1995. Phylogenetic relationships in auriculariaceous basidiomycetes based on 25S ribosomal DNA sequences. *Mycologia* 87: 821–840.

Carling, D.E. 1996. Grouping in *Rhizoctonia solani* by hyphal anastomosis reaction In B.L. Sneh et al., Eds., *Rhizoctonia Species: Taxonomy, Molecular Biology, Ecology, Pathology and Disease Control.* Kluwer Academic, Dordrecht, Netherlands, pp. 37–47.

Christensen, C.M. 1950. *Keys to the Common Fleshy Fungi* (4th ed.). Burgess, Minneapolis, MN, 45 pp.

Fergus, C.L. 1966. *Illustrated Genera of Wood Decay Fungi.* Burgess, Minneapolis, MN, 132 pp.

McLaughlin, D.J., M.E. Berres, and L.J. Szaba. 1995. Molecules and morphology in basidiomycetes phylogeny. *Can. J. Bot.* 73: 684–692.

Sneh, B., L. Burpee, and A. Ogoshi. 1991. *Identification of Rhizoctonia Species.* APS Press, St. Paul, MN, 133 pp.

Swann, E.C. and J.W. Taylor. 1993. Higher taxa of basidiomy-cetes. An 18S rRNA gene perspective. *Mycologia* 85: 923–936.

Vilgalys, R. and M.A. Cubeta. 1994. Molecular systematic and population biology of *Rhizoctonia. Annu. Rev. Phyto-pathol.* 32: 132–155.

Weber, N.S. and A.H. Smith. 1988. *A Field Guide to Southern Mushrooms.* Univ. of Mich. Press, Ann Arbor, MI, 280 pp.

Wilcox, H.E. 1982. Morphology and development of ecto- and endomycorrhizae. In N.C. Schnenk, Ed., *Methods and Principles of Mycorrhizal Research.* APS Press, St. Paul, MN, pp. 103–113.

20 Oomycota
The Fungi-like Organisms

Robert N. Trigiano, Malissa H. Ament, and Kurt H. Lamour

CHAPTER 20 CONCEPTS

- Organisms in the Oomycota were once classified in the kingdom Fungi, but are now included in the kingdom Stramenopila (Chromista).

- The group is characterized by production of oospores, thick-walled resting spores usually resulting from sexual reproduction and capable of surviving adverse environmental conditions.

- Sexual reproduction is by gametangial contact between oogonia and antheridia where meiosis occurs; asexual reproduction is typically by motile, biflagellate sporangiospores (zoospores), although some species do not produce motile cells.

- Hyphae typically are coenocytic (solid septa may be present at the base of reproductive structures or in old and damaged hyphae), are relatively large (6–10 μm diam.), contain cellulose in the wall and have diploid (2N) nuclei.

- Species in this group ecologically range from saprophytes to obligate parasites. Most of the more important plant pathogens are classified in the Peronosporales.

The phylum Oomycota (oomycetes) encompasses an incredibly diverse group of organisms that were, for many years, considered to be fungi or, at the very least, were always included in the study of the fungi. Most mycologists and other biologists recognized that the oomycetes, although being achlorophyllous and having mycelium-like colonies, were somewhat different, and did not fit the conventional and prevailing concept of true fungi. The oomycetes have been variously categorized into different kingdoms and other groups in the past, and some confusion of their taxonomic status still exists today. For the purposes of this chapter, we will follow Alexopoulus et al. (1996) and place them in the kingdom Stramenopila.

The oomycetes include organisms ranging all along a continuum from saprophytic through obligately parasitic. Some are strictly aquatic, whereas others are considered terrestrial—they occupy widely divergent habitats. Whatever their mode of nutrition or ecological niche, all species share some morphological and physiological characteristics. The following is not an exhaustive list, but these primary features readily distinguish the oomycetes from all other plant pathogens.

- The life cycles of oomycetes can be best described as diplontic; that is, for the vast majority of their life history, the ploidy level of the nuclei is diploid (2N). Haploid (N) nuclei are only found in the gametangia where meiosis occurs. Compare this to the true fungi that have typical haplontic life cycles where nuclei are haploid or may exist as a **dikaryon** (N + N), another functional variation of haploid. The zygote is the only diploid structure in most true fungi.

- The hyphae are generally **coenocytic**, but solid (no pores) septa or cross walls do occur at the bases of reproductive structures and in older portions of the mycelium (Figure 20.1). The hyphal walls contain cellulose, which is typically not found in true fungi. Most true fungi have some amount of chitin in their walls.

- Sexual reproduction is accomplished by **gametangial contact** between **oogonia** and **antheridia** (Figure 3.2). The resulting thick-walled **oospore**, which is diploid, is considered to be the resting or overwintering spore for many species (Figure 20.2).

- Asexual reproduction by the vast majority of species is by **zoospores** produced in sporangia. Typically, many more-or-less kidney-shaped zoospores, each with one tinsel flagellum directed forward and one whiplash flagellum facing behind, are cleaved from the cytoplasm in the **sporangium** or vesicle.

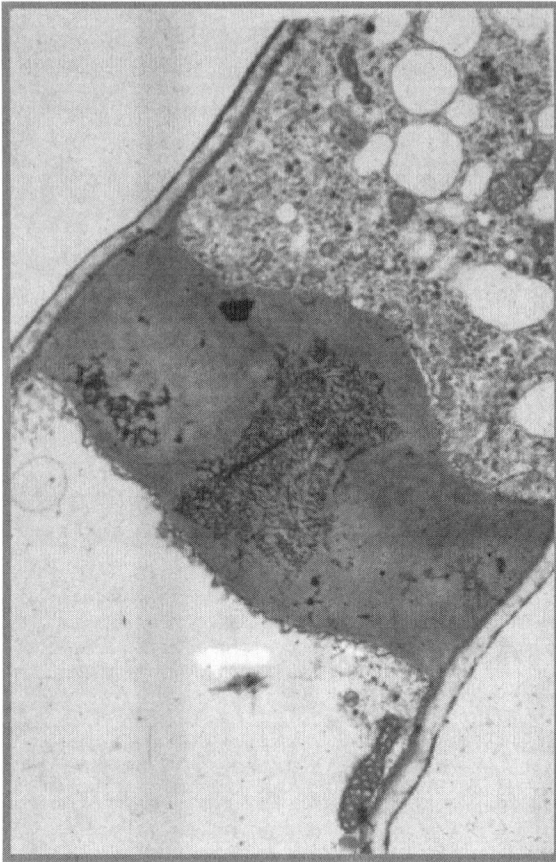

FIGURE 20.1   Septum in an old intracellular hyphae of *Peronospora tabacina*. (Reprinted with permission from Trigiano et al. (1985). *Tob Sci* 29: 116–121.)

- There are a number of biochemical synthetic pathways and storage compounds in the oomycetes that are very different than those present in true fungi.

Just as there is some confusion as to the placement of the members of the Oomycota in a kingdom, there is also some discussion on how the phylum should be divided into orders. Most mycologists and plant pathologists agree that there should be either five or six orders that may sometimes include the order Pythiales. We will again follow the example of Alexopoulos et al. (1996) and recognize six orders. Most of the important plant pathogens fall into a single order—the Peronosporales. Note that, in the Saprolegniales (Saprolegniaceae), there are several species of *Aphanomyces* that can cause root rots of field crops such as peas and sugar beets. These organisms will not be discussed in this chapter. The species in Peronosporales are very diverse and include soil-inhabiting facultative parasites, such as *Pythium* and some *Phytophthora* species, that cause seedling diseases, wilts, and root and seed rots; other *Phytophthora* species are facultative saprophytes that cause aerial blights, root rots, etc.; and the downy mildews and white rust organisms, which are obligate parasites that attack only the aerial portions of plants.

The Peronosporales can be further subdivided rather easily into the following three major families, based primarily on characteristics of sporangia and sporangiophores: Pythiaceae, Peronosporaceae, and Albuginaceae. Species in the Pythiaceae have sporangia that are borne variously, but not in chains, on somatic hyphae or on sporangiophores of indeterminate growth (grows continuously). This family includes the genera *Pythium* and *Phytophthora*. The seven genera classified in the Peronosporaceae (the downy mildews) form sporangia on well-defined, branched sporangiophores of determinate growth (Hawksworth et al., 1995). Some common genera are *Peronospora*, *Plasmopara*, *Bremia*, and *Pseudoperonospora*. The Albuginaceae contains a single genus, *Albugo*, that produces chains of sporangia on club-shaped sporangiophores of indeterminate growth. The species of

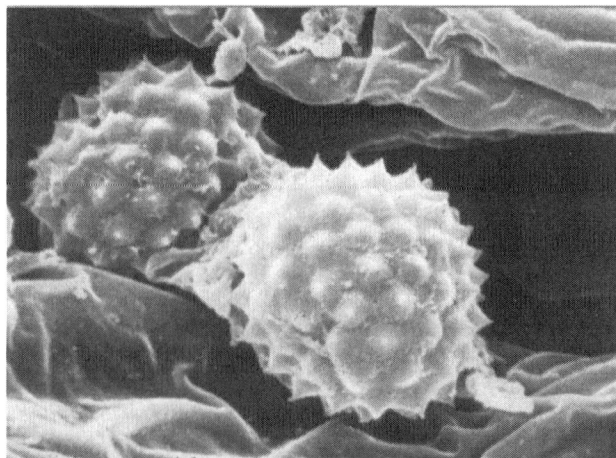

FIGURE 20.2   Oospore of *Peronospora tabacina*.

FIGURE 20.3 Root rot of chrysanthemum caused by *Pythium* species. (Courtesy of Alan S. Windham, University of Tennessee.)

Albugo are collectively known as the white rusts (Alexopolous et al., 1996).

PYTHIACEAE

The family Pythiaceae encompasses ten genera (Hawksworth et al., 1995) and includes *Pythium* and *Phytophthora* species. These two genera contain some of the most destructive and notorious plant pathogens. The Irish famine of 1845 and 1846 caused by *Phytopthora infestans* was largely responsible for the deaths of many who were overly dependent on the potato crop and for the subsequent mass emigration to the United States. A more detailed account of this disease and its influence on human affairs appears in Chapters 1 and 33.

PYTHIUM

There are about 120 species of *Pythium* according to Hawksworth et al. (1995), and many cause seed and root rots and pre- and post-emergence damping off of many greenhouse and field crops. In preemergence damping off, the germinating seedling is attacked, colonized, and killed before breaking through to the soil surface. Many times seedlings will emerge from the soil and immediately begin to wilt (postemergence damping off). In this scenario, the young feeder roots are destroyed, and cortical tissues of the primary root are invaded by the pathogen. Dark, water-soaked lesions develop on stem at the soil line. Extracellular hydrolytic enzymes (pectinase, cellulases, and hemicellulases) (see Chapters 10, 28, and 29) produced by the pathogen degrade cell walls and cause plant tissues to lose their structural integrity. When the seedlings can no longer support themselves, they collapse and fall onto the soil surface (damping off). Often, and especially in high-humidity environments such as greenhouses, white, watery-appearing mycelia can be seen growing on the rotting remains of the shoots. *Pythium* root diseases can also severely damage established plants in the greenhouse and field (Figure 20.3). Roots of the plants often lack the small feeder roots, and secondary roots are discolored (brown or black). The first symptom noticed on the shoots is wilting, and some will possibly exhibit mineral deficiencies. Older plants infected with *Pythium* can also exhibit necrotic, watery-appearing lesions on the stem (Figure 20.4). *Pythium* blight has also become a serious problem of turfgrass, especially on high-value sports fields and golf courses (Figure 20.5).

The life history of *P. debaryanum* (Figure 20.6) is generally representative of most *Pythium* and other oomycete species in that meiosis occurs in gametangia, diploid oospores are the resting or overwintering spore, and biflagellated zoospores are produced. *Pythium* species survive in soil as somatic hyphae or oospores. Zoospores from sporangia and oospores, or from either, are attracted to roots where they encyst and germinate to infect tissue. Somatic hyphae may also directly infect roots. Hyphae grow intercellularly within host tissue, and do not produce **haustoria**. The asexual portion of the life cycle is completed with the formation of sporangia. Sporangia may be indistinguishable from somatic hyphae or quite distinct, depending on the species. The contents of sporangia migrate into **vesicles** where zoospores are differentiated. Vesicles burst, and zoospores are liberated to the surrounding environment. Sexual reproduction is initiated with the differentiation of gametangia (antheridia and oogonia) from somatic hyphae. There are a few **heterothallic** *Pythium* species that require different mating types, but most are **homothallic,** where both gametangia involved in the sexual process may arise from the same hyphae, or they may be formed on different hyphae that lie close together. One or more antheridia contact a single oogonium, and meiosis occurs, reducing the ploidy of the nuclei in the gametangia to haploid. A

FIGURE 20.4   Blackleg of geranium caused by *Pythium ultimum*. (Courtesy of Alan S. Windham, University of Tennessee.)

FIGURE 20.5   Pythium blight of grass. (Courtesy of Alan S. Windham, University of Tennessee.)

nucleus from an antheridium passes into the oosphere (plasmogamy) and eventually fuses with a nucleus in the oosphere (karyogamy) to restore the diploid condition. The oosphere develops a smooth thick wall, which is capable of germinating after a resting period. Oospores may germinate directly via a germ tube or by zoospores, depending on prevailing environmental (primarily temperature) conditions to initiate infection. Some of the more important *Pythium* species include *P. ultimum*, *P. aphanidermatum*, and *P. debaryanum*.

Unfortunately, host resistance to *Pythium* species is generally not available for field and greenhouse crops. Control of these diseases is primarily through cultural practices (Chapter 35) and fungicides (Chapter 36), although there are some biocontrol tactics for some oomycetes (Chapter 37). Good soil drainage and management of soil moisture are important cultural practices used to control diseases caused by *Pythium* species in both field and greenhouse situations. Sound nutritional management of the crop, especially in regard to avoiding excess nitrogen fertilization (Chapter 35), can help to avoid diseases by these organisms. Crop rotation using nonhosts may also be an effective management tactic. In nursery crops, the adoption of mulched hardwood bark as a container medium instead of peat has helped manage root rot diseases. Seeds may be treated with captan or other contact fungicides to prevent root rots. Plants may be treated with systemic fungicides, such as metalaxyl (specific for the oomycetes), either to prevent or control infection.

PHYTOPHTHORA

There are approximately 50 species of *Phytophthora*, according to Alexopoulos et al. (1996), but with the advent of molecular techniques in recent years, this number is subject to change. *Phytophthora* species cause many of

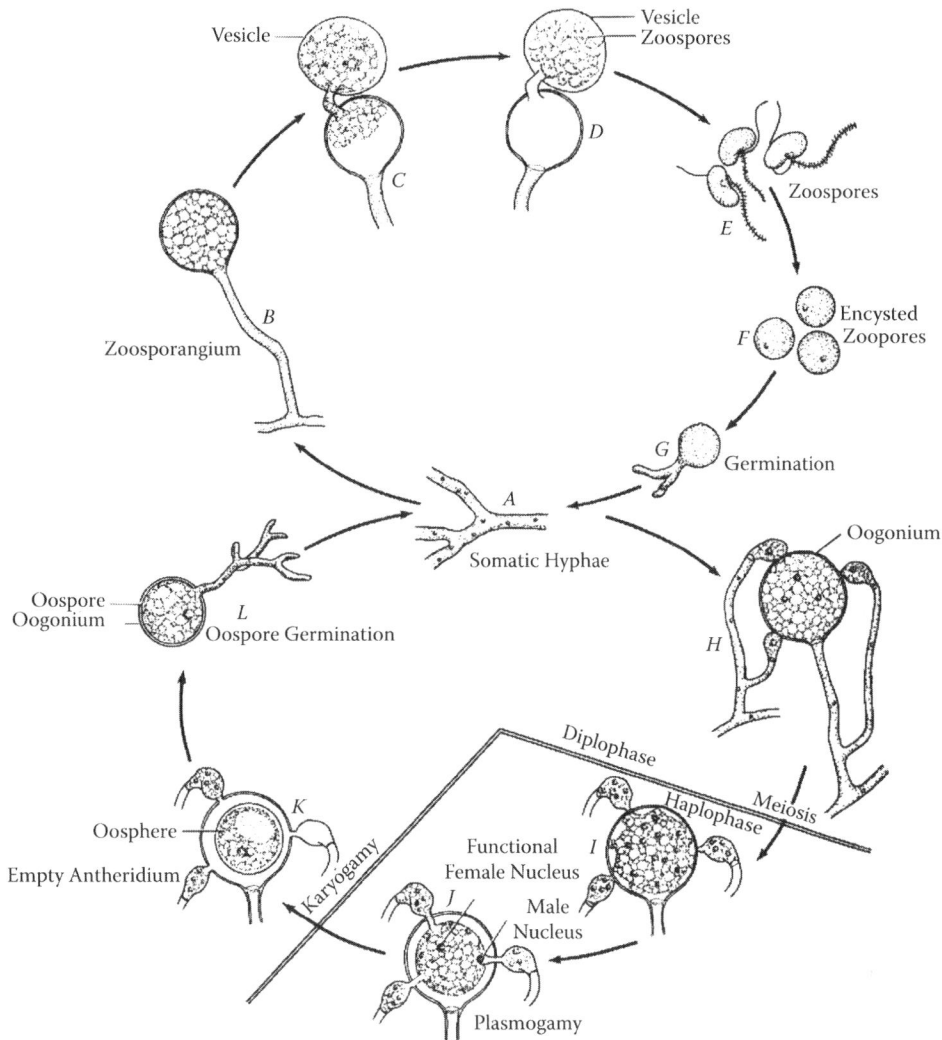

FIGURE 20.6 Life history of *Pythium debaryanum*. (Drawing by R. W. Scheetz from Alexopoulos, C.J., C.W. Mims, and M. Blackwell. 1996. *Introductory Mycology* and reprinted with permission from John Wiley and Sons, New York.)

TABLE 20.1

Some Common Diseases Caused by *Phytophthora* Species

Species	Disease
P. fragariae	Red stele of strawberry
P. megasperma	Root and stem rot of soybean
P. parasitica	Blackshank of tobacco
P. citrophthora	Foot rot of citrus
P. cinnamoni	Root rots of forest and nursery crops and avocado
P. capsica	Cucurbits (squash, pumpkins), tomato, peppers
P. ramorum	Sudden oak death (SOD)
P. infestans	Late blight of potato and tomato
P. palmivora	Black pod of cacao

the same type of diseases as *Pythium* species on the same plants (Table 20.1) and thus may be easily confused with one another when only considering symptomology. In addition, they can cause a number of other diseases, including foliar blights of field, nursery, forest, and greenhouse crops (Figure 20.7).

The life history of *Phytophthora* species differs in detail compared to that exhibited by *Pythium* species. The most notable difference between the two genera is that zoospores are delimited and functional within the sporangium for species of *Phytophthora*, whereas in *Pythium* species, zoospores are formed from the cytoplasm of sporangia that has migrated into a vesicle. *Phytophthora* species also have indeterminate sporangia, but many species have very differentiated sporangiophores and sporangia, or either (Figure 20.8). For example, *P. infestans* produces sympodially branched sporangiophores, which have swollen "nodes" and which produce lemon-shaped, papillate

---

CASE STUDY 20.1

SUDDEN OAK DEATH—NEITHER UNIQUE TO OAKS NOR SUDDEN

- Sudden oak death (SOD), caused by the oomycete *Phytophthora ramorum*, is a new species that attacks a wide variety of woody hosts and is under intense regulation by European and U.S. government agencies.
- *Phytophthora ramorum* was isolated from bleeding stem cankers on tan oak and coast live oak in California in the mid-1990s, which led to the common name "sudden oak death."
- The time from infection to tree death can be from months to years.
- Areas on the west coast of the United States known to have the disease are currently under quarantine, and plant production facilities are routinely inspected and tested for the presence of *P. ramorum*.
- The pathogen is primarily isolated from above-ground plant parts and has been isolated on numerous occasions from plants shipped to many other areas of the United States (Figure 20.9).
- *Phytophthora ramorum* is spread via rain splash, infested soil, and infected plant material.
- *Phytophthora ramorum* is an obligately outcrossing species.
  - Makes abundant thick-walled asexual chlamydospores.
  - Epidemics in the United States are thought to be comprised of one to two clonal lineages of the *P. ramorum*.
  - Does not appear to be completing the sexual stage in natural epidemic populations.
- A whole-genome sequence was completed by the Joint Genome Institute in 2004.
  - Contains a large gene family of rapidly evolving "effector" proteins that are thought to play a key role in overcoming plant defenses.

FIGURE 20.7 Foliar blight of vinca by *Phytophthora* species. (Courtesy of Alan S. Windham, University of Tennessee.)

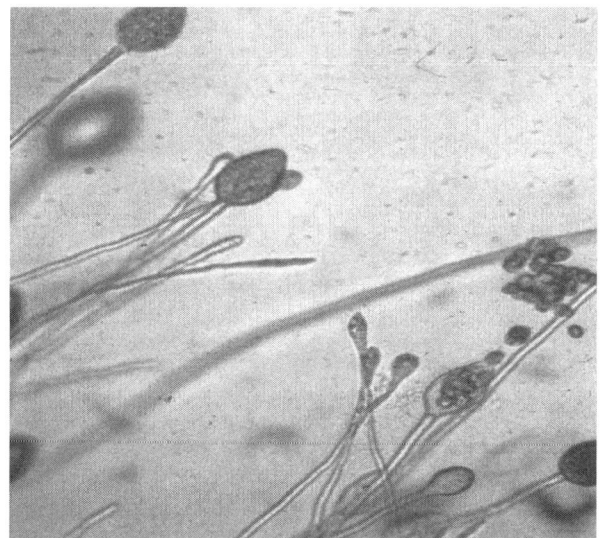

FIGURE 20.8 Sporangiophores and sporangia of a *Phytophthora* species. (Courtesy of Alan S. Windham, University of Tennessee.)

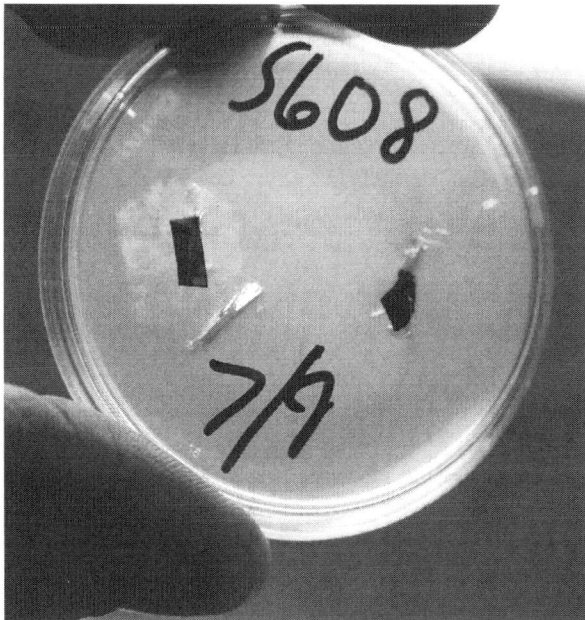

FIGURE 20.9 *Phytophthora ramorum* mycelium expanding from a piece of Rhododendron leaf onto selective agar medium.

sporangia. Some *Phytophthora* species produce haustoria unlike the closely related *Pythium* species.

Phytophthora infestans—Late Blight of Potato (and Tomato)

Phytophthora infestans is probably the best known member of the genus. Along with a few other species, it is also the most extensively studied because of its economic impact on agriculture and history. This organism is heterothallic and requires two mating types for sexual reproduction by gametangia contact. Sexual reproduction is not necessary for the survival of the organism as the mycelium may survive in infected tubers. The disease can be initiated when infected tubers are planted or when volunteer potatoes sprout in the spring. Under favorable environmental conditions (cool and moist), the pathogen grows into the shoot, and after sufficient time, a matter of days, sporangia on sporangiophores will emerge through stomata on the bottom surface of leaves. Sporangia are dispersed by air currents and rain to other plants and deposited on the soil surface. Several (typically less than ten) biflagellated zoospores form and are liberated from each sporangium. Zoospores that land on leaves encyst and then germinate via germ tubes. Germ tubes form **appressoria** that penetrate the leaf surface either directly or indirectly through stomata. Sporangia that are in soil also produce zoospores, which may infect existing or developing tubers. Under favorable conditions, many "generations" or cycles of sporangia may be produced to infect plants—late blight is a **polycyclic** disease (Chapters 2 and 33) and may quickly devastate entire

fields. As the pathogen grows and sporulates, host tissue is destroyed, creating lesions that coalesce. Often the entire shoot portion of the plant is destroyed and yield is dramatically reduced. Infected tubers are often small with sunken lesions, stained purple or brown, and may have an offensive odor due to secondary bacterial and fungal invaders.

Control of late blight of potatoes is accomplished using several tactics. Seed potatoes should be free of the pathogen, volunteer plants should be rouged, and infected debris from any previous potato crop destroyed (sanitation—Chapter 35). Disease-resistant varieties should also be used when possible, although resistance may be broken by environmental conditions favorable to pathogen development, or when new strains of the pathogen develop. Contact fungicides may be applied at prescribed intervals according to disease development models based on temperature and moisture (Chapter 34). Metalaxyl, a systemic fungicide specific for the oomycetes, has been extensively used to protect crops.

Control of other diseases caused by *Phytophthora* species is similar to control tactics used for *Pythium* species. Well-drained soils and careful management of soil moisture will help minimize damage by these pathogens to many crops. For greenhouse crops, growing media and containers should be decontaminated to ensure pathogen propagules are killed. Hardwood bark mulch has been used to control disease in container nursery crops. Unlike for diseases caused by *Pythium*, there are a number of crops species (fruits, etc.) in which there are disease-resistant varieties available. Systemic fungicides such as metalaxyl and fostel-Al are commonly used to protect various crops from infection.

PERONOSPORACEAE— THE DOWNY MILDEWS

The **downy mildews** are obligate biotrophic parasites of aerial portions of flowering plants and belong to the Peronosporaceae. There are seven genera in the family: *Basidiophora, Bremia, Bremiella, Peronospora, Plasmopara, Pseudoperonospora,* and *Sclerospora*. The genera are distinguished by the highly differentiated branching patterns of sporangiophores that always exhibit determinate growth (Figure 20.10). Sporangiophores emerge through stomata on the lower leaf surface, and development is completed before any sporangia are formed. Formation of sporangia on individual sporangiophore is synchronous (Figure 20.11), and mature sporangia are disseminated by air currents. The sporangia of many species germinate by producing biflagellated zoospores characteristic of the Oomycota. However, the sporangia of *Pseudoperonospora, Peronospora,* and some species of *Bremia* germinate via a germ tube and for this reason have been mistakenly termed conidia by some authors.

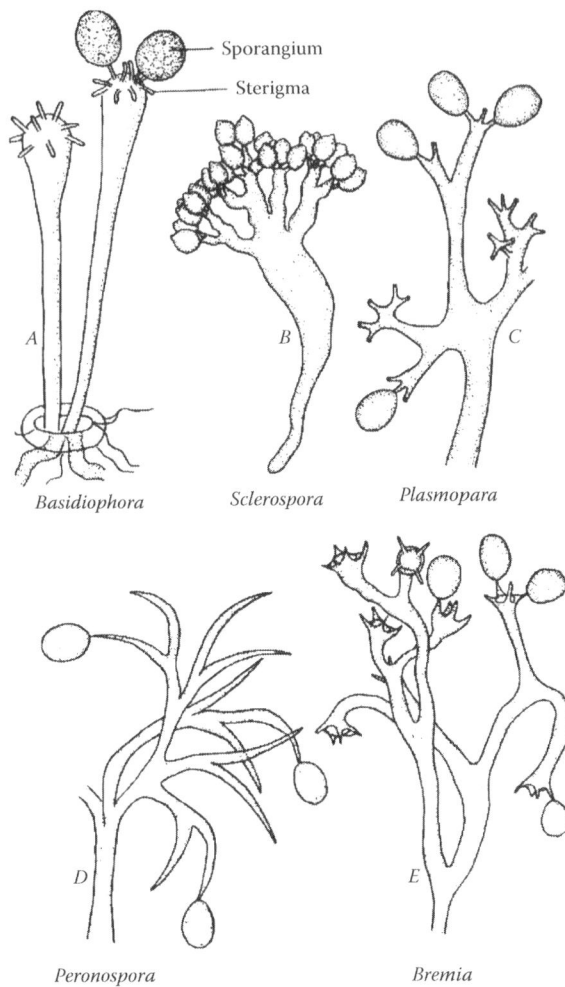

FIGURE 20.10 Sporangiophores characteristic of five genera of Peronosporaceae. (Reprinted courtesy of Alexopoulos, Mims, and Blackwell. 1996. John Wiley and Sons, New York.)

Regardless of the mode of sporangial germination, these propagules serve to infect either healthy plants or healthy tissue remaining on an infected plant. Leaf surfaces are penetrated either directly through the epidermal wall or through stomata. Intercellular hyphae ramify throughout the mesophyll and establish haustoria within host cells (Figure 20.12). The infection may be localized in a leaf or become systemic in the plant. When sufficient energy has become available to the organisms, and when environmental conditions are satisfactory, usually cool temperatures with high humidity and darkness, sporangiophores and sporangia are produced and start the disease cycle over again. Many "crops" of sporangia can be produced in a growing season, and thus diseases caused by these organisms can be considered polycyclic. Sexual reproduction is by gametangial contact, and ornamented oospores are usually produced within host tissue and serve as overwintering structures (Figure 20.2). Oospores usually germinate directly via germ tubes, but sometimes germinate indirectly by zoospores.

Members of the Peronosporaceae infect and cause diseases on many plants (Table 20.2). Probably the most well-known disease in this group is downy mildew of grapes caused by *Plasmopara vitacola*. Briefly, a root aphid native to North America was introduced to France in the mid-1860s with devastating consequences for the wine industry. In an attempt to control the aphid, the wine producers imported American rootstock, which had good tolerance to the insects. Unbeknown to them, they also imported *P. vitacola* with the plants. The French grapes were extremely susceptible to the disease and threatened the continued existence of the industry. In 1882, Alexis Millardet noticed that some vines sprayed with a substance did not have the disease. The owner of the vines had applied a mixture

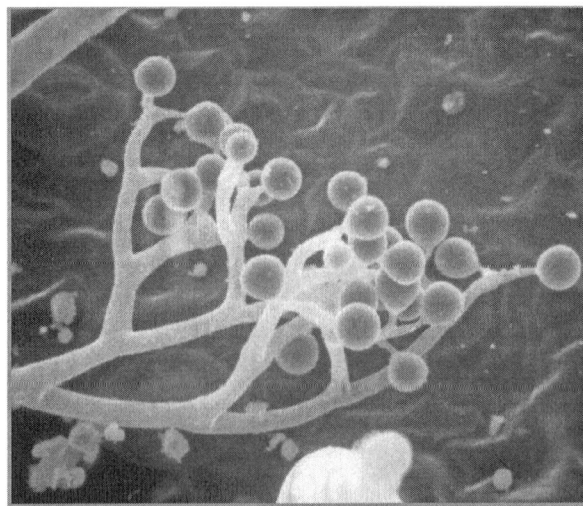

FIGURE 20.11 Synchronous development of sporangia on a sporangiophore of *Peronospora tabacina*. (Reprinted with permission from Trigiano et al. (1985). *Tob Sci* 29: 116–121.)

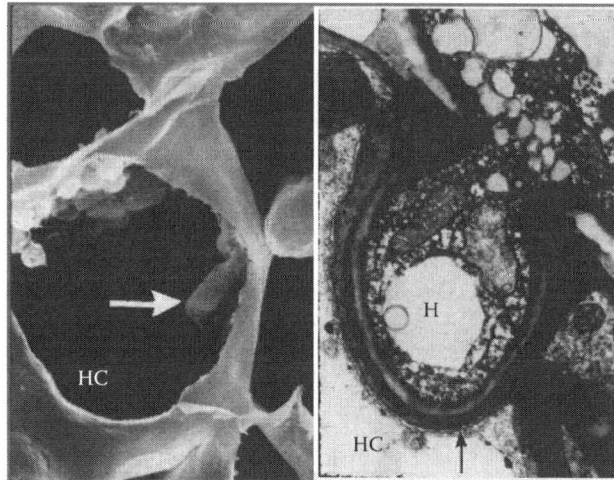

FIGURE 20.12 Haustoria of *Peronospora tabacina* in tobacco. (Left) Scanning electron micrograph of a haustorium (arrow) in a host cell (HC). (Right) Transmission electron micrograph of a haustorium (H) in a host cell (HC). Notice that the plasmalemma (cell membrane) is intact. (Reprinted with permission from Trigiano et al. (1983). *Can. J. Bot.* 61: 3444–3453.)

CASE STUDY 20.2

PERONOSPORA TABACINA—BLUE MOLD OF TOBACCO

Blue mold of tobacco is caused by the oomycete *Peronospora tabacina* (syn. *P. hyoscyami*) and is a member of the Peronosporaceae, the downy mildews. It is only a pathogen of *Nicotiana* (tobacco) species (http://www.ces.ncsu.edu/depts/pp/bluemold/thedisease.php).

- All downy mildews are obligate plant parasites (cannot live and reproduce without a living host), and as such, form an intimate relationship with their host. These relationships are characterized by formation of haustoria (Figure 20.11), which transport nutrients from the host to the intercellular hyphae.
- *Peronospora tabacina* reproduces sexually by oospores (Figure 20.2) and asexually by sporangia (sporangiospores) borne on sporangiophores (Figure 20.11). Sporangia are the primary mode of dissemination for this pathogen.
- The disease may affect seedlings as well as older plants, and derives its name from the blue to grayish-blue mass of sporangia and sporangiophores that emerge from stomata on the undersurface of leaves with chlorotic (yellow) lesions. These lesions will eventually turn brown or necrotic. Leaves or portions of leaves may be "cupped." The pathogen usually produces discrete lesions on leaves, but may also be found as systemic infections that may be common in some regions.
- *Peronspora tabacina* does not overwinter in the United States and Canada. Annual outbreaks and epidemics, or either of the disease are initiated by sporangia arriving on wind currents from tropical zones south of the 30th parallel.
- Rapid disease development and spread is favored by cool, wet, and overcast conditions, whereas hot, dry weather adversely affects disease development and spread.
- Some general strategies for controlling blue mold:
  - Local management of tobacco crops
  - Use more than one control method
  - Make the environment less favorable for the pathogen and for the spread of the disease;
  - Use protective fungicides (see Websites for details)
  - Manage beds, greenhouses, and fields to favor the crop.

See http://www.ces.ncsu.edu/depts/pp/bluemold/control_2006/php and http://www.uky.edu/Ag/kpn/kyblue/kyblu04/cntl0201.htm.

TABLE 20.2

Some Common Downy Mildew Diseases

Species	Crop
Bremia lactucae	Lettuce
Peronospora tabacina (syn. P. hyoscyami)	Tobacco
P. destructor	Onion
P. antirrhini	Snapdragon
Peronosclerospora sorghi	Sorghum
Plasmopara vitacola	Grape
P. halstedii	Sunflower
Pseudoperonospora cubensis	Many cucurbit species
Sclerophthora macrospora	Cereal and grass species

of copper sulfate and lime to persuade passers-by not to eat his crop. Millardet, being a professor at the university, seized upon his observation and developed the first fungicide—Bordeaux mixture (Chapter 36). Thus, the French wine industry had solutions to the aphid and downy mildew problems.

Varieties of some crops resistant to downy mildews are available, but contact fungicides are still used to control diseases. More recently, systemic fungicides, including metalaxyl and fosetyl-Al, which may or may not be combined with contact fungicides, have been used successfully to control downy mildew diseases. However, in the last several years, some of the species that cause downy mildew diseases (e.g., *P. tabacina*—blue mold of tobacco and *Pseudoperonospora cubensis*), have developed resistance to acylanaline-class (metalaxyl) fungicides.

ALBUGINACEAE—THE WHITE RUSTS

The **white rusts** are obligate plant parasites on flowering plants and belong to the genus *Albugo*. Species are differentiated by host and oospore ornamentation. *Albugo candida* can be a serious pathogen of crucifers. Another common species is *A. ipomoeae-panduranae* and is found growing in sweet potatoes and, more frequently, in wild morning glory (Figure 20.13).

The life history of *Albugo* species is similar to that exhibited by many other oomycetes except in the production of sporangiophores and sporangia. The club-shaped sporangiophores are contained within the host tissue, and each produces several sporangia under the epidermis. The pressure exerted by the production of sporangia and mycelium against the epidermis distorts the shape of the sporangia from globose to box- or cube-like. The pressure also ruptures the epidermis, liberating the sporangia and forming a white crust. Old "pustules" on morning glory may appear orange to pink. Sporangia are disseminated by wind and rain and germinate via zoospores that have typical oomycete morphology. Zoospores encyst, form germ tubes, and initiate infections on suitable host plants. Oospores are formed by gametangial contact, and they serve as resting spores.

The plant pathogens classified in the Oomycota are a very interesting and distinct group of organisms. They cause a number of very economically important diseases and have influenced the history of mankind. There is no doubt, that with additional research, we will understand their taxonomic position better and learn more about the diseases they cause and how to control them.

FIGURE 20.13 *Albugo ipomoeae-panduranae* infecting morning glory. (Left) Top of leaf showing chlorotic tissue. (Right) Bottom surface of leaf with white crusty pustules containing sporangia. (Courtesy of Alan S. Windham, University of Tennessee.)

LITERATURE CITED AND SUGGESTED READING

Agrios, G.N. 2005. *Plant Pathology,* 5th ed. Academic Press, New York. 952 pp.

Alexopoulos, C.J., C.W. Mims and M. Blackwell. 1996. *Introductory Mycology.* John Wiley. New York. 868 pp.

Anonymous. 2004. Blue mold management. http://www.uky.edu/Ag/kpn/kyblue/kyblu04/cntl0201.htm.

Hawksworth, D.L., P.M. Kirk, B.C. Sutton and D.N. Pegler. 1995. *Ainsworth and Bisby's Dictionary of the Fungi,* 8th ed. CAB International, Wallingford, U.K. 616 pp.

Ivors, K.L., K. Seebold, C.S. Johnson and A. Mila. 2006. Blue mold control plan. http://www.ces.ncsu.edu/depts/pp/bluemold/control_2006/php.

Main, C.E. 2005. The blue mold disease of tobacco. http://www.ces.ncsu.edu/depts/pp/bluemold/thedisease.php.

Trigiano, R.N., C.G. Van Dyke and H.W. Spurr, Jr. 1983. Haustorial development of *Peronospora tabacina* infecting *Nicotiana tabacum. Can. J. Bot.* 61: 3444–3453.

Trigiano, R.N., C.G. Van Dyke, H.W. Spurr, Jr. and C.E. Main. 1985. Ultrastructure of sporangiophore and sporangium ontogeny of *Peronospora tabacina. Tob. Sci.* 29: 116–121.

21 Laboratory Exercises with the Oomycetes

Robert N. Trigiano, Richard E. Baird, and Steven N. Jeffers

The "water molds" or oomycetes occur worldwide and can be found in diverse ecosystems, including estuaries, lakes, oceans, rivers, and streams. Taxa within the group, however, can also occur in agricultural fields and in forest habitats, growing as saprophytes or parasites. Because of the diversity of oomycetes and their occurrence in many different environments, several representative experiments have been included in this chapter to enable students to learn more about methods to study this unique group.

There are many fascinating and intriguing experiments that can be conducted with the members of the oomycetes, including those that are not plant pathogens; however, we will be limiting consideration to some of the more common species that cause plant diseases, such as *Pythium, Phytophthora, Peronospora,* and *Albugo.* Soil-inhabiting species of *Pythium* and *Phytophthora* are easy to grow and manipulate in axenic cultures, whereas species of *Peronospora* and *Albugo* cannot be grown in culture because they are obligate parasites that require a living host to complete their life cycles. The obligate pathogens are difficult to work with in most experimental systems. Although these obligate pathogens are very challenging to maintain, some educational and research exercises can be completed with relative ease. As has been emphasized throughout this book, please secure the proper permits to obtain and transport pathogens and then destroy all experimental materials by autoclaving or by other means when the experiments are completed.

The following experiments are designed to provide hands-on experiences for students working with *Pythium* species that cause root rots and damping-off of peas and beans, isolation of *Phytophthora* species from diseased plant tissues and directly from soil, sporangia and oospore formation by *Pythium* and *Phytophthora* in culture, *in vitro* coculture of *Peronospora* species, and microscopic observation of *Peronospora* and *Albugo* species in host materials. The latter two experiments are designed as special topics for advanced undergraduate and graduate students.

EXERCISES

EXPERIMENT 1. ROOT ROT OF BEAN AND PEAS CAUSED BY *PYTHIUM* SPECIES

There are a number of *Pythium* species that cause root and seed rots of various crops. These diseases can devastate both field-grown and greenhouse crops. This experiment is designed to demonstrate symptoms and signs of the diseases. For an alternative exercise utilizing a floricultural crop, see Procedure 32.1.

Materials

Each student or team of students will require the following items:

- Culture of *Pythium ultimum* or *P. aphanidermatum*; may obtain cultures from ATCC or colleagues with the proper permits
- Untreated (no fungicides) seeds of any cultivar or cultivars of common edible beans (*Phaseolus vulgaris*) and peas (*Pisum sativum*)
- Six 10-cm diam. plastic pots, pot labels, and permanent marker
- Sand for plastic pots, and pencil or large glass rod for making holes in sand
- Two plastic flats with Promix or other soilless medium and paper towels
- Laboratory blender, scissors, and long transfer forceps
- 250-mL flask containing 125 mL of sterile cornmeal (CM) broth (Difco, Detroit, MI) and two 1000-mL beakers
- Four 10-cm diam. petri dishes containing CM agar (add 15 g agar to the CM formulation)
- Compound microscope and microscope slides

Follow the protocol listed in Procedure 21.1 to complete this experiment.

Procedure 21.1

Root and Seed Rots Caused by *Pythium* Species

Step	Instructions and Comments
1	Approximately 7 to 10 days before the laboratory, plant seeds in flats filled with soilless medium. Grow in a cool greenhouse or laboratory. Each team of students will require at least 12 germinated seeds of each bean and pea. Space seeds so that the roots of individual plants will not grow together, and seedlings can be easily separated.
2	Approximately 1 week before the laboratory, inoculate the sterile CM broth in 250-mL flasks with several plugs from the margin of 5-day-old colonies of either *Pythium ultimum* or *P. aphanidermatum* growing on CM agar dishes. Incubate the liquid cultures at 18–22°C either on a slow (30 rpm) shaker or on a shelf in an incubator or laboratory. Prepare an equal number of uninoculated flasks (medium without *Pythium* species).
3	Autoclave the sand the day before the experiment. This step is not absolutely essential for the success of the exercise.
4	Gently remove the seedlings from the flats and wash the particles of soilless medium from the roots using tap water. Store plants with roots wrapped in moist paper towels on the laboratory bench.
5	Swirl the liquid culture to dislodge hyphal growth from the glass and empty the contents of the flasks into the blender. Add 375 mL of sterile distilled water and homogenize the mixture with short bursts (high speed) of the blender. Pour the suspension into a 1000-mL beaker.
6	Dip the roots of three bean plants into the *Pythium* species suspension. Make three large holes in the sand in each plastic pot using a pencil or glass rod. Be sure that the sand is moist. Very gently plant the beans with as little damage to the roots as possible. Label each pot: Pythium intact roots. Repeat the procedure with pea seedlings and label.
7	Trim about 25% of the root length from another group of three bean plants. Dip the remaining roots in the suspension and plant as in step 6. Label the pots: Pythium cut roots. Repeat the procedure with pea seedlings and label.
8	Plant three bean seeds about 2 cm (0.5–1.0 in.) deep in a pot. Pour about half of the remaining *Pythium* species homogenate onto the surface of the sand in the pot. Label pot: Pythium bean seeds. Repeat the procedure with pea seeds and label.
9	Repeat steps 6–8 using the contents of an uninoculated flask mixed with 375 mL of sterile water for the dip and drench treatments. Label pots according to the treatment.
10	Set the plants on a laboratory bench near a window and observe for symptom development. Water with distilled water and do not allow the sand to dry out.

Anticipated Results

Depending on the temperature at which the inoculated plants are grown and the amount of inoculum applied, symptoms of damping-off should be evident between 3 and 7 days on both bean and pea plants. The inoculated plants should appear wilted at first, and then necrotic and water-soaked lesions will occur on the stem at the soil line. Infected plants will not be able to maintain stature and will fall over. Those plants whose roots were cut will typically display symptoms one or two days earlier than those with uncut roots. Roots of infected plants should appear dark and very soft compared to white and firm roots from uninoculated plants. Seeds treated with *Pythium* will either fail to germinate and rot, or germinate poorly with the seedling succumbing to the disease very quickly. All plants that were not treated with *Pythium*, including those with cut roots, should grow normally unless a contaminating pathogenic *Pythium* species is present in the potting medium. Uninoculated seeds should germinate and the plants should grow normally.

Questions

- What are the controls in this experiment, and why are they necessary?
- How would Koch's postulates (Chapters 39 and 40) be completed for this experiment?

EXPERIMENT 2. ISOLATION OF *PHYTOPHTHORA* SPECIES FROM PLANT TISSUES AND SOIL

Many members of the Pythiaceae, including *Pythium* and *Phytophthora* species, are widely distributed in soils.

They may survive for long periods without a host. This exercise is designed to provide experience isolating *Phytophthora* species from diseased plants and infested soil using a selective medium.

Materials

Each student or group of students will need the following items:

- Roots, stems, or leaves from diseased plants (rhododendron, azalea, soybean, tomato, pepper, or tobacco work well)
- Soil sampling tool (2.5-cm diam. works well)
- Ice chest
- Paper towels and plastic bags
- Scalpel with #10 blade (caution: very sharp)
- Twenty 10-cm diam. plastic petri dishes containing PARP(H) medium (Jeffers and Martin, 1986; Ferguson and Jeffers, 1999) (Table 21.1)
- Incubator set at 20°C without light

TABLE 21.1
Growth/Isolation Media for *Phytophthora* Species

V8 Agar (V8A) Growth and Sporulation Medium	Quantity
V8 juice[a]	200 mL
CaCO₃	2 g
Distilled water	800 mL
Agar	15 g

PARP(H) Isolation Medium	Quantity
Delvocid (50% pimaricin): P	10 mg
Sodium ampicillin: A	250 mg
Rifamycin-SV (sodium salt): R	10 mg
75% PCNB (Terraclor): P	67 mg
Hymexazol[b]: H	50 mg
Clarified V8 juice[c]	50 mL
Agar[c]	15 g
Distilled water	950 mL

Note: All antimicrobial amendments should be added after the base agar medium has been autoclaved and cooled to 50–55°C (autoclaved vessel can be held in hand without much discomfort).

[a] Clarified V8A (cV8A) can be made by stirring 200 mL V8 juice with 2 g CaCO₃ for 15 min and then centrifuging for 10 min at 4000 × g. Supernatant may be stored frozen at −20°C; use 100 mL with 900 mL water and 15 g agar to make cV8A.
[b] Inclusion is optional: hymexazol inhibits *Pythium* species.
[c] Corn Meal Agar (CMA) may be substituted for clarified V8 juice and agar.

- Sieves with 2-mm and 4-mm openings
- Several aliquots of 100 mL of 0.3% water agar (3 g agar in 1 L of water) in 250-mL beakers
- Magnetic stirrer and stir bars
- Wide-bore pipette (1-mL) and bulb or pump
- Top-loading balance and plastic weigh boats
- Compound microscope, glass slides, and coverslips

Follow the protocol outlined in Procedure 21.2 to complete this exercise.

Anticipated Results

Colonies of *Phytophthora* species should develop on PARP(H) from both soil and diseased tissues between 24 and 72 h. Colonies of *Pythium* species may also develop if hymexazol is omitted from the isolation medium (PARP) or if there are hymexazol-tolerant *Pythium* species present in the samples. Soil from sources without infected plants may or may not contain *Phytophthora* and *Pythium* species. *Phytophthora* colonies should not develop on PARP(H) from healthy plant tissues.

Questions

- What morphological characteristics are used to recognize and identify species of *Pythium* and *Phytophthora*?
- What are the characteristics used to distinguish *Pythium* and *Phytophthora* species from each other?
- Can a plant be infected with *Phytophthora* and *Pythium* at the same time? Can soil be infested with both genera?

EXPERIMENT 3. PRODUCTION OF SPORANGIA AND OOSPORES BY *PHYTOPHTHORA* AND *PYTHIUM* SPECIES

Many species of *Phytophthora* and *Pythium* will form sporangia and oospores in culture if provided the proper environmental conditions. These simple experiments are designed to allow students to observe asexual and sexual reproduction in the Pythiaceae.

Materials

Each team of students or class will require the following materials:

- Agar (CM) cultures of several homothallic species of *Pythium* and *Phytophthora* (e.g., *P. cactorum* or *P. citricola*)
- Freshly gathered grass (tall fescue, blue grass, etc.) clippings autoclaved for 20 min on two successive days

Procedure 21.2

Isolation of *Phytophthora* Species from Infected Plant Tissue and Infested Soil

Step	Instructions and Comments
	Isolation of *Phytophthora* Species from Infected Plant Tissues
1	Place roots and stem segments from diseased plants into plastic bags and keep moist using damp towels. Healthy plants should also be sampled. Transport the samples to the laboratory in a cool ice chest. Samples should be kept in the dark.
2	In the laboratory, gently wash the tissues under running tap water for 5–10 min and blot excess moisture with a paper towel.
3	Cut samples into 1-cm segments and place four segments on each of five PARP(H) petri dishes. Each piece should be pushed into the agar so that the tissues are surrounded by agar. Incubate the cultures at 20°C in the dark.
4	Observe colony growth on PARP(H) agar dishes after 48–72 h. Continue to examine dishes for up to 1 week. Transfer small pieces of mycelium from individual whitish-colored *Phytophthora* colonies to fresh dishes containing PARP(H) medium and incubate at 20°C in the dark.
	Isolation of *Phytophthora* Species from Soil
1	Collect soil cores up to 20 cm deep using a sampling tool approximately 2-cm in diameter. Collect 10 core samples within 20 cm of target symptomatic plants and 10 samples of "noninfested" soil from around healthy appearing plants of the same type.
2	Place each set of 10 soil cores into separate and individually labeled plastic bags to make one infested and one noninfested composite soil sample. Return the composite soil samples to the laboratory in a cool ice chest. All samples should be stored in a dark, cool place until the assay begins.
3	Prepare each composite soil for isolation by breaking up the clods. Remove rocks and plant debris by first using a coarse (4 mm) screen followed by a smaller screen (2 mm). Thoroughly mix the soil and return to plastic bags.
4	Add 50 mL of the infested soil to 100 mL 0.3% water agar contained in a 400-mL beaker with a stir bar. Place on a magnetic stirrer at high speed for about 3 min. Pipette 1 mL aliquots of the suspension onto PARP(H) dishes. Evenly spread the suspension across the surface of the agar. Use up to 5 dishes per composite soil sample. Repeat for the "noninfested" soil.
5	Incubate the dishes at 20°C in the dark for 48–72 h. Do not enclose dishes in plastic bags or boxes. Wash soil from dishes under running tap water. Examine dishes with a dissecting microscope (30–50×) and count the colonies. Subculture from these colonies to fresh PARP(H) medium if desired. Subcultured colonies may be saved for later identification. Compare number of colonies from the different locations and soil types.

- Sterile distilled water (To enhance sexual reproduction in *Pythium* cultures, a drop of chloroform containing cholesterol [1 mg in 10 mL chloroform] may be added to 8 mL of distilled water. Allow the chloroform to evaporate under a transfer hood before adding grass or *Pythium*.)
- Sharp scissors to cut grass
- Fine-tipped forceps
- Several 60-mm plastic petri dishes
- V8A and cV8A (Table 21.1)
- Centrifuge and Whatman #1 filter paper
- Incubator or laboratory bench equipped with fluorescent lights
- Sterile plastic drinking straws
- Soil

- Nonsterile soil extract solution (NS-SES) (Stir 15 g of soil in 1 L of distilled water for at least 4 h and allow the suspension to settle overnight. Decant the water and centrifuge for 10 min at 4000 x g, followed by filtration through Whatman #1 filter paper [Jeffers and Aldwinckle, 1987]. Store in the refrigerator.)
- Dissecting and compound microscopes
- Microscope slides and coverslips
- Lactoglycerol solution (lactic acid:glycerol 1:1 v:v) with 0.1% acid fuchsin
- Brightly colored nail polish or vaseline

Follow the protocols in Procedure 21.3 to complete this experiment.

	Procedure 21.3 Production of Sporangia and Oospores by *Pythium* and *Phytophthora* Species
Step	Instructions and Comments
1	Grow cultures of *Pythium* species on CM agar in 60-mm plastic petri dishes at 20°C in the dark for 3 days.
2	Add eight mL of sterile distilled water to a number of empty 60-mm diam. petri dishes. Separate and add five to ten autoclaved blades of grass to the water. Remove several plugs of mycelium from the margin of the *Pythium* species colony with a sterile plastic straw and cut into quarters with a sharp scalpel. Transfer four quarters to the petri dish with the blades of grass. Be sure the agar pieces are in contact with the grass. Incubate at 20°C in the light.
3	Cultures may be observed weekly with either a 40× dissecting scope, or individual pieces of grass from the culture may be mounted in lactoglycerol (with or without acid fuchsin) on microscope slides and viewed with a compound microscope. The slide may be made semipermanent by painting the clean, dry edges of the coverslip with nail polish. Draw and label all structures. To stimulate zoospore release from sporangia of *Pythium* species, chill (2–4°C) the grass cultures for a few hours and then allow them to warm to room temperature. Observe sporangia, vesicle formation, and zoospores with the aid of a dissecting microscope.
4	Transfer *Phytophthora* species cultures to V8A and incubate in the dark at 20–25°C for 48–72 h. Note: colonies should be at least 2 cm in diam. Aseptically remove agar plugs (2 mm) near the margin of the colony with a sterile drinking straw and place five of them into a sterile, empty 60-mm diam. petri dish.
5	Cover the agar plugs with 7–10 mL of NS-SES. Place dishes under continuous fluorescent lights at 20–25°C. Observe plugs with a dissecting microscope after 12–24 h for sporangia, or if not present, after an additional 24 h. Draw and label all structures.
6	To initiate oospore formation in cultures of *Phytophthora* and *Pythium* species, transfer the organisms to 60-mm petri dishes containing clarified V8A and incubate at 20–25°C in the dark. Microscopically examine cultures weekly for up to 6 weeks. Draw and label all structures. This procedures works well for homothallic species of both *Pythium* and *Phytophthora*. Oospores are usually present after 2 weeks and are mature by 4 weeks.

Anticipated Results

Sporangia should form very quickly in all *Pythium* and *Phytophthora* cultures—in as little as 24–48 h for many *Pythium* and *Phytophthora* species in NS-SES. Gametangia and oospores usually take 2–4 weeks to form in the cultures.

Questions

- Why were homothallic species of *Phytophthora* and *Pythium* used?
- Are all species of *Phytophthora* and *Pythium* homothallic?
- What is the taxonomic significance of the origin (same hypha, below oogonium, etc.) of antheridia involved in sexual reproduction?
- What purpose do oospores serve in the life cycle of these organisms?
- Assign a ploidy level (haploid or diploid) to each of the structures seen in the cultures.

EXPERIMENT 4. COCULTURE OF *PERONOSPORA TABACINA* AND TOBACCO CALLUS

Obligate pathogens, such as *P. tabacina* (syn. *P. hyoscyami* or blue mold), require a living host to grow and reproduce.

This is a very specialized and complex relationship influenced strongly by environmental conditions. It is possible to establish the pathogen and host together in culture (see also Chapter 41). *In vitro* cocultures allow active maintenance of isolates of pathogens; studies of physical, physiological, and molecular interactions between pathogen and host; and assessment of resistance or susceptibility, or both, of host cultivars to the pathogen. Coculture of *P. tabacina* and tobacco was first report by Izard et al. (1964) and then by Trigiano et al. (1984). In a very interesting report, Heist et al. (2001) devised a technique for culturing infected tobacco leaf tissue directly and maintaining the association of the organism. As we are more familiar and experienced with the 1984 study, this experiment will be based largely on modifications of the protocols found in Trigiano et al. (1984) and it is intended for advanced undergraduate and graduate students. If you are located in a tobacco-growing area of the country, we suggest that this experiment only be attempted when tobacco is not being grown to prevent accidental escape of the pathogen.

General Considerations

Tobacco cultures. Callus cultures can be established from almost any blue mold-susceptible tobacco cultivar.

Remove the leaves from the stems of seven to nine leaf stage plants, wash the stems in soapy water, and rinse well in tap water. Stem tissue can be surface-sterilized in 20% commercial bleach for 10 min and then rinsed three times in sterile distilled water for 2 min each. Remove and discard the cut ends of the stem damaged by the bleach, and split the remaining stem lengthwise. Excise the pith as small (about 1 cm^3) pieces and place on Murashige and Skoog (MS) medium (1962) modified to contain 2 mg of indolacetic acid (IAA), 1 mg benzylaminopurine (BAP), 30 g sucrose, MS salts (Sigma, St. Louis, MO), and 8 g of purified agar in 1 L of distilled water. The pH of the medium should be adjusted to about 5.7 before sterilization by autoclaving. Medium should be poured into 10-cm diameter dishes, and cultures incubated in a 16-h light/8-h dark photoperiod at 25°C. Callus should form on the excised pith pieces after 2 weeks, and the entire culture may be subdivided and cultured on the same medium every 3 to 4 weeks.

Diseased plants. Obtain sporangia of *P. tabacina* from the ATCC (permit required) or friendly land university plant pathologist (permit required if out of your state). Establish the disease by spraying five to seven leaf stage tobacco plants with a suspension of sporangia (2 × 10^4/mL works well). Place the plants in plastic bags overnight and then remove the next morning. Plants can be grown in a greenhouse or on a laboratory bench. Water the plants only by placing water directly on the soil or potting medium; do not get the leaves wet. Chlorotic spots, which indicate the establishment of the pathogen, should be evident within 7–10 days after inoculation. To induce sporulation, water the soil generously and place pot and plant in a plastic bag overnight or for a minimum of 12 h. Sporangiophores and sporangia should be evident in the morning on the undersides of leaves (blue-white "fuzz" in chlorotic areas). Sporangia can be collected in distilled water and sprayed on new plants to restart the infection cycle.

There is an alternative procedure for producing sporangia to use for inoculation of callus cultures. At the first sign of leaf chlorosis (infection), lightly swab the lower surface of the area with 70% ethanol; do not soak the area. This should be repeated for three successive days. Do not use alcohol the day before sporulation is induced as described. Swabbing with alcohol and maintaining dry leaves dramatically reduces contamination of the cocultures.

Materials

- Autoclaved glass microscope slides and 9-cm diam. filter paper (any grade will work)
- Sterile 10-cm diam. plastic petri dishes
- 100 mL of 1% (10 g agar in 1 L) water agar melted and maintained at 55°C in a water bath

- Laminar flow hood and alcohol lamp
- Scalpel with # 10 blade and long forceps
- Sterile 250-mL beaker containing sterile MS medium as described previously
- Tissue cultures of tobacco (see under General Considerations)
- Tobacco plants infected with *P. tabacina* (see under General Considerations)
- Dissecting microscope and compound microscope equipped with epifluorescence—330–380-nm excitation and 420-nm absorption filters
- 0.05% (w/v) aniline blue in 0.03 M phosphate (K$_3$PO$_4$) buffer (Martin, 1959)
- 100 mL of 0.05 M potassium phosphate buffer, pH = 6.8–7.2
- Sterile distilled water

Follow the protocols listed in Procedure 21.4 to complete this experiment.

Anticipated Results

This procedure should result in coculture of *P. tabacina* and tobacco callus. Sporangia, germ tubes, inter- and intracellular hyphae, appressoria, and haustoria should be observed after 48–96 h. Sporangiophores and sporangia may be produced after 10 to 14 days. Details of the expected results are provided as comments in Procedure 21.4.

Questions

- Is the physical and physiological relationship between the host and pathogen similar in culture and intact leaf tissue?
- If *P. tabacina* sporangia typically produce very short germ tubes on leaf surfaces, why do germ tubes (hyphae) grow for long distances in callus culture?
- What are the benefits to establishing cocultures of pathogens with their hosts?
- Why do haustoria of *P. tabacina* "glow" when stained with aniline blue and viewed with UV light?

EXPERIMENT 5. OBSERVING MORNING GLORY LEAF TISSUE INFECTED WITH *ALBUGO CANDIDA*

White rust of morning glory is very common. Infected morning glory leaves have chlorotic halos on the upper surface and white "pustules" on the lower surfaces of the leaves (Figure 20.13). Pustules may appear pink and very crusty in the fall.

	Procedure 21.4
	Coculture of *Peronospora tabacina* and Tobacco Callus
Step	Instructions and Comments
1	Do this operation in a laminar flow hood and 3–5 days before sporangia will be needed. Aseptically place two pieces of filter paper into a 10-cm diam. sterile petri dish and moisten them with sterile, distilled water. Drain excess water from the dish. Dip a sterile glass microscope slide into sterile MS medium—about 75% of the slide should be immersed in the gel. Withdraw the slide, hold parallel to the surface of the flow hood, and allow agar to solidify. Place slides on top of the moist filter papers in the petri dishes (Hock, 1974). Repeat the procedure.
2	Transfer some friable (breaks into small pieces or individual cells, or both) callus to the agar on the microscope slides. A thin layer of cells or aggregates over the entire surface is best. Seal the dishes with parafilm and incubate as with petri dish cultures of callus. Transfer callus cultures to be used in experiments to fresh MS medium. Try to place the callus on the agar as nearly in the center of the dish as possible.
3	Do this operation on a clean laboratory bench. Melt 1% water agar and cool to about 55°C in a water bath. Place moist filter paper in the bottom portion of several petri dishes and close the dishes. The next two steps must be performed quickly and carefully.
4	Excise several 4 cm^2 (~1 in.^{2}) pieces of leaf tissue with sporangiophores and sporangia and place with the top of the leaf in contact with the filter paper. Close the lid to maintain high relative humidity.
5	Place a drop of 1% molten agar in the center of another dish top and quickly affix an infected leaf piece from step 4 by slightly immersing the top side of the leaf into the agar. Remove the lid of a callus culture or a microscope slide culture and replace with a lid to which the infected leaf tissue is attached. Do not seal with parafilm, but place the dish in the running laminar flow hood. After 4–6 h, remove the lids with attached leaf pieces and replace with new, sterile lids. Incubate cultures in the dark at 25°C. An alternative method for inoculating the tobacco cultures is to transfer the leaf segments to the petri dish lids before sporangiophores and sporangia form. Seal the dishes with parafilm and incubate the cultures in the dark overnight. By the next morning, sporulation will be evident. Remove the parafilm wrapping and place the dish in a running laminar flow hood. This method achieves higher inoculum levels, but generally results in more contaminated cultures.
6	Microscope slide cultures can be examined after 48–96 h directly with a compound microscope (10 and 20× objectives only). Ungerminated and germinated sporangia should be present as well as very long germ tubes and hyphae. Careful examination will reveal appressoria, haustoria, and intracellular hyphae.
7	Cells can be removed from the slides and suspended in either phosphate buffer (Figure 21.1) or aniline blue fluorescent stains. Higher magnification objectives with microscopes equipped with differential interference contrast (Figure 21.2) and epifluorescent (Figure 21.3) microscopy can be used to examine the host–pathogen interaction. Caution: do not look directly at the UV light source. Cells can also be suspended in fixatives for light and electron microscopy (Trigiano et al., 1984).
8	Periodically observe the inoculated callus on agar culture for 2 weeks. Discard any obviously contaminated cultures. Blue mold hyphae are hyaline (glass-like, without color), and may be scanty or abundant on the surface of the calli. Sporulation will not occur in cultures incubated in the dark.
9	Incubate inoculated callus cultures at 25°C with a 14-h light/10-h dark photoperiod. A few sporangiophores with sporangia should develop.

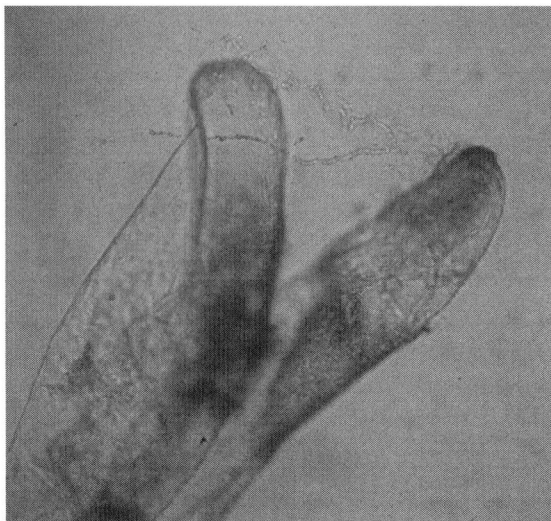

FIGURE 21.1   Whole mount of host cells (HC) with external hyphae. Note fungal growth in the host cells. (Reprinted from Trigiano et al. 1984. *Phytopathology* 74:280–285 with permission from the American Phytopathological Society.)

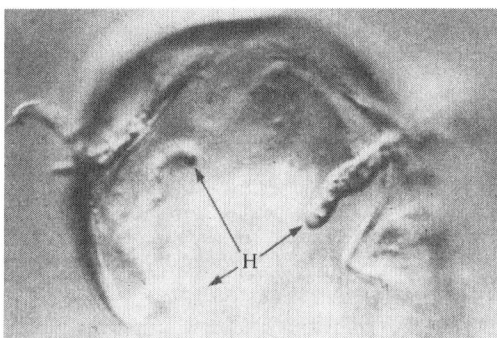

FIGURE 21.2   Differential interference contrast micrograph of a whole mount of three haustoria (H) in a host cell. (Reprinted from Trigiano et al. 1984. *Phytopathology* 74:280–285 with permission from the American Phytopathological Society.)

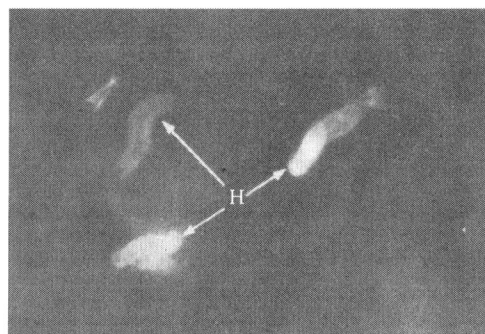

FIGURE 21.3   Fluorescence microscopy of the same haustoria (H) shown in Figure 21.2 stained with aniline blue and viewed with UV light. (Reprinted from Trigiano et al. 1984. *Phytopathology* 74: 280–285 with permission from the American Phytopathological Society.)

Materials

Each student will require the following items:

- Morning glory leaves infected with *A. candida* (see Figure 20.13).
- Compound microscope, glass microscope slides, and coverslips
- Water in a dropper bottle
- Razor blades (caution: very sharp)
- 0.1% calcofluor in water
- Epifluorescence microscope equipped with 395–420-nm excitation filter and 470-nm absorption filter

Follow the protocol in Procedure 21.5 to complete this experiment.

Anticipated Results

Sectioned mounts of pustules should reveal many square- or rectangular-shaped sporangia. Knob-like haustoria may also be seen in host cells. Compare with a prepared slide if these structures cannot be observed in the fresh sections. If the tissue is observed with calcofluor, hyphae, sporangia, and haustoria should "glow" white. Plant cell walls will also "glow."

Questions

- Why is it relatively difficult to observe sporangiophores of *Albugo* compared to *Peronospora*?
- Why are *Albugo* sporangia box-shaped?
- Why does calcofluor stain *Albugo* structures? Would it stain structures of most oomycetes?

Procedure 21.5

Morphological Features of *Albugo candida* in Infected Morning Glory Plants

Step	Instructions and Comments
1	Observe pustules on morning glory leaves using a dissecting microscope (Figure 20.13).
2	With a razor blade, cut cross sections (as thin strips as possible) from morning glory leaves infected with *Albugo candida* and mount in water on a microscope slide. Try to include a portion of a pustule in the section. Draw sporangia and pustule morphology.
3	Cut other thin sections of leaf tissue and pustules and mount in calcofluor solution. View these sections using epifluoresence and draw all pathogen and host structures. Caution: do not look directly at the UV light source.

LITERATURE CITED

Ferguson, A.J. and S.N. Jeffers. 1999. Detecting multiple species of *Phytophthora* in container mixes from ornamental crop nurseries. *Plant Dis* 83: 1129–1136.

Heist, E.P., W.C. Nesmith and C.L. Schardl. 2001. Cocultures of *Peronospora tabacina* and *Nicotiana* species to study host-pathogen interactions. *Phytopathology* 91: 1224–1230.

Hock, H.S. 1974. Preparation of fungal hyphae grown on agar-coated microscope slides for electron microscopy. *Stain Technol* 49: 318–320.

Izard, C., J. Lacharpagne and P. Schiltz. 1964. Comportement de *Peronospora tabacina* dans les cultures de tissus et le role de l'épiderme foliare. SEITA (Serv. Exploit. Ind. Tab. Alumettes). *Ann Dir Etud Equip Sect* 2: 95–99.

Jeffers, S.N. and H.S. Aldwinckle. 1987. Enhancing detection of *Phytophthora cactorum* in naturally infested soil. *Phytopathology* 77: 1475–1482.

Jeffers, S.N. and S.B. Martin. 1986. Comparison of two media selective for *Phytophthora* and *Pythium* species. *Plant Dis* 70: 1038–1043.

Martin, F.W. 1959. Staining and observing pollen tubes in the style by means of fluorescence. *Stain Technol* 34: 125–128.

Murashige, T. and F. Skoog. 1962. A revised medium for rapid growth and bioassays with tobacco tissues. *Physiol Plant* 15: 473–497.

Trigiano, R.N., C.G. van Dyke, H.W. Spurr, Jr. and D.J. Gray. 1984. Infection and colonization of tobacco callus by *Peronospora tabacina*. *Phytopathology* 74: 280–285.

22 Soilborne Plant Pathogens

Bonnie H. Ownley and D. Michael Benson

CHAPTER 22 CONCEPTS

- Soilborne plant pathogens include fungi, bacteria, nematodes, viruses, and parasitic plants that are capable of surviving in soil for extended periods of time in the absence of a host plant.

- Soilborne plant pathogens survive in soil passively by production of dormant propagules, or actively by saprophytic colonization of host debris.

- Management of soilborne pathogens is aimed at reducing the residual amount of inoculum available to infect the next crop, or preventing germination of pathogen propagules in the host infection court.

- Environmental factors in soil, including soil pH, macro- and micro-elements, water, aeration, temperature, and organic matter affect the ability of soilborne plant pathogens to survive and cause disease.

- Disease management strategies to control soilborne plant pathogens include plant resistance, cultural practices, biological controls, conventional fungicides, fumigation, and compost amendments.

Soilborne plant pathogens are a general group of pathogens that are capable of surviving in soil until the host is infected. Examples of soilborne pathogens come from all groups of pathogens including bacteria, fungi, nematodes, parasitic higher plants, and viruses. Soilborne fungal pathogens will be the primary focus of this chapter. Fungal pathogens may be **soil inhabitants**, which have a saprophytic phase that allows for growth and colonization of crop debris in soil; or **soil invaders**, which survive in soil but do not grow saprophytically on crop debris prior to host infection. Soil invaders produce spores or survival structures that may persist in soil only one season or many years depending on the pathogen and survival structure formed. Soil invaders only increase in population by spores or survival structures produced on or in their host.

Life in the soil for soilborne plant pathogens is regulated by environmental factors such as temperature, moisture, oxygen, pH, etc., but foremost, by availability of nutrients for spore germination and growth toward the plant host. It was recognized in the mid-twentieth century that many fungal pathogens survive in soil in a dormant state as spores, chlamydospores, sclerotia, or other survival structures. This phenomenon, termed **soil fungistasis,** was demonstrated to occur in natural soils in all climates, from temperate to tropical to desert to arctic environments. The effect was greatest in the upper few inches of soil where the number and diversity of microorganisms were greatest; the effect diminishes with soil depth. Soil fungistasis is overcome when a source of nutrients, primarily carbon and nitrogen, becomes available to the fungal pathogen. Fungal spores that are "nutrient-dependent"

and "nutrient-independent" are both subject to soil fungistasis. Nutrient-independent spores, once removed from the fungistatic conditions in soil and placed in a drop of water on a microscope slide, are able to germinate without **exogenous** nutrients. Because these spores do not germinate in soil but do germinate in a water droplet on a microscope slide, there must also be inhibitory substances present in soils that prevent spore germination until enough exogenous nutrients are available to overcome soil fungistasis.

Due to the spatial distribution of fungal propagules in the three-dimensional soil environment, most propagules never encounter the nutrients needed to overcome soil fungistasis, and eventually die without ever infecting a host plant. The propagules that will successfully infect the host lie either in the **rhizosphere,** the zone under the influence of the root, or on the **rhizoplane,** the root surface (Figure 22.1). What is it that is unique about the rhizosphere/rhizoplane and root infection? As the root system of a plant develops, **root exudates** in the form of simple sugars (carbon), amino acids (nitrogen), and many other types of compounds found in the plant diffuse into soil from gaps between cortical cells in the **zone of root elongation** just behind the **root cap** (Figure 22.1). Compared to nonrhizosphere soil, i.e., bulk soil, the number of bacteria is 20 to 40 times greater in the rhizosphere, whereas the number of fungi is several folds greater. The large population of microorganisms in the rhizosphere is supported by root exudates. Competition for these nutrients by soil microorganisms, primarily bacteria and fungi, is intense, and thus, the rhizosphere effect is limited to just a millimeter or two from the root in most cases. For propagules of soilborne plant pathogens in the rhizosphere or on

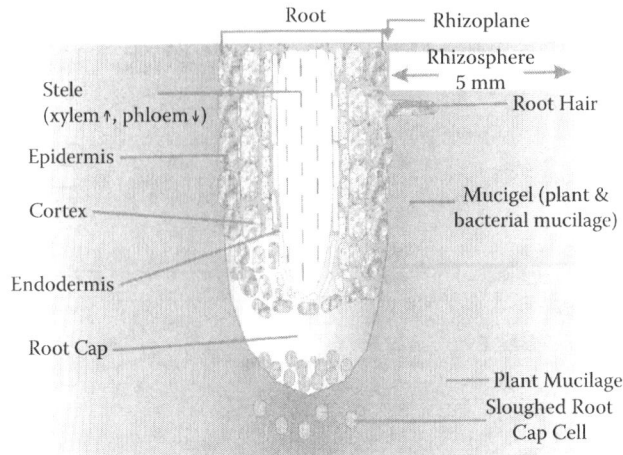

FIGURE 22.1   The rhizosphere. (Reprinted from *Environmental Microbiology*, Pepper, I.L. Beneficial and pathogenic microbes in agriculture, pp. 425–446, Copyright 2000, with permission from Elsevier.)

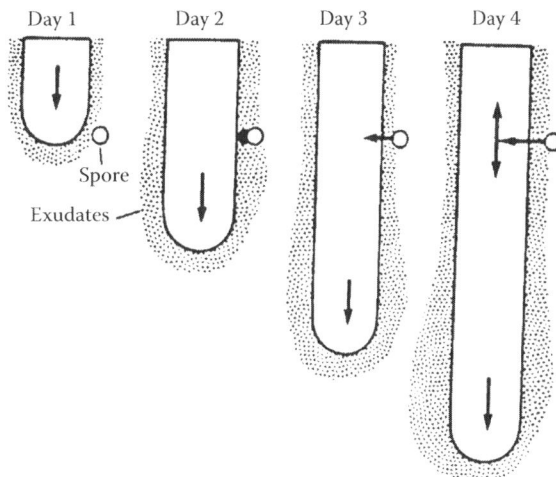

FIGURE 22.2   Exudate from a peanut root tip stimulates germination of a chlamydospore of *Fusarium oxysporum*. The germ tube grew toward the source of the exudate, and penetration occurred behind the root tip. (From Bruehl, G.W. 1987. *Soilborne Plant Pathogens*, Macmillan, New York. With permission from the McGraw-Hill Companies.)

the rhizoplane, these nutrients provide the energy needed for the propagules to overcome soil fungistasis, germinate, and infect the host (Figure 22.2). This same "exudates phenomenon" occurs in the vicinity of hypocotyls of seedlings and the **spermosphere** zone around germinating seeds. For these various host situations, the **infection court** for soilborne pathogens becomes the zone of root elongation, seedling hypocotyl, or seed coat.

The number of propagules of a pathogen in a soil sample, whether in the form of spores, sclerotia, or chlamydospores, is termed **inoculum density** and is expressed as propagules per gram of soil. For example, in a cotton field soil the inoculum density of *Verticillium dahliae*, which causes Verticillium wilt, may range from 20 to 40 microsclerotia per gram of soil. Given a favorable environment

and a susceptible host, this is more than enough inoculum to cause a severe epidemic of Verticillium wilt.

SURVIVAL

MECHANISMS OF SURVIVAL

The success of soilborne plant pathogens from crop to crop and year to year relies exclusively on the ability to survive between crops and years. Soilborne plant pathogens exhibit a variety of survival mechanisms as a group, but most individual pathogens utilize only one or two mechanisms. In general, survival of soilborne plant pathogens may be active or passive. Active forms of survival include parasitic survival on alternate hosts, commensal survival in the rhizosphere of nonhosts, and saprophytic survival. Dormant survival may be imposed by the environment (i.e., soil fungistasis) or inherent in the genetic makeup of the pathogen.

Soilborne plant pathogens like *Rhizoctonia solani* have a highly **competitive saprophytic ability** to survive in soil. Characteristics of these pathogens include rapid germination and growth on crop debris in soil, enzymes to decompose complex carbohydrates, production of antimicrobial compounds, and tolerance of antimicrobial compounds produced by other microorganisms challenging for colonization of the crop debris. Other pathogens such as *V. dahliae* colonize parenchyma cells of host stems and roots by saprophytic growth as the host is dying from the parasitic effects of the pathogen in xylem tissues. In this situation, *V. dahliae* has a "timing" advantage in colonizing the cotton crop debris compared with other microorganisms in soil. *Gaeumannomyces graminis*, which causes take-all of wheat and other cereals and grasses, survives as mycelium on roots and plant debris invaded during the parasitic phase of take-all disease. Survival in wheat debris is shortened in moist, well-aerated soil at moderate to warm temperatures;

FIGURE 22.3 Black rhizomorphs of *Armillaria mellea* on the surface of apple bark. (From APS Digital Image Collections, *Diseases of Orchard Fruit and Nut Crops*. 2002. Photo courtesy of A.L. Jones. With permission.)

these are conditions that hasten the decomposition of crop debris in soil. *Gaeumannomyces graminis* survives longer in dry or compacted soil, conditions that are not favorable for microbial decomposition of organic debris. Pathogens of woody tissues, such as *Armillaria mellea*, are well adapted to survive as mycelium or rhizomorphs on diseased trees or decaying roots. **Rhizomorphs** are thick cordlike threads of mycelium (1 to 3 mm in diameter) with a compact black outer layer of mycelium and an inner white or colorless core (Figure 22.3). Survival of pathogens on woody tissues depends on how long the substrate will last. These pathogens can survive on tissues that are high in carbon and low in nitrogen. Often, they are very tolerant of tannic acid that is naturally found in woody tissues. Tannic acid, which inhibits the growth of most fungi, stimulates growth of *A. mellea* in culture (Cheo, 1982). Generally, successful saprophytic survival requires possession of the plant debris until a host infection court, such as new seed, seedling, hypocotyl, or root, invades the space near the colonized residue.

SURVIVAL STRUCTURES

Soilborne pathogens that depend on dormant survival produce some type of resting propagule such as a thick-walled conidium, **chlamydospore, microsclerotium, oospore,** or **sclerotium** (Table 22.1). *Cochliobolus sativus* causes common root rot in wheat and survives between crops as conidia in soil. Chlamydospores are thick-walled, asexual structures produced by pathogens such as *Fusarium, Phytophthora*, and *Thielaviopsis* that survive in soil for several years. A number of soilborne pathogens like *Verticillium, Rhizoctonia, Macrophomina, Sclerotinia, Sclerotium,* and *Phymatotrichopsis* produce sclerotia or microsclerotia as resting propagules. Sclerotia may be rudimentary or well organized into different tissue layers. Sclerotia of most soilborne pathogens survive many years in soil in the absence of a host. Oospores produced by some spe-

cies of *Phytophthora* and *Pythium* are thick-walled sexual spores formed on or in infected host tissues (Chapter 20). Fertilization of a nucleus in the *oogonium* by antheridial nuclei results in development of the oospore within the oogonium. Oospores may have an inherent dormancy so that some are available for germination over a period of several years.

SURVIVAL OVER TIME

Whether a soilborne plant pathogen survives by active or passive mechanisms, over time the number of surviving propagules of the pathogen will decline in soil. The pattern of decline can be visualized in the form of one of three idealized curves (Figure 22.4) (Benson, 1994). For pathogens that survive as conidia or chlamydospores in soil, the population of those propagules enters a logarithmic death phase wherein most of them die quickly over a relatively short period, such as over winter. (Figure 22.4, Curve A). However, a small proportion of the population of propagules is very successful in survival and persists for a long period, resulting in a residual population. For pathogens with multicellular propagules such as sclerotia, there is an initial lag phase as individual cells in the sclerotium die before the population of sclerotia enters the logarithmic death phase (Figure 22.4, Curve B). For pathogens with high competitive saprophytic ability, the pathogen population may actually increase in soil as the pathogen colonizes new substrates before entering the logarithmic death phase (Figure 22.4, Curve C). In all three curves, however, the number of propagules surviving, i.e., the inoculum density of the pathogen in the residual phase of survival, determines the initial amount of disease that will develop in a subsequent crop. For most soilborne plant pathogens, this residual inoculum density is usually more than adequate to initiate an epidemic. Therefore, plant disease management practices are aimed at reducing this residual inoculum density even lower or by preventing the germination of pathogen propagules in the host infection court.

ENVIRONMENT AND SOILBORNE PLANT PATHOGENS

SOIL WATER AND DISEASE

The amount of water present in soil can have a profound effect on soilborne pathogens and the diseases they cause. Just as plant growth is affected by the amount of water available for root uptake, some soilborne pathogens are uniquely suited to take advantage of soil water availability and cause disease. Whereas many soilborne pathogens cause disease across a normal range of soil moisture for good plant growth, some such as the water molds, i.e., *Phy-*

TABLE 22.1
Characteristics of Selected Soilborne Plant Pathogens

Name of Pathogen	Pathogen Type	Survival	Dispersal	Infection Court	Symptoms	Hosts	Parasitism[a]
Agrobacterium tumefaciens[b]	Prokaryote	Cells in soil, plant galls	Cultivation, infected plants, irrigation water	Wounds on roots and stems	Galls, low vigor, stunting	Many	Hemibiotroph
Armillaria mellea	Basidiomycete	Hyphae and rhizomorphs in infected woody tissues	Root contact between infected and healthy trees, rhizomorph growth in soil, wind-borne basidiospores	Roots	Basal stem cankers, collar rot, root rot	Many woody angiosperms and gymnosperms, some nonwoody plants	Necrotroph
Bipolaris sorokiniana (teleomorph = *Cochliobolus sativus*)	Deuteromycete (teleomorph = Ascomycete)	Thick-walled conidia	Wind-borne conidia	Seeds, coleoptiles, subcrown internode	Root rot, seedling blight, whiteheads	Wheat, barley, rye, other grass species	Hemibiotroph
Cylindrocladium parasiticum (teleomorph = *Calonectria ilicicola*)	Deuteromycete (teleomorph = Ascomycete)	Microsclerotia	Cultivation, wind-borne plant debris	Root tips	Black root rot, leaf chlorosis, stem necrosis	Peanut, soybean, blueberry, tea, hardwoods	Necrotroph
Fusarium oxysporum	Deuteromycete	Chlamydospores	Cultivation, plant debris, infected transplants	Root tips	Stunting, vascular wilt	Many	Hemibiotroph
Fusarium roseum	Deuteromycete	Chlamydospores, mycelia	Cultivation	Basal stems	Foot and crown rot	Many	Necrotroph
Fusarium solani	Deuteromycete	Chlamydospores	Cultivation, plant debris	Hypocotyls	Root rot	Many	Necrotroph
Gaeumannomyces graminis	Ascomycete	Hyphae in crop debris	Cultivation, plant debris	Roots	Chlorosis, low vigor, stunting, whiteheads	Wheat, barley, other grasses	Necrotroph
Hymenula cerealis (synonym *Cephalosporium gramineum*)	Ascomycete	Hyphae in crop debris	Cultivation, crop debris	Wounds on roots and crown	Chlorotic stripes, stunting, vascular wilt, whiteheads	Wheat	Hemibiotroph
Macrophomina phaseolina	Deuteromycete	Microsclerotia	Cultivation, plant debris	Root tips	Charcoal rot of roots and stems	Many	Necrotroph

Pectobacterium carotovorum subsp. *carotovorum* (formerly *Erwinia carotovora* subsp. *carotovora*)	Prokaryote	Cells in soil, potato tuber debris, water	Irrigation water, aerosols, potato seed pieces	Wounds on potato tubers, enlarged lenticels on tubers	Soft rot of tubers	Potato	Necrotroph
Phymatotrichopsis omnivora (formerly *Phymatotrichum omnivorum*)	Deuteromycete	Sclerotia	Growth of hyphal strands in soil	Roots	Bronzing of leaves, permanent wilt	Many dicotyledonous plants (> 2000)	Necrotroph
Phytophthora cinnamomi	Oomycete	Chlamydospores	Cultivation, water, infected plants	Root tips	Root rot, low vigor	Many hosts (> 900)	Hemibiotroph
Pseudocercosporella herpotrichoides	Deuteromycete	Hyphae in crop debris	Rain splash, cultivation	Basal leaf sheaths	Eyespot, foot rot, lodging	Wheat	Hemibiotroph
Pythium spp.	Oomycete	Sporangia, oospores, mycelium	Cultivation, water, infected plants, seeds	Root tips	Root rot, damping-off	Many	Necrotroph
Ralstonia solanacearum	Prokaryote	Cells in soil and plant debris, contaminated water sources	Cultivation, irrigation water	Wounds on roots	Permanent wilt	Solanaceous crops	Hemibiotroph
Rhizoctonia solani (teleomorph = *Thanatephorus cucumeris*)	Deuteromycete (teleomorph = Basidiomycete)	Sclerotia, mycelium in soil	Cultivation, crop debris	Seeds, stems, hypocotyls, roots	Damping-off, stem rot, root rot	Many	Necrotroph
Sclerotinia spp.	Ascomycete	Sclerotia	Cultivation, crop debris	Stem	Blight	Many	Necrotroph
Sclerotium rolfsii	Deuteromycete	Sclerotia	Cultivation, crop debris	Lower stem	Blight	Many	Necrotroph
Thielaviopsis basicola	Deuteromycete	Chlamydospores	Cultivation, infected plants	Root tips	Root rot, chlorosis	Many	Hemibiotroph
Verticillium dahliae	Deuteromycete	Microsclerotia	Cultivation, infected plants	Root tips	Wilt, vascular necrosis	Many	Hemibiotroph

[a] See Chapters 10 and 23.
[b] Proposed name change = *Rhizobium radiobacter* (Young et al., 2001).

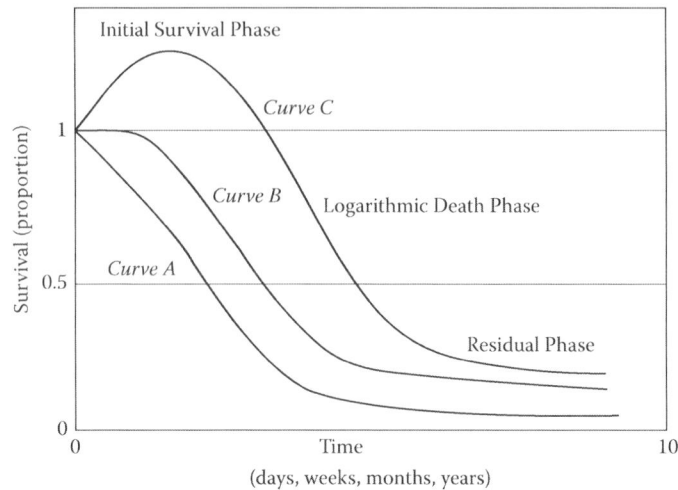

FIGURE 22.4 Idealized survival curves for soilborne plant pathogens. Curve A represents inoculum that enters the logarithmic death phase immediately after dissemination. Curve B represents multicellular inoculum, such as sclerotia, that has a lag phase before entering the logarithmic death phase, and curve C represents saprophytic growth of inoculum after dissemination to soil. (From Campbell, C.L. and D.M. Benson, (Eds.). 1994. *Epidemiology and Management of Root Diseases*, Springer-Verlag, Berlin. With permission).

tophthora and *Pythium* spp., and others such as *Fusarium* spp., cause epidemics under conditions of saturated and drought conditions, respectively.

SOIL WATER POTENTIAL

The energy of soil water is measured in terms of **soil water potential**, which is composed of gravitational, matric, osmotic, and other forces. The amount of energy an organism, e.g., fungus or plant root cell, would expend to take up pure, free water is arbitrarily set at a water potential of 0. However, in soil, there are several forces that bind water to soil particles, and the amount of energy required to overcome these forces is defined in terms of suction or pressure and expressed as –bars or –kilopascals (kPa; the Standard International [SI] unit of pressure). One bar equals 100 kPa. As an analogy, imagine the "energy" or force, i.e., suction in this case, that you would have to exert on a drinking straw to remove water from a glass. **Gravitational potential** is potential energy due to gravity. Gravitational water is water in excess of that which is held by soil particles. Gravitational water passes through soil to deep within the profile and may reach the water table. After gravitational water has drained through the soil profile, the soil is at field capacity with a water potential of about –0.3 bars. For sandy soils field capacity is at –0.1 bars. In most soils, the permanent wilting point for many plants is –15 bars.

   Matric potential is a measure of the force needed to remove adsorption and capillary water held on and between soil particles. Many biological processes in soil are affected by matric potential. For example, as matric potential decreases, the rhizosphere shrinks because water

films become thinner. **Osmotic potential** is the result of ions and molecules dissolved in soil water. The osmotic potential component of total water potential is negligible in all but the most saline soils. Matric potential dominates total water potential in soils and has the greatest impact on water uptake by roots and growth and infection by soilborne pathogens. In plant tissues, osmotic potential dominates and can influence pathogen growth and colonization after infection.

   In soil, water potential can be described in three categories that include gravitational water from 0 to –0.3 bars water potential, which drains through the soil profile after rain or irrigation, **capillary water** from –0.3 to –31 bars, which is held by forces of adsorption on soil particles and capillary forces in pores between soil particles, and **hygroscopic water** with a water potential below –31 bars. Gravitational water is available for root uptake only briefly, as it drains through the soil profile, whereas hygroscopic water is so tightly held on soil particles that it is unavailable to roots. Gravitational water can carry propagules of soilborne pathogens deeper into the soil profile, but in poorly drained soils with slow percolation, oxygen deficiency can harm plant roots and aerobic microbes.

   As gravitational water drains through the soil profile, the soil pores, or space between soil particles, are filled with air when water percolates. The diameter of the pore space determines whether or not the pore can retain water as the soil dries. The distribution of soil pore sizes, i.e., the space between soil particles, varies with soil type. However, the relationship between pore size diameter and water potential has been calculated. At field capacity, i.e., -0.3 bars water potential, pores with a diameter of 8.79 μm would hold water, whereas at –15 bars water potential, the permanent wilting point for most plants, the pore diameter that continues to retain water is only 0.19 μm. Even at –15 bars, however, soil relative humidity is 98.9%.

In soil, capillary water between –0.3 and –15 bars water potential is available for plant growth. Water potential gradients exist in soil as plant roots take up water during the day such that, in bulk soil, the water potential may be –5 bars, dropping to –8 bars in the rhizosphere, and –14 bars at the root surface. At night, the system would reequilibrate to about –5 bars water potential.

SOIL WATER EFFECTS ON DISEASE

Phytophthora root rot caused by a number of *Phytophthora* spp., and Fusarium foot rot caused by *F. culmorum* illustrate two diseases where wet soils and dry soils, respectively, are important in disease development.

PHYTOPHTHORA ROOT ROT

The biology of *Phytophthora* spp. that cause root rot and crown rot is very tightly keyed into soil water and, more importantly, changes in soil water status resulting from rainfall and irrigation. Although chlamydospores and oospores are responsible for long-term survival of *Phytophthora* spp. in soil, once soil temperature is favorable, changes in soil water can trigger sporulation resulting in formation of sporangia. Typically, sporangia of most *Phytophthora* species are formed in soil within 12 h of when matric potential reaches between –0.1 and –0.15 bars; remember that field capacity is near –0.3 bars water potential, so sporangia form under very wet soil conditions. Sporangia can persist in soil for a few days or at most a few weeks; however, when soil reaches 0 bars water potential (saturated conditions), sporangia are triggered to produce and release motile **zoospores**. Exudates of carbon compounds in the rhizosphere serve as a chemotactic gradient that the motile zoospores follow as they swim through water-filled pores to the rhizoplane where they encyst, germinate, and infect the root. Rather than swim in a straight line along the nutrient gradient to the root, zoospores exhibit a helical swimming pattern that requires a turning diameter of 110 μm, in the case of *P. cinnamomi*. This means that soils at matric potentials below about –0.015 bars water potential do not have water-filled pores large enough to accommodate the swimming zoospores. However, soils need to stay saturated (0 bars water potential) for only a few hours for zoospores to find and infect plant roots. For management of Phytophthora root rots, strategies that prevent sporangium formation can be the key to control of this pathogen.

FUSARIUM FOOT ROT

In certain parts of the Pacific Northwest, and other geographical areas with low rainfall (20 to 40 cm annually) and dryland farming practices, Fusarium foot rot of wheat caused by *F. culmorum* can limit crop production. The pathogen survives as chlamydospores in the crop debris mulch layer and infects roots 2 to 3 cm below the soil surface where secondary roots emerge. Under adequate moisture conditions no further disease development occurs, but under water stress conditions (very low water potentials) the pathogen colonizes the crown roots and moves up the stem one to three internodes, causing a chocolate brown discoloration of the inner tissues, although the leaf sheath wrapped around the internodes remains apparently healthy. The symptoms are not apparent until the plant has formed heads, at which time bleached, empty "whiteheads" take the place of seed-filled heads.

In dryland farming, precipitation is collected in the mulch soil profile for one season before the wheat crop is planted in the fall for summer harvest the following year. Fusarium foot rot was first recognized in the Pacific Northwest when semidwarf wheat cultivars were grown on close plant spacings, and high nitrogen fertilization rates were applied to promote yield. These crop production practices resulted in soil and leaf water potentials that stressed the plant and favored infection and colonization of plant tissues by *Fusarium*. As illustrated in Figure 22.5, wheat plants grown under high rates of nitrogen had soil water potentials in the upper 120 cm of the soil profile that were –7 to –8 bars drier than plants grown under low nitrogen regimes. Likewise, the leaf water potential (osmotic potential) was –5 to –8 bars drier in the stems of high nitrogen plants than in low nitrogen plants. This difference in soil and leaf water potential was enough to result in more infections and subsequent severe disease development with corresponding loss of yield. To solve the problem, growers are advised to fertilize at a conservative rate and to use a row spacing width of 40 to 45 cm. These practices will not result in plant water stress later during crop growth.

SOIL AERATION

Organic matter and microorganisms are found mainly in the surface layers of soil because plant roots predominate in the upper layers. The relative abundance of oxygen is the main factor responsible for root distribution because the roots of higher plants are aerobic. In general, soils range from 40 to 60% pore space, and air content of soil varies inversely with water content. If all the soil pores are filled with water, then air is excluded. Oxygen deficiencies often develop when porosity is reduced to 10% or less for an extended time period (Bruehl, 1987).

The rate at which organic matter (such as a dead plant infected with a plant pathogen) decomposes is strongly influenced by aeration and temperature. When the soil is warm, moist, and well aerated, organic matter decomposes quickly. Bringing plant crowns and roots to the sur-

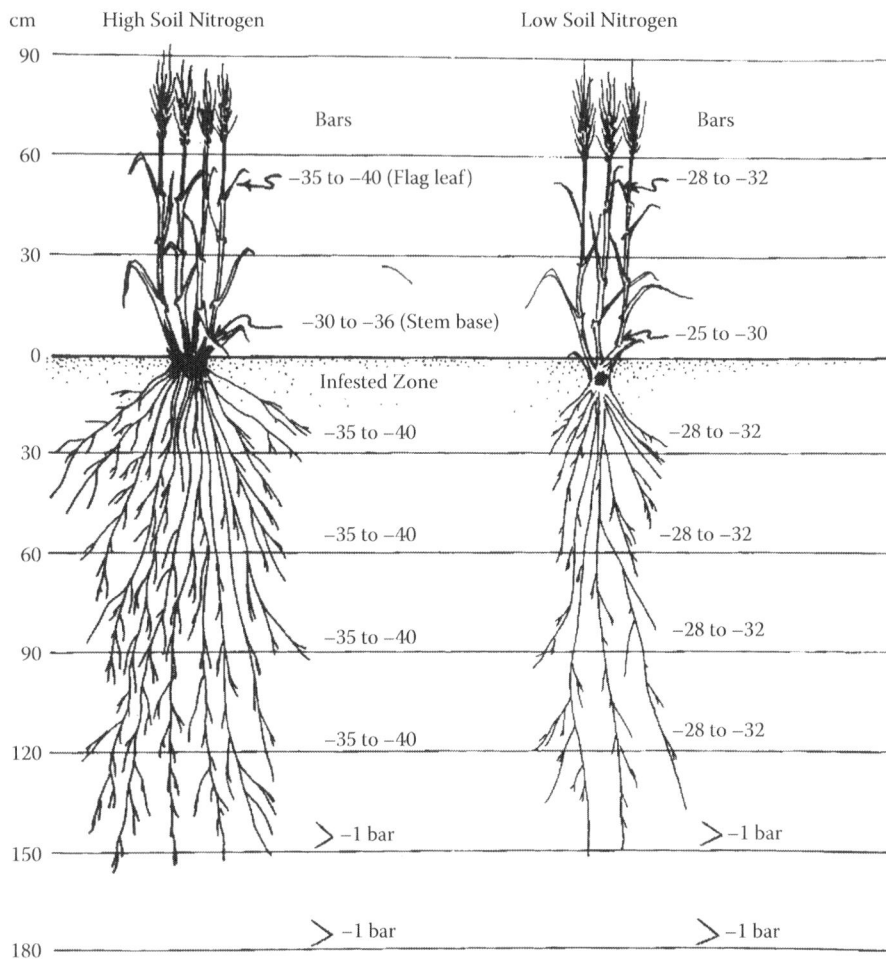

FIGURE 22.5 Typical soil and plant water potentials recorded in the field for *Triticum aestivum* 'Nugaines' (winter wheat) at mid-dough stage, grown under Pacific Northwest dryland conditions with high (60–120 lb N/acre) versus low (20–40 lb N/acre as residual in the profile, none applied) nitrogen fertility, in the presence of soil infestation of *Fusarium roseum* f. sp. *cerealis*. (From Cook, R.J. 1973. *Phytopathology* 63: 451–458. With permission.)

face through cultivation hastens decomposition, mainly by the effect of increased aeration on the activity of soil microbes that are responsible for decomposing the crop debris. Rapid decomposition of diseased plant material can greatly reduce the ability of a plant pathogen to survive as a saprophyte (Bruehl, 1987).

SOIL TEMPERATURE

The metabolism, growth, and interactions of plants, plant pathogens, and other microorganisms in the soil are directly affected by soil temperature. Generally, the lower temperatures that we associate with late fall, winter, and early spring in the Northern Hemisphere are below the minimum required for disease-causing activity of most pathogens. Plant diseases are not usually initiated during that time or, if they are in progress, they often come to a halt. However, pathogens do differ in their preference for higher or lower temperatures; whereas the majority of plant diseases develop more rap-

idly when higher temperatures prevail, others develop best with cooler temperatures.

Vascular wilt disease caused by *V. dahliae* and *V. albo-atrum* can be severe at soil temperatures of 20 to 28°C. *Verticillium albo-atrum* is more important in the cooler, humid, northern regions, and *V. dahliae* causes significant disease problems further south, particularly on irrigated crops. Both fungi are unimportant or absent in the humid tropics and semitropics (Bruehl, 1987).

Verticillium dahliae is more widely distributed because it is adapted to warmer climates and produces abundant, long-lived microsclerotia. *Verticillium albo-atrum* overwinters as microsclerotial-like hyphae, rather than microsclerotia. To illustrate how small differences in response to temperature can be important, consider this example. *Verticillium albo-atrum* is actually more virulent on cotton than *V. dahliae*, but *V. albo-atrum* is found in areas with cool soil temperatures, fails to grow at 30°C, and has not been isolated from the major cotton-producing areas with warm soil temperatures in the United States. On the other hand, *V. dahliae* is an important pathogen of cotton in the southern and southwestern United States. Both species are pathogenic on potatoes from 12 to 28°C, but *V. dahliae* alone is still pathogenic at 30°C (Bruehl, 1987).

In some cases, the optimum temperature for disease development, growth of the pathogen, and growth of the host are quite different. For example, the optimum temperature range for development of black root rot of tobacco caused by *Thielaviopsis basicola* is 17 to 23°C. Tobacco grows best at 28 to 29°C, and plant resistance to black root rot increases as soil temperature rises. Similarly, the pathogen grows fastest in culture at 28 to 30°C. Neither the host nor the pathogen grows well at 17 to 23°C. However, the host is more negatively affected than the pathogen, so that even a weak isolate of *T. basicola* can cause maximum disease development at 17 to 23°C (Bruehl, 1987).

Temperature also affects geographic distribution of soilborne plant pathogens. The fungal pathogen *Phymatotrichopsis omnivora,* which causes root rot of many dicotyledonous hosts, is not found in northern climates. In contrast, temperature plays a minimal role in the geographic distribution of *F. solani* f. sp. *phaseoli,* which can be found wherever bean is grown. After the bean host dies, *F. solani* f. sp. *phaseoli* survives as chlamydospores in soil until favorable conditions return.

Soil pH

Soil pH influences the stability and activity of antibiotics and enzymes, the adsorption of substances and bacteria to soil colloids, the balance between total populations of bacteria and fungi, microbial diversity, and the availability of mineral nutrients to higher plants and microorganisms. Low soil pH makes iron, manganese, zinc, copper, and cobalt more available. When soil pH is less than 5.5, nitrogen, calcium, magnesium, phosphorus, potassium, sulfur, and molybdenum become less available. When soil pH is greater than 7, copper and zinc become less available. Chronic high soil pH often leads to shortages of iron, manganese, cobalt, copper, and zinc. Soils with pH values between 5.6 and 7 are best for most crop plants.

Many soilborne diseases are affected by soil pH. The effect of soil pH on clubroot disease of cabbage, caused by the zoospore-forming obligate parasite *Plasmodiophora brassicae,* is a well known example (Chapter 11). Development of clubroot is strongly favored by acid soils. Clubroot does not occur in heavily limed, moist soils. In fact, adding lime to soil to control clubroot disease has been a management practice for more than 200 years. In agriculture, the term "lime" is generic and refers to several forms of calcium and magnesium. Calcium carbonate ($CaCO_3$) is the most widely distributed naturally occurring form of lime and is found in limestone, chalk, and calcite (Campbell and Greathead, 1990).

Actively growing cabbage roots reduce the rhizosphere pH, which favors infection. In a study by Dobson et al. (1983), in which equivalent amounts of various forms of lime were added to increase the soil pH, the fineness of the material used had an effect on the final pH and, subsequently, the percentage of infected plants. The form of lime, calcium hydroxide, or calcium carbonate was not important. With the addition of the coarser lime materials, acidic microsites could develop and disease was favored.

How does "liming" reduce clubroot disease? The pH increase and the added calcium appear to be important factors. Maturation of zoosporangia and release of zoospores is delayed or prevented by higher levels of calcium. The effect of calcium is enhanced by higher soil pH. There is also an effect on the host–pathogen interaction. With lower calcium levels, infection may occur, but infections are inhibited from further development. Aside from the role of calcium rendering intercellular plant pectins more resistant to enzymatic degradation, calcium may act as a cellular messenger that affects various regulatory mechanisms in higher plants (Campbell and Greathead, 1990).

Take-all of wheat, caused by *Gaeumannomyces graminis* var. *tritici,* is favored by alkaline soils. The severity of disease is also greatly affected by the type of fertilizer applied. Disease is less severe with NH_4-N compared to NO_3-N forms of fertilizer. When wheat plants are supplied with ammonium nitrogen, the pH of rhizosphere soil is lower than bulk soil. In contrast, when nitrate-nitrogen is applied, there is an increase in rhizosphere pH compared to bulk soil (Smiley and Cook, 1973). Apparently, when plants are given nitrate-nitrogen, OH^- and HCO_3^- are excreted from the root, which raises the rhizosphere pH. When plants are given ammonium-nitrogen, H^+ ions are excreted, resulting in a decrease in rhizosphere pH. The rhizosphere pH can deviate from the bulk soil pH by 1.2 units, and this difference can have a large effect on the severity of take-all. A decrease in rhizosphere pH is also associated with increased availability of iron, manganese, phosphorous, and zinc. Increased levels of these elements have been linked to decreases in take-all severity by making the plant less susceptible to disease.

Black root rot of burley tobacco, caused by *T. basicola,* occurs in western North Carolina. However, in some fields the disease does not develop. By comparing the exchangeable Al^{+3} across fields, it was determined that fields with at least 1 milliequivalent (meq) of Al^{+3} per 100 g soil were suppressive to black root rot. Soils with a pH below 5 have soluble Al^{+3} in soil solution that controls soil pH. Amending soil with an alkaline source of calcium, such as calcium hydroxide, raises soil pH as the soluble Al^{+3} is replaced by insoluble $Al(OH)_3$. In an experiment, a black root rot suppressive soil at pH 4.4 with 0.9 (meq) of Al^{+3} per 100 g soil became conducive to disease when calcium hydroxide ($Ca(OH)_2$) was used to raise the pH to 5.8, at the same time lowering Al^{+3} to 0.3 (meq) of Al^{+3} per

100 g soil (Meyer and Shew, 1991). In Brazilian soils with soil pH around 4.8, *P. capsici*, which causes Phytophthora crown and root rot in pepper, is not a disease problem. However, if soils are limed to raise the pH, severe disease develops. In petri dish culture, mycelial growth and sporulation of both *T. basicola* and *P. capsici* is inhibited in the presence of aluminum. The mechanism of aluminum toxicity to fungi is not completely understood, but critical protein synthesis pathways may be inhibited.

CULTURAL PRACTICES

Successful crop production requires integration of a number of cultural practices that result in high yield and elimination or suppression of weeds, insects, and plant pathogens. For soilborne plant pathogens, cultural practices such as avoidance, crop rotation, tillage and residue management, water management, crop management, and sanitation have proven useful to limit diseases caused by these pathogens. Giving cultural practices ample consideration for disease control is taking a holistic approach to plant health. See Chapter 32 for additional information on cultural control of plant diseases.

AVOIDANCE

A common form of avoidance is to delay the planting date. For example, warm-weather crops like corn and cotton are stressed when planted in cool soil early in the spring. Planting after soils have warmed avoids prolonged exposure in cold soils to damping-off pathogens such as *Pythium* and *Thielaviopsis*. In the case of root-knot nematodes, a winter crop of carrots planted in the southeastern United States between November 18 and December 1 has about 90% marketable roots, whereas only 50% are marketable if planted by October 16. As long as the seeding date can be altered economically, a grower has some control over choice of planting date and its corresponding soil temperature, and subsequent potential effects on disease.

CROP ROTATION

Farmers have practiced crop rotation since ancient times. Crop rotation to a nonhost is the most effective cultural practice for crop production because soil nutrients essential to a given crop can be replaced and pathogen populations in soil may decline during the nonhost period of the rotation due to the lack of nutrients. Generally, crop rotation is more effective against soil invaders than soil inhabitants because soil invaders may not persist in soil very long in the absence of a host. The effectiveness of crop rotation is dependent on the longevity of survival structures of soilborne pathogens. For example, 3- to 4-year rotations are sufficient for control of *Phoma lingam,* which causes blackleg of cabbage, whereas soilborne pathogens such as *Rhizoctonia, Sclero-*

tinia, Verticillium, and *Thielaviopsis* produce survival structures that persist in soil for many years and are not likely to be controlled through crop rotation.

Crop rotation can also have preventative value. For example, in Washington State, new lands were brought under irrigation to grow dry beans. For several years, yields were very high, and growers did not practice crop rotation. Eventually, *F. solani* f. sp. *phaseoli* became a serious disease problem, and yields declined. Then the growers started practicing crop rotation to nonhosts; however, by that time, populations of *F. solani* f. sp. *phaseoli* had reached 2000 to 3000 chlamydospores per gram of soil. Once populations were established to this extent, inoculum levels of *F. solani* f. sp. *phaseoli* capable of causing economic loss were maintained, even with crop rotation. In other words, once certain pathogens are increased to very high population levels, they appear to be immortal (Bruehl, 1987). If crop rotation had been followed in the beginning, this problem could have been avoided.

TILLAGE AND RESIDUE MANAGEMENT

The adoption of reduced tillage and conservation tillage has had a tremendous impact on management of soilborne pathogens such as *Sclerotium, Cephalasporium,* and *Pyrenophora tritici-repentis* that survive in crop debris. Unlike conventional tillage where the practice of moldboard plowing would bury crop debris containing these pathogens, reduced tillage provides a favorable microclimate for the pathogens in crop residues on the soil surface. Reduced tillage or no tillage results in moderation of soil temperatures and conservation of soil moisture, factors that favor survival of soilborne pathogens. In addition, crop residues serve as a food base for sporulation and growth of the pathogen prior to infection. Although several soilborne pathogens have become more severe under reduced tillage, the downside of conventional tillage is reduced soil moisture, increased labor and energy costs, and wind and water erosion of bare fields. In Kansas, loss of 2.5 cm of top soil resulted in 100 kg/ha/year reduction in soybean yield presumably due to loss of nutrients.

Fusarium root rot of bean, caused by *F. solani* f. sp. *phaseoli,* is particularly severe in soils with compacted hard pans that restrict rooting of the bean plant. Farmers who used a chisel plow to loosen the compacted soil prior to planting beans enjoyed an almost threefold increase in yield because the root system was able to develop to a greater depth and compensate for the *Fusarium* infection on the shallow roots. On deep soils, the practice of deep plowing to depths of 30 cm, compared with conventional depth of 15 cm, has the potential to bury inoculum beyond the host infection court. *Sclerotinia minor* causes lettuce drop disease and produces sclerotia that survive many years in soil. In lettuce fields that were deep plowed, incidence of lettuce drop was reduced 50% in the first year.

CASE STUDY 22.1

SOUTHERN BLIGHT CAUSED BY *SCLEROTIUM ROLFSII*—BASIC DISEASE-MANAGEMENT STRATEGIES SHOULD NOT BE OVERLOOKED (POTTORFF, 2003)

- *Sclerotium rolfsii* produces sclerotia that enable it to survive in soil.
- *Sclerotium rolfsii* occurs primarily in the warm humid southeastern United States.
- In the late 1990s the pathogen was reported in the Midwest; more recently, it was reported in greenhouses in Colorado.
- Symptoms include wilting, yellowing, and browning of lower leaves of infected plants.
- Signs of the fungus are extensive white mycelial growth starting at the base of plant stems and extending upward; sclerotia (tan spheres the size of mustard seeds) also develop (Figure 22.6).
- Fungicides labeled for control of *S. rolfsii* include flutolanil and pentachloronitrobenzene.
- Basic management practices including immediate removal and destruction of infected plants, pasteurization of potting soil before reuse, ensuring adequate drainage, avoiding overhead irrigation, and using pathogen-free stock can prevent disease outbreaks.

However, the extra cost of deep plowing was not recovered as disease incidence returned to pre-deep plow levels after only 2 years because sclerotia became redistributed throughout the soil profile. On the other hand, *Sclerotium rolfsii*, which causes Southern blight in peanuts and other crops, is effectively controlled with conventional tillage.

SOLARIZATION

Soil solarization is a control strategy for soilborne plant pathogens, initially developed in Israel, whereby soil prepared for planting is covered with a clear plastic tarp during the summer. Each day, as the sun warms the soil under the cover, temperatures in solarized soils may be 10°C warmer at a 10-cm depth than in uncovered soil, reaching near 50°C. Soils are usually left covered for about 6 weeks. Solarization works by direct thermal inactivation of survival propagules of soilborne plant pathogens such as sclerotia, chlamydospores, conidia, and hyphae in crop debris. In general, the higher the soil temperature, the shorter the time needed to kill propagules; however, there is a temperature below which a particular propagule can survive. To reduce the number of pathogen propa-

FIGURE 22.6 Southern blight fungus (*Sclerotium rolfsii*) on soybean. (From Koenning, S.R. 2000. Soybean Disease Information Note 5. Identification and management of mid-to-late season soybean stem and root rots. Department of Plant Pathology, Plant Pathology Extension. North Carolina State University. Photo courtesy D.P. Schmitt. With permission.)

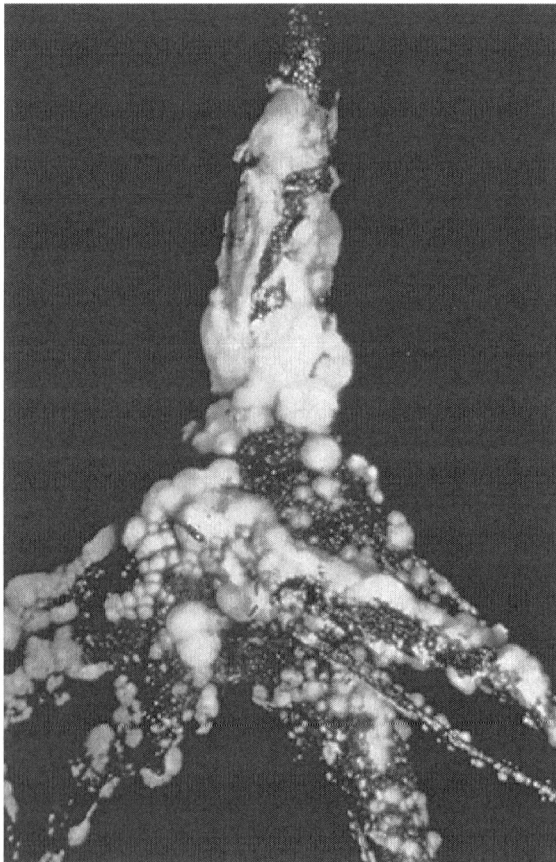

FIGURE 22.7 Dematophora root rot (*Dematophora necatrix*) on grape. (From APS Digital Image Collections, Diseases of Small Fruits. 2000. Photo courtesy A.H. McCain. With permission.)

gules, solarized soils must reach the minimum temperature needed to kill a particular propagule and remain at that temperature long enough for the killing effect to take place. Many diseases caused by soilborne plant pathogens such as *Verticillium* spp., *Fusarium* spp., *Rhizoctonia solani*, and *Sclerotium* spp. have been controlled through soil solarization. Nematodes and weeds are also controlled with this method.

Solarization has several advantages including the following: a nonchemical approach to disease control with no worker exposure concerns; less total costs than fumigation; easy to implement with conventional farm equipment; may provide benefits for multiple years; effective against a range of plant pathogens; and may not kill thermal tolerant beneficial microorganisms. The disadvantages of soil solarization include: the costs of field preparation are the same as fumigation; effective implementation is limited to specific geographic areas with hot, dry summers; it may reduce or eliminate beneficial nonthermal-tolerant microorganisms such as mycorrhizal fungi; costs of plastic disposal; and loss of crop production for 4 to 6 weeks.

BIOFUMIGATION

In the early 1980s, scientists discovered that residues left over from cabbage harvest could be used to control the soilborne pathogen *F. oxysporum* f. sp. *conglutinans* that causes cabbage yellows disease. When dry cabbage residues were amended to soil at 1% (w/w) and covered with a plastic tarp for 8 weeks, the incidence of cabbage yellows in a subsequent cabbage crop was nil. When soil infested with *F. oxysporum* f. sp. *conglutinans* was suspended above moistened cabbage residues in a closed jar, the propagules in the infested soil were killed. Thus, a volatile fungicidal compound was suspected. Later, this volatile was identified as isothiocyanate. Biofumigation

---

CASE STUDY 22.3

BACTERIAL WILT CAUSED BY *RALSTONIA SOLANACEARUM*—BEWARE OF IMPORTING FOREIGN PATHOGENS

- *Ralstonia solanacearum* is a soilborne bacterial pathogen that causes vascular wilt disease in a variety of agricultural and ornamental crops, as well as weedy species.
- The bacterium is categorized into five races (based on plant host) and four biovars (based on metabolic activity).
- *Ralstonia solanacearum* race 1, biovar 1, is commonly found in the southern United States, and has a wide host range.
- *Ralstonia solanacearum* race 3, biovar 2, is tolerant of cold temperatures and is very destructive on potato (*Solanum tuberosum*), other solanaceous crops, and geranium (*Pelargonium × hortorum*). It does not occur naturally in the United States.
- On geranium, symptoms of bacterial wilt include yellowing, wilting, and browning of lower leaves, discoloration of the vascular tissues in the stem, and browning of roots (Figure 22.8). Infection usually occurs through roots. On potato plants, symptoms are similar to those of geranium. In potato tubers, grayish brown discoloration of the vascular tissues and bacterial ooze indicate an established infection (Figure 22.9).
- In 1999, *R. solanacearum* race 3, biovar 2, was imported into the United States on geranium cuttings from Guatemala. Infected plants were destroyed, and industry officials worked with the U.S. Department of Agriculture–Animal and Plant Health Inspection Service (USDA–APHIS) to develop procedures to prevent further introductions.
- To protect important U.S. crops, such as potato, from significant economic loss, *R. solanacearum* race 3, biovar 2, was listed on the select agent list (http://www.aphis.usda.gov/programs/ag_selectagent/ag_bioterr_toxinslist.html) of the Public Health Security and Bioterrorism Preparedness and Response Act of 2002. This list includes many foreign plant and animal pathogens.
- In 2003, *R. solanacearum* race 3, biovar 2, was found in greenhouse geraniums in the United States and Canada that were originally imported from Kenya. After entry into the United States, the geraniums were propagated, and infected cuttings were distributed to growers in nearly 20 states; this led to quarantine of several hundred greenhouse operations by USDA–APHIS officials. Greenhouse growers were instructed to identify and dispose of plants in which the pathogen was confirmed (Daughtrey, 2003; Kuack, 2003).
- Inspection of imported plant materials for diseases is an important component of disease management procedures.

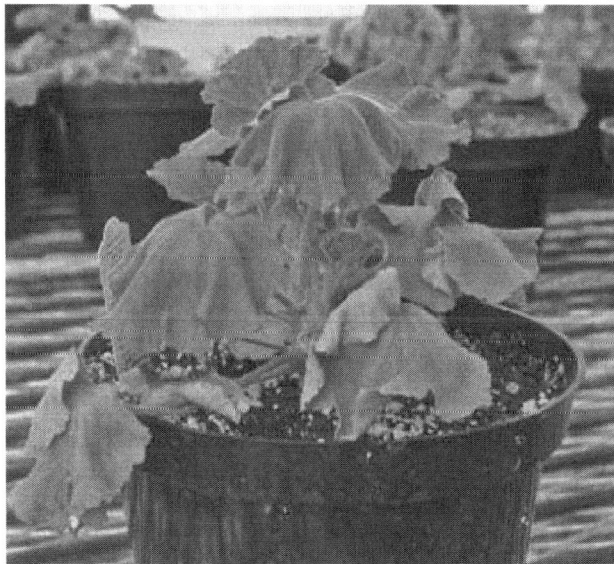

FIGURE 22.8 Bacterial wilt (*Ralstonia solanacearum* race 3, biovar 2) on geranium. (From http://www.mda.state.mn.us/plants/plantdiseases/ralstoniadefault.htm. Minnesota Department of Agriculture. Photo courtesy Caitlyn Allen, University of Wisconsin–Madison.)

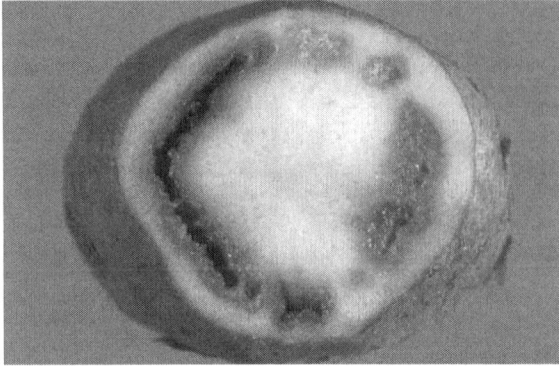

FIGURE 22.9 Brown rot (*Ralstonia solanacearum* race 3, biovar 2) of potato. (From Brown Rot of Potato. 2005. http:// www.defra.gov.uk/planth/brownrot.htm. Photo courtesy of the Department for Environment Food and Rural Affairs, London, U.K. With permission.)

takes advantage of fungitoxic compounds produced during decomposition of crop residues, normally crucifer crops, by holding these volatiles in soil for a period of time with a clear plastic tarp. In addition to the toxic effect of the decomposing residue, plant pathogens are also subject to solarization when biofumigation is carried out during warm periods of the year.

FUMIGATION

Fumigants are biocides that kill cells on contact, whether the cells are bacterial, fungal, nematode, or plant in origin. They include two classes of compounds: the halogenated aliphatic hydrocarbons such as methyl bromide, chloropicrin, and 1-3 dichloropropene (1,3 D) and the methyl isothiocyanate liberators such as metam-sodium, dazomet, and, of course, the cabbage residues mentioned earlier. Because fumigants are biocides, they must be applied before the crop is planted or in such a way to avoid contact with crop roots. Due to cost, fumigants are generally used only on high-value crops such as vegetables and some ornamental crops. A number of fumigants discovered after 1940 have since been banned in the United States due to human health or environmental concerns.

Methyl bromide has been one of the most effective fumigants ever discovered. At one time, virtually all fields in California used to grow strawberries were treated with methyl bromide. However, due to its high volatility, a positive attribute for diffusion through the soil profile, methyl bromide use has been correlated with depletion of the earth's ozone layer. Accordingly, the United States joined with other countries in signing the Montreal Protocol that calls for the elimination of agricultural uses of methyl bromide in developed countries. Although critical use exemptions have been granted by the international governing body for crops where alternatives have not been found, growers will eventually have to rely solely on alternative controls to methyl bromide fumigation.

Once released in soil, fumigants diffuse throughout the soil profile based on their inherent vapor pressure, soil temperature, soil moisture, and soil structure (clay versus sand, clods versus well-tilled soil). Fumigants are less effective in cold, wet, and poorly tilled soils. The

---

CASE STUDY 22.4

RHIZOCTONIA ROOT ROT CAUSED BY *RHIZOCTONIA SOLANI*—PROBLEMS IN POINSETTIA PROPAGATION (BENSON ET AL., 2002)

- Rhizoctonia stem rot is the most significant problem in propagation of poinsettia (*Euphorbia pulcherrima*).
- *Rhizoctonia solani* has a wide host range and survives as sclerotia or mycelium in infested crop debris in greenhouse production facilities.
- The pathogen attacks juvenile host tissue. Lesions from infections caused by *R. solani* develop on the stem at the surface of the rooting cube. Lesions expand rapidly and girdle the stem. Infected cuttings can collapse within 5 to 7 days.
- Root rot develops in the rooting cube or on rooted cuttings transplanted into pots. As lesions expand from stem infections initiated during rooting, crown rot may occur. Stunting is common. Foliar symptoms of infection include chlorosis, leaf necrosis, wilting, defoliation, and plant death (Figure 22.10).
- Several fungicides are registered for use against Rhizoctonia diseases in the greenhouse. Rooting strips can be soaked in fungicide solution to protect cuttings from disease, but care should be taken with selection of fungicide and cultivar of poinsettia because some fungicides can reduce rooting. Fungicide drenches can be used after transplanting to prevent crown and root rot.
- In greenhouse production systems, it is important to practice strict sanitation. All sources of infested crop debris should be removed. During propagation, avoid wetting the foliage of newly made, turgid cuttings. Potted plants should be spaced so that ventilation is adequate to avoid moisture conditions favorable to *R. solani*.

FIGURE 22.10 Poinsettia cutting with stem lesion in leaf abscission area (left) and close-up of *Rhizoctonia* hyphae growing out into rooting cube (right). (From Benson, D.M. et al., 2002. *Plant Health Progress*. DOI: 10.1094/PHP-2002-0212-01-RV. With permission.)

concentration of the toxicant and exposure time for the pathogen propagule determine when death of the propagule occurs. Propagules in the lower portion of the root zone may not be exposed to a sufficient concentration of fumigant to be killed. Fumigants such as chloropicrin are most effective against fungal pathogens, whereas 1,3 D is more effective for nematode control. Methyl bromide is effective against both fungal and nematode pathogens as well as weed seeds.

PLANT RESISTANCE

Selection of disease-resistant cultivars, if available, is an important component of disease management strategies for soilborne plant pathogens. The advantages of plant resistance include reduction or elimination of applications of disease control products, compatibility with other disease management practices, and the long-lasting nature of many disease-resistance genes. For example, use of resistant cultivars is the best management tool for control of wilt diseases caused by *Fusarium* and *Verticillium*. There are also disadvantages associated with host resistance. In some cases, disease-resistant cultivars may lack the more desirable horticultural traits or potential for high yields found in disease-susceptible cultivars; and resistance to one disease does not imply resistance to another. Depending on the host–pathogen system, disease resistance may be shortlived as new strains of the pathogen develop that overcome the plant's resistance genes. Unfortunately, due to the broad host range and large genetic variability of many soilborne plant pathogens, resistant cultivars are not available for all diseases on all crops. Examples of pathogens for which there is little or no genetic resistance available include *Gaeumannomyces*, *Pythium*, *Sclerotinia*, and *Sclerotium*. For additional information on host resistance, see Chapter 34.

BIOLOGICAL AND CHEMICAL CONTROLS

Biological control products (also called biopesticides) and conventional fungicides are both available for control of many soilborne plant pathogens. The commercial products available to growers change often as new research leads to development of more efficacious and environmentally friendly products. Both topics are treated thoroughly elsewhere in this text (Chapters 36 and 37).

ORGANIC CROP PRODUCTION

In October 2002, the U.S. Department of Agriculture (USDA) adopted national certification standards for organic production and processing. In organic crop production, the certification standards do not allow use of conventional synthetic pesticides and fertilizers, and plant growth regulators. Disease management systems developed for organic crops include selection of resistant plants and cultivars (produced by traditional breeding, not genetically modified), selection of planting site, exclusion, many of the cultural practices described earlier, application of biological controls (Chapter 37), and soil amendment with composted organic materials.

Site selection involves consideration of previous soilborne disease problems associated with specific fields, as well as physical features, and soil and water characteristics of the site. For example, well-drained sites reduce the risk of damping-off and root rot caused by soilborne *Pythium* and *Phytophthora* spp. Exclusion, as exemplified in Case Studies 22.1, 22.2, 22.3, and 22.4, involves preventing diseased plants and planting materials, and other production materials contaminated with pathogens, such as soil, potting mix, pots, or trays, from entering the crop production system. Water can also be a source of pathogen contamination. Incorporation of compost into soil is a fundamental practice in organic crop production. Benefits include enhancing soil fertility, increasing organic matter, and increasing the populations and diversity of soil microflora. Depending on the nature of the composted materials, and the combination of plant host and pathogen, specific benefits to management of soilborne diseases have also been reported.

REFERENCES

Benson, D.M. 1994. Inoculum, pp. 1–33 in *Epidemiology and Management of Root Diseases,* C.L. Campbell and D.M. Benson (Eds.), Springer-Verlag, Berlin, 344 pp.

Benson, D.M., J.L. Hall, G.W. Moorman, M.L. Daughtrey, A.R. Chase and K.H. Lamour. 2002. The history and diseases of poinsettia, the Christmas flower. Online. *Plant Health Progress.* DOI: 10.1094/PHP-2002-0212-01-RV.

Bruehl, G.W. 1987. *Soilborne Plant Pathogens.* Macmillan, New York, 368 pp.

Campbell, R.N. and A.S. Greathead. 1990. Control of clubroot of crucifers by liming, pp. 90–101 in *Soilborne Plant Pathogens: Management of Diseases with Macro- and Microelements.* A.W. Engelhard (Ed.), APS Press, St. Paul, MN, 217 pp.

Cheo, P.C. 1982. Effects of tannic acid on rhizomorph production by *Armillaria mellea* obtained from infected roots of California live oak, *Quercus agrifolia. Phytopathology* 72: 676–679.

Cook, R.J. 1973. Influence of low plant and soil water potentials on diseases caused by soilborne fungi. *Phytopathology* 63: 451–458.

Daughtrey, M. 2003. New and re-emerging diseases in 2003, pp. 51–60 in *Proceedings of the 19th Conference on Insect and Disease Management,* Society of American Florists. Alexandria, VA, 129 pp.

Dobson, R.L., R.L. Gabrielson, A.S. Baker and L. Bennet. 1983. Effects of lime particle size and distribution and fertilizer formulation on clubroot disease caused by *Plasmodiophora brassicae. Plant Dis.* 67: 50–52.

Koening, S.R. 2000. Identification and management of mid-to-late season soybean stem and root rots. Soybean Disease Information Note 5. North Carolina State University, Raleigh.

Koike, S.T., M. Gaskell, C. Fouche, R. Smith and J. Mitchell. 2000. Plant disease management for organic crops. University of California, Agriculture and Natural Resources Publication 7252, UC ANR, Oakland, CA, 6 pp.

Kuack, D. 2003. Plants could be dangerous to the health of your business. *Greenhouse Management & Production* 23(5): 2.

Meyer J.R. and H.D. Shew. 1991. Soils suppressive to black root rot of burley tobacco, caused by *Thielaviopsis basicola. Phytopathology* 81: 946-954.

Pepper, I.L. 2000. Beneficial and pathogenic microbes in agriculture, pp. 425-446 in *Environmental Microbiology.* R.M. Maier, I.L. Pepper, and C.P. Gerba (Eds.). Academic Press, San Diego, 585 pp.

Pottorff, L.P. 2003. Foil the soilborne diseases. *Nursery Management & Production* 19 (7): 47–50.

Smiley, R.W. and R.J. Cook. 1973. Relationship between take-all of wheat and rhizosphere pH in soils fertilized with ammonium vs. nitrate-nitrogen. *Phytopathology* 63: 882–890.

Young, J.M., L.D. Kuykendall, E. Martinez-Romero, A. Kerr, and H. Sawada. 2001. A revision of *Rhizobium* Frank 1889, with an emended description of the genus, and the inclusion of all species of *Agrobacterium* Conn 1942 and *Allorhizobium undicola* de Lajudie *et al.* 1998 as new combinations: *Rhizobium radiobacter, R. rhizogenes, R. rubi, R. undicola* and *R. vitis. International Journal of Systematic and Evolutionary Microbiology* 51: 89–103.

SUGGESTED READING

Bockus, W.W. 1998. The impact of reduced tillage on soilborne plant pathogens. *Annu. Rev. Phytopathol.* 36: 485–500.

Cook, R.J. 1980. Fusarium foot rot of wheat and its control in the Pacific Northwest. *Plant Dis.* 64: 1061–1066.

Duniway, J.M. 1975. Formation of sporangia by *Phytophthora drechsleri* in soil at high matric potential. *Can. J. Bot.* 53: 1270–1275.

Duniway, J.M. 1980. Water relations of water molds. *Annu. Rev. Phytopathol.* 17: 431–460.

Engelhard, A.W. (Ed.). 1990. *Soilborne Plant Pathogens: Management of Diseases with Macro- and Microelements.* APS Press, St. Paul, MN, 217 pp.

Gamliel, A. and J.J. Stapleton. 1993. Characterization of antifungal volatile compounds evolved from solarized soil amended with cabbage residues. *Phytopathology* 83: 899–905.

Hall, R. (Ed.). 1996. *Principles and Practice of Managing Soilborne Plant Pathogens.* APS Press, St. Paul, MN, 330 pp.

Katan, J. 1981. Solar heating (solarization) of the soil for control of soilborne pests. *Annu. Rev. Phytopathol.* 19: 211–236.

Katan, J. 2000. Physical and cultural methods for the management of soil-borne pathogens. *Crop Prot.* 19: 25–731.

Lockwood, J.L. 1988. Evolution of concepts associated with soilborne plant pathogens. *Annu. Rev. Phytopathol.* 26: 93–121.

MacDonald, J.D. and Duniway, J.M. 1978. Influence of the matric and osmotic components of water potential on zoospore discharge in *Phytophthora. Phytopathology* 68: 751–757.

Ramirez-Villapuda, J. and D.E. Munnecke. 1987. Control of cabbage yellows (*Fusarium oxysporum* f. sp. *conglutinans*) by solar heating of field soils amended with dry cabbage residues. *Plant Dis.* 71: 217–221.

Rosskopf, E.N., D.O. Chellemi, N. Kokalis-Burelle and G.T. Church. 2005. Alternatives to methyl bromide: A Florida perspective. Online. *Plant Health Progress.* DOI: 10.1094/PHP-2005-1027-01-RV. http://www.plantmanagementnetwork.org/sub/php/review/2005/methylbromide/.

Singleton, L.L., J.D. Mihail and C.M. Rush (Eds.). 1992. *Methods for Research on Soilborne Phytopathogenic Fungi.* APS Press, St. Paul, MN, 265 pp.

Stapleton, J.J. 2000. Soil solarization in various agricultural production systems. *Crop Prot.* 19: 837–841.

Subbarao, K.V., S.T. Koike and J.C. Hubbard. 1996. Effects of deep plowing on the distribution and density of *Sclerotinia minor* and lettuce drop incidence. *Plant Dis.* 80: 28–33.

WEBSITES

North Central IPM Center. Pest Alerts from the North Central Region. http://www.ncpmc.org/alerts/ralstonia.cfm. Updated July 20, 2005.

Agricultural Select Agent List. U.S. Department of Agriculture—Animal and Health Inspection Service. http://www.aphis.usda.gov/programs/ag_selectagent/ag_bioterr_toxinslist.html. Updated May 2005.

Agricultural Select Agent Program. U.S. Department of Agriculture—Animal and Health Inspection Service. http://www.aphis.usda.gov/programs/ag_selectagent/index.html. Updated October 20, 2006.

23 Laboratory Exercises with Soilborne Plant Pathogens

D. Michael Benson and Bonnie H. Ownley

Soilborne plant pathogens are a diverse group of viruses, bacteria, fungi, nematodes, and parasitic higher plants that survive in soil between infection cycles of their hosts. This chapter will concentrate on laboratory exercises for studying four fungal soilborne pathogens. Many soilborne fungal pathogens are either **hemibiotrophs** that invade host cells but keep the cell alive for a period prior to killing the cell, or **necrotrophs** that kill the host cell prior to invading the cell. A few soilborne fungal pathogens such as the club root pathogen, *Plasmodiophora brassicae* (Chapter 11), are **biotrophs** that infect only living host cells.

Soilborne pathogens can be described as **soil invaders** or **soil inhabitants**. Soil invaders do not have an active phase in soil, so the soil environment merely serves as a physical substrate for their survival as various types of propagules, such as **sclerotia**, **chlamydospores**, etc., depending on the pathogen. These propagules serve not only as survival structures but also as inoculum to initiate host infection. Over time, in the absence of a host plant, the population of a soil invader declines. Soil inhabitants not only survive in soil as propagules but also have some saprophytic ability that allows for the colonization of dead organic matter in the soil. Soil inhabitants, in general, are more difficult to control because they can increase in number in the absence of a host.

From the standpoint of plant disease control, it is important to know the type of propagule, i.e., **inoculum**, surviving in soil for a given soilborne pathogen and the number of propagules per unit volume of soil, i.e., **inoculum density**. For many soilborne fungal pathogens, inoculum density is correlated with root infection and subsequent **disease incidence**. For instance, an inoculum density of 0.8 propagules/g soil at 26°C resulted in 50% preemergence damping-off of radish by *Rhizoctonia solani*, whereas 7 propagules/g soil were required to cause the same 50% damping-off at 15°C, a temperature unfavorable for *R. solani* (Benson and Baker, 1974).

EXERCISES

The laboratory exercises will provide an introduction to soilborne pathogens and some methods for detecting and quantifying the diseases they cause. Most of the experiments can be completed in one or two lab periods with minimum preparations.

GENERAL CONSIDERATIONS

The laboratory exercises can be performed by individual students, or teams of two or more students, depending on resources and space. In some exercises, the microorganisms will be isolated directly from natural populations in soil. For other exercises, the microorganisms can be purchased from the American Type Culture Collection (ATCC) (http://www.atcc.org/). An approved U.S. Department of Agriculture, Animal Plant Health Inspection Service, Plant Protection and Quarantine (USDA, APHIS, PPQ) application and permit are required to purchase or for interstate transport of plant pathogens (http://www.aphis.usda.gov/ppq/permits/plantpest/pathogen.html).

EXPERIMENT 1: DETECTION OF *RHIZOCTONIA SOLANI* IN SOIL

Rhizoctonia solani is a common soilborne fungal pathogen with a very large host range that causes pre- and postemergence **damping-off**, root rot, and stem and crown rots in plants. Most agricultural soils contain inoculum of *R. solani* in the form of **hyphae** associated with colonized organic matter and sclerotia. In this experiment you will learn a method to detect and quantify *R. solani* in soil and then later use cultures of the *R. solani* that you have collected for additional experiments on inoculum density and disease.

Materials

Each student or team of students will require the following items:

- Aniline blue (0.5%) or trypan blue (0.05%) in lactophenol solution in dropper bottles as described for nuclei staining in Procedure 17.5
- Bunsen burner or alcohol lamp and ethanol (95%) for sterilizing transfer needle
- Dissecting and light microscopes

TABLE 23.1

Recipes for *Rhizoctonia* Selective Medium, Potato Dextrose Agar, 10% Carrot Agar, and 1.5% Water Agar

Rhizoctonia Selective Medium

Difco™ Agar, granulated, 15 g

Deionized water, 1 L

Combine in 2-L flask and autoclave at 121°C for 15 min

Cool to 50–55°C, then add

Benomyl, 1 mg/L (Note: You can substitute any thiophanate-methyl fungicide for benomyl such as Cleary's 3336, OHP 6672, Fungo Flo. Caution: the 1 ppm rate of benomyl is critical because *Rhizoctonia* is sensitive to benomyl at higher concentrations.)

Chloramphenicol, 100 mg/L

Pours about sixty 10-cm-diameter petri dishes

Potato Dextrose Agar (PDA)

Difco™ Potato Dextrose Agar, 39 g (*Note:* PDA contains the infusion from 200 g of Irish potatoes, 20 g of dextrose, and 15 g of agar)

Deionized water, 1 L

Combine in 2-L flask and autoclave at 121°C for 15 min

Pours about sixty 10-cm-diameter petri dishes

10% Carrot Juice Agar

Difco™ Agar, granulated, 15 g

Fresh carrot juice, 100 mL

Deionized water, 900 mL

Combine in 2-L flask and autoclave at 121°C for 15 min

Pours about forty 10-cm diameter petri dishes

1.5% Water Agar

Difco™ Agar, granulated, 15 g

Deionized water, 1 L

Combine in 2-L flask and autoclave at 121°C for 15 min

Pours about sixty 10-cm-diameter petri dishes

Note: After cooling, poured dishes should be stored in sealed plastic bags in the dark until use.

- Flat wooden toothpicks—birch if possible
- Forceps
- Paper towels
- Parafilm™ strips (1-cm wide) for sealing agar dishes of *Rhizoctonia* pure colonies
- Plastic pots (10-cm diameter), about 450 cm³ capacity
- Plastic wrap
- Pot labels and marking pens
- Potato dextrose agar (PDA) (one agar dish; Table 23.1)
- Resealable plastic bags (3.78-L) for soil samples
- *Rhizoctonia* selective medium (four agar dishes; Table 23.1) per soil assayed per student or team
- Soil samples
- Transfer needle
- Transparent guide (10-cm diameter) with 5-mm squares (*Note*: The guide can be made by photocopying graph paper with 5-mm grids onto transparency sheets, then cutting to fit the size of petri dishes.)

Follow the protocol listed in Procedure 23.1 to complete this experiment.

Anticipated Results

Colonies of *Rhizoctonia solani*, and possibly binucleate *Rhizoctonia* fungi, will grow from the toothpicks within 24 h after culturing on the *Rhizoctonia* selective medium (Figure 23.1). Under the dissecting microscope, *Rhizoctonia* hyphae will appear slightly brownish and glistening. Nuclear staining will be necessary to separate *R. solani* from binucleate *Rhizoctonia*. Isolates of *R. solani* should be recovered from all soil types but not all soils tested, so a large variety of different soil samples should ensure success for the class. Inoculum densities may range from less than one to more than 100 propagules per gram of soil.

Questions

- Why does the toothpick method work for detecting *Rhizoctonia solani* in soil but not a soilborne pathogen like *Phytophthora cinnamomi*?

	Procedure 23.1
	Isolation of *Rhizoctonia* from Soil
Step	Instructions and Comments
1	Prior to lab day, students should collect an assortment of soil samples from various agricultural fields, cultivated gardens, and landscape beds. Collect about 1 L of soil in a resealable plastic bag to a depth of 10 cm from several different locations in the sampling area (field, garden, bed). Break up any clods and mix the soil sample thoroughly by rotating the bag for a few minutes. Store at room temperature, away from direct sunlight. If you have access to a commercial greenhouse, sample potting mix debris left behind on and under the benches and along walkways, or other production areas.
2	If collected soil is moist, use as is. If dry, moisten in the plastic bag by adding small volumes of water and mixing until the soil is moist enough to hold together when squeezed but easily breaks apart and does not form a mud ball.
3	Fill plastic pot with test soil. Mark the pot label with the soil source and your name, and place in pot. Incubate the test soils in pots for 1 or 2 days at 16°C, if possible. It may be necessary to cover the pot surface with plastic wrap to keep the soil from drying out during the incubation period.
4	After the incubation period, insert six toothpicks to a depth of 5 cm into the soil, equidistant around the soil surface. Only the end of the toothpick should be visible so it can be retrieved after incubation in the test soil. Incubate the soil at room temperature for 2 days and cover with a moist paper towel, if needed, to prevent drying.
5	Use the forceps to retrieve the toothpicks after incubation in the test soil. Mark the bottom of the *Rhizoctonia* selective medium dishes with the soil source and your name. Space out three toothpicks per dish of *Rhizoctonia* selective medium. Place the other three toothpicks on a second agar dish. Incubate dishes at room temperature.
6	One day later, examine the agar dishes under a dissecting microscope for colonies of *Rhizoctonia solani* growing from the toothpicks (Figure 23.1). Note: It is critical that the toothpicks be examined after 1 day on the agar dishes because the colonies will grow together after this time.
7	Place the 5-mm grid guide under the petri dish of *Rhizoctonia* selective medium and count the number of 5-mm squares adjacent to the toothpick that have *Rhizoctonia* colonies growing in them. Count the total number of colonies for the six toothpicks. Assume that, in 24 h, only one colony of *Rhizoctonia* will have grown over one 5-mm square.
8	Determine the inoculum density (propagules/gram) of *Rhizoctonia* in the soil sampled. a. Each toothpick has a volume of 0.075 cm^3 so the total volume of soil displaced by the six toothpicks is 0.45 cm^3. b. Therefore, one colony per six toothpicks is 1 propagule/0.45 cm^3 of soil. c. Assuming the soil has a bulk density of 1, then one propagule/0.45 cm^3 of soil is one propagule/0.45 g of soil. Convert to propagules/gram of soil by multiplying propagules per 0.45 g by 2.22. Thus, for six toothpicks, the number of colonies counted multiplied by 2.22 is the number of propagules per gram of soil.
9	Compare the growth of hyphal colonies from the toothpicks under the dissecting microscope to the characteristics of *Rhizoctonia* illustrated in Figure 23.2. Observe right-angle branching with hyphal constriction at the branch hypha.
10	Follow Procedure 17.5 for staining nuclei of *Rhizoctonia* and observe with a light microscope (Chapter 42) to confirm identity of the colony.
11	Use the transfer needle and aseptic technique to transfer a small block of agar containing hyphae of a suspected *Rhizoctonia* colony to the center of a petri dish of PDA. Repeat if the second dish of toothpicks also has a *Rhizoctonia* colony.
12	Store the pure cultures of *Rhizoctonia* for future experiments by sealing the edges with stretched-out Parafilm®.

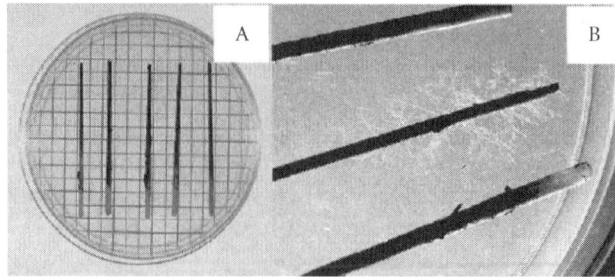

FIGURE 23.1 Toothpick assay. (A) Five toothpicks were removed from soil and placed on a plate of *Rhizoctonia* selective medium, with 5-mm counting grid underneath. (B) Colony of *Rhizoctonia solani* growing out of toothpick after 24 h. (From Paulitz, T.C. and K.L. Schroeder. 2005. *Plant Dis.* 89: 767–772. With permission.)

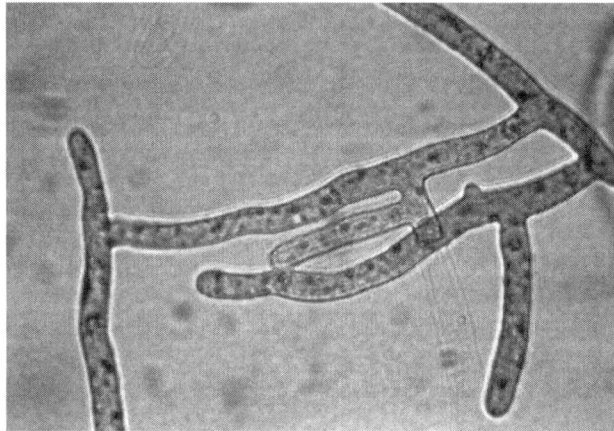

FIGURE 23.2 Right-angle branching of hyphae of *Rhizoctonia solani*. (From Fundamental FUNGI Image Collection & Teaching Resource, APS Press Slide Collections. Photomicrograph courtesy W.E. Batson, Jr. With permission.)

- Were the hyphae of *Rhizoctonia* detected all multinucleate?
- What soils had the highest inoculum density of *R. solani*?

EXPERIMENT 2: RHIZOCTONIA DISEASES

This experiment is designed to demonstrate pre- and post-emergence damping-off as well as stem rot caused by *R. solani*. For the most part, *R. solani* attacks young plant tissues before the host has developed mature tissues that are more resistant. Germinating seeds and seedlings are the prime targets of *R. solani*. Vegetatively propagated plant material can also be severely diseased by this pathogen. In the following exercise, traditional and **ELISA** methods will be used to recover *R. solani* from infected plant tissue. See the chapter on viruses (Chapter 4) for an explanation of how ELISA is able to detect antigens of a pathogen.

Materials

Each student or team of students will require the following items for Procedures 23.2A-C, 23.3, and 23.4:

- Balance (one per class)
- Beaker (500-mL)
- Bedding plant seeds—snapdragons, impatiens, vinca, etc.
- Bent wire, stiff
- Blender, cookie sheet, sieve with 2-mm openings (USA Standard Testing Sieve, No. 10), aluminum foil squares
- Clear plastic bag large enough to hold cuttings in a strip of rooting cubes
- Commercial peat moss potting mix such as Fafard 4P™ or other soilless mix
- ELISA kit for *Rhizoctonia* (Preorder from Neogen Corporation at http://www.adpsa.co.za/Neogen/ NeogFood.htm.)
- Erlenmeyer flask (125 mL) with 50 mL sterile deionized water
- Forceps and scalpel
- Four plastic bags (3.78 L)
- Graph paper (1 sheet)
- Greenhouse flats (four) with drainage sufficient to accommodate 64 seeds on 2.5-cm grid, i.e., about 25 × 25 cm with a depth of 2.5 cm
- Hand-spray bottle for misting

Procedure 23.2A

Pre- and Post-emergence Damping-Off of Vegetable and Bedding Plants Caused by *Rhizoctonia solani*—
Preparation of *R. solani* Inoculum

Step	Instructions and Comments
1	Combine 18 mL deionized water with 25 g rice grains in 250-mL Erlenmeyer flask with aluminum foil cover. Autoclave 40 min.
2	Next day, use stiff bent wire to "stir" up rice grains to break apart the solid mass. Autoclave again for 40 min. Do not add any more water.
3	When cool, add three to five agar disks from *Rhizoctonia* culture to the rice grains. Incubate on lab bench at room temperature.
4	Each day, "bump" rounded bottom edge on thick mouse pad or palm of your hand to keep the mycelium from binding rice grains together.
5	*Rhizoctonia* colonized rice is ready to use after 7 days of incubation.
6	Reserve ten intact rice grains colonized by *R. solani* for use in Procedure 23.3.
7	Add the remainder of the rice grains from one flask to a blender, then pulse blend until rice grains are pulverized.
8	Screen the pulverized rice grains through the 2-mm mesh sieve onto the cookie sheet and store temporarily in folded aluminum foil square.
9	Weigh three aluminum foil packets of rice grains: one packet with 0.05 g of pulverized rice grains, a second packet with 0.5 g, and the third packet with 1.0 g of grains. Note: Once rice grains are removed from the flask, they must be used immediately, i.e., within a couple of hours.

Procedure 23.2B

Pre- and Postemergence Damping-off of Vegetable and Bedding Plants Caused by *Rhizoctonia solani*—
Infesting Soilless Mix and Planting Seeds

Step	Instructions and Comments
1	Transfer 1500 cm^3 of dry soilless potting mix to a 3.78-L plastic bag and rotate the bag slowly 25 times to mix the contents. Pour mix into a greenhouse flat and mark the pot label "Trt 1" (untreated control).
2	To a second lot of 1500 cm^3 of dry potting, mix in a 3.78-L bag, and add 0.05 g of pulverized rice grain inoculum by sprinkling across the surface of the mix. Rotate the bag slowly 25 times, then pour infested mix into a flat. Label "Trt 2" (0.05 g inoculum rate).
3	Repeat step 2 for the 0.5-g rate of inoculum and label "Trt 3" (0.5 g inoculum rate).
4	Repeat step 3 for the 1-g rate of inoculum and label "Trt 4" (1 g inoculum rate).
5	After filling each flat, smooth the soil mix surface. Water gently (use water breaker on hose) to wet the mix and let drain.
6	Make a small depression about 5 mm deep with a pencil or pen and make an 8 × 8 grid at 2.5-cm spacing.
7	Plant one seed per depression and resmooth the surface. Gently rewater the flat to settle the potting mix and place on a greenhouse bench, or shelf in a lab with daylight.
8	Water the flats very lightly every day. Do not create craters in the mix when watering, or seeds will be washed out. On the other hand, the mix must remain moist for good seed germination. A hand-spray bottle set for misting works well.

	Procedure 23.2C Pre- and Post-emergence Damping-Off of Vegetable and Bedding Plants Caused by *Rhizoctonia solani*— Assessing Disease
Step	Instructions and Comments
1	Most vegetable and bedding plant seeds will germinate and emerge in 7 to 10 days. Take stand counts when the first seedlings emerge. Record separate counts for healthy seedlings and seedlings that appeared diseased, i.e., damped-off. Record counts by Trt (= treatment) number.
2	Continue recording stand counts every other day for the next 2 weeks.
3	At the end of the 2-week stand count period, calculate pre- and post-emergence damping-off percentages for each treatment as disease should have reached a maximum level. a. For pre-emergence damping-off, subtract the total number of seedlings emerged (whether healthy or diseased) from the stand count in the untreated control. This value is the numerator to be divided by the stand count in the untreated control. Multiply the resulting fraction by 100 to obtain percentage pre-emergence damping-off. Repeat calculation for each inoculum density of *Rhizoctonia*. b. For post-emergence damping-off, divide the number of diseased seedlings emerged by the total number of seedlings, whether diseased or healthy at that inoculum density. Multiply the fraction by 100; the resulting percentage is post-emergence damping-off. Repeat calculation for each inoculum density of *Rhizoctonia*.
4	Plot the inoculum density (in this case 0.05 g rice/1500 cm^3, 0.5 g rice/1500 cm^3, and 1.0 g rice/1500 cm^3 mix) on the x-axis, and percent damping-off on the y-axis. Use separate symbols for the pre- and postemergence damping-off data. Connect the data points and compare the position of resulting lines. Compare your results with students who used the same or different hosts.
5	Examine diseased seedlings and observe symptoms and signs of disease.
6	Transfer four to six diseased seedlings to a 125-mL Erlenmeyer flask with 50 mL sterile deionized water. Swirl to remove potting mix adhering to the seedlings. Use forceps to transfer the seedlings to paper towels to blot excess moisture. Finally, transfer the seedlings to petri dishes of PDA (two to three seedlings per dish).
7	Observe the culture dishes for mycelial growth characteristic of *Rhizoctonia*.

- Inoculum of *R. solani* on PDA from toothpick isolations described in Procedure 23.1.
- Long grain white rice, 25 g; deionized water, 18 mL; 250-mL Erlenmeyer flask; aluminum foil cover for flask
- Paper towels
- PDA (two agar dishes, Table 23.1)
- Pot labels and marking pens
- *Rhizoctonia* selective medium (two agar dishes, Table 23.1)
- Ruler (15 cm)
- Strip of polyfoam rooting cubes in plastic liner (*Note*: Most horticultural laboratories will have a supply of these for vegetative propagation research or courses.)
- Vegetable seeds—tomato, cucumber, etc.
- Vegetative poinsettia plants to provide five cuttings

Follow the methods outlined in Procedures 23.2A–C to complete the first part of this experiment.

Anticipated Results

As a variety of hosts were used, the percentage of pre- and post-emergence damping-off will vary by crop. In general, the more inoculum that was used, the more disease that should develop up to a limit, i.e., 100% disease. Often the maximum is lower for host and environmental reasons. Pre-emergence damping-off symptoms may be hard to observe unless you can find the diseased seed among the mix and note its decomposing state; otherwise, this disease is noted by the difference in percentages of emergence between the untreated control and the flats with *Rhizoctonia* inoculum. Post-emergence damping-off is easy to observe if you look at the flats every 2 days. Seedlings that were apparently healthy during the last observation period will be collapsed and possibly covered with mycelium of *Rhizoctonia*. Thus, stand count of healthy seedlings in flats with *Rhizoctonia* will decrease over the observation period. Transfer of damped-off seedlings to petri dishes of PDA should result in cultures with brownish mycelium typical of *R. solani*.

<table>
<tr><td colspan="2" align="center">**Procedure 23.3**
Rhizoctonia Stem Rot of Poinsettia Caused by *Rhizoctonia solani*</td></tr>
</table>

Step	Instructions and Comments
1	Make five vegetative cuttings from a poinsettia stock plant. Cuttings should be 8 to 10 cm long with the cut made about 2 cm below the second node of leaves from the terminal. Remove the second node of leaves by breaking off the petiole at the node.
2	Rooting strips may be 13 to 20 cubes long. To save space, break or cut them into five-cube sections and saturate the strip with water.
3	Place one of the cuttings in the hole provided in the strip and proceed to do the same with the rest of the cuttings.
4	With a pair of curved forceps make a small depression about the length and depth of a rice grain in the cube at the junction line between the first and second cubes, then again between the fourth and fifth cubes.
5	Use the forceps to transfer an intact rice grain colonized by *Rhizoctonia solani* (see Procedure 23.2A, Preparation of *R. solani* inoculum) to each of the two depressions. Mark a pot label with your name and date, and stick it in the cube at one end.
6	Prepare a second set of cuttings without rice grain inoculum as an untreated control. Only a few of these are needed for the class.

7	Incubating the cuttings.
	   a. If a mist chamber dedicated to plant pathology research is available, place the rooting strip under a mist system that cycles on for 1 min/h, dawn to dusk.
	   b. If a mist chamber is not available, moisten several layers of paper towels and place them in a plastic bag. Put the strip of cuttings in the bag and seal the bag. With the bag method, it will be necessary to place a second pot label at the opposite end of the strip of cuttings to hold the bag away from the cuttings. Incubate the bags at room temperature on a shelf with daylight but not direct sun.

8	Examine the cuttings weekly for signs and symptoms of *Rhizoctonia* stem rot. Record your observations.
9	With a scalpel, cut a small piece (3- to 5-mm long) of diseased tissue and place in the center of a petri dish of PDA. After 2 to 3 days, check the PDA dish for growth typical of *R. solani*. Record your observations.
10	Excise four pieces of the rooting cube (3- to 5-mm long) at various points along the entire strip and place them on a dish of PDA or *Rhizoctonia* selective medium. After 2 to 3 days, check the cube pieces on the petri dish for growth typical of *R. solani*. Record your observations.

Questions

- Did the percent of pre- and post-emergence damping-off increase with greater amounts of *Rhizoctonia* inoculum?
- Did the different hosts tested vary in their susceptibility to pre- and post-emergence damping-off?

Continue with this experiment by following the protocols listed in Procedure 23.3.

Anticipated Results

Assuming that the toothpick cultures isolated from the soil were indeed *Rhizoctonia solani*, symptoms and signs of Rhizoctonia stem rot will be very obvious in 3 to 21 days, depending on temperature and method of incubation. Cuttings incubated in plastic bags will develop symptoms first, whereas those under a mist system will take 5 to 14 days for initial symptoms. Disease will progress from cuttings nearest the rice grains with *Rhizoctonia* to those furthest from the inoculum. In the bag method, cuttings will be totally collapsed and mushy brown in appearance, especially those nearest the rice grains with *Rhizoctonia*. Hyphae of *R. solani* will be apparent, entwining the diseased tissue and growing from and along the rooting cube. Separating the plastic liner of the rooting strip from the polyfoam rooting cubes will reveal the presence of *Rhizoctonia* hyphae on the sides and bottom of the rooting strip. If a mist chamber is used, initial symptoms of stem rot will be apparent as brownish white lesions on the stems of the cuttings at the cube surface level. Aerial hyphae growing between the strip and the stem will be observed. As the lesions expand and girdle the stem, the cutting will become necrotic and collapse. Leaf lesions that are necrotic and water soaked may develop on leaves that touch the rooting strip and make contact with inoc-

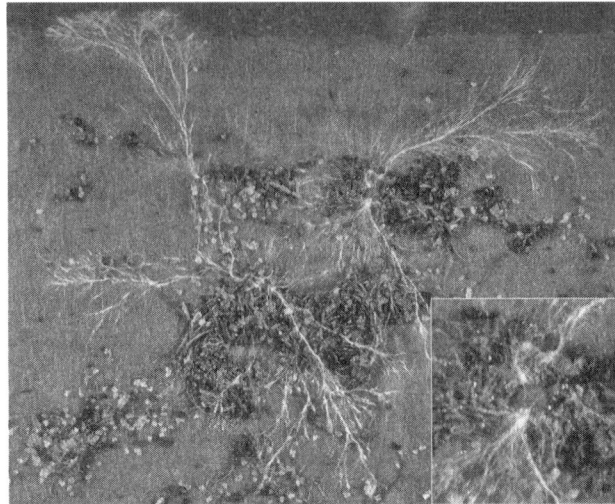

FIGURE 23.3 Eruptive germination of two sclerotia of *Sclerotium rolfsii* during soil assay on moist paper toweling. In eruptive germination, note that, from each central sclerotium, two strands composed of numerous hyphae each have emerged, one above and one below the sclerotium with subsequent diffuse growth. Inset shows close-up of eruptive germination. (Photo courtesy B. Shew, North Carolina State University.)

ulum on the strip. Cultures made from diseased tissue should be typical of *R. solani*, but there may be some fungal and bacterial contamination depending on technique. The control cuttings without *Rhizoctonia*-colonized rice grains should remain healthy. Cuttings under the mist system will root in 3 to 4 weeks.

Questions

- Did you observe hyphae of *Rhizoctonia* growing on the cube?
- Did you recover *Rhizoctonia* on the PDA dish from pieces of the strip that were furthest from the rice grains?
- Did this experiment fulfill Koch's postulates?
- What control methods do you think a grower should adopt to prevent Rhizoctonia stem rot?

Finish this set of experiments by completing the protocol listed in Procedure 23.4.

Anticipated Results

Rhizoctonia-infected tissue will result in a color change to blue after a few minutes.

Questions

- Did you observe a blue positive color change for your sample?
- How long did the ELISA procedure take to run?
- Were the results of the ELISA in agreement with the isolations on culture dishes for *Rhizoctonia*?

EXPERIMENT 3: DETECTION OF *SCLEROTIUM ROLFSII* IN SOIL

Sclerotium rolfsii is a soil inhabitant that attacks the stems of a wide range of host plants at or near the soil surface. The pathogen survives as sclerotia free in soil or associated with crop debris. Only a few sclerotia per kilogram of soil are needed to incite disease in most hosts. A food base is needed for plant infection in addition to a favorable microenvironment in the plant canopy. In row crops, canopy closure results in environmental conditions favorable for infection. Volatile compounds such as methanol produced from remoistened host materials, such as peanut hay, stimulate the germination of sclerotia, and possibly provide an energy source for growth. Thus, a very simple method can be used to detect the pathogen in field soils.

Procedure 23.4
Detection of *Rhizoctonia solani* from Diseased Tissue with ELISA

Step	Instructions and Comments
1	Collect samples of suspected *Rhizoctonia*-diseased material from Procedures 23.1 and 23.2.
2	Follow manufacturer's directions for grinding the tissue sample and extracting the antigen in buffer.
3	Place extract in Alert® test wells per manufacturer's instruction, followed by the various solutions as directed.
4	Observe color change and compare to positive and negative controls.

	Procedure 23.5
	Assay for Detecting *Sclerotium rolfsii*
Step	Instructions and Comments
1	Visit agricultural fields, home gardens, and landscape beds where crops such as ajuga, chrysanthemum, carrots, hosta, melons, tomato, peanut, pepper, pumpkin, corn, wheat, etc., have grown. Note: *Sclerotium rolfsii* is common in southern states and California, but the pathogen is not normally found in regions were the average winter temperature falls below 0°C. However, *S. rolfsii* var. *delphinii* is cold tolerant and found in Iowa and surrounding Midwest states on hosta and other herbaceous ornamentals (Edmunds and Gleason, 2003). Collect a total of 1 L of soil from within the top 5 cm of the soil surface at several locations in a given area and store in a resealable plastic bag. Prior to assay, break up clods. If soil will not be used within the week, it should be air dried for long-term storage.
2	Line the aluminum pan with three to four layers of dry paper towels.
3	Place 250 cm^3 of soil in the bucket and add about 1 L tap water. Suspend the soil in the water by stirring or swirling the bucket, then decant through the sieve slowly so that the sieve does not overflow. It may be necessary to add more water and repeat to transfer all of the soil onto the sieve. Bumping the sieve or gentle rubbing the underneath side of the sieve will speed the flow of soil solution through it. Next, tilt the sieve and wash the soil to the lower side of the sieve. Transfer the soil residue to the aluminum pan by bumping the sieve upside down on one-half of the pan. Use a pot label to spread out the soil residue, then thoroughly moisten the soil residue and paper towels with the methanol solution. Repeat this process with another 250 cm^3 of soil, until the entire 1-L sample has been processed.
4	Use the wash bottle with 1% methanol solution to transfer the organic debris and soil particles on the sieve to the aluminum pan, spreading the debris as evenly as possible.
5	Place the plastic bag around the aluminum pan and seal.
6	In 48 h, count the number of developing colonies of *S. rolfsii*. Colonies will have coarse white mycelium growing several centimeters from the origin (Figure 23.3).
7	Compare the results from the various soils with different crops.
	Optional: Obtaining pure cultures of *S. rolfsii* for other experiments.
1	Use forceps and aseptically transfer individual sclerotia from the aluminum pan to petri dishes of PDA.
2	In 24 to 48 h, aseptically transfer a small block of agar with mycelium from the edge of the developing colony to the second dish of PDA to obtain a pure culture.
3	Seal the PDA dish with Parafilm® or place in resealable plastic bag and store for future use.

Materials

Each student or team of students will require the following items:

- Aluminum baking pan (22 × 30 cm, or larger)
- Bucket
- Collection of field, landscape, and garden soils for detection of *S. rolfsii*
- Paper towels
- Plastic bag sized for aluminum pan
- Plastic beaker (400 mL) for measuring soil and water
- Resealable plastic bags (3.78 L)
- Sieve with 425-µm openings (USA Standard Testing Sieve, No. 40)
- Tap water, 1 L
- Trowel
- Wash bottle with 1% (v/v) aqueous methanol solution
- Wooden stick (ca. 30 cm)

Optional

- Bunsen burner or alcohol burner with ethanol (95%), forceps, transfer needle
- Parafilm® or resealable plastic bag for storage of cultures on PDA
- PDA (two dishes, Table 23.1)

Follow the instructions provided in Procedure 23.5 to complete this experiment.

Anticipated Results

Rapidly spreading colonies of *S. rolfsii* with white coarse mycelium and developing sclerotia will be seen in 48 h.

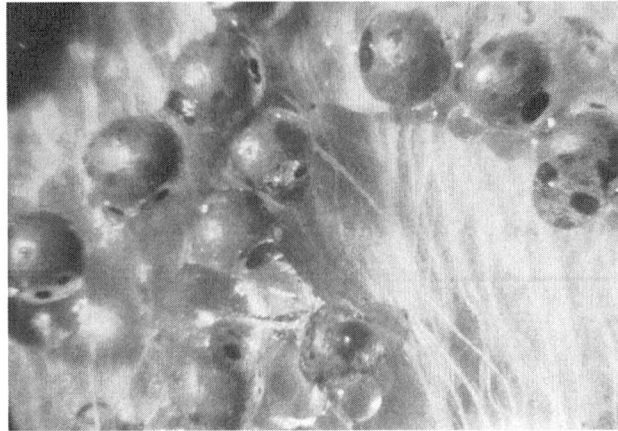

FIGURE 23.4 Brown sclerotia and white mycelia of *Sclerotium rolfsii*. (From Fundamental FUNGI Image Collection & Teaching Resource, APS Press Slide Collections. Photo courtesy L.F. Grand. With permission.)

The mycelium will develop additional tan sclerotia over time. If the sample contained more than two to three sclerotia per liter of soil, the developing colonies may grow together. For *S. rolfsii*, an inoculum density of two to three sclerotia per liter is enough to cause severe epidemics in susceptible crops.

Questions/Discussion

- What is the purpose of the 1% methanol solution?
- Why was a 425-μm mesh sieve used?
- Describe the change in appearance that hyphae of *S. rolfsii* go through as sclerotia develop on the mycelium.

EXPERIMENT 4: STUDENT PET PATHOGEN PROJECT

Most soilborne plant pathogens are microscopic and hard to observe. However, *S. rolfsii* produces coarse mycelium and relatively large sclerotia that can be seen easily with the naked eye (Figure 23.4). Although a plant pathogen, *S. rolfsii* has a very strong saprophytic habit that allows it to grow on a broad range of crop residues whether of host origin or not. Over the course of a few weeks you will be "feeding" your pet culture of *S. rolfsii* various residues to determine just how wide ranging this ability to colonize substrates saprophytically really is.

Materials

Each student or team of students will require the following items:

- Bunsen burner or alcohol burner with ethanol (95%), transfer needle
- Clear plastic box with lid (22 × 30 × 10 cm deep)
- Culture of *S. rolfsii* on PDA
- Paper towels and wash bottle with tap water
- Various sources of organic matter such as broccoli, carrot slices, chewing gum, wood chips, etc.

Follow the protocol outlined in Procedure 23.6 to complete this experiment.

Procedure 23.6
Keeping Your Pet *Sclerotium rolfsii* Culture Alive

Step	Instructions and Comments
1	Line the plastic observation box with several layers of paper towels and moisten.
2	Use the transfer needle to cut a 3 × 3 cm square of agar from the center of the PDA dish with *Sclerotium rolfsii* and place the agar square at one end of the observation box.
3	Place the various types of organic residues throughout the box.
4	Store the box at room temperature on the lab bench away from direct sunlight.
5	Observe the growth and development of *S. rolfsii* after 2 and 7 days, and then for the next few weeks with particular attention to the number and size of sclerotia produced on the various residues and the extent of mycelia growth. It may be necessary to rewet the paper towels weekly.

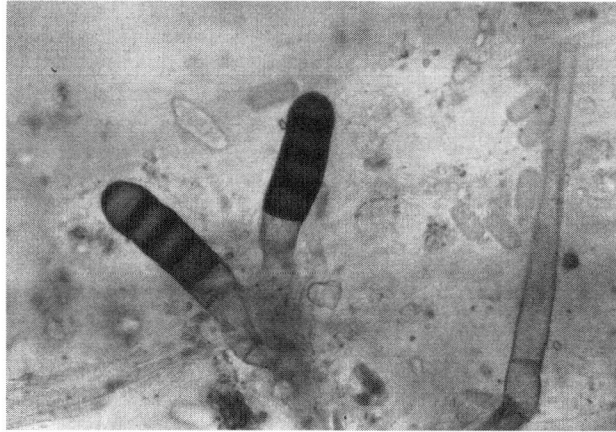

FIGURE 23.5 Darkly pigmented, globose aleuriospores, and hyaline, barrel-shaped endoconidia of *Thielaviopsis basicola*. (From Fundamental FUNGI Image Collection & Teaching Resource, APS Press Slide Collections. Photomicrograph courtesy L.F. Grand. With permission.)

Anticipated Results

The nutrients in the PDA culture of *S. rolfsii* will serve as an initial food base as the fungus starts to grow across the paper towels. Because *S. rolfsii* has a high competitive saprophytic ability, mycelium and sclerotia will eventually colonize the residues and spread throughout the box, including the sidewalls. Size and number of sclerotia on the debris will depend on the nutrient value of that debris to the pathogen.

Questions/Discussion

- Describe the color changes that sclerotia go through from initiation to maturity.
- What type of residues did your pet prefer?

EXPERIMENT 5: THIELAVIOPSIS BLACK ROOT ROT

Thielaviopsis basicola (alternative name *Chalara elegans*), a soilborne fungal pathogen, causes black root rot and damping-off of a wide range of host plants, including vegetables, ornamentals, and field crops such as cotton. The fungus reproduces asexually and produces two very different spore types: **aleuriospores** (also referred to as chlamydospores), and **endoconidia** (also called **phialospores** and **phialoconidia**) (Figure 23.5). Aleuriospores produced by *T. basicola* are darkly pigmented, thick-walled, ovoid to subglobose in shape, and occur in chains of five to eight cells. In contrast, the endoconidia are hyaline (colorless), barrel-shaped or cylindrical, and occur in chains of three to five cells. The endoconidia are produced by large conidia-producing cells called **phialides**. In culture, copious quantities of both spore types can be produced. However, growth on nutrient-poor medium tends to promote production of endoconidia (Weber and Tribe, 2004). Aleuriospores are

the primary overwintering or survival propagules, but both types of spores can incite disease. The following exercises are designed to familiarize students with use of a **hemacytometer** to quantify fungal spore inoculum, signs and symptoms of black root rot caused by *T. basicola*, and a method to recover this pathogen from soil and diseased roots.

Materials

Each student or team of students will require the following items for Procedures 23.7, 23.8, and 23.9:

- Carrot
- Cheesecloth
- Culture of *T. basicola* on 10% carrot juice agar
- Culture tube (20 mL)
- Deionized water
- Dissecting and light microscope; microscope slides and cover slips
- Hand counter
- Hand sprayer for misting
- Hemacytometer (cleaned with ethanol) and glass cover slip
- NaOCl (1.0%)
- Pansies (4-week-old plants in cell pack)
- Plastic pots (10-cm diameter)
- Pot labels and markers
- Potting soil
- Spreader
- Sterile filter paper disk in sterile petri dishes
- Sterile knife
- Streptomycin sulfate (50 µg/mL)
- Transfer pipette

A hemacytometer is a glass counting chamber with etched grids that can be viewed under a microscope. Although

Procedure 23.7

Quantification of *Thielaviopsis basicola* Spore Inoculum Using a Hemacytometer

Step	Instructions and Comments
1	Flood 3-week-old culture of *Thielaviopsis basicola* growing on 10% carrot juice agar (Table 23.1) with 10 mL deionized water.
2	Lightly scrape the culture surface with a spreader to free fungal spores. The inoculum will likely be a mixture of endoconidia and aleuriospores.
3	Remove the spore suspension with a transfer pipette; filter the suspension through four layers of cheese cloth, then transfer the suspension to the 20-mL culture tube.
4	Place the cover glass over the hemacytometer and carefully transfer a small amount of the spore suspension into one of the V-shaped wells with a transfer pipette (Figure 23.6A). Allow the mirrored surface of the hemacytometer to be covered by capillary action. Note: It is important to avoid overflow.
5	The etched grid of the hemacytometer is divided into nine large squares, each with a surface area of 1 mm^2. With the cover slip in place, the chamber depth is 0.1 mm. This represents a total volume of 0.1 mm^3 or 10^{-4} cm^3 or 10^{-4} mL. Place the hemacytometer on the microscope stage and focus under low power. Using the hand counter, count the number of spores in the four large squares in the outer four corners (Figure 23.6B). Count only those spores that are within the boundaries of the four counting squares. Calculate the mean total number of spores in four squares for three different samples of the spore suspension. The average total number of spores in the four squares should be in the range of 60 to 200. Note: If many of the spores are clumped, dilute the suspension and repeat the process. Alternatively, too few spores will be observed if the spores settled to the bottom of the tube before the sample was withdrawn from the suspension.
6	Calculate the number of spores per milliliter of your suspension according to the following: a. Multiply the total number of spores counted in the four squares by 2500 to yield the number of spores per milliliter of suspension. b. If the spore suspension was diluted, multiply the calculation from Step 6a by the dilution factor. For example, if a 1:2 dilution factor was used, multiply by 2.

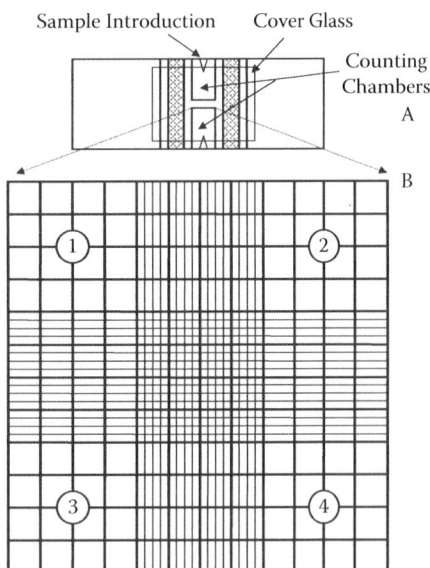

FIGURE 23.6 Hemacytometer. (A) Top view. (B) Enlarged etched grid pattern of hemacytometer (Becker et al., 1996).

originally designed for quantifying red blood cells, it can be used to determine how many fungal spores are in a suspension. *Note*: The procedure described applies to hemacytometers with Neubauer rulings (grid). See Chapter 42 for details on use of microscopes.

Follow the protocols described in Procedure 23.7 to complete this part of the experiment.

Anticipated Results

The sample of spores in the suspension will be primarily endoconidia because filtration through cheesecloth will remove the mycelia and aleuriospores.

Question

- A dilution of 1:10 of a spore suspension was made prior to counting the spores in a hemacytometer. In the four outer squares, a total of 120 spores were counted. Calculate the number of spores per milliliter in the spore suspension.

Follow Procedure 23.8 to complete the experiment with black root rot of pansy.

	Procedure 23.8
	Black Root Rot of Pansy Caused by *Thielaviopsis basicola*
Step	Instructions and Comments
1	The mixture of endoconidia and aleuriospores or the suspension of endoconidia following filtration through cheesecloth (Procedure 23.7) can be used as inoculum. Add the 10 mL spore suspension of inoculum to 90 mL deionized water.
2	Uproot a pansy plant, remove soil to expose the roots, and dip the roots into the spore suspension of *Thielaviopsis basicola*. Uprooting the plant will likely cause minor wounding that will enhance disease. Repot the pansy plant. Pour the remaining spore suspension around the plant roots. Note: To enhance disease severity further, the pH of the growing medium should be 6 or greater, soil temperature should be maintained below 25°C, and planting medium should be kept moist during the incubation period.
3	Write your name and date on a pot label and place it in the pot.
4	Controls should be prepared that do not receive any *T. basicola* inoculum. Only a few controls are needed per class.
5	Observe plants weekly and compare with untreated control plants for foliar symptoms of disease. If different forms or rates of inoculum were used, i.e., mixture of aleuriospores and endoconidia or endoconidia alone, or different rates of endoconidia as determined with a hemacytometer, then compare the rate of disease progress based on development of foliar symptoms.
6	When symptoms appear, gently remove plants from pots, wash potting soil from roots, and observe roots for signs and symptoms of black root rot.
7	Observe roots with a dissecting microscope.
8	Cut a small section of blackened root tissue and prepare a squash mount for viewing with 4×, 10×, or 40× magnification with a light microscope. To prepare a squash mount, place the tissue in a drop of deionized water, cover with a glass coverslip, and press gently with a pencil eraser to flatten the preparation. Observe under the lowest power before proceeding to higher magnification.

Anticipated Results

Foliar symptoms of black root rot of pansy include stunting, wilting, and yellowing (chlorosis). Symptoms on roots will range from dark spots visible on normally white roots to large sections of blackened, water-soaked roots. With magnification, the darkly pigmented aleuriospores of *T. basicola* should be visible. If different rates of inoculum are compared, disease onset should be earlier, and disease should be more severe with increasing rates of inoculum.

Questions

- What was the time span between inoculation and symptom expression?
- Describe the condition and size of diseased roots relative to the untreated control.
- Based on the condition of the roots, do you think that the fungus progressed beyond the cortex into the vascular tissue?

Follow the instructions in Procedure 23.9 to complete the rest of this experiment.

Anticipated Results

After incubation of the carrot disks, *T. basicola* colonies will appear as grayish colonies. The colonies will become black as masses of aleuriospores are produced.

Question/Discussion

- Following incubation, describe the appearance of carrot disks with roots or soil applied relative to the untreated control carrot disks.

EXPERIMENT 6: TAKE-ALL OF WHEAT CAUSED BY *GAEUMANNOMYCES GRAMINIS* VAR. *TRITICI*

Take-all is a severe root and crown rot disease of wheat, barley, and other grasses caused by the soilborne fungal pathogen *Gaeumannomyces graminis* var. *tritici*. Seedlings infected with *G. graminis* var. *tritici* are yellow and stunted, and may be killed by the fungus. Older plants produce very few tillers, and the heads of grain have a bleached appearance (termed whiteheads) and contain little or no grain. Roots of infected plants are brown-black

---

Procedure 23.9

Detection of *Thielaviopsis basicola* in Rotted Pansy Roots and Infested Soil Using a Carrot Root Disk Assay*

Step	Instructions and Comments
1	Surface-disinfest the whole carrot by dipping it in 1.0% NaOCl for 20 min, then rinse with deionized water to eliminate pathogens from the surface of the carrot.
2	Cut the carrot into disks (5-mm thick) with a sterile knife.
3	Place carrot disks onto filter paper disks inside sterile petri dishes. Moisten the filter paper disks with sterile deionized water.
4	Prepare untreated carrot disk controls by incubating moistened disks as described in the following text, but not adding soil or diseased roots.
5	Inoculate the carrot disks with roots or soil.
	a. Using the rotted pansy roots from Procedure 23.8, wash the roots free of potting soil in running tap water, blot dry, and place a clump of roots on the carrot disks.
	b. Using the infested potting soil from Procedure 23.8, or agricultural, garden, or landscape soils, spread soil sample thinly over the surface of the carrot disks. Mist the soil to moisten it, but avoid free water on the surface. Preincubate the carrot disks for 2 to 4 days at 22 to 24°C in the dark. Wash carrot disks free of soil and place in sterile petri dishes with moistened filter paper.
6	Seal the petri dishes containing carrot disks with rotted roots, or disks preincubated with soil, with Parafilm® and incubate for 7 to 10 days at 22 to 24°C in the dark. The carrot disks should be moist but not wet during the incubation period. Note: If bacterial soft rot develops, mist the carrot disks with 50 µg/mL streptomycin sulfate.

Note: Selective culture media for *T. basicola* have been developed (Specht and Griffin, 1985); however, they contain toxic ingredients and may not be suitable for classroom laboratory exercises.

* Adapted from Yarwood, 1946.

---

in color and very brittle. If you tug on an infected wheat plant, the roots break off easily near the crown or soil surface (Figure 23.7).

Gaeumannomyces graminis var *tritici* survives between crops in infected wheat roots and crowns. Most infections occur when mycelia come into contact with the roots of growing plants. **Ectotrophic** runner hyphae of the fungus extend mainly along the root surface (Figure 23.8) and produce short, dark, feeder hyphae (**hyphopodia**), which grow toward the host root. The hyphopodia have a pore through which a hyphal peg grows into the root. The fungus invades the cortex and the vascular tissue of the root through the hyphal pegs. Thick dark strands of runner hyphae on roots are diagnostic for take-all.

The fungus is classified in the Ascomycota (see Chapter 15) and produces sexual **ascospores** in **perithecia** (Figure 23.9). When *G. graminis* var. *tritici* becomes systemic in the plant, it extends upwards to the crown and base of the stem, and to the lower leaves. The necks of the perithecia can be seen as small raised black spots on the

FIGURE 23.7 Wheat roots infected with *Gaeumannomyces graminis* var. *tritici*. (From Bockus, W.W. and N.A. Tisserat. 2000. *The Plant Health Instructor.* DOI: 10.1094/PHI-I-2000-1020-01. Photo courtesy M.V. Weise. With permission.)

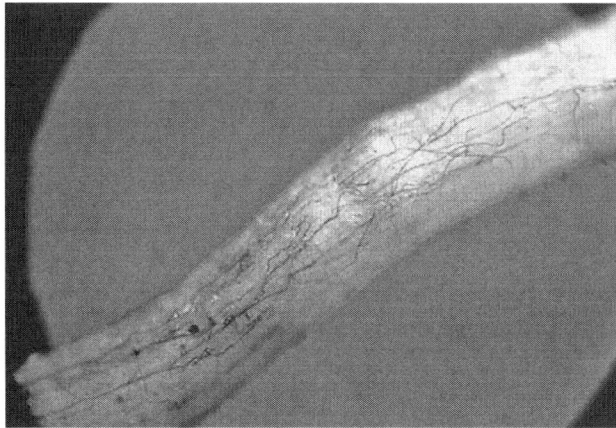

FIGURE 23.8 Runner hyphae of *Gaeumannomyces graminis* var. *tritici* on infected wheat root. (From Mathre, D.E. 2000. *Plant Health Progress*. DOI: 10.1094/PHP-2000-0623-01-DG. Photo courtesy BASF. With permission.)

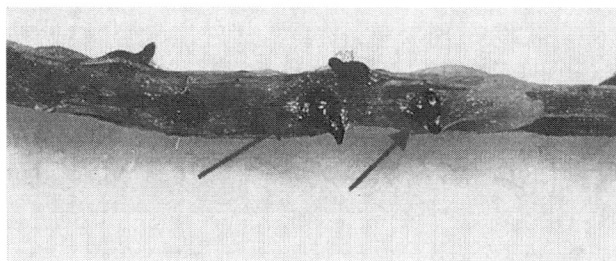

FIGURE 23.9 Perithecia (indicated by arrows) of *Gaeumannomyces graminis*. (From Bockus, W.W. and N.A. Tisserat. 2000. *The Plant Health Instructor*. DOI: 10.1094/PHI-I-2000-1020-01. Photo courtesy R. Kane. With permission.)

leaf sheaths. The purpose of this exercise is to familiarize students with the signs and symptoms of take-all disease.

Materials

Each student or team of students will require the following items to complete Procedures 23.10 and 23.11:

- Agricultural soil (silt loam preferred); coarse grain sand
- Bunsen burner or alcohol lamp with ethanol (95%)
- Culture of *Gaeumannomyces graminis* var. *tritici* on PDA (Table 23.1)
- Dissecting microscope; light microscope, microscope slides, and glass cover slips
- Commercial bleach solution (NaOCl), 10%
- Oat grains, 150 cm³; deionized water, 150 mL; 1-L Erlenmeyer flask; nonabsorbent cotton plug; 10-mm² square of two layers of cheesecloth, 10-cm cotton string; 10-mm² square of aluminum foil
- Paper towels
- Plastic pots (10-cm diameter)
- Pot labels; marking pens
- Sieve with 10-mm openings (USA Standard Testing Sieve, No. 10)

- Sieve with 500-μm mesh openings (USA Standard Testing Sieve, No. 35)
- Sieve with 250-μm mesh openings (USA Standard Testing Sieve, No. 60)
- Sterile deionized water
- Sterile filter paper in sterile petri dishes
- Transfer needle
- Water agar, 1.5% (1 dish, Table 23.1)
- Wheat seeds

Follow the protocols outlined in Procedure 23.10 to complete this part of the experiment.

Anticipated Results

Aboveground symptoms, including stunting and chlorosis, should be apparent in wheat seedlings. Disease caused by *G. graminis* var. *tritici* will appear initially as pinpoint black lesions on wheat roots. As disease progresses, lesions will coalesce, and large segments of the roots will appear black. If the fungus has become systemic in the seedling, there will be brownish black discoloration on the stem.

Questions/Discussion

- Describe appearance of diseased roots.
- Are runner hyphae visible on the root surface?

Procedure 23.10

Lesions and Runner Hyphae Caused by *Gaeumannomyces graminis* var. *tritici* on Wheat Roots

Step	Instructions and Comments
	Oat Grain Inoculum of *Gaeumannomyces graminis* var. *tritici*
1	To prepare oat inoculum, combine 150 cm³ oat grains with 150 mL deionized water per 1-L flask. Allow the oats to soak for 1 h prior to autoclaving. Prepare a plug for the flask by placing nonabsorbent cotton in a 2-layer square of cheesecloth, tie with a string to create the plug, and place in the top of the Erlenmeyer flask; cover the plug with aluminum foil. Autoclave the flask of oats for 30 min at 121°C. In 24 h, autoclave the flask again to ensure sterility.
2	Using sterile technique, cut several small squares from the PDA culture of *Gaeumannomyces graminis* var. *tritici* to seed the flask of oats. Incubate the culture flask at room temperature on a lab bench for 2 weeks. Shake the flask every other day to redistribute the mycelia and avoid clumping.
3	Spread the oat grain inoculum on paper towels and allow to air dry for 1 to 2 days.
4	Pulverize the oat grain inoculum in a blender. Sieve the fragments through a set of three nested sieves (10-mm, 500-µm, and 250-µm mesh openings). Collect particles that are in the size range of 250- to 500-µm diameter for use as inoculum.
	Infesting Soil and Planting Seed
1	Mix an agricultural soil with coarse sand in a 3:1 w/w ratio. Add fragments of oat grain inoculum of *G. graminis* var. *tritici* to the soil mix at 1% w/w. Place soil mix in plastic pots. Note: Increasing the soil pH to 7.0 or greater, and increasing the coarseness of the soil (by adding sand), will increase disease severity.
2	Mark the pot label with your name and date, and place it in the pot.
3	Controls should be prepared that do not receive any *G. graminis* var. *tritici* inoculum. Only a few controls are needed per class.
4	Plant several wheat seed in the pot, water as needed to keep soil moist, and incubate at 15 to 20°C for 3 weeks.
	Assessing Disease
1	Observe wheat seedlings for symptoms of disease.
2	Remove plants from pots, wash soil from roots, and examine wheat roots under a dissecting microscope and low-power light microscope for black necrotic lesions and runner hyphae.
3	Infected wheat stems can be saved for Procedure 23.11.

To complete this experiment, follow the instructions in Procedure 23.11.

Anticipated Results

Small black perithecia should develop on the colonized wheat seedlings and the infected wheat stems.

Questions/Discussion

- Were perithecia formed using both methods?
- Which method produced the most abundant production of perithecia?
- Which method resulted in the most rapid production of perithecia?
- Describe the perithecia produced.

ACKNOWLEDGMENTS

The authors gratefully acknowledge the advice of Barbara Shew and Larry Grand, Department of Plant Pathology, North Carolina State University, in the design of the experiments with *Sclerotium rolfsii*.

Procedure 23.11	
Comparison of Methods for Production of Perithecia by *Gaeumannomyces graminis* var. *tritici*	
Step	Instructions and Comments

Production of Perithecia on Wheat Seedlings

1 Surface-sterilize wheat seeds in 10% commercial bleach for 2 min, rinse with sterile deionized water, and blot the seeds dry on sterile paper towels.

2 Place three wheat seeds on 1.5% water agar (Table 23.1) and allow them to germinate.

3 Cut small squares of *G. graminis* var. *tritici* from PDA dishes and place on 1.5% water agar next to a germinated wheat seed. Prepare control dishes by placing uncolonized squares of PDA next to germinated wheat seeds.

Production of Perithecia on Infected Wheat Stems

1 Cut 3-cm sections of wheat stems from the infected plants in Procedure 23.10.

2 Surface-sterilize stems for 30 s in 10% commercial bleach. Rinse with sterile deionized water and place three stems on moistened sterile filter paper in a sterile petri dish. Additional sterile deionized water may be added to keep the filter paper moist.

Incubation of Test Dishes

1 Seal the dishes with Parafilm® and incubate at 25°C for 3 to 6 weeks with 12 h of light per day.

2 Observe the dishes weekly for development of perithecia by *G. graminis* var. *tritici*.

REFERENCES

Becker, J.M., G.A. Caldwell and E.A. Zachgo. 1996. Determination of cell number, pp. 221 and 222 in *Biotechnology—A Laboratory Course*, Academic Press, New York.

Benson, D.M. and R. Baker. 1974. Epidemiology of *Rhizoctonia solani* preemergence damping-off of radish: inoculum potential and disease potential interaction. *Phytopathology* 64: 957–962.

Bockus, W.W. and N.A. Tisserat. 2000. Take-all root rot. *The Plant Health Instructor*. DOI: 10.1094/PHI-I-2000-1020-01.

Edmunds, B.A. and M.L. Gleason. 2003. Perennation of *Sclerotium rolfsii* var. *delphinii* in Iowa. Online. *Plant Health Progress* DOI: 10.1094/PHP-2003-1201-01-RS.

Mathre, D.E. 2000. Take-all disease on wheat, barley, and oats. Online. *Plant Health Progress* DOI: 10.1094/PHP-2000-0623-01-DG.

Paulitz, T.C. and K.L. Schroeder. 2005. A new method for the quantification of *Rhizoctonia solani* and *R. oryzae* from soil. *Plant Dis.* 89: 767–772.

Specht, L.P. and G.J. Griffin. 1985. A selective medium for enumerating low populations of *Thielaviopsis basicola* in tobacco field soils. *Can. J. Plant Pathol.* 7: 438–441.

Weber, R.W.S. and H.T. Tribe. 2004. Moulds that should be better known: *Thielaviopsis basicola* and *T. thielavioides*, two ubiquitous moulds on carrots sold in shops. *Mycologist* 18: 6–10. DOI: 10.1017/S0269915X04001028.

Yarwood, C.E. 1946. Isolation of *Thielaviopsis basicola* from soil by means of carrot disks. *Phytopathology* 38: 346–348.

SUGGESTED READING

Mathre, D.E. 1992. Gaeumannomyces, pp. 60–63 in *Methods for Research on Soilborne Phytopathogenic Fungi*, Singleton, L.L., J.D. Mihail and C.M. Rush (Eds.), APS Press, St. Paul, MN.

Mims, C.W., W.E. Copes and E.A. Richardson. 2000. Ultrastructure of the penetration and infection of pansy roots by *Thielaviopsis basicola*. *Phytopathology* 90: 843–850.

Rodriquez-Kabana, R., M.K. Beute and P.A. Backman. 1980. A method for estimating number of viable sclerotia of *Sclerotium rolfsii*. *Phytopathology* 70: 917–919.

Shew, B.B. and M.K. Beute. 1984. Effects of crop management on the epidemiology of Southern stem rot of peanut. *Phytopathology* 74: 530–535.

Shew, H.D. and J.R. Meyer. 1992. Thielaviopsis, pp. 171–174 in *Methods for Research on Soilborne Phytopathogenic Fungi*, Singleton, L.L., J.D. Mihail and C.M. Rush (Eds.), APS Press, St. Paul, MN.

24 Parasitic Seed Plants, Protozoa, Algae, and Mosses

Mark T. Windham and Alan S. Windham

CHAPTER 24 CONCEPTS

- Parasitic seed plants can be chlorophyllous or achlorophyllous and can cause significant economic crop loss.

- Parasitic seed plants may range in size from a few centimeters tall, such as Indian pipe, to several meters tall, such as buffalo nut.

- Dodder has a large host range, is entirely aerial, and may destroy entire ornamental plantings of annuals and perennials.

- Dwarf mistletoe is extremely destructive on black spruce in western North America.

- Witchweed is an extremely destructive pest of grass crops such as corn, millet, sorghum, and sugarcane in Africa and Asia. The pest is established in North and South Carolina in the United States.

- Flagellated protozoa can cause serious diseases of coffee and oil palm, cassava, and coffee.

- Algae can cause leaf spots on numerous plants, including southern magnolia and cultivated azaleas in coastal areas of the southern United States.

Although most of the important plant pathogen groups have been covered in previous chapters, a few remaining plant pathogens or organisms that have been associated with disease symptoms are parasitic seed plants, protozoa, algae, and mosses. The most important pathogens among these groups are the parasitic seed plants. Parasitic seed plants have flowers and produce seed, but may be deficient in other typical characters associated with plants such as roots or chlorophyll. Some parasitic seed plants, such as buffalo nut (*Pyrularia pubera*), are shrubs that have green leaves and a root system, whereas others, such as squawroot (*Conopholis americana*), Indian pipe (*Monotropa uniflora*), pine-sap (*M. hypopiths*), witch weed (*Striga lutea*), dodder (*Cuscuta* species), and dwarf mistletoe (*Arceuthobium* species) and leafy mistletoe (*Phoradendron* species), have a very small thallus, which does not include roots and/or green leaves. In this chapter, several examples of parasitic seed plants; tropical protozoa that have been associated with several diseases; an alga whose colonies are associated with leaf spotting of magnolia and azalea; and ball moss that causes death of shrubs and trees in isolated areas of Louisiana, Texas, and Florida will be considered.

PARASITIC SEED PLANTS

DODDER

Dodder, also known by the names strangle weed and love vine, is a small plant with a thin yellow-to-orange stem. Infestations of dodder look like yellow-orange straw spread along roadsides. The stems of dodder are leafless but contain abundant whitish flowers that form in early summer and are produced until frost (Figure 24.1). Seeds are tiny, brown to gray to black, and mature in a few weeks. Dodder has a very wide host range that includes alfalfa, potatoes, numerous herbaceous annuals and perennials, and young shrubs or seedlings of trees in nurseries. Dodder infections can kill young plants. Severe infestations can destroy entire plantings of wild flowers along a roadway, or annuals and perennials in a flowerbed. Dodder can also be used as a bridge (vector) to transmit a number of plant viruses (Chapter 4), but the economic importance of virus transmission by dodder is unknown.

Symptoms and Signs of Dodder Infection

Symptoms of infected plants include stunting, loss of vigor, poor reproduction (flowering), and death. Signs of

FIGURE 24.1 Close-up of dodder on stem of lespedeza. Note knobby protuberances on dodder stems that are haustoria primordia.

dodder include yellow-orange stems of dodder entwined around plant stems, petioles, and foliage. Dodder does not have roots and is entirely aerial. Patches of dodder enlarge by growing from plant to plant and by new dodder plants emerging from seed produced during the growing season. In container nurseries, dodder can spread rapidly throughout the canopy of plants that are tightly packed together.

Disease Cycle

Dodder seed may remain dormant in soil for many years or may be introduced into a field or flower bed at planting. Only a stem is produced when seeds germinate (Figure 24.2), and the young seedlings rotate until they come in contact with a host plant. If a host plant is not available, the seedling eventually dies. However, if a suitable host is found, the stem of the seedling wraps around the host stem and produces haustoria that penetrate the host stem. After successful infections have taken place, the dodder stem begins to grow from host plant to host plant if the host plants are close enough, and produce many small white flowers. Seeds mature in approximately three weeks and may contribute to the current epidemic, or lie dormant until the next growing season. Seeds are spread by water, animals, tillage equipment and/or in mixtures with host seed.

Control

Dodder infestations can be very difficult to control once the pathogen has become established in a field or flower bed. Equipment should be cleaned before moving from an infested area to a noninfested area. Livestock in infested areas should be kept there and not moved to areas that are thought to be free of dodder. Once dodder infestations are established in a field, it is usually controlled with contact herbicides that destroy both the dodder and host plants before the parasite has a chance to flower. Cultivation or fire can be used to destroy dodder if done before flowering. Fumigation is possible in flower beds, but is impractical in fields due to the patchiness of dodder infestations.

DWARF MISTLETOES

Dwarf mistletoes are the single most economically important parasitic seed plant in North America and are the most important disease problem in conifer production in the western United States. In Oregon and Washington, about 13% of the total wood production is lost to dwarf mistletoes annually. In Minnesota, infestations are primary in black spruce (*Picea mariana*), and about 11% is lost due to the disease. Losses in tree production are due to mortality, poor growth and wood quality, reduction in seed production, wind breakage, and predisposition of infected trees to other diseases and

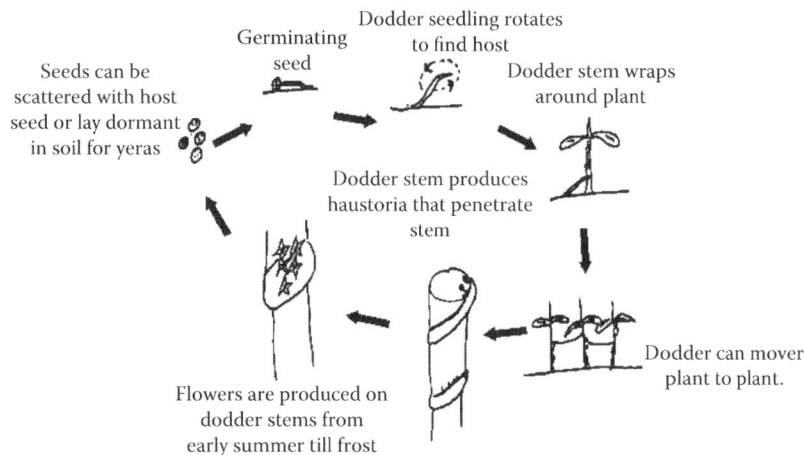

FIGURE 24.2 Generalized life cycle of dodder (*Crucuta* species).

insects. There are nearly 40 species of dwarf mistletoe in the genus *Arceuthobium*. Economically important *Arceuthobium* species and their hosts are listed in Table 24.1.

Symptom and Signs

Infected portions of the tree become swollen, and excessive branching at the infection site leads to a **witches' broom** (Figure 24.3); multiple witches' brooms can occur on the same tree. Although the growth rate of infected tree parts increases, growth in the rest of the tree is retarded, and growing points on the tree often die. Signs of dwarf mistletoe include small yellow-to-orange plants with sessile leaves and white flowers (Figure 24.4A). Developing berries are white, and darken with age (Figure 24.4B).

Disease Cycle

The dwarf mistletoe is dioecious, and flowers are pollinated by insects. The berries contain one seed and are surrounded by **viscin**, a sticky mucilaginous pulp.

The seeds are forcibly discharged for a distance up to 15 m (Figure 24.5). The discharged seeds land directly on foliage of neighboring trees. Long-distance spread is by seed that become stuck to the feet of birds. During periods of rainy weather, the seeds slide down the needle (or off the feet of birds) and land on twigs. Seed usually start to germinate within a few weeks, but germination may be delayed until spring. Once the seeds germinate and the radicle comes in contact with a rough area in the bark surface, a primary haustorium penetrates the limb. After the primary haustorium is successfully formed, the dwarf mistletoe plant develops an extensive absorption system that eventually includes the xylem of the host tree. After three or more years, an aerial shoot of the parasite will emerge from the bark of the infected twig and flower production begins the following year. Dwarf mistletoe plants contain chloroplasts, but fix little carbon and obtain practically all their nourishment from the host. The dwarf mistletoe is most severe in open stands along ridge tops (higher elevations) and seldom is a problem in bottom lands.

TABLE 24.1

Economically Important Species of *Arceuthobium* (Dwarf Mistletoe) and Principal Hosts

Arceuthobium Species	Host
A. abietinum f. sp. *concoloris*	White and grand firs
A. americanum	Jack, lodgepole, and beach pine
A. douglasii	Douglas fir
A. laricis	Western larch
A. pusillum	Alberta and black spruce
A. tsugense	Mountain and western hemlock
A. vaginatum subsp. *vaginatum*	Apache pine, rough bark Mexican pine, and ponderosa pine

Note: Summarized from Tainter and Baker (1996).

(a)

(b)

FIGURE 24.3   (A) Dwarf mistletoe infection of lodgepole pine. (B) Close-up of infected limb.

(a)

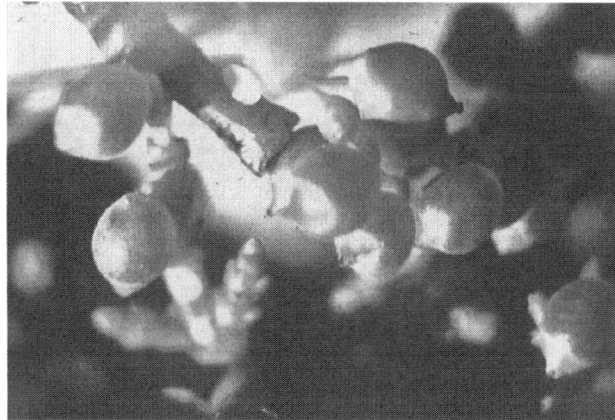

(b)

FIGURE 24.4 (a) Dwarf mistletoe plant growing on conifer. (b) Close-up of dwarf mistletoe plant showing sessile leaves and ripening berries.

Control

Because dwarf mistletoes are obligate parasites, clear-cutting infected trees and prescribed burning to remove any ripe berries and fallen limbs on the forest floor can be an effective control tactics. In many western areas, eradication of dwarf mistletoe is not practical, and control strategies center around reducing the amount of dwarf mistletoe after the current stand is harvested. After timber harvest, heavily infected trees are removed and not used as sources of spruce seed. Plantings are accomplished in a manner to take advantage of natural barriers to dwarf mistletoe such as highways, large streams, nonhost species, etc. In recreational sites, the parasite may be controlled by pruning infected branches before seeds are produced.

Resistance to this disease has not been successfully utilized except in the Rouge River valley in Oregon among the ponderosa pine (*Pinus ponderosa*). This tree has drooping needles, which causes seeds to slide down to the tip of the needles and drop harmlessly to the ground. However, seedlings taken from this valley revert to having erect needles. Thus, the ability of ponderosa pines in this area to escape the disease may be due to environmental conditions instead of a genetic trait.

LEAFY MISTLETOES

Leaf mistletoes occur throughout the world and usually attack hardwoods in forest and landscape areas. Heavily infected trees may begin to decline in vigor, but economic losses are small when compared to losses attributed to dwarf mistletoe infections. In North America, infection by leafy mistletoes is usually due to *Phoradendron* species, but leafy mistletoes of *Viscum* species, common in Europe, are also found in California. Leafy mistletoes are evergreen plants and may be so numerous in a deciduous tree that they may make the tree appear to be an evergreen during the winter months (Figure 24.6). Symptoms of leafy mistletoe infections are similar to those of dwarf mistletoes in that infected branches become swollen and often form a

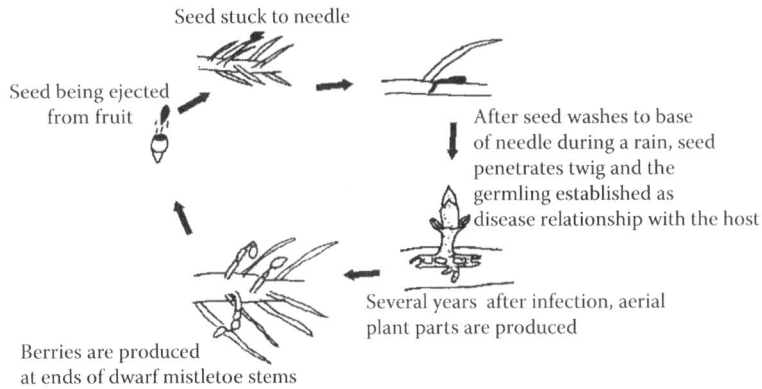

Seed stuck to needle

Seed being ejected from fruit

After seed washes to base of needle during a rain, seed penetrates twig and the germling established as disease relationship with the host

Several years after infection, aerial plant parts are produced

Berries are produced at ends of dwarf mistletoe stems

FIGURE 24.5   Generalized life cycle for dwarf mistletoe (*Arceuthobium* species).

FIGURE 24.6   Hardwood tree with multiple infections by a leafy mistletoe.

witches' broom. Heavily infected trees may have reduced growth and limb death. Leafy mistletoes have white berries that are eaten by birds. The seeds are excreted and stick to the branches where the birds perch. Control of leafy mistletoes is usually not necessary, but infected branches can be pruned. A comparison of dwarf and leaf mistletoes is provided in Table 24.2.

OTHER PARASITIC SEED PLANTS OF ECONOMIC IMPORTANCE

WITCHWEED

Witchweed is an economically serious parasite in Africa, Asia, Australia, and in limited areas in the United States (i.e., limited to a few coastal counties in North and South

Carolina). Witchweed can parasitize hosts such as corn, millet, sorghum, rice, sugarcane, tobacco, and cowpeas. Infected plants are stunted, usually wilted, chlorotic, and often die. Infected roots may contain numerous large haustoria from a single witchweed plant, or haustoria from more than one witchweed plant. Flowers of witchweed are brightly colored with red or yellow petals that are showy. A single plant may produce nearly one-half million tiny seeds.

Witchweed is difficult to control, and avoiding the introduction of the seed into fields is paramount. **Quarantines** have been effective in limiting spread of the parasite in the Carolinas. Eradication of witchweed has reduced the infested area significantly since the disease was discovered in the mid-1950s. Witchweed can also be controlled by using trap or catch crops, and by destroying

TABLE 24.2

A Comparison of Dwarf and Leafy Mistletoes

Dwarf Mistletoes	Leafy Mistletoes
Found in western and northern parts of North America	Found in the southern half of North America
Attacks conifers	Attacks hardwoods
Propagated by forcibly ejected brown-to-gray seeds that stick to needles or to the feet of birds	Propagated by white seeds not forcibly discharged and are eaten and excreted by birds
Dioecious plants	Monoecious plants
Economically important in United States	Seldom of economic importance in the United States
Yellow-orange to brown plants	Green plants

the host plants and parasite with a herbicide before flowering and seedset.

BROOMRAPES

Broomrapes (*Orobanche minor* and *O. ramosa*) attack more than 200 species of plants and occur throughout much of the world. In India, broomrape infections in tobacco may destroy one-half of the crop, whereas yield losses due to broomrapes on tobacco in the United States are rare. Plants attacked early in the season are stunted, whereas plants attacked later in the growing season suffer few effects from the parasite. Broomrape infections are usually clumped in a field. Broomrapes are whitish to yellowish to purplish plants that grow at the base of the host plant. Often, ten or more broomrape plants can be found attacking the same plant. Broomrape can produce more than one million seeds at the base of a single host plant.

Broomrapes, like witchweed, are hard to control in areas where infestations are severe. In India, some control may be obtained by weeding out broomrape plants before seeds are produced. Weeding must be done throughout the growing season because more broomrape plants will emerge after the initial weeding. Soil fumigation is effective in killing the seed.

FLAGELLATE PROTOZOA

Some protozoa in the family Trypanosomatidae are known to parasitize plants in tropical areas. Although thousands of protozoa can be found in the phloem of diseased plants, formal proof of their pathogenicity has not been achieved because infections with pure cultures of the protozoa have not been accomplished (Koch's postulates have not been satisfied—see Chapter 39). Diseases associated with flagellate protozoa include sudden wilt of oil palm, heartrot of coconut palm, empty root of cassava, and phloem necrosis of coffee. Symptoms of these diseases include chlorosis (yellowing of fronds), stunting, and death. These diseases can spread very rapidly. For example, heartrot can spread to thousands of trees in one year. Protozoa that cause phloem necrosis in coffee and empty root of cassava can

be transmitted by root grafts or grafting. Protozoa associated with sudden wilt of oil palm and heartrot of coconut palm are transmitted by pentatomid insects. Control is primarily by avoiding infected stock at transplanting. Control of potential vectors is of questionable value.

ALGAE

Although at least 15 species in three families of algae are known to parasitize plants, only three species of *Cephaleuros* are common worldwide, and *C. virescens* is the only species common in the United States. Parasitic algae affect more than 200 species of plants in the United States along coastal areas of North Carolina extending south and westward to Texas. Economically important hosts that may suffer noticeable damage include southern magnolia (*Magnolia grandiflora*) and cultivated azaleas (*Rhododendron* species hybrids). Algal colonies are only successful if they form in small wounds and develop between the cuticle and epidermis. Superficial colonies on the leaf surface wash away during heavy rains. Host cells directly beneath the colony die causing necrotic leaf spots. Many colonies can be found on a single leaf (Figure 24.7), and colonization is most severe during periods of warm, wet weather. Control is usually not recommended, but multiple sprays of copper are effective if needed.

MOSSES

Ball moss, *Tillandsia recurvata*, is a bromeliad that is closely related to Spanish moss. It is an epiphyte that is found in the southernmost United States (Florida, Louisiana, Texas, and Arizona) and southward to Argentina. Ball moss occurs on many deciduous and evergreen species. Large clusters of ball moss may completely encircle smaller branches and prevent buds from developing, which results in branch death. Other damage to the trees may be due to the weight of the epiphyte populations causing mechanical damage to host plants. Leaf abscission and branch death have also been attributed to abscission substances produced by the epiphyte. Control of ball moss is accomplished by using copper or sodium bicarbonate sprays.

FIGURE 24.7   Colonies of *Cephaleuros virescens* on a leaf of southern magnolia.

Among pathogens covered in this chapter, the most economically important worldwide are dwarf mistletoes and witchweed. Yield losses from these two diseases can destroy plantings over a wide area, and both diseases may cause pandemics. Other pathogens discussed in this chapter can cause high yield losses in specific locations. Flagellate protozoa may destroy plantings of palms, coffee, or cassava in particular locations. Dodder-infested flower beds can lead to severe limitations as to what types of annuals or perennials may be grown in those areas. Broomrapes may cause severe disease losses in tobacco in India. Control of most of these pathogens remains difficult, and little host resistance is known to any of these pathogens. Chemical control of any of the pathogens is almost impossible without also destroying the host plants. Most of these pathogens have been the subject to lesser amounts of research than many fungal, bacterial, or viral pathogens. Until research scientists give pathogens in this group more attention, many questions concerning their life cycles, infection processes, and disease control tactics will remain unanswered.

SUGGESTED READING

Agrios, G.N. 2005. *Plant Pathology*. 5th ed. Academic Press. San Diego. 952 p.

Coyier, D.L. and M.K. Roane (Eds). 1986. *Compendium of Rhododendron and Azalea Diseases*. APS Press. St. Paul, MN. 65 p.

Holcomb, G.E. 1995. Ball moss: an emerging pest on landscape trees in Baton Rogue. *Proc. Louisiana Acad. Sci.* 58: 11–17.

Lucas, G.B. 1975. *Diseases of Tobacco*. 3rd ed. Biological Consulting Associates. Raleigh, NC. 621 p.

Sinclair, W.A., H.H. Lyon, and W.T. Johnson. 1987. *Diseases of Trees and Shrubs*. Comstock Publishing Associates. Ithaca, NY. 575 p.

Tainter, F.H. and F.A. Baker. 1996. *Principles of Forest Pathology*. John Wiley. New York. 805 p.

25 Abiotic Diseases

Alan S. Windham and Mark T. Windham

CHAPTER 25 CONCEPTS

- Abiotic diseases of plants are often caused by cultural practices or environmental factors.

- Abiotic diseases may be difficult to diagnose as the causal agent or factor may have dissipated prior to symptom development.

- Abiotic diseases may predispose plants to infection by plant pathogens.

- Sun scald occurs when leaves acclimated to low levels of light are exposed to full sun.

- Drought stress often predisposes woody plants to infection by canker-causing fungi.

- Bark splitting is often associated with winter injury or a sudden drop in temperature.

- The most common reasons for pesticide injury are the misuse or misapplication of pesticides, movement from the initial point of application due to vaporization, drift or movement in water, and injurious residue left from a previous crop.

- Symptoms associated with herbicide injury may be confused with symptoms produced by certain virus diseases.

Abiotic or noninfectious diseases are caused by cultural practices (Chapter 35) or environmental factors on plants in nature and also on cultivated crops (contrast with Chapter 39 for disease diagnosis). Although not true plant diseases, damage by abiotic extremes, such as light, water, and temperature, can be quite severe under certain circumstances. Most plants grow best at optimum levels of environmental conditions. If, for example, temperature drops below or exceeds the optimum for growth, damage may occur. Also, although water is necessary for normal plant function, excessive amounts of water or insufficient amounts may cause injury.

Damage caused by abiotic diseases may be difficult to diagnose because symptoms may not appear until well after plants were exposed to suboptimum cultural conditions or environmental extremes. Another interesting note is that abiotic diseases may **predispose** plants of infection by plant pathogens. For instance, woody plants exposed to drought stress are more likely to be infected by canker-causing fungi, such as *Botryosphaeria* species. Small grains and turfgrass grown in alkaline soils are at risk of infection by *Gaeumannomyces*, a soilborne fungus associated with take-all patch diseases (Chapters 22 and 23). In many cases, damage from abiotic diseases is compounded by **biotic diseases** that follow.

LIGHT

Low light decreases plant vigor, slows growth, elongates internodes, and may reduce flowering and fruit. Suboptimum levels of light lead to decreased carbohydrate production and damage to the plant's photosynthetic system through reduced chlorophyll production.

It is often difficult to separate the effects of high light levels and high temperature on leaves. Plants grown at low light intensities have leaves with little or no wax and cutin. **Sun scald** occurs when plant material is moved from low light conditions to high light. An example is moving a shade-loving plant such as rhododendron from a lathe house to a landscape bed in full sun. Also, leaves of woody shrubs may develop sun scald after pruning exposes leaves that are acclimated to low light intensities. Occasionally, plants grown at high elevations will develop sun scald when exposed to full sun after several days of cloudy weather.

LOW-TEMPERATURE INJURY

Freeze injury occurs when ice crystals form in intercellular and intracellular spaces. If the cell membrane is ruptured, the cell will die. Cold-temperature injury to young trees often leads to bark splitting. Extremely low winter tem-

FIGURE 25.1 Winter injury to the foliage of southern magnolia (*Magnolia grandiflora*). Note marginal leaf desiccation.

peratures may cause severe injury to woody plants, killing all above-ground plant parts. However, woody shrubs may also be damaged by a sudden drop in temperature, which can lead to **bark splitting** on the lower stem. Fungi that cause **cankers** may cause branch dieback or death. Tender foliage or shoots may be injured by freezing temperature (Figure 25.1).

WATER STRESS

Drought stress and water stress are sometimes used interchangeably. However, water stress can also be used when discussing an excessive amount of water, such as flooding in poorly drained soils. Drought stress is normally used when discussing a shortage of natural rainfall.

DROUGHT STRESS

Wilting, chlorosis, shortened internodes, and poor flower and fruit production are all symptoms that have been associated with drought stress. Leaf scorch, which is a **marginal** or interveinal leaf necrosis, is often seen on deciduous plants exposed to drought stress (Figure 25.2). Evergreen plants, such as conifers, may shed needles in response to a shortage of water. Remember that any biotic disease that affects the root system, the vascular system, or the trunk and branches of a plant may produce symptoms that could be confused with those associated with drought stress.

One of the most common causes of losses of landscape plants is the lack of water. Annual flowering plants are especially sensitive to drought stress; most herbaceous

FIGURE 25.2 Drought injury on Kousa dogwood (*Cornus kousa*). Note marginal leaf scorch.

perennial plants are less sensitive; however this varies by species and even cultivar. It is not uncommon to see native trees in forests die after several years of below-normal rainfall. Many woody plants are more susceptible to canker-causing fungi, such as *Botryosphaeria* and *Seiridium* species, if the plants have been exposed to significant levels of drought stress.

Sometimes drought stress may be localized. Soils with underlying rock or construction debris, sandy soils, or golf greens with hydrophobic areas may have localized areas of plants exhibiting wilting and chlorosis. Also, if pine bark media that is used in container nurseries is not stored properly, it may become infested with fungi, such as *Paecilomyces,* that cause the bark to become hydrophobic. Irrigation water forms channels and is not absorbed by the media in affected containers, simulating drought stress. This sometimes leads to losses once plants are installed in landscape beds as the bark mix continues to repel water.

EXCESSIVE WATER OR FLOODING

Flooding or prolonged periods of saturated soil conditions can lead to the decline or death of many cultivated plants. Chlorosis, wilting, and root necrosis are all symptoms that may be exhibited by plants exposed to seasonal or periodic flooding. In waterlogged soils, low oxygen levels lead to root dysfunction and death. Water molds such as *Pythium* and *Phytophthora* are favored by the conditions found in saturated soils or growth media (Chapter 20).

During periods of cloudy weather, uptake of water may remain high, whereas transpiration rates are slow. When this occurs, leaf tissue saturated with water ruptures, forming corky growth on the underside of leaves. This condition, edema, is fairly common in ornamental flowering plants such as ivy geranium. A similar condition, called intumescence, may be observed on sweet potato and ornamental sweet potato. The symptoms of intumescence can occur on both the upper and lower leaf surface.

FIGURE 25.3 Hail injury to the trunk of flowering dogwood (*Cornus florida*).

LIGHTNING INJURY

The damage caused by lightning strikes is frequently observed in forest and shade trees. Large wounds are often created on the trunk of the tree as the charge moves down the cambium, blowing out strips of bark and sapwood. The wound may be nearly vertical or spiral down the trunk following the grain of the wood. Trees injured by lightning may soon die from the strike or decline after insects and wood decay fungi enter the wound.

Cultivated crops, such as vegetables, are sometimes injured by lightning strikes in fields. Circular areas of damaged plants are often visible within days after the strike. In succulent vegetables, such as tomato, wilting may be visible within hours of a strike. Further damage often follows the initial strike in vegetables such as tomato, cabbage, and potato when necrotic pith tissue develops in the stem of affected plants. On golf greens, lightning sometimes strikes flag poles, and the charge radiates out in a circular pattern across the green in an almost cobweb fashion.

HAIL

Cultivated crops may be ruined within a few minutes during a hailstorm. Hailstones quickly shred the leaves of many plants; the bark of young trees may be severely damaged (Figure 25.3). Bark damage on trees is often on one side only. Wounds from hail damage are often invaded by insect borers, canker-causing fungi, and wood decay fungi. In many cases, nursery stock severely damaged by hail will have to be destroyed.

MINERAL DEFICIENCY AND TOXICITY

Normal plant growth is dependent upon the availability of several mineral elements. Minerals such as nitrogen, phosphorus, potassium, magnesium, calcium, and sulfur are needed in larger amounts (macro elements) than elements such as iron, boron, zinc, copper, manganese, molybdenum, sodium, and chlorine (minor or trace elements). If these elements are not available in sufficient amounts for typical plant growth, a variety of symptoms may develop (Table 25.1). Chlorosis, necrosis, stunted growth, rosette, reddish-purple leaves, and leaf distortion are symptoms that have been associated with nutrient deficiencies. Also, if certain nutrients are available in excessive amounts, plant damage may occur.

Soil **pH** may have a significant effect on nutrient deficiency or toxicity (Figure 25.4). In alkaline soils, ericaceous plants such as blueberry, azalea, and rhododendron may show symptoms of iron deficiency. In contrast, manganese is sometimes toxic to crops such as tobacco growing in acidic soils. In **saline** soils, sodium and chlorine ions may occur at damaging levels.

The diagnosis of deficiency or toxicity of mineral nutrients in cultivated crops is complex. Observe the crop showing symptoms of a nutrient problem. Are the symptoms on young leaves, older leaves, or both? Chlorosis of older leaves may indicate that a mobile nutrient such as nitrogen has moved from older foliage to newly developed leaves. Are affected leaves chlorotic or necrotic? Chlorosis is associated with nitrogen, magnesium, and sulfur deficiencies, whereas marginal necrosis of leaves is often linked to potassium deficiency. Keep in mind that visible foliar symptoms are more often associated with nutrient deficiencies than toxicities. The diagnosis of nutrient problems is often complicated by plant disease symptoms, insect damage, and pesticide injury.

Soil and plant tissue analyses are very helpful in diagnosing fertility problems. Besides revealing soil pH, soil analysis shows the potential availability of mineral nutrients. It is often helpful to collect a soil sample from a problem area and from an adjacent area where plants are growing normally for comparison. Plant tissue analysis gives a "snapshot" of the actual nutritional status of the plant. Note the stage of growth when collecting plant tissue for nutrient analysis. Collecting separate samples of old and young leaves can give you information on nutrients that are mobile within the plant. Finally, be aware

TABLE 25.1

Symptoms of Nutrient Deficiencies in Plants

Nutrient	Symptoms
Nitrogen (N)	Chlorosis in older leaves, leaves smaller-than-normal, stunted plants
Phosphorus (P)	Purple-to-red leaves, smaller-than-normal leaves, limited root growth
Potassium (K)	Chlorosis in leaves, marginal chlorosis in older leaves
Magnesium (Mg)	Chlorosis in older leaves first, chlorosis may be interveinal
Calcium (Ca)	Leaf distortion such as cupping of leaves; fruit of some plants may rot on blossom end
Sulfur (S)	Chlorosis in young leaves
Iron (Fe)	New growth chlorotic, chlorosis often interveinal, major veins may be intensely green
Zinc (Zn)	Alternating bands of chlorosis in corn leaves; rosette or "little leaf"
Manganese (Mn)	Interveinal chlorosis, may progress to necrosis
Boron (B)	Stunting, distorted growth, meristem necrosis
Copper (Cu)	Tips of small grains and turfgrass are chlorotic, rosette of woody plants such as azalea

FIGURE 25.4   Rosette or witches' broom (on left) associated with copper deficiency of azalea (*Rhododendron* species).

that fungicides containing copper, zinc, or manganese may affect the results of plant tissue analysis.

PESTICIDE INJURY

The most common reasons for pesticide injury are the misuse or misapplication of pesticides, movement from the initial point of application due to vaporization, drift or movement in water, and injurious residue left from a previous crop. The inappropriate use of pesticides frequently leads to plant injury. For instance, a nonselective soil sterilant used to edge landscape beds may lead to disastrous results if the roots of desirable plants absorb the herbicide. Other common causes of pesticide injury are the drift of herbicides onto nontarget plants, using the same sprayer for herbicides and other pesticide applications, storing fertilizer and herbicides together, mislabeled pesticide con-

tainers, inappropriate tank mixtures, and planting crops into soil with harmful herbicide residues (Figure 25.5). The amount of damage caused by a particular pesticide is often related to the type of pesticide and the concentration applied. Some plants may recover from the initial damage induced by some herbicides if the concentration applied was low, such as in drift of spray droplets from an adjacent field.

There are numerous symptoms of pesticide injury. Necrosis, chlorosis, witches' broom, or rosette, strap-shaped leaves, cupping, and distinct leaf spots symptoms have been associated with pesticide injury. It is important to remember that the symptoms of many plant diseases caused by plant pathogens may be confused with the symptoms of pesticide injury, so a thorough investigation is important. The symptoms of some viral diseases may be mistaken for symptoms caused by **herbicide injury**.

FIGURE 25.5 Phenoxy herbicide injury of flowering dogwood (*Cornus florida*). Note curled, strap-shaped leaves.

The diagnosis of pesticide injury may be tedious and time consuming. Collect information about pesticide applications within the last two growing seasons as some damage may be the result of a preemergent herbicide applied the previous year. Ask if pesticides were applied to adjacent fields or utility right-of-ways. Look for diagnostic symptoms of injury. Check several plant species for similar symptoms. Question the applicator about the pesticide applied, the rate at which the chemical was applied, and check the calibration of the sprayer. Collect plant and soil samples for analysis by a plant disease clinic, soil test laboratory (for soil mineral analysis and pH), and an analytical laboratory (for pesticide detection). It is helpful to collect samples as soon as possible after the damage occurs (Figure 25.6). Also, it is more economical to ask an analytical lab to assay the soil or plant material for a pesticide suspected of causing the problem rather than an open screen of many pesticides.

Bioassays may also be used to determine if damaging levels of pesticide residues remain in soils. Collect soil from problem areas; place in pots or flats in a green house, and plant several plant species such as a small grain, radish, tomato, and cucumber into the suspect soil. Observe germination rates and look for symptoms development on young seedlings. Solving the cause of pesticide injury may be time consuming, but it is worth the effort for a crop of high value.

AIR POLLUTION INJURY

Air pollutants that originate from humanmade sources or are produced naturally may damage plants. Some of the more common pollutants that cause plant injury are ozone (O_3), sulfur dioxide (SO_2), and ethylene. Ozone is generated naturally during lightning strikes, but may also be produced when nitrogen dioxide from automobile exhaust combines with oxygen in the presence of ultraviolet light. Sulfur dioxide originates from several sources, including automobile exhaust and coal-fired steam plants. Ethylene may be produced by poorly vented furnaces (such as those used to heat greenhouses) and from plant material or fruit stored in poorly ventilated areas.

Ozone is one of the most damaging air pollutants. It may cause chlorotic stippling of the needles of conifers or chlorotic-to-purple discoloration of the leaves of deciduous plants such as shade trees. Sulfur dioxide may also cause chlorosis of foliage. However, in eastern white pine, the tips of the needles of affected trees turn bright red. Not all trees are affected equally. In a pine plantation, it is not unusual to see a low percentage of trees with symptoms associated with sulfur dioxide exposure. Ethylene is a plant hormone. However, plants that are exposed to abnormal levels of ethylene produce distorted foliage that is often confused with

FIGURE 25.6 Sublethal dose of glyphosate herbicide on variegated periwinkle (*Vinca major*). Note dwarfed new growth (left).

symptoms of virus diseases. Ethylene may also reduce fruit and flower production. Ultimately, the damage caused by air pollutants depends on the concentration of the pollutant, the time of exposure, and the plant species. The diagnosis of air pollution injury is very difficult, and is often tentative and based on the distribution of the damage, symptoms, and the plant species affected.

Abiotic plant diseases should not be discounted or overlooked as a cause of plant damage. Nearly half of all plant samples submitted to plant disease clinics exhibit symptoms associated with abiotic diseases such as drought stress, nutrient deficiencies, or pesticide injury. Diagnosis is not easy, and often depends on the experience of the diagnostician (Chapter 39). Abiotic diseases related to cultural practices are easier to solve. Diseases associated with environmental or climatic factors are often harder to diagnose.

SUGGESTED READING

Agrios, G.N. 2005. *Plant Pathology.* 5th ed. Academic Press, New York.

Fitter, A.H. and R.K.M. Hay. 1987. *Environmental Physiology of Plants.* Academic Press, New York.

Kramer, P.J. and T.T. Kozlowski. 1979. *Physiology of Woody Plants.* Academic Press, New York.

Levitt, J. 1980. *Responses of Plants to Environmental Stresses,* Vol. I, Academic Press, New York.

Levitt, J. 1980. *Responses of Plants to Environmental Stresses,* Vol. II, Academic Press, New York.

Marschner, H. 1986. *Mineral Nutrition of Higher Plants.* Academic Press, New York.

Part 3

Molecular Tools for Studying Plant Pathogens

26 Molecular Tools for Studying Plant Pathogens

Timothy A. Rinehart, XinWang Wang, and Robert N. Trigiano

CHAPTER 26 CONCEPTS

- All living organisms contain DNA that can be assayed by a variety of methods to answer important questions about plant pathology, such as relatedness, phylogeny, and pathogen identity, among others.

- DNA fingerprinting techniques, such as DAF, RAPD, and AFLP, utilize arbitrary priming and do not require prior knowledge of the plant pathogen genome, and are useful for investigations of diseases of unknown etiology.

- SSR and SNP methods rely on DNA sequence data and provide more detailed information, including the potential to detect specific pathogens using PCR amplification.

- Gene expression microarrays and differential display answer questions about the genetic underpinnings of pathogenicity and other traits associated with plant pathogens.

- Non-DNA-based approaches, such as ELISA, detect gene products, not the DNA itself.

The ubiquitous nature of DNA is a central theme for all biology. The nucleus of each cell that makes up an organism contains **genomic DNA**, which is the blueprint for life. The differential expression of genes within each cell gives rise to different tissues, organs and, ultimately, different organisms. Changes in genomic DNA give rise to the functional advantages that make some plant pathogens more successful than others. Success, as measured by the ability to reproduce, dictates that organisms that accumulate useful mutations in their genomic DNA will be more likely to pass those changes on to future generations. Heritable mutations are the **genetic variation** that is visualized by molecular tools. In this chapter, we will discuss the many different forms of genetic variation, current molecular methods for characterizing genetic variation, and possible questions about plant pathogens that can be answered with molecular tools.

The methods that we will discuss rely heavily on understanding DNA, so it is worthwhile to review its structure. DNA is made up of four nucleotide bases: adenine (abbreviated A), cytosine (C), guanine (G), and thymine (T), which are covalently linked together by a sugar (deoxyribose)-phosphate backbone into a long polymer. Strands of DNA are linear and directional, that is, they are read from 3′ to 5′ along the sugar-phosphate backbone (Figure 26.1). Within the nucleus, two strands of DNA intertwine to form a double-helix structure where G bonds with C, and A bonds with T, to form **base pairs**.

Complementary strands encode the same information, a redundancy that ensures fidelity during DNA replication. Under most conditions, complementary strands are annealed to each other via hydrogen bonding. During cell division, complementary DNA strands are split apart by DNA polymerase enzymes, and new DNA is synthesized from each of the template strands (Figure 26.1).

Genetic information is determined by the sequence of nucleotides. Regions of DNA that encode functional products are called **genes** and consist of trinucleotide units called **codons**, each of which codes for one of 22 possible amino acids. Genes are transcribed by cellular enzymes into a temporary nucleotide monomer called **messenger RNA (mRNA)**. Proteins are produced by cellular machinery that translates the codon sequence of the mRNA and assembles the corresponding amino acids into linear chains. As amino acids are linked together via peptide bonds, they often fold into complex structures that can be combined with other protein subunits, be transported, embedded in cellular compartments, or modified to become functional units. Functional proteins are the cellular machinery that makes tissues, organs, and organisms what they are. The unidirectional flow of genetic information from DNA to mRNA to protein, often called the central paradigm of molecular biology, is critical to understanding how genetic variation in genomic DNA can produce phenotypes that are subjected to natural selection.

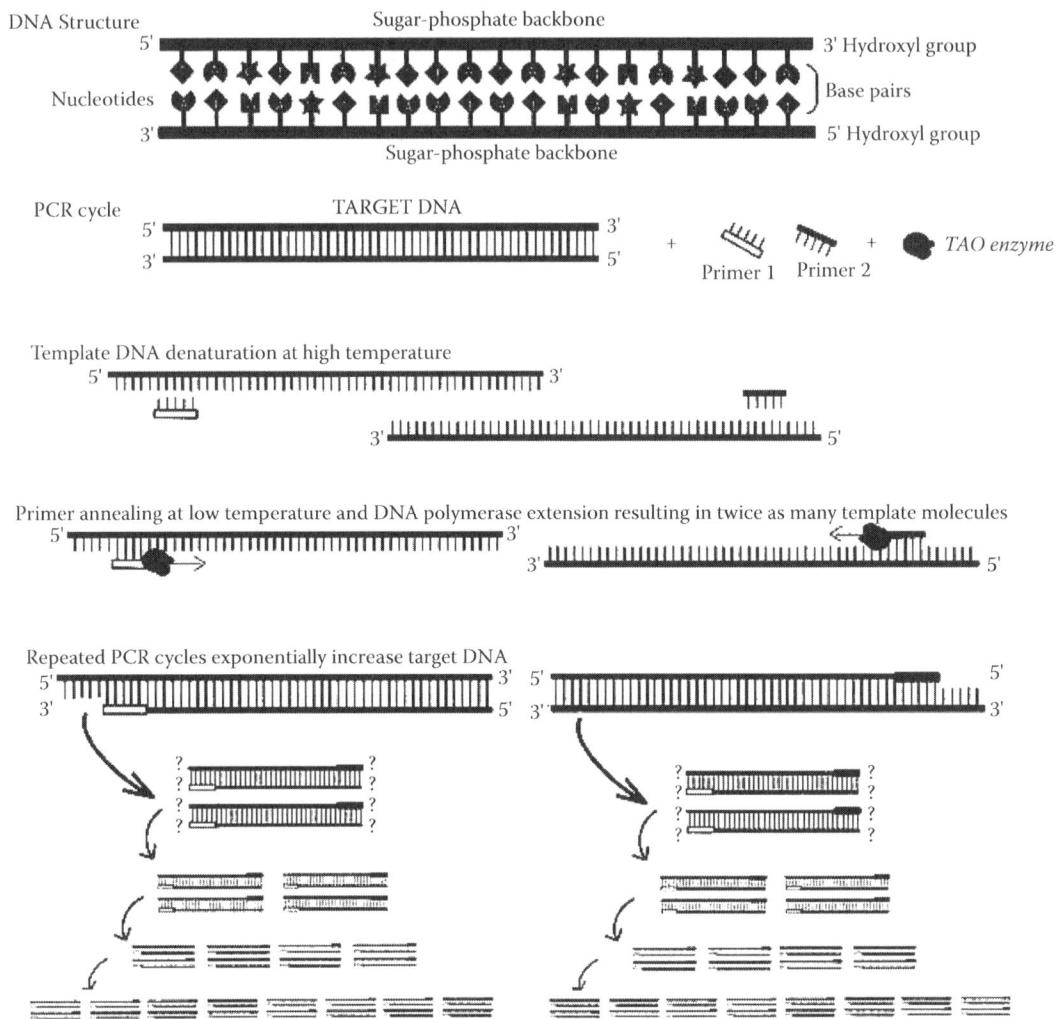

FIGURE 26.1 The complementary structure of DNA and PCR amplification of DNA fragments. Nucleotide bases pair with each to form base pairs with A and T, and G and C, as partners. Sugar phosphate backbone is directional, denoted by 5′ to 3′ end. Primers are short DNA fragments that anneal to specific regions of DNA according to the complementary base pair, which serves as a starting point for DNA polymerase to synthesize more DNA. Synthesis proceeds directionally from 3′ starting point. When PCR primers are designed for both DNA strands facing each other, the resulting cycles of template DNA denaturation, primer annealing, and synthesis of new DNA results in exponential amplification of discreet regions of DNA defined by the 5′ ends of the primers.

The type of genetic variation observed between individuals is highly dependent on where you look within the genome. Genes contain 64 possible combinations of three nucleotide codons that only correspond to 22 possible amino acids. Thus, there is considerable redundancy in codon usage. A single base change mutation can change the genomic DNA sequence, but not necessarily the amino acid sequence of the protein, and are labeled silent mutations because the phenotypic effect is not apparent or subject to natural selection. Mutations that alter the codon so that a different amino acid is incorporated into the peptide sequence are called missence mutations. Mutations can also consist of nucleotide insertions or deletions. One or two nucleotide insertions or deletions can disrupt the trinucleotide codon sequence, and are known as frameshift mutations. These mutations typically disrupt protein syn-

thesis and are not observed as often as silent or missence mutations. This does not necessarily mean that frameshift mutations occur at a lower rate; just that they are less likely to be passed on due to their deleterious effects. Different regions of the genome demonstrate distinct patterns in the type of mutations they accumulate and the rate at which mutations are observed. Where to look for genetic variation and the types of variation observed play an important role in how genetic data are analyzed and conclusions that can be made about plant pathogens.

Not all genomic DNA codes for cellular machinery. Many DNA sequences are associated with packaging the genome into **chromosomes** and serve as recognition sites for DNA-binding proteins. Other DNA is only useful as spacing between genes or as a buffer against DNA loss. Noncoding regions, especially those with no apparent

TABLE 26.1

Summary of DNA Characterization Techniques

Technique	Prior Knowledge	Type of Information	Use/Comparison at Level	Difficulty/Expense
DAF	None, PCR uses arbitrary primers	Anonymous bands or loci, dominant data	DNA profiling of individual isolates from the same population or same species	Easy and inexpensive
RAPD	None, PCR uses arbitrary primers	Anonymous bands or loci, dominant data	DNA profiling of individual isolates from the same population or same species	Easy and inexpensive
AFLP	None, PCR uses known primers annealing to linker sites after restriction digestion	Anonymous bands or loci, dominant and codominant data	DNA profiling of individual isolates from the same population, species, or genera	Moderately easy and inexpensive
SSR	DNA sequence of random SSR loci	Allele size variation of known loci, codominant data	DNA profiling of isolates from the same species, genera, or higher-order taxa	Moderately difficult, expensive to develop
DNA sequence	DNA sequence data of specific gene or genes	DNA base mutations at known locus, codominant data	DNA profiling of isolates from the same species, genera, or higher-order taxa	Moderately difficult, expensive to develop
SNP	DNA sequence data of multiple genes	DNA base mutations at multiple loci, codominant data	DNA profiling of isolates from the same species, genera, or higher-order taxa	Moderately difficult, moderately expensive
Whole genome sequence	DNA sequence data of large genomic regions and/or chromosomes	DNA base mutations at multiple loci, codominant data	DNA profiling of isolates from the same genera or higher-order taxa	Difficult and very expensive

function at all, typically accumulate mutations at a higher rate. The types of mutations observed varies widely among noncoding sites but often includes large insertions and deletions that would not be tolerated in coding regions. Eukaryotic genomes also contain large amounts of selfish DNA, or mobile DNA, which encodes proteins solely for the purpose of copying and inserting additional copies of the DNA, the phenotypic effects of which may not be readily apparent. Regardless, all types of DNA play a vital role when assaying genetic diversity between individuals or populations. In this first part of the chapter, we will focus on methods that can be used to genetically identify and describe a plant pathogen. In the second half, we will focus on molecular techniques that are useful for characterizing the genes expressed in plant pathogens.

GENETIC IDENTIFICATION AND CHARACTERIZATION OF PLANT PATHOGENS

DNA variation can be used to distinguish between all taxonomic levels of plant pathogens including individuals,

populations, strains, species, genera, and families. DNA fingerprinting has significantly accelerated the important task of plant pathogen identification. If a pathogen has been previously described, new isolates can often be identified from field samples without laboratory culturing, which can be difficult for some fungal, microbial, and viral pathogens. Plant pathogen systematics, or the study of pathogen diversity and relationship among pathogens over time, utilizes many different DNA fingerprinting methods to identify and classify plant pathogens and evaluate relationships between pathogens and groups of pathogens. Genetic characterization of plant pathogens can also answer questions regarding population structure, pathogen movement, modes of reproduction, modes of dispersal, and efficacy of chemical and cultural controls.

Some of the most popular molecular tools to identify and genetically characterize unknown pathogens are arbitrarily primed techniques such as DNA Amplification Fingerprinting (DAF), Random Amplified Polymorphic DNA (RAPD), and Amplified Fragment Length Polymorphism (AFLP) (Table 26.1) (also see Chapter 27). These protocols do not require prior knowledge of the pathogen's genome. Both methods make use of polymerase chain reaction (PCR) to amplify large amounts of specific DNA from small amounts of total genomic DNA (Mullis and Faloona, 1987). PCR primers are short segments of manmade single-strand DNA that anneals to specific regions in the genome based on the complementary DNA sequence. DNA polymerase synthesizes new DNA using the genomic DNA as a template. When this reaction is repeated several times, the DNA fragment between the PCR primers is exponentially amplified (Figure 26.1). Thermal-stable DNA polymerase is used so that the template DNA can be melted apart (passive denaturation) with heat without destroying the function of the enzyme. A single PCR cycle consists of several seconds at high temperature to denature the DNA, followed by a low temperature to anneal the primers, and finally a period of optimum temperature for the DNA synthesis, usually 68–72°C. PCR amplification of short regions, typically less than 1 kilobase (1000 nucleotides), is more robust than longer amplifications, which require specialized amplification protocols.

DAF and RAPD utilize arbitrary primers of short length, often 12 bases or less, that anneal throughout the genome (Welsh and McClelland, 1990; Williams et al., 1990). The sequence of the primers is random, and the probability of two primers annealing within 1.5 kilobases of each other can be calculated based on the size of the genome being assayed. However, genomes are not composed of random DNA sequences, so primer pairs must be empirically tested. Size separation of the PCR fragments can be done using either acrylamide or agarose gels, or other inexpensive equipment making optimization of DAF or RAPD affordable, especially because hundreds of discreet loci may be amplified during a single PCR amplification. The banding patterns produced indicate genetic similarities and differences between samples (Figure 26.2). Same-sized PCR fragments produced under identical cycling conditions indicate genetic similarity, whereas fragments that are different between samples, or polymorphic, suggest that a mutation disrupted the PCR amplification. Missing bands can be due to nucleotide changes in the primer annealing site, which would eliminate the production of the polymorphic fragment, or an insertion/deletion mutation in the region to be amplified, which would change the size of the fragment. PCR fragments that are unique to a group of samples suggest inheritance of that mutation and can be used to reconstruct the genetic relatedness of individuals and estimate genetic diversity within and between populations.

Reproducibility is often cited as a disadvantage of DAF and RAPD analyses. This is understandable because slight differences in temperature or reagents might bias the amplification process toward or away from certain PCR fragments. When this bias is multiplied by the enormous number of fragments that can be produced, there is con-

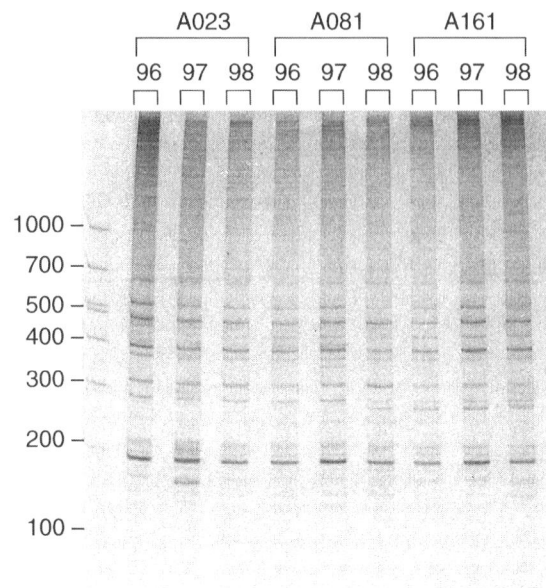

FIGURE 26.2 DAF analysis of *Marasmius oreades*. DNA extracted from the fruiting bodies of several isolates was amplified by using the oligonucleotide primer GTATCGCC, generated fingerprints were separated using polyacrylamide gels, and fragments were stained with silver. Isolates are labeled according to fairy ring and the year they were collected. Molecular markers are given in base pairs, and selected polymorphic bands are indicated. (Reprinted from Abesha, E. and G. Caetano-Anolles. 2004. Studying the ecology, systematics, and evolution of plant pathogens at the molecular level., pp. 209–215. In: Trigiano, R.N., M.T. Windham and A.S. Windham (Eds.). *Plant Pathology Concepts and Laboratory Exercises*, CRC Press, Boca Raton, FL. With permission.)

siderable justification for being cautious about DAF and RAPD results. However, the overall conclusions from DAF and RAPD analyses have proven reliable as long as the necessary controls are observed (Brown, 1996). Because of the low cost and virtually infinite combinations of primers and amplification conditions, researchers with enough dedication and time can compare an exhaustive number of loci between individuals. Randomly sampling genomic DNA includes comparisons between coding, noncoding, and mobile DNA, which increases the chances of finding a difference because some regions of the genome are more prone to accumulating mutations than others. Thus, DAF and RAPD potentially offer greater resolving power between highly related individuals, even when evaluating clonally reproducing pathogens such as the casual agent for Dutch Elm Disease (Temple et al., 2006).

AFLP is based on Restriction Fragment Length Polymorphism (RFLP) in which genomic DNA is cut into small chunks by restriction endonuclease enzymes. The resulting DNA fragments are visualized by radiolabeled probes made from known genes (Vos et al., 1995). AFLP employs the same restriction endonucleases to chop up genomic DNA, but then utilizes PCR to selectively amplify

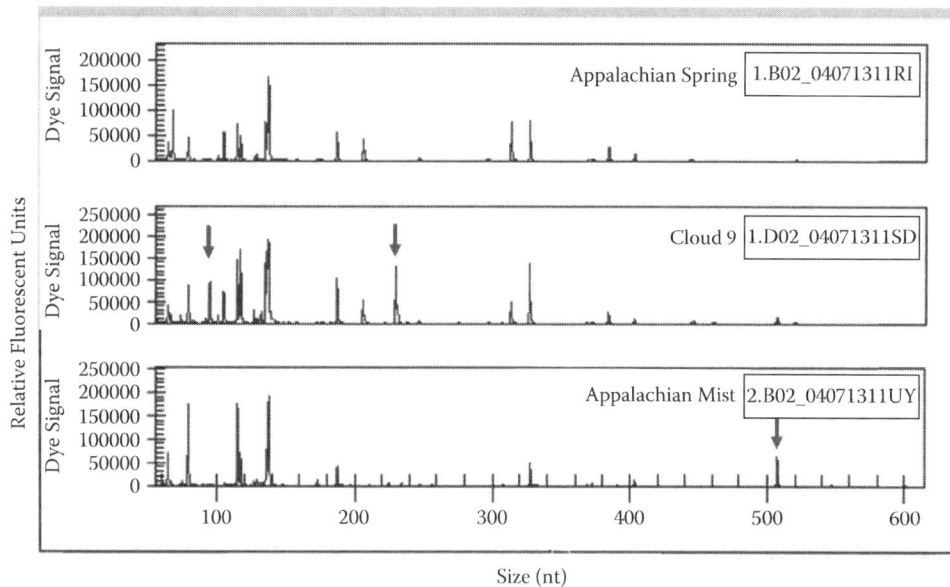

FIGURE 26.3 Amplified fragment length polymorphism (AFLP) fingerprints, or electropherograms, for *Cornus florida* "Appalachian Spring," "Cloud Nine," and "Appalachian Mist." Note the unique markers for "Cloud Nine" and "Appalachian Mist" (arrows). Size (nt) = size of fragment in nucleotides. (Reproduced from Smith, N.R., R.N. Trigiano, M.T. Windham, K.H. Lamour, L.S. Finley, X. Wang, and T.A. Rinehart. 2007. AFLP markers identify *Cornus florida* cultivars and lines. *J. Amer. Soc. Hort. Sci.* 132: 90-96. With permission.)

hundreds of discreet fragments. Unlike DAF and RAPD, which use random primers, AFLP primers are specific sequences designed to anneal to manmade linker DNA that is attached to the ends of the pieces of chopped-up genome. Linkers are annealed to the overhangs left behind by the restriction endonuclease enzymes and attached by ligating them to the sugar-phosphate backbone. The resulting pool of DNA contains small fragments, typically 500–2000 base pairs, which can be PCR amplified using primers designed from the linker DNA sequence. Primers are radioactive or fluorescently labeled so that amplified fragments can be visualized on acrylamide gels, which can separate fragments that differ by a single nucleotide. Because there may be hundreds to thousands of amplifiable and detectable DNA, additional bases are sometimes included in the primer at the 3′ end to further reduce the complexity of the amplified DNA. The resulting DNA fingerprint consists of size-separated fragments for each sample that can be compared side by side. Much like RAPD results, fragments that are present or absent in one sample but not the other suggest a genetic difference. Same-sized fragments produced by both samples indicate genetic similarity. Researchers can choose from a number of restriction endonucleases and primer combinations, which potentially produce hundreds of fragments during a single PCR amplification. Thus, AFLP generates robust DNA fingerprints from loci across the entire genome, but uses specific PCR primers that increase reliability and repeatability. AFLP has been adapted to run on automated capillary array sequencing instruments using fluorescent labels for greater throughput and automated data analysis (Figure 26.3).

AFLP, DAF, and RAPD produce dominant markers. Data is tabulated as DNA fragments that are present (1) or absent (0). The exact nature of the genetic mutation creating differences in the DNA fingerprints is unknown. Genetic similarity between samples and phylogenetic inference regarding shared ancestry are based solely on the mathematical frequency of DNA fragments, not biological models for DNA evolution. Amplified DNA fragments that are unique to a specific individual or population can be purified, ligated into plasmid vectors, and cloned into *E. coli*. The plasmid DNA can then be sequenced to identify the underlying nature of the polymorphism. These sequenced regions are referred to as Sequence Characterized Regions (SCARs), which, once described, can be exploited as codominant markers. PCR primers are designed flanking the mutation such that the amplified DNA fragments can be visualized, either by size variation or DNA sequence differences depending on the nature of the mutation, and recorded as alleles specific to each sample. In a sexually reproducing diploid organism, researchers expect to see two alleles per sample, one corresponding to the paternal chromosome, and another allele from the maternally contributed chromosome. Codominant data is not scored as a binary (absence or presence) but as allele variation. Identical alleles suggest shared ancestry, whereas different alleles indicate genetic divergence. Codominant data is considered more informative because every PCR amplification produces DNA fragments, and a technical failure during PCR amplification cannot be mistakenly scored as absence of a fragment (0).

One technique that generates codominant data is Simple Sequence Repeats (SSR), microsatellites, or SSR markers (Tautz, 1989). Eukaryotic genomes contain many short regions of repeated DNA. Generally, the repeat unit is 1 to 4 base pairs long and repeated 10 to 100 or more times. These repeats have a tendency to change in number when DNA is replicated due to a phenomenon known as DNA polymerase slippage. PCR primers adjacent to an SSR region can amplify the repeat. Size differences in the repeat length can be visualized by radiolabel or fluorescent molecules incorporated into the PCR products during amplification. Because they are uniformly spread around the genome, and some mutate faster than others, SSR are robust molecular markers.

Polyploidy and multinucleonic conditions found in certain pathogens can produce numerous SSR alleles per amplification, making analysis of the data somewhat complex. Because SSR markers are specific types of DNA, they are not present in all genomes. Therefore, viral and bacterial plant pathogens are less likely to produce results. SSR markers are particularly suited for diploid organisms that reproduce sexually and for evaluating genetic differences between species, genera, and higher-order taxa. The main disadvantage of the SSR technique is cost. To develop SSR markers, researchers must locate and sequence SSR regions before they can develop specific primers. However, SSR data are reproducible and easily verified by sequencing the amplified products. There are also specific models for the evolution of SSR regions that can be invoked during data analysis for more accurate conclusions. For example, a trinucleotide repeat consists of three base pair units. Changes in allele size can be weighted during analysis such that a 15 base-pair change is weighted more than a 3 base-pair change because it is likely that multiple slippage events, or more than one mutation, contributed to the 15 base-pair variant.

The ultimate molecular tool for comparing genetic variation between plant pathogens is sequencing the entire genome of each sample. Such an experiment would be cost prohibitive. Most DNA sequencing methods focus on only a few loci. DNA sequence variation can range from highly conserved to highly variable, and researchers often use different regions of the genome to answer different questions. Conserved DNA sequences are more appropriate for evaluating higher-level relationships such as comparing genera or families, whereas more variable regions are appropriate for comparing individuals and populations. Public databases such as GenBank (www.ncbi.nih.nlm.gov) contain more than 100 gigabases of DNA sequence data and computational tools to search for analogous DNA sequences that share a high level of similarity (Benson et al., 2006). Gene sequences, particu-

larly conserved gene sequences, from related taxa can be used to design PCR primers to amplify the same DNA regions in unknown plant pathogens. DNA sequences from the unknown pathogens can then be compared to related organisms in order to estimate genetic diversity and relatedness. Studies using conserved loci for species identification are also cataloged in GenBank, making it possible to classify unknown plant pathogens based solely on DNA sequence comparisons to previously classified organisms (Rinehart et al., 2006). This work builds upon the collective research of others with the expectation that researchers will add their own DNA sequences once studies are published. Genes commonly sequenced include elongation factor genes, tublin genes, and other universally conserved eukaryotic sequences.

When more variation is desirable, which may be necessary for the identification of strains or phylotypes, noncoding regions of the genome are more useful because they accumulate mutations rapidly. The optimum scenario is a short hypervariable region sandwiched between conserved gene sequences such that PCR primers can be designed to anneal to the conserved regions and amplify the more variable internal DNA. For example, Internal Transcribed Spacer (ITS) regions of ribosomal DNA (rDNA) are short sections of hypervariable DNA located adjacent to conserved 5.8s, small subunit (SSU) and large subunit (LSU) ribosomal regions (White et al., 1990) (see Chapter 27). PCR amplification is robust using universal primers, in part because the primers anneal to the conserved regions and because rDNA is repeated in tandem and found in high copy number per cell. ITS variants containing base changes, and nucleotide insertions and deletions, are usually prevalent, sometimes even among individual samples in a population. Verified ITS sequence data for many plant pathogens are available publicly for comparison, which increases the validity of the results and reduces the labor involved.

There is a wide array of genes that could be PCR amplified and sequenced, but the main disadvantage is the cost, which is considerable as this requires genomic DNA extraction, PCR amplification, and then DNA sequencing for each sample to be genetically characterized. Pure isolates are usually necessary, which may limit application to real-world plant pathogen problems. Moreover, valid conclusions require comparison of unknown DNA sequences to additional samples, preferably from related organisms or previously characterized reference isolates. Unless these data exist already, researchers are obligated to repeat the DNA sequencing on many other related pathogens to align the DNA sequences and compare genetic variation. This can rapidly increase the number of samples and escalate costs, particularly if additional isolates must be collected from the wild. One advantage, aside from the detail inherent in DNA sequence variation, is that sophisticated models for DNA evolution can be incorporated into computer analyses of genetic variation, making conclusions based on DNA sequence data statistically testable. Decisions to use DNA sequencing data to answer question about plant pathogens generally come down to how much is already known about the pathogen and how much effort is justified in acquiring DNA sequence data.

If large amounts of DNA sequence data are known or can be generated, the single base differences between samples can be tabulated. These single-nucleotide polymorphisms (SNPs) can be tabulated for many different loci across the genome to generate high-resolution genetic characterization and organism identification. Once an SNP site has been identified as informative, or unique to a particular strain or individual, it can be assayed in new or unknown samples much like SCAR markers generate codominant data. SNP databases are powerful tools because they approximate the strengths of entire genome sequencing by focusing only on polymorphic base pairs in genes and ignoring most noncoding DNA.

DNA sequence data is critical to PCR-based detection of plant pathogens in field-collected samples. All of the molecular tools discussed so far have the potential to uncover DNA fragments specific to a particular plant pathogen. Once characterized, these genetic regions can be exploited as targets for PCR amplification using PCR primers specific to the genetic variation that is unique to species, strain, phylotype, or even individual plant pathogens. Once optimized, PCR amplification can be a robust indicator of the presence or absence of a particular plant pathogen from field samples. Because the assay does not rely on pathogenicity or the presence of visible infection, diagnostic results can be generated during quiescence, latent infection, or among other plant pathogens and host tissue. Advances in biotechnology have made possible real-time PCR (RTPCR) methods, which measure the amount of amplified product during the PCR cycles using DNA-specific dyes or incorporation of fluorescent labeled nucleotides. Regression analysis can be used to plot the starting number of template molecules. In this manner, RTPCR can quantify how much pathogen DNA is present in a sample, offering an estimate of amount of plant pathogen in a given field sample (Heid et al., 1996).

Non-DNA-based approaches such as serology also detect the presence of plant pathogens in field samples. These molecular tools detect gene products, not the nucleic acids (DNA or RNA), and rely on antibodies that bind to proteins found on the outside of specific plant pathogens. These antibodies can be discovered by synthetic generation of a pool of reacting antibodies and subsequent selection for antibodies that only react with the plant pathogen of interest. Antibodies that cross-react with many plant pathogens are usually not as useful as antibodies that only bind to the specific virus, bacteria, or fungus causing disease. Positive detection is generally visualized by enzyme-linked immunosorbent assay, or ELISA (Clark and Adams, 1977; Kohler and Milstein, 1975; Voller et

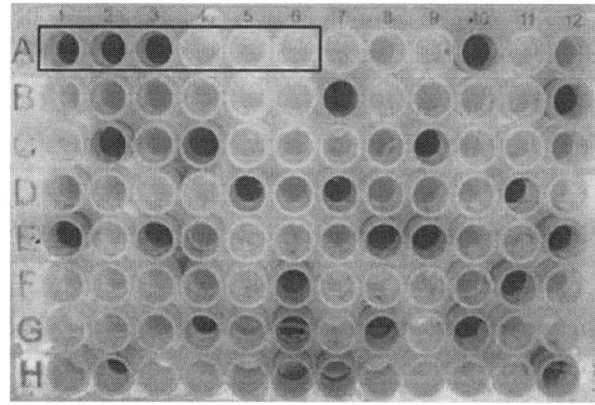

FIGURE 26.4 ELISA-based assay for *Rhizoctonia*. Colorimetric change, or increasing pigment, indicates the presence of *Rhizoctonia* species in freshly rubbed plant tissue samples. Three positive and three negative control reactions are indicated in the boxed region.

al., 1978) (see Chapters 4 and 39). This biochemical technique uses two antibodies: one antibody is specific to the pathogen; the other reacts to the pathogen–antibody complex and is coupled to an enzyme. Activation of the enzyme linked to the second antibody causes a colorimetric or luminescence change, which can be quantified using a luminometer or spectrophotometer. Positive results are rapid and sometimes visible to the naked eye. For this reason, serology-based plant pathogen detection is useful in field applications where laboratory equipment is not available. Depending on the detection protocol, there may be a linear relationship between the amount of pathogen and the amount of color change or light produced such that the amount of pathogen in a sample can be quantified. Unfortunately, ELISA is typically 100 to 1000 times less sensitive than PCR, so pathogens with low titer or in the initial stage of infection may go undetected. New and emerging plant pathogens may also not be detected. However, the development of commercial kits has vastly improved the selection and reliability of immunology-based assays, and many protocols are available in high-throughput formats for rapid, efficient screening of plant tissue for quarantine and disease identification purposes (Figure 26.4).

CHARACTERIZING GENE EXPRESSION IN PLANT PATHOGENS

Looking for the genetic underpinning of a trait, such as pathogenicity, typically requires linking the genomic DNA information with gene expression, protein expression and, finally, protein function. Although the central paradigm describes information going from DNA to RNA to protein, studying this functional pathway does not necessarily have to start with the analysis of genomic DNA. **Expressed Sequence Tag (EST)** libraries provide snapshots of the genes expressed in a pathogen at a particular time, such as initial infection, and can yield a wealth of genetic data regarding the mechanism of plant–pathogen interactions. EST libraries are created from mRNA isolated from the pathogen, so they reflect only those genes being expressed. A typical mRNA library consists of 10,000 or more clones, each corresponding to an mRNA transcript. Once the mRNA has been converted to cDNA by rt-PCR (reverse transcriptase PCR), it is usually cloned into a plasmid vector for propagation inside *Escherichia coli*. These bacterial colonies are easily stored, and the plasmid DNA can be extracted and sequenced. Generally, several EST libraries are made from different life stages of the pathogen or infection so that the genes expressed in one library can be compared to the transcripts contained in the other.

Much like SSR and SNP DNA fingerprinting, the cost of DNA sequencing can be a critical factor when deciding to use EST libraries to understand gene expression in plant pathogens. Libraries are often redundant because any mRNA transcript in high copy number will be cloned multiple times. There are molecular methods for removing high copy number DNA and increasing the complexity of the cDNA before cloning. A normalized EST library is one in which redundant cDNAs have been reduced by a hybridization step. Similarly, EST libraries can be subtracted from one another to enrich for genes that are unique to each mRNA extraction. DNA sequence data from EST libraries are typically deposited in GenBank or other public databases to accelerate research in multiple labs. Oftentimes, researchers working on the same plant pathogen will pool resources to produce and release EST information. This synergistic approach allows researchers to assemble smaller bits of gene expression information into a coherent model of the genes expressed in a plant pathogen. EST sequence data can also be mined for SNP and SSR marker development.

Differential display is a molecular tool that allows researchers to visualize a large number of expressed genes and uncover those genes that are up- or downregulated at a specific time in the plant pathogen's life (Liang and Pardee, 1992). Differential display compares mRNA transcripts from two different time points side by side in a manner similar to AFLP fingerprinting. mRNA transcripts are reverse-transcribed into DNA using a mixture of primers anchored to the polyA tail of mRNA and arbitrary primers. The resulting short, labeled fragments are separated and visualized either by acrylamide gel or capillary array electrophoresis to resolve size differences as small as a single nucleotide. When compared side by side, DNA fragments specific to a particular mRNA extraction can be identified as missing from other samples. Results look very similar to DNA fingerprinting using RAPD or AFLP, except that the starting template is mRNA, not genomic DNA. Thus, differences do not indicate genetic variation between isolates but, rather, spatial or temporal differences in gene expres-

sion. Unique DNA fragments can be excised from the gel, cloned, and the DNA sequenced.

DNA sequence data for differentially expressed genes can be compared to DNA databases to locate analogous sequences. For example, BLAST (Basic Local Alignment Search Tool) searches of GenBank can uncover previously characterized genes, possibly from other organisms. Even novel gene sequences can be analyzed for possible protein function by theoretical reconstruction of the protein itself. Using computer software, triplet codons are translated into amino acids that can be strung together and folded into a three-dimensional model of the protein itself. X-ray crystallography has produced a wealth of information regarding the structural components of proteins and their cellular function. For example, membrane-spanning proteins generally alternate characteristic hydrophilic and hydrophobic regions (Seshadri et al., 1998). These models produce reasonable hypotheses about novel protein function, especially when coupled with any analogous DNA or protein sequence information in public databases. Differentially expressed gene sequences can also be used to design specific PCR primers for real-time PCR. In this way, gene expression patterns visualized by differential display can be quantified, verifying that mRNA transcripts in one sample are increased or decreased in relation to another sample. There are several alternative protocols for visualizing differences in gene expression between two samples, including subtractive hybridization, RNA-arbitrarily primed PCR (RAP-PCR), representational difference analysis (RDA), and serial analysis of gene expression (SAGE).

Biotechnology and database resources can be leveraged in other ways to understand more about what makes a plant pathogen pathogenic. **DNA microarrays** utilize EST information to create multiple probes for each of the gene transcripts even when full-length mRNA sequence is not available, or gene function has not been characterized. Thousands to tens of thousands of these probes are affixed to a small surface, typically a glass slide, and may represent an equal number of expressed genes. Plant pathogen mRNA is extracted and converted to fluorescent labeled cDNA. These labeled fragments are then hybridized to the DNA probes that are fixed to the microarray slide. Non-binding transcripts, or those cDNA that do not share high levels of similarity, are washed off, and the resulting intensity of fluorescence for each microarray probe reflects the number of copies of mRNA transcript present in the sample at that time point. In this manner, the expression level of thousands of genes can be simultaneously compared in a single experiment (Schena et al., 1995). Because microarrays are reusable, additional mRNA extractions and experiments can uncover a coordinated, quantitative picture of the genes that are up- and downregulated (Figure 26.5).

DNA microarrays require advanced knowledge of expressed genes; however, DNA sequence databases may

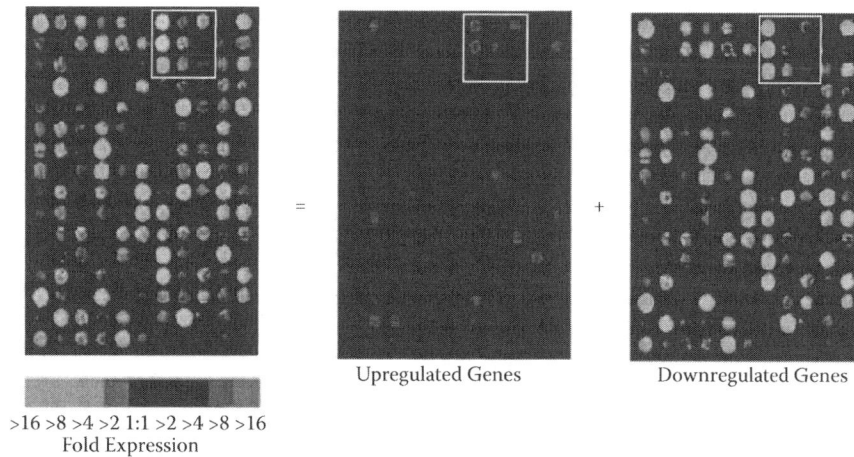

>16 >8 >4 >2 1:1 >2 >4 >8 >16
Fold Expression

FIGURE 26.5 DNA microarray results demonstrating up- and downregulated gene expression. mRNA from an uninfected plant sample was labeled with green fluorescent molecules, and mRNA from a plant undergoing infection was labeled with a red florophore. Samples were mixed and hybridized to a gene array. Dots showing green fluorescence indicate reduced gene expression, whereas red indicates increased expression. Boxed area highlights the gene expression changes in nine genes. (See CD for color figure.)

contain sufficient analogous gene sequences from related genera that the creation and sequencing of new EST libraries is not necessary. For example, the rust pathogen that attacks wheat may be a different species from the rust pathogen that attacks daylily, but the microarray for wheat rust may produce comparable data for daylily rust. Microarrays that are universal to a range of plant pathogens are possible because the DNA sequence variation between genomes, at least the expressed portion of the genome, may be minimal among related pathogens, and multiple probes are generally included for each expressed gene, increasing the chances that hybridization between probes and cDNA will occur for at least some regions of the gene. Thus, investment in DNA microarrays can be spread across several crops or disciplines.

The expanding field of **proteomics** uses computer-assisted analysis of two-dimensional gel-based protein profiles to compare protein fingerprints from different stages or tissues involved in plant pathogen attacks, and then picks unique proteins and analyzes their amino acid composition using mass spectrometry, specifically matrix-assisted laser desorption-ionization time-of-flight (MALDI-TOF) mass spectrometry system. This molecular tool skips genomic DNA and mRNA information and goes straight to the differences in functional protein products. Once proteins are characterized, they can be modeled in three dimensions using computer software. Theoretical mRNA transcripts can also be reverse engineered, and RTPCR can be used to verify differences in gene expression.

Characterizing gene expression in a plant pathogen does not necessarily rely on theoretical interactions or computational modeling. Targeting-induced local lesions in genomes, or TILLING, is the process of systematically mutating a specific gene and observing the mutant phenotypes to better understand protein function (Till et al., 2003). Large numbers of identical plant pathogens, or clones, are exposed to a chemical mutagen called ethyl methanesulfonate (EMS), which generally causes single base mutations in the genomic DNA. Thousands of these mutagenized clones are produced and stored. Genomic sequence data must be available because the randomly mutated pathogens are screened using PCR primers corresponding to expressed genes. The net result is a gigantic collection of mutants that can be characterized at the DNA level for changes in amino acid composition of the protein, possible truncation, or structural changes in the protein folding. Researchers usually place orders for mutations in specific regions of well-characterized genes based on gene function, protein modeling, or gene sequence homology. The collection of mutants is then screened by PCR for base changes in those regions. Any isolate with mutations in the desired region are sent out to the researcher for characterization.

It is also possible to generate specific mutations in well-characterized plant pathogen systems that have transformation systems. The ability to insert genomic DNA into an organism gave rise to the controversial field of genetically modified organisms. For research purposes, the ability to integrate foreign genetic material into the host genome is generally used to turn off specific genes. RNA-mediated interference, abbreviated RNAi, is a molecular tool based on a cellular defense system that is found in most eukaryotic organisms, which shuts down over expressed genes (Fire et al., 1998). The host system targets RNA transcripts via a double-stranded RNA (dsRNA) intermediate that is complementary to the gene being shut down. Once the dsRNA is detected, ribonuclease enzymes cleave the complementary mRNA transcript into useless chunks. Gene expression is effectively stopped before mRNA can be translated into protein. Because RNAi occurs after transcription, synthetic dsRNA that is added to cells can induce RNAi and selectively reduce the production of a specific protein without having to know where in the genome the mRNA is being expressed. In fact, dsRNA based on analogous genes from other organisms can shut down host genes if there is enough similarity between gene sequences. The resulting mutant has reduced production of a single protein, which can yield insights into the protein's role and function. RNAi and other tools described in this chapter are universal to plant pathology as they can be applied to the host plant as well. Because pathogens are involved in an arms race with plants, natural selection acts in concert on genes for disease resistance found in the plant and genes for pathogenicity in the plant pathogen.

Increases in biotechnology drive modern biology and have even impacted the role of science in popular culture from television shows and court cases. It is no surprise that molecular tools have become important techniques in plant pathology research. In this chapter we covered molecular tools to rapidly and unambiguously identify pathogenic organisms using DNA markers or DNA sequence data. Molecular data can also be used for the classification and increased taxonomic understanding of the relationships between different plant pathogens, and the rapid quantification of how many and what types of pathogens are present. These techniques typically require only small amounts of tissue, and new or unknown pathogens may be detected. Advances in molecular genetics rely heavily on public sharing of DNA sequence information and computer software, and offers the possibility of understanding the genetic basis for plant pathogen phenotypes.

LITERATURE CITED

Abesha, E. and G. Caetano-Anollés. (2004). Studying the ecology, systematics, and evolution of plant pathogens at the molecular level. Pp. 209–215. In: Triggiano, R.N., M.T. Windham and A.S. Windham (Eds.). *Plant Pathology Concepts and Laboratory Exercises*, CRC Press, Boca Raton, FL.

Benson, D.A., I. Karsch-Mizrachi, D.J. Lipman, J. Ostell and D.L. Wheeler (2006). GenBank. *Nucleic Acids Res.* 34: 16–20.

Brown, J.K.M. (1996) The choice of molecular marker methods for population genetic studies of plant pathogens. *New Phytologist.* 133: 183–195.

Clark, M.F. and A.N. Adams (1977). Characteristics of the microplate method of enzyme-linked immunosorbent assay (ELISA) for the detection of plant viruses. *J. Gen. Virol.* 34: 475–483.

Fire, A., S.Q. Xu, M.K. Montgomery, S.A. Kostas, S. E. Driver, and C.C. Mello (1998). Potent and specific genetic interference by double-stranded RNA in *Caenorhabditis elegans*. *Nature*. 391: 806–811.

Heid C.A., J. Stevens, K.J. Livak, and P.M. Williams (1996). Real time quantitative PCR. *Genome Res.* 6: 986–94.

Kohler, G., and C. Milstein (1975). Continuous culture of fused cells secreting antibody of predefined specificity. *Nature*. 256: 495–497.

Liang, P. and A.B. Pardee (1992). Differential display of eukaryotic messenger RNA by means of the polymerase chain reaction. *Science*. 257: 967–971.

Mullis, K.B., and F.A. Faloona (1987). Specific synthesis of DNA in vitro via a polymerase-catalyzed chain reaction. *Methods Enzymol.* 155: 335–350.

Rinehart, T.A., C. Copes, T. Toda, and M. Cubeta (2006). Genetic characterization of binucleate *Rhizoctonia* species causing web blight on azalea in Mississippi and Alabama. *Plant Dis.* 91: 616–623.

Schena, M., D. Shalon, R.W. Davis, and P.O. Brown (1995). Quantitative monitoring of gene expression patterns with a complementary DNA microarray. *Science*. 270: 467–470.

Seshadri, K., R. Garemyr, E. Wallin, G. von Heijne, and A. Elofsson (1998). Architecture of beta-barrel membrane proteins: analysis of trimeric porins. *Protein Sci.* 7: 2026–2032.

Smith, N.R., R.N. Trigiano, M.T. Windham, K.H. Lamour, L.S. Finley, X. Wang and T.A. Rinehart. (2007). AFLP markers identify *Cornus florida* cultivars and lines. *J. Amer. Soc. Hort. Sci.* 132: 90–96.

Tautz, D. (1989). Hypervariability of simple sequences as a general source for polymorphic DNA. *Nucl Acids Res.*17: 6463–6471.

Temple, B., P.A. Pines, and W.E. Hintz (2006). A nine-year genetic survey of the causal agent of Dutch elm disease, *Ophiostoma novo-ulmi* in Winnipeg, Canada. *Mycological Res.* 110: 594–600.

Till, B.J., S.H. Reynolds, E.A. Greene, C.A. Codomo, L.C. Enns, J.E. Johnson, C. Burtner, A.R. Odden, K. Young, N.E. Taylor, J.G. Henikoff, L. Comai, and S. Henikoff (2003). Large-scale discovery of induced point mutations with high-throughput tilling. *Genome Res.* 13: 524–530.

Voller, A., A. Bartlett, and D.E. Bidwell (1978). Enzyme immunoassays with special reference to ELISA techniques. *J Clin Pathol.* 31: 507–520.

Vos, P., R. Hogers, M. Bleeker, M. Reijans, T. van de Lee, M. Hornes, A. Frijters, J. Pot, J. Peleman, and M. Kuiper (1995). AFLP: a new technique for DNA fingerprinting. *Nucleic Acids Res.* 23: 4407–4414.

Welsh, J. and M. McClelland (1990). Fingerprinting genomes using PCR with arbitrary primers. *Nucleic Acids Res.* 18: 7213–7218.

Williams, J.G.K., A.R. Kubelik, K.J. Livak, J.A. Rafalski, and S.V. Tingey (1990). DNA polymorphisms amplified by arbitrary primers are useful as genetic markers. *Nucleic Acids Res.* 18: 6531–6535.

27 Molecular Techniques Used for Studying Systematics and Phylogeny of Plant Pathogens

Robert N. Trigiano, Malissa H. Ament, S. Ledare Finley,
Renae E. DeVries, Naomi R. Rowland, and G. Caetano-Anollés

The primary objective of these laboratory exercises is to familiarize undergraduate and graduate students (and instructors) with three very powerful molecular techniques that are used either to characterize **DNA** of plant pathogens and other organisms and/or define relationships between organisms. All of the techniques, which are discussed in Chapter 26, are similar in that they utilize the **polymerase chain reaction (PCR)** to amplify or increase copies of DNA, but differ primarily in the sequences of the **genomic DNA** that are targeted for amplification.

The first technique is **DNA Amplification Fingerprinting or DAF** (Caetano-Anollés et al., 1991). This technique is a DNA profiling protocol that employs relatively short **arbitrary primers** (5 to 10 base pairs) that target anonymous but discrete regions of genomic DNA. Amplification in the DAF technique produces a multitude of products of various sizes, which can be separated and visualized as bands on an acrylamide gel. The DAF procedure is partitioned into four independent laboratory exercises that include DNA isolation, DNA amplification, gel electrophoresis and silver staining, and data collection and analysis. Although the DNA amplification and gel electrophoresis exercises are emphasized, very detailed, easy-to-follow instructions and protocols are provided for all aspects of the DNA fingerprinting process. The procedure is adapted largely from Trigiano and Caetano-Anollés (1998) with permission from the American Society for Horticultural Science, and we recommend that you obtain a copy for your reference.

Amplified fragment length polymorphisms (AFLPs), the second profiling technique included in this chapter, also targets anonymous but discrete regions of genomic DNA (Vos et al., 1995). However, the technique involves several additional steps including a restriction digest of the genomic DNA, and several PCR (preselective and selective) reactions. PCR products can be separated on typically agarose gels and visualized with ethidium bromide or other fluorescent stains, but are more often separated using capillary electrophoresis and detected using fluorescence labeled primers, which have been incorporated into the PCR products. One advantage of using capillary gel electrophoresis is that the data are recorded electronically and can be manipulated into any number of statistical analysis programs. Although the AFLP technique is more difficult than DAF or other arbitrary primer techniques, AFLP data are generally considered to be superior because they are more reproducible within and between laboratories. We have divided the AFLP procedure into seven easy-to-follow steps.

The third technique involves the selective amplification of the internal transcribed spacer (ITS) regions that flank the 5.8S nuclear ribosomal unit (rRNA). The PCR reaction is completed with longer (18–22 base pairs) primers than those used for DAF and, upon amplification, produce a single band or product. The products from individuals are then sequenced (omitted in these exercises), the sequences compared, and relationships among the individuals inferred. These exercises or similar ones have been successfully completed on the first attempt by several classes of novice undergraduate, graduate students, and other researchers.

DNA fingerprinting can be defined functionally as a sampling procedure capable of reducing the extraordinarily complex genetic information contained in DNA to a relatively simple and manageable series of bands or "bar codes," which represent only selected but defined portions of a genome. Comparison of **DNA profiles** or fingerprints from different but closely related organisms can reveal regions with unlike nucleotide sequences (**polymorphisms**) that uniquely identify individuals much like the distinctive patterns of a person's fingerprints. A note of caution about the limitations of arbitrary fingerprinting techniques is in order, especially in determining relationships among a group of organisms. One assumption of the techniques made by inexperienced researchers is that bands appearing at the same position for different samples (locus) in the gel are of the same base pair (sequence) composition. In reality this is not always the situation. An individual "band or locus" from a sample organism may actually either contain several different amplified products of similar weight that comigrate, or the sequence of the product may be very different from products from other organisms in the comparison. One way to minimize these types of errors is to select very closely related organisms such as isolates of a fungus species or cultivars of plants. These limitations notwithstanding, DNA fingerprinting has been used in genetic and physical mapping, map-based cloning, ownership rights, molecular systematics, phylogenetic analysis, marker-assisted breeding, parentage testing, gene expression, and in many other applications in the plant sciences including plant pathology.

Prior to the 1990s, DNA characterization required molecular hybridization (Southern, 1975) or selective DNA amplification (Mullis et al., 1986; Erlich et al., 1991). These techniques demanded prior knowledge of DNA sequence information or clones and/or characterized probes, and often required extensive experimentation (Caetano-Anollés, 1996). Since then, a multitude of techniques (see Caetano-Anollés and Trigiano, 1997) has been developed that employ relatively short (5–20 nucleotides), arbitrary oligonucleotide primers to direct **DNA polymerase**-mediated amplification of discrete but anonymous segments of DNA. Among these techniques are R̲andom A̲mplified P̲olymorphic D̲NA (RAPD) analysis (Williams et al., 1990) and D̲NA A̲mplification F̲ingerprinting (DAF) (Caetano-Anollés et al., 1991; Caetano-Anollés and Gresshoff, 1994). Both methods produce information that characterizes a genome somewhere between the level of the DNA sequence and chromosomes.

Arbitrary oligonucleotide primers amplify multiple genomic regions (**amplicons**), many of which are variant (polymorphic) and represent allelic differences that can be traced in inheritance studies or can be treated as characters that can be used in population or phylogenetic analyses. The amplification reaction occurs through the succession of temperature cycles. Under low stringency conditions (low annealing temperature and ionic environment), the primer with arbitrarily (user)-defined sequence binds to many sites distributed in the genomic DNA template. DNA synthesis is initiated by a thermostable DNA polymerase even in those cases where there is substantial mismatching between primer and template base sequences. Despite perfect or imperfect priming, the DNA polymerase continues the amplification process by the successive addition of template-complementary bases to the 3′ terminus of the primer. Strand elongation is increased by raising the reaction temperature to an optimum level (usually about 72°C), and generally ends when the temperature is high enough to allow for the denaturation (disassociation) of the template DNA and the newly copied DNA strand. The separated strands now serve as template DNA when the reaction temperature is decreased to a point where primer annealing is permitted again (in most applications, usually less than 60°C). Following this initial amplification cycle, successive changes in temperature result in the selective amplification of genomic regions bordered by primer annealing sites occurring in opposite strands and separated by no more than a few thousand nucleotides (bases). The outcome of the amplification reaction is primarily determined by a competition process in which amplicons that are the most stable (efficient) primer annealing sites adjoining the easily amplifiable sequences prevail over those that are inefficiently amplified. A model to explain the amplification of DNA with arbitrary primers was proposed (Caetano-Anollés et al., 1992) and later discussed in detail (Caetano-Anollés, 1993), and is based on the competitive effects of primer–template as well as other interactions established primarily in the first few cycles of the process. Essentially, the rare but stable primer–template duplexes are transformed into accumulating amplification products. The final outcome is the selection of only a small subset (5 to 100) of possible amplification products.

Within the sample's DNA sequences, polymorphisms arise from nucleotide substitutions that either create, abolish, or modify particular primer annealing sites, which may also alter the efficiency of amplification or priming, or deletions or insertions that shorten or extend, respectively, the amplicon length. The resultant polymorphic RAPD or DAF fragments are useful DNA markers in general fingerprinting or mapping applications. These markers have been profusely applied in the study of many prokaryotic and eukaryotic organisms.

Although both RAPD and DAF analyses produce similar types of information, there are some differences between the two techniques that should be noted.

1. DAF uses very short primers, usually 7 or 8 nucleotides in length, whereas RAPD typically utilizes 10 nucleotide primers.
2. DAF products are usually resolved using 5 to 10% polyacrylamide gel electrophoresis and silver staining (Bassam et al., 1991), whereas RAPD products are typically separated electrophoretically in agarose gels, stained with ethidium bromide, and visualized under UV light.
3. The **reaction mixture or cocktail** in DAF contains higher primer-to-template ratios than RAPD and produces relatively complex banding profiles containing 30 to 40 products that are fewer than 700 base pairs in length. In turn, RAPD usually generates more simple patterns of 5 to 10 bands.
4. DAF polyacrylamide gels are backed using polyester films and are amenable to permanent storage, whereas RAPD agarose gels are difficult to store, and a photograph serves as the only permanent record. One could argue the relative merits of each fingerprinting technique, but from our experience, DAF is easier for students and instructors to learn and use and is very tolerant, almost for-

TABLE 27.1

Sources for Laboratory Equipment and Materials

Supplier	Product	Address	Phone, Fax, or URL
Applied Biosystems	DNA polymerase	850 Lincoln Center Drive, Foster City, CA 94404	800.327.3002, www.appliedbiosystems.com
Barnstead/Thermolyne Corp.	Nanopure water	2555 Kerpar Blvd. PO Box 797, Dubuque, IA 52004	800.553.0039, www.barnstead.com
Bio-Rad	Electrophoresis supplies; tips	2000 Alfred Nobel Dr., Hercules, CA 94547	800.424.6723, www.bio-rad.com
BioVentures, Inc.	Biomarkers	PO Box 2561, Murfreesboro, TN 37133	800.235.8938
BioWhittaker Molecular Applications	Gelbond PAG film	101 Thomaston St., Rockland, ME 04841	800.341.1574, www.bmaproducts.com
Electron Microscopy Sciences	16% Formaldehyde	P.O. Box 251 321 Morris Rd., Fort Washington, PA 19034	800.523.5874
Ericomp, Inc.	Thermalcycler	6044 Cornerstone Ct. W., Suite E, San Diego, CA 92121	800.541.8471
Exeter Software	NTSYS-pc, version 2.2	100 North Country Road, Sedtauket, NY 11733	516.689.7838
Fisher	Centrifuge tubes, acrylamide	P.O. Box 4829Norcross, GA 30091	800.766.7000, www.fischersci.com
Gentra Systems, Inc.	DNA isolation kit	15200 25th Ave. N., Suite 104, Minneapolis, MN 55447	800.866.3039, www.gentra.com
Integrated DNA Technologies	Synthesis of primers	1710 Commercial Park, Coralville, IA 52241	800.328.2661, www.idtdna.com
Midwest Scientific	Flat loading tips	280 Vance Rd., Valley Park, MO 63088	800.227.9997, www.midsci.com
MJ Research	Thermalcycler	149 Grove St., Watertown, MA 02172	800.729.2165
Pharmacia Biotech, Inc.	Mini-Fluorometer; plates	800 Contennial Ave., Piscataway, NJ 08855-1327	800.526.3593, www.apbiotech.com
Phenix	Pipette tips, centrifuge tubes	3540 Arden Road, Haywood, CA	800.767.0665, www.phenix1.com
Qiagen	QiaQuick PCR clean-up kits DNA isolation kit	9600 DeSoto Ave. Valencia, CA 91311	800.426.8157, www.quigen.com
Rainin Instrument Company	Pipettes and tips	Rainin Road, Box 4026, Woburn, MA 01888-4026	800.472.4646, www.rainin.com
Sierra-Lablogix, Inc.	Staining trays	1180-C Day Rd., Gilroy, CA 95020	800.522.5624
US Biochemical	dNTPs	PO Box 22400, Cleveland, OH 44122	800.321.9322

giving, of some errors typically made by novices, such as inaccurate pipetting, etc. The data are permanently recorded in the form of a gel instead of a photograph, which is very gratifying to students and facilitates research by allowing repeated and close scrutiny of data. Some research laboratories also have the capability to scan and store data from gels as computer records.

The intention of these exercises is not to fully describe and explore the theoretical aspects of DNA fingerprinting, which may be otherwise obtained by reading the literature cited throughout this paper. The educational objectives of the laboratory exercises are to acquaint students with the general concepts, techniques, and uses of DNA finger-printing and to remove some of the perceived mystique underlying molecular genetics.

A number of products are mentioned throughout the laboratory exercises. Complete information is provided in Table 27.1 should an instructor wish to order from any company mentioned in this paper. These are simply what we normally use and do not constitute product endorsements by either the authors, the publishers, or the University of Tennessee, nor implied criticism of those products not mentioned. There are equally suitable, if not alternative, products and equipment that may be substituted for those listed herein.

Before beginning any of the exercises in the chapter, a few essential generalities applicable to all laboratories are listed.

1. All pipette tips, Eppendorf centrifuge tubes, water, and reagents used to assemble the amplification reactions either should be autoclaved or filter-sterilized (0.22 μm) and made with sterile, high-quality water.
2. Participants should wear either latex or acetonitrile gloves to avoid hazardous materials (acrylamide and silver nitrate) and protect samples from **DNAses** found on the skin (Dragon, 1993).
3. Where possible, use only ACS (American Chemical Society)-certified pure chemicals and double distilled or nanopure water (<16 MΩ/cm) (Barnstead/Thermolyne Corp., Table 27.1), hereafter referred to as "pure." It is not necessary to use HPLC-grade water.

Characterization of genomes using DAF, as well as any other of the arbitrary primer-based techniques, always consists of at least four independent phases, including isolation of DNA, amplification of DNA, **electrophoresis** and visualization of amplified products, and collection and analysis of data. Each of these steps requires between 4 and 6 h to complete. If class and laboratory time is limited, the instructor may opt to complete one or more of the laboratory sessions for the students. In fact, for large classes, the instructor may wish to complete the exercises as a demonstration. However, students will derive the most benefit by fully participating in each of the laboratory sessions. This chapter emphasizes DNA amplification and DAF product separation and visualization, and, to a lesser degree, DNA isolation and data analyses. We recommend that these laboratory experiments and procedures be completed by advanced undergraduate or graduate students working in teams of four or less.

EXPERIMENT 1: DNA AMPLIFICATION FINGERPRINTING

Following an examination of a few journal articles concerning some aspect of DNA analysis, it is evident that there is a multitude of methods to isolate genomic DNA, all of them more or less suitable for the purposes of this laboratory exercise. Fortunately, DAF reactions do not require the high quality or large quantity of DNA necessary, for example, in restriction fragment length polymorphisms (RFLP) analysis. In our lab, we use a DNA isolation kit especially formulated for plants and/or fungi (Puregene®, Gentra Systems, Inc., or Qiagen; Table 27.1) or a procedure developed specifically for fungi (Yoon et al., 1991). Unlike earlier methods for isolation of DNA, most commercially available kits avoid the use of highly toxic materials, such as phenols. Regardless of the technique or kit used to isolate DNA, young, quickly growing cultures of Gram negative bacteria or fungi should be used. Mycelia or bacterial cells should be stored at

−70°C until needed. These laboratory exercises will be illustrated using *Fusarium oxysporum* isolates.

EXERCISE 1: DNA ISOLATION

Materials

The following will be needed for each team of students:

- Puregene plant or fungus DNA isolation kit (one kit for the entire class will be sufficient) or follow Yoon et al. (1991) (Other methods may be substituted.)
- 100% Ethanol or isopropanol
- 70% Ethanol
- Sterile 1.5 and 0.65 mL Eppendorf centrifuge tubes
- Sterile 100- and 1000-μL pipette tips and pipettors
- High-speed tabletop centrifuge
- 60°C water bath
- Liquid nitrogen (wear insulated gloves and eye protection) and Dewar vessel
- Sterile, chilled mortar and pestle for each isolate
- Mycelium (either fresh or frozen at −70°C)
- Sterile, pure water
- Other reagents required by DNA isolation kit (see manufacture's instructions).
- Insoluble polyvinylpolyprolidone (PPVP)

Follow the protocol outlined in Procedure 27.1 to complete DNA isolation.

Determining DNA Concentration

DAF is exceptionally tolerant of both the quality (purity) and quantity of DNA used in the reaction mixture. DNA concentration can be determined spectrophotometrically with a dedicated **fluorometer** (e.g., Mini-Fluorometer, Pharmacia Biotech, Table 27.1), invariably set at 365 nm. The fluorometer only reads DNA; RNA is not detected. Instructions for using the fluorometer are included with the instrument; dye and calf thymus DNA for standard concentrations of DNA may be purchased directly from Pharmacia Biotech (Table 27.1). Newer fluorometers do not require a fluorescent dye. Because most DNA isolations from fungi usually yield concentrations between 10 and 75 ng/μL, we recommend preparing and calibrating the fluorometer with standards of similar concentrations.

If a dedicated fluorometer is not available, DNA content can be determined directly using a spectrophotometer (Procedure 27.2). The 260/280 ratio of a pure double-stranded DNA preparation should be between 1.65 and 1.85. Although this ratio is dependent on the fractional GC content, higher ratios are often due to RNA contamination, and lower values due to protein or

---

Procedure 27.1

Isolation of DNA from Mycelium

Step	Instructions and Comments
1	For each isolate, place 25 mg or less of mycelium with about 25 mg polyvinylpolypyrrolidone (PPVP) (sequesters plant phenols) into a sterile mortar and pestle, add liquid nitrogen, and grind frozen mycelium to a powder. Add 500 μL of extraction buffer.
2	Continue to grind and freeze and thaw at least twice, adding more extraction buffer if necessary. Slurry should be very thin and watery when melted.
3	Load about 400 μL of the slurry into sterile, 1.5-mL centrifuge tubes and float in a 60°C water bath for about 1 h. Centrifuge at maximum rotation (14,000 rpm) for 10 min to deposit (pellet) cellular debris and PPVP. Transfer supernatant to a new, sterile centrifuge tube and complete the kit's instructions except for the RNAse step when using a fluorometer, and include when DNA concentration will be determined using a spectrophotometer (see Procedure 27.2).
4	At the end of the isolation procedure, do not redissolve the DNA in TE buffer; instead use 50 μL of sterile, pure water. Note: A large DNA pellet is unlikely and, in fact, you may not see a distinctive pellet. A little faith is required now—there is DNA in the bottom of the tube. Heat the contents of the tubes in a 60°C water bath for about 2 min to help dissolve the DNA and refrigerate (4°C) overnight. The next morning, centrifuge for 2 min to pellet any undissolved particulate material, then carefully pipette the supernatant containing the DNA into new, sterile, 0.65-mL centrifuge tubes labeled F1–F7 and store at 4°C.

---

phenol contamination. Thus, if determining DNA concentration using this method, it is imperative that the RNA in the sample be eliminated with **RNAase** step in the isolation procedure. Note that low concentrations of DNA are very difficult to read using the spectrophotometer. The approximate DNA concentration in the solution may also be determined using an agarose gel and known concentrations of standards. In this case, the relative intensity of ethidium bromide stained standards is compared to the intensity of the samples. Follow the instructions described in Exercise 5—DNA Isolation for AFLPs—to determine DNA concentration.

Follow the protocols outlined in Procedure 27.2 for this portion of the exercise.

Adjusting the Concentration of DNA

Typically, the concentration of DNA from most isolations is too high to be used directly as the template in the amplification stage of DAF. Optimum concentrations of DNA range from 0.02 to 2.0 ng/μL of the reaction mixture. Therefore, the DNA must be diluted with sterile, pure water to a more functional concentration, such as 5.0 ng/μL (Procedure 27.3). We have stored isolated DNA and dilutions at 4°C for more than 5 years without apparent degradation. DNA stock solutions may also be stored at −20°C.

Follow the protocols outlines in Procedure 27.3 to complete this portion of the exercise.

Questions

- Why is EDTA included in the extraction procedure?
- Why is the extraction solution buffered?
- What is genomic DNA?
- Why is it important to wear gloves and use sterile tubes?

Procedure 27.2

Determining DNA Concentration Using a Spectrophotometer

Step	Instructions and Comments
1	"Zero" spectrophotometer by pipetting 1 mL of distilled water into both sample and reference quartz cuvettes.
2	Pipette and mix by gentle inversion 2 μL of DNA into 1 mL of distilled water in thesample cuvette.
3	Read absorbance (optical density) at 260 nm (e.g., 0.012).
4	Because an O.D. of 1.0 = 50 μg/mL DNA, the entire sample contains 50 μg/mL × 0.012 = 0.6 μg or 600 ng/mL.The amount of DNA in each μL = 600 ng/2 μL or 300 ng/μL.

Procedure 27.3

Diluting DNA for Use in Reaction Mixtures

Step	Instructions and Comments
1	Determine concentration of DNA in isolation from fluorometer or Procedure 27.2, e.g., 79 ng/μL.
2	Make a 5 ng/μL solution using the following formula:

$$C_1 \times V_1 = C_2 \times V_2$$

where: C_1 = concentration of the isolated DNA in nanogram/microliter (ng/μL), V_1 = volume (μL) of the isolated DNA to dilute (arbitrarily used 20 μL), C_2 = concentration of diluted DNA (5 ng/μL), and V_2 = volume of diluted DNA in μL (unknown).

Substitute in the equation and solve for V_2:

$$79 \times 20 = 5 \times V_2 = (79 \times 20)/5 = V_2$$

$$V_2 = 316 \ \mu L = \text{total volume of diluted DNA.}$$

| 3 | Because 20 μL of original DNA was used, the amount of sterile, pure water to add is |

$$316 \ \mu L - 20 \ \mu L = 296 \ \mu L$$

Pipette 20 μL of the original DNA solution into a sterile 0.65-mL tube and add 296 μL of sterile, pure water. Mix throughly by vortexing and centrifuge (14,000 rpm for 5 s) to remove air bubbles. This is the template concentration you will use for the amplification cocktail.

| 4 | Store all DNA stocks at 4°C or –20°C. |

EXERCISE 2: DNA AMPLIFICATION

This laboratory exercise requires careful experimental design and planning, and involves handling of many liquid reagents. By completing this exercise, students will gain experience and confidence with routine procedures in a molecular biology laboratory.

To begin, here are a few helpful hints. Plan ahead and write everything down! It is very easy to forget what has been done and needs to be done; record keeping is an integral part of good laboratory practices. When in doubt, change sterile pipette tips. Do not risk cross contamination of solutions and templates or introduction of DNAses to save a pipette tip. Mix, by vortexing, all stock solutions, except DNA polymerase, and centrifuge tubes to remove large air bubbles from the liquid before opening and to avoid aerosols. Look at the pipette tip to ensure that an appropriate amount of fluid has been taken into the lumen. Finally, always wear gloves to pro-

---

Procedure 27.4

Preparation of Primer Stocks

Step	Instructions and Comments
1	Prepare 300 μM primer stock, e.g., 159 nmoles (on label) provided by supplier.

$$159 \text{ nmoles primer/ } x \ \mu L = 300 \ \mu M \ \ x = 532 \ \mu L$$

Briefly centrifuge tube containing the primer. Now add 532 μL of sterile, pure water to the manufacturer's tube. Mix thoroughly by vortexing and centrifuge briefly. *Note:* Tube may be heated to 65°C briefly to facilitate resuspending the primer.

| 2 | Prepare 30 μM primer stock: |

Pipette 30 μL of 300 μM into a sterile 0.65-mL Eppendorf tube and add 270 μL of sterile, pure water. Mix thoroughly by vortexing and centrifuge briefly. This is the stock to use in the amplification cocktail.

| 3 | Store all stocks at -20°C in a nondefrosting refrigerator. |

Fungal DNA Templates

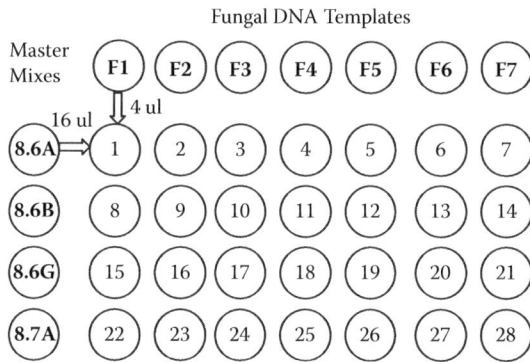

FIGURE 27.1 Scheme for dispensing master mixes and DNA templates. Pipette 16 µL of master mix into each row of seven sterile 0.65-mL reaction centrifuge tubes. Pipette 4 µL of DNA template into each reaction tube. Remember to change tips between tubes.

tect stock solutions and DNA from contamination with either bacteria or DNases from your hands.

Materials

The following materials will be needed for each team of students:

- AmpiTaq Stoffel Fragment DNA polymerase (Applied Biosystems, Table 27.1)
- 66 mM $MgCl_2$ for fungi (do not use the 10X $MgCl_2$ for plants included in the kit; use all other components). Sterilize using a 0.22-µm filter.
- Primers (Integrated DNA Technologies, Inc., Table 27.1) with the sequences (5′ to 3′) GAGCCTGT (8.6A), CCTGTGAG (8.6B), CTA-ACGCC (8.6G), and CCGAGCTG (8.7A). The first number in the primer codes denotes oligonucleotide length, and the second represents the approximate fractional GC content. Follow Procedure 27.4 to prepare the correct primer concentration.
- Deoxynucleoside triphosphates (dNTPs, US Biochemical, Table 27.1) are supplied as a set of four ampoules containing 25 µmol of each dNTP in 250 µL of water (100 mM). Simply combine the four ampoules (1 mL) in a sterile container and add 11.5 mL of sterile, pure water to produce a 2 mM solution containing all the necessary dNTPs. Dispense 250 µL aliquots into 50 sterile centrifuge tubes of each 0.65 mL and store at −20°C. This is the working dNTP concentration for the amplification reaction mixture.
- Heavy mineral oil (no need to sterilize; most **thermalcyclers** do not require oil)
- Assorted sterile pipette tips
- Sterile 0.65-mL Eppendorf tubes

- Sterile, pure water
- Thermalcycler—PCR machine

Follow the protocols in Procedure 27.4 to prepare primers.

Planning the Experiment

DNA from the seven isolates of your organism should be amplified with four primers. A total of 28 sterile 0.65-mL centrifuge tubes are needed. Label the tubes 1–28 according to the scheme depicted in Figure 27.1 and record in a laboratory notebook.

Amplification of DNA

The assembly of the amplification cocktail is the heart of the DAF technique. Each group of students should make "master mixes" (one or more) containing all the ingredients common in each amplification reaction cocktail (Procedure 27.5). The only two variables in the *Fusarium* experiment are the DNA templates from the individual isolates and the primers. Master mixes should therefore contain sterile pure water, stoffel buffer, dNTPs, magnesium chloride, stoffel enzyme, and a single primer; assembly should be in a sterile centrifuge tube labeled with the primer code (e.g., 8.6A). DNA template will be added to individual reaction tubes later. Procedure 27.5 details how to make a master mix and provides a list of reagents and their final concentration in 20 µL of the mixture. Start by removing the ingredients from the freezer and, after thawing, vortex and centrifuge briefly (except the enzyme). Gloves should be worn by all persons involved in making the master mixes.

Follow the protocols in Procedure 27.5 to complete this portion of the experiment.

Thermalcyclers are programmed to establish **annealing** (30–62°C), **extension** (72°C), and **denaturing** (95°C) **temperatures** for prescribed times. This set of temperature regimens constitutes a cycle, which is repeated 30 to 40 times. However, because there may be significant differences in ramping times between thermalcyclers, proceed with caution when searching for a suitable cycle. Ramping time can be thought of as the time necessary for the amplification mixture to go from one designated temperature to the next (e.g., from annealing to extension temperature). The Easy Twin Block System (Ericomp Inc., Table 27.1) has relatively slow ramping times compared to the DNA Engine PTC-200 (MJ Research, Table 27.1). Reproducible, clear profiles can be generated with the Ericomp machine using a cycle of 30 sec at 95°C and 30°C, without an extension step. However, for the DNA Engine, which has faster ramping times, the cycle of 1 min at 95°C and 30°C with an extension step at 72°C for 30 sec works well. A general rule is to increase the annealing, extension, and denaturing times when the

Procedure 27.5

Preparation and Assembly of Master Mixes for Each Primer (Sufficient for Eight Reactions; Always Prepare More Than You Have Samples to Allow for Pipetting Errors)

Step	Instructions and Comments
1	Pipette 65.6 μL (8.2 × 8) of sterile, pure water into a sterile 0.65-mL tube.
2	Add 16 μL (2 × 8) of 10× stoffel buffer provided by the manufacturer (final conc.[a] = 1×).
3	Add 16 μL (2 × 8) of 2 mM dNTPs (final conc. = 200 μM).
4	Add 16 μL (2 × 8) of 30 μM primer stock (final conc. = 3 μM).
5	Add 9.6 μL (1.2 × 8) of 66 mM $MgCl_2$ (final conc. = 4.0 mM).
6	Add 4.8 μL (0.6 × 8) of DNA polymerase provided by manufacturer.
7	Mix thoroughly by vortexing and centrifuge briefly at 14,000 rpm to eliminate air bubbles.
8	Place the master mixes to the left of the four rows of seven reaction tubes (in a plastic flipper rack) labeled 1–7, 8–14, 15–21, and 22–28 as shown in Figure 27.1. Dispense 16 μL of the master mix into each of seven sterile 0.65-mL tubes. Master mix for an individual primer can be distributed to tubes without changing tips. However, a different tip should be used for each of the four master mixes because they contain different primers.
9	Remove the 5.0 ng/μL DNA stocks from the refrigerator, vortex, and briefly centrifuge at high speed. Place the DNA stocks at the top of the flipper rack above those reaction tubes 1, 2, 3, 4, 5, 6, and 7, which correspond to the isolates to be analyzed (Figure 27.1).
10	Pipette 4 μL from F1 into tube 1, then close the reaction tube and discard the tip. Repeat the sequence for tubes 8, 15, and 22, then close stock tube F1. Repeat the procedure for cultivars F2 through F7. Each tube now contains 20 μL of reaction mixture plus template.
11	Mix the contents of the tubes by vortexing and centrifuging briefly at high speed. Open the tubes and add a drop of heavy white mineral oil to each to prevent evaporation and condensation during amplification. Note: Some thermalcyclers do not require oil—see thermalcycler manufacturer's instructions.
12	Place the tubes in a thermalcycler for amplification of the DNA.

[a] Final concentrations of reagents are based on 20 μL reaction volumes after the addition of 4 μL of template DNA.

thermalcycler has short ramping times. The entire amplification process takes between 2 and 6 h depending upon the thermalcycler.

Recovery and Storage of Amplification Products

(You may skip this section if oil was not used in the tubes during amplification.) Before removing the products, prepare several tubes containing 20 μL of blue water [i.e., a drop of xylene cyanol stock: 12 g urea and 2 mL of xylene cyanol stock solution (4 mg xylene cyanol/ 8 mL of pure water) in 20 mL of water] under a drop of mineral oil. Students can easily see the blue color and practice pipetting the water without drawing any oil into the tip. Once the amplification program is completed, the reaction products are removed from beneath the oil. Label 28 sterile, 0.65-mL centrifuge tubes by designating an experiment (e.g., 1) and the tube number (1.1, 1.2, . . . 1.28). A P-100 (Rainin, Table 27.1) pipette or equivalent set on 22 or 23 μL with sterile Prot/Elec tip (Bio-Rad, Table 27.1) can be used. The operation should be performed quickly and the contents of the tip "squirted" into a new sterile, pre-labeled 0.65-mL centrifuge tube without touching the sides to avoid contamination with oil clinging to the tip. When working with the reaction tubes, be sure to change pipette tips between each sample. Samples may be stored at 4°C as is or diluted 1:1 with pure, sterile water.

Questions

- Why is a "low" annealing temperature used in DAF?
- Why are divalent magnesium ions necessary in the reaction mixture?
- When the DNA polymerase enzyme assembles the new DNA strand, are there copy errors?

EXERCISE 3: ELECTROPHORESIS AND STAINING OF DAF PRODUCTS

This laboratory exercise focuses on the electrophoretic separation of amplification products and should expose students to one of several electrophoretic techniques routinely used to characterize biological molecules. Experi-

Procedure 27.6
Composition of 10X TBE Buffer and 10% Polyacrylamide Stock

Step	Instructions and Comments
	10X TBE Buffer
1	Dissolve 121.1 g Tris base, 51.4 g boric acid, and 3.7 g Na$_2$EDTA·2H$_2$O in 800 mL of pure water.
2	Bring the final volume to 1 L with pure water; pH = 8.3. Store at room temperature. *Note:* If room is cool, salts may precipitate. Try making a 5X buffer.
	10% Polyacrylamide Stock (Wear Protective Clothing, Gloves, and Particle Mask)
1	Dissolve 19.6 g acrylamide, 0.4 g PDA (piperazine diacrylamide), and 20.0 g urea in 130 mL of pure water. [Caution: wear protective particle mask, gloves, eyeglasses, and clothing—unpolymerized acrylamide is a potent neurotoxin; skin contact and accidental inhalation of the compound should be avoided.] Do not substitute BIS (N,N′-methylene bis acrylamide) for PDA as it adversely affects staining quality.
2	Add 20.0 mL of 10X TBE buffer and 10.0 mL glycerol to the acrylamide solution.
3	Bring the final volume to 200 mL with pure water. Store at 4°C in a brown bottle and discard unused portion after 8 weeks.

ence with this technique should facilitate understanding and performance by students of similar procedures such as protein electrophoresis (Chapter 30).

Assembling the Gel Apparatus

While wearing gloves, assemble two Protean II (or III) Electrophoresis Cells (Bio-Rad, Table 27.1) a day before the DAF products are to be separated electrophoretically. We recommend using 0.5-mm spacers that can be purchased separately for the Protean II apparatus, and 0.75-mm spacer for the Protean III. Gelbond flexible backing supports (sheets) can be purchased from BioWhittaker Molecular Applications (Table 27.1). Here are a few helpful hints in assembling the rigs. Meticulously clean the glass plates with running distilled water to remove any dust and **acrylamide** from previous experiments. Assemble the rig under running distilled water. Place the hydrophobic surface (the side that water beads on) of the backing film on and toward the large glass plate and rub the hydrophilic surface until all trapped air is evacuated. All gel rig components should be flush at the bottom. Lastly, do not overtighten the knobs (Protean II)—the glass plates will bow and produce a thickened center portion of the gel, which will not stain properly. The assembled apparatus should be examined carefully to ascertain that the glass plates, spacers, and backing film are flush with each other on the bottom. Run a fingernail across the bottom of the apparatus. If the bottom does not feel smooth or if your fingernail gets "hung up," the level of the glass plates, spacers or support film needs to be adjusted. The gel rigs should be allowed to dry overnight in a place that is dark and dust free.

Casting the Gels: Materials

Each team of students will need the following items:

- Acrylamide stock solution—*Caution:* this solution is toxic, wear gloves (Procedure 27.6)
- A 0.22-μm filter and 10-mL syringe
- TEMED—*Caution:* toxic, wear gloves
- 10% ammonium persulfate (100 mg/mL pure water) solution—ammonium persulfate may be made in bulk, 1.0 mL dispensed into each 1.5-mL centrifuge tubes, and stored frozen at −20°C
- Two assembled Protean II or III gel rigs, casting stand, and two 0.5-mm combs
- One 10-mL disposable pipette and pipette pump
- One 25- or 50-mL beaker and stir bar
- Aluminum foil
- Acetonitrile gloves

Follow the protocols listed in Procedures 27.6 and 27.7 to make running buffer and acrylamide stock solution, and to assemble and pour gels. Always wear acetonitrile gloves when working with acrylamide and TEMED.

Preparing Samples for Electrophoresis: Materials

Each team of students needs the following items:

- A microtiter plate
- Loading buffer (0.25% bromophenol blue, 0.25 xylene cyanol, and 15% type 400 Ficoll in water)
- Molecular marker (ladder) solution (1:10 or undiluted)
- P10 micropipette and tips
- Amplification products

Amplification products can be prepared for electrophoresis while the acrylamide is polymerizing. First, carefully pipette 3 μL of loading buffer into the number of wells in

	Procedure 27.7
	Casting Gels
Step	Instructions and Comments
1	Wear gloves! Mount the gel rigs onto the casting stand using the gray rubber gaskets on the bottom. We usually place several equal-length and width strips of parafilm wrap under the gasket to ensure a good seal with the glass plates. A very distinct snap should be heard as the rigs are set into place on the casting stand. Place the casting stand onto a large piece of aluminum foil on which a 10-mL syringe and 0.22-μm filter can be laid.
2	Pipette 10 mL of the 10% polyacrylamide stock (Procedure 27.6) into a 20-mL beaker containing a small magnetic stir bar. Place the pipette tips containing 15 μL of TEMED and 150 μL of 10% ammonium persulfate solutions into the stirring acrylamide solution and dispense. Dispose of the tips in a safe location. Stir for about 10 sec.
3	The following steps in casting the gel should be completed as quickly as possible (usually less than 2 min). Carefully draw the gel solution into a syringe, avoiding introduction of air into the barrel. If air bubbles are present, hold the syringe upside down at 70° away from the body; the air should rise to the top. Slowly depress the plunger until the air is expelled. Mount a nonsterile 0.22-μm filter on the open end of the syringe. Slowly express a small amount of acrylamide to wet the filter and release any trapped air.
4	Place the tip filter in the middle of the ledge formed by the small (short) plate and quickly dispense the acrylamide solution into the space between the glass plates. Rotate the casting stand 180° and fill the second gel rig with acrylamide. If there are bubbles trapped in the gel, gently tap the inner (short) glass plate and, with luck, they will rise to the top.
5	Position the 10- or 15-well combs about half way (level) in each of the rigs and examine for small bubbles residing on the bottom surface of the teeth. For Protean III rigs, completely insert the combs. If bubbles are present, remove and reposition the combs.
6	Allow the acrylamide to polymerize for at least 20 min. If desired, the gels may be cast the day before the laboratory exercise and stored overnight lying flat on the bottom of a plastic container that is lined with wet paper towels. Be careful not to disturb the combs, and store in the dark.

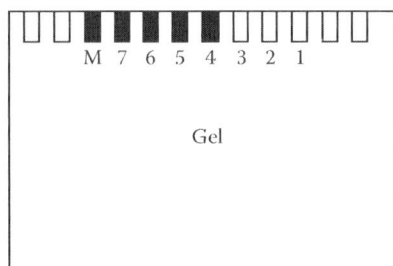

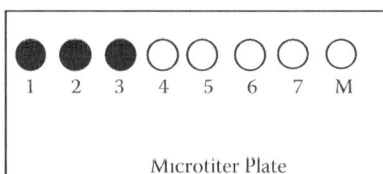

FIGURE 27.2 Loading DNA samples into the gel. Because the gel apparatus was rotated 180° when mounted on the central reservoir stand, the samples must be loaded in reverse order, or from right to left, in the gel. For example, load sample 7 in the fourth well from the left.

Procedure 27.8
Composition of Silver Stain and Developer Solutions

Step	Instructions and Comments
	Silver Stain
1	For 2 gels, dissolve 0.15 g of ACS-certified silver nitrate in 150 mL of pure water.
2	A few minutes before use, add either 750 μL of 16% or 325 μL of 37% formaldehyde.
	Developer
1	For 2 gels, dissolve 4.5 g of ACS-certified sodium carbonate (Na_2CO_3) in 150 mL of pure water and chill to 2–4°C.
2	Add 75 μL of sodium thiosulfate solution (0.2 g/50 mL)
3	Before use, warm the solution to 6–8°C and add either 600 μL of 16% or 260 μL of 37% formaldehyde (open formaldehyde in a fume hood).

a 6 × 10 microtiter plate (Pharmacia Biotech, Inc., Table 27.1). In the case provided in Figure 27.2, each row in the plate would have eight wells filled—seven for the samples and one for the molecular weight marker (M). However, we encourage teams of students to use a separate plate for their DNA. Map the order of the samples in the gel in the laboratory notebook. For instance (from left to right), sample 1.1, 1.2 … 1.7, and M. Now, in the order that the samples will be placed in the gel, carefully pipette 3 μL of each into their respective wells and mix by repipetting the solution several times. Change tips between samples. The last well is the molecular weight marker consisting of a 50- to 1000-base pair ladder (Biomarker Low, BioVentures, Inc., Table 27.2). The working solution is made by mixing 10 μL of biomarker with 90 μL of sterile, pure water. Biomarkers may also be purchased as ready to use—do not dilute. As with the other samples, 3 μL of biomarker are used per well. Be sure to replace the microtiter plate cover to prevent evaporation of the sample preparations.

PRE-RUNNING GELS AND PREPARING STAIN AND DEVELOPER SOLUTIONS: MATERIALS

Each team of students will need the following items:

- Five or ten × TBE running buffer
- 1-L graduated cylinder
- Protean II or III reservoir and central stand
- Tuberculin syringe with 25-gauge needle
- Power supply (two or three teams can share this item)

Make 1 L of 1× TBE buffer by mixing 100 mL of 10X TBE (Procedure 27.6) and 900 mL of pure water in a 1-L graduated cylinder and mix throughly. Alternatively, use 200 mL 5X TBE with 800 mL of pure water to make the 1X TBE. Dismount the two gel rigs from the casting stand and gently remove any polymerized acrylamide from the bottom of the plates with a laboratory tissue. Rotate the rigs 180° and snap into the central stand. Be careful not to touch or disturb the combs. When both rigs are mounted, the small plates of the rigs will be toward the interior, facing each other, and the outer plates will form the top buffer reservoir. Fill inner and outer reservoirs with a total of about 800 mL of 1X TBE. Carefully remove the combs from the gels by gently pulling straight up with equal pressure on both sides; do not damage the wells. Fill a 1-mL (cc) tuberculin syringe, equipped with a 1.5-in.-long, 25-gauge needle, with buffer from the central well. Gently insert the needle tip about one-quarter of the way into the top portion of a well and gently force the buffer into the well. This will flush the accumulated urea and errant bits of acrylamide from the wells. Repeat the process so that all wells of both gels have been cleaned. Connect the apparatus to the power supply and set to a constant 180 to 200 V for 15 to 20 min.

While the gel is prerunning, there will be time to prepare both the silver stain and carbonate developer solutions. Both solutions may be prepared in bulk, including every constituent except formaldehyde and sodium thiosulfate. Silver nitrate solution is light sensitive and should be stored in a brown bottle. The sodium thiosulfate solution should be prepared weekly and stored in the refrigerator at 4°C . If preparing developing and staining solutions for daily use, then plan on 75 mL for each gel.

Follow the protocols in Procedure 27.8 for preparing silver stain and developer solutions.

LOADING SAMPLES AND RUNNING THE GEL

After prerunning the gels, clean the wells in one gel as described previously. With a P10 (or equivalent) pipette adjusted to deliver 6.5 μL, load the samples into the wells using flat tips (Midwest Scientific, Table 27.1). Because the gels were rotated 180°, load the gels in reverse order, or from right to left, as indicated in Figure 27.2. Draw the far right sample in the microtiter plate into a flat pipette tip. Keeping the flat tip parallel to the glass plate, guide

	Procedure 27.9
	Synopsis of Fixing, Staining, and Developing Gels
Step	Instructions and Comments
1	Fix gels in 7.5% acetic acid for 10 min on a rotary shaker (60 rpm).
2	Rinse gels with pure water 3 times each for 2 min on a rotary shaker (60 rpm).
3	Soak gels in silver stain for 20 to 30 min on a rotary shaker (40 rpm) in a fume hood.
4	Rinse gels in pure water for 5 to 10 s.
5	Soak gels in developer for 5 to 8 min (or until bands are dark) on a rotary shaker (40 to 60 rpm) in a fume hood.
6	Fix gels in cold (4°C) 7.5% acetic acid for 5 min on a rotary shaker (60 rpm).
7	Soak gels in pure water 2 times each for 5 min on a rotary shaker (60 rpm).
8	Soak gels in anticracking solution for 5 min on a rotary shaker (60 rpm). This solution may be used over again.
9	Hang gels overnight to dry.

it partially into the third well from the left side with the left index finger and gently dispense the sample into the well. Be careful not to damage the well. Load the next sample with a new flat tip. Load all the samples for this gel and repeat the procedure, including cleaning the wells, for the other gel. *Note:* maintain the same sample loading order in each of the gels. Keeping the same order between gels will greatly facilitate data collection. Reconnect the power supply and run at a constant 180 to 200 V for about 1 to 1.5 h or until the blue tracking dye reaches the level of the bottom electrode.

STAINING AND DEVELOPING GELS

Turn off the power and disconnect the gel apparatus from the power supply. Wearing gloves, disassemble the gels under distilled water by first loosening the four knobs and gently removing the glass plate sandwich from the apparatus. The rigs simply release from the Protean III holder. Holding the "sandwich" with the large glass plate contacting the palm of the left hand and in a stream of or a pan of distilled water, insert the fingernail of the right index finger under the top corner of the small glass plate and gently pry it upward. Let the water do most of the work. The backing film and the gel may now be sepa-rated from the large plate and placed in a staining tray (Sierra-Lablogix, Inc., Table 27.1) or in lids from pipette boxes. Follow the staining and developing procedures outlined in Procedure 27.9. Remember to add formalde-hyde to silver stain and developer solutions (Procedure 27.8) just prior to use. After silver staining is completed, quickly and throughly rinse the gels with pure water to remove all excess silver nitrate solution. Do not pour the cold developing solution directly on the gels; instead, introduce the solution onto the bottom of the staining dish and immediately place on a rotary shaker set at about 40 to 60 rpm. Continue shaking until the bands in the marker and sample lanes are dark and sharp or until the margins and the background of the gel start to dis-color (overdeveloped). Stop with cold 7.5% acetic acid. The gels may be "hung to dry" in a dust-free environ-ment after they are treated with anticracking solution (under a fume hood, add 100 mL of glacial acetic acid, 10 mL of glycerol, and 370 mL of 95% ethanol to 520 mL of pure water).

Label the gels with the experiment number and gel identification number (e.g., 1-A) with a permanent marker after the gels are dry in 12 to 24 h. Gels may be stored indefinitely in photo albums. Table 27.2 describes some

TABLE 27.2
Troubleshooting Gels: Some Common Imperfections and Their Causes

Bands in some lanes but not in others: DNA template missing or degraded in lanes with weak or no products.

No amplification—all lanes blank: missing ingredient in master mix, degraded primer, or, less likely, all DNA templates degraded.

Dark streaks in lanes: old loading buffer or dust particle on the bottom surface of the well.

Lightly staining products in center of gel: developer poured directly on gel; glass plates warped, creating thickened gel in center; developer less than 6–8°C.

Individual bands not straight but jagged: bottom surface of well damaged.

Lanes not straight but deflected: air bubble under support film.

Bubbles in gel: aspirated air from syringe or air bubble adhering to a dirty glass plate.

Light bands or no bands in lanes: not all primers work well with all organisms; try other primers.

Black smudges in gel: incomplete removal of silver stain solution before adding developer.

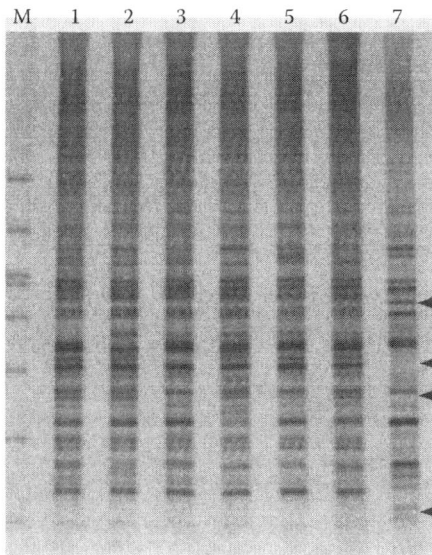

FIGURE 27.3 DNA profiles of *Fusarium oxysporum* isolates (F1–F7: lanes 1–7) using primer 8.6A. Note that some of the many polymorphisms (arrowheads). M = molecular weight markers (from top: 1000, 700, 525, 500, 400, 300, 200, and 100 bp; 50 bp marker not shown).

of the more common gel imperfections, their causes, and remedies.

Questions

- What purpose does the loading buffer serve?
- How does the percentage of acrylamide affect the migration of amplified products?
- Do all amplified products appearing at the same level in the gel (loci) have the same sequence?
- Does a single band in the gel represent a single amplification product?

EXERCISE 4: DATA COLLECTION AND ANALYSES

DAF data will be analyzed using the Numerical Taxonomy and Multivariate Analysis System (NTSYSpc) program, version 2.2 (Exeter Software, Table 27.1). The analyses are easily understood and provide estimates of genetic distances and relationships between isolates.

Data Collection: Materials

- White light box (transilluminator)
- Computer with NTSYSpc version 2.2
- Clear plastic ruler
- Clear, 12 in. × 12 in. glass plate

View the dried gels on a light box and cover with a clear glass plate. Beginning at about 700 bp, align common bands in the different sample lanes with a straight edge and enter the binary data: 1 = product present, 0 = product

TABLE 27.3
An Example of Matrix Definitions and Binary Data for *Fusarium oxysporum* Isolates to Be Analyzed Using the NTSYSpc Program

1	111	7L	9			
F1	F2	F3	F4	F5	F6	F7
0	1	0	0	0	1	1
1	1	1	1	1	1	1
1	1	1	1	1	9	1
0	0	1	0	0	0	0
.	.	.	.	.	.	.
1	1	1	1	1	1	1

This data set (rectangular = 1) includes 111 character loci from 7 isolates (initials) and contains missing data represented by a 9. Rows with all 1s are monomorphic; a row containing a single 1 identifies a unique marker for an isolate.

absent, and 9 = missing data. For example, in Figure 27.3 (primer 8.6A), at about 675 bp, a prominent band appears in all sample lanes. The data for this character locus would be the following: 1 1 1 1 1 1 1. Continue to record data for the entire gel. Combine data from the primers and enter in the computer as shown in Table 27.3. The order in which the data from individual primers are entered into the data set is not important. Be careful to include a hard return after each line of the data set and eliminate any extraneous spaces within the lines. The first line of the data set should begin with the number "1," followed by the number of lines (character loci) in the data set, e.g., "154," then the number of samples (7L), and "1," then "9" if there were any missing values. If there were no missing values, enter "0" instead of "1." The next line contains the abbreviations for the samples—F1, F2, etc. Save the data set as an ASCII file. You may also use an Excel spreadsheet for data entry, which can be imported by NTSyspc.

Data Analyses

After your data is in the correct format, you are now ready to open NTSyspc. At any time during your session, you can click on the "notebook" icon at the top of the screen to view a log of your analyses. This notebook can be saved as text, printed, etc. It could be a useful record of where to find output files for later reference, or to find out why the program will not run on your data.

Follow the instructions in Procedure 27.10 to analyze data.

Questions

- Are the isolates closely related to each other?

| **Procedure 27.10** |
| Cluster and Principal Component Analyses DAF Data Using NTSYSpc |

Step	Instructions and Comments
	Cluster Analysis
1	Click the *Similarity* tab. Select the SimQual button. Double click the entry window to choose the file from the hard drive or diskette, or type the path and name of the data file. At the coefficient line, select *J for Jaccard*. Enter a path and filename for the output. For example, if the data file is "Fusarium," call the output "Fusarsim." Then click on *Compute*. See Table 27.4 for Similarity Matrix.
2	Click the *Clustering* tab. Select the SAHN button. Double click the entry window to choose the file from the hard drive or diskette: "Fusarsim" above, or type the path and filename of the similarity output. Type the path and filename for the output, for example, "Fusarclus." Be sure that UPGMA is selected in the clustering method. Click *Compute*.
3	To view the phenogram (Figure 27.4A), click the icon resembling a phenogram in the lower left of the window. To save the phenogram as an *.emf file, click *File/Save* and name the phenogram. To print, click *File/Print/ OK*. To copy the phenogram into a word processing document, click *Edit/Save bitmap*. Minimize NTSYSpc. Open the word processing program. Right click in the body of the document to insert the phenogram. Click Paste. Save this word processing file.
	Principal Component Analysis
1	Click the *General* tab. Select the *Dcenter* button. Double click the entry window to choose the file from the hard drive or diskette: "Fusarsim" above, or enter the path and filename of the similarity output. Enter the path and filename for the output, for example, "FusarDC." Leave the "*Square Distances*" box checked. Click *Compute*.
2	Click the *Ordination* tab. Select the *Eigen* button. Double click the entry window to choose the file from the hard drive or diskette: "FusarDC" above, or enter the path and filename of the Dcenter output. Enter the path and filename for the output, for example, "FusarEign." Click *Compute*.
3	To view the 3D PCA graph (Figure 27.4B), click the icon resembling a PCA graph in the lower left of the window. Be sure to Click *Options/Plot Options* and click the button by the word *Label* so that the graph is labeled. The view of the 3-D graph can be adjusted by turning it or leveling it. To save the PCA graph as an *.emf file, click *File/Save* and name your graph. To print, click *File/Print/OK*. To copy your PCA graph into a word processing document, click *Edit/Save* bitmap. Minimize NTSYSpc. Open the wordprocessing program. Right click in the body of the document, where you want to insert the PCA graph. Click *Paste*. Save this word processing file.

TABLE 27.4

An Example of a Similarity Matrix Generated Using Jaccard Coefficient for Seven Isolates of *Fusarium oxysporum*

Isolates	F1	F2	F3	F4	F5	F6	F7
F1	1.000						
F2	0.780	1.000					
F3	0.863	0.883	1.000				
F4	0.823	0.822	0.865	1.000			
F5	0.831	0.830	0.895	0.897	1.000		
F6	0.779	0.796	0.817	0.800	0.826	1.000	
F7	0.627	0.624	0.644	0.615	0.618	0.693	1.000

- How many character loci are needed to get an accurate representation of the relationships among isolates?
- Does the choice of coefficient in the similarity measure influence the calculated relationship between isolates?

EXPERIMENT 2. AMPLIFIED FRAGMENT LENGTH POLYMORPHISM (AFLP)

Amplified fragment length polymorphism (AFLP) is a technique that potentially allows for the recognition of genetic relationships between isolates. Even when little information is known about a genome, such as its genetic complexity or specific DNA sequences, the AFLP technique can be applied to determine genetic variability (Vos et al., 1995). AFLP typically produces a large number of loci that can

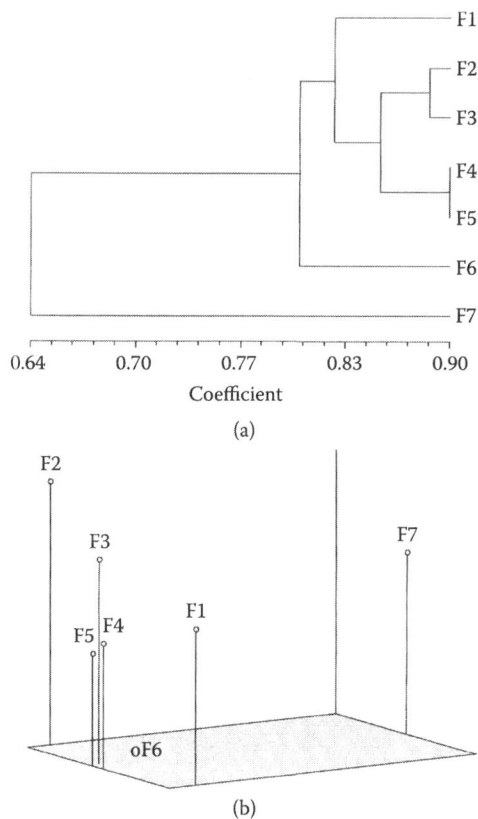

(a)

(b)

FIGURES 27.4 Analysis of *Fusarium* DAF data. A. Diagrammatic representation of UPMGA or cluster analysis of the relationships between seven isolates of *F. oxysporum*. B. Diagrammatic representation of principal component analysis of the relationships between seven isolates of *F. oxysporum*.

be assayed quickly and applied to most genetic variability studies because markers can be detected at almost any taxonomic level (Vos and Kuiper, 1997).

Although AFLP can be used in various applications, as with any widely applied technique, it has limitations. If the organisms being studied share less than 90% homology, few bands may be found that are common to all samples, indicating that even when similar-sized bands are detected, they may be attributed to chance (random). For example, AFLP may not be an appropriate technique to use to study organisms such as bacteria, where genomic changes occur at a high frequency. Homology can decrease the usefulness of AFLP markers in bacterial studies because changes can occur very quickly in the genome. In contrast, if the genome being studied has little sequence variation, AFLP may also produce few polymorphic markers that are useful for genetic variation research (Vos and Kuiper, 1997).

AFLP includes the strategies of **restriction fragment length polymorphism (RFLP)** and polymerase chain reaction (PCR) in concert to detect selectively amplified fragments (Miyashita, 1999) (see Chapter 26). This process includes the following three general steps: (1) digestion of the genomic DNA with restriction **endonucleases** and modification of the fragmented DNA by attachment of synthetic **oligonucleotides** sequences, (2) selective amplification by PCR, and (3) analysis of fragments by gel electrophoresis (Figure 27.5). These steps are described in more detail throughout this section.

Throughout this exercise, we suggest that students work in groups of four.

Exercise 5: DNA Isolation for AFLPs

As stated in the DNA amplification fingerprinting section, there are many different types of fungal DNA extraction methods that can be used to successfully extract high-quality DNA from mycelium. This section describes a method that works particularly well in our lab and produces sufficient quantities of DNA to perform several different types of reactions. To obtain high-quality DNA in high concentration, it is important to ensure that the fungus is cultured in media that does not allow for excessive carbohydrate accumulation. Excessive carbohydrates in the sample can block the membrane in the purification column and therefore hinder DNA elution extraction. It is also very important to keep the sample frozen until it is ground to a powder and the lysis buffer is thoroughly mixed.

Materials

The following materials will be required by each team of students:

- Fungal samples stored at –70°C
- Qiagen DNeasy® Plant Mini Kit (Qiagen, Valencia, CA) (Catalog # 69106)
- Sterile mortar and pestle for each sample
- Liquid nitrogen
- Sterile disposable pipette tips (wide-bore 1-mL, 2–20-μL, and 1–10-μL)
- Pipettes
- Water bath or heating block set at 65°C
- Vortex
- Sterile 1.5 Eppendorf microcentrifuge tubes
- Microcentrifuge with rotor for 2 mL tubes
- Ice bucket with ice
- Ethanol (96–100%)
- Gloves (latex or nitrile)
- 1X TAE buffer: 20 mL of 50× TAE in 980 mL of ddH$_2$O
  - 50X: 242 g Tris base
    57.1 mL glacial acetic acid
    100 mL of 0.5 M NaEDTA, pH 8.0
    fill to 1 L with ddH$_2$O
- Agarose

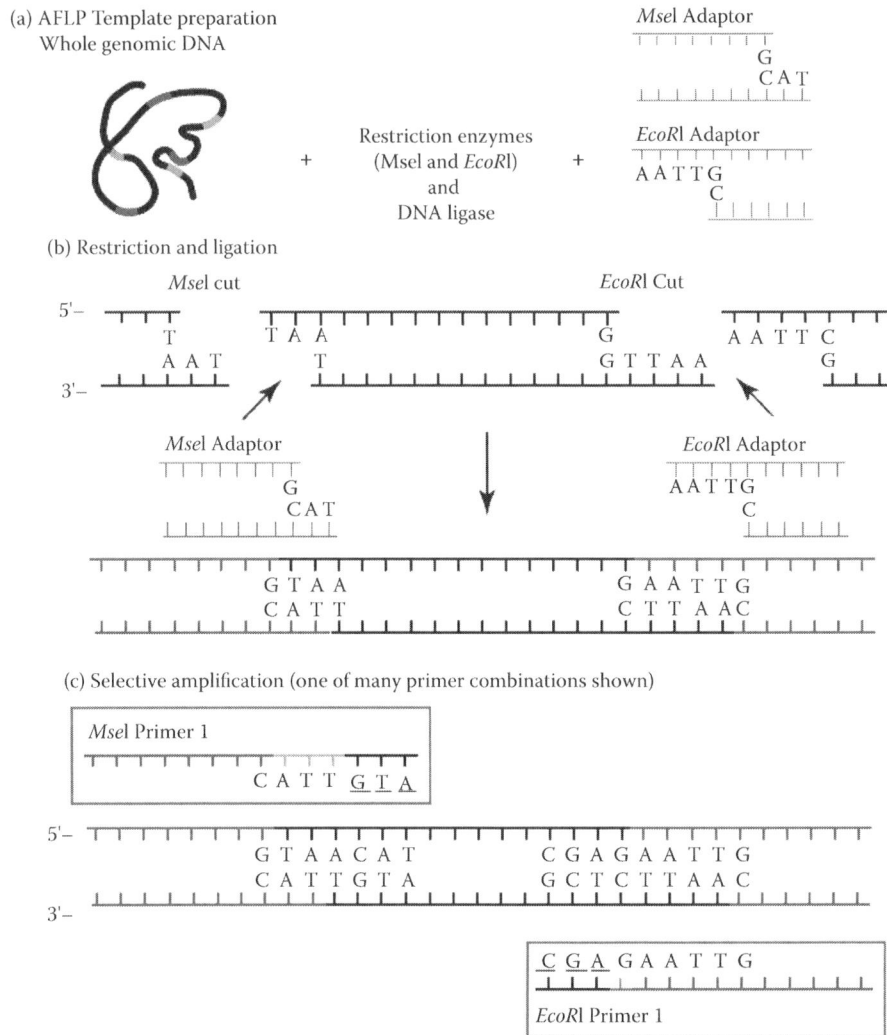

FIGURE 27.5　AFLP begins as the genomic DNA is digested by restriction enzymes. Adaptors can then be ligated to the overhangs created by the restriction enzymes. PCR can then be performed to select for fragments that have adaptors and have nucleotides next to the restriction site that match the selective primer. (Reprinted from Mueller, U.G. and L. LaReesa Wolfegnbarger. 1999. AFLP genotyping and fingerprinting, *Trends Ecol. Evol.*, 14: 389–394. With permission from Elsevier.)

- Ethidium bromide (10 mg/mL)
- Low Mass Molecular Ladder (Invitrogen, Carlsbad, CA)
- UV light box
- **Spectrophotometer** or **fluorometer**
  - If using fluorometer, Hoest dye is necessary. Newer instruments do not require dyes; DNA concentration is determined directly from the sample.
- Electrophoresis apparatus

Follow the instructions in Procedure 27.11 to complete this section.

Determining DNA Concentration

Protective gloves should be worn at all times.

The concentration of DNA in a sample can be determined by agarose gel electrophoresis. First load 5 μL of each sample of genomic DNA on a 1% agarose gel made with 1× TAE buffer (Tris base, acetic acid, and EDTA) amended with 10 mg/mL ethidium bromide (usually approx. 1.2 μL works for a 40 mL 1% agarose gel), and run at 100 V for 60 min. Visualize the samples with UV light and quantify by comparing the relative intensity of the DNA in the gel to the intensity of a Low Mass DNA Ladder (Invitrogen, Carlsbad, CA). High-quality DNA should appear as a crisp band with very little "smearing" or "cloudiness" below the band in the gel.

Once the mass (ng) of the DNA in the gel is known, the concentration (ng/μL) can be determined by dividing the mass by the amount of DNA loaded on the gel (5 μL). For instance, if you determine that a band in the gel has the same relative intensity as the 40 ng band in the stan-

Procedure 27.11	
AFLP—Isolation of DNA from Mycelium	
Step	Instructions and Comments
1	Pour liquid nitrogen into each mortar to chill before adding frozen sample. Immediately pour liquid nitrogen onto sample in mortar and grind to a fine powder with pestle. This may need to be done a couple of times to powder the entire sample. Add 500 µL of AP1 lysis buffer and 4 µL of RNase provided with the Qiagen DNeasy® Plant Mini Kit to each sample. Mix using the pestle to ensure all DNases are inhibited as the sample thaws. Transfer sample to a sterile 1.5-mL Eppendorf tube using a pipette with a wide-bore tip and vortex. Place tube in 65°C water bath or heat block for 10 min.
2	After incubation, continue following the instructions that are supplied with the DNA extraction kit. Elute the DNA with 10 mM Tris, pH 8.0, in the final step and store at –20°C.

dard lane, you can assume that the band in the gel also contains 40 ng. The calculation to determine ng DNA/µL in the sample would be as follows:

$$40 \text{ ng} \div 5 \text{ µL (amount of sample loaded in the gel)}$$
$$= 8 \text{ ng/µL}$$

Other methods for determining DNA concentration can also be used, such as using a fluorometer or spectrophotometer. These methods are described with the DAF procedure.

For the restriction/ligation reaction, the genomic DNA needs to be at 100 ng/µL or 500 ng in 5 µL. If you were to use the sample in the preceding example, the DNA would be too dilute for the restriction/ligation reaction. Concentrate the sample by speed vacuuming or ethanol precipitating. The calculation to determine how much sample to speed vacuum is as follows:

$$C_1 \times V_1 = C_2 \times V_2$$

$$(8 \text{ ng/µL}) (\times) = (100 \text{ ng/µL}) (5 \text{ µL})$$

$$= 62.5 \text{ µL of DNA}$$

Once the sample is nearly dry, 5 µL of sterile distilled water can be used to dissolve (resuspend) the sample. The concentration in the sample should now be at 100 ng/µL or 500 ng/5 µL.

EXERCISE 6: RESTRICTION/LIGATION REACTION

DNA is fragmented (restricted) using one restriction enzyme as a rare cutter and another as a frequent cutter during the AFLP technique. For example, *Eco* RI recognizes a restriction site sequence that is 6 bp long, whereas *Mse* I recognizes a restriction site sequence that is 4 bp long. Statistically speaking, because *Mse* I searches for a shorter DNA sequence, it "cuts" the DNA more frequently than *Eco* RI. *Eco* RI and *Mse* I are conventionally used for

most studies because other enzyme combinations do not produce high-quality fingerprints. Due to the nature of the two endonucleases (*Eco* RI and *Mse* I), a sticky end is left after the palindromic restriction site sequence is identified. A sticky end refers to the single strand overhang produced by the enzyme once it cuts at the restriction site. In this case, *Eco* RI cuts less frequently than *Mse* I, and therefore it is expected that more fragments will be produced with both ends being cut by *Mse* I. The only fragments that are amplified during this process are those that were cut by the rare cutter/frequent cutter endonucleases (Vos et al., 1995). Simultaneous with digestion, restriction fragments are modified by ligation of synthetic DNA adapters to the ends with DNA ligase. This exercise suggests using *Eco* RI and *Mse* I restriction enzymes to digest the DNA, but other frequent cutter and rare cutter combinations can be used. If other enzyme combinations are more applicable, keep in mind that the adaptor sequences may need to be modified to ensure they are complementary to the restriction enzyme site.

Ligation of the adaptors can be accomplished by taking advantage of the sticky ends produced by the two restriction endonucleases. The base pairs of the complementary sequences of the adapters and the restriction site sequence overhang can anneal, and the nick in the phosphodiester backbone can be sealed by the enzymatic activity of T4 DNA ligase and adenosine triphosphate (ATP). The ligated adaptors serve as annealing sites for primers in the subsequent amplification reactions. Taq polymerase then can complete the amplification of the fragment by the addition of deoxynucleotide triphosphates (dNTPs) and primers complementary to the adaptors in the preselective reaction (Vos et al., 1995).

Materials

The following materials will be required by each team of students:

- 0.5 mL thin-walled PCR tubes, one per sample
- Sterile pipette tips

	Procedure 27.12
	Annealing of Adaptors to DNA
Step	Instructions and Comments
1	Upon receiving lyophilized adaptors, reconstitute each in 10 mM Tris, pH 8.0, to reach a concentration of 200 μM. Vortex and allow the adaptors to sit out on the desktop for a few minutes to ensure they are totally resuspended. Vortex again before using. Store at –20°C.
2	In a thin-walled Eppendorf PCR tube, mix 5 μL of each adaptor pair (Mse I adaptors 1 and 2 together; Eco RI adaptors 1 and 2 together) and heat at 95°C for 5 min in a thermalcycler. Adaptor pairs become annealed during the heating stage. The tubes should be left in thermalcycler overnight to allow for slow cooling to room temperature to further ensure complete annealing. Prepare the proper dilutions for the restriction/ligation reaction. Store at –20°C until use.

- Pipettes
- 1.5-mL Eppendorf tubes
- Genomic DNA from each sample (100 ng/μL)
- Ice bucket with ice
- 6-place personal microcentrifuge for 1.5- to 2.0-mL tubes (with adaptors for 0.5-mL microcentrifuge tubes)
- *Eco* RI restriction enzyme (10 U/μL) (Invitrogen, Carlsbad, CA)
- *Mse* I restrction enzyme (5 U/ μL) (Invitrogen, Carlsbad, CA)
- T4 ligase kit (New England Biolabs, Beverly, MA):
  - T4 ligase (3 U/ μL)
  - Ligase buffer (10×)
- NaCl (0.5 M)
- Bovine serum albumin (BSA) (1 mg/mL)
- *Mse* I adaptors (50 μM) (Integrated DNA Technologies, Coralville, IA)
  - Adapter 1- (5′ GAC GAT GAG TCC TGA G 3′)
  - Adapter 2- (5′ TAC TCA GGA CTC AT 3′)
- *Eco* RI adaptors (5 μM) (Integrated DNA Technologies, Coralville, IA)
  - Adapter 1- (5′ CTC GTA GAC TGC GTA CC 3′)
  - Adapter 2- (5′ AAT TGG TAC GCA GTC TAC 3′)
- Tris pH 8 (10 mM)
- Thermalcycler

Keep all reagents and samples on ice before preparing the reaction, especially the enzymes. Promptly return enzymes to the freezer after use. All oligonucleotides (primers or adaptors) should be resuspended in 10X Tris buffer in this reaction and all subsequent reactions.

Follow the instructions in Procedures 27.12 and 27.13 to complete this section.

EXERCISE 7: PRESELECTIVE REACTION

An unmanageable amount of fragments are usually produced from the restriction/ligation step; therefore, it would be difficult to begin data analysis at this step. The preselective reaction helps decrease the number of fragments through amplification of only fragments with adaptors attached. When analyzing large genomes, such as plants, primers in the preselective reaction select for one base within the sequence next to the restriction site. For instance, the primer (5′ CTC GTA GAC TGC GTA CCA ATT C**A** 3′) is annealed to the *Eco* RI side and is complementary to the adapter and the *Eco* RI restriction site. In addition to the complementary sequence, it also has a one base extension, adenine, on the 3′ end. The primer (5′ GAC GAT GAG TCC TGA GTA A**C** 3′) for the *Mse* I side of the fragment is complementary to the adapter and the *Mse* I restriction site and also has a one base extension, but uses cytosine (Vos et al., 1995). When studying smaller genomes such as fungi, primers are used during the preselective reaction that do not contain a one base extension but are only complementary to the adaptor. The one base extension may be too discriminatory, and enough useful fragments may not be produced for analysis (Vos and Kuiper, 1997). For the purposes of this exercise, primers will be named Eco+0 and Mse+0; "0" indicates no base extensions other than the sequence that was complementary to the adaptor.

Materials

The following materials will be required for each group of students:

- 0.5-mL thin-walled PCR tubes, one per sample
- Sterile pipette tips
- Pipettes
- 1.5-mL Eppendorf tubes
- Eco+0 Primer (5′ GAC TGC GTA CCA ATT C 3′) (2.75 mM) (Integrated DNA Technologies, Coralville, IA)

	Procedure 27.13
	Restriction/Ligation Reaction

Step	Instructions and Comments
1	To prepare the master mix for 8 samples, for example, combine the reagents below in a 1.5-mL Eppendorf microcentrifuge tube. Although there are only 8 samples, prepare extra master mix to make up for pipetting errors. It is very important that the enzymes are kept on ice in order to maintain viability. Combine on ice the master mix specifically in this order: **9 reactions (total of 52.02 μL):** 9.9 μL of ligase buffer (10X), 9.9 μL of 0.5 M NaCl, 4.5 μL of BSA (1 mg/mL), 9 μL Mse I adaptors (50 μM), 9 μL Eco RI adaptors (5 μM), 4.5 μL Eco RI (10 U/μL), 2.25 μL Mse I (5 U/μL), 2.97 μL T4 ligase (3 U/μL)
2	While holding the top of the tube with your forefinger and thumb, mix the contents in the tube by gently flicking the bottom. Centrifuge the tube in a microcentrifuge for only a couple of seconds—just long enough to make sure all the liquid is in the bottom of the tube.
3	Add 5.78 μL of the master mix to the bottom of each 0.5-mL thin-walled microcentrifuge tube on ice followed by 5 μL of 100 ng/μL genomic DNA. Be sure to change pipette tips between DNA samples.
4	Mix and centrifuge the DNA/master mix mixture as in step 2.
5	Incubate on the benchtop over night at room temperature (~20°C). In the morning, dilute the entire reaction with 189 μL of 10 mM Tris, pH 8. This dilution deactivated enzymes that could interfere with subsequent reactions. Samples can be stored at –20°C until the preselective reactions can be performed.

- Mse+0 (5′ GAT GAG TCC TGA GTA A 3′) (2.75 mM) (Integrated DNA Technologies, Coralville, IA)
- Diluted restriction/ligation reaction
- Sterile ddH$_2$O
- Eppendorf Master Taq DNA polymerase kit:
  - 10× Taq buffer with 25 mM MgCl$_2$
  - Eppendorf Master Taq DNA polymerase (5 U/μL)
- dNTPs (Eppendorf, Westbury, NY) (2 μM)
- Ice bucket with ice
- Thermalcycler
- Agarose
- Ethidium bromide (10 mg/mL)
- Biomarker Low DNA Ladder (BioVentures, Murfreesboro, TN)
- UV light box
- Electrophoresis apparatus
- Tris pH 8.0 (10 mM)

Follow the instructions in Procedure 27.14 to complete this section.

EXERCISE 8: SELECTIVE REACTION

The selective reaction selects for fragments that contain specified bases next to the restriction site sequence. For example, only fragments that contain 5′ACC 3′ on the *Eco RI* side of the fragment, and 5′CCT 3′ on the *Mse* I side of the fragment will be amplified. This step further reduces the number of fragments produced by PCR and visualized during capillary gel electrophoresis. Just as the preselective reaction, selection of three bases next to the restriction site causes too large a reduction in fragments for analysis when studying smaller genomes, and substitution of two bases next to the restriction site may be necessary.

The preselective and the selective steps are performed separately to increase the selectivity of fragments by the primers. In studies by Vos et al. (1995), the amount of background smears were decreased in the fingerprint patterns, and bands were absent when the preselective step was not performed. These bands were otherwise present when both the preselective and the AFLP amplification were performed. Primers with a three-base extension caused a decreased amount of base mismatching by the sequence adjacent to the restriction site during PCR in comparison to using four base extension primers (Vos et al., 1995). Vos and Kuiper (1997) suggest using two base extensions on both primers for selective reactions in microorganisms, bacteria, or fungal DNA studies. These primer combinations are described in Table 27.5. Primers in this section are exactly in the same sequence as in the preselective reactions, except for the addition of base

Procedure 27.14	
Preselective AFLP Reaction	
Step	Instructions and Comments
1	Preparation of the preselective master mix should always occur on ice. Mix the following reagents in a 1.5-mL Eppendorf microcentrifuge tube: **9 reactions (135 µL):** 61.2 µL sterile ddH$_2$O, 18 µL 10X Taq buffer, 18 µL dNTPs (2 µM), 18 µL Eco+0 primer (2.75 µM), 18 µL Mse+0 (2.75 µM), 1.8 µL of Eppendorf Master Taq DNA polymerase (5 U/µL). Mix the master mix slightly as in the restriction/ligation reaction section. Aliquot 15 µL of the master mix to each 0.5-mL, thin-walled microcentrifuge tube. Pipette 5 µL of diluted restriction/ligation reaction to its corresponding tube on ice. Mix slightly as in the restriction/ligation reaction section.
2	The following thermalcycler program should be used: initial incubation at 72°C for 2 min followed by 20 cycles of 94°C for 20 s, 56°C for 30 s, 72°C for 2 min, final extension at 72°C for 2 min, final incubation at 60°C for 30 min, and 4°C hold.
3	After amplification, separate 5 µL of each PCR product as well as 3 µL of Biomarker Low DNA Ladder by electrophoresis at 100 V for 1 h in a 1%-agarose gel containing ethidium bromide dye. If the restriction/ligation and preselective reactions were successful, smears should be present in the lanes of the gel.
4	Dilute each reaction with 135 µL of 10 mM Tris pH 8.0. This diluted reaction can serve as template for several subsequent selective reactions. Store at –20°C until the selective reaction can be completed.

extensions. Troubleshooting reactions may need to be performed that test different combinations of selective bases to find primer pairs that produce polymorphisms.

Materials

The following materials will be required for each group of students:

- 0.5-mL thin-walled PCR tubes, one per sample
- Sterile pipette tips
- Pipettes
- 1.5-mL microcentrifuge tubes
- Eco+NN[a] Primer (2.75 mM) (Integrated DNA Technologies, Coralville, IA)
- Mse+NN[b] Primer (2.75 mM) (Integrated DNA Technologies, Coralville, IA)
- Diluted preselective reaction
- Sterile ddH$_2$O
- Eppendorf Master Taq DNA polymerase kit (or equivalent):

[a] "N" indicates any nucleotide.
[b] This primer can also be designed to select for Mse+N if Mse+NN primers do not produce polymorphisms.

- 10X Taq buffer with 25 mM MgCl$_2$

- Eppendorf Master Taq DNA polymerase (5 U/µL)
- dNTPs (2 mM) (Eppendorf, Westbury, NY)
- Ice bucket with ice
- Thermalcycler

Follow the instructions in Procedure 27.15 to complete this section.

EXERCISE 9: FLUORESCENT LABELING REACTION

Using fluorescently labeled primers for the various selective reactions can become very costly. In the past, a separate labeled Eco side primer would be needed for each possible selective primer combination. To remedy this issue, an AFLP protocol was developed by Habera et al. (2004) that included a separate and simple PCR reaction to label fragments. This reaction uses a single Eco+0 primer that can be used for all fluorescent labeling reactions (Habera et al., 2004).

During this step the selective reaction products are labeled with WellRED (Sigma-Aldrich, St. Louis, MO) phosphoramidite dye using a labeled Eco+00 primer composed of the same sequence as preselective reaction, and Mse+NN primer from the selective reaction (Habera et al., 2004). Due to the light sensitivity of the fluorescent dyes, the reactants should be maintained on ice and in diminished light conditions (turn off overhead lights in laboratory).

TABLE 27.5

Proposed Primer Combinations for Various Plants, Bacteria, and Fungi

Mse I	Eco RIa	0+	+1 AA, AC, AG, AT, TA, TC, TG, TT	2+ AA, AC, AG, AT, TA, TC, TG, TT	3+ AAC, AAG, ACA, ACC, ACG, ACT, AGC, AGG
0+		Cosmids	PACs		
		PACs	BACs		
		BACs	YACs		
+1	A	PACs	YACs	Yeast	
	C	BACs	Bacteria	Bacteria	
	G	YACs			
	T				
2+	AA		Yeast	Fungi	
	AC		Bacteria	Bacteria	
	AT			Microorganisms	
	AG				
	CA				
	CC				
	CG				
	CT				
3+	CAA			Arabidopsis	Oilseed rape
	CAC			Rice	Sunflower
	CAG			Cucumber	Brassica
	CAT			Peach	Potato
	CTA				Tomato
	CTC				Lettuce
	CTG				Pepper
	CTT				Maize
					Barley

Source: Reprinted from Vos, P. and M. Kuiper. 1997. AFLP analysis. In: G. Caetano-Anollés and P. M. Greshoff (Eds.). *DNA Markers, Protocols, Applications, and Overviews.* Wiley-Liss, New York. With permission from John Wiley & Sons.

Procedure 27.15

Selective AFLP Reaction

Step	Instructions and Comments
1	As in the previous reactions, the master mix should be prepared on ice. Selective reaction master mix contains the following reagents: **9 reactions (135 µL):** 72 µL of sterile ddH$_2$O, 18 µL 10X Taq Buffer, 7.2 µL dNTPs (2 mM), 18 µL Eco+(AA or TG) primer (2.75 µM), 18 µL Mse+C primer (2.75 µM), 1.8 µL of Taq polymerase (5 U/µL), After mixing, aliquot 15 µL of the master mix to each 0.5-mL, thin-walled microcentrifuge tube. Pipette 5 µL of diluted preselective reaction into its respective tube. Mix as in previous reactions.
2	The following touchdown thermalcycler program should be used: initial DNA denaturation temperature of 94°C for 2 min, then 94°C for 20 sec, 66°C for 30 sec, 72°C for 2 min, followed by 9 cycles of 94°C for 20 sec, decrease by 1°C/cycle from 66°C, 72°C, 20 cycles of 94°C for 20 sec, 56°C for 30 sec, 72°C for 2 min, final incubation at 60°C for 30 min, and 4°C hold.
3	After PCR, dilute the preselective reaction 1:500 with 10 mM Tris, pH 8.0. Store plates at −20°C until fluorescent labeling reactions.

	Procedure 27.16
	Fluorescent Labeling Reaction
Step	Instructions and Comments
1	This reaction should also be prepared on ice as well as in diminished light. Combine the following reagents in a 1.2-mL microcentrifuge tube: **9 reactions (135 µL):** 43.2 µL of sterile ddH$_2$O, 18 µL 10X Taq buffer, 18 µL dNTPs (2 µM), 36 µL Eco+0 labeled primer (1.38 µM), 18 µL Mse+C primer (2.75 µM), 1.8 µL Taq polymerase (5 U/µL). After slight mixing, aliquot 15 µL of the master mix to each 0.5-mL, thin-walled microcentrifuge tube after slight mixing. As in other reactions, pipette 5 µL of diluted preselective reaction. Mix and centrifuge as in previous reactions.
2	The same touchdown thermalcycler program can be used as in the selective reaction. Wrap the tubes in foil and store at –20°C until capillary gel electrophoresis.

Materials

Each group or team of students will require the following materials:

- 0.5-mL thin-walled PCR tubes, one per sample
- Sterile pipette tips
- Pipettes
- 1.5-mL microcentrifuge tubes
- Eco+0 primer labeled with phosphoramidite dye (1.38 µM) (Sigma-Aldrich, St. Louis, MO)
- Mse+NN[a,b] primer (2.75 µM) (Integrated DNA Technologies, Coralville, IA)
- Diluted selective reaction
- Sterile ddH$_2$O
- 10× Taq buffer with 25 mM MgCl$_2$
- Taq DNA polymerase (5 U/µL)
- dNTPs (2 µL)
- Aluminum foil
- Ice bucket with ice
- Thermalcycler

[a] "N" indicates any nucleotide.
[b] or Mse+N, if applicable.

Follow the instructions in Procedure 27.16 to complete this section.

EXERCISE 10: CAPILLARY GEL ELECTROPHORESIS

AFLP fragments were often separated by vertical polyacrylamide gel electrophoresis, but with the advent of capillary gel electrophoresis, it is rarely used now. Capillary gel electrophoresis allows for better resolution of bands and is less labor intensive. Software is often available with various capillary gel electrophoresis systems, which streamlines data analysis. The following exercise describes how samples can be prepared for capillary gel electrophoresis on the Beckman Coulter CEQ™ 8000 Genetic Analysis System. This particular system provides software that can convert the AFLP fragments into binary data ("1" representing presence of a fragment or "0" representing absence of a fragment). Binary data can then be easily analyzed by statistical software, which will be described in a later section.

Materials

Each team of students will require the following materials:

- Undiluted fluorescently labeled product
- CEQ™ 8000 Genetic Analysis System
- 96-well plate (must be compatible with CEQ™ 8000 Genetic Analysis System)
- CEQ™ SLS (sample loading solution)
- Fluorescently labeled CEQ™ DNA Size Standard (400- or 600-bp ladder)
- Mineral oil (comes with CEQ™ SLS)
- PCR amplification tape (Fisher, Atlanta, GA)
- Aluminum foil
- 96-well flat-bottom plate (well diameter 6.4 mm) (Corning Glass Company, Corning, NY)
- CEQ™ Separation Buffer
- CEQ™ LPA-1 (Linear Polyacrylamide) Gel Cartridge
- CEQ™ Capillary Array (33 cm)

Follow the instructions in Procedure 27.17 to complete this section.

	Procedure 27.17
	Capillary Gel Electrophoresis
Step	Instructions and Comments
1	Before gel electrophoresis, a mixture of fluorescently labeled size standard and SLS (sample loading solution) should be prepared in the following volumes: 9 samples (274.5 µL): 30 µL of CEQ™ SLS, 0.5 µL of fluorescently labeled CEQ™ DNA Size Standard (400- or 600-bp ladder). The type of size standard used depends on the length of fragments that are anticipated. Add 30.5 µL to each well of the column in a CEQ™ 8000 Genetic Analysis System compatible 96-well plate. In diminished light, add 5 µL of each undiluted fluorescently labeled product into its corresponding well. A CEQ™ 8000 Genetic Analysis System compatible 96-well plate in the dark. Add a drop of mineral oil to the top of each sample and cover with PCR amplification tape. The plate should then be wrapped in foil to protect it from the light.
2	Fill the corresponding wells of a 96-well flat-bottom plate (well diameter 6.4 mm; Corning Glass Company, Corning, NY) with CEQ™ Separation Buffer for gel electrophoresis.
3	Insert the gel cartridge and capillary array, and place the uncovered sample plate and buffer plate into the CEQ™ 8000 Genetic Analysis System. Purge the capillary with 0.5 mL of gel before each run and ensure that the optics are aligned.
4	The CEQ method for capillary gel electrophoresis should be run as follows: 2 min of capillary preparation at 50°C, followed by denaturation at 90°C for 120 sec, injection at 2 kV for 30 sec, and separation at 4.8 kV for 75 min with no pause between samples.

EXERCISE 11: FRAGMENT AND DATA ANALYSIS

After gel electrophoresis raw data (an electropherogram graphing fluorescence against time) is produced by the CEQ™ 8000. This raw data is analyzed and the sizes of AFLP fragments are compared to fragments of known size, or the size standard. If a 600-bp CEQ™ DNA Size Standard is used, a quartic model should be implemented for analysis. Alternatively, if a 400-bp CEQ™ DNA Size Standard is used, a cubic model should be used for data analysis. By looking at the analyzed electropherogram (graphing fluorescence versus size), it is possible to detect if a sample failed, e.g., the sample has low signal, size standard cannot be detected, or the sample does not contain fragments (Figure 27.6). If a sample did not pass the analysis or had too low of signal, troubleshooting reactions should take place starting at the fluorescent reaction and working backwards, repeating previous reactions. If the size standard could not be detected, the sample may simply need to be rerun on the machine.

After the raw data is analyzed by the software, it can be converted to binary data through the AFLP analysis option. Common settings for the AFLP analysis include setting the Y threshold for signal strength at 5000 RFU and bin at 1.5 nucleotides. The signal strength threshold should be determined based on the particular sample type you are working with, the relative fluorescence of the amplicons, and the amount of background noise present. Once binary data is produced, all samples should be manually edited to ensure that the software correctly called the peaks as fragments and for low background

noise. Exporting the data to an Excel file permits you to make changes in the binary data as well as placing it in a format that can be readily imported into statistical software. To edit, the binary data should be compared to the electropherogram. If the software called a fragment at certain bp length as a "1," meaning a fragment was detected, the electropherogram should show a peak at that fragment size.

If more than one AFLP primer set was analyzed, all edited binary data can be combined from the all primer sets, end to end. After importing binary data into NTSYSpc version 2.20b (Exeter Software, 2005), the similarity indices can be calculated by using Jaccard association coefficient. The equation is as follows:

$$GS(ij) = 2a/(2a + b + c),$$

where **GS (ij)** is the measure of genetic similarity between individuals **i** and **j**, **a** is the number of polymorphic bands that are shared by **i** and **j**, **b** is the number of bands present in **i** and absent in **j**, and **c** is the number of bands present in **j** and absent in **i**. In addition, pairwise absolute distances, or the number of bands that were different between two samples, can be determined using PAUP (Swofford, 2001) and combined on a table with similarity indices. Cluster analysis can also be completed from the similarity indices by NTSYSpc version 2.20b (Exeter Software, 2005) using the Unweighted Pair Group Method with Arithmetic Mean (UPGMA) method. UPGMA produces a tree dendrogram that illustrates the genetic relationships of the samples. The dendrogram can further be analyzed by WinBoot (Yap,

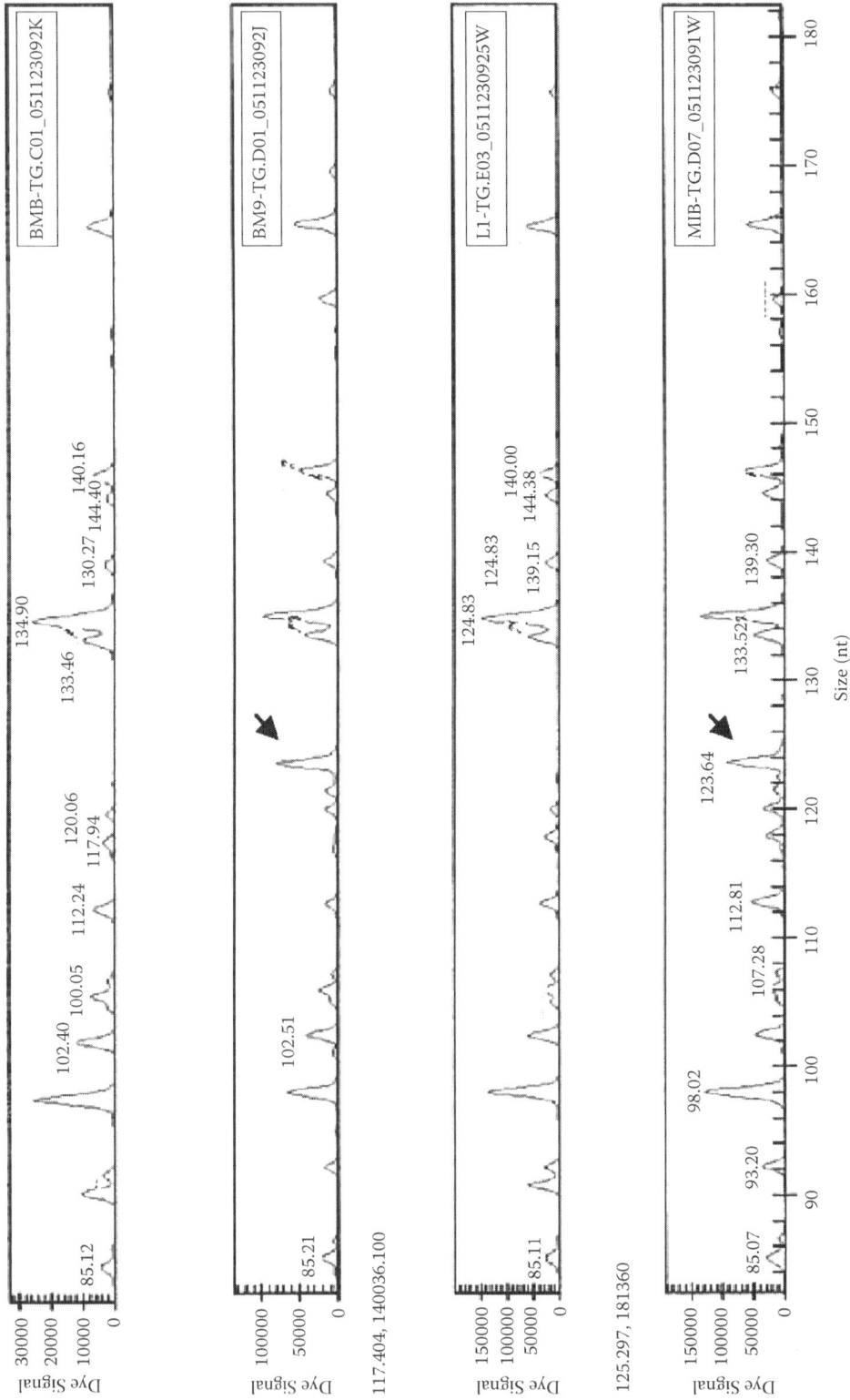

FIGURE 27.6 AFLP electropherogram produced by the Beckman Coulter CEQ™ 8000 Genetic Analysis System. Arrows indicate an example of a polymorphism.

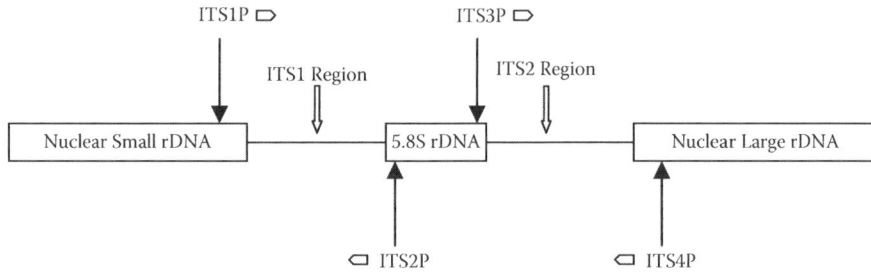

FIGURE 27.7 Diagrammatic representation of ITS and ribosomal regions. (Modified from White, T.J., T. Burns, S. Lee and J. Taylor. 1990. pp. 315-321. In *PCR Protocols: A Guide to Methods and Applications*. Innis, M.A., J. Gelford, J. Sninky, and T.J. White (Eds.) Academic Press, San Diego). ITS1P and ITS2P are primers for ITS1 Region, and ITS3P and ITS4P are primers for ITS2 region.

1991), which calculates bootstrap values for cluster support. Bootstrapping tests the reliability of nodes on a tree by resampling. Higher bootstrap values give confidence that a particular node of a tree appears the same after each time the data set is resampled. Normally, values below 50 are not labeled on the tree (Trigiano et al., 2004).

Questions

- What advantages does AFLP offer over other fingerprinting techniques such as DAF or RAPD? Disadvantages?
- Can AFLP data be used to construct a probable phylogeny between species of the same genera? Same family? Why or why not?
- What are some of the other uses for AFLP data?

EXPERIMENT 3: AMPLIFICATION OF INTERNAL TRANSCRIBED SPACERS (ITS) OF rRNA

Eukaryotic ribosomal genes are arranged in tandem repeats with the 5.8S coding region flanked by internal transcribed spacers (ITS) (Figure 27.7). Although these regions are important for maturation of nuclear ribosomal RNA (rRNA), ITS regions are usually considered to be under low evolutionary pressure and therefore typically treated as nonfunctional sequences. The sequences of the ITS regions have been used in many phylogenetic studies and molecular systematics of fungi. This experiment will allow students to amplify one or both of the ITS regions (1 and 2) and is largely based on Caetano-Anollés et al. (2001) and White et al. (1990). The experiment will not describe DNA sequencing (usually completed at a university center) or the analysis of sequence data. Instructors and students are encouraged to seek expert help on their campus.

MATERIALS

Each team or group of students will require the following items:

- AmpliTaq Gold DNA polymerase kit with buffer II (Applied Biosystems, Table 27.1)
- All materials under Experiment 2—DNA amplification except for the primers
- Use the following primers (5′-3′) in 3 μM concentrations:
    for ITS1 region use: TCCGTAGGTGAACCT-GCGG (ITS1) and
    GCTGCGTTCTTCATCGATGC (ITS2)
    for ITS2 region use: GCATCGATGAAGAAC-GCAGC (ITS3) and TCCTCCGCTTATT-GATATGC (ITS4)
- Stock fungal DNA from Experiment 1 diluted 1:9 or 0.5 ng/μL
- QiaQuick PCR Purification Kit (Qiagen, Table 27.1)

Follow the protocol as outlined in Procedure 27.18 to amplify the ITS region.

The thermalcycler should be programmed as follows:

Step 1: 95°C for 9 min—95°C activates the DNA polymerase enzyme that the Stoffel fragment polymerase did not require. This step is also termed a "hot start."
Step 2: 96°C for 1 min—Denatures DNA
Step 3: 56°C for 1 min—Primers anneal to sites
Step 4: 72°C for 1 min—Extension
Steps 2–4 should be repeated for 35 cycles
Step 5: 72°C for 7 min

Recovery, electrophoresis, staining, and developing of the amplified product are the same as described under Experiment 1. Sometimes, more than one product appears in the gel (Figure 27.8). The lighter product is an artifact and will not affect sequencing—the vast majority of product is the complete ITS region. If you were to run the product on an agarose gel and stain with ethidium bromide, the second lighter product would not be visible. You may or may not be able to detect visually differences in absolute number of base pairs in the ITS regions between *Fusarium* isolates.

Procedure 27.18

Amplification of the ITS1 and ITS2 Regions

Step	Instructions and Comments
1	Pipette 49.6 μL (8 × 6.2) of pure water into a sterile, 0.65-mL Eppendorf tube.
2	Pipette 16.0 μL (8 × 2) of 2 mM nucleotides into the tube.
3	Pipette 16.0 μL (8 × 2) of 10× Buffer II (without magnesium chloride) into the tube.
4	Pipette 9.6 μL (8 × 1.2) of 25 mM magnesium chloride into the tube.
5	Pipette 16 μL (8 × 2) of **each** of either ITS1 and ITS2 or ITS3 or ITS4 primers. (*Note:* as an alternative, use ITS1 and ITS4 primers and amplify both ITS1 and 2 regions plus the gene.)
6	Pipette 4.8 μL (8 × 0.6) of AmpliTaq Gold DNA polymerase.
7	Vortex and centrifuge briefly. Dispense 16 μL of the reaction mixture into seven sterile 0.65-mL Eppendorf tubes. Now add 4 μL of fungal DNA template (0.5 ng/μL) to each of the seven as indicated Figure 27.1. Be sure to change pipette tips between different isolates. Vortex and centrifuge briefly.
8	Add drop of heavy, white mineral oil to each tube and place in thermalcycler (most thermalcyclers do not require oil—follow the manufacturer's instructions).

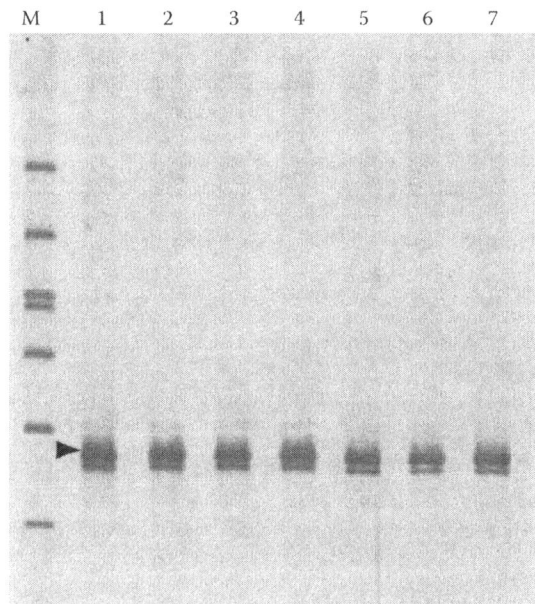

FIGURE 27.8   Gel showing single product (arrowhead at about 275 bp) of the ITS1 region from seven *Fusarium oxysporum* isolates (lanes 1–7) produced by using primers ITS1 and ITS2. The lighter band below is probably an artifact. M = molecular-weight markers (from top: 1000, 700, 525, 500, 400, 300 and 200 bp).

Before the product can be sequenced, other remaining components of the reaction mixture must be removed. We use a QiaQuick PCR Purification Kit (Qiagen). The instructions in the kit are excellent and easy to follow. Enquire at your sequencing center whether the DNA should be eluted in pure, sterile water or buffer. It is imperative to use sterile water that is between pH 7.0 and 8.5 for the QiaQuick kit. If you use distilled, pure water (pH = ~5.5), the DNA bound to the membrane will not be released. Most sequencing centers require the DNA concentration in the samples to be between 5 and 10 ng/μL,

and you must also supply some of the primers used in the amplification reaction.

QUESTIONS

- What are insertions and deletions?
- Why would you not send DNA in TE (Tris–EDTA) buffer to the sequencing center?
- Would any product be formed using only one primer in the amplification cocktail?

LITERATURE CITED

Bassam, B.J., G. Caetano-Anollés, and P.M. Gresshoff. 1991. Fast and sensitive silver staining of DNA in polyacrylamide gels. *Anal. Biochem.* 196: 80–83.

Caetano-Anollés, G. 1993. Amplifying DNA with arbitrary oligonucleotide primers. *PCR Methods Appl.* 3: 85–94.

Caetano-Anollés, G. 1996. Scanning of nucleic acids by in vitro amplification: new developments and applications. *Nat. Biotechnol.* 14: 1668–1674.

Caetano-Anollés, G., B.J. Bassam, and P.M. Gresshoff. 1991. DNA amplification fingerprinting using very short arbitrary oligonucleotide primers. *Bio/Technology* 9: 553–557.

Caetano-Anollés, B.J. Bassam, and P.M. Gresshoff. 1992. Primer-template interactions during DNA amplification fingerprinting with single arbitrary oligonucleotides. *Mol. Gen. Genet.* 235: 157–165.

Caetano-Anollés, G. and P.M. Gresshoff. 1994. DNA amplification fingerprinting of plant genomes. *Methods Mol. Cell. Biol.* 5: 62–70.

Caetano-Anollés, G and RN Trigiano. 1997. Nucleic acid markers in agricultural biotechnology. *AgBiotech: News and Information* 9: 235N–242N.

Caetano-Anollés, G., R.N. Trigiano and M.T. Windham. 2001. Patterns of evolution in Discula fungi and origin of dogwood anthracnose in North America, studied using arbitrarily amplified and ribosomal DNA. *Curr. Genet.* 39: 346–354.

Dragon, E.A. 1993. Handling reagents in the PCR laboratory. *PCR Meth Appl.* 3: S8–S9.

Erlich, H.A., D. Gelfand, and J.J. Sninsky. 1991. Recent advances in the polymerase chain reaction. *Science* 252: 1643–1651.

Exeter Software, 2005. NTSYSpc version 2.20b. E. Setauket, NY.

Habera, L.H., K.H. Lamour, R. Donahoo, and N.R. Smith. 2004. A single primer strategy to fluorescently label selective AFLP reactions. *Biotechniques* 37: 902–904.

Miyashita, N.T., A. Kawabe, and H. Innan. 1999. DNA variation in the wild plant *Arabidopsis thaliana* revealed by amplified length polymorphism analysis. *Genetics* 52: 1723–1231.

Mueller, U.G. and L. LaReesa Wolfenbarger. 1999. AFLP genotyping and finger printing. *Trends Ecol. Evol.* 14: 389–394.

Mullis, K.B., F.A. Faloona, S. Scharf, R. Saiki, G. Horn, and H. Erlich. 1986. Specific enzymes amplification of DNA *in vitro*: the polymerase chain reaction. *Cold Spring Harbor Symp. Quant. Biol.* 51: 263–273.

Southern, E.M. 1975. Detection of specific sequences among DNA fragments by gel electrophoresis. *J. Molec. Biol.* 98: 503–517.

Swofford, D.L. 2001. PAUP*. Phylogenetic Analysis Using Parsimony (*and Other Methods). Version 4.0b 10 for 32-bit Microsoft Windows. Sinauer Associates, Sunderland, MA.

Trigiano, R.N., M.H. Ament, M.T. Windham, and J.K. Moulton. 2004. Genetic profiling of red-bracted *Cornus kousa* cultivars indicates significant cultivar synonymy. *Hortscience* 39: 489–492.

Trigiano, R.N. and G. Caetano-Anollés. 1998. Laboratory exercises on DNA amplification fingerprinting for evaluating the molecular diversity of horticultural species. *HortTechnology* 8: 143–423.

Trigiano, R.N., G. Caetano-Anollés, B.J. Bassam and M.T. Windham. 1995. DNA amplification fingerprinting provides evidence that *Discula destructiva*, the cause of dogwood anthracnose in North America, is an introduced pathogen. *Mycologia* 87: 490–500.

Vos, P., R. Hogers, M. Bleeker, M. Reijans, T. van de Lee, M. Hornes, A. Frijeters, J. Pot, J. Peleman, M. Kuiper, and M. Zabeau. 1995. AFLP: a new technique for DNA fingerprinting. *Nucleic Acids Res.* 23: 21, 4407–4414.

Vos, P. and M. Kuiper. 1997. AFLP analysis. In: G. Caetano-Anollés and P. M. Greshoff (Eds.). *DNA Markers, Protocols, Applications, and Overviews.* Wiley-Liss, New York.

White, T.J., T. Bruns, S. Lee and J. Taylor. 1990. Amplification and direct sequencing of fungal ribosomal DNA genes for phylogenetics. In: M.A. Innis, J. Gelfand, J. Sninky and T.J. White (Eds.). *PCR Protocols: A Guide to Methods And Applications*, pp. 315–322. Academic Press, San Diego.

Williams, J.K., A.R. Kubelik, K.J. Livak, J.A. Rafalski, and S.V. Tingey. 1990. DNA polymorphisms amplified by arbitrary primers are useful as genetic markers. *Nucleic Acid Res.* 18: 6531–6535.

Yap, I.V. 1991. WinBoot UPGMA bootstrapping for binary data. University of Washington and Joseph Felsenstein.

Yoon, C.-K., D.A. Glawe and P.D. Shaw. 1991. A method for rapid small-scale preparation of fungal DNA. *Mycologia* 83: 835–838.

Part 4

Plant–Pathogen Interactions

28 Plant–Fungal Interactions at the Molecular Level*

Ricardo B. Ferreira, Sara Monteiro, Regina Freitas, Cláudia N. Santos, Zhenjia Chen, Luís M. Batista, João Duarte, Alexandre Borges, and Artur R. Teixeira

CHAPTER 28 CONCEPTS

- Plants produce an astonishing plethora of defense reactions that arose from a multimillion-year, ping-pong-type coevolution, in which plant and pathogen successively added new chemical weapons in this perpetual battle. As each defensive innovation was established in the host, new ways to circumvent it evolved in the pathogen.

- A clear-cut border line between nonhost resistance, incompatible host–pathogen interaction, and host-specific resistance is difficult to establish. Often, plants cannot be considered either 100% resistant or 100% susceptible to a particular pathogen. Even susceptible plants mount a (insufficient) defense response upon recognition of pathogen-elicited molecular signals.

- The challenging battle to infection involves recognition of host by the pathogen, followed by penetration, i.e., transition from extracellular to invasive growth. Central to the activation of defense responses is timely perception of the pathogen by the plant. A crucial role is played here by elicitors, either of pathogen origin (exogenous elicitors) or host components released or modified by pathogen effectors (endogenous elicitors).

- The attempted infection of a plant by a pathogen, such as a fungus or an Oomycete, may be regarded as a battle whose major weapons are proteins and smaller chemical compounds produced by both organisms.

INTRODUCTORY CONCEPTS IN PLANT–PATHOGEN INTERACTIONS

Fungi, forming a kingdom that is evolutionarily related to animals, are an extremely diverse group of organisms, with about 250,000 species widely distributed in essentially every ecosystem. Oomycetes, which were long considered as a class within the kingdom Fungi, are now classified in the kingdom Stramenopila and are related to heterokont biflagellate, golden-brown algae. Together with the fungi, they comprise the two most important groups of eukaryotic plant pathogens, exhibiting many similarities in what concerns their lifestyles. For example, they both show filamentous growth in their vegetative stage and produce mycelium and form spores for asexual and sexual reproduction. In addition, there is no distinction between the response of plants to fungi or to Oomycetes. They both include biotrophic (which grow and reproduce in living plant tissues, e.g., in the genera *Erysiphe* and *Plasmopara*, respectively), necrotrophic (which feed on dead plant cells, e.g., in the genera *Botrytis* and *Pythium*), and hemibiotrophic species (which initially establish a bio-

trophic relationship with their host, but subsequently, the host cells die as the infection proceeds, e.g., in the genera *Colletotrichum* and *Phytophthora*). Therefore, convergent evolution seems to have forced the development of similar infection strategies in these two types of plant pathogens, which elicit similar responses from the host plants.

Plants constitute an excellent ecosystem for microorganisms because they offer a wide diversity of habitats, including the phyllosphere (aerial plant part), the rhizosphere (zone of influence of the root system), and the endosphere (internal transport system). In contrast to the situation in animals, in which fungal diseases are less frequent, many plant diseases are caused by fungi. Thus, 120 genera of fungi, 30 types of viruses, and 8 genera of bacteria are responsible for the 11,000 diseases that have been described in plants. Therefore, plant organs, both above and below the ground, are continuously and permanently exposed to a vast range of potential pathogens. However, as the normal state of a plant is to be healthy, the development of disease requires the coincidence of a susceptible host, a virulent pathogen, and a favorable environment (Chapter 33). In other words, not all pathogens can attack

---

* Adapted from Ferreira et al. 2006. Fungal pathogens: The battle for plant infection. *Crit. Rev. Plant Sci.* 25: 505–524.

CASE STUDY 28.1

AN EXAMPLE OF A HIGH DEGREE OF SPECIFICITY BETWEEN PLANT AND MICROBE: AVENACIN

* The preformed saponin glycoside avenacin constitutes a good example of the high-level specialization some-times achieved between plant and pathogen.
* The growth of the wheat (*Triticum aestivum*) root pathogen *Gaeumannomyces graminis* var. *tritici* is inhibited by avenacin. Therefore, oat (*Avena sativa*) plants producing the compound exhibit resistance to the pathogen.
* The related oat root pathogen *G. graminis* var. *avenae* produces a glycosidase that removes the sugar residue from avenacin, effectively detoxifying it. Therefore, this strain is not inhibited by avenacin, and oat plants are susceptible to it (Keen, 1999; Thordal-Christensen, 2003).
* A mutation in the glycosidase gene rendered *G. graminis* var. *avenae* sensitive to avenacin and incapable of infecting oat plants (Bowyer et al., 1995).
* Engineered oat plants that do not express avenacin are susceptible to *G. graminis* var. *tritici* (Papadopoulou et al., 1999).
* These results indicate that *G. graminis* var. *tritici* is a nonhost pathogen for oats because it lacks the capacity to metabolize avenacin, considered a resistance factor unless a pathogen can deal with it.

all plants, and a single plant is not susceptible to the whole plethora of phytopathogens. In fact, only a very small proportion of these pathogens are capable of successfully invading each plant host and causing disease. In addition, pathogen population and host plant change during their life cycle according to the development stage, affecting pathogen virulence and host susceptibility. Disease is also strongly dependent on environmental conditions, such as water availability, temperature, and plant surface wetness. For all these reasons, development of disease is usually less frequent than one would expect.

In their long association with pathogens, plants evolved an intricate and elaborate array of defensive tools. At the same time, those very same pathogens developed means to overcome plant resistance mechanisms in what must have been a multimillion-year evolutionary game of ping-pong. As each defensive innovation was established in the host, new ways to circumvent it evolved in the pathogen. More recently, phytopathogens overcame plant resistance through the acquisition of virulence factors. These newly evolved pathogen race-specific factors drove the coevolution of plant resistance genes and the development of phylogenetically related, pathogen race/plant cultivar–specific disease resistance. Over time, the coevolutionary struggles between would-be pathogens and their erstwhile hosts have generated some of the most complex and interesting interactions known to biology. This complex coevolution process has probably led to redundancy or ineffectiveness of some defensive genes, which often encode numerous enzymes with overlapping activities, and to the exquisite specificity between pathogen and host that is commonly observed.

The existence of redundant and eventually inoperative chemical defense mechanisms against pathogen attack raises inevitably the question of the role they ful-fill in plants. Supposedly, these mechanisms appeared casually during evolution, meaning that their presence may be of value, may be indifferent, or may harm the host plant. Let us consider, for example, a casual mutation that induces the accumulation of an antifungal protein upon infection by a fungal pathogen. If the protein biosynthetic process consumes a significant proportion of the plant metabolic potential, two possible outcomes are expected: (1) The mutation does not add value to the host plant, which will be unable to compete as efficiently as the wild-type for the available resources. Under these circumstances, the selective pressure will presumably eliminate promptly the modified plant and its useless antifungal protein. (2) The new antifungal protein provides the plant with a selective advantage over the wild-type, allowing it to be more successful in a certain competitive habitat. Under these conditions, the modified plant may also become the dominant form in that habitat. However, if the new defense mechanism does not use up a considerable fraction of the plant assimilatory power, the modified plant may subsist for a long time even with a useless or redundant defense mechanism. Over millions of years, one expects an ordinary plant to have acquired or lost a number of redundant or inoperative defense mechanisms. Nevertheless, it seems imprudent to conclude that a given protein is useless just because it has no obvious function. This protein may be worthless today, but may have fulfilled an important defense role against pathogens that existed long ago.

Case Studies 28.1 and 28.2 provide examples of the high degree of specificity sometimes observed between plant and pathogen and of the evolution of a plant–pathogen interaction into a highly intricate complex mechanism.

Resistance to any given pathogen is due to a combi-nation of defense mechanisms that include passive or pre-

---

CASE STUDY 28.2

Sometimes Plant–pathogen Interactions Evolved into Highly Intricate
Complex Mechanisms: Cyanogenesis Reconstitution in Barley

- Appressoria and appressorial hook formation of *Blumeria graminis* f. sp. *hordei*, the causal agent of barley (*Hordeum vulgare*) powdery mildew, are stimulated *in vitro* by epiheterodendrin, a leucine-derived cyanogenic beta-D-glucoside that accumulates specifically in the epidermis of barley leaves.
- A (1,4)-β-D-glucan exohydrolase located in the starchy endosperm of barley grains was found to hydrolyze epiheterodendrin *in vitro*, releasing hydrogen cyanide.
- Barley leaves are not cyanogenic, i.e., they do not possess the ability to release hydrogen cyanide, because they lack a cyanide-releasing β-D-glucosidase.
- Cyanogenesis was reconstituted in barley leaf epidermal cells by expression of a cDNA encoding a cyanogenic β-D-glucosidase from sorghum (*Sorghum bicolor*), which resulted in a 35 to 60% reduction in colonization rate by *B. graminis* f. sp. *hordei*.
- These observations suggest that loss of cyanogenesis in barley leaves has enabled the fungus to utilize the presence of epiheterodendrin to facilitate host recognition and to establish infection.

---

formed mechanical and chemical barriers, which provide nonspecific protection against a wide range of organisms, and pathogen-induced or active structural (i.e., physical) and chemical barriers that supply host- or cultivar-specific resistance. Preformed defenses comprise waxy cuticles, cell-wall components, and antimicrobial compounds, such as secondary metabolites and antifungal proteins. Upon pathogen detection, plants activate a number of defenses such as changes in plasma membrane permeability preceding Ca^{2+} and H^+ influx and K^+ and Cl^- efflux; rapid oxidative burst, leading to production of reactive oxygen species (ROS) such as superoxide, hydrogen peroxide, and hydroxyl free radicals; hypersensitive response at the site of infection; cell wall reinforcement by oxidative cross-linking of cell wall components; apposition of callose and lignin; increased synthesis of antimicrobial secondary metabolites, such as phytoalexins; transcriptional activation of defense-related genes encoding pathogenesis-related (PR) proteins (chitinases, glucanases, and proteases) and other antimicrobial proteins (e.g., defensins); and induction of phospholipases (PLPs), which act on lipid-bound unsaturated fatty acids within the membrane, resulting in the release of jasmonate, methyl jasmonate, and related molecules. In contrast, disease results either from the failure of the recognition event or from the ability of the pathogen to avoid or overcome the resistance response.

Stability of species immunity is likely due to the multiple and intertwined functionally redundant layers of protective mechanisms. Recognition of multiple signals derived from each pathogen may result in finer-tuned response or may induce redundant recognition systems that act as independent backup systems. Therefore, impairment of individual defense responses may not alter significantly a particular plant–pathogen interaction, rendering it difficult to genetically dissect plant disease resistance.

Unlike animals, plants have no circulating cells capable of sensing the presence of microbial nonself. Therefore, defense responses of plant cells are autonomous with pathogen recognition and response occurring at the level of each single cell. Host resistance is the form of plant disease resistance that is cultivar or accession specific. On the other hand, nonhost resistance provides resistance against pathogens throughout all members of a plant species. A pathogen that cannot cause disease on a nonhost plant is referred to as a nonhost pathogen. In contrast to host resistance, nonhost resistance operates under less-understood mechanisms.

PLANT RESISTANCE

Host Resistance

The genetic basis determining race–cultivar specificity in some plant–pathogen interactions was initially and tentatively explained in 1955 by the gene-for-gene theory of H. H. Flor, following his studies on the interaction between flax (*Linum usitatissimum*) and the fungus *Melampsora lini*, a rust (Chapter 18). In this theory, host plant disease resistance (*R*) genes (Chapter 34) mediate the recognition of specific pathogen-derived components, the products of avirulence (*avr*) genes. This gene-for-gene interaction subsequently was complemented with the receptor–ligand model, in which the avirulence protein binds to the corresponding resistance protein, which in turn triggers the plant surveillance system. However, there is also accumulating evidence that R proteins are not the primary receptor for the avr proteins. When complementary pairs of *R* and *avr* genes are present, the interaction between the host and the pathogen is incompatible, triggering host cell defense mechanisms and resulting in resistance to disease. Lack of, or nonfunctional products of, either gene establishes a compatible plant–pathogen interaction, resulting in disease. According to this terminology, the pathogen is said to be avirulent in incompatible interactions and virulent in compatible interactions.

Mutational analysis demonstrated in several cases that *avr* genes provide a selective advantage to the pathogen in the absence of the corresponding *R* gene, explaining their persistence in pathogen populations. In other words, in plants lacking the appropriate *R* gene, *avr* genes are often required for maximal virulence of particular pathogen strains. Most avr proteins are therefore considered virulence factors that contribute to host infection and are required for the colonization of host plants. In the course of plant–pathogen coevolution, avr proteins were targeted by *R* genes present in the resistant host plant cultivars. The identity and biochemical function of most avr proteins remains unknown. One *Phytophthora avr* gene, *avr1b* from *P. sojae*, has been cloned and characterized. It encodes a secreted elicitor that triggers a systemwide defense response in soybean (*Glycine max*) plants carrying the cognate *R* gene, *Rps 1b*. Interestingly, these virulence effectors originally were identified not by their promotion of virulence but rather by their "avirulence" activity.

Corresponding pairs of *R* genes in the host and *avr* genes in the invading microbe often determine the hypersensitive response (HR), a pathogen-induced, rapid and spatially confined cell suicide at the spot of infection. As a result, the pathogen remains confined to necrotic lesions near the site of infection. Localized acquired resistance in a ring of cells surrounding necrotic lesions ensure that they become fully refractory to subsequent infections.

Local HR often triggers a systemic signal that transduces nonspecific resistance throughout the plant, leading to systemic acquired resistance (SAR), which confers long-lasting immunity against subsequent infections by a broad spectrum of pathogens. In contrast to HR, the development of SAR is slow and gradual. However, SAR shares many features with local resistance, including the involvement of salicylic acid (SA) and the oxidative burst. In addition, pathogen infection appears to establish new sinks in remote plant organs that compete with physiological sink organs, highlighting the systemic effects of pathogen infection.

Nonhost Resistance

Plants are permanently exposed to many different pathogens in the environment. However, only a few of these pathogens are capable of causing disease in each plant species. Conversely, each pathogen has a limited range of plants on which it causes disease. All other plants are by definition nonhost plants, and the attacking microbes are nonhost pathogens. If a particular pathogen is unable to attack a plant, that means the plant is resistant to the pathogen, i.e., it cannot host the pathogen. This durable type of immunity of a plant to a pathogen is called nonhost resistance and consists of a multistep defense system. Therefore, nonhost resistance stops almost all parasite attacks on plants in a broadly effective and durable manner. Although it has been the subject of little research, the mechanisms of nonhost resistance may be exploited to introduce resistance traits in crop plants.

Nonhost resistance, the most common form of disease resistance exhibited by plants against the majority of potential pathogens, may be classified into the following two types:

- Type I nonhost resistance, which does not produce any visible symptoms (necrosis), is probably the most common type of nonhost resistance. In this case, the pathogen is not able to get past the first or second obstacles of plant defense, which include preformed and inducible resistance mechanisms, completely arresting its multiplication and penetration into the plant cells.
- Type II nonhost resistance is always associated with rapid localized necrosis (HR). Phenotypically, it is more similar to an incompatible gene-for-gene interaction and is a more sophisticated plant defense mechanism than type I nonhost resistance. After overcoming the preformed and inducible defense mechanisms, the pathogen can directly penetrate the plant cells. The extracellular hyphal proteins and pathogen-secreted proteins are recognized by the plant surveillance system, and the defense reaction leading to HR is activated to prevent further spread of the pathogen.

The type of nonhost resistance triggered in a nonhost plant depends on the plant–pathogen pair, so that a nonhost plant may exhibit type I nonhost resistance toward one pathogen species, and type II against another. Conversely, a single pathogen species can trigger both type I and type II nonhost resistances in different plant species.

Nonhost resistance against pathogens is considered to be associated with the penetration process or, in other words, with the incapacity of fungal invasive growth. Until recently, it was assumed to be mostly based on passive defense mechanisms, such as preformed structural components of the cell wall, antimicrobial compounds and other secondary metabolites on the surface of the plant, or simply a lack of molecular entry sites for the pathogen. However, recent evidence has shown that active immune responses may also be key components of nonhost resistance mechanisms.

Host Resistance versus Nonhost Resistance

A potential plant pathogen must overcome many barriers to become an actual pathogen. Although detection of nonspecific, pathogen-elicited molecular signals often plays a

CASE STUDY 28.3

AN EXAMPLE OF PATHOGEN–HOST INTERACTION THAT BRIDGES THE GAP BETWEEN NON-HOST AND HOST-SPECIFIC RESISTANCE: FLAGELLIN PERCEPTION

- Flagellin perception determines the host range of the bacteria *Acidovorax avenae* and *Pseudomonas syringae*. The *FLS2* gene, which encodes an R protein, controls perception of the bacterial protein flagellin in *Arabidopsis*. Flg22, a conserved 22-amino acid residue subdomain of flagellin, is necessary and sufficient to induce the flagellin response, including accumulation of PR proteins, but not an HR (Gomez-Gomez and Boller, 2002).
- Different strains of *A. avenae* produce flagellins with distinct primary structures, which correlate with their capacity to induce resistance responses in plants. Thus, flagellin from the millet (*Setaria italica*) strain, but not the rice (*Oryza sativa*) strain, induces resistance response in rice cells, including an HR and expression of PR proteins (Che et al., 2000).
- A flagellin deletion mutant of *A. avenae* is incapable of overcoming resistance in rice, indicating that flagellin perception is not the only determinant of resistance in this interaction (Tanaka et al., 2003).
- Different strains of *P. syringae* produce flagellins with differential glycosylation patterns, which also correlate with their ability to induce resistance responses in plants. Thus, flagellin from *P. syringae* pv. *glycinea*, but not *P. syringae* pv. *tabaci*, induces resistance response in tobacco (*Nicotiana tabacum*) (Taguchi et al., 2003). Glycosylation of flagellin in *P. syringae* pv. *tabaci* has been shown to be required for bacterial virulence (Taguchi et al., 2006).
- However, a flagellin deletion mutant of *P. syringae* pv. *tabaci* becomes pathogenic on tomato (*Lycopersicon esculentum*), indicating that the flagellin perception is the major determinant of nonhost resistance in this interaction (Shimizu et al., 2003).
- Using rice cDNA microarrays, cultured rice cells, and compatible and incompatible strains of *A. avenae*, Fujiwara et al. (2004) demonstrated the following:
  - Differentially expressed between incompatible and compatible interactions are 131 genes.
  - Of these, 94 genes are upregulated and 32 genes are downregulated during incompatible interactions, but only 5 genes are upregulated during compatible interactions.
  - Among the 126 genes that are up- or downregulated during incompatible interactions, expression of 46 genes is decreased when the cells are inoculated with a flagellin-deficient incompatible strain, indicating that approximately 37% of the 126 genes are directly controlled by flagellin perception.

role in nonhost resistance, detection of specific molecular signals is required to resist pathogens that are capable of overcoming all the nonspecific barriers.

Nonhost and host-specific resistance often share signaling components and defense responses, in the same way nonhost and host pathogens may be recognized by similar mechanisms. Indeed, following recognition of an intruder by R proteins, subsequent defense responses are remarkably similar, including rapid ion fluxes, production of ROS, and accumulation of antimicrobial compounds and HR. Therefore, it was suggested that the difference between nonhost and host resistance may reside in the solidity of the recognition leading to resistance. For example, differentiation of a *Blumeria graminis* f. sp. *hordei* appressorium depends on the composition of the plant surface wax, determining whether plants are hosts or nonhosts.

However, it is still questionable if nonhost and host resistance involve the same signal transduction pathways. It was proposed that nonhost and host resistance involve separate signal transduction pathways, with significant amount of cross-talk between them, but the two pathways

may converge at a later time. An example of a pathogen–host interaction that bridges the gap between nonhost and host-specific resistance is given in Case Study 28.3.

A clear-cut boarder line among nonhost resistance, incompatible host–pathogen interaction, and host-specific resistance is sometimes difficult to establish. Often, plants cannot be considered either 100% resistant or 100% susceptible to a particular pathogen. It is well known that even susceptible plants mount a (insufficient) defense response upon recognition of pathogen-elicited molecular signals. The identification of many *Arabidopsis* host mutants that shift phenotypes of powdery mildew infection from susceptible to super-susceptible strongly implicates the existence of active, though inefficient, defense responses even in compatible interactions.

THE CASE EXAMPLE OF *VITIS* SPP.

Grapevines are classified into the genus *Vitis*. Because of its high quality for wine, fresh fruit, and dried grape production, a single *Vitis* species, the European *V. vinifera*, is by far the major species cultivated, with over 5000 cultivars and varieties currently in use worldwide. American *Vitis* species, such as *V. riparia*, *V. rupestris*, *V. berlandieri*, *V. labruscana*, and their hybrids, do not produce good quality grapes but possess root systems resistant to phylloxera (*Daktulosphaira vitifoliae*), a small homopterous insect that normally feeds on vine roots. This insect is considered the worst pest worldwide affecting vines and justifies the need for grafting and the use of American *Vitis* species as rootstocks. For this reason, the individual plants that compose a vineyard nowadays are chimeras, with the aerial part belonging to one species (*V. vinifera*—susceptible to phylloxera but producing superior quality grapes) and the root system to another (an American *Vitis* species or hybrid—resistant to phylloxera but incapable of producing good quality berries). However, besides phylloxera, *V. vinifera* is particularly susceptible to an array of fungal diseases.

American and European vines differ widely in their resistance to fungal pathogens. American *Vitis* species are generally regarded as resistant against *Erysiphe necator*, *Plasmopara viticola*, and *Phomopsis viticola*, and partially insusceptible to *Botrytis cinerea*. On the contrary, *V. vinifera* is extremely susceptible to all the American pests and diseases, including phylloxera, *E. necator*, *Pl. viticola*, *Ph. viticola*, and *B. cinerea*.

From a historical and evolutionary perspective, it is relatively easy to explain the marked differences between American *Vitis* and *V. vinifera* in what concerns their resistance to disease. The causal agents of the powdery (i.e., *E. necator*) and downy (i.e., *Pl. viticola*) mildews coevolved with the American *Vitis* on the American continent, where they are endemic, over millions of years, which resulted in the development of natural plant resistance. However, the introduction in Europe of *E. necator* and *Pl. viticola* in the 19th century, together with accessions of the American wild *Vitis* species, means that *V. vinifera* evolved in the absence of those obligate pathogens. Since then, not enough time has elapsed for *V. vinifera* to evolve natural resistance against those parasites. Even if there were, such development would have been blocked by the exclusive, vegetative reproductive system currently in use for all *V. vinifera* varieties.

Traditional breeding programs involving crosses between American *Vitis* species and *V. vinifera* were successful in the sense that disease resistance traits were introduced in some hybrids, but they were ineffective because the high quality of grapes was lost. It is therefore increasingly important to use genetic manipulation tools, in an attempt to transfer the disease resistance mechanisms exhibited by the American *Vitis* species into *V. vinifera* genome, with the aim of accelerating the process of disease resistance acquisition by the European vine. The molecular mechanisms responsible for the observed resistance of American *Vitis* to pathogens have not yet been elucidated.

American *Vitis* species exhibit different levels of resistance toward the powdery and downy mildews, economi-

cally the most important diseases affecting grapevines worldwide. Thus, whereas *V. labruscana* is moderately resistant to *E. necator* and *Pl. viticola*, *V. berlandieri* is resistant to those pathogens, although in some years powdery and downy mildews have been observed on the foliage, but not in significant amounts. *Vitis riparia* and *V. rupestris* are among the most recalcitrant American species to these pathogens, but even in these cases, defense mechanisms do not confer 100% resistance. Germination of *E. necator* spores, as well as fungal penetration, mycelia growth, and conidia formation onto *V. riparia* leaves can easily be observed, although the frequency and size of the powdery spots are much smaller than on *V. vinifera* leaves. *Vitis rupestris* exhibits an even higher degree of resistance. Nevertheless, under laboratory conditions, a few powdery spots may still be observed on young leaves a number of hours after inoculation, indicating some degree of appressoria development, penetration, mycelial growth, and conidia formation (Figure 28.1).

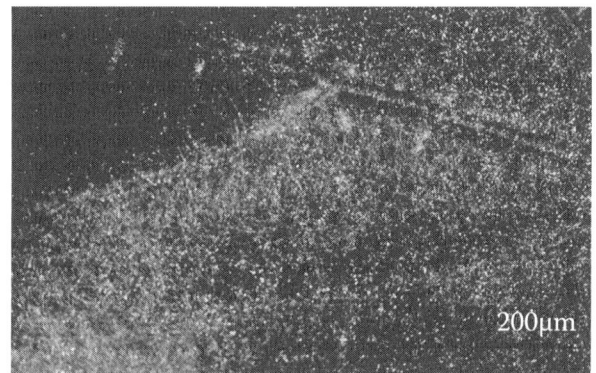

(a)

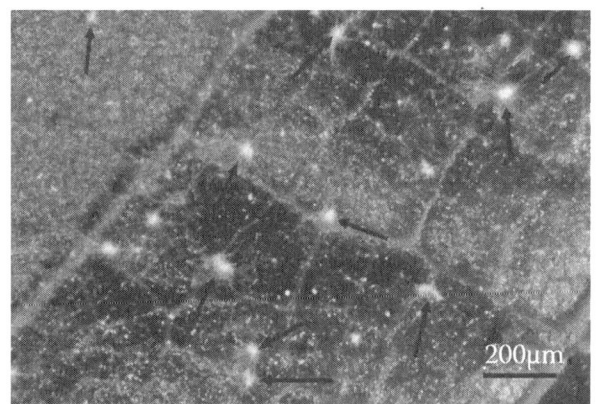

(b)

FIGURE 28.1 A—Detached *Vitis vinifera* cv Dona Maria leaf inoculated with *Erysiphe necator* conidia, producing white mycelia on the upper surface 5 days after inoculation. B—Detached *V. rupestris* leaf inoculated with *E. necator* conidia, producing white mycelia on the upper surface 10 days after inoculation.

CASE STUDY 28.4

EXAMPLES OF ELICITORS

A. Botrycin and cinerein
- Botrycin and cinerein, for example, are two elicitors that have been isolated from the crude mycelial cell wall and from culture filtrate preparations, respectively, of the necrotrophic fungus *Botrytis cinerea*.
- Both elicitors cause formation of necrotic lesions in grapevine that mimic a typical HR.
- Infiltration of minute amounts of these elicitors into leaves activates specific and distinct MAP kinases, stimulates a rapid transcriptional activation of genes encoding enzymes of the phenylpropanoid pathway, and induces ion fluxes across the plasma membrane and the production of ROS (Repka, 2006).

B. Microbe-associated hydrolytic enzymes
- Microbe-associated hydrolytic enzymes, which release fragments from structural plant cell wall components through limited degradation, are often essential virulence factors and provide the attacking pathogen with nutrients.
- It is the case, for example, of pectic fragments (oligogalacturonides [OGAs]) (Ebel and Cosio, 1994; Ebel and Scheel, 1997; Shibuya and Minami, 2001).
- Besides their enzymatic products, the cell wall-degrading enzymes themselves, including endopolygalacturonase and xylanase, are ubiquitous virulence factors among plant pathogens that can also function as elicitors.
- In this way, plants must not only discriminate self from nonself but also respond to some endogenous self-derived structures.

C. Fungal cell-wall bits
- Fungal cell wall bits liberated by the action of plant-derived hydrolases may also function as elicitors in a highly specific mode.
- Thus, for example, the chitooligosaccharides (i.e., *N*-acetly-D-glucosamine oligomers) released from pathogen cell walls by apoplastic chitinases are involved in oxidative burst, protein phosphorylation, phytoalexin biosynthesis, transcriptional activation of defense genes, and cell division stimulation (Kim et al., 2000; van der Holst et al., 2001; Malinowski and Filipecki, 2002).
- Specific receptors located in the plant cell membrane are responsible for chitooligosaccharide perception (Felix et al., 1998). The binding affinity increases with the degree of chitooligomer polymerization: chitotetraose is a minimal length oligomer, whereas the 8-mer produces an intensive cellular response (Day et al., 2001).
- A putative chitooligosaccharide-binding membrane receptor, termed chitinase-related receptorlike kinase (CHRK1), contains an extracellular enzymatically inactive, chitinase-related, and chitooligosaccharide-binding domain, linked via a hydrophobic transmembrane region to an intracellular serine/threonine kinase domain, which triggers signal transduction (Kim et al., 2000).

Vitis vinifera cultivars exhibit differential degrees of sensitivity to the powdery and downy mildews. In addition, the level of susceptibility of any given *V. vinifera* plant to *E. necator* infection varies along time not only in a development-controlled manner but also in an apparently random manner. This may be due to synthesis and accumulation of PR proteins, phytoalexins, and other antifungal compounds as a result of biotic or abiotic stresses (Chapter 25).

PERCEPTION OF PATHOGENS BY PLANT CELLS

Central to the activation of defense responses is timely perception of the pathogen by the plant. Although originally employed to designate any compound capable of inducing the production of phytoalexins, the term elicitor is now commonly used for any molecule that triggers a plant defense reaction. Included in this broader definition of elicitor are molecules of pathogen origin (exogenous elicitors) or host components released or modified by pathogen effectors (endogenous elicitors). Listed in Case Study 28.4 are several examples of elicitors.

Although the first elicitors characterized were predominantly oligosaccharides, research over recent years revealed that elicitors do not belong to any common chemical group but encompass a wide range of chemical structures, including oligosaccharides, peptides, proteins, glycoproteins, and lipids. This indicates that plants are capable of recognizing a tremendous diversity of more or

---

CASE STUDY 28.5

UNIFYING FEATURES OF PATHOGEN-ASSOCIATED MOLECULAR PATTERNS (PAMPs)

- Unifying features of PAMPs are their highly conserved structures, their indispensable nature for microbial fitness, their presence in many microorganisms—pathogenic or not—and their apparent absence in potential host organisms (Nurnberger et al., 2004). They are typically shared by large groups of pathogens.
- The Pep13 motif, for example, is a surface exposed, internal peptide sequence present within a cell-wall transglutaminase that has been highly conserved among different *Phytophthora* species, and it activates resistance responses in solanaceous plants (Brunner et al., 2002) but could not be detected in plants (Parker, 2003).
- The extreme conservation of the Pep13 motif suggests that it cannot easily be mutated to avoid detection by plants.
- Mutational analysis within the Pep13 sequence identified amino acid residues indispensable for both transglutaminase activity and the activation of plant defense responses (Nurnberger et al., 2004).
- In cells of parsley (*Petroselinum crispum*), a nonhost for *P. sojae*, Pep13 derived from this Oomycete triggers a defense reaction (Nurnberger et al., 1994; Hahlbrock et al., 1995).
- Besides Pep13, other pathogen-derived elicitors indispensable for the microbial pathogen may be considered as PAMPs, such as NPP1, fungal chitin, and Oomycete glucans (Fellbrich et al., 2002; Nurnberger et al., 2004).

---

less specific structures from a large number of potential pathogens as signals for pathogen defenses. This problem is compounded by the tendency of pathogens to mutate. The challenge prompted the evolution of an enormous arsenal of perception systems that recognize pathogen-conserved motifs not found in plants. These motifs often play essential roles in the biology of the invading pathogen and are therefore not subjected to high mutation rates. Only in a very few cases has the cognate receptor for a particular elicitor been identified. Nevertheless, the elicitor does not necessarily need to interact with a receptor, but may, for example, form membrane pores or bilayer discontinuities that elicit responses.

Elicitors, which trigger plant defense reactions at low concentrations, are usually clearly set apart from toxins, which operate only at higher concentrations and affect the plant detrimentally. However, the fungal compound fumonisin B1 acts both as a phytotoxin in the interaction of the necrotrophic pathogen *Fusarium verticillioides* with its host maize (*Zea mays*) and as an elicitor in *Arabidopsis*, switching on defense genes and programmed cell death (PCD).

The biological role of elicitor detection in host–pathogen interactions is not fully understood because susceptible plants can support the growth of elicitor-producing pathogens with triggering defense responses. Based on these observations, Jones and Takemoto (2004) suggested that pathogens possess mechanisms to suppress elicitor-induced plant defense signaling or ways to avoid their detection by host plants.

Depending on their mode of action, two major types of elicitors may be considered: general or nonspecific elicitors and race-specific elicitors. General elicitors, which include proteins, glycoproteins, peptides, carbohydrates, and lipids, are apparently involved in the activation of the first line of plant defenses. They are released during attacks by both host and nonhost pathogens and signal the presence of the potential pathogens, in both host and nonhost plants, in a non-cultivar-specific manner. It is becoming increasingly evident that general elicitors play a crucial role in rejection of nonhost pathogens. Their presence is often constitutive and essential for the pathogen lifestyle and, upon receptor-mediated perception, they inevitably betray the invader to the plant's surveillance system. A few of them can be regarded as molecular signatures of microbial invaders and may be considered physiologically equivalent to the invariant pathogen-associated molecular patterns (PAMPs) described for mammals and *Drosophila* (Case Study 28.5). As a result, they are conserved within a class of microbes. For example, general elicitors are released during the enzymatic degradation of pathogen cell-wall polymers when the pathogen is making its way into the host. Such elicitors include oligomers of chitin and glucans, are often indispensable to the microbe, and plants are suggested to exploit them during recognition. The nonspecific nature of general elicitors is relative, however, because some of them are recognized by only a restricted number of plants.

In contrast to general elicitors, specific elicitors are encoded by *avr* genes and trigger cultivar-specific responses, often accompanied by HR. This specific recognition conforms to the gene-for-gene hypothesis and is determined by direct or indirect interaction of host R proteins and cognate pathogen-derived avr effectors. In those cases where avr factors are recognized by the complementary *R*-gene-encoded protein counterparts, they act as specific elicitors of plant defense rather than virulence or pathogenic factors.

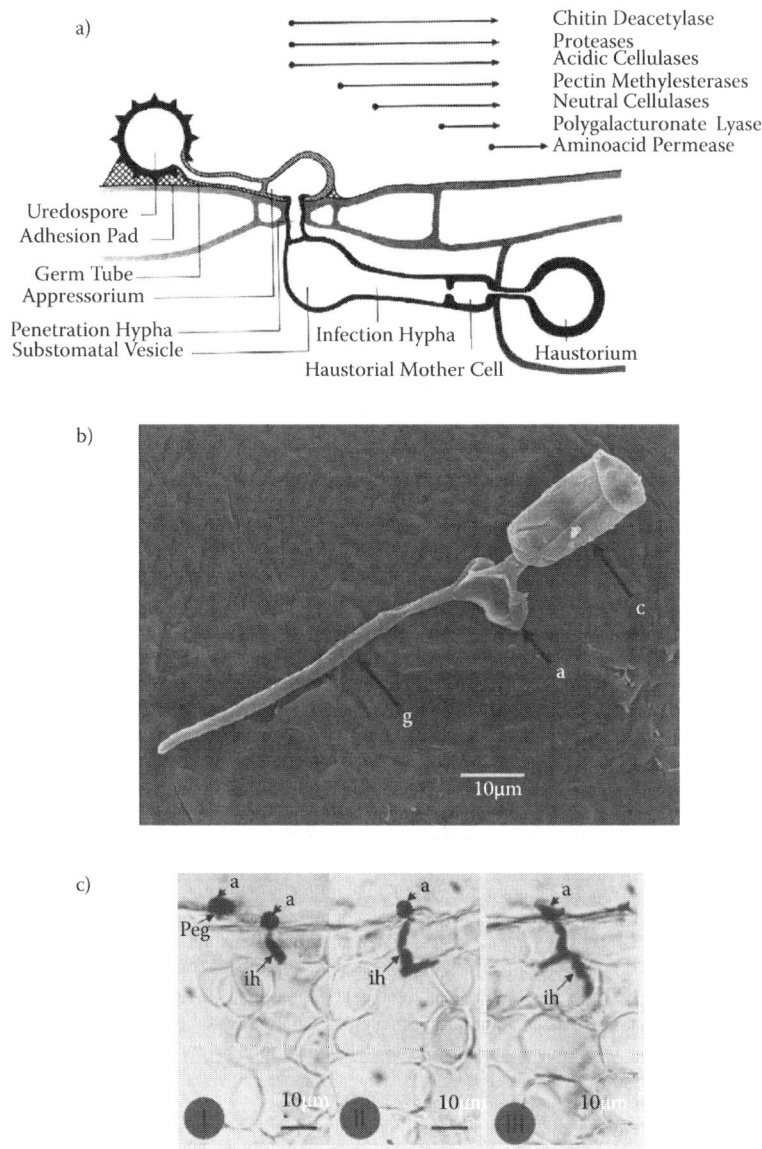

FIGURE 28.2 (a) Concerted differentiation of infection structures and production of enzymes by *Uromyces viciae-fabae*. Penetration occurs through stomata. (b) *Erysiphe necator* conidium (c), 16 h after inoculation on a *Vitis vinifera* cv. Dona Maria leaf, producing appressorium (a) and infecting germ tube (g). Penetration occurs through the intact leaf surface. (c) The infection process of *Colletotrichum gloeosporioides*: infected heat shocked (55°C, 30 sec) coffee leaves at 36 (I), 48 (II), and 72 h (III) after inoculation. a: appressoria; peg: infection peg; ih: infection hyphae.

THE CHALLENGING BATTLE TO INFECTION

RECOGNITION OF HOST BY THE PATHOGEN

A host pathogen must overcome a series of obstacles before it succeeds in causing disease. Usually, if a pathogen fails to survive one of the first obstacles, it is most likely a non-host pathogen. The same obstacles are also serious challenges for host pathogens and often reduce their success rate significantly. For successful infection, host-cell entry or penetration through the cuticle and the plant cell wall represents a critical and crucial step in the pathogenesis of most biotrophic fungi and marks a lifestyle transition from extracellular to invasive growth. Figure 28.2 represents the major steps in developing pathogenesis by the biotrophic fungi *Uromyces viciae-fabae* and *Erysiphe necator,* and the hemibiotrophic fungus *Colletotrichum gloeosporioides*.

The earliest step in any typical plant–pathogen interaction involves recognition of host by the pathogen followed by perception of the pathogen by the host defense machinery. Pathogens require specific signals from the plant surface to induce cell differentiation and express essential pathogenic genes. The major signals provided by the host are hydrophobicity, surface hardness, components of the plant surface, and topographical properties. Thig-

---

CASE STUDY 28.6

NEW MOLECULAR GENETIC DATA CHALLENGE THE TRADITIONAL VIEW OF THE PLANT CELL WALL AS A PASSIVE STRUCTURAL BARRIER TO PATHOGEN INVASION

- Apparently, plants are able to sense perturbation of the cell wall by monitoring the integrity of its structure. Biotrophic fungi may manipulate this surveillance system for the establishment of biotrophy, subverting the interconnected plant defense signaling pathways or the underlying resistance mechanisms (Jones and Takemoto, 2004; Schulze-Lefert, 2004).
- For example, mutations in the *Arabidopsis* cellulose synthase gene *CESA3* exhibit constitutive activation of jasmonate- and ethylene-mediated defense gene expression and enhanced resistance to powdery mildew pathogens (Cano-Delgado et al., 2003; Nishimura et al., 2003).

CASE STUDY 28.7

RECENT OBSERVATIONS REVEAL THAT MOLECULAR PROCESSES OCCURRING AT AND IN PLANT CELL WALLS MAY ALSO FUNCTION IN FUNGAL PATHOGENESIS

- The papillae are localized cell-wall fortifications formed on the inner side of plant cell walls at the site of pathogen penetration that are generally regarded as an inducible structural barrier.
- In the case of powdery mildews, the papillae have been reported to play an important role against fungal invasive growth (Thordal-Christensen et al., 2000; Zeyen et al., 2002).
- The β-1,3-D-glucan callose is rapidly synthesized and deposited at plant cell walls upon microbial attack.
- *Arabidopsis* mutants in the gene encoding the single glucan synthase responsible for papillary callose synthesis exhibit broad-spectrum enhanced resistance to powdery mildew fungi, suggesting a role for the wild-type gene in fungal colonization of the host cells. Callose may facilitate penetration of pathogens into host cells by providing a structural collar for the intruder. The glucan also accumulates after successful invasion of the pathogen into host cells at haustorial complexes (Jacobs et al., 2003; Nishimura et al., 2003).
- Glucan synthase participates in the containment of pathogen-derived elicitors at infection sites, thereby preventing their perception by the plant, or in the protection of the invading pathogen against plant-derived antimicrobial compounds (Gomez-Gomez and Boller, 2002; Schulze-Lefert, 2004).

---

motropic responses direct germ tubes on leaf surfaces, allowing some plant pathogens to recognize an array of anticlinal walls or stomatal openings.

Passive or nonmetabolic adhesion of fungal structures to the host surfaces often involves hydrophobic interactions between spores and hyphae and the cuticle. A second stage follows involving secretion of a protein, glycoprotein, or carbohydrate sheath around the germ tube and parts of the cuticle to enable hyphae to sense the plant surface and to differentiate infection structures. Some spores release cutinases or esterases (Chapter 10) that erode the cuticle and contribute to fungal adhesion.

The cuticle and the plant cell wall are preformed physical barriers often claimed to constitute the first line of plant defense by protecting against pathogen penetration. In addition, they are also a source of signals used by the invading pathogen to activate pathogenic responses or by plants to induce defense mechanisms. Nevertheless, new molecular genetic data challenge the traditional view of the plant cell wall as a passive structural barrier to pathogen invasion (see, for example, Case Studies 28.6 and 28.7).

PENETRATION: TRANSITION FROM EXTRACELLULAR TO INVASIVE GROWTH

Fungi have evolved an astonishing array of invasive strategies. Targeted and regulated secretion of cutinases and a variety of cell wall–degrading enzymes (depolymerases) (see Chapters 4 and 30), production of phytotoxic metabolites such as mycotoxins and oxalic acid, and, as shown in Case Study 28.8, an increase in pressure within the fungal infection structures support penetration of the cuticle and plant cell walls.

The expression of fungal pathogen–induced, cell wall–associated defense responses has been shown to be dependent on adhesion between the plant cell wall and the plasma membrane. Peptides containing an Arg-Gly-Asp

CASE STUDY 28.8

FUNGAL PENETRATION OF HOST CUTICLE AND PLANT CELL WALLS

- Although capable of synthesizing cutinases and cell wall–degrading enzymes, fungi with melanized appressoria, such as *Magnaporthe* and *Colletotrichum*, penetrate the host cuticle and cell wall mainly by means of turgor pressure (Chen et al., 2004).
- An enormous turgor pressure of 8.0 MPa (80 bar; compare with the normal air pressure of 0.2 MPa (2 bar) used in a typical car tire), apparently due to molar concentrations of glycerol, was measured in appressoria of *M. grisea* (Howard et al., 1991).
- In the case of *Blumeria graminis* f. sp. *hordei*, penetration is thought to result from combined cellulase activity and a lower turgor pressure of about 2 to 4 MPa. In this study, the presence of fungal cellulase at the appressorial germ tube tip was found to correlate in time with penetration of the host-cell wall (Pryce-Jones et al., 1999).
- An elegant study demonstrated that unlike forward movement and shape of the hyphal tip, penetration of the water mold *Saprolegnia ferax* is mainly mediated by turgor pressure (Money, 1995).
- Hyphae of *S. ferax* grown at full turgor pressure (0.44 MPa) are able to penetrate into the agar medium.
- A reduction of turgor pressure to less than 0.02 MPa by increased concentrations of osmolytes in the medium decreased the capability of the hyphae to penetrate the agar, but not their ability to grow forward on the agar surface.

(RGD) motif, which interferes with plasma membrane–cell wall adhesion as shown by the loss of the thin plasma membrane–cell wall connections known as Hechtian strands, reduce the expression of cell wall–associated defense responses during the penetration of nonhost plants by biotrophic fungal pathogens. This reduction is associated with increased fungal penetration efficiency. The same study demonstrated that disruption of plant microfilaments has no effect on Hechtian strands but mimics the effect of RGD peptides on wall defenses, suggesting that the expression of cell wall associated defenses involves communication between the plant cell wall and the cytosol across the plasma membrane. In this respect, it was suggested that rust fungi may induce a decrease in plasma membrane–cell wall adhesion as a means of disrupting the expression of nonspecific defense responses during penetration of host cells.

Both *Magnaporthe grisea* and *Colletotrichum* species incorporate a melanin layer into their appressorial walls. The essential role of melanin for penetration is provided by the chemically induced melanin-deficient strains of *C. kahawae* and in melanin-deficient mutant strains of *M. grisea* that are unable to form melanized appressoria, fail to generate turgor, and are nonpathogenic (Figure 28.3). Addition of appropriate precursors of melanin biosynthesis restored melanin incorporation into apressorium and pathogeny of the mutants.

Pathogenic fungi that do not differentiate a fully developed appressorium rely basically on cutinase and cell wall–degrading enzymes for penetration (Chapter 30). Plant cell-wall fragments and polymers such as pectin, polygalacturonic acid, cellulose, and xylans activate fungal transcription factors capable of inducing the expression of genes encoding an array of cell-degrading enzymes required to degrade those carbon sources. However, whether or not enzymes, such as cutinases, laccases, polygalacturonases, pectin, and pectate lyases and pectin methylesterases, are required for penetration of plant cell walls or only for saprophytic growth of the fungus remains to be elucidated. Determination of the role played by these enzymes, frequently exhibiting overlapping activities, in penetration is hampered by their redundancy and variable regulation. Functional redundancy is often indicative of processes with vital importance to an organism. Alternatively, it may derive from the multimillion-year, ping-pong-type coevolution, in which plant and pathogen successfully added new weapons in this perpetual protein battle.

THE PROTEIN BATTLE FOR PENETRATION

The attempted infection of a plant by a pathogen involves a battle whose major weapons are essentially proteins produced by both organisms. The outcome of this complex confrontation determines in large part the success or failure of the attempted pathogenesis. Figure 28.4 summarizes the major known steps in this warfare.

Plant cell walls, which consist mainly of polysaccharides (i.e., cellulose, hemicelluloses, and pectins) and proteins, play an important role in defending plants against pathogens. Many pathogenic fungi release an array of cell wall degrading, hydrolytic enzymes to fragment the plant cell wall polymers, including proteases and glycanases (e.g., galacturonases, xylanases, and glucanases), thus facilitating the colonization of the host cells. Several pathogenic fungi secrete

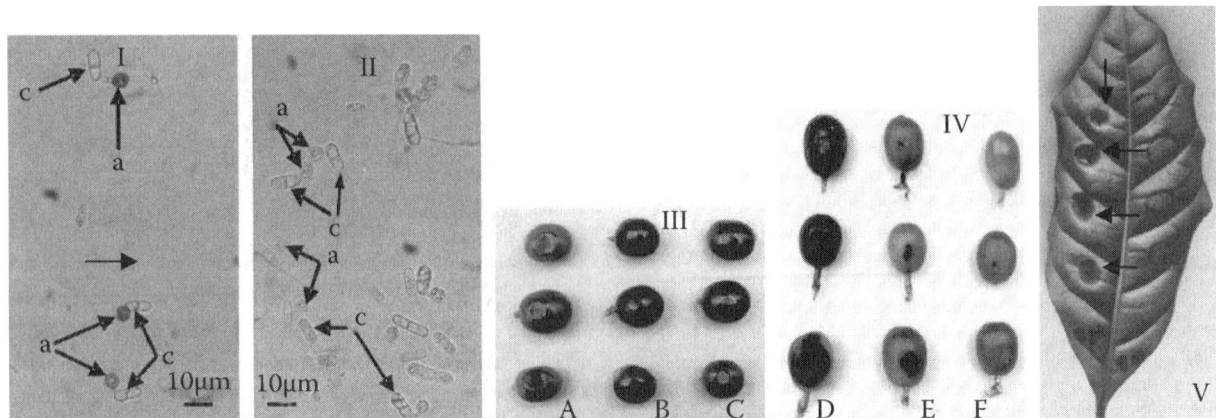

FIGURE 28.3 Influence of epicatechin on the appressorial melanization of *Colletotrichum kahawae* and on the symptom expression on ripe and green coffee berries. I: conidia germinated in distilled H_2O (control; a: appressorium; c: conidium); II: Inoculated conidia in a suspension containing 1.2 mg epicatechin/mL; III and IV: Symptoms expression on ripe and green coffee berries, respectively; inoculated conidia containing 0 (A and D), 1.2 (B and E), and 2.3 mg (C and F) epicatechin/mL. Symptoms were evaluated at day 7 (ripe berries) and at day 10 (green berries) after inoculation. Inoculated zones were circled with a black pen on ripe berries. Similar results were obtained on coffee leaves (V) inoculated with a conidial suspension of *C. kahawae* in tricyclazole at 10 μ/mL (right half; spots circled with a black pen) or in H_2O (control; left half). Note normal necrotic lesions in control (arrowheads in left half) and no lesions in the right half; picture taken at day 7 after inoculation.

proteases whose importance in pathogenesis is highlighted by the observation that protease-deficient mutants lose the ability to induce lesions in plants. Also, during appressorium formation, several fungal proteases exhibit specificity toward fibrous hydroxyproline-rich proteins.

Pectin, a major component of the cell walls in many plants, is cleaved by fungal endopolygalacturonases (EPGs) with the transient formation of elicitor-active oligogalacturonides (OGAs) with degrees of polymerization between 9 and 15. Thus, the oligosaccharides generated by glucanases not only provide the fungus with a carbon source but are also perceived by, and elicit, the plant defense mechanisms. For this reason, OGAs are rapidly converted to smaller, biologically inactive fragments by the EPGs.

To increase the lifetime of the biologically active oligosaccharides, plants release inhibitors of the fungal glycanases. These include inhibitors of pectin-degrading enzymes such as polygalacturonases, pectin methyl esterases and pectin lyases, and cross-linking glycan (known earlier as hemicelluloses) degrading enzymes such as endoxylanases and xyloglucan endoglucanases. For example, plant polygalacturonase-inhibiting proteins (PGIPs) are glycoproteins present in the apoplast of many plants that form reversible high-affinity complexes with fungal EPGs, reducing their catalytic activity by one to two orders of magnitude. By limiting EPG activity, the lifetime and concentration of OGAs are increased, prolonging or enhancing plant defense responses. PGIPs limit the growth of plant pathogens and also elicit defense responses in plants. They belong to the superfamily of leucine-rich repeat (LRR) proteins, which also include the products of several plant resistance genes.

Fragmentation of the fungal cell wall by plant-derived chitinases and β-1,3-glucanases also generates oligosaccharides that induce plant defense responses. In response, fungi produce glucanase inhibitor proteins (GIPs) that prevent degradation of their own cell wall, thus limiting their perception by the plants.

Once inside the host tissues, many biotrophic fungi, such as rust and powdery mildew fungi, form a specialized infection structure called haustorium with a presumed role in nutrient uptake. Interestingly, an increase in chitin deacetylase activity is observed after appressoria development in *C. lindemuthianum* and *U. viciae-fabae*. Chitin deacetylation on the surface of the penetration hypha has been suggested to protect the fungal cell wall from the action of plant chitinases, which have both a direct damaging effect and an elicitor-releasing activity.

The tremendous threat imposed by pathogenic fungi on crops, the development of fungal resistance mechanisms against the fungicides currently available in the market, and the increasing public concern over the use of toxic chemical fungicides exert a continuous pressure on the search for nontoxic and environment-friendly new fungicides. In the era of "omics," recognition of potential targets may be achieved by studying the changes occurring in the genome, proteome, and metabolome of both host plants and pathogens. Identification of elicitor receptors in the near future and of the action mechanisms of the increasing number of known avr/R effectors will certainly contribute to a complete understanding of the processes underlying pathogenesis and host resistance and lead the way to the development of suitable tools to engineer plants into crops resistant to pathogen attack. The ultimate and challenging goal will be

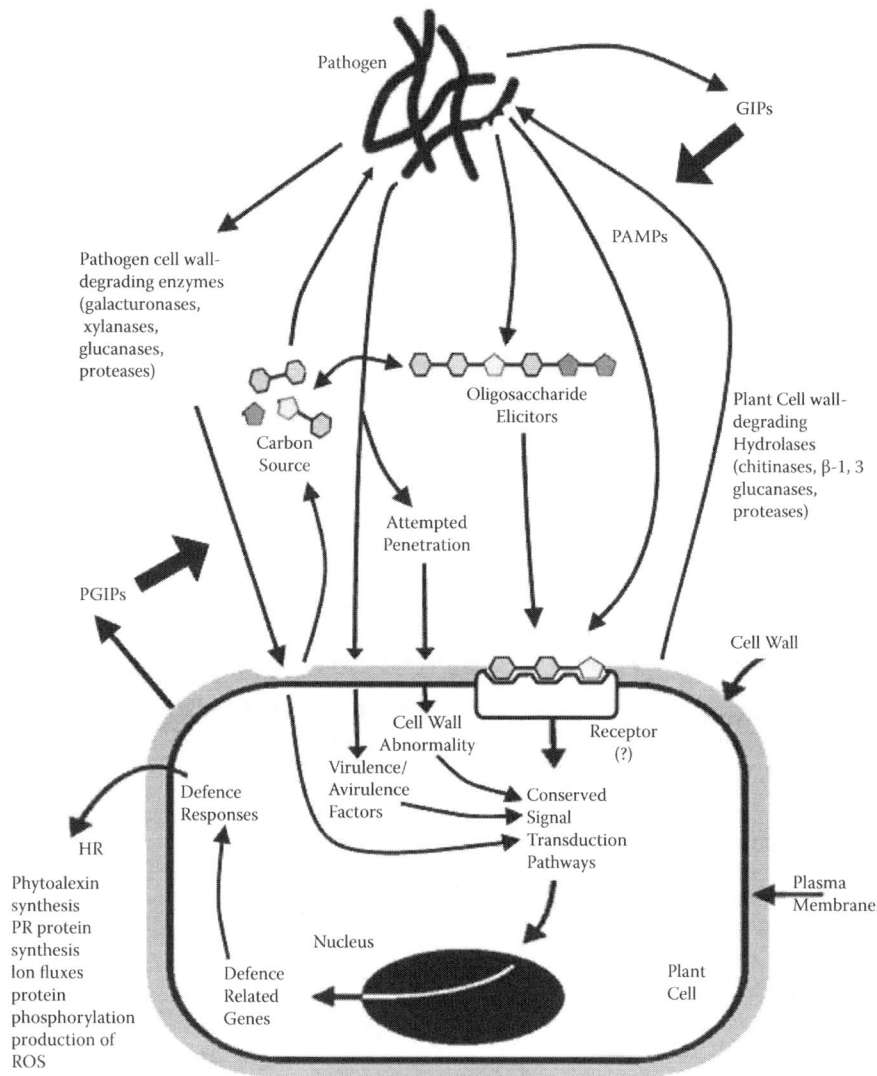

FIGURE 28.4 The protein warfare going on between plant host and attacking pathogen during attempted infection, and the path leading to plant defense response. Pathogen refers to fungal or Oomycete invaders. Abbreviations: PAMP, pathogen-associated molecular pattern; GIP, glucanase inhibitor protein; PGIP, polygalacturonase-inhibiting protein; HR, hypersensitive response; PR, pathogenesis-related; ROS, reactive oxygen species.

the unveiling of the plant signal transduction pathways following perception of pathogen by the host and the complex ways in which they interact. This will culminate in an academic achievement of excellence with potential practical applications that only time will unfold.

LITERATURE CITED

Bowyer, P., Clarke, B.R., Lunness, P., Daniels, M.J. and Osbourn, A.E. 1995. Host range of a plant pathogenic fungus determined by a saponin detoxifying enzyme. *Science*, 267: 371–374.

Brunner, F., Rosahl, S., Lee, J., Rudd, J.J., Geiler, C., Kauppinen, S., Rasmussen, G., Scheel, D. and Nurnberger, T. 2002. Pep-13, a plant defense-inducing pathogen-associated pattern from *Phytophthora* transglutaminases. *EMBO J.*, 21: 6681–6688.

Cano-Delgado, A., Penfield, S., Smith, C., Catley, M. and Bevan, M. 2003. Reduced cellulose synthesis invokes lignification and defense responses in *Arabidopsis thaliana*. *Plant J.*, 34: 351–362.

Che, F.S., Nakajima, Y., Tanaka, N., Iwano, M., Yoshida, T., Takayama, S., Kadota, I. and Isogai, A. 2000. Flagellin from an incompatible strain of *Pseudomonas avenae* induces a resistance response in cultured rice cells. *J. Biol. Chem.*, 275: 32347–32356.

Chen, Z., Nunes, M.A., Silva, M.C. and Rodrigues, C.J., Jr. 2004. Appressorium turgor pressure of *Colletotrichum kahawae* might have a role in coffee cuticle penetration. *Mycologia*, 96: 1199–1208.

Day, R.B., Okada, M., Ito, Y., Tsukada, K., Zaghouani, H., Shibuya, N. and Stacey, G. 2001. Binding site for chitin oligosaccharides in the soybean plasma membrane. *Plant Physiol.*, 126: 1162–1173.

Ebel, J. and Cosio, E.G. 1994. Elicitors of plant defense responses. *Intl. Rev. Cytol.*, 148: 1–36.

Ebel, J. and Scheel, D. 1997. *Signals in host-parasite interactions*, Springer-Verlag, Heidelberg.

Felix, G., Baureithel, K. and Boller, T. 1998. Desensitization of the perception system for chitin fragments in tomato cells. *Plant Physiol.*, 117: 643–650.

Fellbrich, G., Romanski, A., Varet, A., Blume, B., Brunner, F., Engelhardt, S., Felix, G., Kemmerling, B., Krzymowska, M. and Nurnberger, T. 2002. NPP1, a *Phytophthora*-associated trigger of plant defense in parsley and *Arabidopsis*. *Plant J.*, 32: 375–390.

Fujiwara, S., Tanaka, N., Kaneda, T., Takayama, S., Isogai, A. and Che, F.S. 2004. Rice cDNA microarray-based gene expression profiling of the response to flagellin perception in cultured rice cells. *Mol. Plant Microbe Interact.*, 17: 986–998.

Gomez-Gomez, L. and Boller, T. 2002. Flagellin perception: a paradigm for innate immunity. *Trends Plant Sci.*, 7: 251–256.

Hahlbrock, K., Scheel, D., Logemann, E., Nurnberger, T., Pariske, M., Reinold, S., Sacks, W.R. and Schmelzer, E. 1995. Oligopeptide elicitor-mediated defense gene activation in cultured parsley cells. *Proc. Natl. Acad. Sci. USA*, 92: 4150–4157.

Howard, R.J., Ferrari, M.A., Roach, D.H. and Money, N.P. 1991. Penetration of hard substrates by a fungus employing enormous turgor pressures. *Proc. Natl. Acad. Sci. USA*, 88: 11281–11284.

Jacobs, A.K., Lipka, V., Burton, R.A., Panstruga, R., Strizhov, N., Schulze-Lefert, P. and Fincher, G.B. 2003. An *Arabidopsis* callose synthase, GSL5, is required for wound and papillary callose formation. *Plant Cell*, 15: 2503–2513.

Jones, D.A. and Takemoto, D. 2004. Plant innate immunity— direct and indirect recognition of general and specific pathogen-associated molecules. *Curr. Opin. Immunol.*, 16: 48–62.

Keen, N.T. 1999. Plant disease resistance: Progress in basic understanding and practical application. In *Advances in Botanical Research Incorporating Advances in Plant Pathology*, 30: 291–328.

Kim, Y.S., Lee, J.H., Yoon, G.M., Cho, H.S., Park, S.W., Suh, M.C., Choi, D., Ha, H.J., Liu, J.R. and Pai, H.S. 2000. CHRK1, a chitinase-related receptor-like kinase in tobacco. *Plant Physiol.*, 123: 905–915.

Malinowski, R. and Filipecki, M. 2002. The role of cell wall in plant embryogenesis. *Cell Mol. Biol. Lett.*, 7: 1137–1151.

Money, N.P. 1995. Turgor pressure and the mechanics of fungal penetration. *Can. J. Bot.*, 73: S96–S102.

Nielsen, K.A., Hrmova, M., Nielsen, J.N., Forslund, K., Ebert, S., Olsen, C.E., Fincher, G.B. and Moller, B.L. 2006. Reconstitution of cyanogenesis in barley (*Hordeum vulgare* L.) and its implications for resistance against the barley powdery mildew fungus. *Planta*, 223: 1010–1023.

Nishimura, M.T., Stein, M., Hou, B.H., Vogel, J.P., Edwards, H. and Somerville, S.C. 2003. Loss of a callose synthase results in salicylic acid-dependent disease resistance. *Science*, 301: 969–972.

Nurnberger, T., Brunner, F., Kemmerling, B. and Piater, L. 2004. Innate immunity in plants and animals: striking similarities and obvious differences. *Immunol. Rev.*, 198: 249–266.

Nurnberger, T., Nennstiel, D., Jabs, T., Sacks, W.R., Hahlbrock, K. and Scheel, D. 1994. High affinity binding of a fungal oligopeptide elicitor to parsley plasma membranes triggers multiple defense responses. *Cell*, 78: 449–460.

Papadopoulou, K., Melton, R., Leggett, M., Daniels, M. and Osbourn, A. 1999. Compromised disease resistance in saponin-deficient plants. *Proc. Natl. Acad. Sci. USA*, 96: 12923–12928.

Parker, J.E. 2003. Plant recognition of microbial patterns. *Trends Plant Sci.*, 8: 245–247.

Pryce-Jones, E., Carves, T. and Gurr, S.J. 1999. The roles of cellulase enzymes and mechanical force in host penetration by *Erysiphe graminis* f.sp. hordei. *Physiol. Mol. Plant Pathol.*, 55: 175–182.

Repka, V. 2006. Early defense responses induced by two distinct elicitors derived from a *Botrytis cinerea* in grapevine leaves and cell suspensions. *Biol. Plant.*, 50: 94–106.

Schulze-Lefert, P. 2004. Knocking on the heaven's wall: pathogenesis of and resistance to biotrophic fungi at the cell wall. *Curr. Opin. Plant Biol.*, 7: 377–383.

Shibuya, N. and Minami, E. 2001. Oligosaccharide signalling for defense responses in plant. *Physiol. Mol. Plant Pathol.*, 59: 223–233.

Shimizu, R., Taguchi, F., Marutani, M., Mukaihara, T., Inagaki, Y., Toyoda, K., Shiraishi, T. and Ichinose, Y. 2003. The DeltafliD mutant of *Pseudomonas syringae* pv. tabaci, which secretes flagellin monomers, induces a strong hypersensitive reaction (HR) in non-host tomato cells. *Mol. Genet. Genomics*, 269: 21–30.

Taguchi, F., Shimizu, R., Inagaki, Y., Toyoda, K., Shiraishi, T. and Ichinose, Y. 2003. Post-translational modification of flagellin determines the specificity of HR induction. *Plant Cell Physiol.*, 44: 342–349.

Taguchi, F., Takeuchi, K., Katoh, E., Murata, K., Suzuki, T., Marutani, M., Kawasaki, T., Eguchi, M., Katoh, S., Kaku, H., Yasuda, C., Inagaki, Y., Toyoda, K., Shiraishi, T. and Ichinose, Y. 2006. Identification of glycosylation genes and glycosylated amino acids of flagellin in *Pseudomonas syringae* pv. tabaci. *Cell Microbiol.*, 8: 923–938.

Tanaka, N., Che, F.S., Watanabe, N., Fujiwara, S., Takayama, S. and Isogai, A. 2003. Flagellin from an incompatible strain of *Acidovorax avenae* mediates H2O2 generation accompanying hypersensitive cell death and expression of PAL, Cht-1, and PBZ1, but not of Lox in rice. *Mol. Plant Microbe Interact.*, 16: 422–428.

Thordal-Christensen, H. 2003. Fresh insights into processes of nonhost resistance. *Curr. Opin. Plant Biol.*, 6: 351–357.

Thordal-Christensen, H., Gregersen, P. and Collinge, D. 2000. The barley/*Blumeria* (syn. Erysiphe) *graminis* interaction. In *Mechanisms of Resistance to Plant Diseases*. 77. Slusarenko, A. and L.C. van Loon, (Eds.), Kluwer Academic Publishers, Dordrecht, The Netherlands.

van der Holst, P.P., Schlaman, H.R. and Spaink, H.P. 2001. Proteins involved in the production and perception of oligosaccharides in relation to plant and animal development. *Curr. Opin. Struct. Biol.*, 11: 608–616.

Zeyen, R., Carver, T. and Lyngkjaer, M. 2002. Epidermal cell papillae. In *The Powdery Mildews: A Comprehensive Treatise*. Bushnell, W. (Ed.), APS Press, St. Paul, MN.

29 Testing Blad, a Potent Antifungal Polypeptide

Sara Monteiro and Ricardo B. Ferreira

The control of pathogenic fungi constitutes a serious problem worldwide with respect to the most important crops. Control is generally achieved by massive applications of chemical fungicides (Chapter 36). The economic costs, the negative environmental impact, and the increasing public concern associated with such applications led to a search for alternative strategies.

A promising alternative in the fight against phytopathogens is the identification and purification of substances of biological origin with potent antifungal activity (Chapter 37). The detection of such compounds involves screening a variety of organisms, such as plants and microorganisms, for substances that are subsequently tested in antifungal assays and finally isolated and characterized. In this way, many classes of antifungal proteins have already been isolated, including chitinases, cysteine-rich proteins that bind strongly to chitin, β-1,3-glucanases, permeatins, thionins, and lipid transfer proteins. These proteins are thought to play a fundamental role in the natural defenses of plants against the attack of microbial pathogens.

Recently, a novel protein, termed Blad, was discovered and identified. This protein occurs in germinating cotyledons of plants belonging to the genus *Lupinus* and exhibits the following unusual characteristics:

1. Potent antifungal activity, due to the breakdown of several cell wall components that confers great potential as a fungicide
2. Extreme resistance to denaturation, which makes it possible to utilize it under field conditions
3. Very high susceptibility to proteolysis, which makes it harmless to the environment and nontoxic to man

Therefore, Blad is a promising target, with great potential to be developed into an efficient method to control the fungi that affect plants. This potential has been protected under patent application number PCT/IB2006/052403.

The fungus *Botrytis cinerea* Pers. is one of the most destructive plant pathogens. It attacks over 200 plant species, and infects flowers, fruits, and vegetative tissues. This fungus causes gray mold, or *Botrytis* blight in grapevine plants, one of the important damaging diseases affecting this crop (Hausbeck, 1993). The disease symptoms are characterized by gray sporulating lesions that are commonly observed under humid conditions. These lesions produce masses of conidia that become airborne and are the primary means by which the fungus is disseminated. A wet, humid, greenhouse environment is ideal for the rapid growth and prolific sporulation of *B. cinerea* (Hausbeck and Moorman, 1996). Control of *B. cinerea* is challenging because of its abilities to survive as a saprophyte, rapidly invading host tissue and quickly producing abundant conidia that are easily distributed by air currents. Moreover, the fungus is capable of growing within a wide range of temperatures. The optimum temperature for *B. cinerea* growth is 24 to 28°C, but fungal growth occurs between 0 and 35°C. Numerous fungicides, such as benzimidazole and dicarboximide, are used to control gray mold. Many of them are preventives and must be repeatedly applied to prevent the appearance of infection symptoms. However, this is becoming unacceptable because it conflicts with public concern about fungicide residues and also increases the potential for the build-up of resistance in *B. cinerea* populations to fungicides. In this respect, resistances to these chemicals have already been reported in natural populations of *B. cinerea* (Leroux et al., 1999). Biological control, on the other hand, has advantages over fungicides, but its efficacy varies depending on timing and environmental conditions. Finally, traditional breeding for resistant cultivars is a difficult problem, mainly because of a lack of host resistance to *B. cinerea*. Consequently, finding specific compounds exhibiting antifungal properties against *B. cinerea* is a requisite for disease control and/or for creating cultivars with improved resistance to this pathogen.

In this laboratory exercise, the ability of *Lupinus albus* Blad protein to inhibit *B. cinerea* growth is investigated. *In vitro* antifungal assays (see also Chapter 41) are performed to determine the potential of Blad to inhibit mycelial growth as well as conidial germination of *B. cinerea*. This information is useful to design strategies to enhance crop protection against *Botrytis* diseases.

EXERCISES

The following experiments are designed to provide hands-on training for students working in phytopathology to give

TABLE 29.1

Mixture for Preparation of 12.5% (w/v) Acrylamide Slab Gels (SDS-PAGE)

	Separating Gel		Stacking Gel	
Stock solutions/Number of gels	1	2	1	2
Acrylamide 30% (w/v) (mL)	16.7	33.4	1.67	3.34
Bis-acrylamide 1% (w/v) (mL)	4.1	8.2	1.3	2.6
H2O (double-distilled) (mL)	4.3	8.6	6.1	12.2
1M Tris-HCl (pH 8.8) (mL)	14.9	29.8	—	—
1M Tris-HCl (pH 6.8) (mL)	—	—	1.25	2.5
TEMED (μL)	13.3	26.6	5.0	10.0
PSA 10% (w/v) (μL)	130	260	100	200

them experience with proteins having antifungal properties and applying these in the control of pathogenic agents that attack plants.

EXPERIMENT 1: GERMINATING *LUPINUS* SEEDS AND BLAD PURIFICATION

Blad is a novel protein with potent antifungal properties that exhibits a powerful inhibitory activity on the germination and development of spores from fungal pathogens of plants.

Blad is extracted from germinated seedlings of the genus *Lupinus* and is a phosphorylated and nonglycosylated 20-kDa polypeptide that is part of a larger 210-kDa protein. This protein occurs naturally between the 4th and 12th day after the onset of germination. β-conglutin (accession number AY500372), the major seed storage protein from *Lupinus* genus, is the biosynthetic precursor of Blad. Indeed, Blad is a highly processed polypeptide that has undergone several levels of chemical modification. The initial steps in the catabolism of β-conglutin involve proteolytic cleavage of all or most of its constituent subunits, resulting in the accumulation of Blad.

Due to its intrinsic antifungal properties, which are naturally exploited by the host plant, this protein is maintained in very high concentrations in the cotyledons of the developing plants during a life stage in which the plant is most sensitive to fungal and insect attack. After a few days, the protein is degraded and its amino acids used in the growth of the young plant, fulfilling its additional role as a seed storage protein.

The DNA nucleotide sequence of the gene fragment that encodes Blad (accession number DQ142920) does not share any significant homology with any other antifungal protein that has been isolated from plants or other organisms. Blad constitutes a novel type of protein among the antifungal proteins that have been described to date.

In this experiment, seeds from *Lupinus albus* L. are germinated and grown for a period of 8 d and the total protein from the resulting cotyledons extracted. The amount of protein obtained is quantified and the presence of Blad detected by polyacrylamide gel electrophoresis.

Materials

Each student or team of students will require the following items:

- Dry seeds of white lupin (*Lupinus albus* L.) obtained from a seed supplier
- Plastic pots
- Sterilized soil
- Water
- Liquid nitrogen
- Mortar and pestle
- Extraction buffer: 100 mM Tris-HCl buffer, pH 7.5, containing 10% (w/v) sodium chloride (NaCl), 10 mM ethylenediaminetetraacetic acid (EDTA), and 10 mM ethyleneglycol-bis(β-aminoethyl-ether)-*N,N,N',N'*-tetraacetic acid (EGTA)
- Refrigerated centrifuge
- Desalting PD-10 columns (GE Heathcare)
- Spectrophotometer
- Vertical electrophoresis apparatus
- Sodium dodecyl sulfate-polyacrylamide gel electrophoresis (SDS-PAGE) (Table 29.1)
- Electrophoresis buffer for SDS-PAGE (stock 10× concentrated): 0.25 M Tris-HCl, 1.92 M Glycine, pH 8.3, 20% (w/v) SDS
- Sample buffer (4× concentrated): 0.8 M Tris-HCl buffer, pH 6.8, containing 15% (v/v) glycerol, 2% (w/v) SDS, 0.01% (w/v) *m*-cresol purple (tracking dye), and 0.1 M ß-mercaptoethanol
- Molecular mass standards for SDS-PAGE: bovine serum albumin (BSA) 66 kDa, ovalbumin 45 kDa, carbonic anhydrase 29 kDa, trypsinogen

---

Procedure 29.1

Extraction and Electrophoresis of the Blad Protein

Step	Instructions and Comments

Germination and Seedling Growth

1 Hand sort the dry seeds of Lupinus albus to remove damaged seeds and imbibe the seeds under aerated, running tap water for 2 days.

2 Transplant the seedlings into sand (depth 0.5 cm) and grow for 6 days at 25°C with a 16 h/8 h light/dark photoperiod.

3 Harvest both cotyledons from each plant. Freeze each individual 1 g of tissue in liquid nitrogen and store frozen until required.

Total Protein Extraction

4 Grind 1 to 3 g of frozen material under liquid nitrogen to a fine powder using a chilled mortar and pestle (5 mL of extraction buffer for 1 g of ground material). Incubate at 4°C for 30 min. Centrifuge for 1 h at 30,000 g and 4°C.

5 Desalt the resulting supernatant on PD-10 columns previously equilibrated in double-distilled water, pH adjusted to 7.5. This desalted extract containing Blad is used for the antifungal assays.

6 Quantify the total protein extracted, assuming an extinction coefficient (Ecoef.) of 1.85 mL.mg^{-1}.cm^{-1} for this fraction. Read in a spectrophotometer at 280 nm the sample absorbance. Using the formula

$$\text{Abs.} = [\text{sample protein concentration}] \times \text{Ecoef.} \times 1 \text{ cm},$$

calculate the protein concentration of the sample.

SDS-PAGE Electrophoresis

7 Sample preparation for SDS-PAGE electrophoresis: Use 80 µg of the protein sample to run electrophoresis. Add sample buffer to the sample (1:3) and boil for 3 min. Freeze the electrophoretic sample at −20°C until needed.

8 Prepare the SDS-PAGE gel according to Table 29.1. Prepare the electrophoresis buffer 1×. Add to the anode buffer 0.1 M sodium acetate to resolve polypeptides with molecular masses ranging from 2.5 kDa to greater than 200 kDa.

9 Apply the sample to the gel using the molecular markers as control. Run electrophoresis at 200 V and 70 mA for 2.75 h or until the tracking dye has migrated through the entire gel.

10 Dismantle the electrophoretic apparatus. Transfer the gel into a container with 12% (w/v) TCA, and incubate for 10 min with gentle agitation.

11 Remove the TCA from the container, add the staining solution, and incubate under gentle agitation. Staining will take at least 2 h.

12 Destain the gel with destaining solution for several hours. This solution should be renewed several times until the stained polypeptides can be visualized against a pale blue background.

13 Store the gel in 10% (v/v) acetic acid.

---

24 kDa, trypsin inhibitor 20.1 kDa, and α-lactalbumin 14.2 kDa. This mixture may be obtained from Sigma, St. Louis

- 12% (w/v) trichloroacetic acid (TCA)
- Staining solution: 0.25% (w/v) Coomassie Brilliant Blue R-250, 25% (v/v) 2-propanol, and 10% (v/v) acetic acid
- Destaining solution: 25% (v/v) 2-propanol and 10% (v/v) acetic acid
- Storing solution: 10% (v/v) acetic acid

Follow the protocol listed in Procedure 29.1 to complete this experiment.

Anticipated Results

Under the conditions studied, the cotyledons emerge from the sand and become photosynthetic (green) 3 to 4 days after the onset of germination. The analysis of the electrophoretic gel shown in Figure 29.1 indicates the presence of an abundant 20-kDa polypeptide (Blad) as well as a number of other minor polypeptides.

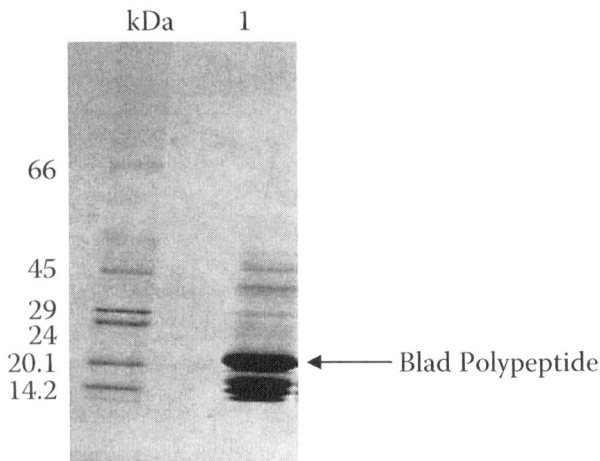

FIGURE 29.1 SDS-PAGE of the total protein fraction extracted from 8-days-old *Lupinus albus* cotyledons (lane 1). The presence of Blad is highlighted. The molecular masses of standards are indicated (lane kDa).

Questions/Discussion

- Why does the *Lupinus* polypeptide migrate downwards during electrophoresis, i.e., from the cathode to the anode?
- Estimate Blad apparent molecular mass (*mm*) from the SDS-PAGE gel obtained. (Remember that the distance migrated by a polypeptide in SDS-PAGE is a linear function of $\log_{10}$ *mm*.)
- Using the molecular mass calculated, estimate the number of amino acid residues that compose Blad.

EXPERIMENT 2: ANTIFUNGAL EFFECT OF BLAD ON THE GERMINATION OF *BOTRYTIS CINEREA* CONIDIA

For this experiment, any pathogen that is easily cultured on solid media may be used. However, as explained earlier, *Botrytis* is a very important pathogenic fungus for several crops and difficult to control under field conditions. This illustrates the importance of finding specific compounds exhibiting antifungal activity against this fungus.

Materials

Each group of students will need the following materials:

- Culture of *B. cinerea* isolated from infected grape tissues or obtained from other sources
- Potato dextrose medium (PDA, Difco Laboratories, Detroit), 15 g/L of agar
- Potato dextrose broth (PDB, Difco Laboratories, Detroit)
- Petri dishes

- Ringer solution ¼ strength (Oxoid, Ltd., Basingstoke, England)
- Sterile kimwipes
- Neubauer hemacytomete chamber (Brand, Germany)
- 96 well-microtiter plates
- Microplate reader
- BSA (bovine serum albumin)
- Optical microscope
- Growth chamber
- Sterile transfer hood

Follow the protocol listed in Procedure 29.2 to complete this experiment. All steps must be performed under sterile conditions.

Anticipated Results

This experiment allows calculation of the minimal inhibitory concentration (MIC) value, defined as the lowest protein concentration that prevents any detectable *B. cinerea* growth.

The aim of this experiment is to test the potential antifungal activity of Blad against *B. cinerea*, the fungus that causes gray mold. Blad was purified from *Lupinus albus* germinated cotyledons and used in the antifungal assays. The concentrations required for 50% inhibition (inhibitory concentration, IC_{50}) and for complete inhibition of fungal growth (MIC) in the microtiter plate assays are taken as a measure of the inhibitory potency of Blad on the *B. cinerea* isolate. At Blad concentrations below MIC, the antifungal potency tends to decrease with time. Fungal cultures treated with BSA are expected to display a normal growth behavior.

Questions/Discussion

- With the data obtained, draw a graphic similar to the one depicted in Figure 29.2. Estimate the values of IC_{50} and MIC.
- Why are the controls in this experiment necessary?
- Why does the absorbance at 595 nm increases with time under control conditions?

EXPERIMENT 3: ANTIFUNGAL EFFECT OF BLAD ON THE GROWTH OF *BOTRYTIS CINEREA* MYCELIA

Botrytis cinerea rapidly invades host tissues or culture media and quickly produces abundant conidia that are easily distributed. To diminish this fast growth rate that could eventually compromise the experimental results, this laboratory exercise will use a nutrient-poor culture medium.

	Procedure 29.2
	Antifungal Effect of Blad *Botrytis cinerea* Conidial Germination
Step	Instructions and Comments
1	The strain of *Botrytis cinerea* must be maintained on PDA at 25°C under a 16 h/8 h light/dark cycle and fresh subcultures must be prepared every 2 weeks. Collect the conidia suspension by adding sterile ringer solution onto the surface of the mycelia
2	Vortex the conidia suspension for 1.5 min and filter through sterile kimwipes to remove hyphal fragments. Adjust the conidia suspension concentration to 5×10^4 spores per mL with the improved Neubauer hemacytometer chamber.
3	Add to each well of the microtiter plate 150 μL of PDB and mix with 50 μL of conidia suspension. Allow *Botrytis conidia* to pregerminate for 6 h at 26 to 28 °C.
4	Read the absorbance in a microplate reader at 595 nm, using PDB as the blank.
5	Add four different concentrations of desalted extract containing Blad to each well of the microplate: 0 μL/mL (control-1), 50 μL/mL, 150 μL/mL, and 250 μL/mL. As control-2, use 250 μL/mL of a stock solution containing 1 mg BSA per mL.
6	Incubate the microplates for 24 h at 26 to 28°C.
7	Read the absorbance at 595 nm. Confirm the spectrophotometric data, take some conidia suspension from each well, and check under the optical microscope (Figure 29.2).

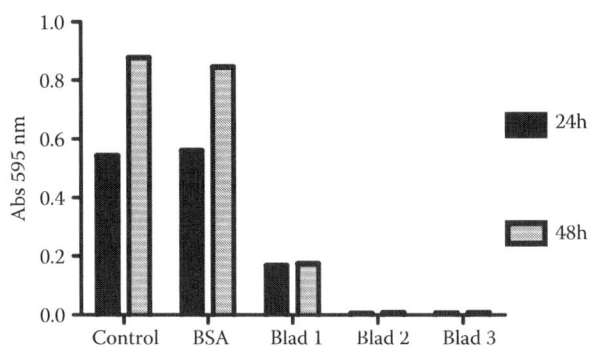

FIGURE 29.2 Antifungal effect of Blad on *Botrytis cinerea* conidial germination after 24 and 48 h of incubation. Control, no added protein; BSA (250 μL/mL); Blad 1 (50 μL/mL); Blad 2 (150 μL/mL); and Blad 3 (250 μL/mL).

Materials

Each team of students will need the following items:

- Culture of *B. cinerea* isolated from infected grape tissues or obtained from other sources
- Solid malt extract medium: 3% (w/v) malt extract (Biokar Diagnostics), 2% (w/v) agar
- Petri dishes
- 8 mm discs of sterile paper (Advantec, Toyo Roshi Kaisha)
- Cork borer, 0.7 cm diameter

- 50 μL micropipette
- Ruler
- Forceps

Follow the protocol listed in Procedure 29.3 to complete this experiment. Students should perform this procedure under aseptic conditions; working in a sterile environment, such as a laminar flow hood, will minimize contamination.

Anticipated Results

Figure 29.3 illustrates an example of growth inhibition of *B. cinerea* grown in malt extract (a nutrient-poor medium) by Blad. As expected, no fungal inhibition is detected in the presence of water (control) or heat-denatured Blad. However, the growth of *B. cinerea* hyphae is heavily restricted in the vicinity of native Blad.

Discussion

- Explain why *B. cinerea* hyphae stop growing at a considerable distance from the filter paper containing Blad. (Hint: Structurally, the growth medium may be considered as a mesh of large pore size.)
- Provide an explanation for the negative result obtained with denatured Blad.

Procedure 29.3

Antifungal Effect of Blad on *Botrytis cinerea* Mycelial Growth

Step	Instructions and Comments
1	Cut 0.7-cm diameter mycelial discs from the edge of an actively growing colony of *Botrytis cinerea* using a sterile cork borer. Place one mycelial disc in the center of a layer of solid fungal growth medium in a petri dish; the mycelial side of the disc should be in contact with the surface of the medium. Incubate for 48 h at 26 to 28°C or until the colony diameter reaches 1.5 cm.
2	Using forceps, place three discs of sterile filter paper 2 cm away from the edge of the phytopathogen. Using a micropipette, apply 50 µL of water (control), 50 µL of the solution containing native Blad, and 50 µL of the solution containing denatured Blad (boiled for 4 min) to each filter paper and replace the cover on the petri dish. Incubate at room temperature.
3	The growth rate of the fungus will determine the length of the incubation period; for *Botrytis* growing in this medium, this period usually lasts 3 to 5 days. However, cultures should be checked daily and the diameter of the colonies measured before hyphae in the control reach the edge of the plate.
4	Percentage inhibition is given by the following equation:

$$\text{Inhibition (\%)} = [(\text{growth diameter in untreated control} - \text{growth diameter in treatment}) \times 100] / \text{growth diameter in untreated control}.$$

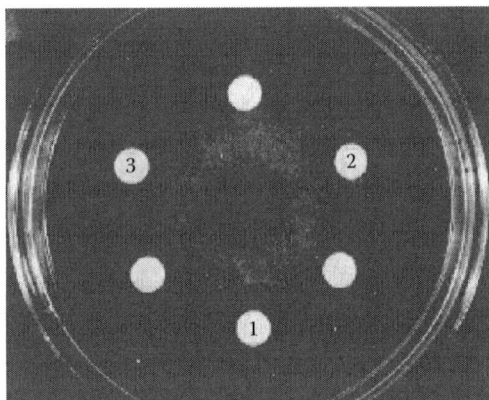

FIGURE 29.3 Comparison among the antifungal activity of native Blad (1), denatured Blad (2), and water (control) (3) on *Botrytis cinerea* mycelium growth. The picture was taken 48 h after inoculation.

REFERENCES

Hausbeck, M.K. 1993. Botrytis blight. Pages 223–228 in *Geranium IV*. J. White, Ed. Ball Publishing, Geneva.

Hausbeck, M.K. and G.W. Moorman, 1996. Managing *Botrytis* in Greenhouse-grown flower crops. *Plant Dis.* 80: 1212–1219.

Leroux, P., Chapeland, F., Giraud, T., Brygoo, T., and Gredt, M. 1999. Resistance to sterol biosynthesis inhibitors and various other fungicides in *Botrytis cinerea*. Pages 297–303 in *Modern Fungicides and Antifungal Compounds II*. H. Lyr, P.E. Russell, H.W. Dehne, and H.D. Sisler, Eds. Intercept, Andover, UK.

30 Detecting and Measuring Extracellular Enzymes Produced by Fungi and Bacteria

Robert N. Trigiano and Malissa H. Ament

Fungi, fungilike organisms, and bacteria do not "eat" or obtain nutrition in the ways that many animals do—they lack organized digestive systems in the traditional sense. Instead, they absorb all of the essentials for life directly from the environment. However, many of the simple molecules, such as sugars that are easily transported into the organism, are present only as complex polymers, such as cellulose or starch, in the environment. Complex carbohydrates and other classes of polymers cannot be used by the microorganisms unmodified. Therefore, these organisms must have a means of degrading the carbohydrate polymers, proteins, etc., into their constituent smaller molecules.

Regardless of their relationship to substratum (pathogenic or saprophytic), these classes of organisms produce **extracellular enzymes**, which interact with the environment outside of the cell or hypha. These enzymes are complex proteins that are manufactured in the bacterial cell or fungal hypha, transported across the plasmalemma (cell membrane) and cell wall to the outside environment. These are typically hydrolytic or oxidative, and systematically degrade or break down very complex plant (and animal) polymers, such as cellulose, lipids, starch, lignin, pectin, protein, etc., to simple molecules. In turn, the simple molecules such as sugars, amino acids, fatty acids, etc., resulting from enzymatic actions are absorbed through the cell wall, transported across the plasmalemma, and used for growth, energy, reproduction, and other life processes. Some enzymes that degrade plant cell walls may also be involved in plant pathogenesis (Chapters 10 and 28). For example, the fungus *Rhizopus stolonifera* and the bacterium *Pectobacterium* (*Erwinia*) *carotovora* both produce a battery of **pectinolytic** enzymes (pectinases) that degrade the middle lamella, which is composed chiefly of pectin and lies between plant cells. These enzymes cause soft rots. Many obligate parasites, such as *Peronospora tabacina* (blue mold of tobacco), utilize cell-wall-degrading enzymes (e.g., cellulases) to establish contact between haustoria and plant cell membranes.

The laboratory exercises in this chapter will be concerned with detection of the activities of extracellular enzymes produced by microorganisms. There are three basic methods used to measure enzyme activities: detection of products of the enzymatic reaction (e.g., reducing sugars, acids, change in pH and viscosity, etc.); depletion of enzyme substrate (phenolic compounds, etc.); and exhaustion of a cofactor (ATP and others) in the reaction. The enzymes discussed in the exercises generally fall into the following two broad classes: constitutive and inducible. Constitutive enzymes are produced by the organisms at all times, albeit at a low level perhaps. Inducible enzymes are produced only when the microorganism is grown in the presence of the enzyme substrate. The time for the enzyme to be induced and produced to levels of detection will depend on the organism and the specific enzyme of interest.

In the first exercise designed for undergraduate students, **amylase**, **lipase**, and **polyphenol oxidase** activities will be qualitatively determined using solid agar media in which the enzyme substrate has been incorporated. These three enzymes are usually considered to be constitutive, but the quantity of enzyme produced may be influenced by the presence of the substrate. The second exercise devised for graduate students or special projects for advanced undergraduates will quantitatively determine **cellulolytic** (cellulase) activity of fungi grown in liquid medium. Cellulases are usually considered inducible, and the test organisms will require contact with a cellulose or modified cellulose substrate, such as carboxymethylcellulose (CMC) before any appreciable enzyme activity can be detected. The third exercise is designed for use by either undergraduate or graduate students and involves detection of endopectinolytic (pectinase) isozymes using **polyacrylamide gel electrophoresis**. Pectinases are generally inducible enzymes and will require a specialized liquid medium containing pectin for enzyme production.

EXERCISES

EXPERIMENT 1. QUALITATIVE DETERMINATION OF SOME ENZYME ACTIVITIES

Often the only information desired from an experiment is whether or not the fungus produces a specific enzyme. For

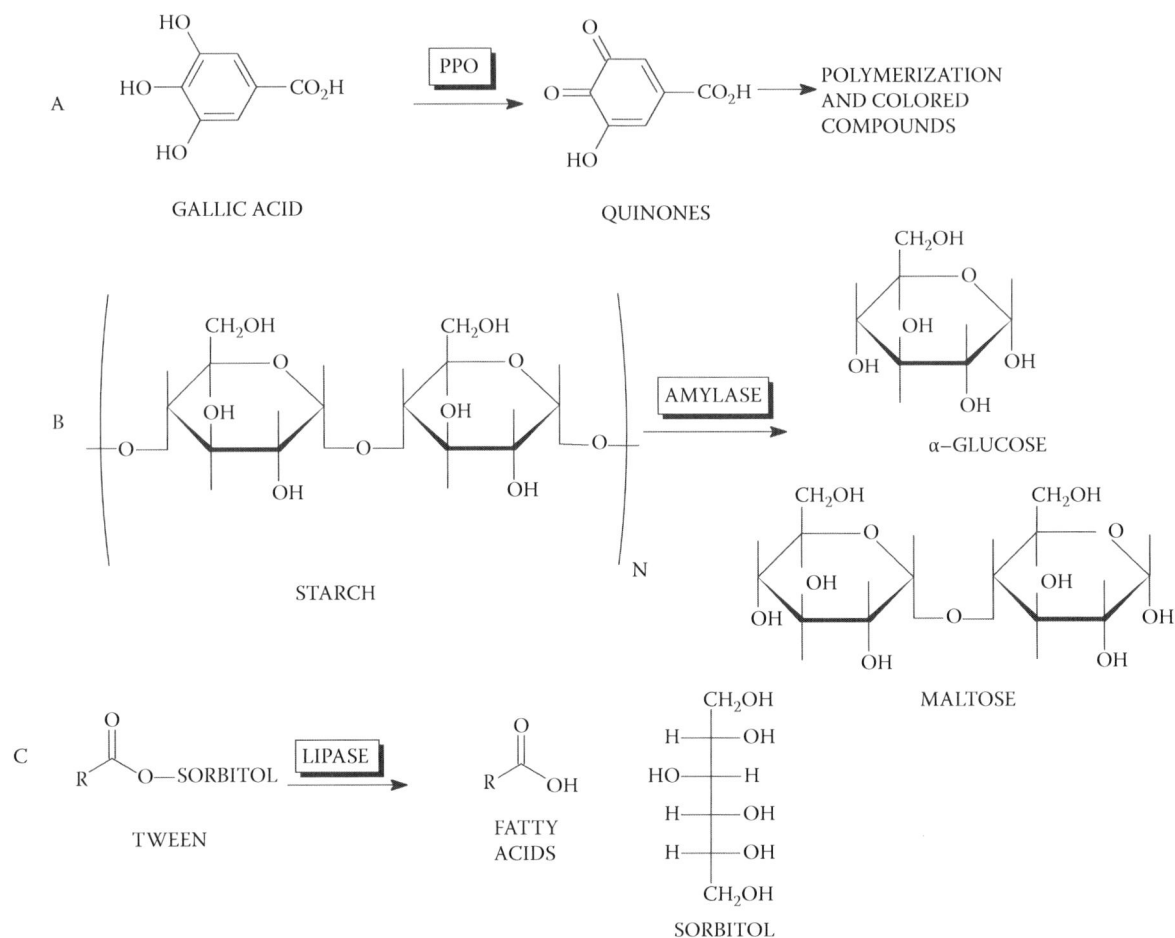

FIGURE 30.1 Substrates and products of polyphenol oxidase, amylase and lipase. A. Polyphenol oxidase (PPO) oxidizes adjacent hydroxyl groups on gallic acid (3,4,5- trihydroxybenzoic acid) and creates quinones, which are unstable and spontaneously polymerize to form pigmented (colored) products. B. Starch is a α 1-4 polymer of D-glucose residues. α-amylase hydrolyzes the bond between adjacent glucose units to form a random mixture of D-glucose and maltose (two D-glucose molecules linked α1-4) residues. C. The Tweens are synthetic fats with a sorbitol (a sugar alcohol with five carbons instead of glycerol, which has three carbons) backbone esterified to various fatty acids (R: e.g., lauric or oleic acid). Note only carbon one is shown. Lipase hydrolyzes the ester bond between the carbon in sorbitol and the carbonyl carbon of the fatty acid to form sorbitol and a free fatty acid. Changes in pH and calcium bonding with the free fatty acids combine to produce the white, flocculent precipitate suspended in the medium. (Drawing provided by Dr. James Green, University of Tennessee.)

example, the fungi that cause dogwood anthracnose, *Discula destructiva* and an undescribed species of *Discula*, can be distinguished from each other by their ability to produce polyphenol oxidases, which is a presumptive test of the ability to degrade lignin. Qualitative techniques are ideally suited to achieve this goal of detection only when quantification is neither needed nor desired. Typically, evaluation of the ability to produce an enzyme or battery of enzymes can be accomplished using an agar medium in which the substrate for the enzymes has been incorporated. The fungus is allowed to grow on the agar and positive enzyme activity is indicated by a color change in the medium or precipitation of a product. For some other qualitative methods, such as some of the pectinolytic enzymes, staining for the original substrate can be used.

In this experiment, we will qualitatively evaluate different fungi for the ability to produce amylase (AMY), lipase (LIP), and polyphenol oxidase (PPO). The substrates and enzymatic products for these three enzymes are shown in Figure 30.1.

Materials

The following items are needed for each student or team of students:

• Duplicates of 10-cm-diam. petri dishes of five to eight cultures of different fungi grown on the appropriate inoculum medium (Table 30.1) (We suggest not using slow-growing fungi such as *Geotrichum* or *Acremonium* or prolific spore- or

TABLE 30.1

Medium for Growing Inoculum

Amylases and Lipases	Polyphenol Oxidase
Nutrient Agar (NA)	Malt Extract–Yeast Extract (MEYE)
Nutrient broth (Difco Lab, Detroit, MI) 8 g	Malt extract (Difco Lab, Detroit, MI) 20 g
Agar 20 g	Yeast extract (Difco Lab, Detroit, MI) 1 g
Distilled water 1 L	Agar 20 g
	Distilled water 1 L

Note: Combine all ingredients and autoclave at 121°C for 20 min. Dispense media into sterile 10-cm diam. petri dishes when cooled, but not hardened.

sporangia-producing fungi, such as *Penicillium* or *Rhizopus*, respectively.)

- One dish of uninoculated (without fungi) NA and YEME to use as controls (Table 30.1)
- Three 60-mm-diam. petri dishes containing enzyme assay medium for each fungus
- A supply of autoclaved plastic 5-mm to 7-mm-diam. drinking straws
- Laminar flow hood (optional)
- Alcohol burner
- Scalpel and number 11 blade or stainless steel spatula
- Growth room or incubator
- Iodine reagent consisting of 15-g KI and 3-g I$_2$ per liter of distilled water
- Parafilm
- Large forceps
- Aluminum foil

Follow the protocols outlined in Procedure 30.1 to complete the experiment.

Anticipated Results

Most species of fungi produce both amylase (Figure 30.3A) and lipase (Figure 30.3B), and the enzymes are easily detectable after a short period (5 to 10 d) in culture. Amylase-positive cultures are indicated by clear zones in the agar after staining with iodine solution. A positive test for lipase is a white flocculent precipitate in the medium. It is possible that not all species classified in the same genus will produce these enzymes.

Polyphenol oxidase activity should be expressed as either darkened inoculum plugs or darkened assay medium within 24 to 48 h. Generally, the darker and more widespread the discoloration of the medium, the greater the amount of PPO that has been produced (Figure 30.4). In some cultures, the plugs may be very slightly discolored, and it may be difficult to determine if the fungal isolate produces PPO. Many fungi produce PPO and, generally, isolates of the same species are all capable of producing the enzyme. However, there may be great variance within

species of the same genus, and this information may be useful in taxonomic considerations. Some fungi grow on PPO assay medium, whereas others do not. Growth or lack of growth of individual species is probably influenced by low pH of the medium and the toxicity of gallic acid more than the ability to oxidize the substrate.

Questions

- What is an enzyme?
- If a fungus does not grow on PPO assay medium, then how can the assay medium turn dark?
- Besides participating in lignin degradation, what other advantage might PPO confer on a plant pathogenic fungus?
- How do enzymes move through the assay medium and what does it mean if enzyme activity is apparent well beyond the perimeter of the fungal colony?
- What role does calcium chloride play in the LIP assay medium? (Hint: What is precipitated in the medium?)

EXPERIMENT 2. QUANTITATIVE MEASUREMENT OF CELLULOLYTIC ACTIVITY

Quantitative measurement allows more accurate assessment of enzyme activities of individual isolates and comparison of activities between organisms. The procedures described in this experiment are designed to detect products of cellulolytic activity colorimetrically. However, this experiment may be adapted to any enzyme system, such as pectinases and xylanases, in which **reducing sugars** (or other compounds with **anomeric** carbons) are produced.

Crystalline cellulose, 1,4-β linkages of D-glucose residues, is very insoluble in water at neutral or slightly acid pH typically found in growth media. An individual polymer consists of 500 to 15,000 D-glucose units and has a molecular weight in the range of about 5×10^4 to 2.5×10^6 Da. Therefore, carboxymethylcelluloses (CMC), which are far more soluble in water, are used as a substrate in most media. CMC has methyl groups esterified through

Procedure 30.1

Detection of Extracellular Amylase (AMY), Lipase (LIP), and Polyphenol Oxidase (PPO) Activity

Step	Instructions and Comments
1	Choose any number of fungi (we suggest five to eight species or include some isolates of the same species) and grow cultures for inoculum in 10-cm-diam. petri dishes using the appropriate medium for each of the enzymes (see Table 30.1). Two cultures of each fungus will be adequate for each student or team of students. Also provide a "blank" petri dish with medium (without fungus growth) for each enzyme.
2	Cut 10 to 12 equal diam. plugs of inoculum from the perimeter of the mycelium with a sterile plastic straw for each fungus (Figure 30.2). If the plugs should get stuck in the straw, squeeze them out into the dish with sterile forceps. Repeat the process for each fungus and the "blank" agar using different straws.
3	Using a sterile scalpel or spatula, transfer plugs with mycelium to the center of three dishes containing AMY enzyme assay medium. The mycelium side of the plug should be in contact with the medium. In a fourth dish, place a "blank" agar plug. Be sure to flame-sterilize the scalpel between transferring isolates of fungi. Repeat the process for LIP and PPO assay media. Seal each dish with parafilm, and place the cultures in an incubator at 25°C. Wrap the PPO cultures in aluminum foil to prevent exposure to light.
4	Examine the AMY cultures periodically for mycelial growth. After the mycelium has covered 50–75% of the agar surface, flood one of the cultures with iodine reagent; there is no need to maintain axenic cultures at this time. Using a scalpel, make numerous cuts through the mycelium in one dish, and observe any color changes in the assay medium. Cleared zones underneath or in advance of the mycelium indicate amylase activity, whereas blue coloration denotes intact starch polymers and lack of amylase activity (Figure 30.3A). If the test is negative for amylase at this time, allow the fungus in the other dishes to grow for an additional week and retest using the iodine reagent. Record observations.
5	Observe the LIP cultures for white flocculent inclusions either beneath and/or in advance of the mycelial mat (Figure 30.3B). Observation of the medium with a dissecting microscope may help see the inclusions. Precipitation is a positive indication of lipase activity. As with the amylase assay, if the cultures are negative for lipase at this time, allow another week for growth and look for precipitation. Note where the precipitation occurs in the medium and record observations.
6	Examine the PPO cultures for darkened plugs and/or assay medium after 24 h and again after 48 h (Figure 30.4). Be sure to compare to the "blanks" or control cultures to inoculated cultures. Dark coloration indicates PPO activity and polymerization of the oxidized gallic acid. Measure the diameter of the discolored area, and record observations.

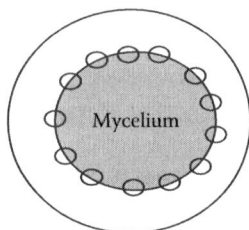

FIGURE 30.2 Harvesting mycelial plugs from cultures. Push the open end of a sterile plastic straw through the periphery of the colony as designated by the circles. Transfer the plugs to the center of the assay medium with the mycelium side of the plug in contact with the medium.

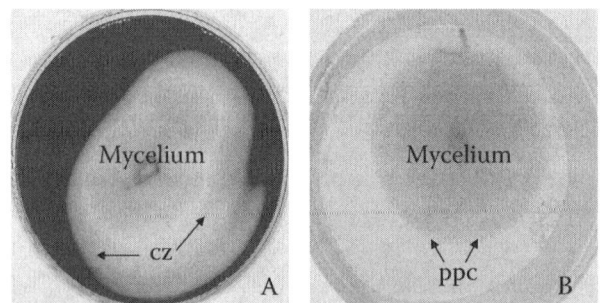

FIGURE 30.3 Assay media that are positive for enzymatic activity. A. Amylase-positive dish. Cleared zone (cz) indicates the lack of starch (amylase positive), whereas the dark (blue) zone has intact starch polymers. B. Lipase positive dish. Note the flocculent (white) precipitation (ppc) in the medium that indicates lipase activity.

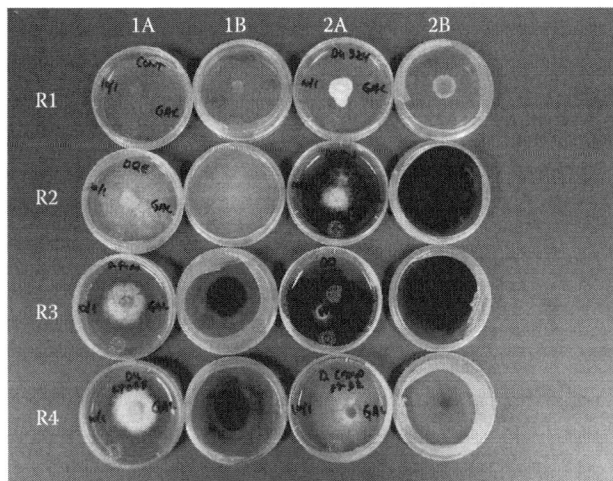

FIGURE 30.4 Solid medium test for polyphenol oxidase (PPO) using gallic acid as the substrate. All fungi were *Discula* spp., and cultures were 10 days post inoculation. Columns 1A and 2A are views of the top surface of the petri dish, and columns 2A and 2B are the corresponding bottom views of the medium. Row 1 (R1) columns 1A and 1B represent the medium inoculated with an agar plug only. Note that there is no discoloration of the medium indicating no PPO activity. The dishes labeled DQE (R2, C1A,B), DU324 (R1, C2A,B) and D. Camp (R4, C2A,B) also show no discoloration of the medium, which indicates no PPO activity. In contrast, all other test organisms produced PPO and corresponding discoloration of the growth medium. Note that most of the fungi grew on the medium, and of those that were positive, the discoloration was limited to directly beneath the mycelium (D. frax: R3, C1A,B and LT068: R4, C1A,B) or the discoloration extended well beyond the mycelium margin (DDTN12: R2, 2A,B and DDAH2: R3, 2A,B). The area of discoloration probably indicates relative abundance of the PPO enzyme, although it is very difficult to quantify.

TABLE 30.2

Enzyme Assay Media

Amylase (Amy)	Lipase (Lip)[a]	Polyphenol Oxidase (PPO)[b]
Soluble starch 2 g	Peptone 8 g	Solution A:
Nutrient broth 8 g	$CaCl_2 \cdot H_2O$ 0.1 g	Gallic acid 5 g
Agar 20 g	Agar 20 g	Distilled water 250 mL
Distilled water 1 L	Distilled water 990 mL	Solution B:
		Malt extract 15 g
	Tween 20 or 80 10 mL	Agar 20 g
		Distilled water 750 mL
(Society of American Bacteriologists, 1957)	(Sierra, 1957)	(Davidson et al., 1938)

[a] Autoclave Tween [polyoxyethylene sorbitan monolaurate (20) or monooleate (80)] and base medium separately at 121°C; then aseptically add Tween to base medium after cooled, but not hardened. Swirl to mix and dispense into sterile, plastic 60-mm diam. petri dishes.

[b] Autoclave Solutions A and B separately at 121°C; then combine after cooled, but not hardened. Swirl to mix and dispense into sterile, plastic 60-mm diam. petri dishes. This medium should be stored in the dark or wrapped in aluminum foil to exclude light.

CARBOXYMETHYLCELLULOSE (CMC)

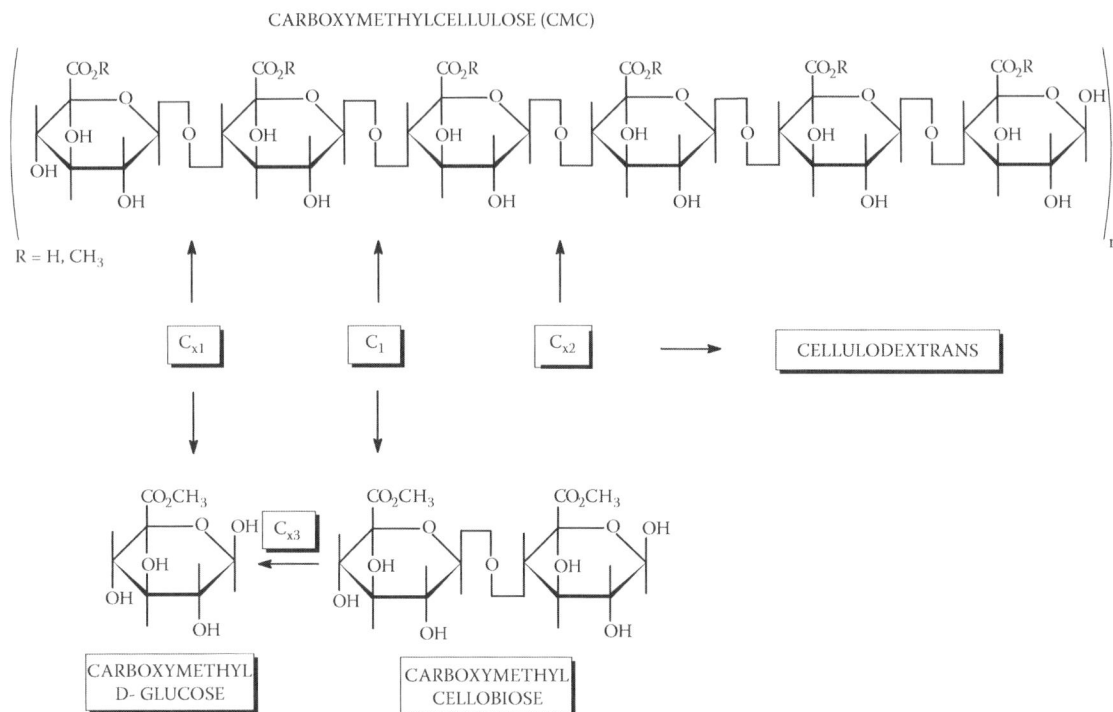

FIGURE 30.5 Diagrammatic representation of the action of cellulolytic enzymes on the substrate carboxymethyl cellulose (CMC—OH and C=O groups not shown). The C_1 enzyme (1,4-β-glucan cellobiohydrolase) hydrolyzes the 1-4 β bond on the nonreducing end of the polymer to yield carboxymethyl cellobiose molecules. C_{x3} or cellobiase (1,4-β-glucosidase) hydrolyzes cellobiose to yield carboxymethyl D-glucose residues, which is a reducing sugar. C_{x1} is exocellulase (1,4-β exoglucanase) that cleaves one carboxylmethyl D-glucose residue from the non-reducing end of the CMC molecule, whereas C_{x2} is an endocellulase (1,4-β endoglucanase) that internally hydrolyzes bonds to produce various-length cellulodextrans. All of the cellulodextrans can be substrates for C_1 or C_{x1} enzymes and, thus, C_{x2} may increase the apparent activity of the other enzymes by providing more substrate (if substrate is a limiting factor in the reaction mix). (Drawing provided by Dr. James Green, University of Tennessee.)

the carboxyl group on carbon six. Solubility in water is achieved through the partial positive charge imparted by the methyl groups to the molecule. Not all of the available carboxyl groups are methylated, and CMC is available in a wide range of percentage substitutions.

Cellulases are inducible enzymes and for organisms to produce them, cellulose or substituted cellulose compounds must be present in the growth medium. Cellulases are hydrolytic (adding a water molecule between individual units) and actually consist of a number or battery of enzymes. The C_1 enzyme or 1,4-β-glucan cellobiohydrolase enzymatically hydrolyzes crystalline cellulose to form cellibiose, a disaccharide composed of two D-glucose residues. The C_{x1} enzyme (1,4-β-exoglucanase), cleaves off individual glucose residues from the end of the chain, whereas the C_{x2} enzyme (1,4-β-endoglucanase) randomly hydrolyzes the cellulose polymer into a mixture of cellulodexans of various lengths. The C_{x3} enzyme is a 1,4-β-glucosidase (cellibiosase) that hydrolyzes cellobiose into two D-glucose residues (Figure 30.5).

Experiment 2 is divided into the following three parts: fungal growth and protein isolation; developing standard curves and measuring enzyme activity; and calculating cellulolytic activity. Fungal cultures will take about 2 to 3 weeks to grow. Two 4- to 6-h laboratory periods should be scheduled to complete the remaining tasks. If this experiment is used as a class activity, we suggest that the instructor initiate the cultures and if time is a limiting factor, isolate total protein from the cultures. Teams of students may also be assigned portions of the experiment. For example, a team of four students can help isolate the protein. Two of the students can work on determining protein concentration, and the other two measure reducing sugars.

Fungal Growth and Protein Isolation

Materials

Each student or team of students will require the following materials and cultures:

- Five to eight fungal isolates grown on nutrient medium described in Table 30.1
- Liquid cellulolytic induction medium (Table 30.3) (Reese and Mandels, 1963)

TABLE 30.3

Composition of Liquid Cellulase Induction and Assay Medium

Cellulase Induction Medium		Cellulase Assay Medium	
KH_2PO_4	2.0 g	Citric acid · H_2O	10.5 g
$(NH_4)_2SO_4$	2.5 g	K_2HPO_4	17.4 g
Urea	0.3 g	Water to	1 L
$MgSO_4 \cdot 7 H_2O$	0.3 g	CMC[a]	5.5 g
$CaCl_2 \cdot 2 H_2O$	0.5 g	pH = 5.0–5.2	
Peptone	1.0 g	Do not autoclave; store at 4°C	
Water to	1 L		
CMC[a]	10.0g		
pH = 5.5[b] Autoclave 121°C for 15 min			

[a] pH medium before adding CMC (carboxylmethyl cellulose). May require extensive stirring to dissolve CMC.

[b] CMC may precipitate during autoclaving.

- Three 125-mL Erlenmeyer flasks of the above medium for each fungal isolate
- Alcohol lamp
- Sterile 6-mm to 8-mm-diam. straws
- Scalpels fitted with number 11 blades
- Incubator at 25°C (other temperatures may be required for specific organisms)
- Büchner funnel, side-arm flask, and vacuum source
- Whatman #1 filter papers and a 250-mL graduated cylinder
- Ammonium sulfate—about 750–1000 g
- 250-mL beakers, magnetic stir bars, and stir plates
- Dialysis tubing (Spectrum Laboratories, Inc. Rancho Dominguez, CA; 800-445-7330) 3500 DA exclusion and clamps (suture string will also work)
- Ice bath
- Table top centrifuge (3,000–10,000 g—does not need to be refrigerated)
- 50-mL centrifuge tubes (do not need to be sterile)
- Top loading balance
- Squeeze bottle of distilled water
- Distilled water
- Refrigerator

Follow the protocols outlined in Procedure 30.2 to complete this part of the exercise.

Anticipated Results

Most fungi will grow well in cellulase-inducing medium, and the enzymes should be present in the medium between 1 and 3 weeks after inoculation. Ammonium sulfate yields three osmotically active particles when dissolved in water and will salt out most proteins (e.g., cellulases) at 90% saturation. It is most important that undissolved salt crystals are not present when starting the centrifugation step (salt crystals will cause excess water to enter the dialysis tube and dilute the protein). A mat of other insoluble material, mostly carbohydrates, may be present at the top of the centrifuge tubes after centrifugation. This is normal and the mat should be discarded. After overnight dialysis, the contents of the tubes should be increased—some tubes may be very swollen and turgid with water. The quantity of water in the tube is dependent on the amount of salt that was carried over after centrifugation. In many experiments, each tube will yield about 4 mL of an aqueous solution of various proteins and other compounds. Some pigments may have coprecipitated with the proteins, which is not a cause for concern. Salting out and dialysis concentrates the protein from 15 to 25 times that found in the inoculated medium.

DEVELOPING STANDARD CURVES AND MEASURING ENZYME ACTIVITY

In order to determine enzyme activity, it is necessary to have solutions of known concentration of total protein and enzyme product (in this case, reducing sugars) to compare to solutions of unknown concentrations. Standard solutions are prepared in several concentrations and equations calculated to predict intermediate values of unknowns. The equations are formulated by regressing absorbance on concentration and usually follow Beer's law, especially in the midrange of the standards; that is, the relationship between known concentrations and absorbance is linear, except at very low or high concentration of product, in

	Procedure 30.2
	Growth of Fungi and Isolation of Protein
Step	Instructions and Comments

Step	Instructions and Comments
1	Grow 5 to 8 isolates of fungi (can be different isolates of the same species) in petri dishes containing nutrient agar as described in Table 30.1.
2	Prepare sufficient liquid cellulase induction medium (about 150 mL per isolate) to dispense 35 mL into each of four capped 125-mL flasks for all isolates included in the experiment. After autoclaving the medium in the flask, some of the CMC may be precipitated as fine white threads at the bottom of the flask, but this will not affect the induction of cellulases. Prepare inoculum plugs using sterile straws as shown in Figure 30.2, and dissect the plugs into quarters using a scalpel and number 11 blade. Transfer four quarters to each of three flasks per isolate. Prepare a control (cut from an uninoculated agar dish—no mycelium) to inoculate the fourth flask. Label the flask with the species name, isolate, and date. Repeat the inoculation step for each species using a different straw each time. Place all flasks in a 25°C incubator for 14–21 days depending on the growth rates of the fungi selected for the experiment.
3	Examine all flasks for contamination, especially for bacteria. The medium should be clear; cloudiness usually indicates bacterial contamination. Place a piece of Whatman's #1 filter paper in the Büchner funnel, and attach to the side arm flask. Moisten the paper with distilled water and apply a gentle (not greater than 15 psi) vacuum. Pour the contents of a flask into the funnel and express the medium through the paper into the flask. Repeat this procedure for the remaining two inoculated flasks for the same species or isolate. The paper may need to be changed between flasks depending on the nature of the mycelium. Pour the contents of the side arm flask into a graduated cylinder, record the volume, and transfer the liquid to a 250-mL beaker. Rinse the Büchner funnel, side arm flask, and graduated cylinder with distilled water between species or isolates. Repeat entire procedure for the remaining fungal isolates and the uninoculated flasks. Place used filter papers in a biohazard bag and autoclave before disposal.
3a	Alternatively, the liquid from the cultures may be poured into 50-mL centrifuge tubes, taking care to not to transfer the mycelial mat to the tube. Balance tubes (two tubes of the same weight), place opposite of each other in the centrifuge, and centrifuge for 10 min at 10,000 g. Pour the liquid into a graduate cylinder and record the volume. Discard the pellet in a biohazard bag, and autoclave before disposal.
4	Add 66 g of ammonium sulfate for every 100 mL of induction medium in the 250-mL beakers. For example, if 110 mL of medium was collected, then add $$66 \text{ g} \times (110 \div 100) = 66 \text{ g} \times 1.1 \text{ or } 72.6 \text{ g of ammonium sulfate.}$$ Stir until all the ammonium sulfate is dissolved, and place beakers on ice for at least 30 min.
5	Stir briefly, dispense the liquid into 50-mL centrifuge tubes (do not add more than about 45 mL to each tube), and label. Be sure to balance opposite tubes by weighing on a balance (within 1 g is okay) and centrifuge for 10 min at about 3000–5000 g. Decant and discard the supernatant ,and invert the tube on a paper towel. A pellet may or may not be present in the bottom of the centrifuge tubes. After the excess liquid has drained, dissolve the total protein by adding 0.5 mL of distilled water per tube and shake or vortex. Allow to stand for 10 min. Combine the contents of the tubes for each species or isolate.
6	Cut 10- to 15-cm lengths of dialysis tubing, and soak in distilled water overnight or autoclave for 3 min. Tubing should be rinsed several times in distilled water before use. Express all of the water from the tube, and clamp one end. Load the total protein solution from one species into each of the tubes. Clamp the other end of the tube taking care not to introduce air into the lumen. Immerse the loaded dialysis tubing into chilled, distilled water, and place in the refrigerator at 2–4°C overnight. See also Procedure 29.1 for column desalting of proteins.
7	The next morning, aliquot 0.5 mL from the dialysis tubes to several 1.5-mL labeled Eppendorf tubes. The number of tubes needed per isolate will vary. The Eppendorf tubes may be stored at −70°C for future use, or placed on ice for use later this day. Note that the samples can only be thawed once from −70°C without permanently denaturing the enzymes or significantly decreasing the activity.

this case protein or reducing sugars. In this portion of the experiment, the students will develop standard curves for total protein using the method developed by Lowry et al. (1951) and for reducing sugar with dinitrosalicylic acid (DNSA) reagent (Miller, 1959).

The following materials will be required for completion of this portion of the experiment:

- D-glucose and bovine serum albumin (BSA)
- Several 1-, 5-, and 10-mL pipettes
- Visible light spectrophotometer
- Centrifuge that can accommodate 50-mL plastic tubes and develop 3,000–10,000 g
- Fumehood
- Glass test tubes (greater than 10-mL capacity) and racks
- 50-mL centrifuge tubes (not necessarily sterile)
- Disposable plastic cuvettes and transfer pipettes
- Folin–Ciocalteau reagent (Fisher Scientific Company) (Caution: This reagent is caustic; wear hand and eye protection.)
- 2% (w/v) sodium carbonate in 0.1N sodium hydroxide solution
- 2% (w/v) aqueous sodium tartrate solution
- 1% (w/v) aqueous cupric sulfate
- Acetonitrile gloves and safety glasses
- DNSA reagent (Table 30.4) (Caution: These reagents are caustic; wear hand and eye protection and only heat under fume hood.)
- 100°C water bath or hot, freshly autoclaved water
- 40°C water bath
- Ice
- Vortex
- Computer with programming capable of performing linear regression

Follow the protocols provided in Procedure 30.3, Procedure 30.4, and Procedure 30.5 to complete this section of the experiment.

Anticipated Results

Measurement of protein concentration is quick, easy and accurate using Lowry's (1951) method. The blank or "0" protein mixture will develop a very slight blue tint and, with increasing protein concentrations, the mixture will be progressively darker. The highest standard concentration of protein will be almost opaque (Figure 30.6). Absorbance measurements of the unknown concentrations should lie within the range of absorbances of the standards. A general rule is not to extrapolate the line beyond the range of standards. Most unknown preparations will fall between 0.1 and 0.3 mg protein/mL. However, if any unknown is not within the range, additional standard concentrations, especially above 0.5 mg/mL, may be prepared and the regression

TABLE 30.4

Dinitrosalicylic Acid (DNSA) Reagent for Detecting Reducing Sugars[a] (Caution: Wear Gloves and Eye Protection. Prepare Solution under Fume Hood)

3,5-dinitrosalicylic acid	8.0 g
Crystalline phenol	6.9 g
Sodium bisulfite	6.9 g
Sodium-potassium tartrate	2.55 g
Sodium hydroxide	15.0 g
Water	Bring to 1 L

Note: Prepare at least 24 h before use, and store in brown bottle at 4°C.

[a] To begin making the DNSA solution, add 15 g of sodium hydroxide to 900 mL of water and stir until dissolved. Sodium hydroxide generates heat (exothermic) as it dissolves in water. After the sodium hydroxide has dissolved, add all other chemicals and bring the final volume of the solution to 1 L with water.

Source: Adapted from Miller, 1957.

line recalculated. Alternatively, the unknown sample can be diluted 1:1 or 1:2 with water and reassessed. If the concentration is low, the sample can be concentrated by evaporation of water and then remeasured. Two reminders: always use the equation of the line to compute the protein concentration of the unknown, and dispose of all reagents properly.

The color of the DNSA becomes increasingly more red to dark red with higher concentrations of reducing sugars (Figure 30.7). The reducing sugar content from unknown assays will usually fall between 0.3 and 0.5 mg D-glucose/mL. The same considerations for determining proteins apply to reducing sugars. If the absorbance of the unknown sample is beyond that of the most concentrated standard, dilute the sample 1:1 or 1:2 with assay mixture and reassess the concentration with DNSA. Once again, use the regression equation to compute reducing sugars in the unknown samples. Also, heat DNSA and samples under the fume hood and dispose of all reagents properly.

Questions

- Why are three replications of each standard concentration used to estimate the regression line?
- Is the total protein isolated composed entirely of cellulases? What other types of enzymes and/or proteins may be present?
- The activities of which cellulases are measured with this D-glucose (reducing sugar) assay procedure? How can the activity of C_{x2} be measured?
- Why is it necessary to include a boiled enzyme preparation in the assay?

Procedure 30.3

Developing a Standard Curve for Protein

Step	Instructions and Comments
1	Dissolve 100 mg of bovine serum albumin (BSA: protein) in 100 mL of distilled water. This makes a 1 mg BSA/mL solution. Make the following standard solutions using dilutions: 0.1, 0.2, 0.3, and 0.5 mg BSA/mL. Use the following formula: $$C_1 \times V_1 = C_2 \times V_2$$ where C = concentration of protein and V = volume. For example, to make a 0.5 mg BSA/mL, start with 10 mL of 1.0 mg BSA/mL. $$1.0 \text{ mg BSA/mL} \times 10 \text{ mL} = 0.5 \text{ mg BSA/mL} \times V_2.$$ Solve the equation for V_2, which is the total volume of the desired solution. $$10 \text{ mg} \div 0.5 \text{ mg/mL or } 20 \text{ mL} = V_2.$$ Because 10 mL volume was initially used: $$V_2 - V_1 = 10 \text{ mL}.$$ Add 10 mL of distilled water to the original 10 mL of 1.0 mg BSA/mL. There should be 20 mL of 0.5-mg BSA/mL. The following amounts of distilled water should be added to 10-mL of 1.0 mg BSA/mL to make the appropriate standards: 90 mL water for 0.1 mg BSA/mL; 40 mL of water for 0.2 mg BSA/mL; and 23.4 mL of water for 0.3 mg BSA/mL. Mix all solutions well, transfer to 50-mL centrifuge labeled tubes, and if not used within the day, store at –20°C.
2	Prepare 200 mL of a 0.1 N NaOH solution by dissolving 0.8 g of NaOH in 200 mL of distilled water. NaOH can be difficult to weigh accurately. Alternatively, dissolve 4 g NaOH in 100 mL (1 = N) and dilute 20 mL of this solution with 180 mL of distilled water to make a 0. 1 = N solution of NaOH. To 200 mL of 0. 1 = N NaOH add 4 g of sodium carbonate. Stir until the sodium carbonate is completely dissolved, and pipette 1.0 mL of each 1.0% (w/v) cupric sulfate and 2.0% (w/v) sodium tartrate into the solution and mix thoroughly. At this time, dilute 20 mL of Folin–Ciocalteau reagent with an equal amount of distilled water. Wear gloves and eye protection when making this solution.
3	Carefully pipette 1 mL of each of the BSA standard concentrations including a water control (0 mg BSA/mL) control into glass test tubes. Prepare three replications of each BSA concentration. For each fungal species or isolate, pipette 1 mL of the total protein isolated from induction medium into a glass test tube. Add 5 mL of the cupric sulfate solution (step 2) to each of the test tubes, and mix well using a vortex mixer at low speed. After 10 min at room temperature, quickly add 0.5 mL of the diluted Folin–Ciocalteau reagent (step 2) to each of the tubes and mix. Set aside at room temperature for 30 min.
4	After the 30 min, pipette the standards and unknown samples into disposable plastic cuvettes. Determine the absorbance values in a spectrophotometer set at 500 nm and record. Establish the baseline using the "0" BSA samples as a reference. After reading, dispose of the liquid (and cuvettes) in an approved, labeled, hazardous waste container.
5	Calculation of standard curves using linear regression and determination of protein in unknowns are discussed in Procedure 30.6.

Procedure 30.4

Developing a Standard Curve for Reducing Sugars—DNSA Reagent

Step	Instructions and Comments
1	Follow the instructions for preparation of standard BSA solutions in Procedure 30.3, step 1, but instead use D-glucose. Prepare the following standard concentrations: 0.1, 0.2, 0.3, 0.4, and 0.6 mg D-glucose/ mL.
2	Pipette 1 mL of each standard d-glucose solution (including 0 mg d-glucose/mL) into each of three glass test tubes. Reducing sugars (RS) in the cellulolytic assay medium samples can be assayed at this time if Procedure 30.5 has been completed. Wearing gloves and eye protection, pipette 3 mL of DNSA reagent (Table 30.4) into each of the tubes.
3	In a fume hood, place the test tube rack with tubes into a water bath containing 100°C water for 15 min. If a water bath is unavailable, the rack can be placed in a plastic dishwashing tub and freshly autoclaved water added.
4	Remove the test tubes and rack from the heated water, and allow to cool to room temperature (about 30 min).
5	Transfer D-glucose standards and unknowns to disposable plastic cuvettes, read on a spectrophotometer set at 550 nm, and record the absorbance values. If a dual beam spectrophotometer is available, establish the baseline using the "0" D-glucose samples as a reference. After reading, dispose of the liquid and cuvettes in an approved, labeled, hazardous waste container.
6	Calculation of standard curves using linear regression and determination of unknowns are discussed in Procedure 30.7. However, for illustrative purposes, the equation Y = 2.34 X + (-0.27) for reducing sugars was calculated from actual data (not shown).

Procedure 30.5

Assaying Isolated Proteins for Cellulase Activity

Step	Instructions and Comments
1	Before starting the procedure, set a water bath to 40°C. Select two 1.5-mL Eppendorf tubes containing 0.5 mL of protein for each fungal species or isolate. If the samples were frozen at −70°C, allow to come to room temperature—do not heat to defrost. Ensure the tube is labeled and securely closed. Place one of the tubes into boiling water for 15 min to irreversibly denature any enzymes. This will serve as a control.
2	Prepare cellulase liquid assay medium the day before as described in Table 30.3, and store in the refrigerator (4°C). Allow the assay medium to come to room temperature, and pipette 4.5 mL into test tubes. Depending on the volume of the protein isolation, additional replicates can be prepared for unknown samples. Dispense 0.5 mL of nondenatured isolated protein (not boiled) into test tubes for each of the unknowns. Dispense 0.5 mL of boiled preparations from step 1 into separate tubes. Also, prepare at least one tube to which 0.5 mL of sterile water has been added to reaction mixture. Mix the contents of all the tubes thoroughly, and cap with aluminum foil to prevent water loss.
3	Incubate the test tubes in a 40°C water bath for 1 h. *Note:* If after the reducing sugar assay, some samples are negative or very low, incubate the reaction assay medium and enzymes for an additional hour. Note the additional time used for the assay will be considered when calculating enzyme activity.
4	Complete the assay for reducing sugars using the DNSA procedure described in Procedure 30.4. Note that the unknowns may be assayed for reducing sugar at the same time that the standard concentrations of D-glucose are determined.

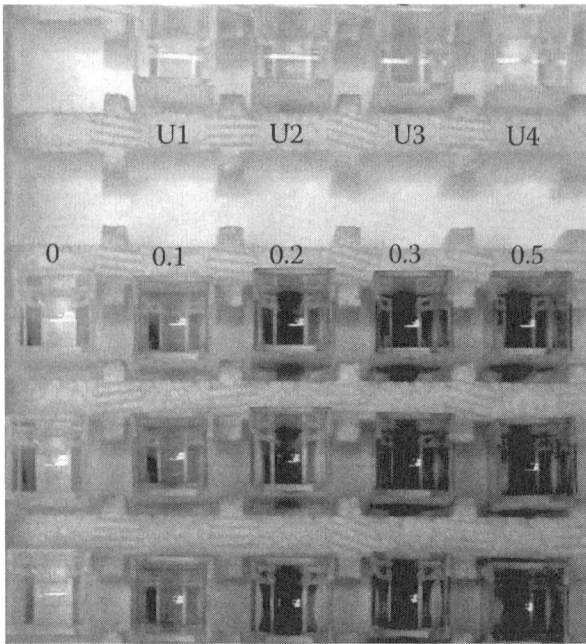

FIGURE 30.6 Protein standards and determination of protein in isolations. The amount of protein in the standards and "unknowns" is linearly related to the amount of blue color developed in the tubes. Three replications of known concentrations of protein (0, 0.1, 0.2, 0.3, and 0.5 mg BSA/mL) are regressed on absorbance and a standard curve developed. Protein concentrations from isolations (U1–U4) can then be determined. (See CD for reaction colors.)

CALCULATION OF ENZYME ACTIVITY

All of the essential information for calculating enzyme activity is now available. The hypothetical example will use the standard curves for the data as calculated by Excel. The final expression of data will be in micromoles (μM) of reducing sugars (as D-glucose equivalents)/mg of crude protein/hour. Follow the steps outlined in Procedure 30.6 to calculate the regression equations for protein and reducing sugars and Procedure 30.7 to determine enzyme activity of the unknown samples.

Questions

- Why is it important to calculate a rate of enzyme activity for each fungus? Why not compare only the reducing sugar produced in each enzyme assay mixture?
- Would you expect the enzyme activity rate to remain constant over time (i.e., after 1, 2, and 4 weeks of culture)?
- How would the enzyme activity change if the incubation temperature were decreased to 30°C or 20°C or increased to 50°C or 60°C?

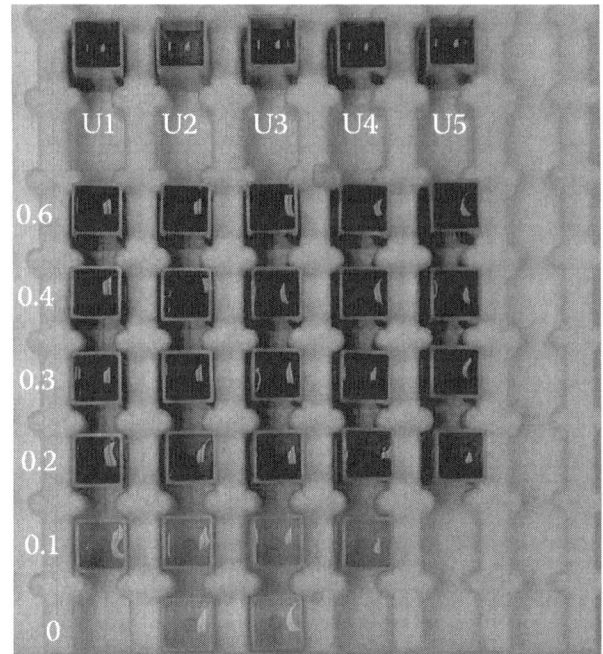

FIGURE 30.7 Reducing sugar standards and determination of reducing sugars produced by the action of cellulases. Using the dinitrosalacylic acid test, the absorbance of three replicates of each standard concentration (0.0, 0.1, 0.2. 0.3, 0.4, and 0.6 mg D-glucose/mL) are determined. Absorbance is regressed (linearly) on D-glucose concentration and a line calculated. The amount of reducing sugar from the action of cellulases in the "unknown tubes" (U1–U5) can then be determined and the activity of the enzyme assessed. (See CD for actual colors.)

EXPERIMENT 3: DETECTION OF ENDOPECTINASE ISOZYMES USING POLYACRYLAMIDE GEL ELECTROPHORESIS (PAGE)

Pectin polymers are composed of galacturonic acid residues (a hexose sugar acid) linked α1-4. In some forms of pectin, methyl groups (CH_3) may be esterified to the carboxyl group on carbon 6. There are several classes of pectin-degrading enzymes, some which demethylate pectin and others that are only act at either low (5.0) or high (9.0) pH. Some of the enzymes remove single galacturonic acid units from the end of the pectin polymer (exoenzymes), whereas other enzymes will act at random along the interior length of the pectin polymer, reducing the polymer to a series of variable-length smaller chains (endoenzymes). In the previous experiment with cellulase, activity of exocellulases was measured primarily by detecting reducing sugars with DNSA reagent. In contrast, this experiment is designed to detect the activities of individual **isozymes** of endopectin methyl glacturonase (endo PMG) and/or endopolygalacturonase (endo PG). Both of these enzymes typically have high activity at pH 5.0.

	Procedure 30.6
	Construction of Standard Curves and Determination of Unknowns
Step	Instructions and Comments
1	For calculating protein and reducing sugars standard curves, data should be entered in columns, as shown in Table 30.5. Column A contains the independent variable (concentration of the standards) and Column B the dependent variable (absorbance values recorded from the spectrophotometer). To build a data set that can be properly analyzed, you should have multiple samples and corresponding readings for each independent variable. The first measurement taken was to zero the spectrometer. Because there were not many samples to read, the zero measurement was not repeated. For each of the protein concentrations (0, 0.1, 0.2, 0.3, and 0.5 mg BSA/mL, for example), three samples were analyzed, and their absorbance measurements were recorded. Note that data generated by the class may be similar, but not exactly the same as sample data presented in Table 30.5. The data may be graphed, but it is not necessary to calculate either protein or reducing sugar concentration. There are other programs available that allow more control over the appearance of the graph than either Microsoft Excel or Corel Quattro Pro. To use the Regression tool in Excel, the version must have the Analysis Tool Pack installed.
2	For linear regression, enter your data using Microsoft Excel, click Tools/Data Analysis/Regression/OK. In the appropriate box, select the range of Y values. These values are the dependent variables (absorbance values) in the data set. Select the range of X values in the next box. These are independent variables (concentration). Select an area in the spreadsheet in which Excel can place the output of the regression calculations. Click OK. In the results displayed in Table 30.6, the X coefficient (or slope of the line, m) is labeled "0" under the "coefficients" heading, and the Y-axis intercept (or b) is labeled "intercept" under the "coefficients" heading. The R^2 value is found at "R Square" under "Regression Statistics" and is a measure of how well the data agree with the estimated line. From these calculated values, you can construct the equation of the regression line, $y = mx + b$; where m is the slope of the line and b is the Y-intercept. The equation from this analysis is: $$Y = 0.99\ X + 0.03.$$ Use this equation to determine the unknown concentrations. Remember Y is the absorbance value for the unknown; solve for X, the concentration. Microsoft Excel provides a tool to graph the data and apply a trendline to complete regression analysis, but that method will not be addressed here. The regression equation is the most accurate way to determine concentration. Trying to determine it using the graph provides only a good estimate of the value.
3	Corel Quattro Pro can also be used to calculate linear regression. Data should be typed in the format shown in Table 30.5. Click Tools/Numeric Tools/Regression. In the appropriate boxes, select the independent (concentration) and dependent (absorbance) values, and an area on the spreadsheet in which the output of the regression analysis can appear. Be sure that the Y-intercept radio button is set to "Compute" and Click OK. The results from analysis of the data listed in Table 30.5 are shown in Table 30.6, and the calculated regression equation is $$Y = 1.01\ X + 0.02.$$ There are minor differences in the values of m and b depending on which program (Excel or Quattro Pro) was used to estimate the regression line. Discrepancies are because of variations in the algorithms used in the programs. Very slight differences in the regression equations will not cause major differences in the calculations in either protein or reducing sugar concentrations. Choose one spreadsheet program and consistently use it to avoid introducing minor errors into the statistical analysis.

TABLE 30.5

Sample Data for Protein Standard Curve Entered in Excel Spreadsheet

Column A	Column B
Protein Concentration (mg/mL)	Absorbance (a.u.)
0	0.000
0.1	0.115
0.1	0.116
0.1	0.125
0.2	0.239
0.2	0.233
0.2	0.235
0.3	0.340
0.3	0.335
0.3	0.334
0.5	0.518
0.5	0.521
0.5	0.519

TABLE 30.6

Regression Output for Protein Data Using Excel and Quattro Pro

	Excel	Quattro Pro
Constant (Y-intercept or b)	0.029	0.022
Standard error of Y estimate	0.006	0.012
R-squared	0.996	0.995
No. of observations	13	13
Degrees of freedom	11	11
X coefficient (slope or m)	0.992	1.015
Standard error of coefficient	0.019	0.021
Estimated equation	Y = 0.99 X + 0.029	Y = 1.02 X + 0.022

Isozymes (iso = the same or equal; zyme = enzyme) can be conveniently defined as molecular weight variants of the same functional enzyme that can occur in the same isolate or between different isolates of fungi, bacteria, or other organisms, including plants and animals. Although these enzymes are of different weights, they essentially perform the same task (e.g., degradation of pectin), although the specific activities (e.g., how "fast" they work under specified conditions) may be quite different when compared to one another. Isozymes are usually detected by gel electrophoresis, a technique used to separate different molecular weight compounds, such as protein or isoenzymes. A gel is a porous support medium composed of either agarose or polyacrylamide (potato starch may also be used) that permits the migration of charged molecules in an electric (DC) field. Migration of proteins (in our case) through the gel is dependent on the strength of the electrical field (volts/cm²), the composition of the gel (% agarose or acrylamide), which dictates the pore size, and the charge and size of the isozymes. The charge on the isozymes is controlled by the pH of the buffer, and when all other parameters of electrophoresis remain constant, migration through the gel is dependent on size of the molecules. Typically, most of the proteins have a net negative charge and migrate toward the anode or positive pole. However, one must be aware of the fact that some proteins, and potentially some of interest, might have a net positive charge in the buffer, move in the opposite direction or toward the anode and eventually be lost from the gel. The smaller (lower-molecular-weight) proteins migrate more rapidly and thus end up closer to the anode (positive pole)

at the bottom of gel compared to large (higher-molecular-weight) proteins.

The activity of many of the endoenzymes is typically measured by noting the change in viscosity of a standard solution containing the appropriate polymer (or modified polymer) over time. The viscosity of the solution becomes less as the endopectinases cleave the polymer into smaller units. However, this methodology determines the total activity of all the enzymes and nothing can be said about how many isozymes are present in the mixture. Another way of detecting activity of endopectinases is to take advantage of the fact that intact pectin or modified pectin polymers incorporated into acrylamide gels stain red-pink with ruthenium red. Electrophoresis of the crude protein will deposit the pectinolytic isozymes at various locations in the gel. If the gel is incubated under the proper conditions (pH, temperature, etc.), then the endoenzymes will degrade the pectin incorporated into the gel so that it will no longer stain with ruthenium red. Each "cleared" or "nonstained" band in each of the lanes represents an isozyme or comigrating enzymes of various endopectinases. Isozyme production from various isolates of the same species or species of different genera can be compared and physiological relatedness, for example, estimated.

In this experiment, endopectinases found in 5- and 10-d-old cultural filtrates of various fungi or bacteria will be isolated with total proteins using the same methods that were used to isolate cellulolytic enzymes. Instructors may opt to complete the experiment up to and including isolating the protein from the cultures and/or pouring the gels.

The class will require the following materials to complete the laboratory activity:

- Five to eight fungal isolates grown on nutrient medium described in Table 30.1
- Liquid pectinolytic induction medium (Table 30.7; modified from Cruickshank and Pitt, 1987)
- Three 125-mL Erlenmeyer flasks of the above medium for each fungal isolate

Procedure 30.7

Calculation of Enzyme Activity

Step	Instructions and Comments
1	Calculate total protein in unknown samples. The estimated equation of the regression line given in Procedure 30.6 step 3 is:

$$Y = 0.99\ X + 0.03$$

where Y = absorbance measurement in arbitrary units (a.u.) and X = concentration of protein in mg/mL. If the absorbance of an unknown sample (e.g. #1) is 0.33 a.u., the following substitution into the preceding equation should be made:

$$0.33 = 0.99\ X + 0.03.$$

Solving for X, the concentration yields:

$$(0.33 - 0.03)/\ 0.99 = X\ \text{or } 0.30\ \text{mg total protein/mL.}$$

Note: Not all of the proteins in the sample are cellulases.

2   Calculate reducing sugar (RS)/mL in the reaction mixture. The example of estimated equation for the regression line given in Procedure 30.4, step 6 is:

$$Y = 2.34\ X + (-0.27)$$

where Y = absorbance measurement in a.u. and X = concentration of RS in mg/mL. If the absorbance of an unknown RS sample (e.g., #1) is 0.52 a.u., the following substitution into the above equation should be made:

$$0.52 = 2.34\ X + (-0.27).$$

Solving for X, the concentration yields:

$$(0.52 + 0.27)/2.34 = X\ \text{or } 0.34\ \text{mg RS/mL of reaction assay medium.}$$

3   Calculate total RS in the 5 mL of reaction assay medium using the following equation:

$$\text{Total RS} = \text{mg RS/mL} \times 5\ \text{mL (total volume of the reaction mixture).}$$

Substituting in the equation provides:

$$0.34\ \text{mg RS/mL} \times 5\ \text{mL reaction mixture} = 1.7\ \text{mg RS.}$$

Remember to incorporate any dilutions of the assay volume (e.g., if the assay was diluted 1:1 with assay medium, then the total volume would equal 10 mL).

4   Calculate mg of RS/mg of crude protein in the assay. Remember that the assay protocol used only 0.5 mL of protein extract (divide concentration by two to determine the amount of protein incubated with the assay mixture). From step 1 above, the sample contained 0.30 mg total protein/mL.

$$0.30\ \text{mg total protein/mL} \times 0.5\ \text{mL} = 0.15\ \text{mg of total protein.}$$

From step 3 above, the total RS in the reaction mixture was 1.7 mg. Divide the RS by the mg of total protein used:

$$1.7\ \text{mg RS/0.15 mg total protein} = 11.33\ \text{mg RS/mg total protein.}$$

This calculation provides a measure of the amount of RS each mg of crude protein can produce.

Continued

Procedure 30.7 (Continued)	
Calculation of Enzyme Activity	
Step	Instructions and Comments
5	Calculate mg RS /mg crude protein/ hour that the assay medium was incubated at 40°C.
	11.33 mg RS/mg total protein/1 h = 11.33mg RS/mg total protein/h.
	Most incubation times are 1 h, but occasionally 2 h is used. This calculation provides a rate for formation of the product.
6	Calculate micromoles of RS using the molecular weight of D-glucose (180.2 g). First, convert grams to milligrams.
	180.2 g D-glucose/mole $\times$ 1000 mg/g = 180,200 mg D-glucose/ mol.
	Now, calculate the number of moles of RS produce in the reaction:
	11.33 mg RS $\div$ 180,200 mg/mole = 6.29×10^{-5} mol RS.
	Lastly, convert moles to micromoles:
	0.0000629 moles $\times$ 1,000,000 µmoles/mole = 62.9 µmol.
	The final expression of enzyme activity becomes:
	62.9 µmol RS/mg total protein/h.

TABLE 30.7

Pectinase Induction Medium

Component	Amount
KCl	1.0g
$(NH_4)_2SO_4$	2.5 g
$MgSO_4 \cdot 7 H_2O$	0.2 g
$CaCl_2 \cdot 2 H_2O$	0.5 g
Water	1 liter
Citrus pectin	5.0 g

Note: Add all components of the medium except citrus pectin to 900 mL of water and adjust the pH to 5.3. Chill to 4°C in refrigerator, then with constant vigorous stirring, slowly add citrus pectin. May need to break clumps with a glass rod. Stir until dissolved, autoclave, and dispense 35 mL into 125-mL flasks.

- Alcohol lamp
- Sterile 6-mm to 8-mm-diam. straws
- Scalpels fitted with number 11 blades
- Incubator at 25°C (other temperatures may be required for specific organisms)
- Büchner funnel, side-arm flask and vacuum source
- Whatman #1 filter papers and a 250-mL graduated cylinder
- Ammonium sulfate—about 750–1000 g
- 250-mL beakers, magnetic stir bars and stir plates
- 25-mL beakers and magnetic stir bars
- Dialysis tubing (Spectrum Laboratories, Inc. Rancho Dominguez, CA; 800-445-7330) 3500 exclusion and clamps (suture string or binder clips will also work)
- Ice bath
- Table top centrifuge (3,000–10,000 g) (does not need to be refrigerated)
- 50-mL centrifuge tubes (do not need to be sterile)
- Top loading balance
- Squeeze bottle of distilled water
- Large 4-L beaker with distilled water at 4°C
- Refrigerator

- Pectin, glycerol (37°C) and 0.01% (10 mg/100 mL) aqueous solution of bromophenol blue-tracking dye
- Acetonitrile gloves, goggles, and lab coats
- 2X loading buffer. 80 mL distilled water; 40 mL 0.5M Tris-HCl (7.85 g Tris-HCl in 70 mL distilled water; pH to 6.8 with NaOH, and bring to 100 mL with distilled water); 32 mL glycerol (much easier to measure if at 37°C), and 1–4 mL 0.01% bromophenol blue. Mix thoroughly and store at room temperature.
- 4X resolving gel (1.5 M Tris) buffer. Dissolve 18.2 g Tris Base in 70 mL of water. Adjust pH with 6 N HCl to 8.8. Bring to 100 mL with water. Store at room temperature.
- 30% acrylamide solution (29.2 acrylamide dissolved in 60 mL of water; add 0.8 g BIS (N,N'-Methylene-bis-acrylamide) and bring to 100 mL in a graduated cylinder—store in refrigerator at 4°C). Caution: Wear gloves, particle mask, and laboratory coat. Clean up any spills immediately.
- Pectinase (commercial preparation) <0.1 mg/1 mL water. Suggestion: Pectolase Y-23 Seishin Corporation, Toyko, Japan, and/or Macerozyme R-10 (contains both pectinases and cellulases), Sigma. Dispense 25 μL of the protein and 25 μL of loading buffer into each of a number of 0.65-mL Eppendorf tubes.
- Running buffer (dissolve 9.8 g Tris-base and 43.2 g glycine in 700 mL of water bring to 1 L). To use, dilute 300 mL of this stock with 600-mL distilled water. Store at room temperature.
- Horizontal electrophoresis unit (Protean III) with either 0.75-mm or 1.0-mm combs and power supply
- TEMED and 10% ammonium persulfate (1 g in 10 mL of water: aliquot 1 mL into each of ten, 1.5-mL Eppendorf tubes and freeze at −20°C until use).
- 10-mL disposable plastic pipette with pump or bulb
- 0.1 M malic acid solution (dissolve 13.4 g of malic acid in 700 mL of water and bring to a final volume of 1 L). Store at room temperature.
- 0.01% (100 mg in 1 L of water) ruthenium red aqueous solution. Store at room temperature.
- Micropipettes (two 20 μL and 200 μL or equivalent) and disposable tips
- Plastic staining trays
- Hot plate with boiling water

Follow the instructions in Chapter 29, Procedures 29.2, to grow the fungi and isolate protein; however, substitute pectin for cellulase. Also in step 29.7, pipette 25 μL of the isolated protein into a number of 0.65-mL Eppendorf tubes and 25 μL of loading buffer. Place the samples to be used during the laboratory period on ice and freeze the remaining samples at −70°C. These samples can be thawed only once before the enzymes lose substantial activity. Now, complete Procedure 30.8 to finish this experiment.

Anticipated Results

See the anticipated results for the growth of organisms and isolation of protein with cellulases in Experiment 2. Often, the medium containing pectin is cloudy after autoclaving, but will become crystal-clear after 7 to 10 d of fungal growth (the uninoculated medium remains cloudy). Clearing of the medium probably signifies that enodpectinases were produced by the fungi.

Electrophoresis of the proteins isolated from the pectinase medium generally takes about an hour. After soaking the gels in 0.1 M malic acid and then staining overnight, cleared or colorless areas in the gel should be apparent (Figure 30.8). Some of these areas will appear as discrete lines or bands, whereas other will be diffuse and run into one another. Some areas will appear to be extended outside of the lane, which generally indicates that the specific enzymes are relatively abundant or the lane was "overload." The "cure" for this condition is to load less protein (less volume). However, if less protein is loaded there is a danger that less abundant or rare isozymes will not be detected. A compromise is to load the gel as in our exercise, but additionally, load and run other gels. The other gels could be loaded with a series of less volume (e.g., 5, 2.5, and 1 μL) or the protein content of all of the protein preparations could be determined after dialysis. Generally, when using the smaller gels in the Protean III, 1–2 μg of protein per lane would be sufficient to detect most isozymes. If a larger gel is used, then each lane should be loaded with 3–5 μg of protein.

Questions

- What effect would changing the pH of the incubation solution have on endopectinase activity?
- Do you think that the isozyme "profile" would change over time? For example, if you harvested proteins after 1 and after 3 weeks, would the "profiles" be the same or different?
- What are the advantages (or disadvantages) for producing isozymes?
- How could you adapt the procedure of detecting endopectinases to detect endoamylases?
- If the liquid medium remains cloudy throughout the incubation period, do you think there is any endopectinase activity? Why might the medium appear cloudy despite enzyme activity?

	Procedure 30.8
	Electrophoresis of Proteins and Detection of Endopectinolytic Isozymes
Step	Instructions and Comments
1	Assemble the horizontal electrophoresis units (Protean III) (2) onto the casting stand. Be sure that the plates are free of dust and dried acrylamide—the gel will stick to the glass surfaces if contaminants are present on the glass plates. Also, it is helpful if a few pieces of parafilm are cut to fit under the gray gaskets at the bottom of the casting stand. This ensures a good seal between the plates and gasket and will prevent leaks.
2	Wear gloves and lab coat. Weigh 8–10 mg of pectin and pour into a 25-mL beaker with a small stir bar. Pipette 4.0 mL of distilled water, 2.5 mL of resolving gel 4X buffer, and 3.3 mL of 30% acrylamide solution into the beaker—this solution should be cool (4°C). Stir until all of the pectin dissolves. Essentially, this is a 10% acrylamide-pectin gel.
3	Pipette 100 µL of 10% ammonium persulfate and 10 µL of TEMED into the acrylamide solution while constantly stirring. Stir for 10 sec, draw the solution into a 10-mL syringe, and place a 0.22-um filter on the end of the syringe barrel. Dispense the acrylamide solution between the sets of glass plates. Each gel unit will require about 3.5–4.0 mL of the acrylamide solution—fill to the top of the 0.75-mm gel rig; about 4.5 mL is needed for the 1.0-mm gel rig. Tap the outside glass of the gel units to remove any air bubbles. Insert the 10-tooth combs into the top slots of each gel unit—ensure that there are no air bubbles adhering to the teeth—reset the comb if necessary. Allow to stand for about 30 min or until the acrylamide is gelled. You can determine when the gel has set by observing the remaining solution in the beaker.
4	Clean excess acrylamide from the bottom of the unit with a laboratory tissue. Mount the gels on the central stand, and place in the running buffer reservoir (clear plastic box). Fill the inner and outer spaces with running buffer, taking care not to overfill the central reservoir. Remember to dilute the running buffer stock 300-mL stock to 600 mL distilled water. Gently remove the combs by lifting directly upward, and rinse them in water. Clean the individual wells with a tuberculin syringe (without a needle) using the running buffer. Place the top on the apparatus and run at 100 V for 10 min.
5	Remove protein samples, including the commercial pectinase (cp) preparation, from the freezer, and allow to thaw on ice. Mix well. Each sample should contain 25 µL of protein solution and 25 µL of loading buffer. Turn off the power to the electrophoresis unit, remove the top, and clean the wells again using running buffer and the tuberculin syringe. Load 2.5–10 µL of each sample into the wells—you may also "skip" wells between the samples. The solution should sink to the bottom. There is really no need to change tips between samples. Remember that you are actually loading backward. So, if you are planning to load five samples + cp, you should begin with sample cp and load it on the left-hand side, continue with #5, then #4, and so on to #1. This will ensure that when the electrophoresis is completed, and the gel is removed and rotated 180°, that the #1 sample will be on the left side and the cp sample on the right side as it faces you. (see Chapter 25; Figure 25.2). Try not to use the outer lanes of the gels.
5a	Note: You may also complete the experiment using a commercial preparation of pectinase (<0.1 mg/mL). Dispense 100 µL into two 0.65-mL Eppendorf tubes. Boil the first tube for 15 min to denature the pectinases and allow to cool to room temperature. To both tubes add 100 µL of loading buffer and vortex to mix. Add 5 µL of the boiled preparation to one lane. Change the tip, skip a well (lane), and add 10, 8, 6, 4, and 2 µL of the undenatured (not boiled) sample to the five wells to the right. The order should be as follows: blank, boiled sample, blank, 10, 8, 6, 4, and 2 uL.

Continued

6   Replace the top and set the power pack for constant 100 V. Allow the tracking dye to migrate to within 1 cm or less of the bottom of the gel. Turn off power, remove the inner glass plate (the smaller one) with the aid of running water, and cut a small piece of acrylamide at 45° to mark the right side of the gel. This will help you with keeping track of the sample order. Remove the gels from the outer or large glass plates with running water and place in a staining tray (one gel per tray). Pour off any water and cover with 0.1 *M* malic acid solution at room temperature for 1 h.

7   Rinse briefly with distilled water, pour ruthenium red solution over gels and allow to stand at room temperature overnight, and then rinse with distilled water. The stain solution may be reused several times and may be stored for a month or more at room temperature in the dark.

8   Note cleared zones (no stain), which may be easier to see with transmitted white light. These are areas where endopectinases migrated during electrophoresis. Draw the patterns for each lane or organism.

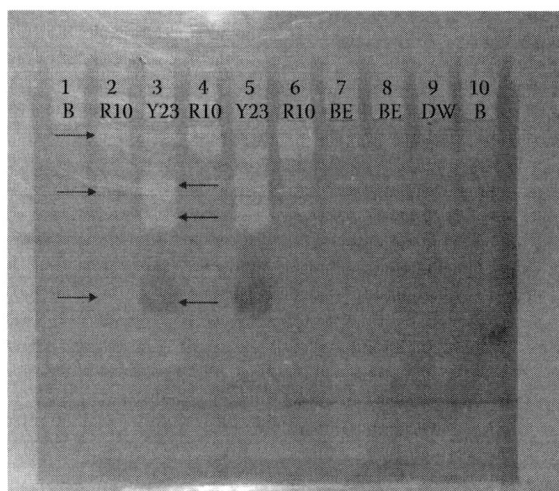

FIGURE 30.8 Polyacrylamide gel electrophoresis of endo-pectinolytic isozymes from Y-23 and Macrozyme R-10 pectinases. Proteins were electrophoresed for 1 h and then incubated in 0.1-*M* malic acid. Gels were stained in 0.01% ruthenium red overnight at 4°C, and then washed with water for 1 to 2 h. Gels were then viewed on a light box. Arrows indicate cleared (non-stained) areas in the gel and correspond to endopectinase activity. B = Blank, R10 (lane 2) = 5 µl of R10, Y23 (lane 3) = 5 µl of Y23, R10 (lane 4) = 2.5 µl R10, Y23 (lane 5) = 2.5 µl Y23, R10 (lane 6) = 1 µl of R10, BE (lane 7) = boiled R10 (5 ul), BE (lane 8) = boiled Y23 (5 ul), DW (lane 9) 5 µl of distilled water, B (lane 10) = blank. (See CD for color figure.)

LITERATURE CITED

Cruickshank, R.H. and J.I. Pitt. 1987. Identification of species in *Penicillium* subgenus *Penicillium* by enzyme electrophoresis. *Mycologia* 79:614–620.

Davidson, R.W., W.A. Campbell, and D.J. Blaisdell. 1938. Differentiation of wood decaying fungi by their reaction on gallic or tannic acid medium. *J. Agric. Res.* 57:682–695.

Lowry, O.H., N.J. Rosenbrough., A.L .Farr, and R.J. Randell. 1951. Protein measurement with the Folin phenol reagent. *J. Biol. Chem.* 193:265–275.

Miller, G.L. 1959. Use of dinitrosalicylic acid reagent for determination of reducing sugar. *Anal. Chem.* 31:426–428.

Nelson, N. 1944. A photometric adaption of the Somogyi method for the determination of glucose. *J. Biol. Chem.* 153:375–380.

Reese, E.T., and M. Mandels. 1963. Enzymic hydrolysis of cellulose and its derivatives. pp. 139–143. In: *Methods in carbohydrate chemistry*. Ed., Whisler, D.L. Academic Press, New York.

Sierra, G. 1957. A simple method for the detection of lipolytic activity of microorganisms and some observations on the influence of the contact between cells and fatty substrates. Antonie *Van Leeuwenhoek Ned. Tijdschr. Hyg.* 23:15–22.

Society of American Bacteriologists. 1957. *Manual of Microbiological Methods*. McGraw-Hill Book Company, New York. 315 p.

31 Host Defenses
A Physical and Physiological Approach

Kimberly D. Gwinn, Sharon E. Greene, James F. Green, and David Trently

CHAPTER 31 CONCEPTS

- Most plants are resistant to most pathogens.

- Resistant plants produce passive barriers that the pathogen cannot overcome or they must be able to activate successful defense(s) that arrest pathogen development.

- Although both active and passive defense can be important at any stage of the disease cycle, passive defense is usually more important during pre-penetration and penetration, and active defense is usually most important in the infection stage.

- Active disease responses require signal recognition/transduction followed by gene activation.

- Examples of passive defenses are inhibitory plant surface chemicals, thick cuticle, lignified tissues, and phyto-anticipins. Examples of active defenses are antimicrobial proteins, phytoalexins, the hypersensitive response, and systemic resistance.

Healthy plants grow in an atmosphere crowded with fungal spores, bacterial cells, and viruses; in soil, healthy roots predominate despite the high numbers of fungal spores, bacterial cells, and nematodes that thrive in the rhizosphere (soil immediately around the roots of the plant). In the face of this onslaught of potential pathogens, plants defend themselves with an arsenal of weapons, and as a result, most plants are resistant to most pathogens. Plants have developed defense strategies that successful pathogens must overcome. Although plants defend themselves against potential pathogens in different ways, scientists have categorized strategies into the following two basic categories: **passive** (present before pathogen recognition) or **active** (induced after pathogen recognition by the host). Knowledge and exploitation of host defenses can lead to new pathogen control strategies (e.g., "fungicides" that turn on resistance, or transgenic plants that can silence viruses, plants that overproduce bioactive natural products, or plants that produce antimicrobial proteins).

Successful host defenses disrupt the disease cycle (Chapter 2), primarily in the pre-penetration, penetration, or infection phases. In general, passive defenses against pathogen attack are more prevalent in the prepenetration and penetration phases, and active defenses are more important in the infection phase.

PREPENETRATION

The barriers most important during prepenetration are passive (see Chapter 28). Plant surfaces are colonized by a numerous microorganisms, both pathogenic and nonpathogenic. The extent to which an organism can colonize the plant surface is related to the chemical nature and topography of the surface. If the host does not have the necessary physical or chemical attributes, then the pathogen cannot recognize, attach, or colonize the surface of the plant. The nature and quantity of available nutrients may increase the fitness of the pathogen. Pathogens must attach to the surface and out-compete existing nonpathogenic organisms. All of these organisms compete for nutrients (carbon, nitrogen, and essential inorganic molecules) that are present on the surface as well as for physical sites (e.g., depressions on the leaf surfaces). When some pathogens contact host chemicals, they form structures for attachment and penetration. For some fungi, formation of penetration structures is induced by physical contact with the ridges and valleys of the host surface. Changes in host surface chemistry or topography can limit the pathogen.

Most pathogens change or multiply when they contact the plant surface. Fungal spores germinate, nematode eggs hatch and bacteria multiply. The chemical and physical attributes of the plant surface can affect these processes. Lipopolysaccharides (LPS), which are cell surface components of Gram negative bacteria, allow the bacterium to exclude antimicrobial substances including those on the plant surface and those induced by active defense by the plant. Bacteria that lack LPS are less effective pathogens.

The thickness of physical barriers plays an important role in host defense during prepenetration, penetration, and infection. In prepenetration, waxes on the surface of many plant parts limit the availability of the free water that many pathogens need to multiply (bacteria) or spores to germinate (fungi). The role of cuticle and cell wall thickness are discussed in the section on penetration.

PENETRATION

PASSIVE

Physical barriers can limit pathogen penetration. The thickness of the cuticle plays an essential role in host defense against some pathogens; a thin cuticle layer is more easily penetrated by some fungi. The cuticle is composed primarily of cutin, a lipid. Suberin, a similar compound associated with cork cells, can be a passive defense or formed in an active response. Cell wall thickness plays a role both in the epidermal cells and in the cells being invaded by a pathogen. Many fungi require enzymes to penetrate the cuticle and cell walls (Chapters 10 and 30), but some plants contain compounds that inhibit these enzymes.

ACTIVE

Plants have developed a number of active mechanisms to limit the entry of pathogens that evade preestablished barriers. The key to active defense is host recognition of the pathogen. Plants react to the presence of a pathogen by signal recognition, followed by signal transduction, and finally gene activation. Gene activation results in end products that contain, inhibit, or kill the invader. The response to invasive organisms, whether they are pathogens or mutualists, is fundamentally similar (Andrews and Harris, 2000). The following three categories of active defense have been described: primary, secondary, and systemic. Primary responses are those limited to the cell in contact with the pathogen. Secondary responses are induced in cells adjacent to the cell in contact with the pathogen. Systemic responses are induced throughout the entire plant.

An understanding of **signal recognition/transduction** is essential to understanding active defense mechanisms. In response to pathogen penetration, a signal recognition/transduction pathway is activated. This involves ion fluxes, oxidative bursts, protein phosphorylation, and signaling molecules. An ion flux is an efflux of Cl^- and K^+ ions coupled with an influx of H^+ and Ca^{++}; Ca^{++} is believed to be the central mediating component of early plant defense responses, although the mechanism by which it controls sig-

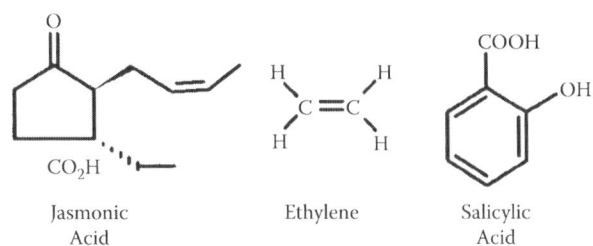

FIGURE 31.1 Some examples of signal compounds. The active defense systems of all plants studied thus far are induced by these molecules.

naling is not known. An oxidative burst is the rapid production of hydrogen peroxide (H_2O_2) and is believed to signal cell death, overwhelm antioxidant cell protection, lead to rapid cell wall reinforcement and induce gene expression. Protein phosphorylation is a primary means for the regulation of transcription. Phosphorylations can increase binding to DNA or signal the protein to move to the nucleus. Signal molecules are the active defense systems of all plants studied thus far and are induced by the same molecules—salicylic acid, jasmonic acid, and ethylene (Stuiver and Custers, 2001) (Figure 31.1). Jasmonic acid and its methyl ester, methyl jasmonate, increase in response to pathogen attack both locally and systemically and are both preexisting and induced compounds. These molecules are produced at penetration (for some fungal and nematode diseases) and in early infection for bacterial, viral, and some fungal and nematode diseases. An *Arabidopsis* mutant impaired in the jasmonic acid response has increased susceptibility to necrotrophic fungi but not to biotrophic fungi. Ethylene, a gaseous plant growth regulator, is synthesized in both compatible and incompatible reactions. Ethylene appears to mediate resistance against necrotrophic fungal pathogens and also nonhost resistance. Treatment of plants with salicylic acid, the active component in aspirin, induces systemic resistance in many plants. Salicylic acid or its methyl ester, methyl salicylate, were once believed to be the long-distance signals in systemic acquired resistance; however, several recent studies have shown that this is unlikely.

In response to these signals, genes are activated, resulting in increased translation (reading the DNA code) and transcription (producing the protein sequences coded by the DNA). Enzymes that are directly important in defense or in the biosynthesis of defense compounds are produced.

Plants can limit penetration of many pathogens to the first cell that is attacked or to cells in the immediate vicinity (i.e., primary and secondary responses). Papillae are cell wall appositions deposited between the plasma membrane and the cell wall. Papilla formation is generally preceded by the aggregation of the cytoplasm in the cell under attack. They consist primarily of callose, a polymer of β-1,3 D-glucose residues, but also may contain cellulose, phenolic compounds, and lignin. The penetration peg of the fungal pathogen is either stopped by the papilla or grows through it. Halos, locally modified regions of the host cell wall around a penetration site, are rich in substances such as phenolics, silicon, lipids, proteins (peroxidases), and lignin. Halos are often larger at sites of unsuccessful penetration than at sites of successful penetration. Epidermal cells adjacent to the papilla or adjacent to an appressorium are often lignified. This induced lignin may be chemically different from the lignin in nonchallenged tissue. Suberization, deposition of insoluble polymers at the cell wall, also increases resistance to some fungal pathogens.

INFECTION

PASSIVE

Plants are capable of synthesizing several thousand different low-molecular weight compounds (secondary metabolites), and many of these products are present at concentrations that affect pathogens. Antimicrobial natural products produced by plants are grouped into the following two classes: the **phytoanticipins** and the **phytoalexins**. Phytoanticipins are preestablished small molecular weight compounds that are stored in the plant cell or are released from a glucoside. Phytoalexins are formed in response to the pathogen and so are discussed in the section on active defense responses. Some compounds are phytoalexins in one species and phytoanticipins in another. Most antimicrobial natural products are broad-spectrum; specificity, if it exists, is determined by the pathogen's ability to break down the compounds. In a few studies, plants with lower amounts of antimicrobial compounds have been shown to be more susceptible to disease. Crops may be susceptible to many pathogens because selective breeding for other characteristics has decreased fitness by reducing numbers and concentrations of natural products (Dixon, 2001). Many phytoanticipins are stored within vesicles, whereas others stored in the glycosylated form until cells are damaged. Glycosylation (chemically bonding to a glucose molecule) converts a reactive and toxic phytoanticipin to a stable, nonreactive storage form that is more likely to be water soluble. Transgenic plants have been engineered that make glycosidic phytoanticipins and demonstrate increased disease resistance.

ACTIVE

Phtyoalexins

Phytoalexins are antimicrobial natural products that are produced after infection or elicitation by abiotic agents. Plants produce natural products by a number of metabolic pathways; phytoalexins (Figure 31.2) are chemically diverse because they are manufactured by a number (or combination) of biosynthetic pathways. However, species in the same plant families tend to make phytoalexins that are derived from the same pathway and therefore are very similar in chemical structure. Although it is generally accepted that phytoalexins play a role in host defense, they are likely only one part of an overall defense strategy. However, phytoalexin production may simply be a response that is correlated with the expression of defense. Several lines of evidence support a role for phytoalexins in host defense and include the following: (1) in several gene for-gene systems, resistance is associated with phytoalexin production; (2) phytoalexins accumulate rapidly to inhibitory concentrations at the site of pathogen development; (3) pathogens can overcome host resistance by phytoalexin detoxification; and (4) plants genetically transformed to overproduce phytoalexins are more resistant to disease (Hammerschmidt, 1999). Phytoalexins are produced by the cell under attack (primary response) as well in the adjacent cells (secondary response).

Antimicrobial Proteins

Antimicrobial proteins have been detected in many plant species and tissues. The widespread localization of antimicrobial proteins (e.g., chitinases and glucanases) in plants, coupled with their activity against pathogens *in vitro,* suggests that these enzymes may serve a protective role. Plants transformed with genes that code for these antimicrobial proteins can be more resistant to pathogens.

Hypersensitive Response

The active responses described above are nonspecific; they occur in response to pathogens and other organisms. The hypersensitive response, however, is highly specific and occurs only when the product of a pathogen avirulence gene interacts with the product of a plant resistance gene. Activation of this gene in gene-for-gene resistance results in a cascade of reactions within the cell. The **hypersensitive response (HR)**, a rapid death of a few host cells that limits the progression of the infection, is a manifestation of recognition of the pathogen avirulence gene product. Typically, an HR includes signal transduction, programmed cell death, increased activation of defense-related genes (e.g., synthesis of phytoalexins, salicylic acid, and antimicrobial proteins), and a distant induction of general defense mechanisms that serve to protect the plant (i.e., systemic acquired resistance [SAR]).

Elicitation of the primary responses results from the recognition of a pathogen effector protein by the host receptor protein. Most effector proteins have no apparent enzymatic activity, but are capable of binding to the receptor protein. When the two proteins interact, a signal transduction pathway is activated and the primary response is initiated. A rapid burst of oxidative metabolism leads to the production of superoxide and subsequent production

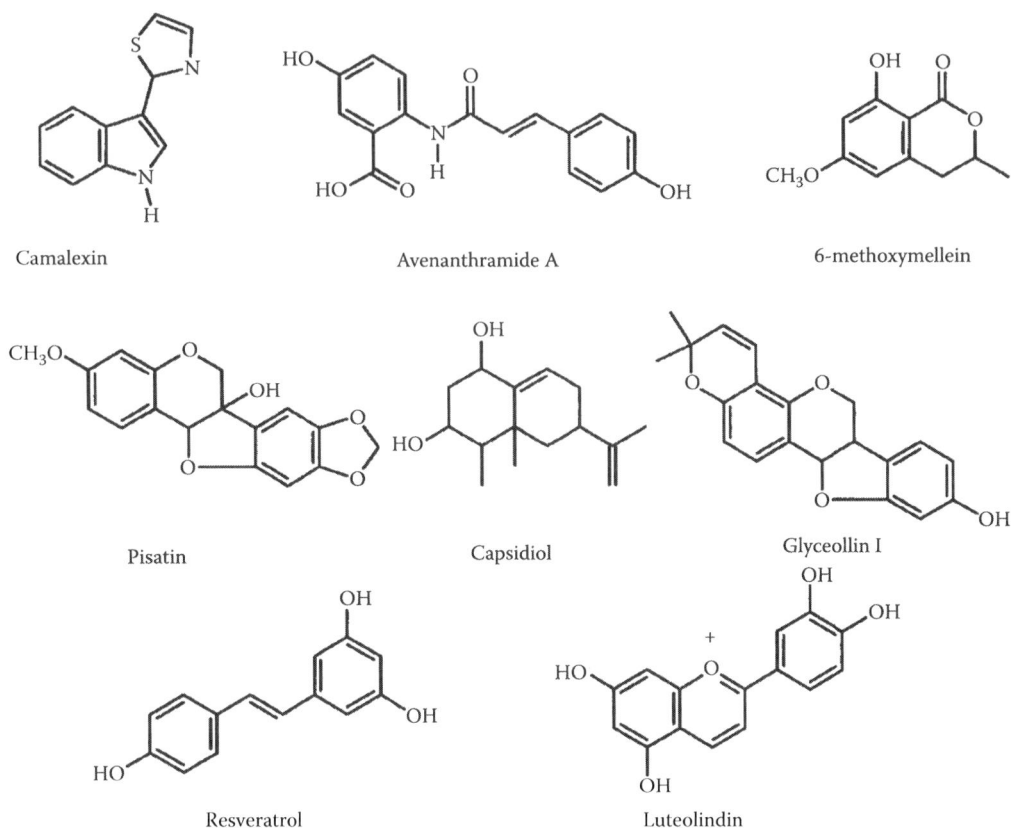

FIGURE 31.2 Structures of some selected phytoalexins.

of hydrogen peroxide that precedes the development of visual symptoms of an HR.

Plants appear to have adapted **programmed cell death**, a general process commonly associated with reproductive and xylem tissue development, as a host defense response. The attacked cell and several cells around it die in response to chemical signals. The sacrifice of these cells isolates the pathogen and is a particularly good resistance mechanism for biotrophic pathogens. The hypersensitive response is not limited to biotrophic pathogens. Pathogens may be killed by the reactive oxygen species generated by the oxidative burst or by the antimicrobial compounds and enzymes formed in response to gene activation.

SYSTEMIC RESISTANCE

Although plant systems do not truly mimic the immune system of mammals, they do have the ability to better resist pathogens after exposure to other organisms. Infection or colonization of the plant by one organism can induce host resistance to other pathogens. There are two major types of systemic resistance, **systemically-acquired resistance (SAR)** and **induced systemic resistance (ISR)**. In SAR, the attacking organism is

a necrotrophic pathogen. Some bacteria that colonize the roots, but do not cause disease, may induce ISR. Although both result in systemic host resistance, methods and mechanisms differ significantly. In SAR, salicylic acid or methyl salicylate are produced as a primary and secondary response. An unknown compound then moves into the phloem and transports the signal to distal portions of the plant. Salicylic acid then accumulates in the distal portions of the plant and is converted to methyl salicylate, which is volatile and may serve as a signal to neighboring plants. Antimicrobial proteins appear in the distal portions of the plant. In ISR, both jasmonic acid and ethylene signal distal portions of the plant; antimicrobial proteins are not produced.

The summaries (Case Studies 31.1–31.3) of resistance mechanisms in well-known pathogens illustrate the complexity of host resistance. Whenever possible, examples of active and passive defenses are used.

EXERCISES

The following laboratory exercises are designed to demonstrate the effects of essential oils (commonly used as candy flavorings) on the growth of plant pathogenic fungi and to demonstrate the impact of wound healing on disease.

CASE STUDY 31.1

TOBACCO MOSAIC VIRUS

- Viral pathogen: tobacco mosaic virus. In some cultivars of solanaceous crops, symptoms of TMV are mild to severe mottling, chlorosis, and dwarfing of leaves. Tobacco cultivars with the N gene produce an HR.
- All viruses are obligate biotrophic organisms.
- Methyl salicylate, released from necrotic lesions on tobacco caused by TMV, increases the resistance of uninfected neighboring plants. The PR1 gene is also induced in the neighboring plants.
- Concentration of a protein located in the plastid can control host reaction to TMV. Overexpression of this protein increases the rapidity of the HR; low concentrations lead to a suppression of the HR.
- Rapid cell death occurs in plants with an N gene except when the plants are transformed with viral genes that slow down the HR and cause the plant to become susceptible.
- At least three protein products have been demonstrated to function as avirulence determinants: the replicase protein (in lines carrying the N gene), the coat protein (in tobacco lines carrying the N′ gene), and the movement protein (in tomato lines carrying the Tm-2 and the Tm-22).

CASE STUDY 31.2

BACTERIAL WILT

- Bacterial pathogen: *Ralstonia solanacearum*. This bacterium causes a bacterial wilt of many hosts. The vascular tissue of stems, roots, and tubers turn brown and when cut, ooze a stream of bacteria. *Ralstonia solanacearum* colonizes both the soil, which is nutrient-poor, and the inside of a plant, which is nutrient-rich but well-defended.
- Bacteria cells attach to the plant roots and form microcolonies at two sites on the root that are vulnerable in two major passive defense systems: the sites of lateral root emergence and the root elongation zone.
- While entering the plant cortex and the vascular system, the pathogen causes minimal tissue/cell damage and either avoids or suppresses host recognition. The exopolysaccharide (EPS1) may be responsible for masking bacterial structures that are targets of host recognition because strains of the bacterium that do not produce EPS1 simply agglutinate and degenerate in the cortex, perhaps due to defense responses.
- Bacteria sense that they have arrived in the plant cell and induce genes for host infection, the so-called hrp genes. The pathogen has greater than 20 hrp genes. Inactivation of the hrp genes causes nearly complete loss of ability to cause disease and incites a hypersensitive response on resistant plants. The hrp genes encode proteins that produce a secretion apparatus that allows translocation of avirulence factors across the bacterial membrane for delivery to the host cell. One protein (PopA) encoded by the hrp genes causes a HR-like response on plant tissues; another protein (PopC) is similar in structure to host receptor proteins and may interfere with host defenses.
- In plants that have the corresponding resistance genes, the plant recognizes the pathogen proteins and responds with cell wall thickening, separation of the cell membrane from the wall surface, and forms wall appositions (Schell, 2000).
- The lipopolysaccharide produced by *R. solanacearum* prevents the HR induction in many dicots. Similar to ISR, salicylic acid does not act as a signal in this localized induced response and it does not require necrosis. Addition of the lipopolysaccharide to plants can induce production of defense-related proteins (e.g., peroxidase in tobacco and synthesis of phytoalexins) and changes in the plant cell surface. The lipopolysaccharide does not induce the oxidative burst typically associated with an HR (Dow et al., 2000).

CASE STUDY 31.3

POWDERY MILDEW

- *Blumeria graminis* causes powdery mildew fungus of wheat and barley.
- Although host cuticle plays an important role in resistance to many pathogens, *B. graminis* effectively degrades the barley cuticle using esterase enzymes that break the ester bonds that hold together the cutin molecule. Esterases are produced within 2 h of the conidia landing on the host
- Host genes govern resistance at different stages of the interaction. In plants containing the mlo gene, penetration of the fungus is stopped by the formation of effective papillae; this type of resistance is effective against all races of the pathogen. Occasionally a few mildew colonies appear in mlo resistant barley. Silicon-rich halos are observed in wheat and barley infected with *B. graminis*. Silicon levels are higher in regions of failed penetration attempts than in sites of successful penetration. Peroxidases and hydrolytic enzymes are also found in the halos.
- In plants containing the Mlg gene, fungal growth is arrested within papillae of cells that subsequently undergo an HR.
- In barley plants that have single gene controlled, race specific resistance to *B. graminis*, failure of the papillae response leads to successful penetration of the cell; when the pathogen race is incompatible with the host, then an HR is initiated (Hückelhoven and Kogel, 1998).

EXPERIMENT 1. EFFECT OF VOLATILE COMPOUNDS FROM CANDY FLAVORINGS ON THE GROWTH OF PLANT PATHOGENIC FUNGI

Essential oils are highly volatile substances isolated from an odiferous plant; the term essential was used because these oils were thought to contain the essence of odor and flavor (Linskens and Jackson, 1991). The oil bears the name of the genus or common name of the plant from which it is derived and can be somewhat misleading because chemistry can be highly variable within a genus or species. Since antiquity, essential oils have been used as perfumes, medicines, and flavorings. Essential oils are well known for their antibacterial, antifungal, antiherbivore, and antioxidant activities, and have been proposed as natural, safe pesticides. Essential oils of many plants (e.g., sage, oregano, citrus, and various mints) contain antifungal phytochemicals (compounds produced by plants). Effects of essential oils on postharvest and grain spoilage pathogens (or similar species) have been studied most often.

Candy flavorings are easily manipulated, nontoxic sources of phytochemicals. Many flavorings (clove, lemon, orange, peppermint, and spearmint) are extracted without solvents as pure oils. These contain several antifungal phytochemicals, most notably various monoterpenes.

General Considerations

Candy flavorings that are essential oils can be purchased at local grocery or specialty shops, but care should be taken to avoid flavorings that contain alcohols or other solvents because these will confound the results. Clove, lemon, orange, peppermint, and spearmint oils produced by Lorann Gourmet (Lorann Oils, Inc., 4518 Aurelius, Lansing, MI 48910) can be used with good results.

Any pathogen that is easily cultured on solid media can be used in these experiments. For example, *Alternaria*, *Fusarium*, and *Sclerotinia* species are inhibited by compounds found in the candy flavorings.

These experiments are designed for teams of students. Each student should have an opportunity to transfer the mycelial plug to the culture medium, pipette oils onto the filter paper, and measure fungal colony diameter.

Materials

The following items are needed for each team of students:

- For each pathogen to be tested, fourteen 10-cm diam. petri dishes containing a fungal growth medium are needed (This is based on three oils at two concentrations run in duplicate plus dishes for non-amended controls. The choice of medium should be dictated by the choice of pathogen. Potato dextrose agar works well for most pathogens.)
- Two actively growing cultures of each pathogen
- 5- to 7-mm diam. cork borers, sterile plastic straws or other tool for cutting similar size mycelial plugs
- Bunsen burner and 70% ethanol
- Dissecting probe
- Ten 100-μL pipettes and tips
- Parafilm
- Filter paper
- Plastic ruler

Follow the protocols listed in Procedure 31.1 to complete this experiment.

	Procedure 31.1
	Effect of Essential Oils on Fungal Growth
Step	Instructions and Comments
1	Label dishes with your name, genus of pathogen, amount and type of flavoring, and date.
2	Cut several disks (ca. 5 mm) from the edge of an actively growing culture using a cork borer or sterile plastic straw.
3	Place one agar disk, mycelium side down, in the center of the culture medium. If available, place a single sclerotium in the center of the dish. Inoculate all dishes in the same manner.
4	Layer two pieces of filter paper inside the lid of each dish. They should fit snugly and the bottom layer should be in contact with the inside of the lid.
5	Pipette a drop (0, 5 or 50 µL) of candy flavoring in the center of the filter paper. Invert the lid and fit the bottom of the dish into the lid. The petri dish should be upside-down with the mycelium plug above the filter paper. Each treatment should be repeated in a separate dish.
6	Incubate the cultures at room temperature for one week. Dishes containing the different flavorings should be stored in separate areas. Observe growth of the fungi daily. Control treatments (0 µL) should not be allowed to grow to the edge of the petri dish. Data should be recorded when the mycelial growth of the controls nears the edge of the dishes or after one week. All data should be recorded during the same class period.
7	Measure and record the diameter of each fungal colony. Determine the mean diameter for each treatment. Calculate comparative growth index (CGI) from the following equation: $$CGI = 100\left(\frac{D_c - D_t}{D_c}\right) - 100$$ where Dc = mean colony diameter for controls and Dt = mean colony diameter for each treatment. Values will be greater than zero if mean growth in the treatment exceeds mean growth of control. Values will be less than zero if mean growth in the treatment is less than the control.

Anticipated Results

Treatment with essential oils should reduce the growth of all fungi, i.e., comparative growth index should be less than zero for all compounds. Growth should be inhibited more by the higher concentration (50 µL) of the oil than the lower concentration (5 µL). Some oils will reduce growth more than others. For example, fungal growth in an atmosphere of lemon oil should be less than growth in an atmosphere of orange oil; both should be less than controls.

Questions

- What role might antimicrobial phytochemicals play in plant defense?
- Would a pathogen of the orange rind (e.g., *Penicillium*) be more or less sensitive or have the same sensitivity to volatile compounds produced by orange than *Fusarium*, which is usually considered a non-pathogen of orange? Why or why not?

EXPERIMENT 2. PHYSICAL DEFENSES OF POTATO AGAINST SOFT ROTTING BACTERIA (M. WINDHAM, PERSONAL COMMUNICATION)

Plants can defend themselves in many ways when they are wounded or when a pathogen tries to attack. These methods include induced structural and biochemical defenses. In this experiment, potato pieces that have had or have not had time to induce defensive mechanisms will be challenged with the soft rot bacterial pathogen, *Erwinia carotovora*. Host tissues will be examined for induced defensive mechanisms and how these mechanisms affect infection by the pathogen.

Materials

The following supplies are needed for each student, student team or class:

- *Erwinia carotovora* isolate
- Several 250-mL flasks
- Nutrient broth (Difco)

Procedure 31.2

Physical Defenses of Potatoes Against Soft Rotting Bacteria

Step	Instructions and Comments
1	One week before lab, start cultures of *Erwinia carotovora* in 250-mL flasks containing 75 mL of nutrient broth. Incubate at room temperature on a rotary shaker set for 100 rpm.
2	Two days before lab, pour the contents of two flasks into the soil of each flat and mix well.
3	One day before lab, take half the potatoes and cut them into pieces that are about 5 cm^2.
4	The day of the experiment, students should cut the remaining potatoes into pieces that are about 5 cm^2 in size.
5	Using the scalpel, make thin sections of a potato cut the day before and from a freshly cut piece of potato. Place thin sections of both on a microscope slide with a drop of water and place a cover glass on top of the specimens. Examine the thin sections microscopically and record any differences in the sections.
6	Take pieces of freshly cut potato and pieces that were cut the previous day and lay them down (wounded side down) on the surface of the soil. Place flats in the greenhouse for one week. Keep flats moist, but not flooded.
7	After a one week incubation, remove the potato pieces and examine them for soft rot symptoms. Record your results.

- Rotary shaker
- Flats of soil and labels
- Potatoes
- Compound microscope
- Microscope glass slides and coverslips
- Water
- Eyedropper
- Knife to cut potatoes
- Scalpel to make thin sections
- Greenhouse or incubator

Follow the protocols listed in Procedure 31.2 to complete the experiment.

Anticipated Results

The response of the potatoes to wounding can be observed microscopically. In the sections from the potato cut a day prior to the experiment, a white, milky film should be evident. The film should be missing from the freshly sliced potatoes. There should also be differences in disease incidence and disease severity between the 1-d-old pieces of potato and the freshly wounded pieces of potato after they have incubated for one week on the surface of the soil infested with *E. carotovora*.

Questions

- What type of defense mechanisms had developed when you examined the potato pieces microscopically?
- What is suberin?
- What would you recommend to potato growers after conducting this experiment?

LITERATURE CITED/SUGGESTED READING

Andrew, J.H. and R.F. Harris. 2000. The ecology and biogeography to microorganism on plant surface. *Annu. Rev. Phytopathol.* 38: 145–80.

Dixon, R.A. 2001. Natural products and plant disease resistance. *Nature* 411: 643–647.

Dow, M., M.-A. Newman and von Roepenack, E. 2000. The induction and modulation of plant defense responses by bacterial lipopolysaccharides. *Annu. Rev. Phytopathol.* 38: 241–261.

Hammerschmidt, R. 1999. Phytoalexins: What have we learned after 60 years? *Annu. Rev. Phytopathol.* 37: 285–306.

Hückelhoven, R. and Kogel, K.-H. 1998. Tissue-specific generation at interaction sites in resistant and susceptible near isogenic barley lines attacked by the powdery mildew fungus (*Erysiphe graminis* f. sp. *hordei*). *Molec. Plant Microbe Interact.* 11: 292–300.

Hutcheson, S.W. 1998. Current concepts of active defense in plants. *Annu. Rev. Phytopathol.* 36: 59–90.

Kurita, N., Miyaji, M., Kurane, R., and Takahara, Y. 1981. Antifungal activity of components of essential oils. *Agric. Biol. Chem.* 45: 945–952.

Linskens, H.F. and Jackson, J.F. 1991. *Essential Oils and Waxes. Vol. 12 in Modern Methods of Plant Analysis.* Springer Verlag, New York.

Mercier, J. and S.E. Lindow, S.E. 2000. Role of leaf sugars in colonization of plants by bacterial epiphytes. *Appl. Environ. Microbiol.*, 66: 369–374.

Petrini, O. and G.B. Ouellette (Ed.) 1994. *Host Wall Alterations by Parasitic Fungi.* APS Press. St. Paul, MN.

Schell, M.A. 2000. Control of virulence and pathogenicity genes of *Ralstonia solanacearu-m* by an elaborate sensory network. *Annu. Rev. Phytopathol.* 38: 263–292.

Vanacker, H., T.L.W. Carver, and C.H. Foyer. 2000. Early H_2O_2 accumulation in mesophyll cells leads to induction of glutathione during the hypersensitive response in the barley-powdery mildew interaction. *Plant Physiol.* 123: 1289–1300.

32 Disruption of Plant Function

Melissa B. Riley

CHAPTER 32 CONCEPTS

- Plant disease symptoms often result from plant pathogens affecting the normal physiological activities of the plant.

- Effects of a pathogen on the normal physiology of the plant can be at the macroscopic level (i.e., production of galls) or at the microscopic level (i.e., alternation of cell membrane permeability).

- Major plant physiological activities affected by plant pathogens include photosynthesis, respiration, production of plant growth hormones, absorption/translocation of water and nutrients, and transcription and translation.

- Study of plant pathogens led to the discovery of important compounds associated with normal plant activity—gibberellins, which are plant growth hormones produced by fungal plant pathogens.

- Reduced photosynthesis and increased respiration induced by plant pathogens ultimately result in reduced plant growth and yield.

- Disruptions of absorption/translocation within a plant can result in mineral deficiency symptoms.

- One of first responses of a plant to plant pathogen attack is alteration in cell membrane permeability.

Since the beginning of time, humans have observed the effects of plant pathogens on plants, although they did not know about plant pathogens or how they caused the observed effects. In many cases today, the processes that result in symptom development are still not completely understood. Plant disease has been defined as alteration in the normal **physiology** of a plant, but this definition obviously requires extensive knowledge of normal plant physiology. Normal plant physiology includes many processes such as **photosynthesis**, **respiration**, **absorption** and **translocation** of water, transport of photosynthetic products, production of compounds, such as plant growth **hormones**, **enzymes**, proteins, carbohydrates, lipids, and nucleic acids by the plant; and movement of materials between individual cells. Changes in any of these processes may have an effect on the overall appearance of the plant, which is observed as symptoms of disease. How do plant pathogens interrupt these processes? Responses may be something as small as the alteration of the metabolism within a cell or the movement of materials across a cell membrane to an overall plant response such as reduction in crop yield. We have learned much about the effects of plant pathogens on the normal physiology of plants, but still many questions remain, especially related to the exact sequence of events that results in the expression of symptoms.

Some common symptoms observed in response to plant pathogens and their possible relationship to normal plant physiology, and examples of plant pathogens are provided in Table 32.1. This chapter will examine how plant pathogens can disrupt the normal functions of a plant. Specifically, the effects of these pathogens on the major activities of the plant, including photosynthesis; respiration; production of plant growth hormones; absorption and translocation of water and nutrients; protein, carbohydrate and lipid production; and cell **permeability** will be considered.

PHOTOSYNTHESIS

Essentially all organisms on Earth depend on the process of photosynthesis in which plants absorb solar energy and convert it into carbohydrates that can be further utilized as energy sources. Photosynthesis occurs in the **chloroplasts** of plant cells where carbon dioxide and water in the presence of solar energy and chlorophyll are ultimately converted to carbohydrates. This process is generally expressed by the following formula:

$$6 \, CO_2 + 6 \, H_2O + \text{Light} + \text{Chlorophyll} \rightarrow C_6H_{12}O_6 + 6 \, O_2$$

The process can be divided into a light and a dark portion. During the light phase, solar energy produces reduced chemical compounds in the form of NADPH (reduced nicotinamide adenine dinucleotide phosphate) and ATP

TABLE 32.1

Common Plant Disease Symptoms, the Physiological Processes Affected, and Examples of a Disease and the Associated Plant Pathogen Affecting the Process

Symptom	Physiological Function	Example Disease/Pathogen
Chlorosis	Photosynthesis	*Tobacco mosaic virus*
Wilting	Xylem transport	Bacterial wilt/tomato and tobacco—*Ralstonia solanacearum*
Hyperplasia—cell division	Growth hormone regulation	Crown gall—*Agrobacterium tumefaciens*
		Black knot/plum—*Dibotryon morbosum*
Necrosis	Many different functions	Fire blight/apple—*Erwinia amylovora*
Hypertrophy—cell enlargement	Growth hormone regulation	Root knots—*Meloidogyne incognita* (root-knot nematode)
Leaf abscission	Growth hormone regulation	Coffee rust—*Hemileia vastatrix*
Etiolation	Growth hormone regulation	Bakanae "foolish seedling" disease of rice—*Gibberella fujikuroi*
Stunting	Many different functions	Many different viral diseases
Abnormal leaf formation	Growth hormone regulation, Respiration	*Cucumber mosaic virus*—ornamentals

(adenosine 5′ triphosphate). During the dark phase, energy captured in the NADPH and ATP is utilized to convert carbon dioxide into carbohydrates. These reactions can occur through two major pathways: (1) the Calvin cycle in C_3 plants and (2) the C_4 cycle in C_4 plants. Alteration or inhibition in the overall photosynthetic activity of the plant can result in obvious gross symptoms in the plant, such as chlorosis or reduced growth and yield. Reduction in the photosynthetic activity of the plant can be accomplished in many different ways. One way is from a reduction in total leaf area due to leaf necrosis and thereby destruction of photosynthetic tissue. A study of the effects of rust and anthracnose diseases on the photosynthetic competence of diseased bean leaves revealed that rust infections reduced photosynthetic rates in direct proportion to visible lesion area, whereas the reduction in photosynthetic activity in anthracnose-infected leaves was greater than could be attributed to the visibly infected area (Lopes and Berger, 2001). Photosynthesis in the green areas beyond the necrotic symptoms was severely impaired in anthracnose-infected leaves, indicating that the pathogen was affecting photosynthesis beyond the development of necrotic areas. Chlorosis is a common symptom observed in response to plant pathogens and can result from inhibition of chlorophyll synthesis (Almási et al., 2000), increases in the rate of chlorophyll degradation (Strelkov et al., 1998), and reduction in the chloroplast size and number (Kema et al., 1996).

Blumeriella jaapii, the causal agent of cherry leaf spot, seems to mainly interfere with the enzymatic process of the Calvin cycle, and these effects were observed prior to the occurrence of visible symptoms (Niederleitner and Knoppik, 1997). The activity of three specific enzymes of the Calvin cycle, including ribulose-1,5-bisphosphate carboxylase/oxygenase (RUBISCO), decreased in powdery mildew-infected wheat leaves (Wright et al., 1995). RUBISCO activity was also decreased in *Arabidopsis thaliana* in response to *Albugo candida* (white blister rust)

(Chapter 20) (Tang et al., 1996), in sugar beet infected with beet curly top virus (Swiech et al., 2001), and *Parthenocissus quinquefolia* (liana) infected with *Xyllela fastidiosa* (McElrone and Forseth, 2004) indicating that different types of pathogens can have similar effects on their hosts. An interesting report to note is that carbon fixation associated with the dark phase of photosynthesis in uninfected bean leaves on plants infected with *Uromyces phaseoli* (bean rust) (Chapter 18) increased compared to bean leaves of uninfected controls (Murray and Walters, 1992). These results indicated that the pathogen can affect the photosynthetic activity in an area of the plant that is not infected. This can be a result of the plant trying to compensate for the loss of activity in infected leaves. In studies of the wheat cultivar Miriam, which is susceptible but tolerant to *Septoria tritici* blotch (STB), the rate of carbon fixation per unit chlorophyll and per green leaf area was higher than that observed in healthy plants. Some plants exhibit enhanced photosynthesis in remaining green tissue to compensate for loss of photosynthetic tissue due to the pathogen (Zuckerman et al., 1997).

The photosynthetic system of plants can be affected in many ways by pathogens. *Seiridium caridinale*–infected cypress contained less total chlorophyll and carotenoids, and reduced RUBSCO and nitrate activity, but more importantly the infection resulted in an inactivation of the donor side of Photosystem II (Muthuchelian et al., 2005). Examples such as this illustrate how the application of functional genomics techniques can permit more complete descriptions of disrupted photosynthetic systems. Microarrays and advanced protein analysis can be utilized to assess the effects of a pathogen on each of the enzymes in the photosynthetic process. Regardless of the details of disease process, the overall effect of plant pathogens that interfere with the photosynthetic potential is less carbohydrate production. Ultimately disease is manifested in reduced growth and yield. In a study on the effects of widespread viral infections of orchids, specifically *Oncidium*, healthy plants following virus elimination had a 17% increase in plant height, 65% increase in the inflorescence size, and 21% increase in photosynthetic capacity (Chia and He, 1999).

RESPIRATION

Respiration in plants involves oxidative processes in which complex molecules, such as carbohydrates produced during photosynthesis, are broken down into carbon dioxide, water, and energy. Energy is transferred and available to life processes when ADP is converted into ATP. The conversion reaction of ADP to ATP is referred to as **oxidative phosphorylation** and ATP serves as the energy component needed for almost all the operations in the plant cell. There are numerous metabolic pathways that serve to generate the substrates needed for oxidative phosphorylation and include the following: **glycolysis** (often referred to as Embden-Meyerhof pathway), tricarboxylic acid cycle (TCA), and oxidation of lipids. Another pathway that may be important, but that is not directly connected to oxidative phosphorylation, is the oxidative pentose phosphate pathway in which NADPH is generated along with pentose phosphates that can enter the glycolysis pathway. Respiration in plant pathogens (except viruses) has metabolic pathways similar to those of the plant host. Because of this it is difficult to differentiate between increases in respiration in plants in response to a plant pathogen or to determine if increases in respiration are due to the activity of the pathogen. Increases in respiration have generally been measured by determining increases in oxygen consumption. Because viral pathogens do not have respiratory pathways, they have been used to determine the effect of plant pathogens on respiration. All oxygen consumption in a viral infection can be directly correlated with plant host respiration.

Respiration usually increases in response to an invasion of a plant by a pathogen. Respiration increases more rapidly in plants exhibiting a resistance reaction due to the requirement for energy to rapidly produce and mobilize defense mechanisms. The rapid rise in respiration is followed by a decrease back to normal rates. In plants exhibiting a susceptible response to the pathogen, respiration levels increase more slowly, but continue to rise. Elevated respiration rates can be the result of increased activity of many different enzymes, faster breakdown of carbohydrates, including starch, and an uncoupling of oxidative phosphorylation. Uncoupling (ADP is not converted to ATP) results in increased concentration of ADP, which stimulates respiration. The activity of six enzymes associated with glycolysis and mitochondrial respiration in *Cucurbita pepo* infected with *Cucumber mosaic virus* (CMV) increased within lesions, whereas levels of RUBISCO involved in photosynthesis decreased only slightly (Técsi et al., 1996). Instead of using viruses as a nonrespiring pathogen, heat-killed bacteria have been employed as the elicitor of host responses. Oxygen uptake

in tobacco cells treated with dead bacteria increased within 4 min and lasted for 10 min; thereafter, respiratory rates returned to a steady state that was approximately twice the initial rate (Baker et al., 2000).

ABSORPTION AND TRANSLOCATION OF WATER AND NUTRIENTS

Many pathogens, including viruses, fungi, bacteria, and nematodes, disrupt the absorption and translocation of nutrients in plants. In some cases this may be on a small scale, such as within a single leaf, but in other cases the pathogen can affect the absorption and translocation of water and nutrients throughout the entire plant. This disruption can result from destruction of the root system, blockage of movement of water and nutrients in the xylem or phloem, or the redirection of host nutrients.

Some of the most obvious symptoms noted in response to plant pathogens that attack the plant roots are wilting, chlorosis, and general decline of the host plant. Root systems plants infected with *Phytophthora*, *Pythium*, or *Thielaviopsis* are often nonfunctional. Rotting and destruction of roots make it difficult, if not impossible, for the plant to absorb and transport water and nutrients normally obtained by the uninfected root system. Many nematodes (Chapter 8) cause disruption of absorption and translocation of water and nutrients in the host plant through the destruction of the small feeder roots. Ectoparasitic nematodes, such as ring nematodes (*Criconemoides* sp.) (Chapter 8) in turf grasses and peach trees, can lead to a significant reduction of feeder roots that can lead to chlorosis as well plant mineral deficiency symptoms.

Other plant pathogens are able to affect the absorption and translocation of water and nutrients within the plant because of their presence in the xylem and phloem tissues. *Ralstonia (Pseudomonas) solanacearum*, the causal agent of bacterial wilt in tobacco, tomato, eggplant, and potato, and Moko disease in bananas, invades the xylem through wounds. The bacteria spreads through the plant, rapidly multiplies within the xylem, and produce copious amounts of exopolysaccharides (EPSs) that essentially dam or block water flow. Loss of water translocation results in rapid wilting of young plants, wilting of a portion of more mature plants, chlorosis, and general decline. Bacterial streaming can be observed when the stem of a young infected tobacco or tomato plant exhibiting wilt symptoms due to *R. solanacearum* is cut and immediately placed in water. Fungal agents such as *Fusarium oxysporum* f. *specialis lycopersici* (wilt in tomato) and *Ophiostoma novo-ulmi* (Dutch elm disease) decrease water flow through the xylem. This decrease in water flow can be attributed to physical obstruction of vessels with hyphae, secretion of polysaccharides and pectolytic enzymes, production of gums/mucilages and the formation of **tyloses**

TABLE 32.2

Major Groups of Plant Growth Hormones and Their Associated Physiological Activities within the Plant

Hormone Class	Physiological Activity
Auxin (indole-3-acetic acid or IAA)	Promote cell elongation
	Induce high levels of ethylene formation resulting in growth inhibition
	Induce cambial cell division
	Initiate root formation
Gibberellins (GA: many forms)	Promote stem elongation by cell elongation
	Stimulate α-amylase production in seeds
	Stimulate flower production
	Retard leaf and fruit senescence
Cytokinins (Zeatin (Z), Z riboside (ZR), and ZP phosphate)	Promote cell division and differentiation
	Inhibit senescence of plant organs
	Induce stomatal opening
	Suppress auxin-induced apical dominance
Ethylene	Stimulate fruit ripening
	Stimulate leaf and flower abscission
Abscisic acid (ABA)	Induce maturation of seeds including development of desiccation tolerance
	Induces water stress associated stomatal closure

in the xylem by the host plant. The most obvious result in blockage of the vascular elements is the reduction in water movement, but it also diminishes the transport of essential minerals throughout the plant.

The presence of various plant pathogens can also result in the redirection of nutrients and resources within the plant. Pathogens divert resources such as carbohydrates and amino acids of the host for their own for growth and reproduction. Leaves of *A. thaliana* infected with *Albuga candida* showed increased levels of both soluble carbohydrates and starch when compared to uninfected leaves. Activities of both wall-bound and soluble invertases—enzymes involved in sucrose hydrolysis—were higher in infected leaves compared to those in control leaves. Additionally, an invertase isozyme was present in infected leaves and not found in healthy tissues (Tang et al., 1996). In studies of CMV infection in cotyledons of marrow plants, virus replication and synthesis of viral protein created a strong sink within lesions, resulting in increased photosynthesis and starch accumulation. There was more than twice as much starch hydrolase activity within lesions when compared to that in areas outside the lesions and to healthy leaves (Técsi et al., 1996).

Healthy melon plants have stachyose as a major component of sugars in phloem. Absolute levels of stachyose in phloem sap were similar in CMV-infected leaves, but sucrose levels dramatically increased, and the sucrose to stachyose ratio was 15- to 40-fold higher compared to that of healthy plants. Elevated sucrose concentrations did not appear to result from stachyose hydrolysis because there was no corresponding increase in galactose. Therefore, it is possible that CMV infection affected the movement of sucrose and was not due to metabolism (Shalitin and Wolf, 2000).

PLANT GROWTH HORMONES

Normal plant growth is controlled by various plant growth hormones and regulators, which have numerous effects and interactions within the plant, especially related to plant–plant pathogen interactions (Table 32.2). Alterations in the levels of these compounds can have significant effects on the plant's growth. Many different plant pathogens cause the host to change the hormone concentration in the plant or in some cases they actually produce these growth-controlling substances themselves. Symptoms observed in response to plant pathogens such as galls or tumor growth, excessive branching, leaf **abscission** or **epinasty**, abnormal leaf shape, and abnormal fruit ripening can be attributed to alterations in the levels of various hormones.

The pathogen *Gibberella fujikuroi*, causal agent of Bakanae or "foolish seedling" disease in rice, produces gibberellins in the host plant and causes elongation of the internodes. The disease results in spindly plants that often **lodge**. Prior to the identification of gibberellins as an actual plant growth hormone, a Japanese scientist identified gibberellins from culture filtrates of *G. fujikuroi*. This is an example of how the study of plant pathogens and their activities has been important to other areas of plant research. Conversely, Fusarium wilt of oil palm, which results in stunting, may be caused by an inhibition of gibberellin synthesis, because similar symptoms were observed following the application of an inhibitor of gibberellin synthesis. Application of gibberellin to infected palms resulted in a partial, but not complete, elimination of symptoms (Mepsted et al., 1995).

Some of the best and most complete evidence to demonstrate production of plant growth hormones by a plant pathogen is with *Agrobacterium tumefaciens*, the causal agent of crown gall. This bacterium has a self-replicating circular piece of DNA referred to as the Ti (tumor inducing) **plasmid**. The plasmid has the ability to transfer a section of its DNA, referred to as the T-DNA, into the host cell, where it becomes incorporated into a host cell chromosome within the nucleus. T-DNA contains genes for the production of opines, cytokinin, and auxin. Opines (amino acids) are an unusual nutrient source that cannot be utilized by the plant cell or by many other organisms except *Agrobacterium*, which has genes for opine metabolism. The production of cytokinin and auxin resulting from the T-DNA incorporated into the plant cell is not under the normal control mechanisms of the plant. The overproduction of these hormones results in gall development due to uncontrolled cell division (**hyperplasia**) and enlargement (**hypertrophy**). The ultimate result is the observed symptom, which gives the disease its common name, crown gall.

Pseudomonas savastanoi, the causal agent of olive knot disease, contains genes for auxin production on a plasmid as well as on its bacterial chromosome. In this case, the bacteria rather than the plant host produce the auxin that induces gall formation. In studies of *Erwinia herbicola* pv. *gypsophilae*, cytokinin and auxin genes are present in a pathogenicity-associated plasmid, which induces the production of a gall. The size of root galls observed in response to *Plasmodiophora brassicae*, the causal agent of clubroot of cabbage (Chapter 11), was correlated with the free indole acetic acid content in the clubs, indicating the role of auxins in symptom development (Grsic-Rausch et al., 2000). Increases in cytokinins and the expression of four or five putative cytokinin synthase genes from *Brassica rapa* were also correlated with development of clubroot (Ando et al., 2005)

Ethylene is a gaseous plant growth hormone. It has many different effects in the plant as shown in Table 32.2. Early in the 1900s it was noted that oranges shipped with bananas promoted banana ripening. Ethylene, which caused fruit ripening, was not produced by the oranges, but by the fungus *Penicillium digitatum*, the cause of green mold on oranges. A nonethylene-producing mutant showed that ethylene production has no significant role in the pathogenicity of *P. digitatum* even though low levels of ethylene predispose fruit to postharvest diseases. Many plant pathogens are known to induce ethylene production in the host as a response to infection. Ethylene then induces many other responses within the plant such as the production of various pathogenesis-related (PR) proteins (Chapters 28 and 31) and defense-related compounds. Ethylene production by the host or pathogen can result in premature ripening of fruit, leaf epinasty (downward bending), leaf abscission, and chlorosis.

Abscisic acid (ABA) in general is viewed as a growth inhibitor, and increases in its production have been noted in some plant diseases. ABA production associated with these plant–pathogen interactions resulted from a plant host response to the plant pathogen rather than being produced by the plant pathogen. ABA-producing strains of plant pathogenic fungi have been identified, but the involvement of ABA in pathogenesis has not been thoroughly investigated.

Many times plant pathogens can cause multiple changes in plant growth regulators. In a study of premature fruit drop in citrus caused by *Colletotrichum acutatum* ethylene evolution increased threefold, IAA increased 140-fold, ABA showed no change, both *trans*- and *cis*-12-oxo-phytodienoic acid (precursors of jasmonic acid (JA)) increased 8- to 10-fold, *trans*-JA was unchanged, *cis*-JA increased fivefold, and salicylic acid (a signaling compound) (Chapters 28 and 31) increased twofold. Many of the enzymes associated with the production of ethylene, IAA, and JA were also shown to be upregulated in infected plants (Lahey et al., 2004).

ALTERATION OF CELL PERMEABILITY

The plant cell membrane consists of a bilayer of phospholipids with embedded proteins. Membranes exist within the cell wall and surrounding plant organelles such as the chloroplasts and mitochondria. Proteins within the cell membrane are important in the regulation and movement of materials across the membrane and are extremely important to the integrity of the cell. When this integrity is altered, the membrane is unable to control the movement of materials across the membrane and substances may be lost from the cell. When pathogens attack plants, one of the first responses that can generally be detected is the alteration of membrane **permeability**. Membrane integrity is measured by determining electrolyte (charged ions, such as K^+ or Cl^-) leakage from cells. Several toxins associated with plant pathogens have been shown to affect cell membrane permeability or integrity. Some examples of these toxins include T-toxin (*Cochliobolus heterostrophus*: Southern corn leaf blight), HC-toxin (*Cochliobolus carbonum*: leaf spot disease in maize), AM-toxin (*Alternaria alternata*: alternaria leaf blotch of apple), and syringomycein E, syringotoxin and syringopeptin (*Pseudomonas syringae* pv. *syringae*: necrotic lesions in a broad range of monocot and dicot species, resulting in tip dieback, bud and flower blast, spots and blisters on fruit, stem canker, and leaf blight). Many of these toxins are cytotoxic in plant cells at nanomole (10^{-9} M) concentrations and cause necrosis by forming ion channels, which freely allow the passage of divalent cations (e.g., calcium and magnesium). In a study of the cultivar-specific necrosis toxin produced by *Pyrenophora tritici-repentis* (tan spot, a widespread foliar disease of wheat), electrolyte leakage was observed

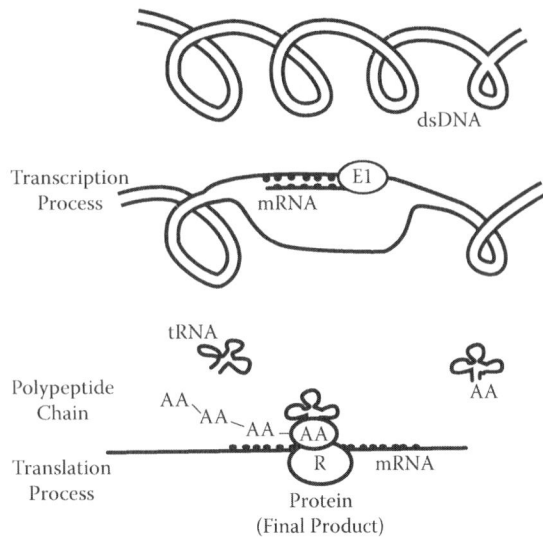

FIGURE 32.1 Transcription and translation process for the formation of proteins based on information located on dsDNA. dsDNA = double stranded DNA, mRNA = messenger RNA, tRNA = transfer RNA, AA = amino acid, E1 = RNA polymerase, R = ribosome.

in a toxin-sensitive cultivar after 4 h, but was not observed in a toxin-insensitive cultivar (Kwon et al., 1996).

TRANSCRIPTION AND TRANSLATION

The processes of **transcription** and **translation** are often affected by plant pathogens, but are most clearly observed with plant viruses. Transcription is the process whereby information associated with a DNA sequence or gene is copied into a complementary piece of RNA referred to as messenger RNA (mRNA) (Figure 32.1). RNA polymerase recognizes the start of a gene on the DNA, combines with the DNA separating the two nucleotide strands, and then progresses down the DNA section making the complementary strand of RNA until a stop code is encountered. Translation is the process by which this piece of mRNA is copied into a protein whose amino acid sequence is determined by the sequence of nucleotides in the mRNA. This process involves a ribosome binding to the mRNA and essentially reading three nucleotides at a time that correspond to specific amino acids, which are brought to the ribosome active site by a transfer RNA (tRNA). The amino acids are connected by peptide bonds forming a protein. Some processing of the protein may be required after the amino acid sequence is complete. Included in the proteins produced by this process are the enzymes that are responsible for the production of other materials within the plant cell, such as the polysaccharides and lipids.

Virus particles that infect plants consist of genetic information in the form of ssDNA, dsRNA, or ssRNA surrounded by a protein coat (Chapter 4). Viruses do not have the ability to self-replicate, but instead utilize the cellular machinery of the host. The DNA or RNA from the virus moves into the host cell, where it undergoes transcription and translation, allowing the genetic material to be produced along with the protein coat. The viral genome serves as the template for transcription for the production of the nucleic acid component of the virus and translation serves for the production of the components of the viral protein coat. The total protein content of 20 cultivars of alfalfa infected with *Phoma medicaginis* was determined following infection. Infected leaves had lower protein content in 19 of 20 cultivars with an average reduction of 22%. Five cultivars had protein levels reduced from 35 to 65% indicating a significant variability associated with different cultivars (Hwang et al., 2006).

Much of the previous work concerning the effects of plant pathogens on the physiology of its host has involved the detection of changes in the production of specific enzymes, such as the enzymes involved in glycolysis or chlorophyll biosynthesis and degradation, the measurement of chlorophyll levels, hormone concentrations, and membrane permeability. With the advancement of techniques to measure changes in gene expression (Golem and Culver, 2003; Li et al., 2006; Whitman et al., 2006) and the resulting protein production as well as protein identification (Devos et al., 2006), a better and more complete understanding of the effects of a pathogen on the host physiology is becoming possible and is advancing rapidly.

LABORATORY EXERCISES

"Seeing is believing" and "Prove it to me" are common phases that are used in all kinds of situations. These ideas are particularly important in research—we always want to know what the effect of the plant pathogen is on the plant. The overall objective of the following exercises is to illustrate some of the effects that plant pathogens have on the physiological processes of the plant. We see the results when we observe symptoms, but in these exercises we will be looking at the effects of plant pathogens on specific physiological activities of the plant. The first exercise can be conducted with many different plants. The only requirement is that the plant pathogen causes symptoms resulting in a chlorotic or mosaic appearance on leaves. Healthy plants with uninfected leaves are used for comparison to the concentration of chlorophyll present in the infected leaf. The second exercise was developed to be used with the bacterial pathogen *Ralstonia solanacearum* and tomato plants. This exercise illustrates the blockage of the xylem system by bacteria and how these bacteria can also be isolated from the xylem stream on nutrient agar plates. The final exercise was developed to illustrate how a plant pathogen, *Agrobacterium tumefaciens*, can alter the normal growth patterns of plant tissue.

Procedure 32.1

Determination of Total Chlorophyll Present in Leaf Area for Comparison of Healthy and Infected Leaf Areas and to Compare Effects of Different Pathogens

Step	Instruction and Comments
1	Chose leaves from a plant that is exhibiting symptoms such as chlorosis or mosaic patterns. Nonsymptomatic leaves must also be obtained from the same plant or from the same general area of a healthy plant. Leaves associated with different diseased plants can be compared in order to compare the effects of different plant pathogens.
2	Cut 10 discs from each type of leaf area (normal and symptomatic) with a # 5 cork borer (1 cm diam.). Calculate the area of the leaf disks by the following formula: $$\text{Area} = \pi r^2 \times 10$$ where $\pi = 3.14$ and $r = 0.5 \times$ diameter, in this case 0.5 cm. Obtain weight of leaf discs prior to proceeding to the next step.
3	Weigh the leaf disks and place disks in 5 mL cold dimethylformamide (DMF). Incubate in the dark at 10°C for a minimum of 48 h.
4	Determine absorbance of the DMF at 664 nm and 647 nm and calculate the total chlorophyll using the following formula: $$\text{Total chlorophyll (µg chlorophyll/mL)} = 7.04\ \text{Abs}_{664} + 20.27\ \text{Abs}_{647}$$
5	Convert the µg/mL to µg/cm^2 by multiplying the total chlorophyll by 5 mL (DMF solution volume) and dividing by the total leaf area associated with the disks as calculated in Step 2. Convert the µg/mL to µg/gm fresh weight by multiplying the chlorophyll total by 5 mL and dividing by the weight in grams obtained in Step 3.

Experiment 1. Effects of a Plant Pathogen on Chlorophyll

Chlorophyll is probably one of the most important molecules on earth. In combination with other compounds and enzymes in the plant, chlorophyll is able to harvest light energy and transfer it into a usable form of chemical energy that the plant utilizes for growth and reproduction. This stored chemical energy also supplies energy for many other forms of life on earth, including humans and plant pathogens. Any time we eat food, we are utilizing this energy either directly or indirectly. The amount of chlorophyll associated with cells has a direct effect on the amount of the stored energy that a plant ultimately produces. A reduction in the chlorophyll can lead to effects such as stunting due to the loss of the energy conversion potential. Chlorophyll extractions such as the one outlined here (Moran, 1982) can be easily done and provide a quick measurement of the effects of plant pathogens on chlorophyll content in plants.

Materials

Each student or team of students will require the following items:

- Leaves from a plant exhibiting symptoms chlorotic or mosaic symptoms (The specific plant is not that important, but it is vital to have a healthy plant that can be used for comparison. Examples of plants that can be used include tobacco plants infected with *Tobacco mosaic virus* or cucumber plants infected with *Cucumber mosaic virus*.)
- Balance to determine the weight of individual leaves
- #5 cork borer
- Ruler
- Test tubes
- Dimethylformamide (DMF) (*Caution*: Use in a chemical hood only and wear rubber gloves to prevent absorption through skin. Obtain information on material safety data sheet prior to use.)
- Disposable rubber gloves
- Spectrophotometer to measure absorbance at 664 and 647 nm
- Disposable or quartz cuvettes for spectrophotometer
- Pipettes for transferring DMF
- Incubator set at 10°C without lighting
- Calculator

Follow the instructions in Procedure 32.1 to complete the experiment.

Anticipated Results

The results of this experiment provide a basis for suggesting that one pathogen or strain of the pathogen causes more severe reactions in the plant than another. Reductions in chlorophyll have a direct effect on plant growth. If the chlorophyll is reduced by 10% by one pathogen and 40% by a second pathogen, it is clear that this second pathogen is more severely affecting the plant, based on the effect on chlorophyll content. In many cases, however, reductions in chlorophyll are not the only effect a plant pathogen has, so other tests may be needed to quantify the total effect that the plant pathogen has on the plant. This experiment should have definite results. An additional component to the experiment may be to determine the variability of results for a specific group illustrating the purpose of having replicates in experiments.

Questions

- What is the percent reduction in the levels of chlorophyll associated with the infected leaves compared to the healthy leaves?
- What was the variation in chlorophyll values within the laboratory for the infected leaves and healthy leaves? Minimum value? Maximum value? Were there similar levels of variation for the healthy and infected groups or was there more variation associated with the infected leaves? How can you explain the variation within these groups?
- Why would it be important to know the variation in chlorophyll levels of plants prior to setting up a research project investigating the effects of a plant pathogen on chlorophyll levels of a plant?
- How can plant pathogens affect the total level of chlorophyll in a plant?

EXPERIMENT 2. DISRUPTION OF WATER TRANSLOCATION

Disruption of the absorption and translocation of water within the plant can occur when plant pathogens cause a destruction of the root system, block the movement of water and nutrients in the xylem and phloem, or redirect the movement of host nutrients. The blockage of the water flow within plants is a common mechanism for plant pathogens that results in wilting of the plant and ultimately may kill the plant. Many of the pathogens that cause these symptoms are very good saprophytes in the soil and survive for many years even in the absence of a host. *Ralstonia solanacearum* utilized in this exercise is common in tropical and subtropical areas and causes severe diseases in tobacco, tomato, potato, and eggplant in warm areas outside of the tropics. Several races of

this organism have various host ranges, and a race which attacks tomato plants is utilized in this exercise.

Materials

Each student or team of students will require the following items:

- 2- to 4-week-old tomato plants ("Marion" or "Rutgers" varieties that are susceptible and various other varieties may be included to compare susceptibility of different varieties.)
- 24- to 48-h-old culture of *R. solanacearum* (race pathogenic to tomato plants) grown in nutrient broth or trypicase soy broth (Difco, Lansing, MI)
- 1-mL sterile syringes with ½ in. needles
- Trypicase soy broth or nutrient broth
- 30°C growth chamber for tomato plants
- Glass microscope slides and cover slips
- Sterile distilled water in dropper bottle
- Nutrient agar or trypticase soy agar in 10 cm diameter Petri dishes
- Shaker
- 28°C to 30°C bacterial culture incubator
- Bacterial transfer loop that can be sterilized by flaming
- Alcohol burner
- Safranin O (0.25% solution in 10% ethanol, 90% water; should be made up initially in ethanol)
- Disposable nitrile gloves
- Paper towels
- Sink with water
- Compound microscope with oil immersion lens

Follow the instructions in Procedure 32.2 to complete the experiment.

Anticipated Results

Tomato plants infected with *R. solanacearum* will generally wilt within 2 to 4 d, depending on the strain and the age of the tomato plant. Older plants will take longer to wilt. The temperature of the growth chamber is vital. If the temperature is less than 30°C the wilting of the plant may be delayed or may not occur. Bacterial colonies should appear within 48 h on the nutrient agar plates on which the bacteria were transferred from the slide. Bacteria should be obvious on the stained slide made from the xylem exudates–water mix.

Questions

- Is the wilting associated with *R. solanacearum* infection of tomato strictly due to the presence

---

Procedure 32.2

Disruption of Vascular Flow in Tomato Infected with *Ralstonia solanacearum*

Step	Instruction and Comments
1	Obtain tomato plants approximately 2 to 4 weeks old. ("Marion" and "Rutgers" varieties are extremely susceptible, but other varieties can be included for comparison of resistance to *Ralstonia solanacearum*.)
2	Grow a virulent strain of *R. solanacearum* in nutrient broth or trypicase soy broth on the shaker for 24 h or until medium is turbid.
3	Using a 1 mL (cc) sterile syringe with ½-in. needle (smallest diameter possible), draw up 0.2 mL of bacterial suspension.
4	Inject the bacterial suspension just below the surface and into the lower stem of the tomato. Incubate the inoculated plants at approximately 30°C until wilt symptoms develop. This usually occurs within 2 to 7 d. Generally, the older the tomato plant, the longer it will take for the plant to wilt.
5	After the plants wilt, place a couple of drops of sterile distilled water on a glass slide. Cut the tomato stem across the area above the site of inoculation and immediately place the cut stem in the water on the slide to observe bacteria streaming from the cut surface. (Water on slide should become turbid due to the presence of bacteria).
6	Take a sterile loop and streak the water-bacteria suspension from the slide onto either nutrient agar or trypticase soy agar in petri dishes and seal with parafilm. After incubating the sealed dishes at 28°C to 30°C for 24 h observe bacterial growth.
7	Touch loop to the bacteria—water mix on the glass slide prepared in Step 5 and transfer a drop to a clean slide. Spread the suspension and air day.
8	Heat-fix the bacteria to the slide by passing it quickly through the flame of an alcohol burner. Be careful not to get the slide too hot.
9	Add a couple of drops of safranin stain to the slide for 30 to 60 s; then rinse with water (wear gloves).
10	Blot, do not rub, the slide dry with a paper towel. Rubbing will remove the bacteria.
11	Observe the stained bacteria using an oil immersion lens on a compound microscope.

of bacteria in the xylem stream blocking the passage of water?

- What compounds do some bacteria produce that may also be important in the blockage of water flow in a plant?
- How are bacteria such as *R. solanacearum* disseminated in nature? What can be done to manage diseases such as *R. solanacearum*? What are the limitations?
- Were there any observable differences between varieties of tomato used in the experiment?
- How many different types of bacterial colonies were isolated from the xylem exudate? Are these other organisms involved in the wilting observed? What might you need to do to determine their involvement?

EXPERIMENT 3. EFFECT OF *AGROBACTERIUM* SPECIES ON NORMAL PLANT GROWTH

Plant pathogens are known to either produce normal plant growth hormones or have the ability to increase or alter the production of plant growth hormones by the plant. Virulent stains of *Agrobacterium* contain plasmids that contain genes for the production of various plant growth hormones. The genes associated with the plant growth hormones are transferred to the plant during the infection process and as a result more plant hormones are produced that interferes with normal plant function and development. The presence of these extra levels of hormones causes various abnormalities in the normal growth of the plant. Tomato plants serve as experimental subjects and can rapidly produce callus or root growth following inoculation with *Agrobacterium*.

Materials

Each student or team of students requires the following items:

- Tomato plants 4 to 6 weeks old (Various varieties can be used for comparison to "Marion" and "Rutgers" varieties, which are susceptible.)
- 24- to 48-h-old culture of virulent strain of *A. tumefaciens* (induces gall formation) and *A. rhizogenes* (induces root formation) growing on nutrient agar or trypticase soy agar

Procedure 32.3
Induction of Gall or Root Formation by Inoculation of Tomato Plants with *Agrobacterium tumefaciens* or *A. rhizogenes*

Step	Instruction and Comments
1	Obtain tomato plants approximately 4 to 6 weeks old. ("Marion" and "Rutgers" varieties are susceptible, but other varieties can be included for comparison of gall or root formation following inoculation.)
2	Grow a virulent strain of *Agrobacterium tumefaciens* (for induction of gall formation) and *A. rhizogenes* (for induction of root formation) on nutrient agar or trypticase soy agar for 24 h or until growth is observable on the dish.
3	Take a sterile scalpel and touch it to the bacterial growth in the petri dish.
4	Cut the surface of the tomato stem with the scalpel coated with the bacteria, and then wrap a piece of parafilm around the cut. Be careful not to cut completely through the tomato stem.
5	For the control treatment take a sterile scalpel and touch it to the surface of a sterile agar dish; then cut the surface of a tomato stem with the scalpel followed by wrapping with parafilm.
6	Remove the parafilm after one day. Observe the tomato plants weekly for the production of callus tissue (galls) or roots. Abnormal growth should be observed after approximately 2 to 3 weeks. Plants can be incubated in a window, greenhouse, or growth chamber.
7	Compare callus/root growth in different tomato varieties by measuring the size/number of abnormal growths. Cut the abnormal growth associated with the plant and weigh the tumor/root growth.

- Sterile scalpel with #10 blade
- Shaker
- Parafilm
- Balance
- Dissecting microscope
- Growth chamber, greenhouse, or window where tomato plants can be maintained for several weeks

Follow the instructions in Procedure 32.3 to complete the experiment.

Anticipated Results

The tomato plants should be observed on a weekly basis. The plants should begin to show symptoms of gall or root formation within approximately 2 to 3 weeks. The first indication of abnormal growth will be a roughness on the surface of the stem where the plant was inoculated. Plant height should be measured for all plants prior to inoculation and at the end of the experiment. Control plants that are cut with the scalpel but are not inoculated with *Agrobacterium* should be included as controls. The total weight of abnormal callus or root growth at the end of the experiment (after approximately 6 weeks) should be determined and should be related to the resistance of the various tomato varieties to *Agrobacterium*.

Questions

- What was the average callus or abnormal root growth for each of the varieties of tomatoes used in the experiment? Does the average weight of callus or root growth correspond to the resistance to *Agrobacterium*? Why or why not?
- Was there any response on the control plants? If so, what was that response? Why did it occur?
- Did the infection with *Agrobacterium* have any effect on the overall height increase of the infected plant compared to that of the control? Would you expect it to have an effect? Why or why not?
- Why is parafilm placed around the plant after it is inoculated?
- What are the methods used for the management of *Agrobacterium* in nature?

LITERATURE CITED

Almási, A., D. Apatini, K. Bóka, B. Böddi, and R. Gáborjányi. 2000. BSMV infection inhibits chlorophyll biosynthesis in barley plants. *Physiol. Mol. Plant Pathol.* 56: 227–233.

Ando, S., T. Asano, S. Tsushima, S. Kamachi, T. Hagio, and Y. Tabei. 2005. Changes in gene expression of putative isopentenyltransferase during clubroot development in Chinese cabbage (*Brassica rapa* L.). *Physiol. Mol. Plant Pathol.* 67: 59–67.

Baker, C.J., E.W. Orlandi, and K.L. Deahl. 2000. Oxygen metabolism in plant/bacteria interactions: characterization of the oxygen uptake response of plant suspension cells. *Physiol. Mol. Plant Pathol.* 57: 159–167.

Chia, T.-F. and J. He. 1999. Photosynthetic capacity in *Oncidium* (Orchidaceae) plants after virus eradication. *Environ. Exp. Bot.* 42: 11–16.

Devos, S., K. Laukens, P. Deckers, D. Van Der Straeten, T. Beeckman, D. Inzé, H. Van Onckelen, E. Witters, and E. Prinsen. 2006. A hormone and proteome approach to picturing the initial metabolic events during *Plasmodiophora brassicae* infection on *Arabidopsis*. *MPMI* 19: 1431–1443.

Golem, S. and J.N. Culver. 2003. *Tobacco mosaic virus* induced alteration in the gene expression profile of *Arabidopsis thaliana*. *MPMI* 16: 681–688.

Grsic-Rausch, S., P. Kobelt, J.M. Siemens, M. Bischoff, and J. Ludwig-Müller. 2000. Expression and localization of nitrilase during symptom development of the clubroot disease in *Arabidopsis*. *Plant Physiol.* 122: 369–378.

Hwang, S.-F., H. Wang, B.D. Gossen, K.-F. Chang, G.D. Thurbill, and R.J. Howard. 2006. Impact of foliar diseases on photosynthesis, protein content and seed yield of alfalfa and efficacy of fungicide application. *Eur. J. Pl. Path.* 115: 389–399.

Kema, G.H. J., D. Yu, F.H.J. Rijkenberg, M.W. Shaw, and R.P. Baayen. 1996. Histology of the pathogenesis of *Mycosphaerella graminicola* in wheat. *Phytopathology* 86: 777–786.

Kwon, C.Y., J.B. Rasmussen, L.J. Fancl, and S.W. Meinhardt. 1996. A quantitative bioassay for necrosis toxin from *Pyrenophora tritici-repentis* based on electrolyte leakage. *Phytopathology* 86: 1360–1363.

Lahey, K.A., R. Yuan, J.K. Burns, P.P. Ueng, L.W. Timmer, and K.-R. Chung. 2004. Induction of phytohormones and differential gene expression in citrus flowers infected by the fungus *Colletotrichum acutatum*. *MPMI* 17: 1394–1401.

Li, C.Y. Bai, E. Jacobsen, R. Visser, P. Lindhout, and G. Bonnema. 2006. Tomato defense to the powdery mildew fungus: differences in expression of genes in susceptible, monogenic- and polygenic resistance responses are mainly in timing. *Plant Mol. Biol.* 62: 127–140.

Lopes, D.B. and R.D. Berger. 2001. The effects of rust and anthracnose on the phytosynthetic competence of diseased bean leaves. *Phytopathology* 91: 212–220.

McElrone, A.J. and I.N. Forseth. 2004. Photosynthetic responses of a temperate liana to *Xylella fastidiosa* infection and water stress. *J. Phytopath.* 152: 9–20.

Mepsted, R., J. Flood, and R. M. Cooper. 1995. Fusarium wilt of oil palm I. possible causes of stunting. *Physiol. Mol. Plant Pathol.* 46: 361–372.

Moran, R. 1982. Formulae for determination of chlophyllous pigments extracted with N,N-dimethylformamide. *Plant Physiol.* 69: 1376–1381.

Murray, D.C. and D.R. Walters. 1992. Increased photosynthesis and resistance to rust infection in upper, uninfected leaves of rusted broad bean (*Vicia faba* L.). *New Phytol.* 120: 235–242.

Muthuchelian, K., N. La Porta, M. Bertamini, and N. Nedunchezhian. 2005. Cypress canker induced inhibition of photosynthesis in field grown cypress (*Cupressus sempervirens* L.) needles. *Physiol. Mol. Plant Pathol.* 67: 33–39.

Niederleitner, S. and D. Knoppik. 1997. Effects of the cherry leaf spot pathogen *Blumeriella jaapii* on gas exchange before and after expression of symptoms on cherry leaves. *Physiol. Mol. Plant Pathol.* 51: 145–153.

Shalitin, D. and S. Wolf. 2000. Cucumber mosaic virus infection affects sugar transport in melon plants. *Plant Physiol.* 123: 597–604.

Strelkov, S.E., L. Lamari, and G.M. Ballance. 1998. Induced chlorophyll degradation by a chlorosis toxin from *Pyrenophora tritici-repentis*. *Can. J. Plant Pathol.* 20: 428–435.

Swiech, R., S. Browning, D. Molsen, D.C. Stenger, and G.P. Holbrook. 2001. Photosynthetic responses of sugar beet and *Nicotiana benthamiana* Domin, infected with beet curly top virus. *Physiol. Mol. Plant Pathol.* 58: 43–52.

Tang, X., S.A. Rolfe, and J.D. Scholes. 1996. The effect of *Albugo candida* (white blister rust) on the photosynthetic and carbohydrate metabolism of leaves of *Arabidopsis thaliana*. *Plant Cell Environ.* 19: 967–975.

Técsi, L.I., A.M. Smith, A.J. Maule, and R.C. Leegood. 1996. A spacial analysis of physiological changes associated with infection of cotyledons of marrow plants with cucumber mosaic virus. *Plant Physiol.* 111: 975–985.

Whitman, S.A., C. Yang, and M.M. Goodin. 2006. Global impact: elucidating plant responses to viral infection. *MPMI* 19: 1207–1215.

Wright, D.P., B.C. Baldwin, M.C. Shephard, and J.D. Scholes. 1995. Source-sink relationships in wheat leaves infected with powdery mildew. II. Changes in the regulation of the Calvin cycle. *Physiol. Mol. Plant Pathol.* 47: 255–267.

Zuckerman, E., A.E. Eshel and Z. Eyal. 1997. Physiological aspects related to tolerance of spring wheat cultivars to *Septoria tritici* blotch. *Phytopathology* 87: 60–65.

Part 5

Epidemiology and Disease Control

33 Plant Disease Epidemiology

Kira L. Bowen

CHAPTER 33 CONCEPTS

- Epidemiology is the study of properties of pathogens, hosts and the environment that lead to an increase in disease in a population.

- Polycylic pathogens that disperse readily usually cause the most damaging epidemics.

- Temperature, moisture, wind, soil properties, radiation, and other components can comprise a conducive environment.

- Host developmental stage and population uniformity can affect epidemic development.

- Disease spreads in space as it increases in incidence (numbers of infected plants).

- Mathematical models that allow disease predictions also contribute to disease management.

Epidemiology is the study of factors that lead to an increase in disease in a population. Understanding the reasons why diseases increase in populations of plants and what can influence those increases can contribute to decisions concerning plant disease management. Man's ability to produce food and fiber is, in part, limited by his ability to manage plant diseases. Thus, plant disease epidemiology has contributed to the highly technological culture in which we currently live. Concepts related to plant disease epidemiology will be discussed in this chapter.

COMPONENTS OF AN EPIDEMIC

Disease occurs on a plant when a virulent pathogen and a susceptible host interact in a conducive environment. In order for disease to increase, these three components—pathogen, host, and environment—must continue to interact over time. The interaction of these components might be thought of as a three-sided pyramid, with each side representing each component of the epidemic, and the height of the pyramid representing the disease level. Thus, as the susceptibility of the host population increases or with greater conduciveness of the environment, the greater the resultant disease level and the higher up the pyramid. In a population of plants, disease becomes important when the damage caused by disease increases to the extent that there are social and economic impacts. There are numerous examples of plant disease epidemics that have had profound impacts on human history and these illustrate the interaction needed over time between the pathogen, host, and environment. Well-known examples include the

potato late blight epidemic of 1845, chestnut blight of the early 20th century, and the Southern corn leaf blight epidemic of 1970. Some more recent examples of epidemics affecting U.S. citizens include sudden oak death, citrus canker, and head blight of wheat and barley.

Potato Late Blight. The potato, *Solanum tuberosum* L., native to the American continent, did well as a crop in the cool damp climate of Ireland. Prior to the 1840s, the farmers of Ireland had become dependent upon the potato for food, and the crop was grown throughout that island. Therefore, two of the three necessary components for a disease epidemic were present—the susceptible crop and an environment that was conducive to disease. *Phytophthora infestans* (Mont.) De Bary, the fungal-like oomycete (Chapter 20) that causes potato late blight, was introduced into Ireland by early 1844, providing the third element for an epidemic. It is probable that potato late blight had become widely established by the end of the 1844 growing season (Andrivon, 1996), but had remained at low levels. In 1845, excessively cool and wet weather prevailed throughout Europe. This weather, which was highly conducive to the development of *P. infestans*, allowed rapid increase in the severity of late blight and resulted in the loss of the entire potato crop by the end of the 1845 growing season. Because farmers were almost entirely reliant on the potato for food, this loss led to the Irish potato famine, with many of the Irish emigrating to other countries.

Chestnut Blight. The American chestnut, *Castanea dentata* (Marsh.) Borkh., was an important and predominant tree species in native forests of the eastern United States through the 18th century. The rot-resistant wood was important in the timber industry, and the bark was a source of tannins for the leather industry. American chestnut also provided food in the form of nuts, for wildlife as well as humans. In 1904, American chestnut trees in the New York Zoological Park began wilting and suddenly dying, apparently from a canker disease. During the next year, the disease, called chestnut blight, was noticed on American chestnuts in other areas of New York. By 1911, chestnut blight had spread through New Jersey, south to Virginia, and north into Massachusetts. Communities in the Appalachians depended on the American chestnut tree for their livelihood, so the spread of this disease caused much concern. Attempts to stop the spread of chestnut blight were unsuccessful, and mature chestnuts trees died throughout the eastern United States, decimating Appalachian communities.

Chestnut blight is caused by the pathogen, *Cryphonectria parasitica* (Murrill) Barr (previously *Endothia parasitica*), and is spread when sticky conidia exuding from cankers adhere to birds and insects. This pathogen was brought to the United States with botanical specimens from China. In China, *C. parasitica* is **endemic** where it affects native Chinese chestnuts causing low levels of disease. American chestnuts had no resistance to *C. parasitica* and there were no elements of the environment in the United States that effectively limited its spread. *C. parasitica* continued to spread and kill mature American chestnuts throughout the eastern United States. The chestnut blight epidemic changed the make-up of American forests in less than 50 years, but only after introduction of the virulent pathogen into the United States.

Southern Corn Leaf Blight. Southern corn leaf blight, caused by *Cochliobolus heterostrophus* (Drechs.) Drechs. (formerly *Bipolaris maydis*), is endemic to the United States. Through the 1960s, hybrid field corn cultivars (*Zea mays* L.) were generally resistant to this disease. However, this reaction changed due to a change in the way hybrid seed was produced. The change in seed production involved the use of a type of cytoplasmic male sterility, *Tcms*, which allowed the production of hybrid seed without the labor-intensive removal of pollen-producing tassels. Hybrid seed is produced by planting several rows of a desired "female parent" corn inbred (A) between one or two rows of the "male parent" corn inbred (B). Pollen-producing tassels are removed from the female parents in these hybrid seed production fields. Because all pollen in the field is from the male parent, all seed produced on the female parent is a result of the cross (A x B) and is hybrid seed. Sterile pollen is produced by corn plants with cytoplasmic male sterility, and this sterility was incorporated, through selection, into inbred lines used as female parents. *Tcms* was incorporated into numerous corn breeding lines used as female parents in hybrid seed production and the hybrid plants produced from these crosses still contained traces of *Tcms* cytoplasm.

By 1970, 80% of hybrid corn in the United States was produced using *Tcms*, creating a genetic uniformity in corn throughout corn-producing regions. About that time, corn cultivars that had previously been resistant to Southern corn leaf blight were becoming diseased. As *Tcms* had been increasingly incorporated into corn germplasm, a new race of *C. heterostrophus* had developed. This race, dubbed Race T, was more aggressive than the former dominant Race O of the pathogen and was highly virulent to *Tcms* corn. *Cochliobolus heterostrophus* Race T developed quickly on the susceptible host populations of *Tcms* plants in the hot moist conditions prevailing in the southern corn-producing regions of the United States (Figure 33.1). Losses due to this disease were 100% in some fields in the South, where the epidemic apparently initiated. While the moist weather of the south contributed to disease development, the Southern corn leaf blight epidemic of 1970 was a result of man's creation of the susceptible host population.

These examples of historically important plant disease epidemics illustrate the interaction of each of the components of the disease triangle over time. American chestnut trees were an important part of eastern U.S. forests until introduction of the virulent pathogen *C. parasitica* to the United States where the environment was suitable for disease development. The environment, host and pathogen interaction allowing Southern corn leaf blight, did exist before the 1970s but had been adequately managed with plant resistance. This disease only became a problem when the host population was changed to a susceptible population. If the weather had not been highly conducive to the rapid development of *P. infestans* in Ireland in 1845, the Irish potato famine may never have happened. These plant disease epidemics also illustrate additional concepts relating to plant disease development and epidemiology.

PATHOGEN

Chestnut blight arose in the United States after the importation of infected botanical specimens for a collection prior to 1907. Similarly, potato late blight in Ireland probably traces back to a shipment of tubers, with at least one infected tuber, into Belgium from the Americas. These infected specimens and tubers provided the **initial inoculum** of virulent pathogens for these plant disease epidemics. Given that the favorable environment and susceptible hosts were present upon introduction of the initial inoculum, the pathogen increased, causing more disease.

In recent years, several pathogens have been newly found in the United States, including *Phytophthora ramorum,* the cause of sudden oak death (Chapter 20) (Davidson et al., 2003), *Phakopsora pachyrhizi* Syd., causing Asian soybean rust (Chapter 18) (Schneider et al., 2005), and *Candidatus* L. americanus, the cause of Huanglongbing disease, otherwise known as citrus greening (USDA, 2005). Initial inoculum of

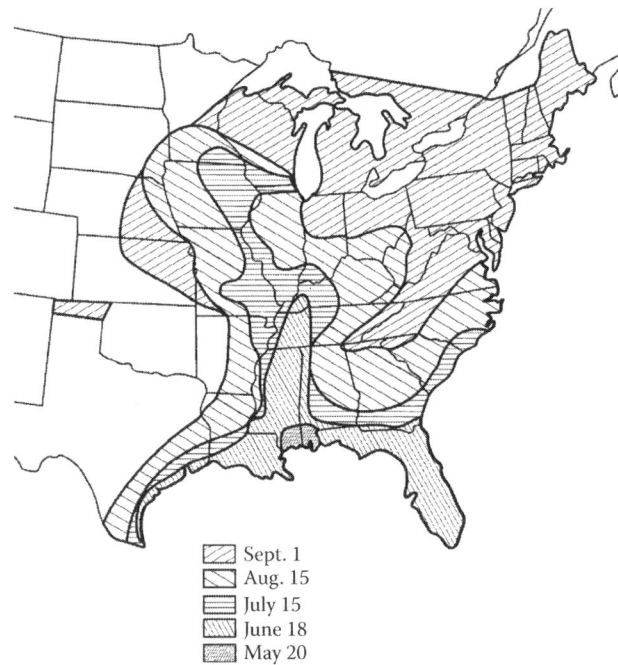

FIGURE 33.1 Disease progress of southern corn leaf blight during the 1970 growing season (adapted from Moore, 1970).

these newly introduced pathogens came to the United States in varying ways. It is possible that *P. ramorum* arose spontaneously from other *Phytophthora* species, which are genetically flexible organisms. Hurricane winds may have carried spores of *P. pachyrhizi* to U.S. Gulf Coast states (Pan et al., 2006) from South America. The pathogen responsible for Huanglongbing disease is vectored by specific species of psyllid insects that have been found in the Florida since 1998.

While plant disease epidemics can be initiated by the introduction of an exotic pathogen to a new location, more commonly, initial inoculum for an epidemic is the quantity of the pathogen that survives a period without a host, as through a winter (also called primary inoculum), or the quantity that arrives at a location after a dispersal event. The initial inoculum for a plant disease epidemic may come from infected seed, or may be the viable propagules of a soilborne pathogen (Chapter 22) at planting time. Initial inoculum can be from a single infected plant brought into a greenhouse, or even from a vector for many viral and bacterial diseases of plants. Of course, the greater the quantity of initial inoculum, the greater the disease level at the start of an epidemic.

Potato late blight and Southern corn leaf blight disease levels were observed to increase in a matter of days or weeks, whereas the increase in chestnut blight was observed over years. Chestnut blight increased over years, not because the tree has a long life span, but because of inherent characteristics of the pathogen, *C. parasitica*. *Cryphonectria parasitica* is considered **monocyclic**, that is, this pathogen reproduces only once in a growing season. There are pathogens of annual plants that are considered monocyclic, such as species of *Sclerotium* that cause diseases on numerous plants including stem rot of peanut, white rot of onion, and Southern stem blight of tomato. *Sclerotium rolfsii* survives as sclerotia, and these germinate and cause plant infections during a growing season. At the end of the growing season, new sclerotia are produced on infected plants to serve as inoculum for the following growing season. Like chestnut blight, many canker diseases are monocyclic. Azalea gall, caused by *Exobasidium vaccinii* (Fuckel) Woronin, common in the spring in Southern landscapes, is also a monocyclic disease. Certain microcyclic rusts that have no repeating urediniospore stage, such as cedar-apple rust caused by *Gymnosporagnium juniperi-virginianae* Schwein (Chapter 18), are also monocyclic diseases. Plant disease epidemics that are due to monocyclic diseases will only increase over years, not a few months.

Southern corn leaf blight and potato late blight are both **polycyclic**. Pathogens that cause polycyclic diseases multiply several times in a growing season. It is likely that only one or a few potato tubers with a viable infection of *P. infestans* survived the trip from the Americas, to Europe, then Ireland by early 1844. That infected tuber produced a plant with foliage infections from which sporangia and/or zoospores were produced. Sporangia are easily spread on wind currents, infecting more plants in moist conditions and producing new sporangia in as little as 5 d when weather conditions are optimal. Thus, in a 90-day growing season, this disease cycle could have repeated 17 times, increasing disease intensity with each cycle. Most plant disease epidemics that occur unexpectedly are polycyclic because of the rapidity with which they can develop.

Polycyclic diseases occur on perennial plants, just as they do on annual crops such as potatoes or corn. Apple scab, caused by *Venturia inaequalis* (Cooke) G. Wint., for example, is a polycyclic disease of apple trees, as is fire blight. *Erwinia amylovora*, the bacterial cause of fire blight, will reproduce and cause increasing numbers of infections as long as moist conditions and appropriate temperatures prevail. Many rust diseases that have the repeating urediniospore stage, including zoysia rust and leaf and stem rusts of cereals, are polycyclic, as are powdery mildew diseases. Conidia of *Erysiphepulchra* (syn. *Microsphaera pulchra*) and *Uncinula necator* (Schwein.) Burrill, causing powdery mildew of dogwood and grape, respectively, can infect their hosts and produce more conidia in 5 to 8 d under favorable conditions. Diseases of plants caused by viruses are also often polycyclic, especially when caused by viruses that are vectored by insects as long as the insects are moving among plants. Cucumber mosaic and tomato spotted wilt are examples of polycyclic diseases caused by viruses (Chapter 4). Diseases caused by nematodes (Chapter 8) can also be polycyclic. Root-knot nematodes (*Meloidogyne* spp.) complete a life cycle in 3 to 4 weeks when soil temperatures are 25°–30°C (77°–86°F). In the southern United States, these nematodes go through three to four reproductive cycles in a growing season.

There are also pathogens that cause diseases that are **polyetic**. Polyetic diseases are those that take several years from the time of infection until symptoms develop and the pathogen reproduces. Because the pathogen *Cronartium ribicola* Fisch. can take several years to grow into the main stem of the tree before sporulating, white pine blister rust is a polyetic disease. Mistletoe parasites are also polyetic because they reproduce only after a few years of plant growth.

Another aspect of the pathogen that affects epidemic development is the diversity within its population. Different individuals or isolates of the same pathogen may vary in the host cultivar that they can infect, and these are called **races** of a pathogen. Races have been identified in a number of plant pathogens, including *Phytophthora sojae* M.J. Kaufman & J.W. Gerdemann, cause of soybean root rot, and *Xanthomonas campestris* pv. *vesicatoria*, cause of bacterial leaf spot of pepper. Differences in a pathogen population may also affect the pathogen aggressiveness, as seen with Race T of *C. heterostrophus*. When a new race of a pathogen develops, plant epidemics can become severe in crops previously thought resistant to that pathogen. This is what happened in wheat with *Puccinia graminis* Pers.:Pers. f. sp. *tritici* Eriks. and E. Henn. through the early part of the 20th century. Wheat cultivars were developed and released to growers with improved resistance to stem rust, then, within a few years, the disease once again became severe on the crop (see also Chapter 18).

ENVIRONMENT

Many of the most destructive diseases of plants are polycyclic diseases, and each of them differs in some way. Whereas the disease cycles for Southern corn leaf blight and potato leaf blight can reoccur very rapidly when conditions are optimum, many disease cycles are somewhat longer. Disease cycles, from infection by inoculum through production of more inoculum, lengthen in time with less than optimum conditions. Temperature and moisture, and the interaction of these two components of weather, are often the most important factors affecting the length of a disease cycle. For example, the Southern corn leaf blight disease cycle can be completed in 3 d when moist conditions prevail and temperatures are between 20 and 30°C (68° and 86°F). If drying occurs, especially if relative humidity remains below 90% for 18 or more hours in a 24-h period, the disease cycle lengthens. Urediniospores of *P. graminis* f.sp. *tritici*, causing stem rust of wheat, cause infection and produce more urediniospores in as little as 7 d in ideal conditions at 30°C (86°F). However, under normal field conditions, this cycle repeats every 14–21 d (Roelfs, 1985).

In the field, temperature and moisture fluctuations detract from ideal conditions for pathogen development. Such fluctuations may be second by second as wind currents vary, but we are usually most aware of those that occur diurnally, in a 24-h cycle, or due to weather systems moving through a region. Often, unfavorable weather conditions limit disease development substantially, as can be seen with decreasing levels of powdery mildew in winter wheat as daytime temperatures warm or minimal black spot on roses during dry seasons. Diseases that seem to disappear during the summer, such as brown patch of warm-season turfgrasses caused by *Rhizoctonia* spp. are actually limited by high temperatures and reinitiate in late summer as temperatures decrease.

Temperature and moisture are readily understood as influences on pathogen development, but there are other aspects of the environment that can affect pathogen development. Wind, for example, plays several important roles. Wind currents are one of the most important means by which plant pathogen inoculum is **dispersed** or becomes spread over a geographical area. It is easy to imagine how a fungal spore can be carried on the wind from one plant to another in a field, but it is possible for inoculum to move even greater distances as might have been the case with hurricanes carrying *P. pachyrhizi* (causing Asian soybean rust) to the United States. Generally, faster wind currents carry inoculum propagules further, and smaller propagules are carried further than larger propagules. Urediniospores of *P. graminis* f. sp. *tritici*, which are relatively small spores, have been carried by wind currents up to 680 km (> 420 miles) from their source. Wind currents also have direct effects on temperature and moisture since they can cool and dry surfaces, thus affecting the length of the disease cycle.

Solar radiation, the energy from the sun, is another aspect of the environment that influences disease development. Radiation can inhibit or stimulate germination of a number of fungal spores. Most bacterial cells quickly lose viability with exposure to radiation. Dogwood anthracnose lesions develop more slowly and have reduced sporulation on foliage in full sun than in full shade. Conversely, spores of *Botrytis* spp., causing gray mold of flowering plants, have been found to germinate only with exposure to particular UV wavelengths. Thus, the prevalence of overcast versus clear skies can contribute to the development of an epidemic.

Other aspects of the environment can affect disease development. Soil, for example, can have physical and chemical characteristics that can profoundly affect disease development, particularly diseases by soilborne pathogens. Nematodes develop and move most readily in moist, but not waterlogged soil, and near the soil surface where oxygen availability is not limiting. Species of *Sclerotium* also have an oxygen requirement, so root rots by *Sclerotium* tend to begin near the soil surface. Soil acidity is one of the chemical properties that can affect disease development, as seen with take-all of wheat and turfgrasses by *Gaeumannomyces graminis* (Sacc.) Arx & D. Olivier, which is favored in more neutral (higher pH) soils. Soil nutrients can also affect the development of both soilborne and airborne pathogens in causing foliar diseases. For example, nitrogen fertilization has been associated with increased potato late blight severity. High N is also conducive to brown patch development in turf caused by *Rhizoctonia* species and excessive fertilization can initiate a severe outbreak of brown patch. In a very broad sense, the economic and political environment that led to reduced tillage practices over substantial acreage in the north central states may be responsible for recent severe epidemics of Fusarium scab of wheat (McMullen et al., 1997).

HOST

In addition to influences of the environment, the genetic makeup of a host population can affect an epidemic. A higher degree of genetic uniformity among plants is likely to allow more rapid and more severe disease development than lower uniformity. In any population of plants, there will be some differences between individual plants that can affect the amount of disease resulting from an infection. However, plants that are self-pollinated, like wheat, will have less variability between individuals in a population than those that are cross-pollinated, such as corn. Plants that are propagated vegetatively, such as potatoes and many ornamentals, have lower variability between individuals than even self-pollinated plants. Thus, epidemics would be expected to develop most rapidly across a population of clonally- or vegetatively-propagated plants, less rapidly in

self-pollinating plants, and more slowly in cross-pollinated populations. In wild or natural plant populations, where plant species are intermingled and different ages and origins of plants exist together, disease epidemics are considered rare. This rarity of plant disease epidemics in mixed species populations is primarily due to the probability of pathogen propagules landing on an appropriate host.

Humans, of course, are responsible for creating massive populations of identical plants that can readily succumb to disease problems. We have also discovered that certain plants have resistance or the inherent ability to withstand pathogen infection or disease development (Chapter 34). Two types of resistance are recognized, one is easy to select for or breed into plants because it is usually encoded for by a single gene. In addition, this **single gene resistance** frequently prevents the development of any symptoms of disease. The disadvantage of this type of resistance is that a change in the pathogen can overcome that single gene and allow as much disease as if the plant were susceptible. Thus, adaptation of the pathogen to this single gene resistance can result in the development of a severe plant disease epidemic. In wheat, for example, numerous genes have been identified that encode for resistance to stem rust, and these genes can be found in different cultivars of wheat. Because races of *P. graminis* f. sp. *tritici* shift from year to year, different cultivars need to be planted from one year to the next to avoid a devastating disease epidemic. Another type of resistance is **multigenic resistance**, and this is encoded by several to many genes. Multigenic resistance is difficult to incorporate into a single cultivar, but is not as readily overcome as single gene resistance. However, multigene resistance usually does allow some disease development, which is why this is also called partial resistance. An advantage to this type of resistance is that it is effective on most or all races of a pathogen. Epidemics do develop in crops with multigenic resistance, but disease severity generally stays low (Vanderplank, 1963).

Tissue age or age of host plants can also influence the rate of disease development. Many pathogens, such as *Venturia inaequalis,* the cause of apple scab, will infect young tissues more readily, so disease develops more quickly early in the growing season. There are also plant diseases, such as Sclerotinia blight of peanut caused by *Sclerotinia minor* Jagger, charcoal rot of soybean (caused by *Macrophomina phaseolina* (Tassi) Goidanich) (see Case Study 16.2), and chestnut blight, that do not develop until plants have entered their reproductive stages. Most often, plants are more susceptible to diseases when they are young, or when the plant or particular plant parts are actively growing. Young tissue can be more susceptible to pathogen infection because natural barriers (e.g., cuticular coats) have yet to develop.

Plant growth provides more tissue with the potential to become diseased. If disease is not increasing as rapidly as the plant is growing, the proportion of diseased tissue on that plant will appear to decrease. Similarly, if defoliation occurs due to disease, as happens with peanut leaf spot diseases and black spot of rose, the proportion of the plant that is diseased may appear to decrease. Many plants have limited or finite growth during a season, so that disease can eventually affect the entire plant if conditions remain appropriate. Thus, cultivars of plants that have longer growing seasons may suffer heavier damage due to a particular disease than other, shorter-season cultivars because an epidemic developed for a longer period of time.

As plants grow, they also go through various developmental stages that involve physiological changes, and these changes can influence disease development. Though it might be tedious work, it is easy to understand that various aspects of plant size can be measured. Plant height, numbers of leaves, total area of foliage, root depth, root volume, and root mass, if monitored over time, would reflect plant growth. Host development, too, can be monitored, and for many plants distinct **host growth stage** keys or diagrams are available (Campbell and Madden, 1990). In corn, for example, distinct stages were delimited by Hanway (1963) and include emergence, tassel formation, silk development, and several stages in kernel maturation. Each of these developmental stages is separated in time, but not necessarily by equal time increments. Growth stage keys have been developed for most agronomic crops, including cereals, soybeans, and peanuts.

DISEASE

Prior to the complete devastation of their crops in 1845, Irish farmers did not notice late blight on their potatoes even though it had likely occurred in the preceding year. Late blight that may have affected potato plants in 1844 would have been easy to overlook, in part because the concept of plant disease was unknown then. However, even today, disease is difficult to find in any plant population when damage levels are low. Most people who work regularly with plants will not notice symptoms of a disease until it has increased to about 1% intensity. This initial observation of disease is called **onset**, and onset is often preceded by several to many reproductive cycles by the pathogen following introduction of the initial inoculum. Yet the difference between onset at ~1% disease level and severely diseased plants at 80% damage might only be another one or few reproductive cycles. Thus, it appears that disease development is slow at the beginning of an epidemic. This can be illustrated with a disease that may occur, for example, in an acre of 20,000 plants. A single lesion of this disease might occupy 0.1% of the foliage of a single plant and account for about 0.000005% of all foliage in the acre of plants. If that single lesion produces inoculum that results in a twentyfold increase in numbers of lesions, then the second "generation" of disease lesions would occupy 0.0001% of foliage

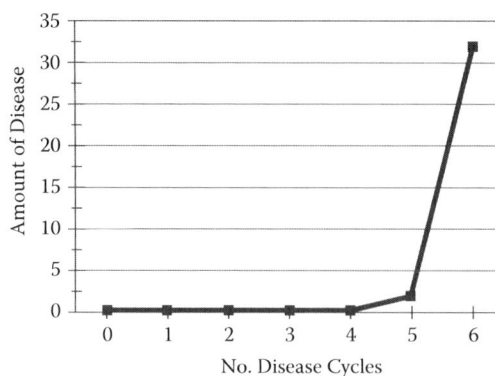

FIGURE 33.2 Graph showing increase in disease that starts from a single lesion affecting 0.000005% of foliage on one plant among 20,000 plants, when each disease cycle allows the development of 20 new lesions from each older lesion.

in the acre, and so on. It would take more than four disease cycles before disease at ~1% is readily noticed, and only 1.6 more cycles before 100% of foliage is diseased, if that disease increase continued at this rate (Figure 33.2).

Epidemics are not always rapid developments of severe disease; epidemics occur when there is **any** increase in disease over time. Thus, even the increase of one diseased plant to two disease plants, reflects an epidemic. An increase in **disease incidence**, the number or proportion of plants in a population that are affected by disease, is due to spread or dispersal of disease-causing inoculum. When formerly healthy plants become diseased due to inoculum spread from another plant, this is termed **alloinfection**. Some inoculum spreads no further than to leaves of the originally diseased plant. However, even this **autoinfection** of the same plant leads to a greater disease level through an increase in severity. **Disease severity** represents the proportion of tissue of a single plant affected by disease. It is probable that potato late blight was widely established through Ireland at the end of the 1844 growing season and that disease incidence was high, with more than 50% of individual plants affected. However, the level of disease on individual plants was probably fairly low, perhaps less than 5% disease severity and thus not very noticeable.

Disease severity and disease incidence are usually positively related to one another. That is, as disease severity increases, so does disease incidence. However, there are a number of diseases that increase only in incidence because a single infection systemically affects the entire plant, frequently causing plant death. Bacterial wilt on tomato by *Pseudomonas solanacearum* (Smith) Smith and Verticillium wilts of many plants including redbud, snapdragon, and most vegetables are just some of the diseases that only increase in incidence over time in a plant population. Whereas there may be differences in the stage of disease progress at some point in time among systemically infected plants in a population, these differences are

TABLE 33.1

Two Rating Scales for Assessing Disease—The Horsfall-Barratt Scale Based on the Weber-Fechner Law of Visual Acuity and an Alternate Scale Based on 1/5th of the Plant

	Horsfall-Barratt		Alternate Rating Scale
Grade	Disease Ranges (% Severity or Incidence)	Grade	Disease Severity Range (%)
0	0	0	0
1	0–3	1	0–20
2	3–6	2	20–40
3	6–12	3	40–60
4	12–25	4	60–80
5	25–50	5	80–100
6	50–75		
7	75–88		
8	88–94		
9	94–97		
10	97–100		
11	100		

more likely due to when infection occurred or genetic differences between the infected plants.

Disease rating scales. Disease severity, or the proportion of the plant that is affected by disease, can range from 0% to 100%. However, the entire range of disease severity is difficult to distinguish with precision, especially when disease levels are moderate or between 20 and 80%. The Weber-Fechner law of visual discrimination relates to this ability to distinguish moderate levels of plant disease severity. This law states that man's ability to see differences decreases by the logarithm of the intensity of the stimulus. In other words, when the stimulus is a disease lesion on a leaf amid healthy tissue, it is easier to distinguish 1% from 3% or 5% disease severity than to tell 20% from 40% severity. Similarly, when a leaf is nearly completely diseased, as happens with Southern corn leaf blight, it is relatively easy to distinguish 95% from 99% disease severity, but more difficult to determine 65% from 85% disease. It is easier to distinguish between high levels of disease as severity approaches 100% because the visual stimulus becomes the healthy tissue amidst diseased tissue.

Horsfall and Barratt developed a disease rating scale based on the Weber-Fechner law and the difficulty of distinguishing among moderate levels of disease severity. This rating scale consists of 12 levels or grades, each of which include disease severity ranges that decrease as disease approaches 0% or 100% (Table 33.1). In addition, disease severity ranges are symmetrical around 50%. The Horsfall-Barratt disease rating scale is only one example of many aids that can be used in rating disease severity in order to monitor increase in disease over time. For example, in evaluating disease on a

plant, it is relatively easy to think of whether the severity encompasses 1/5, 2/5, 3/5, etc., of the plant, and this consideration can be implemented as a rating scale with 0 for no disease and 5 additional levels (Table 33.1). One advantage of the Horsfall-Barratt scale over one based on fifths of a plant is that there are more grades or levels in the scale. The more restricted ranges of disease at low and high levels allow greater accuracy in keeping track of an increase in disease over time. Rating systems with fewer grades may be more useful for comparing the effectiveness of fungicidal products or the susceptibility of a number of cultivars.

Disease diagrams. As an aid in assessing disease severity on plants, disease diagrams have been developed for a number of diseases on several important crops (e.g., James, 1971). Such diagrams provide pictorial aids for determining particular disease levels (Figure 33.3). Although frequently used in training, such pictorial aids improve the precision of disease ratings, even by experienced personnel. In addition, these aids provide consistency in disease evaluations made at different times and among different evaluators. Current technology allows us to easily develop customized diagrams for specific diseases on any host.

Complications. Disease assessment, especially with the help of rating scales and disease diagrams, is not always straightforward. Often, disease symptoms vary in color, and these colors may not be distinct from one another. Similarly, more than one disease may be affecting a single leaf or plant at any point in time.

Disease in space. As illustrated by the southern corn leaf blight and chestnut blight epidemics, increases in

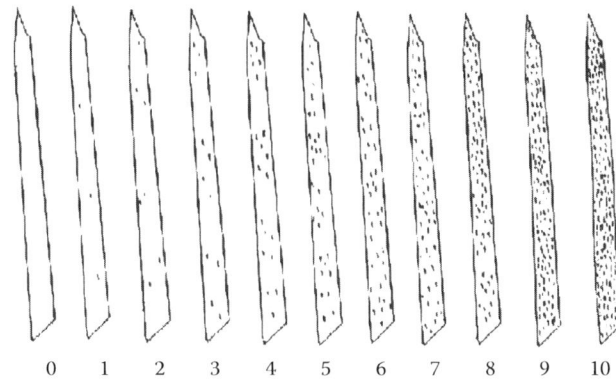

FIGURE 33.3 Disease severity diagram of rating scale used for leaf rust of wheat.

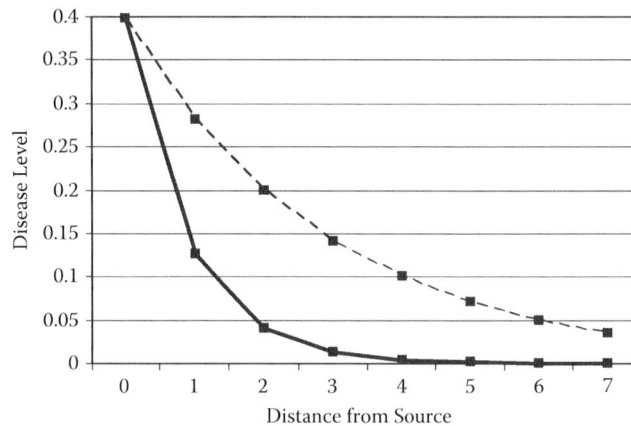

FIGURE 33.4 Graphical representation of disease gradients or the decrease of disease with increasing distance from the source. Solid line represents "steeper" disease gradient.

disease incidence over time also leads to an increase in the size of the area in which a disease can be found (Figure 33.1). At a single point in time, severity on individual plants is expected to decrease as distance increases from the original infection. This is called a **disease gradient**. Gradients can be due to gradual changes in the microclimate or soil type, but are most often seen with increasing distances from the first infection and are due to dispersal. Those pathogens that are most easily dispersed, like powdery mildew fungi and *P. infestans*, will infect plants further and more quickly from an original infection site than pathogens with larger propagules that are dispersed with more difficulty. *Phytophthora infestans* is said to have a "flat" dispersal gradient because of its ease of spread (Figure 33.4). Conversely, nematodes and *G. graminis*, causal organism of take-all of grasses, are considered to have "steep" dispersal gradients because their spread is due to their own movement and growth through soil, which is limited. Actually, dispersal gradients are not usually applicable to diseases caused by nematodes. Rather, because the affected area is very limited, diseases caused by nematodes and other soilborne pathogens (Chapter 22) are considered "focal diseases."

A number of diseases not only cause tissue damage, but also cause defoliation or premature loss of foliage. In addition to defoliation, disease development can cause lodging or the falling of plants, stunting or a decrease in size, wilting, and diminished quality (e.g., sugar, protein content, or toxin contamination). Any of these disease characteristics can increase in intensity over time.

DISEASE ANALYSIS OR MODELS FOR EPIDEMICS

In trying to understand plant disease epidemics, mathematical models have been used as a means of simplification. These mathematical models provide a quantitative way to represent the sometimes complicated processes of an epidemic. Mathematical models have also allowed comparisons of epidemics from different times or places and provide a framework for forecasting future disease levels.

Exponential model. One model that has been applied to plant disease development over time is the exponential model (sometimes called the logarithmic model) (Figure 33.5):

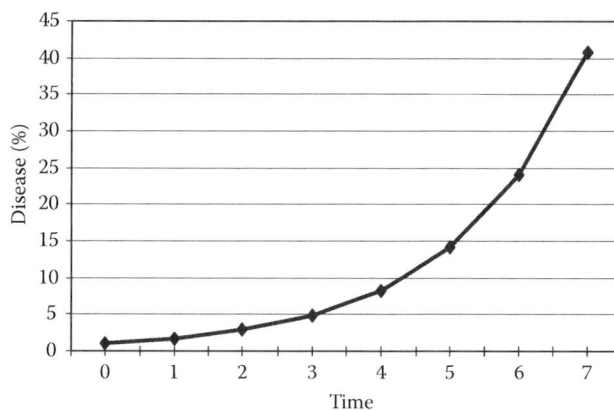

FIGURE 33.5 Exponential model of disease development over time.

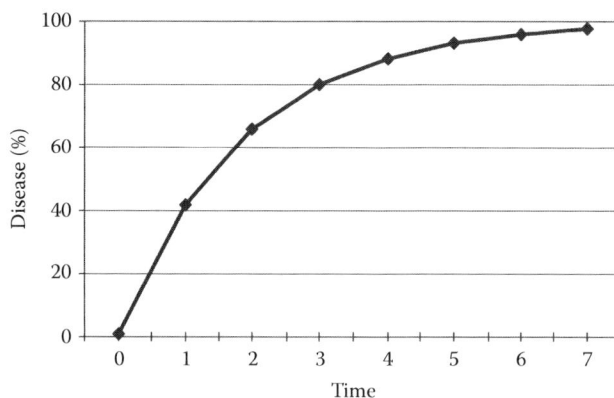

FIGURE 33.6 Increase of disease over time according to the monomolecular model.

$$y_t = y_0 e^{rt} \qquad (33.1)$$

where y_t = the amount of affected tissue after t time intervals, given a starting disease level = y_0 and a rate parameter (r), where e is the mathematical constant 2.718. Note that the graph of this model looks similar to one presented previously (Figure 33.2) in that disease stays low during the first few cycles, then rapidly increases. The linear form of this model is frequently easier to understand:

$$\ln y_t = \ln y_0 + rt \qquad (33.2)$$

Terminology. In applying this and similar models to plant disease, several things should be noted. For example, the variable t can represent different time intervals, depending on the pathosystem. As noted earlier in this chapter, the time interval applicable to late blight of potato would be a day, while for chestnut blight a year might be more applicable. Similarly, the infection rate r differs among various pathosystems. Calculations done on a number of plant disease epidemics indicate that $0.10 < r < 0.56$ for diseases of annual plants. The starting disease level is y_0 and represents the initial inoculum level.

Monomolecular model. Although the exponential model allows for infinite increase in the value of y, usually there is some finite limit to the amount of plant tissue that can become diseased. This is especially true when dealing with disease severity, when the proportion of disease cannot become greater than 100%, or when considering the proportion of plants in a finite population. Therefore, the exponential model is not often applicable to disease increase in plant populations. The monomolecular model is another model for simplifying disease increase over time, and with this model (Figure 33.6), disease levels are limited to 100%:

$$y_t = 1 - (1 - y_0)e^{-rt} \qquad (33.3)$$

The linearized form of the monomolecular model is:

$$\ln y_t = \ln (1/(1 - y_0)) \qquad (33.4)$$

Logistic model. A third model for simplifying disease increase in time is the logistic model (Figure 33.7). This model also limits disease to a maximum of 100%:

$$y_t = 1/(1 + e^{-[(\ln y_0/(1-y_0))+rt]}) \qquad (33.5)$$

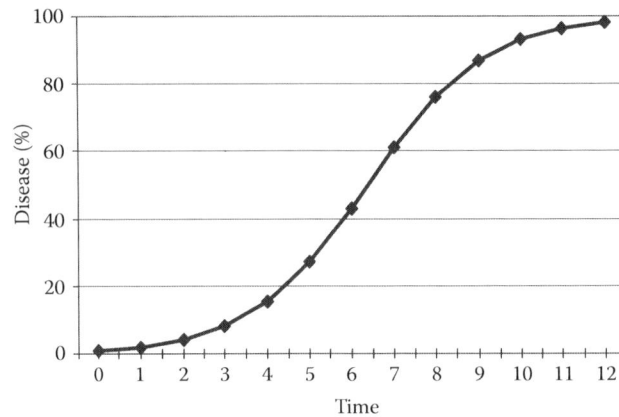

FIGURE 33.7   Graph of increase of disease in time as described by the logistic model.

The linearized form of the logistic model is:

$$\ln(y_t/(1 - y_t)) = \ln(y_0/(1 - y_0)) + rt \qquad (33.6)$$

It should be noted that when disease levels are low, less than 10%, the exponential model is very similar to the logistic model. Thus, either the exponential model or the logistic will apply during early stages of an epidemic. When disease levels are high, approaching 100%, the monomolecular model is very similar to the logistic model. Thus, either the monomolecular or the logistic model will apply during later stages of many epidemics.

Simple assumptions. These models idealize disease progress as they assume constant and uniform values for each of the variables. They also simplify the processes of disease development and have been fit to numerous plant disease epidemics, from potato late blight development to the development of dollar spot by *Sclerotinia homeocarpa* F.T. Bennett on creeping bentgrass. Vanderplank, in his 1963 book *Plant Diseases: Epidemics and Control*, was the first to quantitatively analyze plant disease epidemics. These initial analyses were criticized and built upon, and ultimately led to the plant disease epidemiology as an autonomous area of study.

The infection rate of these models is where most of the simplification comes in relative to understanding epidemics. As described earlier in this chapter, there are numerous factors that influence the rate of epidemic development and these are incorporated into the value of *r*. In real life, an increase in plant disease over time appears to be somewhat erratic—sometimes increasing rapidly, sometimes seeming to decrease. Quantitative models smooth out these apparent stops and starts of disease increase, as seen in the difference between Figures 33.2 and 33.5. In addition, these models for plant disease increase over time do not account for any **latent period**, but assume that new infections are immediately visible and instantaneously infectious. All of these fluctuations and stops and starts in disease development are incorporated into the infection rate *r*.

CONTROL APPLICATIONS

Using the linearized forms (Equations 33.2, 33.4, and 33.6) of the models for disease increase in time, it is easy to see how changes on the right side of each equation affect the left side. For example, a decrease in the value of y_0 will decrease y_t if the values of other parameters stay the same. Thus, these equations help in understanding the effects of many control measures.

Effects on y_0. Crop rotation (Chapter 35), for example, decreases the amount of viable inoculum of a soil-borne pathogen, thus decreasing y_0. Initial inoculum is also decreased through sanitation, which is the removal of inoculum, and through the use of protectant fungicides, which are fatal to inoculum. Single gene resistance is also considered as affecting y_0. A decrease in initial inoculum, if substantial enough, can result in lower disease at the end of a finite time period, such as a growing season, compared to no decrease in y_0 (Figure 33.8). The figure shows that with $y_0 = 0.01$ (Line A), by $t = 9$ disease is 100%. However, when initial inoculum is decreased such that $y_0 = 0.005$, disease at $t = 9$ is 92% (Line B). A further decrease to $y_0 = 0.0025$ results in disease at 72% at $t = 9$ (Line C). Lower disease levels are desirable because lower losses are associated with lower disease.

Effects on t. Control practices can also affect the *t* variable of models for disease increase. For example, early harvest or a short-season cultivar would end an epidemic earlier, perhaps at $t = 8$ rather than $t = 9$. For example, with Line C in Figure 33.8, an earlier harvest at $t = 8$ rather than at $t = 9$ could mean disease severity of 46% instead of 64%.

Effects on r. Control measures also can affect the infection rate, *r*, of these models. For example, multigene resistance in a plant is considered to reduce *r* compared to susceptibility. The use of sterol biosynthesis-inhibiting fungicides, which delay fungal development and reduce inoculum production, reduce the rate of disease increase. Cultural methods such as pruning a tree or shrub to increase air flow, thereby decreasing duration of moist periods, and

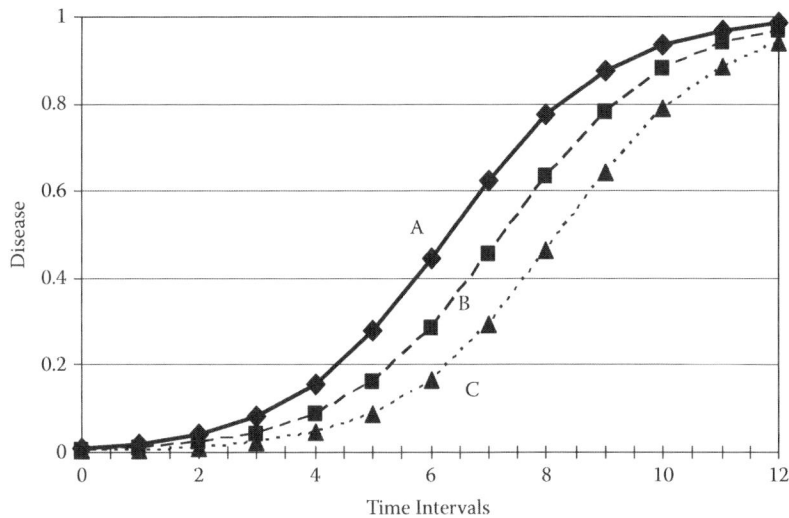

FIGURE 33.8 Graphs of disease in time according to the logistic model with three levels of initial inoculum: (A) $y_0 = 0.01$; (B) $y_0 = 0.005$ (50% of A); and (C) $y_0 = 0.0025$ (50% of B).

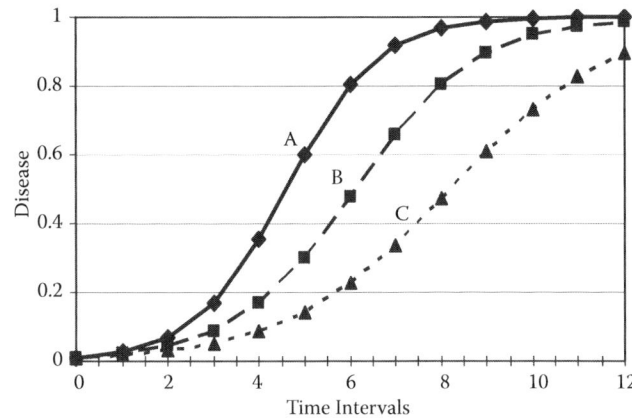

FIGURE 33.9 Graphs of disease in time according to the logistic model with three levels for the rate parameter. (A) $r = 1.0$; (B) $r = 0.75$ (25% less than for A); and (C) $r = 0.56$ (25% less than B).

use of drip irrigation instead of sprinkler irrigation, also reduce the value of r in these models (Figure 33.9).

Each of the components of these models act together and distinctively for each pathosystem. In general, polycyclic diseases have low y_0 but high r values. Because polycyclic diseases can develop very quickly, the level of y_0 has less influence on the epidemic than does r. Thus, control strategies for diseases with high r values should be aimed at decreasing r. Black spot disease of rose is a polycyclic disease with a high r value in the southern United States. Reduction of y_0 through sanitation by removal of fallen infected foliage is recommended for minimizing black spot. However, on many rose plants, fungicide applications are still needed in order to manage this disease. Similarly, peanut leaf spot diseases, caused by *Cercospora arachidicola* S. Hori and *Cercosporidium personatum* (Berk. & M.A. Curtis) Deighton, are polycyclic diseases with high values for r. Initial inoculum for these leaf spot diseases is believed to originate from plant debris remaining in the field. As these fungal pathogens only infect peanut, the only source of inoculum is from previously infected tissue. However, if peanuts are grown in a field not previously planted to this crop, leaf spot diseases develop rapidly from the small quantities of inoculum that are carried to these new fields by wind. Powdery mildew diseases of many plants are additional examples of polycyclic diseases with high r values.

Conversely, most monocyclic diseases have low r values, but relatively high y_0 values. Most effective control strategies for monocyclic diseases would be to reduce y_0. As presented earlier, Southern blight of tomato is a monocyclic disease. If planting is done in a field that had not been previously cropped to tomato, the incidence of Southern blight is likely to remain low through the growing season. Thus, crop rotation is a recommended and effective strategy for minimizing Southern blight of tomato, a monocyclic disease.

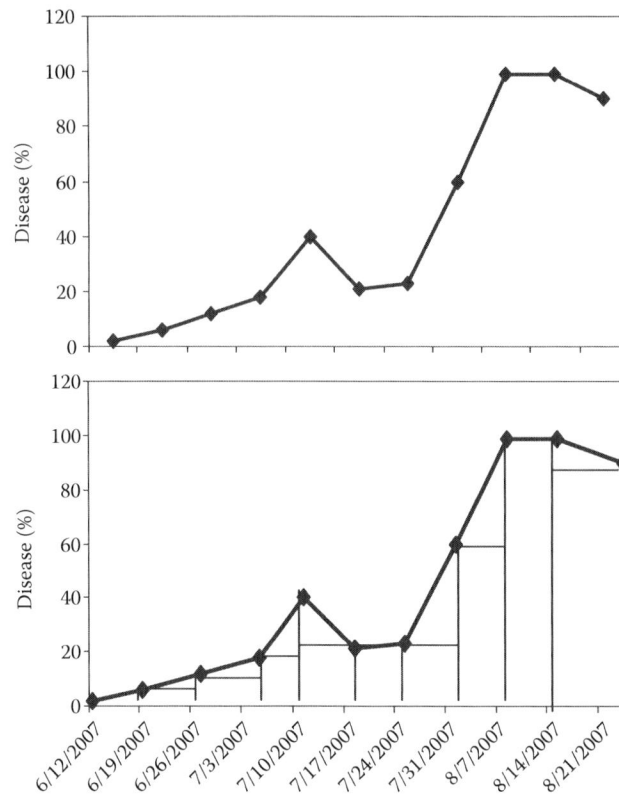

FIGURE 33.10 Geometric approach to calculating area under the disease progress curve (AUDPC) by summing all areas of rectangles and triangles that fit under disease curve. Area of rectangle = (width * height) or (difference between adjacent time intervals * difference between adjacent disease levels). Area of triangle = ½ (width * height) or ½ (difference between adjacent time intervals * difference between adjacent disease levels).

Although increase of disease severity over time may be fit to the logistic model, a simpler means of comparing different epidemics is frequently desired. In these situations, the total area under the graphed curve of the epidemic might be used. The "area under the curve" or "area under disease progress curve" (AUDPC) is relatively easy to calculate and provides a single numerical value for quantifying an epidemic. This area might be quantified by using calculus, or more simply with geometry, where the curve of the epidemic is broken into rectangles and triangles, and the areas of these figures are summed (Figure 33.10). Several epidemics can be compared using AUDPC values, as long as the monitored duration of those epidemics is the same.

Spatial modeling. As has been mentioned, when there is disease increase in a population, the increase occurs over time as well as over geographical area. Spatial aspects of epidemics can also be described using mathematical models. Such spatial models can be used to determine how far a disease can spread. One such model is an exponential model with different parameters than the model above:

$$y = ae^{-bx} \qquad (33.7)$$

The linear form being:

$$\log y = \log a - b(\log x) \qquad (33.8)$$

where y is disease at some distance x from an infection source, a represents the amount of viable inoculum, and b reflects the difficulty with which the pathogen is dispersed. As the value of b approaches zero there is decreasing difficulty in dispersal, or the disease gradient is steeper (Figure 33.4). Thus, diseases caused by pathogens that disperse readily, such as *Botrytis* spp. with its small conidia, have small values for b and a gradient that is flattened. Values for b for soilborne diseases are high, approaching 1, since the pathogens causing many of these diseases spread primarily by means of their own growth or movement.

Spatial analyses of plant diseases can contribute to improved management of those problems. For example, spatial analyses have been used with geostatistics and provided documentation for the role of weeds as alternate hosts in tomato virus management in the Del Fuerte Valley of Mexico (Nelson et al., 1999). These analyses have also reinforced the need for management strategies that reduce insect vectors and pathogen reservoirs when dealing with Pierce's disease of grape (Tubajika et al., 2004).

CROP LOSSES

In each of the three examples starting this chapter, plants and crops were killed or yields were substantially reduced due to plant disease epidemics. Potato late blight killed potato plants, and any tubers that did develop and eventually harvested were likely to rot in storage due to infection by *P. infestans*. Similarly, mature American chestnut trees died from infection with *C. parasitica* throughout eastern U.S. forests. Although not all plant pathogens kill their hosts, any disease will detrimentally affect plant growth and development. In the vast majority of diseases, leaf spots or even a few rotten roots will not negatively impact human lives. However, when disease incidence or severity increases to the point that most of a plant's foliage becomes necrotic or limbs of trees do not produce fruit, this is cause for concern.

One of the primary goals in gaining knowledge about plant diseases is to reduce crop losses to an acceptable level. An acceptable level of loss may be that which cannot be controlled with affordable management strategies. As the value of the yield of a crop or an ornamental plant increases, more can be done to manage any disease that is detrimental to the yield or the aesthetics of plants such as ornamentals. However, not all levels of disease have noticeably detrimental effects on yield or plant development. So, first the relationship between disease and yield must be determined and understood.

The relationship between disease and yield (or losses in yield) has been determined for numerous pathosystems. With some plants, such as wheat, specific growth stages affect different components of the productivity of that plant and any disease occurring during that growth stage can subsequently affect yield. For example, the flag leaf, or the leaf just below the grain head of wheat, has been documented as contributing most to grain fill. For this reason, yield loss models for diseases of cereals often use the percent disease severity on the flag leaf (e.g., Burleigh et al., 1972) for estimating losses due to disease. Of course, in order to use such estimates of loss, disease severity must be determined with some accuracy, which justifies the use of standardized rating scales as previously presented in this chapter. Another type of loss relationship can be illustrated with Southern stem rot in peanuts where each locus of a *S. rolfsii* infection at plant maturity, up to 30 cm or 1 foot in length, causes 0.9 to 2.9% loss in yield quantity (Bowen et al., 1992). Knowledge of this relationship can aid a grower in determining how yields could change if disease incidence increased or decreased, and may provide justification for implementation of control measures in the following year. More useful, perhaps

would be a means of estimating crop loss due to disease before the occurrence of disease. Such knowledge would allow the use of control measures before damage and losses have developed.

DISEASE FORECASTING AND DISEASE MANAGEMENT

One of the uses of the mathematical models that describe plant disease development over time is for predicting or forecasting future disease levels. Knowledge about a future disease level can assist in making disease management decisions.

With monocyclic diseases, disease levels might be forecast by a direct assessment of the amount of initial inoculum at the beginning of the season. Indeed, this has been done with white rot of onion, caused by *Sclerotium cepivorum* Berk., in New Jersey (Adams, 1981) and for sclerotium rot of sugar beets, caused by *S. rolfsii* in Uruguay (Backman et al., 1981). With both of these diseases, sclerotia in the soil provide the initial inoculum, which is enumerated for a forecast. Similarly, nematode populations are determined prior to planting to estimate the potential damage in a crop. When initial inoculum levels are high, substantial disease is predicted, and the recommendation is made to plant the crop elsewhere or to take protective measures. Another disease for which disease is predicted from initial inoculum is Stewart's wilt of corn (Chapter 34). The assessment of the initial inoculum, however, is indirect because it is the corn flea beetle vector of the pathogen, *Erwinia stewartii* (Smith) Dye that is considered rather than the pathogen itself. The corn flea beetle population is adversely affected by cold temperatures during winter months. If the average winter temperature is at or below freezing, few flea beetles survive, and the risk of high disease is substantially reduced. Knowledge of the corn flea beetle survival helps growers determine which cultivar they should plant, especially when cultivars differ in their resistance to Stewart's wilt (Pataky et al., 1995).

Initial inoculum also plays a role in the forecast for apple scab. However, the actual amount of initial inoculum is not quantified, rather the weather that is conducive for release of spores and infection by initial inoculum is analyzed. When infection is favored, the forecast recommends the initiation of a fungicidal spray program. Thus, regular fungicidal sprays are used to manage this polycyclic disease, but the forecast is largely based on weather that is favorable for initial inoculum.

Most polycyclic diseases are best controlled by reducing the infection rate. As presented earlier, weather is one of the primary influences on the rate of development of an epidemic. Thus, some of the best forecasts of polycyclic diseases are based on weather conditions that are favorable for pathogen development. In Alabama, for example, the AU-Pnut leaf spot advisory provides a recommendation for fungicide applications on peanuts when weather is predicted as favorable for infection. This advisory is based on the accumulation of rain events beginning at planting or following a period of protection after a fungicide spray. Further north, in Virginia and North Carolina, temperature can be a limiting factor for the development of peanut leaf spot diseases, so the Virginia peanut leaf spot advisory system includes temperature as well as moisture in predicting favorable conditions for disease increase. These two forecast systems, based on reducing the infection rate of an epidemic, have been shown to save growers 1 to 2 fungicide applications per year compared to applications made every 14 d through a growing season.

Moisture and temperature are also used in predicting potato late blight using BLITECAST. This forecasting system computes daily disease severity values which are based on numbers of hours during which relative humidity (RH) > 90% and the average temperature during moist periods. Greater disease severity values are predicted with longer periods of high RH, and higher temperatures allow some shortening of high RH periods. Disease severity values for BLITECAST are accumulated over a 7-d interval, and along with numbers of rain days, are used in recommending whether to spray. In addition to the recommendation to spray or not spray, BLITECAST suggests spray intervals depending on the accumulated disease severity values. A similar forecast system that uses moisture and temperature for determining disease severity values is TOM-CAST, a management aid for control of early blight (caused by *Alternaria solani* Sorauer), Septoria leaf spot, and fruit anthracnose (caused by *Colletotrichum coccodes* (Wallr.) S.J. Hughes) in tomatoes. Disease severity values in TOM-CAST are greatest with wet weather and temperatures between 21° and 27°C (70° and 80°F). Fungicide applications are recommended by TOM-CAST when disease severity values have accumulated to particular levels. Each of these forecasting systems provides recommendations in order to decrease the number of fungicide applications when the environment is not conducive for disease development.

In addition to the four plant disease forecasting systems presented here, numerous additional systems have been developed. Many of these systems have been implemented with some success. The reasons that forecasting systems for plant diseases have not been more widely used are varied. One reason that growers do not adapt to using a plant disease forecast system is that they consider the possible reduction in pesticide applications as too risky. Adding to the perception of risk is that pesticide labels do not address the possible use of forecasting systems. Another reason for a lack of implementation has been the limited availability of appropriate data for a forecast. However, this is changing with increased access to the Internet and ease of obtaining customized local and accurate weather information.

CONCLUSION

Plant disease forecasts and advisory systems are tools that have been developed through epidemiological studies. Plant disease epidemiology, or the study of the factors that allow disease increase, provides information that can improve the efficiency of plant and crop production. Numerous opportunities exist for continued application of knowledge gained through epidemiology for improving our ability to manage plant diseases.

EXERCISES

EXERCISE 1. DISEASE ASSESSMENT

Horsfall and Barratt developed a disease rating scale based on the Weber-Fechner law and the difficulty of distinguishing among moderate levels of disease severity. This rating scale consists of 12 levels or grades, each of which includes disease severity ranges that decrease as disease approaches 0% or 100% (Table 33.1). In addition, disease severity ranges are symmetrical around 50%. The Horsfall-Barratt disease rating scale is only one example of many that can be used in rating disease severity in order to monitor increase in disease over time. For example, in evaluating disease on a plant, it is relatively easy to think of whether the severity encompasses 1/5, 2/5, 3/5, etc., of the plant and this consideration can be implemented as a rating scale with 0 for no disease and 5 additional levels (Table 33.1). One advantage of the Horsfall-Barratt scale over one based on fifths of a plant is that there are more grades or levels in the scale. The more restricted ranges of disease at low and high levels allow for greater accuracy in keeping track of an increase in disease over time. Rating systems with fewer grades may be more useful for comparing the effectiveness of fungicidal products or the susceptibility of a number of cultivars.

Materials

The following items are required for each student.

- Four collections of diseased leaves or plants, at least 5 samples per set (Samples may be collected at different times, but from the same plant population. Alternatively, copies of disease diagrams can be used [e.g., James, 1971].)
- Calculator

Follow the experimental outline provided in Procedure 33.1.

Procedure 33.1

Disease Assessment

Step	Instructions and Comments
1	Using Table 33.1, assign a rating to each sample based on the Horsfall-Barratt scale; then rate each sample using the alternative scale. Do this for each collection of samples.
2	Calculate an average disease rating from each of the collections and for each rating scale.
3	Compare average disease ratings for each collection.

Procedure 33.2

Disease Progress

Step	Instructions and Comments
1	By using Equation 33.5 or Equation 33.6, and $y_0 = 0.01$, $r = 0.48$, calculate y_t values for $t = 0, 1, 2, ..., 20$.
2	Graph the data, calculated in Step 1, on graph paper. Connect data point to create Line 1.
3	Repeat Step 1 and Step 2, using $y_0 = 0.01$ and $r = 0.24$ for Line 2.
4	Repeat Steps 1 and 2, using $y_0 = 0.005$ and $r = 0.48$ for Line 3.
5	Compare plotted lines.

Anticipated Results

Students will gain an appreciation for differential disease levels, and for the "art" of assessing disease severity. Ideally, using *in vivo* materials, students can see how disease severity might develop over time. If used, copies of disease diagrams can be organized by the instructor to simulate disease development or to compare student improvement with practice.

Questions

- What levels of disease were the most difficult to rate?
- Were there complicating factors affecting the symptoms that were assessed?
- What was taken into consideration when deciding a value to assign to each sample?

EXERCISE 2. DISEASE PROGRESS

In trying to understand plant disease epidemics, mathematical models have been used as a means of simplification. These mathematical models provide a quantitative way to represent the sometimes complicated processes that comprise an epidemic. The logistic model is often used to depict plant disease development:

$$y_t = 1/(1 + e^{-[(\ln y_0/(1-y_0))+rt]})$$

In this model, y_t is the amount of affected tissue after t time intervals, given a starting disease level y_0 and a rate parameter (r), where e is the mathematical constant = 2.718.

Materials

Each student or team of students will require the following items:

- Calculator
- Graph paper

Follow the experimental provided in Procedure 33.2 to complete this part of the exercise.

Anticipated Results

Students will generate three disease progress curves from the models. From the plots of these progress curves, effects of changes in initial inoculum or rate of infection, or both, can be seen.

Questions

- How did the plotted lines differ?
- What plant disease management method might have been taken to reduce y_0, as shown in differences between Lines 1 and 3?
- What approach might be taken to reduce r, as demonstrated by Lines 1 and 2?
- What management strategies might be more effective to reduce the theoretical disease that is modeled?

LITERATURE CITED

Adams, P.B. 1981. Forecasting onion white rot disease. *Phytopathology* 71: 1178–1181.

Andrivon, D. 1996. The origin of *Phytophthora infestans* populations present in Europe in the 1840s: a critical review of historical and scientific evidence. *Plant Pathol.* 45: 1027–1035.

Backman, P.A., R. Rodriguez-Kabana, M.C. Caulin, E. Beltramini, and N. Ziliani. 1981. Using the soil-tray technique to predict the incidence of *Sclerotium* rot in sugarbeets. *Plant Dis.* 65: 419–421.

Bowen, K.L., A.K. Hagan, and R. Weeks. 1992. Seven years of *Sclerotium rolfsii* in peanut fields: yield losses and means of minimization. *Plant Dis.* 76: 982–985.

Burleigh, J.R., M.G. Eversmeyer, and A.E. Roelfs. 1972. Linear equations for predicting wheat leaf rust. *Phytopathology* 62: 947–953.

Campbell, C.L., and L.V. Madden. 1990. *Introduction to Plant Disease Epidemiology.* John Wiley and Sons, New York.

Davidson, J.M., S. Werres, M. Garbelotto, E.M. Hansen, and D. M. Rizzo. 2003. Sudden oak death and associated diseases caused by *Phytophthora ramorum*. *Plant Health Progress* doi:10.1094/PHP-2003-0707-01-DG.

Hanway, J.J. 1963. Growth stages of corn (*Zea mays* L.). *Agron. J.* 55: 487–492.

James, C. 1971. A manual of assessment for plant diseases. *Am. Phytopathol. Soc.*, St. Paul, MN.

McMullen, M., R. Jones, and D. Gallenberg. 1997. Scab of wheat and barley: A re-emerging disease of devastating impact. *Plant Dis.* 81: 1340–1348.

Moore, W.F. 1970. Origin and spread of southern corn leaf blight in 1970. *Plant Dis. Rept.* 54: 1104–1108.

Nelson, M.R., T.V. Orum, and R. Jaime-Garcia. 1999. Applications of geographic information systems and geostatistics in plant disease epidemiology and management. *Plant Dis.* 83: 308–319.

Pan, Z, X. B. Yang, S. Pivonia, L. Xue, R. Pasken, and J. Roads. 2006. Long-term prediction of soybean rust entry into the continental United States. *Plant Dis.* 90: 840–846.

Pataky, J.K., J.A. Hawk, T. Weldekidan, and P. Fallah Moghaddam. 1995. Incidence and severity of Stewart's bacterial wilt on sequential plants of resistant and susceptible sweet corn hybrids. *Plant Dis.* 79: 1202–1207.

Roelfs, A.P. 1985. Wheat and Rye Stem Rust. In Roelfs, A.P., and Bushnell, W.R. (Eds.). *The Cereal Rusts*, Vol. II. Academic Press, New York, pp. 3–37.

Schneider, R. W., C.A. Hollier, and H.K. Whitam, J. M. McKemy, and L. Levy, and R. DeVries-Paterson. 2005. First report of soybean rust caused by *Phakopsora pachyrhizi* in the continental United States. *Plant Dis.* 89: 774.

Tubajika, K.M., E.L. Civerolo, M.A. Ciomperlik, D.A. Luvisi, and J.M. Hashim. 2004. Analysis of the spatial patterns of Pierce's disease incidence in the lower San Joaquin Valley in California. *Phytopathology* 94: 1136–1144.

USDA. 2005. U.S. Department of Agriculture and Florida Department of Agriculture confirm detection of citrus greening. Dept. Press Release. On-line publication.

Vanderplank, J.E. 1963. *Plant Diseases: Epidemics and Control.* Academic Press, New York.

34 Host Resistance

Jerald K. Pataky and Martin L. Carson

CHAPTER 34 CONCEPTS

- A resistant phenotype occurs when growth, reproduction, and/or disease-producing activities of the pathogen are reduced, and fewer symptoms of disease are observed.

- Because plant disease results from the interaction of a complex series of events and pathways in both the host and pathogen, it is easy to envision multiple ways the process can be interrupted or altered, giving rise to a resistant reaction.

- Among a plant species, there often is a range of responses to pathogens that varies continuously from a small amount of disease (resistant) to a substantial amount of disease (susceptible). Discontinuous categories of resistant reactions also occur.

- Populations of pathogens vary genetically both for their ability to infect hosts with specific resistance genes (virulence), and for the degree of damage they cause (aggressiveness).

- Host resistance places selection pressure on pathogen populations for genotypes with virulence against that resistance (i.e., races).

- Although all isolates within a race have common virulence, they may be genetically diverse for other traits.

- To develop a disease-resistant cultivar, sources of resistance must be identified, and resistance must be combined in cultivars that have desirable traits.

- The greatest genetic diversity for resistance usually occurs where the host and pathogen have co-evolved, which often is the center of origin of host species.

- The benefit of genetic uniformity for most traits is considerably greater than the risk of potentially severe epidemics; however, genetic uniformity for disease resistance carries a substantial risk even though it may be greatly beneficial when it works.

- Breeders and pathologists have tried to cope with variable pathogens by constantly searching for new sources of resistance and by selecting combinations of resistance genes that provide resistance to the spectrum of races present in a growing region.

Planting resistant cultivars is one of the most efficient and effective methods of disease control. Host resistance eliminates or minimizes losses due to diseases and reduces the need for and cost of other controls. Resistance is compatible with other methods of disease and pest control and it can be integrated easily in pest management programs (Chapter 38).

Disease resistance is one of the most important factors contributing to the long-term stability of crop production. For example, wheat stem rust (*Puccinia graminis* f. sp. *tritici*) periodically devastated the U.S. wheat crop during the first half of the 20th century, but there has not been a major epidemic since the early 1950s largely due to the use of durable combinations of resistance genes in modern wheat varieties. Resistant varieties produce higher and more consistent yields and have improved the quality of the world's major food crops. Cultivars that are resistant to many diseases are among the major achievements of plant breeders and pathologists in the past century. Resistance will continue to play an integral role in the success of agriculture well into the 21st century.

TYPES OF RESISTANCE

Resistance encompasses a wide variety of host-pathogen interactions. Resistant reactions vary in both degree and kind. A resistant phenotype reduces the growth, reproduction, and/or disease-producing activities of the pathogen.

Disease symptoms are less severe on resistant hosts than on susceptible hosts. Susceptible phenotypes are unable to restrict the growth, reproduction, and/or disease-producing activities of the pathogen, and symptoms are severe. Among a plant species, reactions to a specific disease often vary continuously from highly resistant to highly susceptible much like height of people varies continuously from tall to short. When variation is continuous, host reactions differ only in degree of symptom severity and resistance and **susceptibility** represent the two extremes. In other cases, host reactions can be placed into discrete categories as a result of hypersensitive reactions (Chapters 28 and 31) or resistant reactions that result in distinctly different kinds of phenotypes, much like humans may be either "normal" height or genetically dwarf.

Placing host reactions into broad categories (e.g., highly resistant, resistant, moderately resistant, moderately susceptible, susceptible) can be very useful because this type of classification gives an indication of how a cultivar will respond when grown in an area that is favorable for disease development. Nevertheless, categorizing host reactions can be a source of confusion and disagreement, especially when trying to separate continuous reactions. This problem is somewhat analogous to trying to classify height among people. For example, some people would say that a 6 ft man is tall, whereas others (maybe a basketball coach) would consider 6 feet to be average height or even short. In this example, the relative height of a 6 ft man depends on the population that is being sampled. Compared to professional basketball players, 6 ft is relatively short, but compared to all men in the United States, 6 ft is above average height. The same problem exists when trying to compare host reactions to disease. For example, if the reaction of a corn hybrid to Stewart's bacterial wilt is given a score of 4 on a 1 to 9 scale (i.e., *Pantoea stewartii* spreads nonsystemically more than 30 cm from the site of infection), the hybrid may be classified differently among three different groups of corn germplasm (Pataky et al., 2000). Among all dent corn hybrids that are grown widely in the corn belt of the United States, a hybrid with a Stewart's wilt rating of 4 is worse than average (e.g., "moderately susceptible"). Among a representative sample of corn lines collected from throughout the world, a variety with a Stewart's wilt rating of 4 is close to average (e.g., "moderate"). Among early-maturing sweet corn, a hybrid with a Stewart's wilt rating of 4 is better than average (e.g., "moderately resistant"). Thus, resistance is relative to the population that is being sampled.

The arbitrary nature of boundaries is another problem that occurs when trying to classify levels of resistance among populations with continuous distributions of disease reactions. For example, if cultivars with <5% disease severity are classified as resistant and those with 5 to 15% disease severity are classified as moderately resistant, a cultivar with 7% leaf area infected would be classified as moderately resistant even though it is more similar to a resistant cultivar than it is to a moderately resistant cultivar with 14% severity.

Plant breeders and pathologists use different approaches and statistical procedures to deal with these types of classification problems and consequently, disease reactions are reported in a variety of ways. Numerical scales (such as those from 1 to 9 that correspond to resistant to susceptible) and ordinal classes (such as HR—highly resistant, R—resistant, MR—moderately resistant, M—moderate, MS—moderately susceptible, and S—susceptible) are just a few of the many approaches used to describe host reactions to diseases. Although a uniform method of reporting host reactions does not exist, resistance ratings should correspond to different degrees of protection from crop damage.

The expression of resistance genes requires that plants interact with the pathogen in an environment that is suitable for disease development. In order to separate host reactions to diseases, host lines must be compared under equivalent conditions. For many diseases, reliable inoculation methods can be used to screen for resistance. Inoculation helps ensure that all lines being evaluated are treated similarly. For some diseases, host reactions are evaluated in nurseries that are infected naturally. If inoculation procedures or disease nurseries are not uniform, extra care must be taken to ensure that all lines in an experiment are subjected to the same "disease pressure." For example, when disease nurseries are used to evaluate reactions to soilborne diseases, differences in inoculum density frequently occur as a result of the clustered spatial arrangement of soilborne pathogens. In such cases, moderately susceptible lines evaluated at low inoculum densities can have phenotypes (e.g., amount of disease symptoms) similar to moderately resistant lines evaluated at high inoculum densities (Figure 34.1). Separation of reactions also can be affected by environmental conditions. For example, a group of resistant maize lines that could not be separated for reactions to northern corn leaf blight (NCLB) during a dry season in Wooster, OH, displayed a range of reactions with intermediate classes when evaluated at two locations in Uganda where the environment was more conducive to the development of NCLB (Table 34.1). The aggressiveness of pathogen isolates or strains used to inoculate plants in a nursery also can affect the ability to separate disease reactions. In general, more aggressive isolates separate cultivar reactions more effectively than less aggressive isolates. In order to ensure that the results of disease nurseries are accurate, a set of standard cultivars with known reactions to diseases usually are included.

ADJECTIVES THAT DESCRIBE RESISTANCE

Many adjectives are used to describe various types of host resistance. Frequently, pairs or groups of adjectives are used for specific characteristics. Misuse or haphaz-

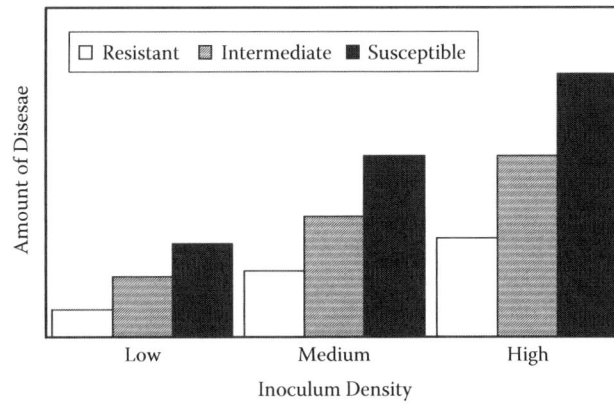

FIGURE 34.1 Amount of disease on cultivars with resistant, intermediate or susceptible reactions when evaluated in field plots with low, medium, or high inoculum density. The phenotype (e.g., amount of disease) may be similar when a resistant cultivar evaluated at a high inoculum density is compared to a susceptible cultivar evaluated at a low inoculum density.

ard use of some adjectives has resulted in their meaning becoming confounded or confused. Confusion occurs sometimes because different people have different meanings for terms. Confusion also occurs when an adjective used to describe one characteristic is thought to imply something about another characteristic. Although these implied associations frequently are correct, there are several examples of resistance where the implications are not correct. For example, monogenic is an adjective used to describe resistance that is inherited as a single gene. Frequently, monogenic resistance is "race-specific" (i.e., virulent and avirulent races of the pathogen occur), but there are several examples of monogenic resistance for which virulent races of a pathogen are not known (i.e., nonrace specific). Many of the most common character-

istics of resistance and their corresponding adjectives are discussed below.

Because resistance is a phenotype, there is no reason that a resistant phenotype cannot be conditioned genetically by any mechanism that conditions the expression of any other plant trait. **Monogenic**, **oligogenic**, and **polygenic** are used to describe the number of nuclear genes involved in the inheritance of resistance. Monogenic resistance is conditioned by a single gene, and is sometimes called **single-gene** resistance. Oligogenic is used to describe resistance that is conditioned by a few genes. If resistance is inherited in a monogenic or an oligogenic manner, resistant and susceptible phenotypes occur in genetically segregating populations; and these phenotypes can be placed in discrete categories that can be fit to Men-

TABLE 34.1
Severity[a] of Northern Corn Leaf Blight (NCLB) on Maize Lines Evaluated in Uganda and Ohio

	Location		
Maize Line	Kabanyolo, Uganda	Nmuloye, Uganda	Wooster, Ohio
B73	75	59	28
Mo17	40	20	1
EV8428-SR	21	20	3
EV8429-SR	16	17	2
Gysau TZB-SR	14	16	1
Jos	8	11	1
KWCA-SR	6	7	1
EV8349-SR	5	10	1
Population 42	2	4	0
Babungo 3	1	2	0
LSD (0.05)	6.8	6.1	2.1

Note: Resistant lines were differentiated more easily in Uganda where the environment was extremely conducive for development of NCLB.

[a] Percent leaf area infected with NCLB (Adipala et al., 1993).

delian ratios. Polygenic refers to resistance that is inherited from several genes. Although the inheritance of polygenic resistance is more complex than oligogenic or monogenic resistance, polygenic resistance usually involves fewer genes than other traits that are inherited polygenically, such as yield. Most experimental estimates of the numbers of loci controlling polygenic resistance are less than ten and commonly only two to four. In populations segregating for polygenic resistance, disease reactions exhibit continuous variation and cannot be classified into discrete Mendelian classes. In a few instances, resistance or susceptibility is conditioned by cytoplasmic genes. Because cytoplasmic genes are transmitted from the female parent of a cross, these traits are said to be inherited maternally.

Resistance is dominant (R) if heterozygotes (Rr) express the resistant reaction completely. Resistance is recessive (r) if heterozygotes (Rr) have a completely susceptible reaction. If there is a complete lack of dominance (i.e., additive gene action), heterozygotes have a reaction that is intermediate between the resistant and susceptible parents. Partial dominance occurs when heterozygotes have a reaction that is more resistant than the mid-parent value between the resistant and susceptible parents.

Qualitative and **quantitative** are adjectives used to differentiate kinds of disease resistance or to describe expression of traits. Qualitative resistance refers to host reactions that can be placed in distinct categories that form the basis of Mendelian ratios. One or a few genes condition qualitative traits, hence qualitative resistance is monogenic or oligogenic. Quantitative resistance refers to host reactions that have no distinct classes because reactions vary continuously from resistant to susceptible. A few or many genes are involved in the expression of a quantitative trait and the effects of individual genes usually are small and hard to detect. The difference between qualitative and quantitative resistance is obvious if we compare progeny in the F_1 and F_2 populations from a cross of a resistant and a susceptible inbred parent. In this example, we assume a phenotype with 10% disease severity for the resistant parent and 70% disease severity for the susceptible parent; gene expression is completely dominant for qualitative resistance and additive for quantitative resistance, and one gene conditions qualitative resistance while three independent genes are involved in quantitative resistance. For qualitative resistance, progeny in the F_1 generation have the same value as the resistant parent due to dominance. In the F_2, there is a 3:1 ratio of resistant and susceptible progeny (Figure 34.2).

For quantitative resistance, reactions of F_1 progeny are intermediate (40%) between the resistant and susceptible parents due to the lack of dominance, but progeny in the F_2 display a continuous range of phenotypes between the reactions of the resistant and susceptible parents (Figure 34.3; Table 34.2). If environmental conditions create slight variation in reactions, the phenotypes in the

F_2 population cannot be separated into distinct classes. Thus, a continuous range of host reactions that appears to be associated with quantitative resistance could be conditioned by as few as three genes with additive gene action. Other factors, such as interactions among genes and partial dominance, may affect the expression of resistant or susceptible reactions. Resistance can be expressed and controlled genetically in as many ways as any other phenotype.

Resistance frequently is divided into two types, **general** and **specific**, to differentiate host-pathogen interactions. General resistance is effective against all biotypes of the pathogen and is sometimes called race nonspecific resistance. Durable resistance is defined as resistance that remains effective during its prolonged and widespread use in an environment favorable to the disease. General resistance is sometimes called durable resistance because changes in frequency of virulence genes in the pathogen population do not affect this resistance. Specific or race-specific resistance is effective against some races of a pathogen (i.e., avirulent races), but ineffective against virulent races. The adjectives vertical and horizontal also are used to describe race-specific and race-nonspecific resistance. Vertical and horizontal refer to the appearance of the histograms that depict reaction of hosts to different races of a pathogen (Figure 34.4). Vertical and horizontal also have additional meanings. Vertical resistance is considered to be inherited monogenically, qualitatively-expressed, and effective against initial inoculum of a polycyclic disease. Horizontal resistance is thought to be polygenic in inheritance, quantitative, and effective at reducing the rate at which polycyclic diseases develop. Since all race-specific resistance does not share the other characteristics of vertical resistance, and all race nonspecific resistance does not share the other characteristics of horizontal resistance, the terms vertical and horizontal should not be used synonymously for race-specific and race-nonspecific even though they frequently are used this way. Similarly, all qualitative and quantitative resistance does not share all of the characteristics of vertical and horizontal resistance so these adjectives also are not synonymous. Caution should be exercised in describing resistance as vertical or horizontal because of the confounded meanings of these adjectives.

Several other adjectives are used to describe specific aspects of certain types of resistance. Complete resistance describes reactions in which some aspect of disease development (usually symptom development or pathogen reproduction) is stopped entirely. In some instances, complete resistance is called immunity, even though plants do not have immune systems. Some types of complete resistance are due to **hypersensitive responses** (Chapters 28 and 31) in which rapid biochemical and histological changes elicited by the pathogen result in necrosis or death of the invaded host cells thereby localizing infection. As

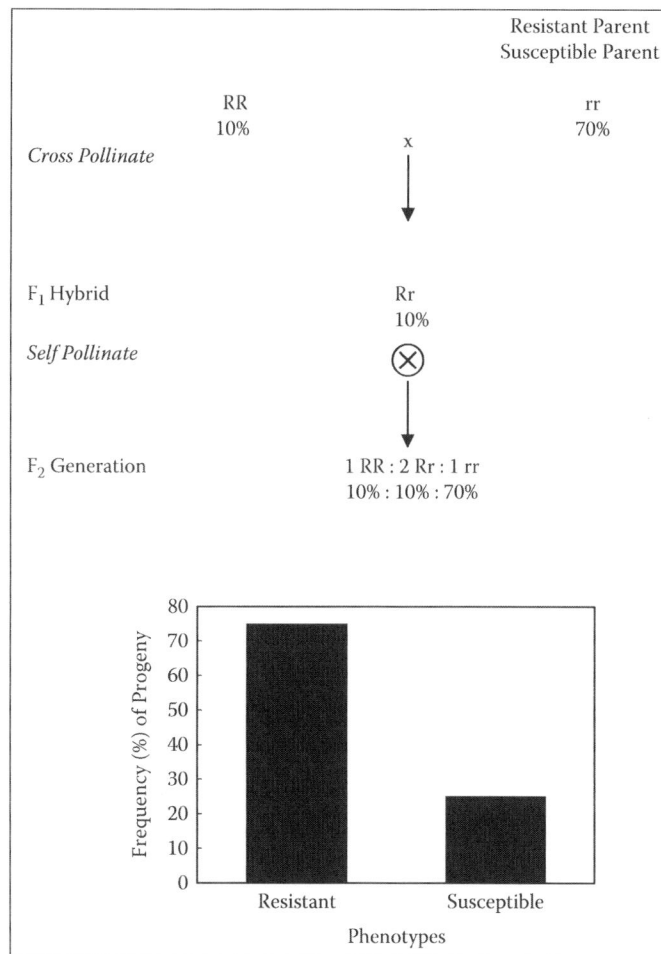

FIGURE 34.2 Qualitative resistance. Progeny in F_1 and F_2 populations of a cross of disease resistant and susceptible parents when resistance is inherited as a completely dominant, single gene. The resistant parent has a phenotype with 10% disease severity and the susceptible parent has 70% severity. The F_2 segregates 3:1 for resistant and susceptible reactions.

single, dominant genes frequently condition complete resistance or hypersensitive reactions, the gene responsible for the reaction can be used as the adjective to describe this type of resistance. For example, resistance in soybean to Phytophthora root rot is conditioned by several different single genes designated *Rps* (resistant to *Phytophthora sojae*). Phytophthora resistance in some soybean cultivars is identified by the specific resistance allele, e.g., "1k" refers to resistance conveyed by the *k* allele at the *Rp1* locus; similarly, "1c" refers to resistance conveyed by the *c* allele at the *Rp1* locus. Several variations of this system are used to identify simply-inherited resistance. For example, in corn, Rp resistance refers to resistance to the common rust fungus, *Puccinia sorghi*. Rp resistance is conveyed by one of nearly 25 different single *Rp* genes. Resistance to the northern corn leaf blight fungus, *Exserohilum turcicum* (syn. *Helminthosporim turcicum*), is conveyed by one of at least four different *Ht* genes ("Ht" = resistance to *H. turcicum*), therefore, this resistance often is referred to as Ht resistance. A variation of using host genes to describe

resistance is to identify resistance based on races of the pathogen that are controlled by the resistance. For example, resistance to Fusarium wilt in tomato is designated by its effectiveness against race 1 or race 2.

The term **partial** resistance is used by some pathologists to describe resistant reactions that are not complete. Some disease symptoms or pathogen reproduction occurs on partially resistant plants. Partial resistance does not imply anything about the genetics, inheritance, or race-specificity of the resistance. Partial resistance is similar to quantitative resistance except that there is no direct implication that the resistance is inherited as a quantitative trait. Field resistance is sometimes used to describe host lines that have less disease than other lines under field conditions, but may or may not exhibit resistance when grown in a greenhouse or growth chamber. Slow-rusting or rate-reducing are used to describe resistance in which rust or other diseases develop at a slower rate on resistant lines than on susceptible lines. Various components of the infection cycle (e.g., reduced infection

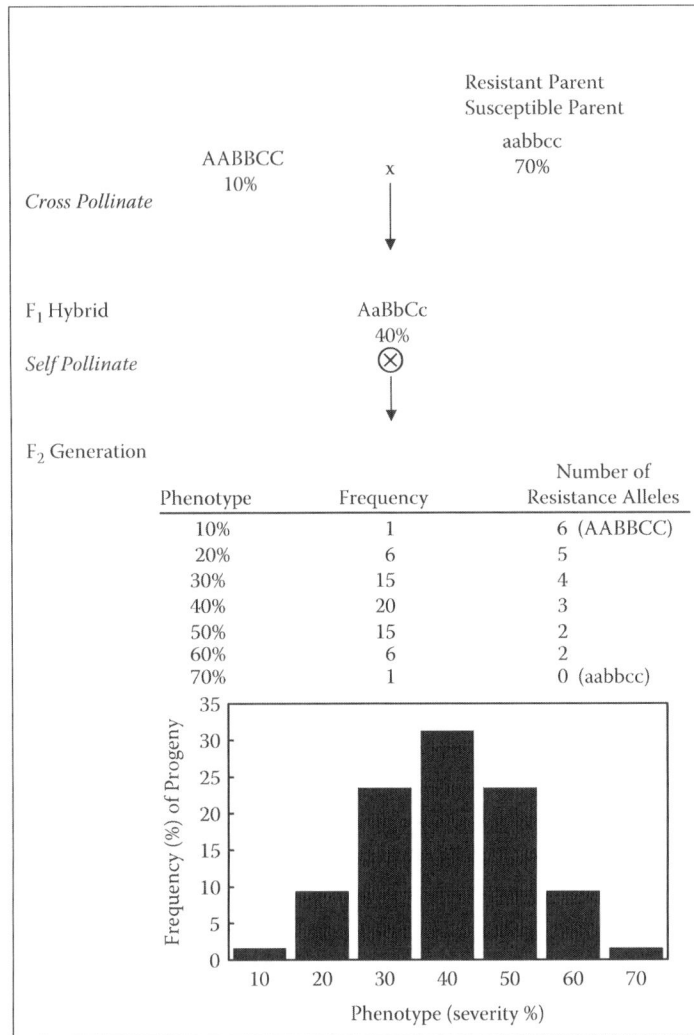

FIGURE 34.3 Quantitative resistance. Progeny in F_1 and F_2 populations of a cross of disease resistant and susceptible parents when resistance is inherited from three genes with additive gene action. The resistant parent has a phenotype with 10% disease severity and the susceptible parent has 70% severity. Each allele for resistance decreases severity by 10%. See Table 34.2 for genotypes in each phenotypic class.

frequency, longer incubation, and latent periods, shorter infectious period, and reduced lesion numbers, lesion size, or sporulation) have been shown to be associated with rate-reducing resistance. Adult-plant resistance refers to resistant reactions that do not occur in seedlings. In some cases, there are documented biological explanations for differences in host responses at different growth stages. For example, juvenile corn leaves are more susceptible to infection by the common rust fungus, *P. sorghi,* than adult leaves. As leaves of a corn plant change from juvenile to adult, at about the fourth-to-seventh internode, leaves differ in various traits such as types and amounts of cuticular waxes, thickness of cuticle, cell shape, and presence of trichomes. Severity of rust infection probably is affected by these morphological differences. All corn displays some adult-plant resistance to common rust regardless of the level of general resistance or susceptibil-

ity. In other cases, differences in seedling and adult-plant reactions to disease are the result of disease development. As adult plants are exposed to several cycles of infection by polycyclic pathogens, slight differences in partial resistance that might be difficult to detect from a single cycle of infection of seedlings are sometimes quite apparent among adult plants.

The diversity in types of resistance and the adjectives used to describe these characteristics can be confusing even for plant pathologists and breeders who work on disease resistance. Nevertheless, it is important to have a general understanding of the adjectives used to describe types of resistance in order to know what kind of resistance you are dealing with. More importantly, recognize that all resistance is not alike even for a single pathogen of one host species. Many different types of resistace reactions may occur within a host species because resistant phenotypes

TABLE 34.2

Phenotypes Frequency of Genotypes, and Number of Resistance Alleles in an F$_2$ Population Segregating for Disease Resistance at Three Loci with Strictly Additive Gene Action

F2 Phenotype[a]	Genotype(s) and Frequency	Number of Resistance Alleles
10	*AABBCC (1/64)*	6
20	*AABBCc, AABbCC , AaBBCC (6/64)*	5
30	*AABBcc, AABbCc, AAbbCC, AaBBCc, AaBbCC, aaBBCC (15/64)*	4
40	*AABbcc, AAbbCc, AaBBcc, AaBbCc, AabbCC, aaBBCc, aaBbCC (20/64)*	3
50	*AAbbcc, AaBbcc, AabbCc, aaBBcc, aaBbCc, aabbCC (15/64)*	2
60	*Aabbcc, aaBbcc, aabbCc (6/64)*	1
70	*aabbcc (1/64)*	0

Note: Distribution of phenotypes presented in Figure 34.3.

[a] Based on percentage disease severity.

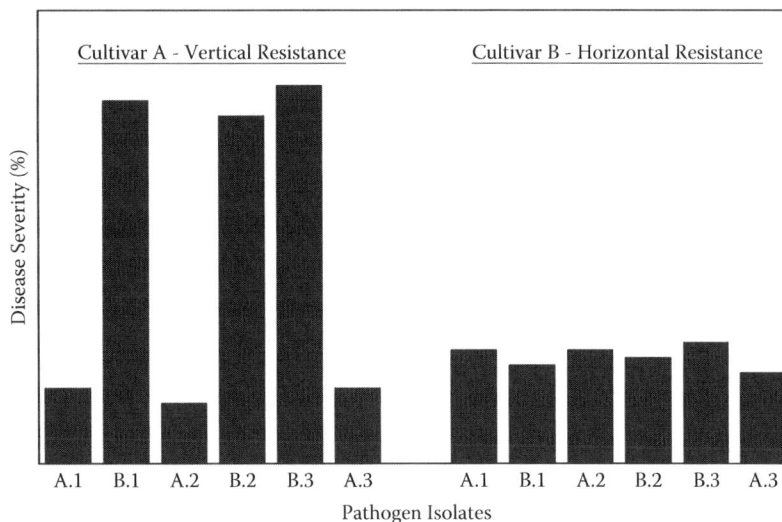

FIGURE 34.4 Vertical and horizontal resistance. Race-specific resistance (cultivar A) is depicted by "vertical" differences in the height of histograms that illustrate disease reactions in response to different pathogen isolates. Race-non-specific resistance (cultivar B) is depicted by a "horizontal" decrease in the height of all histograms that illustrate disease reactions in response to different pathogen isolates.

may arise in many ways. Because plant disease results from the interaction of a complex series of events and pathways in both the host and pathogen, it is easy to envision multiple ways the process can be interrupted or altered.

PATHOGEN VARIATION

It would be neglectful to discuss host resistance without briefly mentioning genetic variation among pathogens. Just as hosts vary genetically for resistant and susceptible reactions to diseases, pathogens display a wide array of genetic variability. Terminology associated with pathogen variation and ability to infect hosts can be confusing. Shaner et al. (1992) presented an overview of uses of various terms in a review of the nomenclature and concepts of vir-

ulence and pathogenicity. For our purposes, we will focus on the following three characteristics of pathogens: (1) the ability to infect a host species, (2) host-pathogen interactions, and (3) the relative amount of damage on a host.

Pathogenicity is the ability of the pathogen to cause disease. For example, *Ustilago maydis* is pathogenic on corn because it causes the disease known as common smut (Chapter 18). This fungus in nonpathogenic on oak trees because the host–pathogen interaction is incompatible and disease does not develop. Another way of thinking about pathogenicity is that corn is a host for *U. maydis* and oak is a nonhost.

Virulence is the ability of a pathogen to cause a compatible (susceptible) reaction on a host cultivar with genetic

TABLE 34.3

Race Identification of Isolates of *Exserohilum turcicum* Based on Reactions of a Set of Differential Host Lines Carrying Ht Resistance Genes

	Ht Resistance Genes				Virulence Formula Effective/Ineffective	
Old Race[a]	None	Ht1	Ht2	Ht3	HtN	Genes	New Race
1	S	R	R	R	R	123N/0	0
2	S	S	R	R	R	23N/1	1
3	S	R	S	S	R	1N/23	23
4	S	R	S	S	S	1/23N	23N

Note: Old race designation based on chronological order in which isolates were identified. New race designation based on ineffective host genes.

[a] Sixteen races could exist based on all possible combinations of resistant and susceptible reactions on four independent host resistance genes.

resistance. **Avirulence** is the inability of a pathogen to cause a compatible (susceptible) reaction on a host cultivar with genetic resistance. In order for a pathogen to be virulent or avirulent, it must first be pathogenic on a host species. Nonpathogens cannot be virulent or avirulent. A **race** of a pathogen is a population of isolates that have the same virulence. Although all isolates within a race have common virulence, they may be genetically diverse for other traits.

Races of pathogens are typically identified by their virulence on a set of differential varieties with known resistance genes. For example, races of *E. turcicum,* which causes northern corn leaf blight of maize, are determined by their reaction on a differential set of inbred lines that carry Ht resistance genes (Table 34.3). Each inbred line contains one of the four different known Ht genes for resistance.

Just as height varies in a population of people, populations of virulent pathogens may have continuous variation in their ability to cause different amounts of disease on a specific host. **Aggressiveness** is the amount of disease caused by an isolate of the pathogen. Aggressiveness is sometimes used synonymously with parasitic fitness, as the amount of reproduction of the pathogen on the host usually is directly proportional to the amount of disease. The same components of the infection cycle that may be affected by partial resistance in the host may also be affected by the aggressiveness of a pathogen strain. For example, when compared to weakly aggressive isolates, extremely aggressive isolates may have increased infection frequency, shorter incubation and latent periods, a longer infectious period, and increased lesion numbers, lesion size, or sporulation.

Using these definitions, virulence and aggressiveness in the pathogen are complementary to race-specific resistance and general resistance in the host. Therefore, some people prefer the terms **specific virulence** and **general virulence** in place of virulence and aggressiveness.

Although a wide range of host-pathogen interactions occur, "resistance" is used most frequently to refer to host–pathogen interactions that involve specific resistance and virulence. In this case, resistant reactions are expressed as qualitative traits that occur when hosts with a specific resistance gene are attacked by isolates of the pathogen that have the corresponding avirulence gene. Susceptible reactions occur when either the host lacks the resistance gene or the pathogen lacks the corresponding gene for avirulence. We now know that resistance genes and avirulence genes are functional genes in the host and pathogen, respectively. Susceptibility and virulence arise when these genes are either absent or rendered nonfunctional by mutations. These interactions have been described as the "quadratic check" that illustrates four possible combinations of host–pathogen interactions and two races of the pathogen (Figure 34.5). The gene-for-gene concept refers to genes for avirulence in the pathogen that correspond to genes for specific resistance in the host. If the host has more than one gene for specific resistance, the number of potential races of the pathogen increases exponentially, i.e., the number of races = 2^N, where N = number of host resistance genes. If the host has two different independent genes for specific resistance, there is a potential for four pathogen races. Three resistance genes could result in eight races, and so on.

	Host	
Pathogen	R_	rr
Avirulent	**R**	**S**	Race 0
Virulent	**S**	**S**	Race 1

FIGURE 34.5 Quadratic check. Four possible combinations of host-pathogen interactions among an avirulent and a virulent race of a pathogen and hosts with and without a dominant, race-specific resistance gene (R).

TABLE 34.4

Indicator Lines (i.e., Host Differentials) Used to Identify Hg Types for Populations of Soybean Cyst Nematode

HG Type	Indicator Line (Host Differential)
1	PI 54840 (Peking)
2	PI 88788
3	PI 90763
4	PI 437654
5	PI 209332
6	PI 89772
7	PI 548316 (Cloud)

When several genes or sources of resistance occur, the number of pathogen races can be extremely large. To avoid confusion, races of some pathogens are designated by the ineffective host genes. The **virulence formula** for an isolate of a pathogen lists the effective host genes and ineffective host genes. For example, in corn there are at least four single dominant genes (Ht_1, Ht_2, Ht_3, and Ht_N) that independently convey resistance against the northern corn leaf blight fungus, *E. turcicum*. As four different host resistance genes exist, there are 16 potential races of the pathogen (i.e., $2^4 = 16$). The old racial designations (races 1, 2, 3, 4, etc.) have been replaced by racial designations based on ineffective host genes from the virulence formula (Table 34.3). Old race 1 is now race 0 because these isolates do not have any specific virulence (i.e., the virulence formula is 123N/0 = effective host genes/ineffective host genes). Old race 3 is now race 23 because these isolates are virulent against the resistance genes Ht_2 and Ht_3 (i.e., the virulence formula is 1N/23).

A similar system has been proposed to describe populations of soybean cyst nematode, SCN (*Heterodera glycines*) (see Chapter 8). HG-types are used to identify populations of SCN, based on their ability of efficiently reproduce on certain host differentials (Table 34.4). If reproduction of a population of SCN on any one of seven soybean indicator lines (i.e., a set of host differentials) is at least 10% of the reproduction on the standard susceptible, Lee 74, then that indicator line is included in the HG-type for that population of SCN just like ineffective host genes from a virulence formula are included in the racial designation of a fungal isolate. For example, a SCN population for which reproduction on PI 88788 is at least than 10% of that on Lee 74 is given an HG type of 2 (Table 34.5). A population for which reproduction on PI 88788 and PI 90763 are greater than 10% of that on Lee 74 is given an HG type of 23; a population for which reproduction on PI 88788, PI 90763, and PI 209332 are greater than 10% of that on Lee 74 is given an HG type of 235; and so on (Table 34.5).

Quantitative or partial resistance usually appears to be race-nonspecific, so virulence and avirulence are not applicable; however, a few researchers have reported small, race-specific effects of isolates against cultivars with quantitative resistance. For example, isolates of barley rust that were allowed to reproduce for several generations on a host cultivar with partial resistance caused more disease on the partially resistant cultivar than on other cultivars (Clifford and Clothier, 1974). In some cases, both specific virulence and general virulence (aggressiveness) were increased for the rust isolates that reproduced on the partially resistant cultivar.

From this brief overview of pathogen variation, it should be obvious that the amount of variation in the pathogen for disease-producing traits is similar to variation in the host for resistant or susceptible reactions. Consequently, plant pathologists and breeders must continually

TABLE 34.5

Examples of HG Types: Five Populations of SCN and Their Corresponding HG Type Based on an Index of Reproduction above 10 (Reproduction above 10% of that on Lee 74)

		Index of Reproduction						
HG Type[a]	Lee 74	(1) PI 54840	(2) PI 88788	(3) PI 90763	(4) PI 437654	(5) PI 209332	(6) PI 89772	(7) PI 548316
0	100	0	4	1	0	0	1	1
2	100	1	17	3	0	1	7	3
3	100	2	6	19	0	0	0	1
23	100	2	42	35	0	5	3	2
235	100	1	45	31	0	14	3	3

[a] HG types designate indicator lines on which the reproduction of a population is at least 10% of that on Lee 74.

identify new sources of resistance and incorporate new resistance into cultivars to compete with constant changes and adaptation in the genetic composition of pathogen populations.

BREEDING FOR DISEASE RESISTANCE

Two steps are required to develop a disease-resistant cultivar. Sources of resistance must be identified, and resistance must be combined in cultivars that have desirable traits necessary for production.

Sources of disease resistance. When a new disease or a new race of a common pathogen occurs, when an old disease increases in prevalence, or when the disease reaction of a particular variety needs to be improved, plant breeders and pathologists look for sources of resistance. Resistance often occurs in germplasm that has been maintained for a long period of time in an area where the host and pathogen have co-evolved. Frequently, this is near the center of origin of the host. For example, resistance to many diseases of peanut are found in germplasm that originated from South America where peanut is native (Wynne and Beute, 1991). On the other hand, resistance sometimes is found in germplasm that was introduced and selected in an area where the pathogen is endemic. For example, the best sources of resistance in maize to Stewart's bacterial wilt are found in germplasm originating from the Ohio River Valley of the United States (Pataky et al., 2000). Germplasm from Central America, the center of origin of maize, is not highly resistant to Stewart's wilt probably because the disease rarely occurs there.

Breeders and pathologists use whatever genetic variation for resistance is available, but resistance can be incorporated more easily from sources that are closely related to adapted, elite cultivars than from unrelated sources. Hence, there is a relatively straight-forward strategy of where to begin looking for resistance.

Adapted cultivars. Resistance found in adapted cultivars is extremely useful because the cultivar already possess other traits necessary for production. In some cases, an existing resistant cultivar may simply replace a susceptible cultivar in commercial production. More important for resistance breeding, an adapted cultivar probably does not have excessively deleterious traits that must be improved when incorporating resistance into new, adapted cultivars. Hence, most types of resistance that are found in adapted cultivars can be incorporated into new cultivars with relative ease. This is particularly useful when trying to improve levels of quantitative resistance in elite cultivars.

- Breeders' stock—Breeders stock are lines that are relatively well-adapted, but have not been released as cultivars because they are deficient in one or a few important traits. Resistance that is found in breeders' stock is used in breeding

programs nearly as easily as resistance found in adapted cultivars.

- Cultivated germplasm—Accessions maintained at plant introduction stations and other collections of germplasm usually represent the greatest amount of easily accessible genetic diversity within a species. Many sources of resistance that are in use today were initially discovered among these collections. For example, soybean cyst nematode resistance originally incorporated into many commercially available soybean cultivars was derived from one of three sources of cultivated germplasm: PI88788, PI90763, or the cultivars Pickett and Peking. Subsequently, other sources of resistance were found among cultivated germplasm (Table 34.4). The ease with which resistance from germplasm is incorporated into adapted cultivars depends on whether inheritance of resistance is simple or complex and whether or not the source of resistance is relatively well adapted. Simply-inherited resistance (e.g., monogenic resistance) can be incorporated from relatively unadapted germplasm, but resistance that is more complex in inheritance (e.g., quantitative resistance) is more difficult to acquire from exotic or unadapted germplasm. The use of molecular markers may improve our ability to incorporate quantitative resistance into adapted germplasm from exotic sources.

- Wild or related species—It is much more difficult to incorporate resistance from wild or related species than from germplasm of the same species; nevertheless, wild or related species can be extremely valuable sources of resistance. The use of wild species for improving disease resistance in food crops has had great success in wheat, potato, and tomato (Jones et al., 1995; Lenne and Wood, 1991). For example, stem rust resistance in wheat is one of the most important instances in the past century where resistance was transferred from related species. Usually, the ability to transfer resistance from related species into an adapted cultivar is limited to simply inherited resistance. This limitation may change in the future if breeders and pathologists are able to use molecular techniques to identify multiple genes for resistance in related species and to transfer those genes through marker-assisted selection.

- Artificially induced variation within a plant species—When genetic variation for resistance does not exist within a species or related species, breeders and pathologist may try to find new genetic variation through alternative approaches such as mutation breeding, somaclonal variation and genetic engineering. Mutation breeding and somaclonal variation have not been highly successful techniques from which new sources of resistance have been incorporated into cultivars (Daub, 1986). Some new sources of resistant have been produced through genetic engineering, such as viral coat protein-mediated resistance to viruses. Similar to resistance from wild species, artificially induced variation must be relatively simply inherited to be incorporate easily into adapted cultivars.

At first, it may seem that there is little need to continue to search for new sources of resistance once reliable sources of resistance are identified and incorporated into adapted cultivars. However, the need for additional sources of resistance is never-ending because genetic variation in the pathogen allows for the occurrence of and selection for virulent isolates (i.e., new races). Therefore, in each crop species, it should be the primary objective of at least one breeder/pathologist to identify diverse sources and types of resistance to each disease of importance and to transfer those resistances into elite, adapted lines that can be used to develop commercial cultivars.

Incorporating resistance. Methods of breeding for disease resistance are no different than those used to breed for any other trait of importance. Backcross methods, pedigree methods, population improvement, or any other breeding technique that might be used for another trait can be used for resistance depending on the host and the type of resistance. Adequate disease pressure usually is required to select disease resistant phenotypes. Typically, this requires environmental conditions that are adequate for disease development and a reliable inoculation method to ensure that the pathogen is present. As pathogens can vary in virulence and aggressiveness, it is important to use a collection of pathogen isolates or biotypes that are representative of the area in which the cultivar is likely to be grown. Similarly, it is important to develop cultivars with resistance to each of the diseases that are likely to be important in an area. For pathogens that are difficult to manipulate (e.g., insect-vectored viruses) or diseases that occur sporadically, marker assisted selection may play an increasingly important role in resistance breeding. Even so, reliable phenotypic data is absolutely necessary to correctly identify markers associated with resistance genes.

Simply inherited resistance usually is incorporated into elite, susceptible lines through backcrossing (Figure 34.6). An elite, susceptible line is crossed with a source of resistance. The resulting F_1 is then crossed "back" to the elite susceptible line (recurrent parent) to produce the first backcross generation. In each backcross generation, progeny are screened for resistant and susceptible phentoypes (1:1 segregation if resistance is conveyed by a single, dominant gene), and resistant progeny are backcrossed with the recurrent parent into which the resistance is being transferred. The exact methods and number of generations of backcrossing vary depending on the specific objective of the program. Following the last generation of backcrossing, the resistant progeny must be self-pollinated and progeny from those pollinations must be screened to identify lines that are homozygous for resistance. Backcross breeding is an easy way to transfer simply inherited forms of resistance, but it is inherently conservative. Often, the recurrent parent may be considered obsolete in its agronomic performance by the time the backcross derived cultivar is released. Although used most commonly to transfer single dominant genes, backcrossing also can be used successfully with appropriate modifications to transfer recessive genes, multiple dominant genes, and even some types of partial resistance.

Usually, breeding for disease resistance is part of an overall program of cultivar development. Disease resistance often is considered just one of many traits, such as yield, quality, lodging resistance, maturity, insect resistance, etc., that must be considered when developing breeding or source populations that are subjected to selection in the breeding process. When synthesizing a breeding population from which to initiate selection, the breeder usually tries to select parents with complementary traits. The breeder hopes to select individuals that have the proper combination of favorable traits from the parents. In crops where the desired end product consists of a pure line cultivar or inbred line for making F_1 hybrids, the source population may consist simply of the F_2 population from a cross of two parent lines. Conversely, the source population may be a much more complicated three, four, or multi-way cross of selected parents. Usually, the less agronomically desirable or poorly adapted the source of disease resistance, the lower the percentage of parentage it represents in the source population. Selection for disease resistance often is initiated early in the inbreeding process to eliminate the most susceptible individuals. Selection may be simply on an individual plant basis if the resistance is oligogenically inherited or heritability is high. Alternatively, selection may be on a family (F_3 or F_4, for example) basis if heritabilities are lower. A reasonably large population should be maintained in order to have a sufficient number of resistant families remaining for agronomic testing in later generations.

The two basic methods used to develop pure line cultivars or inbred lines through the process of continued self-pollination are the bulk method and the pedigree method. In the pedigree method, the identity (i.e., pedigree) of individual families is maintained throughout the inbreeding process, so that any family (e.g., F_5) may be traced back to an individual plant in the original source population. The pedigree method allows selection of individual families at each stage of the inbreeding process and can be an effective means of selecting disease resistance of lower heritability. It requires substantial record keeping

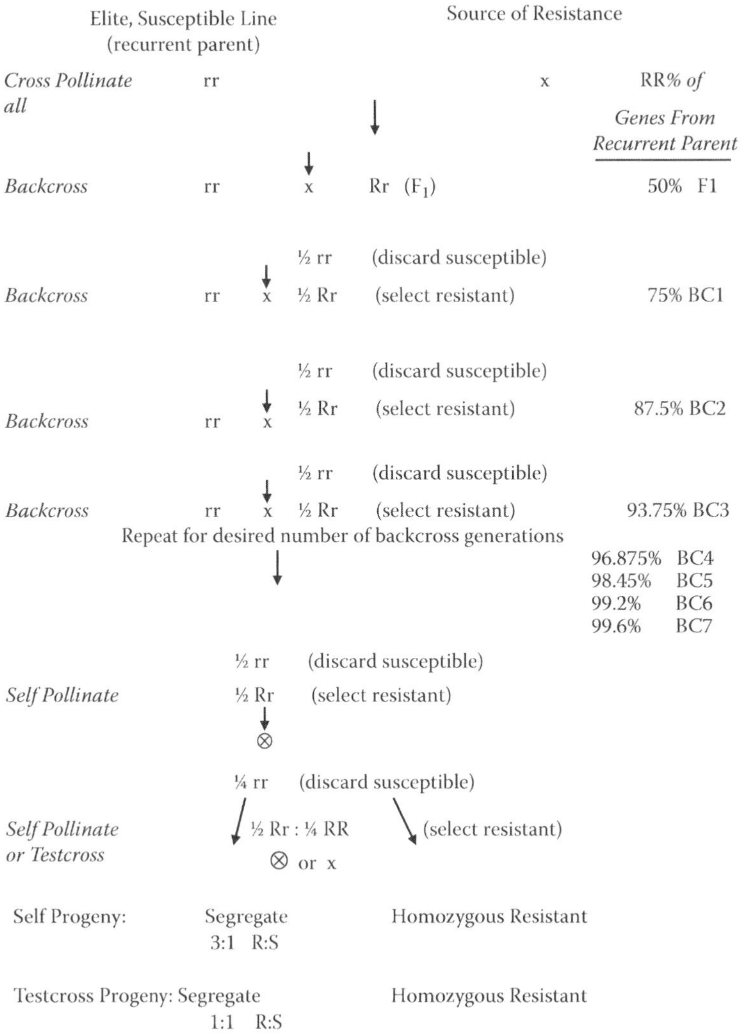

Elite, Susceptible Line   Source of Resistance
(recurrent parent)

Cross Pollinate   rr   x   RR% of
all

                                         Genes From
                                    Recurrent Parent

Backcross   rr   x   Rr (F_1)   50% F1

                      ½ rr (discard susceptible)

Backcross   rr   x ½ Rr (select resistant)   75% BC1

                      ½ rr (discard susceptible)

Backcross   rr   x ½ Rr (select resistant)   87.5% BC2

                      ½ rr (discard susceptible)

Backcross   rr   x ½ Rr (select resistant)   93.75% BC3

Repeat for desired number of backcross generations

                                      96.875% BC4
                                      98.45% BC5
                                      99.2% BC6
                                      99.6% BC7

            ½ rr (discard susceptible)

Self Pollinate ½ Rr (select resistant)
                    ⊗

            ¼ rr (discard susceptible)

Self Pollinate ½ Rr : ¼ RR (select resistant)
or Testcross   ⊗ or x

Self Progeny: Segregate   Homozygous Resistant
                 3:1 R:S

Testcross Progeny: Segregate   Homozygous Resistant
                1:1 R:S

FIGURE 34.6 A general scheme for backcrossing a single dominant resistance gene (R) into a susceptible recurrent parent.

and may use more field space than the bulk method. In the bulk method, seeds from individual plants in each generation are bulked or combined each generation such that it is impossible to trace the pedigree of any plant. Bulk populations usually are self-pollinated for several generations until they consist of a mixture of homozygous pure lines. Seed from individual plants are then planted in progeny rows for replicated testing and selection. The bulk method allows selection on an individual plant basis during early stages of the inbreeding process that can be an effective means of eliminating extremely susceptible individuals if the heritability of disease resistance is high. The bulk method eliminates the need for extensive record keeping and allows a breeder to work with a large number of breeding populations at once since each population takes a minimal amount of space in the nursery.

In crops where the final product is an F_1 hybrid between two inbred lines, the parental lines are commonly developed using the pedigree method that also is used frequently in self-pollinated crops such as small grains and beans. In most crops where hybrids are grown, there are known heterotic patterns or breeding groups that exhibit hybrid vigor when crossed. For example, in maize hybrids grown in the United States, almost all hybrids can trace the pedigree of their parental inbred lines to either the Iowa Stiff Stalk synthetic (typically used as female parents of the hybrid) or the open-pollinated variety Lancaster Surecrop (typically used as the male parent). Disease resistance and other traits of medium-to-high heritability usually are selected during the inbreeding process. Yield can only be assessed in hybrid combination with an appropriate tester, which may be an inbred line, a set of inbred lines, or a cross of two lines from the opposite heterotic group. Although the stage of inbreeding at which testcross performance is assessed varies greatly among hybrid crops and individual breeders, the value or worth of an inbred line is judged by its performance in hybrid combination and not by its performance per se. Thus, in developing

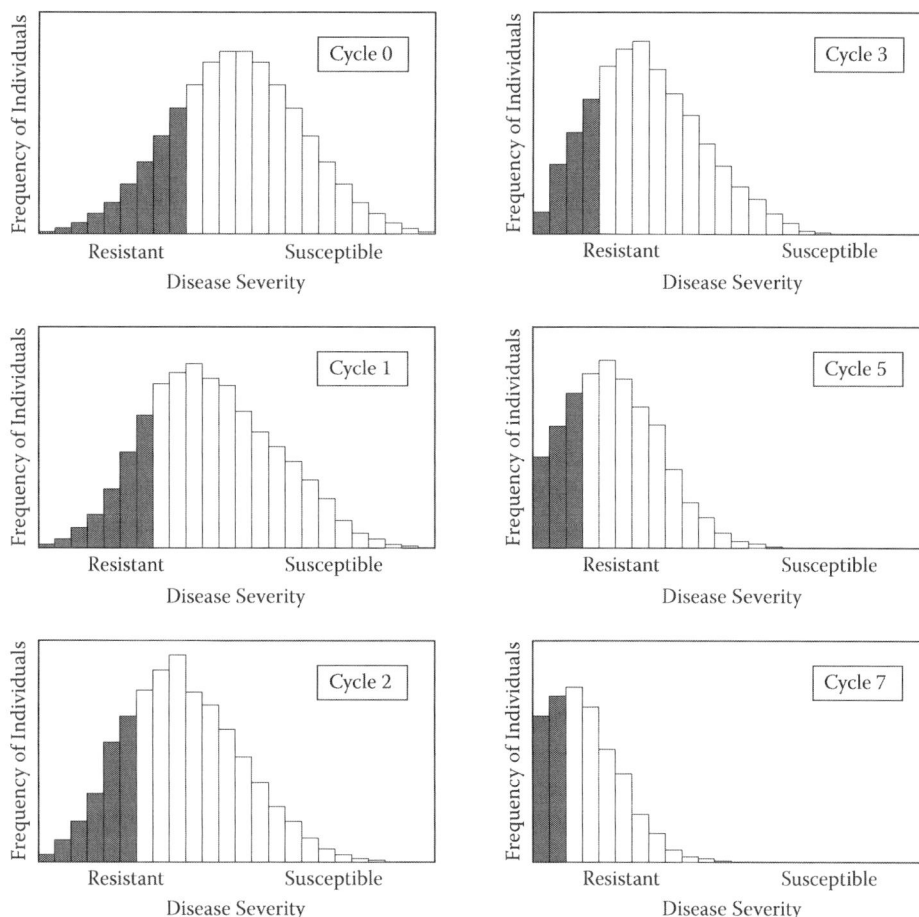

FIGURE 34.7 Population improvement. The frequency of lines with high levels of resistance is greater in advanced cycles of selection because the distribution and mean of the population is continuously moving toward a more resistant phentoype. Shaded area represents plants selected to form the next cycle.

disease resistant hybrids, it may be sufficient for only one parent of an F₁ hybrid to carry resistance if that parent transfers a sufficient level of resistance to the hybrid. Similarly, a hybrid with resistance to multiple diseases may be developed more easily or efficiently if the resistances of the two parental inbred lines are complementary.

Lines with high levels of quantitative resistance can be derived if the source population first undergoes some method of cyclical population improvement. A source population may be created by crossing elite lines with sources of quantitative resistance. After an adequate amount of random mating, the population is screened for disease reactions and other important traits. A certain proportion of the most resistant progeny are selected and inter-mated to advance the population to the next cycle of selection. The exact selection intensity and methods of population improvement vary depending on several factors including heritability of resistance. During any cycle of selection, breeders may try to extract resistant lines from the population. The frequency of lines with high levels of resistance should be greater in the advanced cycles of selection than in the early cycles of selection because the distribu-

tion and mean of the population is continuously moving toward a more resistant phenotype (Figure 34.7).

A phenomenon that is sometimes referred to as the "Vertifola effect" can occur when qualitative resistance prevents selection for quantitative resistance. "Vertifolia" was a cultivar of potato that had a single, dominant gene for qualitative resistance to late blight. This resistance was effective until a virulent race of *Phytophthora infestans* occurred. When the virulent race became prominent, "Vertifolia" was more susceptible than potato cultivars that had been developed without the R gene because the R gene prevented potato breeders from selecting quantitative resistance while developing "Vertifolia."

GENETIC DIVERSITY, SELECTION PRESSURE, AND MONITORING

People often mistakenly believe that genetic diversity is necessary for all of the important traits in a crop in order to avoid significant risks of severe epidemics of diseases or pests. Although any genetic uniformity poses a miniscule risk because of unknown pleiotrophic effects of uniform genes, most crops are uniform for many traits. For example, all sweet corn grown in the United States is uniform for the *su1* gene or the *sh2* gene that increase levels of kernel sugars. The benefit of uniformity for most traits is considerably greater than the risk of potentially severe epidemics. Host resistance to diseases is an exception to this rule. Genetic uniformity for disease resistance carries a substantial risk even though it is greatly beneficial when it works.

Host resistance places selection pressure on populations of pathogens. If genetic variation within the pathogen population allows some biotypes to reproduce on resistant hosts more rapidly than other biotypes, the frequency of biotypes that are more fit will increase in the population and a new, virulent race will occur. This interaction is similar to the response of other populations to selection. For example, antibiotics place selection pressure on populations of bacterial pathogens that can result, by elimination of susceptible strains, in an increased frequency of antibiotic-resistant strains of bacteria. Similarly, synthetic and naturally produced insecticides place selection pressure on populations of insects that can result in an increased frequency of insecticide-resistant insects. Fungicides (Chapter 36) place selection pressure on populations of fungi that can result in an increased frequency of fungicide-resistant isolates of fungi. Therefore, it should not be surprising that host resistance places selection pressure on populations of plant pathogens that can result in an increased frequency of virulent (resistance-resistant) pathogens (i.e., races).

In each example described in the previous paragraph, the occurrence of resistant populations (e.g., virulent races) can be avoided or reduced by altering selection pressure. For example, in human medicine, doctors prescribe, in order to prevent the development of an antibiotic-resistance strain, a variety of antibiotics for patients who are seriously ill with bacterial infections. Similarly, a reduced number of applications of insecticides and fungicides and alternating applications of compounds with different modes of action or mixtures of these compounds are recommended to prevent insecticide-resistant pests and fungicide-resistant pathogens. A plan to delay the occurrence of Bt-resistant insects is based on the maintenance of Bt-susceptible, wild type populations in refugia where selection pressure has been removed completely. In order to prevent the occurrence of virulent pathogens in response to selection pressure created by host resistance to disease, disease resistance must be genetically diverse.

The durability associated with quantitative resistance is related to genetic diversity. As most quantitative resistance is inherited polygenically, or at least oligogenically, it is unlikely that all quantitatively resistant cultivars carry all of the same genes for resistance. Therefore, each quantitatively resistant cultivar exerts slightly different selection pressure on pathogen populations. Also, because most forms

of quantitative resistance allow at least some reproduction of the pathogen, quantitative resistance exerts less selection pressure on the pathogen population than qualitative resistance that often totally suppresses pathogen reproduction. Genetic diversity inherently associated with quantitative resistance also may be due to a different mechanism of resistance or modes of actions of resistance genes. A quantitatively resistant phenotype may be the result of several different factors such as reduced infection frequency, longer incubation and latent periods, reduced lesion size, and reduced sporulation or shorter infectious periods. Each of these factors may be affected by more than one physiological or physical mechanism in the host. Pathogen populations must become adapted to each of these factors in order to become highly virulent. In this case, general virulence (aggressiveness) of the pathogen population probably would change as well as specific virulence.

Monogenic resistance is more likely to result in genetic uniformity in crops because the same gene may be used widely in many or most resistant cultivars. In spite of the potential risk associated with this uniformity, monogenic resistance sometimes is used for a prolonged period of time before virulence becomes frequent. For example, the Ht_1 gene that conveyed chlorotic-lesion resistance to northern corn leaf blight was used widely in dent corn grown in the United States from the mid-1960s through the late-1970s before virulence (i.e., race 1) became frequent in the North American population of *E. turcicum*. Similarly, in the mid-1980s, the *Rp1-D* gene that conveys resistance to common rust was incorporated into most sweet corn hybrids grown for processing in the midwestern United States. Virulence against this gene was not observed in midwestern populations of *P. sorghi* until 1999. During the 15 years that the *Rp1-D* gene was used effectively without the occurrence of virulence, Rp-resistant hybrids increased the on-farm value of processing sweet corn in the Midwest by at least $60 million, based on conservative estimates of yield losses that did not occur and costs of fungicides that were not applied. Clearly, the economic benefits of monogenic resistance can be substantial in certain situations even though experience proves that this type of resistance frequently has selected for new, virulent races of pathogens.

From the very outset of modern efforts at breeding for disease resistance, man has had to cope with the problem of "shifty pathogens." With some diseases such as the fusarium wilts of various crops, resistance has remained durable for many years. For others such as the cereal rusts, resistance genes often are ineffective before they are deployed in a cultivar. Even when effective cereal rust resistance gene is deployed, it is not uncommon for it to remain effective for only 5 years or less. Breeders and pathologists have tried to cope with variable pathogens by constantly searching for new sources of resistance and by selecting combinations of resistance genes that provide resistance to the spectrum of races present in a growing region. These efforts depend on knowledge of the relative frequency, distribution, and virulence of races present in areas where the crop is being grown.

Two basic approaches of monitoring pathogen populations are race surveys and monitor plots. There are many examples of race surveys being used to document diversity in pathogen populations. Usually, scientists in a region cooperate to collect isolates that are then assayed for virulence characteristics (i.e., races) using a set of differential varieties. For example, the USDA Cereal Disease Laboratory in St. Paul, MN, annually collects samples of leaf and stem rust from wheat grown throughout North America and race-types these collections on differential varieties. Wheat breeders use this information to determine which combinations of resistance genes will be effective against prevalent races. The rust survey also serves as an early warning system to alert breeders of the presence of new races virulent on existing cultivars, thus allowing breeders to respond before significant losses have occurred. Data from the rust survey also have been useful as a historical resource from which to examine the influence of resistance gene deployment on pathogen populations.

The second approach, monitor plots, may be used alone or in conjunction with race surveys. Monitor plots consist of field plots of differential varieties planted at various strategically selected locations in a region. The occurrence and severity of disease is monitored in these plots throughout the growing season. By itself, information on disease development on differential varieties may give a rough estimate of the frequency of virulence to specific resistance genes, but it may not provide information on specific virulence combinations present in races. However, samples of isolates usually are collected from monitor plots, and they may be used later to identify races as well as for other purposes. Monitor plots also can serve as an effective early warning system for new diseases, new races of a pathogen, or an increase in prevalence of a previously less-important disease. The Dekalb Plant Genetics Maize Pathogen Monitoring Project is an excellent example of how a private plant breeding company used monitor plots to gather useful information on variation and distribution of maize pathogens throughout North America (Smith, 1977; 1984).

DEPLOYMENT OF RESISTANCE GENES

Several approaches are either being used or have been proposed to increase the durability of disease resistance. Probably, the most common means employed to prolong the useful life of a resistant cultivar is to combine more than one effective specific resistance gene into a single cultivar. The rationale behind this approach is that if virulence toward a specific resistance gene occurs as a very rare random mutational event, then the probability of multiple mutations conferring virulence to multiple resistance genes is so low that the resistance gene combination will have prolonged durability. This "probability" explanation for the apparent durability of certain combinations of stem rust resistance genes in wheat has been criticized. Often it is only certain combinations of resistance genes that exhibit durability. For example, certain combinations of particular seedling *Sr* genes with *Sr2* (an adult plant resistance gene) appear to confer durable resistance to stem rust of wheat. As a general rule, the combination of multiple resistance genes can be effective, but experience will dictate which combinations appear to be durable and which are not.

A related approach to achieve durability is gene pyramiding, i.e., putting as many resistance genes as possible into a cultivar. Proponents of this approach often do not make the distinction between race-specific single genes and polygenes for partial resistance. Some proponents of gene pyramiding consider polygenes for partial resistance merely to be defeated race-specific genes that now have a small, incomplete effect on resistance. This alleged residual effect of defeated race-specific genes is sometimes referred to as "ghost resistance." Evidence for ghost genes is contradictory at best, and does not appear to be of much significance in most pathosystems. The value of arbitrarily pyramiding resistance genes without regard to their specificity or record of durability is questionable. Perhaps a better and more efficient approach would be to pyramid combinations of genes for partial resistance with race-specific genes that appear to be durable in combination.

Multiline cultivars or cultivar mixtures also have been promoted as a means to achieve durable resistance to highly variable pathogens such as the cereal rusts. A multiline cultivar consists of a set of near-isogenic, backcross-derived lines, each of which contains a different race-specific single gene for disease resistance. Because only a fraction of the pathogen population is virulent on a single component line of a multiline variety, the net effect is a reduction in disease development on the multiline as a whole. It is assumed that stabilizing selection (selection against unnecessary genes for virulence) will prevent selection for complex races with virulence on several components of the multiline. Multiline varieties have been developed and appear to have resistance in field tests, but their durability when grown over a large area for a prolonged period of time remains to be demonstrated. Theoretical simulations have shown that stabilizing selection must be relatively strong to prevent the buildup of complex races virulent on multilines. Experimental evidence for strong stabilizing selection operating in agroecosystems is inconsistent, so it is not clearly evident that multilines will be as durable as proposed. High yielding, uniform, and easily maintained multilines have been difficult to develop. These problems have limited the widescale acceptance of this approach.

Regional deployment of race-specific resistance genes in a given epidemiological area is another strategy that has been proposed to prolong the durability of specific resistance. Specifically, it has been proposed that oat varieties grown in different regions of the central North American "Puccinia path" (Chapter 18) carry different genes for crown rust resistance. The crown rust fungus, *Puccinia coronata*, and other cereal rust fungi overwinter in the southern United States and in northern Mexico. These pathogens spread northward each summer with the developing crop. Proponents of regional deployment of resistance believe that this annual northward spread of pathogens could be prevented or delayed by using different resistance genes in the southern states from those used in the Midwest and Canada. In theory, races selected for virulence to resistance genes deployed in the southern states would be avirulent on resistance genes used in the northern states. While this approach is appealing, it requires cooperation among breeders from several states and three countries. Breeders and growers in southern states and Mexico must agree not to use certain resistance genes. That decision could be detrimental to producers in the South if virulent races develop on southern varieties, but not on the varieties used in the North.

Temporal deployment of resistance genes is a variation of the strategy of regional deployment. In this case, cultivars with different resistance genes are planted each growing season in order to alter selection pressure on pathogen populations with each crop. Deploying different resistance genes in time has been proposed as a useful way to prolong the durability of resistance to many soilborne pathogens that are not easily disseminated, such as soybean cyst nematode and Phytophthora root rot.

Polygenically controlled, partial disease resistance has been assumed to be more durable than monogenic forms of resistance. Presumably, this durability is the result of the polygenic nature of the resistance and the incomplete nature of resistance that allows some pathogen reproduction and, thus, reduces selection pressure on the pathogen population. Although polygenic resistance appears to be controlled by several genes, each of which contributes a small portion to the total resistance, most experimental estimates of the numbers of loci controlling polygenic resistance are less than ten and commonly only two to four. Polygenic resistance appears to be found in a wide array of crops to diverse pathogens, but its use depends on the crop and the ease with which it can be manipulated in a breeding program. Perhaps the best example of the successful use of polygenic forms of resistance is in maize where hybrids rely almost exclusively on polygenic, partial resistance for control of various leaf blights. This situation reflects the ease with which maize can be manipulated by an array of breeding methods and the widespread availability of resistant germplasm. In other pathosystems such as the cereal rusts, polygenic resis-

tance has not been widely exploited due to difficulties in effectively selecting for it in traditional pedigree breeding programs, and the availability of easily scored and transferred monogenic forms of resistance. Once useable levels of polygenic resistance have been transferred into adapted, elite germplasm, there are fewer barriers to its use in many breeding programs.

Although polygenic resistance has been effectively used for many years in crops such as maize without any evidence that resistance has significantly eroded, it would be wrong to assume that pathogens cannot eventually adapt to polygenic resistance. There is some experimental evidence to suggest that pathogens have the capacity to at least partially overcome polygenic forms of resistance. However, this adaptation probably would be a slow, gradual process. If breeders continually locate and incorporate new sources of polygenic resistance (and presumably new polygenes) into new cultivars, adaptation by the pathogen to a given set of polygenes would be to little avail because new genes would continue to be deployed in the host.

TOLERANCE

Disease **tolerance** is a term that often evokes different responses from individuals because this term has different meanings, depending upon who is using it. Plant pathologists, particularly those in academia, often define tolerance or "true tolerance" as the ability of a cultivar or variety to sustain less damage when the amount of infection (i.e., disease severity) is the same as on a susceptible cultivar. In other words, the slope of the relationship between yield and disease severity is less negative for a tolerant cultivar than for a susceptible or resistant

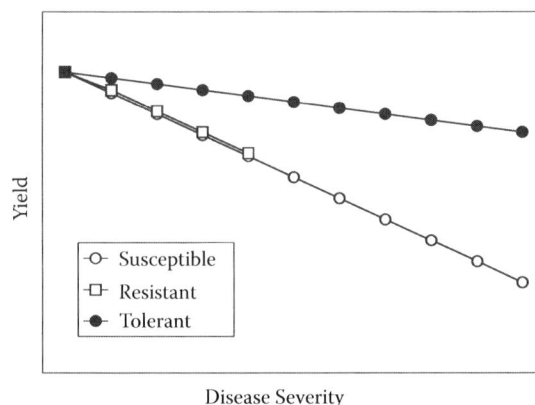

FIGURE 34.8 Tolerance. Based on the academic definition of "true tolerance," the slope of the relationship between yield and disease severity is less negative for a tolerant cultivar than for a susceptible or resistant cultivar because a tolerant cultivar sustains less damage under the same level of infection. Note that the slope of the yield-disease relationship is the same for the resistant and susceptible cultivars, but disease is less severe on the resistant cultivar.

cultivar (Figure 34.8). Although this is not difficult as a concept, it is impractical to select for "true tolerance" in a breeding program because tolerance is difficult to measure experimentally. Due to variability inherent in measuring traits such as yield, yield loss, and disease severity, it is hard to show that two cultivars have equivalent amounts of disease but suffer significantly different amounts of yield losses. In many cases, tolerance can be demonstrated in one environment, but not in another. The practical value of disease tolerance also has been criticized because tolerance often is confounded with low yield potential. That is, low-yielding cultivars are affected less by disease than high-yielding cultivars simply because low-yielding cultivars do not need as much healthy tissue to reach their yield potential as do high-yielding cultivars. Nevertheless, there are examples of susceptible cultivars that sustain less damage from a particular disease than other susceptible cultivars, and these cultivars have "true tolerance."

Tolerance frequently has a somewhat different meaning when used in seed catalogs and other literature produced by the commercial seed industry. Often, disease tolerance is used to identify cultivars that perform well in spite of adverse conditions, including severe disease pressure. When a cultivar yields well in spite of being infected as severely as a poor-yielding, susceptible cultivar, the cultivar may have true tolerance. More likely, the cultivar has a level of quantitative resistance that allows it to perform better than the susceptible cultivar under disease-conducive conditions because it is less severely infected than the susceptible cultivar (Figure 34.9). In this case, "tolerance" is due to partial resistance and the two terms are being used synonymously.

Definitions of resistance, susceptibility, and tolerance vary somewhat among seed catalogs from different companies and among other materials that report on disease reactions of cultivars. Similarly, methods of reporting the disease reactions of cultivars vary among different sources and even among different crops within the same company. Some people view this as confusing and problematic. However, it should be obvious that the types and degree of host resistance to diseases can be as varied as responses for all other host traits. It would be unreasonable to think that all resistant or tolerant reactions could be categorized easily using a single system. Hence, definitions and descriptions differ to some extent because all types of resistance are not the same. Nevertheless, a cultivar that is resistance or tolerant should sustain less damage than a susceptible cultivar grown under similar conditions.

The disease reaction of the cultivar being grown is one of the primary factors that determines whether or not disease will be severe and economically important. The more information that is known about the type and degree of resistance or tolerance in a cultivar, the easier

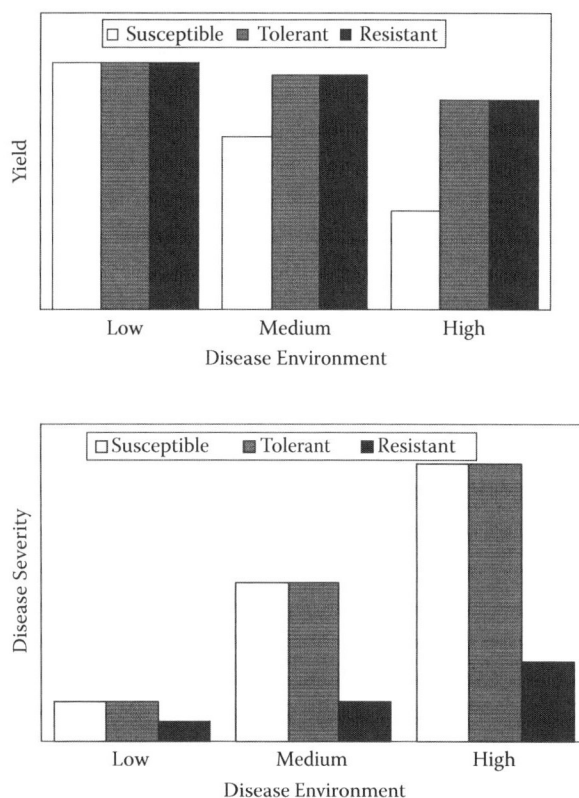

FIGURE 34.9 Susceptible, tolerant, and resistant cultivars yield similarly in environments with low disease pressure because disease severity is low. Under moderate or high disease environments, disease is equally severe on susceptible and tolerant cultivars but yield is affected less adversely on tolerant cultivars. Under moderate or high disease environments, yield of resistant and tolerant cultivars are similar but disease is less severe on resistant cultivars.

it is to predict the performance of the cultivar under various conditions. Under similar conditions, a cultivar that is moderately resistant will be less severely infected than one that is moderately susceptible, and a tolerant cultivar will perform better than a susceptible cultivar. Knowing the reaction of the cultivar being grown allows resistance to be integrated more effectively with other disease control tactics.

LITERATURE CITED/SUGGESTED READING

Adipala, E., P.E. Lipps, and L.V. Madden. 1993. Reaction of maize cultivars from Uganda to *Exserohilum turcicum*. *Phytopathology* 83: 217–223.

Clifford, B.C., and R.B. Clothier, 1974. Physiologic specialization of *Puccinia hordei* on barley. *Trans. Br. Mycol. Soc.* 63: 421–430.

Daub, M.E. 1986. Tissue culture and the selection of resistance to pathogens. *Annu. Rev. Phytopathol.* 24: 159–186.

Hooker, A.L. 1977. A plant pathologist's view of germplasm evaluation and utilization. *Crop Sci.* 17: 689–694.

Hulbert, S.H. 1997. Structure and evolution of the *rp1* complex conferring rust resistance in maize. *Annu. Rev. Phtyopathol.* 35: 293–310.

Jones, S.S., T.D. Murray, and R.E. Allan. 1995. Use of alien genes for the development of disease resistance in wheat. *Annu. Rev. Phytopathol.* 33: 429–443.

Lenne, J.M., and D. Wood. 1991. Plant diseases and the use of wild germplasm. *Annu. Rev. Phytopathol.* 29: 35–63.

Leonard, K.J., Y. Levy, and D.R. Smith. 1989. Proposed nomenclature for pathogenic races of *Exserohilum turcicum* on corn. *Plant Dis.* 73: 776–777.

National Academy of Sciences. 1972. Genetic vulnerability of major crops. NAS, Washington, D.C. 307 pp.

Parlevliet, J.E., and A. von Ommeren. 1985. Race-specific effects in major-gene and polygenic resistance of barley to barley leaf rust in the field and how to distinguish them. *Euphytica* 34: 689–695.

Pataky, J.K., L.J. du Toit and N.D. Freeman. 2000. Stewart's wilt reactions of an international collection of *Zea mays* germplasm inoculated with *Erwinia stewartii*. *Plant Dis.* 84: 901–906.

Schafer, J.F. 1971. Tolerance to plant disease. *Annu. Rev. Phytopathol.* 9: 235–252.

Shaner, G., E.L. Stromberg, G.H. Lacy, K.R. Barker, and T.P. Pirone. 1992. Nomenclature and concepts of pathogenicity and virulence. *Annu. Rev. Phytopathol.* 30: 47–66.

Smith, D.R. 1977. Monitoring corn pathogens. In H.D. Loden and D. Wilkinson (Ed.) Proc. Annu. Corn Sorghum Res. Conf., 32, Chicago. December 6–8. ASTA, Washington, D.C., pp. 106–121.

Smith, D.R. 1984. Monitoring corn pathogens nationally. In D.E. Alexander (Ed.). Proc. 20 Annu. Illinois Corn Breeders School, Univ. of Illinois, Champaign. March 6–8. Univ. of Illinois, Champaign, pp. 101–136.

Vanderplank, J.E. 1963. *Plant Diseases: Epidemics and Control*. Academic Press. 349 pp.

Wynne, J.C., M.K. Beute, and S.N. Nigam. 1991. Breeding for disease resistance in peanut (*Arachis hypogea*). *Annu. Rev. Phytopathol.* 29: 279–303.

35 Cultural Control of Plant Diseases

Gary Moorman and Kimberly D. Gwinn

CHAPTER 35 CONCEPTS

- The way plants are grown and maintained can influence whether or not disease develops and how severe it becomes. Where and when plants are grown, their proximity to other hosts of certain pathogens, and the removal of infected plants or plant parts all have an effect on disease spread.

- Some things can be done to prevent pathogens from coming into contact with susceptible plants. Pathogens can be reduced or eliminated before or during crop production without the use of fungicides, bactericides, or nematicides. In other words, the way plants are grown can be either manipulated to reduce or avoid some plant diseases or make it very likely that disease will develop and become severe.

Cultural practices for growing plants include soil preparation methods, propagation techniques, fertilization and irrigation regimes, where or how seeds or transplants are planted, and how plants are harvested. Each cultural practice may influence the amount of disease that occurs during the growing season. Sometimes, cultural practices can be chosen or modified in ways that reduce the amount of disease. To be effective, a cultural control practice must fit into the production system used by the farmer or the system must be modified to accommodate the practice. Because changing a cultural practice can be expensive or time-consuming, it must be effective enough in suppressing disease to justify its use. In other words, the cost of a cultural control practice has to be recovered through an increase in yield or a suppression of crop loss. In that sense, cultural control practices are no different than using chemicals or biological controls. An advantage of cultural controls is that they may reduce or eliminate the need for other types of control measures. The pros and cons of each cultural practice selected must be weighed, not only in terms of their expense and ease of implementation, but also in terms of how they will influence the most important pathogens that could affect the crop.

IRRIGATION (WATER MANAGEMENT)

Irrigation methods, amounts and frequency can have profound effects on disease incidence. Most plant pathogenic fungi, almost all plant pathogenic bacteria, and foliar nematodes require a film or droplet of water on the plant surface in order to invade the plant. Spores landing on a dry leaf surface do not, in most cases, germinate or penetrate the leaf. Irrigation water applied directly to the soil via furrow flooding or drip and trickle irrigation, and kept off the plant surface, creates an environment unfavorable to many plant pathogens. Overhead irrigation applied through sprinklers, center pivot arrangements, or water cannons wet the foliage of plants and create a favorable environment for spore germination and infection. Many pathogens are splashed from leaf to leaf during overhead irrigation. In dry climates, overhead irrigation may provide moisture necessary for spore germination and infection not available to pathogens through rainfall. An advantage of drip irrigation is that water is placed near a plant's root system without wetting the foliage. This is useful for preventing foliar diseases of horticultural crops.

Some pathogens are inhibited by free water on foliage, and these include the fungi that cause powdery mildew of roses and other plants. When the spores of powdery mildew land on a wet leaf, they do not germinate. Before fungicides were readily available, greenhouse rose growers routinely misted or syringed the leaves during the day in order to suppress spore germination. Thus, the cultural practice of misting helped control powdery mildew on roses. Eventually, fungicides effective for rose powdery mildew became widely available and misting was stopped. One adverse side effect of this change in cultural practice was an increase in mite problems because mites thrive under dry conditions and are suppressed under wet conditions. Also, some fungicides used for powdery mildew management may kill naturally occurring beneficial fungi that parasitize mites. So, all aspects of a cultural practice should be examined before implementation.

Irrigation methods that place water on the soil rather than on the plant tend to inhibit most foliar pathogens but can favor soilborne pathogens. For example **ebb and flow** or flood and drain irrigation systems in greenhouses and nurseries recycle unused water back to a reservoir. If pathogens such as *Pythium* or *Phytophthora* are already in the crop and enter the returning water or contaminate the reservoir from elsewhere, then recycling the water may result in inoculation of a large portion of the crop with each watering.

Regardless of the method of irrigation employed, care must be taken in the amount and frequency of watering. Excessively wet soil is low in oxygen. Roots can be damaged directly by this. *Pythium* and *Phytophthora* are termed "water molds" because their activity is favored by excessive moisture and low oxygen content in soil (Chapter 20). Most species of these pathogens produce a swimming spore stage (zoospore) when soil moisture is high. Zoospores are attracted to roots by chemicals, such as simple sugars, exuded from roots. Roots damaged by excessive soil moisture, excess fertilizer, and other factors tend to leak more such chemicals than healthy roots.

POTTING MEDIA

Over the years, greenhouse operators and nurserymen have moved from using field soil as the growing medium for plants in pots and other containers to **soilless** components. Field soil may harbor plant pathogenic fungi, bacteria, and nematodes as well as weed seeds, and therefore should be treated with heat or chemicals before use. This coupled with the facts that field soil is a nonrenewable resource, and it makes containerized plants heavy to move within the operation as well as heavy to ship long distances, has made the use of lighter, pathogen-free potting media attractive. Perlite, vermiculite, rockwool, and styrafoam are free of pathogens because of the way they are made. These materials can be used in combination or to amend peat moss or composts to produce a potting medium with the desired physical characteristics needed for healthy root development. Their major role, however, is to physically support the plant and they provide few or no nutrients. All nutrients must be applied in a solid or liquid form. It should be noted that while peat moss is usually pathogen-free, it sometimes harbors *Pythium* and another root-rotting fungus, *Thielaviopsis*, and should be treated as if it were field soil. In general, though, the use of soilless potting media has resulted in fewer crop losses in containerized plants than was the case when field soil was used extensively.

Composted tree bark and other composted organic materials, if prepared properly, will be free of pathogens and weed seeds, and their microbial communities will greatly suppress root infecting pathogens. Compost preparation must be monitored carefully because all parts of the pile being composted must reach the proper temperature and must be kept at the proper moisture level so that the organic matter is thoroughly colonized by the desired organisms. If parts of the pile are not composted properly or if the pile becomes too hot or too dry, the beneficial bacteria and fungi may not colonize the organic matter or may be killed.

MULCHING

Mulching can also be used to create an environment favorable to the plant and unfavorable to the pathogen. For example, plastic sheeting can be applied to the soil after initial cultivation and seedlings transplanted through it. Soil covered with black, but especially clear plastic, reaches a higher temperature during the day than bare soil. Verticillium wilt of eggplant is suppressed when plastic mulch is used. The fungus is not killed, however. By the end of the season, many plants exhibit symptoms, and infected plants yield less than healthy plants. However, mulched plants yield more fruit and show symptoms less rapidly than infected plants grown on bare ground.

Organic mulches, such as straw, protect roots from drying and keep rain and overhead irrigation systems from splashing soil onto the plant. When soil harboring a pathogen, such as *Phytophthora parasitica,* that attacks tomatoes is splashed onto leaves, stems, and fruits, rot can occur. Care must be taken in selecting the organic mulch material so that unwanted crop (oats, wheat, etc.) or weed seeds are not introduced into the growing area. Some growers use pine needles as mulch for that reason.

SOLARIZATION

When moderately moist soil (neither excessively wet nor dry) is heated, disease causing organisms are killed at various temperatures. For example, many "water molds" (e.g., species classified in the Oomycota: Chapter 20) are killed at 115°F (46°C), nematodes at 120°F (49°C), and many pathogenic fungi and bacteria at 140°F (60°C). At these temperatures, many beneficial organisms survive and provide competition to any pathogens that may arrive after treatment. For many decades, greenhouse operators have used steam, aerated steam (steam from which liquid water is removed), and electric systems to heat potting soil in order to eliminate pathogens. In some areas, steam and open flames have been used outdoors to treat soil. The heat of the sun, **solarization**, can also be used to treat soil. How well pathogens are reduced in the treated soil is determined by the actual temperature in the soil and the length of time the soil is held at a particular temperature. Some of the pathogen death is caused by the heat itself, some from chemicals released from decomposing organic matter, and some from the enhanced activity of beneficial organisms at elevated temperatures.

To solarize soil, clear plastic mulch is used because soil temperatures reach higher levels than with black plastic. Plant pathogens in dry soil are generally in a dormant, inactive state that is very resistant to heat, cold, or chemicals. Therefore, the cultivated soil should be moistened to a level that would support excellent seed germination. It must not be excessively wet or dry. The soil should then

---

CASE STUDY 35.1

PHYTOPHTHORA BLIGHT IN PUMPKINS—NO EASY ANSWERS

- Phytophthora blight, caused by the oomycete *Phytophthora capsici*, has become one of the most serious threats to production of pumpkins in the United States.
- Phytophthora blight affects the plant at most growth stages (Figure 35.1).
- No resistant cultivars are available.
- Pathogen readily develops resistance to chemical controls.
- **Crop rotation** is not a viable alternative.
  - Pathogen survives in the soil for long periods of time
  - Pathogen has a very wide host range. Hosts include many important food crops (e.g., peppers, cucurbits, tomatoes).
- **Sanitation** would reduce the amount of inoculum, thereby reducing disease losses.
  - Farmers do not typically remove infected plants from the field (**rouging**).
  - Crop residues are typically tilled into the fields, but this does not affect pathogen survival (**tillage**).
  - Why do farmers not use sanitation?
    - Expensive and difficult
    - Does not fit into overall crop production scheme
    - Problems identifying diseased plants in time for effective rouging
- Farmers in many pumpkin producing areas are now unable to grow pumpkins.

---

be covered with the clear plastic sheeting in the summer and kept covered for as long as possible. Several weeks of treatment are better than one week. After the solarization, the plastic can be left in place and slits cut into it for the placement of transplants; or the plastic can be removed. Pathogens will eventually return to the treated area through the movement of pathogen-containing soil on tools, shoes, or by wind and rain.

SANITATION

Sanitation practices are cultural methods that can play a role in suppressing disease outbreaks, especially initial disease development. For example if potato tubers infected with *Phytophthora* are left in the field or are discarded into piles near the field, those tubers act as reservoirs for the initial inoculum of late blight. Initial inoculum may be reduced if the infected tubers are buried or destroyed.

FIGURE 35.1 Phytophthora rot of pumpkins. Inset: pumpkins with white, cottony mycelium. Note the collapse (rot) of the pumpkins. Photos courtesy of Kurt Lamour, University of Tennessee.

Once disease has begun, removing infected plant parts or entire plants can reduce the amount of inoculum present and lessen the possibility that nearby plants will become infected. For example, removing dead flowers from ornamentals can help to reduce the amount of gray mold (*Botrytis*) inoculum. Similarly, removal of branches infected with fire blight (*Erwinia amylovora*) on crabapples, hawthorns, and other rosaecous plants helps to alleviate the disease on those plants. The removal from the greenhouse of entire plants (**roguing**) infected with *Impatiens necrotic spot virus* (INSV; see Chapter 4) is a crucial step in managing that pathogen in potted crops. Roguing of weeds in the greenhouse is important because weeds can act as a reservoir for viruses and other pathogens and for the insects, particularly thrips and aphids, known to move (vector) viruses from plant to plant.

DISINFECTION

Tools used to prune trees and shrubs may become contaminated with fungal spores or bacterial cells and spread the pathogen with each subsequent pruning cut. Thorough disinfection of tools between cuts or between trees being pruned can lessen this means of pathogen spread. Contaminated pruning tools can spread the bacteria that causes fire blight from cankers on infected trees to healthy trees. Likewise the canker stain fungus, *Ceratocystis*, on sycamore trees is readily spread from tree to tree on contaminated pruning tools and other equipment. Tobacco mosaic virus is spread very efficiently in sap from infected plants on almost anything that comes in contact with that plant. Another bacterial disease spread via contaminated tools is crown gall, caused by *Agrobacterium*. This bacterium survives in plant tissue and in soil. In crops where grafting is done, knives must be disinfected on a regular basis so that *Agrobacterium* is not introduced into the cuts made for grafting.

In addition to sanitation measures applied to plant material, pots, flats, and other containers, the greenhouse structures such as walkways and anything that may come in contact with the plants or places where plants may be grown should be disinfected in order to eliminate pathogens in the greenhouse and nursery. Floors and benches that are flooded during ebb-and-flow irrigation and the reservoirs where this water is stored between irrigations should be thoroughly dried, swept clean, and disinfected between crops. Such measures are particularly important in managing the soilborne pathogens *Pythium*, *Phytophthora* (Chapter 20), and *Thielaviopsis* (Chapter 22). Outdoors, it may be important to power wash farm or forestry equipment and vehicles after working in an area known to harbor soil-borne pathogens and before moving that equipment elsewhere. Therefore, sanitation of all tools and equipment coming in contact with infected plants or pathogen-harboring soil is an important cultural control technique used to reduce the spread of certain pathogens.

CROP ROTATION

Crop rotation is used to suppress or avoid certain diseases. Crop rotation is most effective for soilborne pathogens that reside in soil or plant residue left in and on the soil after harvest. One rotation strategy is to plant a crop that does not support the development of a pathogen that is a problem on the next crop to be grown in that particular field. For example, neither wheat nor corn is susceptible to *Sclerotium rolfsii*, the cause of Southern blight of soybeans. Therefore, soybeans can be rotated with grain crops. Sometimes, a noncrop species is used between years of planting the crop. For example, tall fescue planted in rotation with tobacco can result in reduced losses of tobacco to blackshank, caused by *P. nicotianae*. However, the land used for such a rotation will yield little or no revenue during the fescue growing year.

Plant parasitic nematodes obtain their nutrients from live plants. If the crop in the field is not a suitable host for the nematode, many nematodes die and their population level plummets. This is typically the case when lesion nematode (*Pratylenchus*) is present (Chapter 8). The following season, the initial nematode population may be so low that nematodes do not adversely affect the new crop even if it is a suitable host. However, the nematode population may rebound on the suitable crop to a level that would make it unwise to plant that same crop again the following year. Crop rotation can allow you to continue to grow a crop off and on over a period of years despite the presence of serious pathogens at that particular site. Rotation is a very important strategy in managing diseases caused by the root knot nematode (*Meloidogyne* species) on vegetables and the cyst nematode (*Heterodera* species) on soybeans and some other field crops (Chapter 8). Usually the farmer continues to grow the susceptible crop each year, but in a different field each year so that at least part of the farm is producing the main revenue generating crop.

In some cases, it is best to continuously grow a particular crop because doing so tends to create an environment in the soil that inhibits certain pathogens. For example, the continuous culture of wheat may suppress take-all disease caused by *Gaeumannomyces graminis* (Chapter 22). Although during the first few years of continuous wheat production the amount of take-all increases, disease incidence and severity decreases for the next several years. Apparently there is a buildup of beneficial organisms in the soil associated with wheat residue that suppress the pathogen.

TILLAGE PRACTICES

The way soil is prepared for planting can increase or decrease the chances that disease may occur on the new crop. Many disease-causing organisms stay associated with the plant tissue they infected and survive there until the next growing season. If the same crop, or another crop

---

CASE STUDY 35.2

CULTURAL CONTROL—COULD IT HAVE CHANGED HISTORY?

- Ingestion of sclerotia produced by *Claviceps purpurea* causes a toxicosis known as St. Anthony's fire or ergotism in humans and other mammals.
- Ergotism caused problems in populations that relied on rye as the primary grain.
  - Populations that were devastated by ergotism were readily conquered by those who were not suffering; modern geographic borders were shaped by this disease.
  - The Salem witch trials may have been initiated by girls suffering from ergotism (Figure 35.2).
  - Peter the Great was stopped on the brink of conquering the Ottoman Empire because his army suffered from ergotism.
- Ergot disease of rye is easily controlled by two cultural control methods:
  - **Crop rotation**
    - Sclerotia are only viable for 1 year.
    - Rotation into a nonhost for 1 year controls the disease.
  - **Tillage**
    - Deep plowing buries sclerotia so that spores are not discharged into the air.
    - Without spore discharge, the pathogen cannot spread to new hosts.

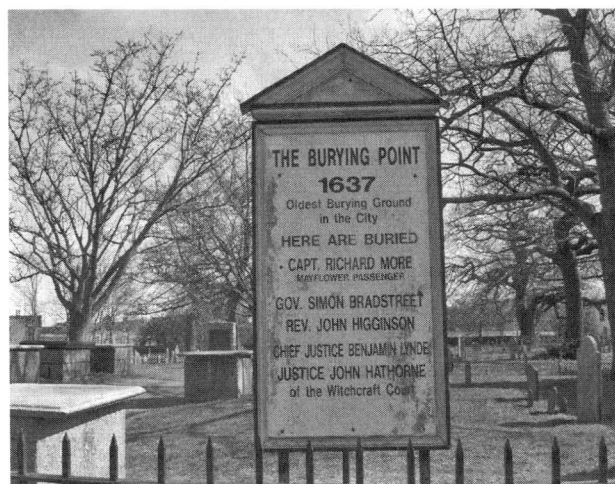

FIGURE 35.2 Graveyard in Salem, MA. Burial point of Benjamin Lynde, the judge in the Salem witchcraft trials.

susceptible to that same organism, is planted in that field then disease may begin again during the new growing season from that inoculum. For example, several different leaf spot fungi that attack wheat and barley survive the winter on the stubble left in the field after harvest. When the stubble and other crop residue are plowed under during soil preparation for the next planting, the fungal spores and other inoculum are buried. The inoculum is not available to infect newly planted seedlings. On the other hand if no-till practices are employed (the soil is not plowed) and the new crop is planted among the stubble left over from the previous year, then inoculum may move from the stubble to the new plants. Similarly, most bacteria that cause plant diseases survive well on plant parts and do not survive well when they are in soil. Thus, tillage practices that result in the decay of bacteria-infected plant parts reduce the amount of bacterial inoculum in the field. Similarly, most viruses die or become inactive as the infected plant part dies, dries, or decays. Whereas almost all viruses will survive in living plant tissue, soil preparation methods that kill any infected plants remaining in a field generally reduce viral inoculum. When selecting tillage practices and when designing crop rotations for a particular field, the possibility of carrying over inoculum from one season to the next must be considered.

CROP RESIDUE BURNING

In addition to plowing under crop residue and thus reducing plant pathogen populations before the next growing

---

CASE STUDY 35.3

DAYLILY DEATH

- Daylilies are susceptible to root rots caused by fungi such as *Fusarium, Phytophthora, Sclerotium, Rhizoctonia*, and *Pythium,* as well as bacterial soft rot caused by *Erwinia.*
- Overwatering, overfertilization, poor soil (or soilless media), and plant stress favor root diseases (Figures 35.3–35.5).
- Disease can be reduced or eliminated by good cultivation, clean soilless media, and correct growing conditions for the plant.
- A grower, who had successfully grown daylilies for many years, potted several thousand daylilies into a soilless medium that he bought from a new vendor.
  - The daylilies began to die due to root rots caused by *Phytophthora* and *Fusarium.*
  - The grower discovered that the tree bark in the soilless medium had not been properly composted and likely contained pathogenic agents.
- Although it cannot be proven conclusively that use of this medium caused the losses, the grower no longer uses soilless media from the vendor.

FIGURE 35.3 Root rot disease of daylilies. Notice small, unhealthy plants in foreground.

season, crop residue can be burned. This method has been used to manage pathogens in grass seed production systems where it is impractical or prohibitively expensive to use other methods of control. Over years of continuous grass seed production, leaf spots, rusts, and smuts build up on the straw and stubble and spread from those harbors to the new stems, leaves, and seed heads. Burning the dormant grass greatly reduces these pathogens and can have the side benefits of killing weeds and destructive insects. Pollution from smoke can be locally controlled by initiating the fire along the entire perimeter of the field so that the fire sweeps from the edges toward the center of the field, thus creating a massive rising air current. The smoke is carried into the upper reaches of the atmosphere away from the field but may cause problems far downwind.

REMOVAL OF NONCROP HOSTS AND ALTERNATE HOSTS

Another cultural control method is the removal of noncrop host plants of the pathogen. Black knot on plum and cherry trees, caused by the fungus *Apiosporina*, is often prevalent on wild plum and wild cherry trees along the edges of the forest and in hedgerows. By removing these trees, the amount of inoculum available to attack orchard trees is reduced. Similarly, the removal of alternate hosts, required by the pathogen to complete its life cycle, is important such as in the case of cedar apple rust on apple trees. The fungus *Gymnosporanium juniperi-virginianae* (Chapter 18) overwinters in galls on *Juniperus* species that are often found in abandoned pastures, in hedgerows, and along roadsides. By removing junipers from the vicinity of the orchard,

FIGURE 35.4   Daylily root rot. A. Healthy root system. B. Diseased roots of several plants.

Procedure 35.1

Effect of Fertilization Rate on Pythium Root Rot of Geranium (*Pelargonium X hortorum*)

Step	Instructions and Comments
1	Plant geranium seed in a flat or pot indoors or transplant poinsettia rooted cuttings in 6-in. diameter pots.
2	Transplant (approximately 10 to 14 days after planting) 120 geranium seedlings into 4-in. diameter pots filled with a potting mix that does not contain compost. (Many composts have Trichoderma in them as a natural part of the flora. *Trichoderma* can be a very efficient biocontrol of *Pythium*. Alternatively, the potting mix can be steamed or autoclaved to eliminate *Trichoderma*).
3	To 35 plants, apply a liquid fertilizer (such as 15-15-15) at a rate of 150 ppm N. Apply the fertilizer to a second 35 plants at a rate of 300 ppm N and at 600 ppm N to the remaining 35 plants. Directions for mixing the fertilizer to achieve the desired concentrations are printed on the product container. Apply the fertilizer to the soil surface to thoroughly soak the potting mix so that about 25% of the moisture added leaches out the bottom of the pot. Repeat this each day for 4 or 5 d in order to establish three different fertilizer levels. After that, fertilizer-amended water should be used each time irrigation is necessary and enough should be applied each time so that about 10% of what is applied leaches through the potting mix. Such leaching helps to maintain the desired fertilizer level and lessens the buildup of salts to higher than desired levels. Measure the conductivity of each fertilizer that was prepared and use these readings as a guide to maintaining the three separate levels desired (see step 5).
4	Transfer a mycelial plug of *Pythium* to each standard 10-cm diameter petri dish containing corn meal agar, V8 juice agar, or potato dextrose agar.
5	At 4 or 5 d after beginning fertilization, take one pot from each fertilizer level and hold it over a beaker while pouring 100–200 mL of distilled water onto the soil surface. Measure the soluble salt level of the leachate using a conductivity meter to be certain three different levels have been established. If the levels seem too low or too high as compared to step 3, apply fertilizer twice a day for 2 d and repeat this step.
6	At 7 to 10 d after transplanting and beginning the fertilizer treatment, inoculate 15 plants at each fertilizer level as follows. Scoop the entire contents of colonized petri dishes of agar into a blender and add 100 mL of tap water per dish. Very briefly homogenize this mixture in order to obtain a suspension of mycelium and agar. Apply 10 mL of suspension to the surface of each of the 15 pots in each fertilizer level. Label these plants "inoculated." Label 15 additional plants "not inoculated." Use one of the remaining five plants per treatment to check the fertilizer level in each of the three treatments as noted in step 5 each week.
7	Each week, note the number of days it has been since inoculation and record the number of dead plants in each inoculated and not-inoculated treatment. Discard the dead plants. Calculate the percentage of plants that have died.
8	Graph the total number of dead plants (Y-axis) versus days after inoculation (X-axis) for each treatment.

FIGURE 35.5 Dissected crown of daylily. Note the discolored roots. Disease was caused by either *Phytophthora* or *Fusarium* species.

the number of spores that spread to infect apple fruit and leaves in the spring is reduced. An example of a forest tree disease that is managed through the removal of the alternate host is white pine blister rust, caused by *Cronartium ribicola*. By removing the alternate host, *Ribes* species, the life cycle of the fungus is disrupted and pines are less likely to become infected.

PLANTING DATE

How plants are established can greatly influence subsequent disease development. Tomatoes and many other high value crops can be seeded directly into soil or can be started in a greenhouse and then transplanted into the field. Seeds and young seedlings are susceptible to damping-off caused by *Pythium* species, especially in cold, wet soils. To have tomatoes for sale early in the season, farmers may plant when soil temperatures favor development of **damping off** due to *P. ultimum*. Because more mature plants tend to be much less susceptible to *Pythium* than young plants, using transplants avoids exposing seeds and small seedlings to the fungus. Seeds can be planted into pathogen-free soilless media in containers in a greenhouse where soil temperature and moisture can be adjusted. More mature plants can be transplanted to the field after the field soil has reached temperature and moisture levels less conducive to damping-off.

INDEXING

Plants propagated vegetatively via cuttings, tubers, or rhizomes are generally less susceptible to disease than young seedlings. However, plant material propagated vegetatively is likely to be infected by pathogens from the mother (stock) plant. For example, bacterial blight of geraniums caused by *Xanthomonas* is carried in practically all cuttings taken from an infected mother plant. To avoid this problem,

specialty propagators of geraniums thoroughly test each mother plant to ensure that it is free of blight bacteria. Most propagators that specialize in geranium production also test for viruses that affect geraniums. This process of testing for specific pathogens is called **indexing**. "Culture indexing" refers to testing for pathogens that can be grown in media (agar or broth), separate from the plant such as bacteria and most fungi. Indexed plants are not necessarily **disease-free** or **pathogen-free**. They may harbor pathogens other than the ones for which they are being tested. Plant indexing and propagating from indexed plants is an important cultural control method for strawberries, geraniums, carnations, chrysanthemums, and other high value crops.

CROP PLACEMENT

Cucumber mosaic virus (Chapter 4), vectored by aphids, is an important pathogen of fresh market bell peppers in some regions. If infection occurs during flowering, severe fruit spotting develops. As few symptoms develop on fruit if infection occurs after flowering, strategies to delay infection can suppress disease losses. The initial onset of disease can be influenced by the crops placement in relation to where aphids can be expected to enter the field. When a very CMV-susceptible crop is planted near a hedgerow or other obstruction that can slow the wind, in the same manner as a snow fence, CMV-carrying aphids may settle out of the air and onto the crop at that location. Aphids are not strong fliers. The aphids may have over-wintered on woody plants in the hedgerow, acquired CMV from nearby weeds in the spring, and then moved to the adjacent crop. Once in the field, the viruliferous aphids move from plant to plant, down the row. By planting the CMV-susceptible crop at a distance from where aphids first enter a field, the onset of CMV can be delayed and at least some crop losses avoided.

PLANT NUTRITION

Plant nutrition can have a dramatic impact on disease susceptibility and development. However, very few generalities can be stated concerning fertilizer type and rate in relation to subsequent disease development. Each crop and pathogen combination must be examined separately. High rates of nitrogen fertilization encourage succulent growth of pear trees that are very susceptible to fire blight. Similarly, high fertilization rates tend to favor verticillium wilt development in trees and shrubs in the landscape. On the other hand, take-all disease severity increases in plants deficient in any of the major elements. Pythium root rot of the ornamental foliage plant, *Peperomia*, is most severe when fertilizer rates are low.

In some host and pathogen combinations, the form of the nutrient plays a role in disease susceptibility. For example, high nitrate fertilization favors verticillium wilt in snapdragons and tomatoes, whereas high ammonium fertilization suppresses verticillium wilt. In other diseases, nitrate may suppress disease.

Soil pH can be used to suppress disease, but results may be erratic. Soil chemistry is very complex. Organic and inorganic compounds, living organisms, soil moisture and temperature, and many other factors all interact to influence soil chemistry. The result of altering the pH of soil with chemicals varies with each soil being amended. The suppression of clubroot of crucifers, caused by *Plasmodiophora brassicae*, through the addition of lime to raise the soil pH has been practiced for well over 200 years. In certain soils where the pH can be adjusted to 6.9 or higher, control can be very good. However, control may be poor at the same pH level in different soils. Many factors in the soil are influenced by pH adjustments and it is difficult to predict what will occur in a particular location without trying the technique.

LOW TEMPERATURE STORAGE

Most decays of fruits, vegetables, and flowers observed in stores actually begin before the plant is harvested. For example, gray mold on cut flowers and strawberries begins when flowers are infected by *Botrytis* outdoors or in the greenhouse. After the fruit is packaged or the flowers are boxed for shipping, the humidity and temperatures rise in the containers to favor the growth and spore formation by the fungus. If left unchecked, fruits and flowers become covered with *Botrytis*, unacceptable to the consumer, and decay rapidly. By holding the plant material at reduced temperatures during shipping, storage, and while on the display shelf, the activity of such pathogens can be greatly inhibited.

Tree and shrub nurseries in regions where spring conditions make it very difficult to dig and prepare the plants for sale and shipping perform those tasks early in the dormant season and then store the plants in a cooler or other location where temperatures can be kept low. Not only do the plants remain dormant until sold the following season, the activity of any plant pathogens on the plants will be slowed or stopped. The risk in using this method is that should the cooling system fail or temperatures be allowed to rise in the storage area, tremendous losses due to disease can result.

SUMMARY

From the examples provided in this chapter, it is clear that each pathogen/host combination needs to be looked at as a separate case when considering the use of cultural practices for disease control. A cultural practice that works with one host and pathogen may not work for a different host of that same pathogen. However, there are many cultural practices that can be used to avoid or reduce disease development. The practice will only be used if it fits into the overall crop production system, is not too expensive to implement, and is effective.

EXERCISE

The effects that cultural practices have on disease development in a crop can be dramatic. For example, the concentration of fertilizer used on geraniums or poinsettias exposed to the root rotting fungus-like organism *Pythium*, determines disease severity. Little disease develops at low fertilizer rates, yet good plant quality is obtained. A high fertilizer rate increases pythium root rot losses. The presence of high amounts of nitrogen (N), phosphorus (P), or both N and P predispose plants to attack by *Pythium*. The effect of fertilizer rate on disease severity is not merely due to damage from the high concentration of salts and associated high conductivity of the soil moisture.

EXPERIMENT 1. EFFECTS OF FERTILIZATION RATE ON PYTHIUM ROOT ROT OF GERANIUMS (*PELARGONIUM* X *HORTORUM*) OR POINSETTIAS (*EUPHORBIA PULCHERRIMA*)

This experiment will require 7–10 weeks to complete.

Materials

The following materials will be required for the entire class:

- Geranium seeds (or 120 poinsettia rooted cuttings)
- A culture of either *Pythium ultimum* or *P. aphanidermatum* (These can be obtained from the American Type Culture Collection. However, APHIS Permit No. 526 must be obtained in order for ATCC to ship the culture. Alternatively, a plant pathologist at a Land Grant university within your state may be able to supply a culture, in which case no permit is required.)

- Plastic pots (4-in. diameter; 6-in. pots for poinsettias)
- Commercial soilless potting mix (preferably one that does not contain composted components)
- Fertilizer to be applied as a liquid with each irrigation
- Conductivity meter for measuring soluble salts
- 250-mL beaker for collecting leachate
- Corn meal agar, V8 agar, or potato dextrose agar in 10 cm-diameter petri dishes

Follow the protocols listed in Procedure 35.1 to complete the experiment.

Anticipated Results

Pythium root rot kills 600 ppm N-fertilized plants at a faster rate than at 300 ppm N. Usually no plants are killed in the 100 ppm N treatment. Some noninoculated plants in both the 300 and 600 ppm N treatments will die as a result of the high salt levels. Although other types of plants can be used, note that begonias and impatiens generally do not react as dramatically as geraniums or poinsettias. Some plants, such as *Peperomia*, may have fewer losses due to root rot at high fertilizer levels than at low levels.

Questions

- What is the advantage of using lower fertilizer amounts in greenhouse plant production?
- How might *Pythium* get into a crop of geraniums in a greenhouse? From where would it have come?
- What can be done to prevent *Pythium* from getting into a crop in the greenhouse?
- What can be done to a greenhouse crop of potted plants once *Pythium* is found on the roots of some of the plants?

SUGGESTED READING

Duczek, L.J., Sutherland, K.A., Reed, S.L., Bailey, K.L., and Lafond, G.P. 1999. Survival of leaf spot pathogens on crop residues of wheat and barley in Saskatchewan. *Canadian Journal of Plant Pathology* 21: 165–173.

Engelhard, A.W. (Ed.). 1989. *Soilborne Plant Pathogens: Management of Diseases with Macro- and Microelements.* APS Press, St. Paul, MN, 217 pp.

Hardison, J.R. 1980. Role of fire for disease control in grass seed production. *Plant Disease* 64: 641–645.

Horst, R.K. 1983. *Compendium of Rose Diseases.* APS Press, St. Paul, MN, 50 pp.

36 Chemical Control of Plant Diseases

Alan S. Windham and Mark T. Windham

CHAPTER 36 CONCEPTS

- A fungicide is a chemical compound that kills or inhibits the growth of fungi.

- A bactericide is a chemical compound that kills or inhibits the growth of bacteria.

- A nematicide is a chemical compound that kills or disrupts the feeding or reproductive behavior of nematodes.

- A disinfectant is a chemical agent that is used to eliminate plant pathogens from the surface of plants, seed, or inanimate objects such as greenhouse benches, tools, pots, and flats.

Agricultural chemicals have been used to control diseases for centuries, even before microorganisms were associated with various plant maladies. Most pesticides targeted at plant diseases are used to minimize losses and are applied prior to infection. Although fungicides can be very efficacious against fungal pathogens, bacterial plant pathogens are much more difficult to control. Few fungicides have the capacity to halt the infection process once the pathogen is firmly established inside a plant.

Many farmers and gardeners equate disease control with the use of pesticides, such as **fungicides**, **bactericides**, **nematicides** or **disinfectants**. Chemical control is one of several strategies used in plant disease management. The use of pesticides for disease control should be a component of an integrated system (Chapter 38) that uses cultural (Chapter 35) and biological controls (Chapter 37). In many cases, chemical control should be the strategy reserved as last when all other methods have been utilized.

PESTICIDE NAMES

Fungicides, as all pesticides, have the following three names: a chemical name of the active ingredient, a common name, and a trade name. Although the chemical and common names remain constant, the fungicide may be marketed under several trade names, depending on the target crop or manufacturer. For instance, a fungicide may be marketed under one trade name to specialty crops such as turf and ornamentals and under a different trade name to fruit, vegetable, and field crop markets.

An example:

Chemical name: Ethyl hydrogen phosphonate
Common name: Fosetyl-al
Trade name: Aliette

PESTICIDE LABELS

Pesticide labels list valuable use and safety information for the person mixing and applying the pesticide. At the top of each label, the trade, common, and chemical names are listed with the percentage of the active ingredient. **Signal words** such as "Caution," "Warning," or "Danger" appear in bold print, depending on the toxicity of the product (Table 36.1). An EPA registration number, which ensures that the product has met all regulatory standards, should be prominently displayed on the label. The precautionary statement lists hazards to humans and wildlife; worker protection standards include necessary personal protection equipment and reentry intervals for workers that may come into contact with fungicide residue. The label will also include directions for use, instructions for mixing and application, and potential environmental hazards. Pesticide labels generally contain information on pesticide storage and disposal.

MATERIAL SAFETY DATA SHEET (MSDS)

The **material safety data sheet** (MSDS) contains important information about pesticides. These sheets should be available for inspection by employees mixing, loading, applying, and working in areas where pesticides have been used. At the top of each MSDS, the pesticide trade or product name, chemical name, and common name are listed. Also listed on these sheets is information about adverse effects of overexposure, first aid measures, fire fighting measures, how to handle a spill or leak, proper storage conditions, and information about worker protection standards, including required personal protection equipment. MSDS also list emergency numbers to call in the event of a spill or accidental poisoning.

TABLE 36.1

Toxicity Classes for Pesticides

Hazard Indicators	Categories			
	I	**II**	**III**	**IV**
Signal Word	**Danger/Poison**	**Warning**	**Caution**	**Caution**
Oral LD50	0–50 mg/kg	50–500 mg/kg	500–5000 mg/kg	+5000 mg/kg
Inhalation LC50	0–0.2 mg/l	0.2–2 mg/l	2–20 mg/l	+20 mg/l
Dermal LD50	0–200 mg/kg	200–2000 mg/kg	2000–20,000 mg/kg	+20,000 mg/kg
Skin effects	Corrosive	Severe irritation at 72 h	Moderate irritation at 72 h	Mild or slight irritation at 72 h
Lethal dose to human adults (oral)	A few drops to 50 mL (1 teaspoon)	5–30 mL (1 teaspoon to 2 tablespoons)	30–486 mL (1 ounce to 1 pint)	7486 mL (> 1 pint)

FUNGICIDES

HISTORY OF FUNGICIDES

In 1802, lime sulfur was first used to control mildew on fruit trees. A few years later in 1807, Prevost used copper sulfate as a seed treatment to prevent bunt of wheat. In the latter part of the century, **Alexis Millardet**, a professor of botany at Bordeaux University, observed that grapes treated with a mixture of copper sulfate and lime to make them unappetizing to travelers were free of downy mildew (Chapter 20). In 1885, he demonstrated the effectiveness of this "Bordeaux Mixture" in controlling downy mildew of grape.

Organomercurial fungicides were used as seed treatments to control bunt of wheat in 1913. The use of mercury and other heavy metals is now restricted by legislation due to animal toxicity. In the 1930s, the organic fungicides called dithiocarbamates were developed. Dithiocarbamates, such as maneb, ziram, and thiram, were used to prevent foliar diseases of plants. Although some of the original fungicides of this group are no longer available, dithiocarbamates are currently some of the most widely used fungicides around the world. By 1950, antibiotics such as streptomycin sulfate and cyclohexamide were introduced to control bacterial or fungal diseases of plants. In a period from 1960 to 1970, widely used protectant fungicides, such as captan and chlorothalonil, were first used for foliar diseases of plants.

One of the first systemic fungicide groups, the benzimadiazoles, was introduced to combat a wide variety of diseases, such as powdery mildew, stem rots, and leaf spot, around 1970. Other systemic fungicides were introduced in the following decade. The SBI (sterol biosynthesis inhibitors) were true systemic fungicides that controlled diseases such as powdery mildew and leaf spots. The acylalanines were the first systemic fungicides used to combat root rot diseases caused by species of *Pythium* and *Phytophthora*. By the end of the 20th century, a new group of fungicides, the strobilurins, were introduced to prevent a long list of diseases including mildews, leaf spots, and stem and root rots. The strobilurins are similar to antifungal compounds produced in nature by wood decay fungi. They are widely used in row and specialty crops.

FUNGICIDE FORMULATIONS

Fungicide **formulations** consist of an active ingredient (a.i.) and inert ingredients, and may contain wetting agents, emulsifiers, or stickers. Wetting agents and stickers help to distribute the product over the crop and slow the weathering process, respectively. Prepackaged mixtures of two a.i.'s may be used to broaden the spectrum of fungicidal activity or to prevent or slow the development of resistance. For example, mancozeb is often mixed with systemic fungicides at risk to resistance. Different formulations of a fungicide may have different concentrations of the a.i., depending on its use and target organisms.

Wettable powders (WP). A wettable powder is a finely ground fungicide powder that does not dissolve when added to water, but remains suspended in the spray tank. Agitation of the spray solution is normally needed to prevent settling. A disadvantage of a wettable powder is dust during measuring and weighing. Many of the older fungicides are formulated as a wettable powder.

Water dispersible granules (WDG). Dispersible granules have larger particles than a wettable powder. These particles break up rapidly when added to water and go into suspension. An advantage of a dispersible granule is less dust during measuring and weighing.

Water soluble pouch (WSP). Water soluble pouches make mixing pesticides easier since a premeasured amount of pesticide is sealed in a bag that dissolves upon contact with water. These pouches are usually designed to be added to 50 to 100 gallons of water. An advantage of this formulation is that it minimizes exposure of the person mixing the pesticide. A disadvantage is that the pouches are preweighed for large volumes of spray and are not convenient for small quantities.

Flowable (F). Flowable formulations have concentrated amounts of fungicide particles suspended in a liquid form. They are convenient since they are measured by volume rather than by weighing as in the dry formulations. A disadvantage of flowable formulations is that settling may occur if they are stored for long periods.

Emulsifiable concentrate (EC). An emulsifiable concentrate, like a flowable fungicide, is a liquid formulation. The active ingredient of this formulation is dissolved in an organic solvent. Emulsifiable concentrates are measured by volume.

Granule (G). A granular formulation is made up of dry pellets or granules of a fungicide, often in low concentrations. These granules may break down or release the fungicide after coming into contact with water after application by a broadcast or drop spreader. Granular formulations are generally used for soilborne pathogens (Chapter 22).

Dust (D). Older fungicides such as sulfur are sometimes applied to foliage of plants as a fine dust. Dust formulations are often shaken or blown onto foliage and redistributed by dew or rainfall.

FUNGICIDE APPLICATION

Fungicides may be applied in many different ways. The application technique depends on the target pest, the crop and available equipment. Most pesticides are mixed with water and applied by hand-held sprayers for small areas or by hydraulic sprayers or mist blowers for larger areas. Hydraulic sprayers are used where larger volumes of water are used to apply pesticides to the canopy of crops. Hydraulic sprayers may use as much as 100 to 200 gallons of water per acre to apply pesticides to the foliage of crops. Air blast or mist blowers use much less water to apply the same amount of pesticide per acre. An air blast sprayer in a nursery or orchard may use as little as 15 to 20 gallons of water per acre to apply pesticides to the canopy of crops.

Fungicides are often applied as seed treatments or in-furrow applications to prevent seed rots and damping off. Vegetable and field crop seed are often treated with fungicides to prevent seed rots caused by fungi such as species of *Pythium* and *Fusarium* species or damping-off caused by *Rhizoctonia solani*. A brightly colored dye is added to the treatment to indicate that the seed has been treated with a fungicide and should not be used for food or animal feed.

Drench applications of pesticides are used in specialty crops for certain soilborne pathogens. Fungicides labeled for stem or root rot diseases are applied in large volumes of water to containers or soil beds, often through a proportioner connected to the hose or irrigation system. Drench applications are intended to saturate the media or soil and may take the place of an irrigation cycle.

Broadcast application of fungicide granules are used when drench applications are impractical. Granules may be applied over the top of containers, propagation beds or landscape beds with a rotary spreader. Drop spreaders are often used for precise application of granules to turfgrass. Irrigation or rainfall is needed after granules are applied to release the active ingredient.

Fungicide granules are sometimes incorporated into bulk soilless media prior to planting. Granules are thoroughly mixed into the media along with other amendments such as fertilizer or lime. Fungicides incorporated into media generally combat damping off, and stem and root rot diseases.

TYPES OF FUNGICIDES

Fungicides are often grouped as contact or eradicants. **Contact fungicides** are most often applied as foliar sprays to protect above-ground plant parts from infection; they may also be used as seed treatments. Contact fungicides must be applied uniformly over the leaf surface. As these fungicides are on the outer plant surface, they are subject to weathering and photodegradation. Contact fungicides do not protect new growth, so they must be applied frequently. **Eradicant fungicides** may be applied as seed treatments or on growing plants. They are usually systemic and move upward though the the plant xylem tissue. Eradicant fungicides, like contact fungicides, work best if they are in place prior to infection; however they will halt fungal growth if applied shortly after infection.

Systemic fungicides usually have one or more of the following characteristics: the ability to enter through roots or leaves, water solubility to enhance movement in the vascular system and stability within the plant. Systemic fungicides may move varying distances in plants after being applied. An advantage of systemic fungicides is that they provide control away from the site of application. They are inside the plant and are not affected by weathering. Locally systemic (mesostemic) fungicides are translaminar and will protect the undersides of leaves after being applied to the upper leaf surface. Locally systemic fungicides may provide protection for short distances on the leaf surface by vapor action. Most truly systemic fungicides move upward in the apoplast (xylem) to protect new plant growth from infection. Movement in the apoplast is usually passive as the fungicide moves upward in the transpiration stream. Arborists take advantage of this flow when they inject fungicides into the trunk of an elm to protect the tree from Dutch elm disease. An example of a fungicide that moves in the apoplast is azoxystrobin, a strobilurin fungicide. Few systemic fungicides move downward in the symplast (phloem) after they are applied to foliage. Systemic fungicides that move in the symplast are transported in phloem sieve tubes to the root system along with carbohydrates manufactured during photosynthesis. One of the few fungicides documented to move in the symplast is fosetyl-al.

TABLE 36.2

Chemical Classes of Fungicides

Fungicide Class	Common Name
Inorganic fungicides	Sulfur, lime sulfur
Sulfur	
Copper	Copper sulfate, copper pentahydrate, Bordeaux mixture
Dithiocarbamate	Mancozeb
Aromatic compounds	Pentachloronitrobenzene, chlorothalonil
Oxathiins	Carboxin, oxycarboxin
Benzimidazole	Benomyl, thiophanate methyl
Dicarboximide	Iprodione, vinclozolin
Acylalanine	Metalaxyl, mefenoxam
Organophosphate	Fosetyl-al
Sterol biosynthesis inhibitor	Fenarimol, myclobutanil, propiconazole, triadimefon, triflumizole
Strobilurin	Azoxystrobin, kresoxim-methyl, trifloxystrobin
SAR stumulant	Acidbenzolar-s-methyl

FUNGICIDE RESISTANCE

If a fungicide is used continuously, there is always the possibility that certain individuals of a fungal population may become less sensitive to the fungicide. This decrease in sensitivity or resistance of a fungal population may be the result of a genetic mutation either present or induced in the population and the subsequent selection and multiplication of resistant individuals. Environmental conditions, disease pressure and the fungicide application frequency all affect resistance development.

Fungicides that attack specific sites in the fungal cell may become vulnerable if the fungus becomes less sensitive to the fungicide with one mutation. Fungi that produce a large number of spores are more likely to become resistant to fungicides. The best chance for resistance development occurs when thousands or millions of fungal spores representing numerous individuals are exposed to one fungicide continuously. A single gene mutation in a few spores may lead to a lack of control if the individual spores are also highly pathogenic. Fungi that cause diseases of floral or vegetable crops in greenhouses such as powdery (Chapter 14) and downy mildews (Chapter 20) and botrytis blight have been reported to acquire resistance to some fungicides.

Cross resistance may occur when a fungus becomes resistant to a particular fungicide active ingredient. For instance, *Botrytis* (gray mold) isolates that are insensitive to iprodione are often tolerant of the fungicide vinclozolin, which is in the same chemical class and targets the same site. Also, powdery mildews that are resistant to benomyl are often resistant to thiophanate methyl, both benzimidazole fungicides. If a resistant strain of a fungus is present, it is often appropriate to choose a fungicide of a different chemical class (Table 36.2).

The development of fungicide resistance can be slowed by the following strategies:

- Using labeled rates for pesticide applications. The use of less than labeled rates may accelerate the development of resistance.
- Alternate or mix a fungicide of a different chemical class to slow resistance. Fungicides vulnerable to resistance should be used sparingly.
- Use fungicides as a part of an integrated disease management program that includes biological (Chapter 37), cultural control (Chapter 35), and host resistance (Chapter 34) when available.

NONTARGET EFFECTS OF FUNGICIDES

The use or repeated use of certain fungicides may have unexpected results. The application of some fungicides may inhibit the growth of some fungi while having no effect on others. Fungicides applied to soil may reduce the growth of soil fungi, such as *Trichoderma* species, which are important competitors of nutrients and potential parasites of plant pathogens. The result is the increased incidence of diseases caused by plant pathogens that are tolerant of the fungicide. Some SBI fungicides are closely related to plant growth regulators. Use at higher than labeled rates and/or at shortened intervals may lead to shortened internodes and stunting of turfgrass; the use of high rates of SBI fungicides as seed treatments may reduce the growth of small grains such as barley or wheat. Broad spectrum fungicides that inhibit the growth of saprophytic fungi in turf areas may lead to increased thatch layers.

BACTERICIDES

Bactericides are used to protect plants from bacterial plant pathogens. Antibiotics such as streptomycin sulfate and inorganic metal compounds such as basic copper sulfate and copper hydroxide are used to combat bacterial diseases. Streptomycin is produced by the actinomycete *Streptomyces griseus*. It is most commonly used to prevent fire blight of pome fruit trees (Chapter 6) and some ornamental hosts during bloom and is often ineffective during the shoot blight phase of fire blight. Copper compounds are used as protectant treatments for foliar bacterial diseases. The copper ion is toxic to bacterial cells. Fixed copper compounds, such as copper hydroxide, are relatively insoluble in water, but release enough copper ions to inhibit the growth of bacteria. Overuse of streptomycin or copper compounds may lead to resistance.

DISINFECTANTS

Disinfectants are used to eliminate bacteria, fungi, and algae from the surfaces of greenhouse benches, pots, flats, tools, and equipment. Common disinfectants include the following: phenolic compounds, alcohol, sodium hypochlorite, quarternary ammonium compounds, and hydrogen peroxide. Disinfectants have little residual activity and dissipate shortly after use, especially if they are exposed to organic matter. Items to be disinfected should be relatively free of soil or organic matter.

CHEMICAL-INDUCED SYSTEMIC ACQUIRED RESISTANCE (SAR)

Plants, unlike animals, do not have an immune system to deter disease. Plants do have defense mechanisms that may be triggered by infection by plant pathogens. Pathogens that cause localized lesions on leaves may activate a systemic defense reaction that protects the entire plant from infection by other plant pathogens. This systemic-acquired resistance (SAR) (Chapters 28 and 31) may also be activated by chemicals, such as acidbenzolar-S-methyl (ASM). Chemicals that induce SAR are generally not toxic to plant pathogens, but work by stimulating the defense system of the plant. As ASM is transported systemically, it is distributed uniformly throughout the plant.

Chemicals that induce SAR have been used successfully to protect plants from diseases such as blue mold (tobacco), downy mildew (lettuce), powdery mildew (wheat), and bacterial speck (tomato). Activity varies depending on disease and plant species. In general, chemicals that induce SAR should be applied several times at weekly intervals prior to possible infection. Traditional fungicides provide short term protection, whereas the chemical that induces SAR provides long term control. Chemically induced SAR may be used to reduce the frequency and quantity of fungicides applied to crops. Since these chemicals do not act directly on plant pathogens, they are at lower risk for inducing pathogen resistance.

Chemical control of plant diseases is an important management strategy that works well when integrated into an overall program that includes host resistance, and biological and cultural controls. The use of chemicals to control plant diseases generally works well if they are applied according to label directions as preventative treatments. This doesn't mean that fungicides should be applied indiscriminately. Consider the cropping history of the field, landscape bed, or turf area. Have certain diseases been observed causing significant damage each year for several years? Have you mapped areas of a field or noted certain golf greens that are disease prone? Can you change the environment in a greenhouse or increase air circulation on a golf green? What cultural practices or inputs could you modify to decrease the incidence or severity of disease? Many factors can affect the use of chemicals in the management of plant diseases including societal and environmental concerns.

Fungicides are valuable tools in crop protection, but should only be used to complement other disease control tactics. Fungicides should not be used to cure diseased plants, but to keep healthy plants healthy or to minimize the damage of plant diseases. They are not a panacea or a "silver bullet"; however fungicides are useful tools for plant protection.

EXERCISES

EXPERIMENT 1. READING AND COMPREHENDING A PESTICIDE LABEL AND MATERIAL SAFETY DATA SHEET

Pesticide labels contain essential information about the safe use of pesticides. Additional facts about mixing, application, and storage may be found on the label. Information on the label should be reviewed before purchasing, mixing, applying, storing, or disposing of the pesticide. Pesticide applicators or farm workers that are exposed to pesticides should be familiar with information on how to handle an accidental poisoning or spill prior to the incident. The goal of this exercise is to familiarize students with the safety and use information found on a pesticide label and material safety data sheet (MSDS).

Materials

- Provide each student with a sample pesticide label from a product guide or agrichemical company Website.
- MSDS sheets may be found in the back of product guides or agrichemical company Website.

Follow the instructions in Procedure 36.1 to complete the study.

Questions

- Who would you call for questions about pesticide use or safety?
- What would you do if a pesticide spill occurred at your business?

| **Procedure 36.1** | |
| Reading Pesticide Labels and MSDS Sheets | |
Step	Instructions and Comments
1	Find the following information on a sample pesticide label:
	Trade name:
	Manufacturer:
	Common name:
	Formulation:
	Percent active ingredient:
	EPA registration number:
	Signal word:
	Precautionary statements: (Are there any hazards to humans or wildlife?)
	Does the pesticide list worker protection standards (WPS)?
	What is the reentry interval (REI)?
	What personal protective equipment (PPE) is required?
	List any environmental hazards:
	List any restrictions:
	How is the pesticide to be applied:
	List emergency numbers (Chemtrec, etc.):
2	Find the following information on the MSDS sheet:
	Pesticide trade name:
	Common name:
	Toxicological information:
	Acute effects of exposure:
	Chronic effects of exposure:
	Carcinogen (yes or no):
	Teratogenicity (birth defects):
	Reproductive effects:
	Neurotoxicity:
	Mutagenicity (genetic effects):
	Accidental spill or leak information:
	Ecological information:
	Handling and storage requirements:

- How should pesticides be stored?
- What does it mean if a pesticide is a carcinogen? Mutagenic?

EXPERIMENT 2. EFFECT OF NOZZLE SIZE ON WATER OUTPUT OF A SMALL, HAND SPRAYER

Sprayer calibration is essential for the safe, economical and effective application of pesticides. Hand sprayers are used to spray small areas of lawns, shrubs, or small trees and may be used to apply fungicides, insecticides, and herbicides. The volume of water applied to a given area is often determined by the pesticide and target pest. For example, many turf herbicides are applied in 1 gal water/1000 sq ft, whereas most turf fungicides are applied in 3 gal water/1000 sq ft since good coverage is necessary to protect the foliage.

The amount of water output of most agricultural sprayers over a given area is determined by travel time (in the case of hand sprayers, walking speed), pressure, and nozzle type and size. With most hand sprayers, output is controlled by nozzle size as pressure is constant if the sprayer is continually pumped. Constant pressure is easier to maintain with backpack sprayers that can be pumped during application. Calibrate the sprayer in a parking lot or other area where the spray pattern of the nozzle on the ground is easily observed.

Procedure 36.2

The Effect of Nozzle Size on Output of a Small, Hand Sprayer

Step	Instructions and Comments
1	Measure a test area of 500 sq ft (e.g., 20 × 25 ft) on a parking lot (where it is easy to see the coverage and spray pattern) or a grassy area.
2	Fill sprayer with a measured volume of water or to a marked level on the sprayer.
3	Uniformly spray the test area. Make sure that the surface is evenly wet.
4	Release compressed air and determine the amount of water used by measuring the water remaining in the sprayer or the amount of water needed to raise the water level to the initial level.
5	Calculate the application rate of the sprayer.
	Method 2: 128th of an acre calibration method.
6	Measure a test area of 340 sq ft (c.a. 18.5 × 18. 5ft).
7	Repeat steps 2–4 from above.
8	Another method for determining water output is to record the amount of time that it takes to spray the test plot. Spray the test plot three times and average the spray times.
9	Fill sprayer with water, pressurize, and catch the output in the 1-L graduated cylinder for the average time required to spray the test plot. The output in fluid ounces is equal to the number of gallons the sprayer would apply to an acre.

Materials

The following items are required for each student or team of students:

- Carboy of water or easily accessible source of water
- A 1 to 3 gal compressed air hand sprayer or a Solo backpack sprayer
- Flat fan nozzles (Delevan D1, D3, D5, or Spraying Systems TK1, TK3, TK5)
- 1-L graduated cylinder

Follow the protocols listed in Procedure 36.2 to complete the exercise.

Anticipated Results

Flat fan nozzles should give uniform coverage. Water output should have increased as nozzles with larger orifice sizes were tested. If adequate coverage is not provided at a normal walking speed, then nozzle size should be increased or walking speed decreased.

Questions

- How would you adjust water output of a compressed air hand sprayer? A tractor-mounted hydraulic sprayer?
- What types of nozzles are used to apply pesticides?

SUGGESTED READING

Agrios, G.N. 2005. *Plant Pathology*. 5th ed. Academic Press, New York. pp. 952.

Bohmont, B.L. 1990. *The Standard Pesticide User's Guide*. Prentice Hall, Englewood Cliffs, NJ.

Hassall, K.A. 1990. *The Biochemistry and Uses of Pesticides*. VCH Publishers, New York.

Hopkins, W.L. 1996. *Global Fungicide Directory*. Ag Chem Information Services, Indianapolis.

Jeffers, S.N., R.W. Miller, and C.C. Powell, Jr. 2001. Fungicides for ornamental crops in the nursery. In Jones, R.K. and D.M. Benson (Eds.). *Diseases of Woody Ornamentals and Trees in Nurseries*. APS Press, St. Paul Minnesota, pp. 409–416.

Köller, W. 1999. Chemical approaches to managing plant pathogens in J.R. Ruberson (Ed.). *Handbook of Pest Management*. Marcel Dekker, New York.

Matthews, G.A. 2000. *Pesticide Application Methods*. 3rd ed. Blackwell Science, London.

Marsh, R.W. 1977. *Systemic Fungicides*. Longman Press, New York.

Parry, D.W. 1990. *Plant Pathology in Agriculture*. Cambridge University Press, Cambridge.

Simone, G.W. 2001. Bactericides and disinfectants. In Jones, R.K. and D.M. Benson (Eds.). *Diseases of Woody Ornamentals and Trees in Nurseries*. APS Press, St. Paul, Minnesota. pp. 417–422.

Staub, T., Kunz, W., and Oostendorp, M. 2001. Chemical activators of disease resistance in crop protection. In *Encyclopedia of Agrichemicals*, John Wiley & Sons, New York.

Waller, J.M., J.M. Lenne', S.J. Waller. 2002. *Plant Pathologist Pocketbook*. CABI Publishing, New York.

37 Biological Control of Plant Pathogens

Bonnie H. Ownley and Mark T. Windham

CHAPTER 37 CONCEPTS

- Plant diseases can be controlled by living microorganisms that are antagonistic to plant pathogens.

- The phenomenon of suppressive soils involves an increase in populations of microorganisms that are antagonistic to disease-causing plant pathogens.

- Mechanisms of biological control of plant pathogens include antibiosis, parasitism, competition, induced systemic resistance, cross protection, and hypovirulence.

- Biological control of plant diseases often involves multiple mechanisms.

Biological control is the use of natural or modified organisms, genes, or gene products, to reduce the effects of undesirable organisms such as plant pathogens, and to favor desirable organisms such as crops (Research Briefings, 1987). This definition is broad and includes genetic modification (genetic resistance) of the host plant. However, the main focus of this chapter will be on natural and modified organisms as biological control agents of plant pathogens. Biocontrol agents are known also as **antagonists**, and **antagonism** is the generalized mechanism that they use to reduce the survival or disease-causing activities of plant pathogens. Antagonism is actively expressed opposition and includes antibiosis, competition, and parasitism. Biological control of plant diseases with antagonists is accomplished by destroying existing pathogen inoculum, excluding the pathogen from the host plant, or suppressing or displacing the pathogen after infection has occurred (Cook and Baker, 1983).

DISEASE SUPPRESSIVE SOILS

Research on the biological control of plant pathogens followed the discovery of **disease suppressive soils**. These are soils in which pathogens either cannot be established, can be established but fail to produce disease, or can be established and cause disease at first, but disease becomes less important with continued culture of the crop (Baker and Cook, 1974). Disease suppression can be characterized as general or specific. General suppression results from the activity of the total microbial biomass in soil and is not transferable between soils; specific suppression results from the activity of individual or select groups of microorganisms and is transferable (Weller et al., 2002). One of the best documented examples of disease suppressive soils is the take-all decline phenomenon. Take-all root disease,

caused by the soilborne fungus *Gaeumannomyces graminis* var. *tritici*, is a root and crown (basal stem) rot disease of wheat (*Triticum aestivum*) and barley (*Hordeum vulgare*) that occurs worldwide in temperate regions. In some soils, take-all decline occurs naturally following 5 to 7 years of growing continuous wheat with severe take-all disease and poor yields. Then the disease becomes less severe and yields recover. Increases in specific populations of microorganisms have been associated with take-all decline. In the United States in the Pacific Northwest, an increase in populations of antibiotic-producing pseudomonad bacteria has been correlated with take-all decline. In Australia, increases in populations of the fungus *Trichoderma* are thought to be responsible for take-all suppressive soils.

WHY IS BIOLOGICAL CONTROL POPULAR?

Scientific interest in biological control of plant pathogens has been spurred by growing public concerns over the potentially harmful effects that some chemical pesticides pose to human health and the environment. There is also a need to control various diseases for which there are currently no controls or only partial control because there is little or no genetic resistance in the host, crop rotation is impractical or not economically feasible or reliable, or economical chemical controls are not available. For example, no practical or economical chemical control for crown gall disease was replaced when biological control with *Agrobacterium radiobacter* K84 was developed (Cook, 1993). In addition, registered biological controls are generally labeled with shorter reentry times and preharvest intervals than conventional chemical pesticides. This gives growers greater flexibility in balancing their operational and pest management procedures (McSpadden Gardener and Fravel, 2002).

PROBLEMS WITH BIOLOGICAL CONTROL

Although an array of microorganisms has been shown to protect crop plants from disease under experimental conditions, commercial development of many antagonists has been hampered due to inconsistent performance between field locations and seasons. Variation in performance of biological control agents has been attributed to many factors. These include compatibility of the host plant and the biocontrol agent due to host plant genotype, agricultural practices, mutation of the biocontrol organism resulting in a loss of effectiveness, resistance of the pathogen to biocontrol mechanisms, vulnerability of the biocontrol agent to defense mechanisms of the pathogen, and effects of the environment on survival and effectiveness of the biocontrol agent.

Antagonists are living organisms and whether applied directly to the host plant, to field soil, or to planting medium in a greenhouse, the antagonists will only be active if their growth and reproduction are favored by the environment. A change in environmental conditions during the growing season may have profound effects on the ability of a biocontrol agent to control a plant pathogen, whereas the same environmental changes may have little effect on the ability of a chemical pesticide to control the pathogen. Preparation and storage of antagonist inocula and application of the antagonist often have exacting requirements. Inocula of biocontrol agents usually cannot be stored at extreme temperatures that might be satisfactory for storing a wettable powder fungicide. In addition, shelf life of antagonist inocula is not as long as that of a chemical pesticide, so growers cannot stockpile large quantities of antagonist inocula for later use.

MECHANISMS OF ANTAGONISM

Antagonists used for biological control of plant pathogens include bacteria, fungi, nematodes, protozoa, and viruses. With some exceptions, antagonists often are not pathogen specific; instead their effect on plant pathogens is coincidental (Cook and Baker, 1983). For example, an antagonist that aggressively colonizes roots and inhibits a wide range of microorganisms may protect roots against pathogens but the effect against any particular pathogen is probably coincidental. However, there are also antagonists that have a true parasitic relationship with their microbial host.

Antagonists interfere with plant pathogens through antibiosis, competition, and parasitism. These mechanisms are not mutually exclusive. An antagonist may utilize multiple mechanisms to detrimentally affect a plant pathogen or may use one mechanism against one type of pathogen and a different mechanism against another. For example, control of *Botrytis* on grapes (*Vitis*) with the fungal antagonist *Trichoderma* involves competition for nutrients and parasitism of **sclerotia**. Both mechanisms

contribute to the suppression of the pathogen's capability to cause and perpetuate disease (Dubos, 1987).

ANTIBIOSIS

Antibiosis is the inhibition or destruction of one organism by a metabolite produced by another organism. Antagonists may produce powerful growth inhibitory compounds that are effective against a wide array of microorganisms. Such compounds are referred to as broad spectrum **antibiotics**. On the other hand, some metabolites, such as **bacteriocins**, are effective only against a specific group of microorganisms. The bacterial antagonist *Agrobacterium radiobacter* K1026 produces agrocin 84, a bacteriocin that is only effective against bacteria that are closely related to *A. radiobacter*, such as the crown gall pathogen *A. tumefaciens*. Antagonists that produce antibiotics have a competitive advantage in occupying a particular niche and securing substrates as food sources because their antibiotics suppress the growth or germination of other microorganisms.

Antibiosis can be an effective mechanism for protecting germinating seeds. For example, the bacterial antagonist *Pseudomonas fluorescens* Q2-87 can protect wheat roots against the take-all pathogen, *Gaeumannomyces graminis* var. *tritici*, when coated onto seed. As the seeds germinate, the bacteria multiply in the rhizosphere, using exudates from the roots as a food source. The **rhizosphere** is the thin layer of soil that adheres to the root after loose soil has been removed by shaking and is directly influenced by substances that are exuded by the root into soil solution. The antibiotic, 2,4-diacetylphloroglucinol, which is produced by *P. fluorescens* Q2-87, is effective against the take-all pathogen in minute quantities and has been recovered from the wheat rhizosphere (Bonsall et al., 1997). However, the effectiveness of antibiotics in soil can be variable as they can become bound to charged clay particles, degraded by microbial activity or leached away from the rhizosphere by water.

ANTIBIOTICS

Many antibiotics synthesized by microbial antagonists are not produced to target a specific pathogen, and a specific antagonist may not produce the same antibiotics under different environmental conditions. Some antagonists produce several bioactive compounds that are effective against different plant pathogens. The bacterial antagonist *P. fluorescens* Pf-5 produces multiple antibiotics, including pyoluteorin, pyrrolnitrin and 2,4-diacetylphloroglucinol. Pyoluteorin inhibits *Pythium ultimum*, a common cause of seedling disease in cotton (*Gossypium hirsutum*), however it has little effect on other cotton seedling pathogens, such as *Rhizoctonia solani*, *Thielaviopsis basicola,* and *Verticillium dahliae* (Howell and Stipanovic, 1980). Pyrrolnitrin inhibits *R. solani, T. basicola,* and *V. dahliae,* but is not active against *P. ultimum* (Howell and Stipanovic, 1979). Examples of bacterial biological control agents that produce antibiotics include *Bacillus, Pseudomonas,* and *Streptomyces*. Fungal antagonists that produce antibiotics include *Gliocladium* and *Trichoderma*.

VOLATILE COMPOUNDS AND ENZYMES

Several volatile substances have a role in biocontrol of plant pathogens. These include ammonia (produced by the bacterial antagonist *Enterobacter cloacae* against the plant pathogens *P. ultimum*, *R. solani*, and *V. dahliae*), alkyl pyrones (produced by *T. harzianum* against *R. solani*), and hydrogen cyanide (produced by *P. fluorescens* against *T. basicola*, which causes black root rot). A mixture of extremely bioactive volatile organic compounds (VOCs) is produced by the biological control agent *Muscodor albus*, which is a fungal **endophyte**. The VOCs produced by *M. albus* include various alcohols, acids, esters, ketones, and lipids (Strobel, 2006). The synergistic effects of the VOCs produced by *M. albus* are lethal to a variety of fungi and bacteria (Strobel et al., 2001).

Although the biocontrol mechanism of parasitism involves many enzymes, there are enzymes that are only involved in antibiosis. For example, the fungal biocontrol agent *Talaromyces flavus* Tf1 is effective against verticillium wilt of eggplant (*Solanum melongena*). *Talaromyces flavus* produces the enzyme glucose oxidase; hydrogen peroxide is a product of glucose oxidase activity, and hydrogen peroxide kills microsclerotia of *Verticillium* in soil (Fravel, 1988). The enzyme alone does not kill microsclerotia.

COMPETITION

Competition is the result of two or more organisms trying to utilize the same food (carbon and nitrogen) or mineral source, or occupy the same niche or infection site. The successful competitor excludes the others due to more rapid growth or reproductive rate, or is more efficient in obtaining nutrients from food sources. *Pseudomonas fluorescens* produces a **siderophore**, pseudobactin, which deprives pathogens such as *Fusarium oxysporum* of iron. Siderophores are extracellular, low molecular weight compounds of microbial origin with a very high affinity (attraction) for ferric iron. Chlamydospores of *F. oxysporum* require an exogenous source of iron to germinate. Although *F. oxysporum* also produces siderophores, the siderophores of *P. fluorescens* are more efficient in binding iron. If *P. fluorescens* is active in soil, chlamydospores of *F. oxysporum* remain dormant and cannot germinate due to low iron conditions.

Biological control of annosus root rot of conifers is another example of using competition to control a plant disease. Annosus root rot is caused by the fungus *Heterobasidion annosum*, which can survive for many years in stumps and logs, and causes extensive damage in managed forests or plantations of pure stands. In Europe, freshly cut stumps are inoculated with the fungal antagonist *Phanerochaete* (= *Phlebia*) *gigantea* to control annosus root rot. The mycelium of *P. gigantea* physically prevents *H. annosum* from colonizing stumps that it would use as a food base for attacking young pine (*Pinus*) trees (Cook and Baker, 1983).

Cross protection is a form of competition in which an avirulent or weakly virulent strain of a pathogen is used to protect against infection from a more virulent strain of the same or closely related pathogen. This term originated in virology to describe cases where infection of a cell by one virus reduces the likelihood that a second virus would damage the cell (Chapter 4). There is no host response in cross protection (see induced resistance discussion below) and it is not transmissible (see hypovirulence discussion below). Inoculation with a small satellite-like self-replicating RNA molecule, known as CARNA 5, can be used to reduce virulence of *Cucumber mosaic virus* (CMV) in some vegetable crops. However, it is important to remember that weakly virulent strains can become more virulent or there may be unexpected synergistic effects that enhance disease. For example, CARNA 5 reduces disease caused by CMV in squash (*Cucurbita pepo*) and sweet corn (*Zea mays*), but will enhance disease severity caused by CMV in tomato (*Lycopersicon esculentum*) (Cook and Baker, 1983).

PARASITISM

Parasitism is the feeding of one organism on another organism. As a mechanism of biocontrol, parasitism can be successfully used to reduce inoculum of sclerotia-forming fungi or prevent root rots, but may be less effective in protecting germinating seeds as establishing a parasitic relationship between the antagonist and pathogen may take more time than the time needed for the seed to become infected by the pathogen.

Mycoparasitism is the term used when one fungus parasitizes another. Parasitism by the fungal antagonist *Trichoderma* often begins with detection of the fungal host (plant pathogen) from a distance. The hyphae of *Trichoderma* grow toward a chemical stimulus given by the pathogen. This is followed by recognition, which is physical or chemical in nature, and attachment of *Trichoderma* hyphae to the host fungus as evidenced by coiling (Figure 37.1). *Trichoderma* produces lytic enzymes that degrade fungal cell walls. In some cases, cell-wall degrading enzymes and antibiotics act synergistically in the biocontrol process (Chet et al., 1998). Another well-known biological control mycoparasite is *Coniothyrium minitans* (current taxonomic name = *Paraconiothyrium minitans*), which effectively parasitizes **sclerotia** of *Sclerotinia sclerotiorum* and *S. minor* (Figure 37.2). Other mycoparasites include *Pythium nunn* and *P. oligandrum*.

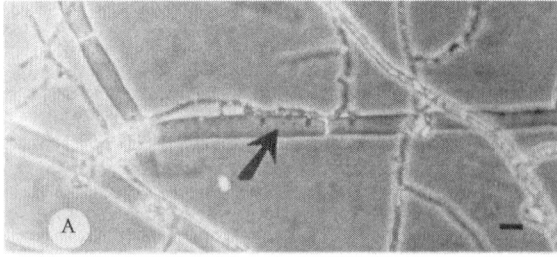

FIGURE 37.1 The smaller hypha of *Trichoderma virens* has penetrated the surface of the *Rhizoctonia solani* hypha at the arrow. Black line on lower right = 10 μm. (Photomicrograph reprinted from Howell, C.R. 1982. *Phytopathology* 72: 496-498.)

OTHER MECHANISMS OF BIOLOGICAL CONTROL

HYPOVIRULENCE

Biological control by **hypovirulence** occurs when a hypovirulent (weakly virulent) strain of a fungal pathogen fuses (**anastomosis**) with a virulent strain of the pathogen and transmits the hypovirulent condition to the virulent strain. Anastomosis is fusion of touching hyphae and represents vegetative compatibility. Transmissible hypovirulence results from the infection of the pathogen with one or more dsRNA of viral origin. The classic example of hypovirulence is biocontrol of chestnut blight caused by the fungal pathogen, *Cryphonectria parasitica,* with hypovirulent strains of the fungus. In Europe, it was noted that cankers on chestnut (*Castanea*) trees were healing or were only superficial. Atypical strains of *C. parasitica* were isolated from the healing cankers. They had reduced pigmentation and sporulation. These atypical strains were also less virulent than strains from cankers that did not heal. In addition, when hyphae from a virulent strain were allowed to fuse with hyphae of a strain from a healing

canker, the virulent strain became hypovirulent. Essentially, the hypovirulent strain converted the virulent strain to hypovirulent by the transfer of dsRNA via hyphal anastomosis (Heiniger and Rigling, 1994).

INDUCED SYSTEMIC RESISTANCE

Colonization of plants with nonpathogenic plant-growth-promoting rhizobacteria (PGPR) can cause **induced systemic resistance** (ISR) in the host plant. (For additional information on ISR see Chapters 28 and 31). ISR is a plant-mediated mechanism of biocontrol where the biocontrol agent and the pathogen do not come in contact with one another. In ISR, host plant defenses are stimulated and plants are protected systemically. The level of host response is modulated by jasmonic acid (JA) and ethylene (van Loon et al., 1998). As many PGPR also produce antibiotics, iron-chelating siderophores or lytic enzymes, their ability to suppress disease may involve more than one mechanism. PGPR may mediate biological disease control or plant growth promotion or both. Several strains or combinations of strains are available as commercial products for crop protection (McSpadden Gardener and Fravel, 2002). Induction of systemic resistance via the JA/ethylene signaling pathway has also been reported for some biocontrol fungi, including *Trichoderma hamatum* strain 382 (Han et al., 2000) and *T. asperellum* strain T203 (Shoresh et al., 2004).

INCREASED GROWTH RESPONSE

Microorganisms utilized in biological control, including PGPR and certain fungal biocontrol agents, can be associated with enhanced plant growth. In some cases, increased growth of the host plant is due to a reduction of viable inoculum of undetected pathogens, such as root-infecting *Pythium* species, which cause only slight reductions

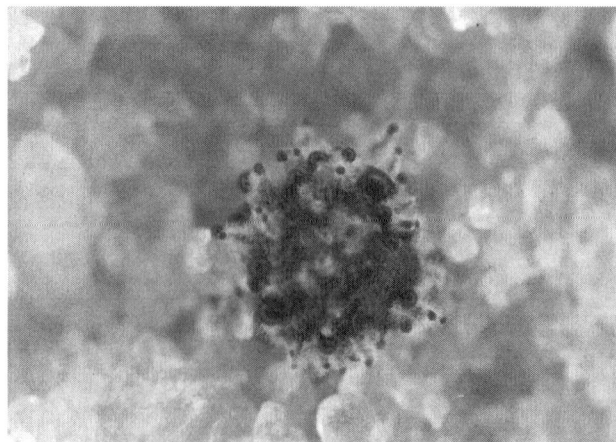

FIGURE 37.2 Sclerotium of *Sclerotinia sclerotiorum* parasitized by *Coniothyrium minitans* (current taxonomic name = *Paraconiothyrium minitans*). (Photo reprinted from Clarkson, J. and J. Whipps. Revised February, 11, 2005. http://www2.warwick.ac.uk/fac/sci/whri/research/biodiseasecontrol/conio/biocontrol/. With permission).

in vigor or yield. In other cases, enhanced plant growth, particularly in the absence of pathogens, may be due to plant growth-promoting compounds of microbial origin (Chet 1993).

DELIVERY, APPLICATION, AND FORMULATION OF BIOCONTROL PRODUCTS FOR PLANT DISEASES

There are various delivery and application methods for applying biocontrol products, including foliar sprays, soil or soilless mix treatments, seed treatments, root dips, and postharvest applications to fruit as a drench, drip, or spray. Formulations of commercially available products include dusts, dry and wettable powders, dry and water-dispersible granules, and liquids (Desai et al., 2002). Foliar sprays are most often used to protect aboveground plant parts from a foliar pathogen. Incorporating inocula of an antagonist into soil is a common approach to biological control of soilborne pathogens. Usually the antagonist is added in a dormant condition, often with a food base so that it can become established quickly in the soil. Commercially available soilless mixes have been marketed in the horticultural industry, that contain one or more biological control agents established in the mixes. The mixes are suppressive to pathogen populations under a prescribed set of environmental conditions. Adding organic amendments such as ground shrimp or crab shells or bean leaf powder to greenhouse mixes or field soil has been used also to stimulate the growth of antagonistic organisms at the expense of plant pathogens.

One of the more successful methods of introducing biocontrol agents into an agricultural system involves treating seeds with antagonists. This approach delivers antagonists of soilborne pathogens as close to the target as possible. Seed treatments place the antagonist in the infection court (the seed coat surface) at planting, before the seed can be attacked by the pathogen. Formulated products can be applied directly at the time of planting as powders or liquids without stickers. Alternatively, antagonists can be pre-coated onto seed as dry powders or liquid-based formulations. Additives are used to prolong the survival of biocontrol agents on seeds. Once the seeds dry, they can be stored with the antagonists until planting. In some biocontrol products, bacterial antagonists and chemical fungicides are applied in combination to seeds (Warrior et al., 2002).

COMMERCIALLY AVAILABLE PRODUCTS

In the past fifteen years, there has been a significant increase in the number of commercially available biological control products for plant disease management. In 1993, there were only seven products registered with the U.S. Environmental Protection Agency (Cook, 1993); currently (2007) there are 35 products registered with the U.S. Environmental Protection Agency (http://www.epa.gov/pesticides/biopesticides/ingredients/index.htm). On the worldwide market, there are approximately 50 biological control products for plant diseases (Desai et al., 2002). Increased understanding of the complex interactions that occur within the microbial community of agricultural soils and improved formulations of biocontrol agents to ensure their survival between production and use are needed for future development of more efficacious, cost-effective products.

CASE STUDIES

The following examples of biological control agents illustrate the diversity and complexity of the interactions between biocontrol agents and their target plant pathogens (Case Studies 37.1–37.7).

CASE STUDY 37.1

AGROBACTERIUM RADIOBACTER STRAIN K1026

Biocontrol agent—*Agrobacterium radiobacter* K1026, a Gram-negative bacterium.

- *Target pathogens*—*Agrobacterium tumefaciens* and *A. radiobacter*, Gram-negative soilborne bacteria.
- *Disease*—Crown gall of fruits (apricot, cherry, nectarine, peach, plum, prune), nuts (almond, pecan, walnut), and ornamental nursery stock (euonymus, rose).
- *Mechanisms of biocontrol*—*A. radiobacter* strain K1026 produces the bacteriocin agrocin 84, which has antimicrobial activity against *A. tumefaciens*. It also competes with the crown gall pathogen for nutrients and attachment sites on plant roots.
- *Application and formulation*—Cells of *A. radiobacter* produced in culture medium are suspended in nonchlorinated water and applied to seeds, seedlings, cuttings, roots, stems, and as a soil drench (McSpadden Gardener and Fravel, 2002; U.S. Environmental Protection Agency, 2007).

CASE STUDY 37.2

Bacillus subtilis GB03

Biocontrol agent—*Bacillus subtilis*, a Gram-positive bacterium.

- *Target pathogens*—Fungi that attack plant roots, including *Rhizoctonia*, *Fusarium*, *Alternaria*, *Aspergillus*, and *Phytophthora*.
- *Diseases*—Root and seed rots, and seedling diseases of many agricultural and ornamental crops.
- *Mechanism of biocontrol*—*B. subtilis* GB03 colonizes plant roots and promotes plant growth. Strain GB03 also produces volatiles that induce a systemic resistance response in plants (Ryo et al., 2004) and antibiotics.
- *Application and formulation*—Seed treatment either by (1) mixing the powdery *Bacillus subtilis* product with seeds in a planter box at the time of planting, or by (2) preparing a slurry mix that includes *Bacillus* product, seeds, insecticides, and/or fungicides, and water. The slurry is continuously agitated and used within 72 h (McSpadden Gardener and Fravel, 2002; U.S. Environmental Protection Agency, 2007).

CASE STUDY 37.3

Coniothyrium minitans CON/M/91-08 (Current Taxonomic Name = *Paraconiothyrium minitans*)

Biocontrol agent—*Coniothyrium minitans*, a deuteromycete.

- *Target pathogens*—*Sclerotinia sclerotiorum* and *S. minor*, soilborne ascomycete fungi.
- *Diseases*—Sclerotinia diseases of many agronomic and ornamental crops including white mold of snap beans, pink mold of celery, stem rot of sunflower and soybean, lettuce drop, white mold of canola, and Sclerotinia blight of peanut.
- *Mechanism of biocontrol*—*C. minitans* is a parasite of *Sclerotinia*. It attacks and parasitizes the sclerotia, or survival stage of the pathogen (Figure 37.2). This leads to a reduction in the amount of *Sclerotinia* inoculum available to cause disease in the following crop.
- *Application and formulation*—*C. minitans* is formulated as a water dispersible granule and sprayed onto agricultural soils, followed by mechanical mixing into the first 1 to 2 in. of the topsoil layer. *C. minitans* is applied after harvest or before planting (McSpadden Gardener and Fravel, 2002; U.S. Environmental Protection Agency, 2007).

CASE STUDY 37.4

Pseudomonas syringae Strains ESC-10 and ESC-11

Biocontrol agent—*Pseudomonas syringae* strains ESC-10 and ESC-11, Gram-negative bacteria.

- *Target pathogens*—Postharvest pathogens, including *Botrytis cinerea*, *Penicillium* spp., *Mucor pyroformis*, and *Geotrichum candidum*; and potato pathogens *Fusarium sambucinum* and *Helminthosporium solani*.
- *Diseases*—Postharvest diseases of citrus, cherries, and pome fruits; and dry rot and silver scurf of potato (Figure 37.4).
- *Mechanism of biocontrol*—Competition is the primary mechanism; however, *P. syringae* strains ESC-10 and ESC-11 also produce the antibiotic syringomycin E, thus antibiosis may play a minor role (Stockwell and Stack, 2007).
- *Application and formulation*—Commodities are harvested and cleaned, then a solution containing *P. syringae* is applied by spraying or by dipping the produce into the solution (McSpadden Gardener and Fravel, 2002; U.S. Environmental Protection Agency, 2007).

CASE STUDY 37.5

Muscodor albus QST 20799

Biocontrol agent—*Muscodor albus* QST 20799, non–spore-forming fungus.

- *Target pathogens*—Bacteria, fungi, and nematodes.
- *Diseases*—Various diseases of field, greenhouse, and stored crops, as well as cut flowers.
- *Mechanism of biocontrol*—Produces volatile organic compounds that are antimicrobial.
- *Application and formulation*—Fungus is incorporated into soil or is added to containers where it does not contact treated commodities (U.S. Environmental Protection Agency, 2007).

CASE STUDY 37.6

Streptomyces lydicus Strain WYEC 108

Biocontrol agent—*Streptomyces lydicus* strain WYEC 108, a streptomycete.

- *Target pathogens*—Root decay fungi such as *Aphanomyces, Armillaria, Fusarium, Geotrichum, Monosporascus, Phymatotrichopsis, Phytophthora, Postia, Pythium, Rhizoctonia, Sclerotinia,* and *Verticillium.* Other target pathogens attack plant foliage, including powdery mildews.
- *Diseases*—Various root and foliar diseases of greenhouse and nursery crops, and turf.
- *Mechanism of biocontrol*—*S. lydicus* colonizes the growing root tips of plants and parasitizes root decay fungi. It may also produce antibiotics that act against these fungi.
- *Application and formulation*—*S. lydicus* is formulated in a water dispersible granule which is mixed with water and applied to soil mix or as a drench to turf grass or potted plants. It can also be applied to plant foliage in greenhouses (McSpadden Gardener and Fravel, 2002; U.S. Environmental Protection Agency, 2007).

CASE STUDY 37.7

Trichoderma harzianum KRL-AG2 (T-22)

Biocontrol agent—*Trichoderma harzianum* KRL-AG2 (T-22), a deuteromycete.

- *Target pathogens*—Several soilborne fungal pathogens, including *Cylindrocladium, Fusarium, Pythium, Rhizoctonia solani, Sclerotinia,* and *Thielaviopsis basicola.*
- *Diseases*—Damping-off and root rot diseases of many different plants, including ornamentals, food crops, and trees.
- *Mechanisms of biocontrol*—*T. harzianum* is a parasite of other fungi and can rapidly colonize plant roots, thereby out-competing pathogens for nutrients and space. *Trichoderma harzianum* also promotes plant growth in the absence of pathogens.
- *Application and formulation*—*T. harzianum* is formulated as a granule that can be mixed with seed, or incorporated into soil or potting mix prior to planting, or mixed with water and applied as a drench (McSpadden Gardener and Fravel, 2002).

EXERCISES

The laboratory exercises are designed to demonstrate the biocontrol mechanisms of antibiosis and parasitism, and the suppressive effect of biological control agents on disease caused by plant pathogens.

GENERAL CONSIDERATIONS

These experiments are designed for teams of two or more students, depending on space and resources. The bacteria and fungi used in these exercises can be purchased from the American Type Culture Collection (ATCC) (http://www.atcc.org/). An approved permit from the Plant Protection Quarantine section of the U.S. Department of Agriculture, Animal Plant Health Inspection Service is required to purchase or for interstate transport of *Rhizoctonia solani* (http://www.aphis.usda.gov/forms/). Permits are not required for nonpathogens, such as *Pseudomonas fluorescens* and *Trichoderma virens*. It is very important to maintain aseptic conditions when conducting Experiments 1 and 2, and preparing cultures of microorganisms.

EXPERIMENT 1. SUPPRESSION OF MYCELIAL GROWTH OF *RHIZOCTONIA SOLANI* BY ANTIBIOTIC-PRODUCING *PSEUDOMONAS FLUORESCENS* PF-5

Pseudomonas fluorescens Pf-5 was originally isolated from the surface of a cotton root. It produces several antibiotics, including pyrrolnitrin, pyoluteorin and 2, 4-diacetylphloroglucinol. It also produces the toxin hydrogen cyanide, and two siderophores, pyochelin and pyoverdin, which are involved in competition for iron in low-iron environments. *Pseudomonas fluorescens* Pf-5 is an antagonist of the soilborne fungal pathogen *R. solani*, which causes seedling diseases of cotton and many other plants. Antibiosis plays a major role in the biocontrol of *R. solani* on cotton seedlings with *P. fluorescens* Pf-5.

Materials

Each team of students will require the following items:

- Microbiological media: Dehydrated media. Follow the manufacturer's instructions for preparation of Difco tryptic soy agar (TSA) and Difco nutrient broth (NB). Prepare dilute potato dextrose agar (dilute PDA) with 4.4 g Difco potato dextrose broth and 20 g granulated agar. Follow manufacturer's instructions for sterilization of media.
  - Four 10-cm petri dishes with TSA.
  - Sterile noninoculated tube with 5 mL of NB.
- 5-day-old culture of *R. solani*. Soil isolates obtained following Procedure 23.1 (Chapter 23)

can be used or isolates from cotton can be purchased from ATCC. The ATCC numbers of these isolates are: 14011, 18184, 28268, 38922, 60734, 90869, and MYA-986.
- 24-hour-old cultures of *P. fluorescens* Pf-5, one tube with a 5-mL culture in nutrient broth. This bacterium is available from the ATCC (ATCC number BAA-477).
- Bunsen burner or alcohol lamp and 95% ethanol for sterilizing the transfer needle
- Four 2 cm × 10 cm sections Parafilm®
- Metric ruler
- Sharpie® marker
- Two sterile, disposable 1-mL pipettes and bulb or pump
- Transfer needle with stiff bent (90° angle) wire to cut 5-mm^2 mycelial plugs of *R. solani*

Follow the protocols listed in Procedure 37.1 to complete Experiment 1.

Anticipated Results

A clear measurable zone of inhibition should be evident between the bacterial colony and the leading edge of the *R. solani* mycelium (Figure 37.3). Antibiotics, such as pyrrolnitrin, produced by *P. fluorescens* Pf-5 diffuse into the TSA medium and inhibit mycelial growth of *R. solani*. A zone of inhibition should not be seen between the drop of nutrient broth and the mycelium of *R. solani*.

Question

- What environmental factors may affect suppression of a soilborne pathogen, like *R. solani*, by an antibiotic-producing antagonist?

EXPERIMENT 2. PARASITISM OF *RHIZOCTONIA SOLANI* BY THE FUNGAL ANTAGONIST *TRICHODERMA VIRENS*

Applications of the fungal antagonist, *Trichoderma virens* (formerly named *Gliocladium virens* GL-21), can protect plants against damping-off and root rot pathogens, such as *R. solani*. *Trichoderma virens* is the active ingredient of the commercially available biological control product, SoilGard™. To control *R. solani*, SoilGard, a granular formulation of *T. virens*, is incorporated into soil or soilless potting mix prior to seeding. The biocontrol mechanisms involved in the interaction between *T. virens* and *R. solani* include parasitism (through production of enzymes and toxins), antibiosis (production of the broad-spectrum antibiotic gliotoxin) and competition for space and nutrients. Seed treatment of cotton with another isolate of *T. virens* has been reported to induce a plant host resistance response against *R. solani* (Howell et al., 2000).

	Procedure 37.1
	Suppression of Growth of *Rhizoctonia solani* by Antibiotic-Producing *Pseudomonas fluorescens*
Step	Instructions and Comments
1	Label the bottom of the TSA dishes with your name, date, and the words "antibiosis assay."
2	Maintain aseptic techniques throughout this experiment.
3	Cut four 5-mm² plugs of mycelium from the actively growing edge of the *Rhizoctonia solani* culture on dilute PDA with a flame-sterilized and cooled transfer needle.
4	Remove lid from TSA dish and use the transfer needle to place one mycelial plug of *R. solani* in the center of the plate. Repeat for the three remaining TSA dishes. Incubate *R. solani* at room temperature (25° to 28°C) for 24 h.
5	On the bottom side of the four TSA dishes, mark two spots about 10 mm from the edge of the dish, on opposite sides of the *R. solani* plug (Figure 37.3). In two of the TSA dishes, use the sterile 1.0-mL pipette to place a 0.1-mL drop of *Pseudomonas fluorescens* broth culture on the agar over each marked spot. In the two remaining TSA dishes, use another sterile pipette to place a 0.1-mL drop of noninoculated sterile nutrient broth over the marked spots. The latter dishes will serve as controls.
6	Allow the liquid drop of bacteria or broth to dry on the TSA before moving the dishes.
7	Seal the top and bottom edges of each TSA dish together with Parafilm® to prevent moisture loss and maintain sterility.
8	Incubate the "antibiosis assay" dishes at 25° to 28°C for 4 to 5 d. Observe daily. Make measurements before the fungal colony reaches the edge of the petri dish in the controls (as in Step 9).
9	Measure and record the width of the inhibition zone between the leading edge of the *R. solani* colony and the *P. fluorescens* colony or the sterile broth droplet.

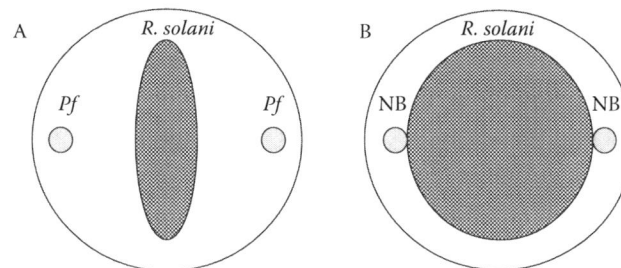

FIGURE 37.3 Schematic of antibiosis assay. (A) Inhibition of *Rhizoctonia solani* by *Pseudomonas fluorescens* Pf-5 (*Pf*) on tryptic soy agar. (B) No inhibition of mycelial growth of *R. solani* by droplet of sterile nutrient broth (NB).

Materials

Each team of students will require the following items:

- Microbiological media: Dehydrated Difco PDA; prepare dilute PDA as described for Experiment 1. Four 10-cm petri dishes with dilute PDA.
- 2-day-old cultures of *T. virens* (*Tv*) and *R. solani* (*Rs*), one culture of each organism on dilute PDA. Isolates of *R. solani* are available from the ATCC (see previously); *T. virens* GL-21 can be cultured on dilute PDA from granules of the commercial product SoilGard.

- Transfer needle with stiff bent (90° angle) wire to cut 5-mm² mycelial plugs of *R. solani* and *T. virens*
- Bunsen burner or alcohol lamp and 95% ethanol for sterilizing the transfer needle
- Eight glass microscope slides (25 mm × 75 mm)
- Four glass microscope cover slips (24 mm × 50 mm)
- Sharpie® marker
- Metric ruler
- Four 2 cm × 10 cm sections of Parafilm®
- Compound microscope with 10× ocular lens and 10× and 40× objective lenses

Procedure 37.2

Parasitism of *Rhizoctonia solani* by the Fungal Antagonist *Trichoderma virens*

Step	Instructions and Comments
1	Label the bottom of dilute PDA dishes with your name, date, and "parasitism assay."
2	Maintain aseptic techniques throughout this experiment.
3	Remove lid from dilute PDA dish and place a flame-sterilized, cooled microscope slide on the agar surface in the center of the dish. Repeat for remaining three dishes of dilute PDA.
4	Cut four 10-mm² plugs of mycelium from the actively growing edge of the *Rhizoctonia solani* culture with a flame-sterilized, cooled transfer needle.
5	Remove the lid from two of the dilute PDA dishes and use a transfer needle to place one mycelial plug of R. solani on the microscope slide, approximately 20 mm from one edge of the slide (Figure 37.5). For the two remaining dilute PDA dishes, place a plug of *R. solani* in the center of the microscope slide.
6	Flame-sterilize and cool the transfer needle; cut two 10-mm² plugs of mycelium from the actively growing edge of the *Trichoderma virens* culture.
7	Remove lids from the two dilute PDA dishes that have a plug of *R. solani* 20 mm from one edge of the microscope slide. Use the transfer needle to place a mycelial plug of *T. virens* on the microscope slide, 20 mm from the opposite edge of the slide from the *R. solani* plug. Place a flame-sterilized, cooled glass cover slip over the pair of plugs of fungal mycelium in both dishes.
8	In the two remaining dilute PDA dishes that have only one plug of *R. solani*, place a flame-sterilized, cooled glass cover slip over the plug.
9	Seal the top and bottom edges of each dilute PDA dish together with Parafilm® to prevent moisture loss and maintain sterility.
10	Incubate the "parasitism assay" dishes at 25° to 28°C for 3 to 5 d.
11	Observe assay dishes daily to determine when hyphae of *T. virens* and *R. solani* have contacted one another. About 48 h after contact, remove the glass cover slip from each "parasitism assay" and place it fungal side down on a few drops of immersion oil onto a clean microscope slide.
12	Observe slides with light microscopy at 100× and 400× magnification. Make observations on the appearance and size of the fungal hyphae on slides.

Follow the protocols listed in Procedure 37.2 to complete Experiment 2.

Anticipated Results

Hyphae of *R. solani* are significantly larger than those of *T. virens* and typically form right angles at the branches. Smaller hyphae of *T. virens* penetrate the surface and parasitize hyphae of *R. solani* (Figure 37.1). The smaller hyphae of *T. virens* are not observed in the control slides.

Question

- Which enzyme would be important for penetration and parasitism of *R. solani* hyphae by *T. virens*?

EXPERIMENT 3. BIOCONTROL OF DAMPING-OFF OF COTTON SEEDLINGS CAUSED BY *RHIZOCTONIA SOLANI* WITH *BACILLUS SUBTILIS* GB03

Bacillus subtilis GB03 is a PGPR and is the active ingredient in the commercially available biological control product Kodiak® that is used to control seedling diseases caused by soilborne pathogens such as *R. solani* in cotton, legumes, wheat, and barley. The biocontrol activity of *B. subtilis* involves induced systemic resistance (ISR) in the host plant, competition for space and nutrients and production of antibiotics. In the absence of pathogens, *B. subtilis* GB03 promotes plant growth.

Materials

Each team of students requires the following items:

- Cornmeal:sand inoculum of *R. solani* prepared by instructor. Isolates of *R. solani* are available from the ATCC (see Experiment 1). Prepare cultures of *R. solani* on Difco dilute PDA (see Experiment 1) and incubate for 5 d at 25° to 28°C. Prepare cornmeal:sand medium using 97% quartz sand, 3% cornmeal, and 15% deionized water by weight. Place mixture into 500-mL Erlenmeyer flasks and cover opening with nonabsorbent cotton plug and aluminum foil. Auto-

FIGURE 37.4 Postharvest biological control of blue mold and gray mold on Golden Delicious apples. (Left) Wounded apples were sprayed with conidia of the postharvest pathogens *Penicillium expansum* and *Botrytis cinerea* only. (Right) Wounded apples were sprayed with conidia of the pathogens and *Pseudomonas syringae* ESC-11. (Photo reprinted from Janisiewicz, W., Revised November 28, 2006. http://www.nysaes.cornell.edu/ent/biocontrol/pathogens/pseudomonas_s.html. With permission).

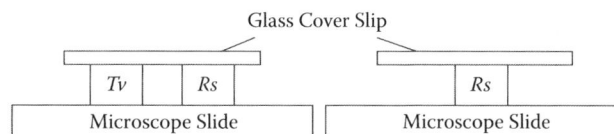

FIGURE 37.5 Schematic of parasitism assay. (Left) Paired mycelial plugs of *Trichoderma virens* (*Tv*) and *Rhizoctonia solani* (*Rs*). (Right) Control slide with one plug of *Rs*. Both slides are placed inside petri dishes containing dilute PDA.

clave the slurry for 90 min on two consecutive days, then inoculate each flask with 20 10-mm^2 mycelial plugs from the edge of the *R. solani* cultures growing on dilute PDA. Incubate the flasks at 25° to 28°C for 14 d. Shake flasks periodically to redistribute the mycelia of *R. solani* in the cornmeal:sand medium.

- Twenty cotton seeds treated with *B. subtilis* (Kodiak® concentrate) per team. Instructor should treat the seeds using manufacturer's instructions.
- Forty untreated cotton seeds
- Three plastic labels for greenhouse flats
- Three plastic greenhouse flats
- Potting soil to fill three greenhouse flats
- Sharpie® marker

Follow the protocols listed in Procedure 37.3 to complete Experiment 3.

Anticipated Results

The untreated healthy control (HC flat) is expected to have nearly 100% seedling survival. The untreated infested control (IC flat) is expected to have the lowest percentage seedling survival, typically about 50% or less. The percentage of diseased seedlings will vary with environmental conditions in the greenhouse. The percentage of surviving seedlings from seed treated with *B. subtilis* GB03 (T flat) is expected to be significantly greater than the IC flat.

Question/Activity

- Additional information could have been gained from including more controls in this experiment. Suggest two additional controls that would help a grower make a decision on whether or not to use this product.

Procedure 37.3

Biocontrol of Damping-Off of Cotton Seedlings Caused by *Rhizoctonia solani* with *Bacillus subtilis*

Step	Instructions and Comments
1	Fill one plastic greenhouse flat with moist potting soil and label with your name, date, and the words "biocontrol assay" and "HC." This flat serves as the untreated healthy control and does not receive either *Rhizoctonia solani* or seed treated with Kodiak™. Plant 20 untreated cotton seeds, in four rows of five seeds in the "HC" flat.
2	Fill two greenhouse flats with moist potting soil. Remove the soil and weigh. Place the potting soil in a plastic bag and thoroughly mix with the cornmeal:sand inoculum of *R. solani* at the rate of 2% cornmeal:sand inoculum by weight. For example, if the potting soil weighs 1000 g, then add 20 g of *R. solani* inoculum.
3	Refill one greenhouse flat with half of the infested potting soil. Label the flat with your name, date, "biocontrol assay," and "IC." This flat serves as the untreated infested control. Plant 20 untreated cotton seeds, in four rows of five seeds in the "IC" flat.
4	Fill the third greenhouse flat with the remaining infested potting soil. Label the flat with your name, date, "biocontrol assay," and "T." This serves as the biocontrol treatment with treated seeds planted into infested potting soil. Plant 20 Kodiak®-treated cotton seeds, in four rows of five seeds in the "T" flat.
5	Place flats on a greenhouse bench and water as needed for 14 d.
6	At 5, 10, and 14 d after planting, observe the flats and record the number of living seedlings emerged.
7	Calculate the percentage of living seedlings emerged from each flat at each date using the equation: (Number of surviving seedlings/20) × 100.
8	Make a line graph with three lines, each line representing the data from one flat. Plot "Days after planting" (X-axis) versus "Percentage of surviving seedlings" (Y-axis) at 0, 5, 10, and 14 d after planting.

LITERATURE CITED

Baker, K.F. and R.J. Cook. 1974. *Biological Control of Plant Pathogens.* W.H. Freeman, San Francisco, CA. 433 pp.

Bonsall, R.F., D.M. Weller and L.S. Thomashow. 1997. Quantification of 2,4-diacetylphloroglucinol produced by fluorescent *Pseudomonas* spp. *in vitro* and in the rhizosphere of wheat. *Appl. Environ. Microbiol.* 63: 951–955.

Chet, I. (Ed.). 1993. *Biotechnology in Plant Disease Control.* Wiley-Liss, New York. 373 pp.

Chet, I., N. Benhamou and S. Haran. 1998. Mycoparasitism and lytic enzymes, in *Trichoderma and Gliocladium, Vol. 2: Enzymes, Biological Control and Commercial Applications,* Harman, G.E., and C.P. Kubicek (Eds.), Taylor and Francis, London, pp. 153–172.

Clarkson, J. and J. Whipps. Biocontrol of soilborne plant diseases: use of mycoparasites that destroy sclerotia of plant pathogens. Warwick HRI, http://www2.warwick.ac.uk/fac/sci/whri/research/biodiseasecontrol/conio/biocontrol/. Revised February 11, 2005.

Cook, R.J. 1993. Making greater use of introduced microorganisms for biological control of plant pathogens. *Annu. Rev. Phytopathol.* 31: 53–80.

Cook, R.J. and K.F. Baker. 1983. *The Nature and Practice of Biological Control of Plant Pathogens.* American Phytopathological Society, St. Paul, MN, 539 pp.

Desai, S., M.S. Reddy and J.W. Kloepper. 2002. Comprehensive testing of biocontrol agents, in *Biological Control of Crop Diseases.* Gnanamanickam, S.S. (Ed.), Marcel Dekker, New York, pp. 387–420.

Dubos, B. 1987. Fungal antagonism in aerial agrobiocenosesin *Innovative Approaches to Plant Disease Control,* Chet, I. (Ed.), Wiley-Liss, New York, pp. 107–135.

Fravel, D.R. 1988. Role of antibiosis in the biocontrol of plant diseases. *Annu. Rev. Phytopathol.* 26: 75–91.

Han, D.Y., D.L. Coplin, W.D. Bauer, and A.J. Hoitink. 2000. A rapid bioassay for screening rhizosphere microorganisms for their ability to induce systemic resistance. *Phytopathology* 90:27–332.

Heiniger, U. and D. Rigling. 1994. Biological control of chestnut blight in Europe. *Annu. Rev. Phytopathol.* 32: 581–599.

Howell, C.R. 1982. Effect of *Gliocladium virens* on *Pythium ultimum, Rhizoctonia solani*, and damping-off of cotton seedlings. *Phytopathology* 72: 496–498.

Howell, C.R., L.E. Hanson, R.D. Stipanovic and L.S. Puckhaber. 2000. Induction of terpenoid synthesis in cotton roots and control of *Rhizoctonia solani* by seed treatment with *Trichoderma virens. Phytopathology* 90: 248–252.

Howell, C.R. and R.D. Stipanovic. 1979. Control of *Rhizoctonia solani* on cotton seedlings with *Pseudomonas fluorescens* and with an antibiotic produced by the bacterium. *Phytopathology* 69: 480–482.

Howell, C.R. and R.D. Stipanovic. 1980. Suppression of *Pythium ultimum*-induced damping-off of cotton seedlings by *Pseudomonas fluorescens* Pf-5 and its antibiotic, pyoluteorin. *Phytopathology* 70: 712–715.

Janisiewicz, W. *Pseudomonas syringae* (saprophytic strain) and "Fruit Yeasts." *Biological Control: A Guide to Natural Enemies in North America.* C. Weeden, A. Shelton, and

M. Hoffman, (Eds.), Cornell University (http://www.nysaes.cornell.edu/ent/biocontrol/pathogens/pseudomonas_s.html). Revised November 28, 2006.

McSpadden Gardener, B.B. and D.R. Fravel. 2002. Biological control of plant pathogens: Research, commercialization, and application in the USA. Online. *Plant Health Progress* DOI: 10.1094/PHP-2002-0510-01-RV.

Research Briefings. 1987. *Report of the Research Briefing Panel on Biological Control in Managed Ecosystems.* National Academy Press, Washington, D.C.

Ryo, C.-M., M.A. Farag, C.-H. Hu, M.S. Reddy, J.W. Kloepper and P.W. Paré. 2004. Bacterial volatiles induce systemic resistance in *Arabidopsis. Plant Physiol.* 134: 1017–1026. DOI: 10.1104/pp.103.026583.

Shoresh, M., I. Yedidia and I. Chet. 2004. Involvement of jasmonic acid/ethylene signaling pathway in the systemic resistance induced in cucumber by *Trichoderma asperellum* T203. *Phytopathology* 95: 76–84.

Stockwell, V.O. and J.P. Stack. 2007. Using *Pseudomonas* spp. for integrated biological control. *Phytopathology* 97: 244–249.

Strobel, G.A. 2006. *Muscodor albus* and its biological promise. *J. Ind. Microbiol. Biotechnol.* 33: 514–522.

Strobel, G.A., E. Dirkse, J. Sears and C. Markworth. 2001. Volatile antimicrobials from *Muscodor albus*, a novel endophytic fungus. *Microbiology* 147: 2943–2950.

U.S. Environmental Protection Agency. EPA: Biopesticide Active Ingredient Fact Sheets. (http://www.epa.gov/pesticides/biopesticides/ingredients/index.htm). Updated February 20, 2007.

van Loon, L.C., P.A.H.M. Bakker and C.M.J. Pieterse. 1998. Systemic resistance induced by rhizosphere bacteria. *Annu. Rev. Phytopathol.* 36: 453–483.

Warrior, P., K. Konduru and P. Vasudevan. 2002. Formulation of biological control agents for pest and disease management, *Biological Control of Crop Diseases*, Gnanamanickam, S.S. (Ed.), Marcel Dekker, New York, pp. 421–441.

Weller, D.M., J.M. Raaijmakers, B.B. McSpadden Gardener and L.S. Thomashow. 2002. Microbial populations responsible for specific soil suppressiveness to plant pathogens. *Annu. Rev. Phytopathol.* 40: 309–348.

WEBSITES

American Type Culture Collection. http://www.atcc.org/.

United States Department of Agriculture, Animal Plant Health Inspection Service, Plant Protection, and Quarantine. http://www.aphis.usda.gov/forms/.

United States Environmental Protection Agency: Biopesticide Active Ingredient Fact Sheets. http://www.epa.gov/pesticides/biopesticides/ingredients/index.htm.

38 Integrated Pest Management

Clayton A. Hollier and Donald E. Hershman

CHAPTER 38 CONCEPTS

- IPM has been defined as, "A sustainable approach to managing pests by combining biological, cultural, physical, and chemical tools in a way that minimizes economic, health, and environmental risks" (Anonymous, 1994).

- There are many definitions of IPM. Most make the following assumptions: (1) the presence of a pest does not necessarily indicate a problem, (2) a cropping system is part of an ecosystem, and (3) all pest management options should be considered before choosing a pest management strategy (Dufor and Bachmann, 2001).

- The first step in IPM implementation is to identify the pest.

- Analysis of information collected in the scouting and identification process is essential to the determination of pest management needs.

- A successful IPM program is one that has been planned and is never considered a "quick fix."

- Biocontrol tactics fit very well in the management of ecologically sensitive systems.

Integrated pest management (IPM) is not a new concept. In reality, IPM was used by the ancient peoples as a means to grow crops for food and fiber. Although their cropping methods were determined by trial and error, they realized that crops should be cultivated in certain ways to protect against pests. Some of their techniques seem foreign to us today, such as the offering of sacrifices to the "rust god" in order to have grains free of diseases. Other techniques are more understandable in a modern context, such as planting particular crops in certain soil types at a certain time of year.

DEFINITION AND IMPORTANCE

IPM has been defined as "... a sustainable approach to managing pests by combining biological, cultural, physical, and chemical tools in a way that minimizes economic, health, and environmental risks" (Anonymous, 1994). Although modern concepts and practices of IPM began in the 1950s in response to the pest management needs in apple production, it expanded into cotton production in the 1960s, and today plays a role in all major crop production systems in the United States (McCarty et al., 1994). In this chapter "**pest**" will be used to describe an insect pest or **vector** and plant **pathogens**.

Pest management represents a significant portion of a grower's crop production costs. Even with current IPM strategies there are still 10 to 30% crop losses (Jacobson, 2001). In the mid-1940s, pesticides (Chapter 36) were seen as tools needed for all existing and future pest problems.

Dependence on pesticides was so heavy that many alternate nonchemical strategies were abandoned (McCarty et al., 1994). Even today, pesticides play an important role in managing pests; however, consumer concerns, environmental regulation, pest resistance to pesticides, and development costs of products have reduced the availability and use of new pesticides. Thus, producers and society in general are seeking new and adequate pest management strategies to protect production and the environment (Jacobson, 2001).

COMPONENTS OF AN IPM PROGRAM

There are many definitions of IPM. Most definitions make the following assumptions: (1) the presence of a pest does not necessarily indicate a problem, (2) a cropping system is part of an ecosystem, and (3) all pest management options should be considered before choosing a pest management strategy (Dufor and Bachmann, 2001). There are many strategies that can be used as part of IPM programs.

SCOUTING

Scouting is the key to successful IPM programs. It involves a comprehensive, systematic checking of fields at regular intervals to gather information on crop progress, pests present and pest levels. Plant pathologists use scouting as an essential part of the IPM process for managing plant diseases. Depending on the crop/pest combination, scouting may help to identify diseases that can be controlled in the same season (e.g., buy applying a foliar fungicide). For most crop/disease combinations, however, scouting provides the information needed to manage a disease in subsequent seasons. Examples of each of these will be discussed later in this chapter.

Scouts that monitor crops for diseases look for the presence of pathogen signs (spore masses, bacterial oozing), symptoms (wilting, leaf spots, damping off, etc.), and pattern of symptoms (low or hill areas, margins of fields, random appearance, clumped, etc.). Various patterns are used to scout fields to make certain that a representative sample of the crop is examined. Scouts often walk fields in an S-, V-, W-, X-, or Z-pattern and look for pests or crop damage; they take the samples they collect to a plant disease diagnostic laboratory for assays. Just as pattern and techniques vary depending on the pest, so too scouting methods vary by crop. The more frequently a crop is monitored, the more information is collected for IPM decision making. The frequency of scouting depends on the crop, the pest, and the environment that exists during the crop/pest growth cycle. A more disease-conducive environment, for example, may necessitate more frequent scouting for disease.

PEST IDENTIFICATION

The strategies for IPM implementation vary greatly in complexity. The first step in IPM implementation is to identify the pest. Most land-grant universities have disease and insect clinics that specialize in plant pest identification (Chapter 39). Many of these clinics use digital imaging to speed pest identification. Rather than shipping a plant or insect specimen to the laboratory, county Extension agents or other practitioners post digital images of the pest or plant damage on the clinic Web page for specialists to view, diagnose, and make management recommendations. Digital diagnostics may allow farmers to react faster to pest outbreaks as the pests are often identified the same day as the images are posted on the Website. Some crop-disease combinations are less conducive to digital diagnostics because of equipment limitations and/or operator knowledge and skill. In these cases, a digital image may still be helpful for assessing the field conditions supporting the disease and disease distribution, but a sample will still need to be submitted to a qualified plant disease diagnostician for positive identification. Once the pest is identified, useful information can be gathered by researching the life cycle of the insect or pathogen. Pest identification is essential before selecting appropriate pest management tactics. Misidentification can lead to the use of inappropriate strategies and lead to further loss of time and money.

PEST SITUATION ASSESSMENT

The **damage threshold** (DT) is the amount of crop damage that is greater than the cost of management measures.

As growers want to protect their crop before a pest population reaches DT, IPM programs use the concept of an **action threshold** (AT). These two levels are related, but the AT is the point at which management tactics should be applied to prevent pest population increases and keep crops from reaching the DT (Zadoks and Schein, 1979). The value of the crop and the cost of likely control measures determine the AT.

Analysis of information collected in the scouting and identification processes is essential to the determination of pest management needs. Is the AT likely to be reached, has it been reached, or has the DT been reached? These are critical questions that producers must be able to answer for the best IPM program to be implemented. However, DTs and ATs have not been determined for many crop and pest combinations. Those that have are generally developed for average conditions and for some, but not all, pests. In times of stress, such as drought, thresholds may have to be altered frequently due to lower crop yield potential (McMullen, 2001).

IMPLEMENTATION AND EVALUATION

Once appropriate management strategies have been determined, they should be implemented effectively and in timely applications. Late implementation usually will result in ineffective management and a loss of time and money. To determine success, the IPM strategy chosen must be evaluated. This is accomplished by comparing the pest activity before and after implementation of the IPM program. Evaluation for some crop-pest combinations can be facilitated by monitoring and yielding areas of fields that have been specifically left untreated with pesticides. Data from these areas are then compared with portions of the field that have received the prescribed IPM tactic. To review: (1) accurately identify the pest, (2) use field sampling procedures that eliminate bias and ensure appropriate representation, (3) choose the most appropriate IPM appropriate strategy based on the AT, if available, and (4) evaluate what was done and make changes, if needed, to make the system better.

DEVELOPING AN IPM PROGRAM

A successful IPM program is one that has been planned and is never considered a "quick fix." Quick fixes and IPM are not compatible strategies. In developing an IPM program, there are several technical and informational needs that must be met (McMullen, 2001). Before developing a program, consider if an IPM program tailored to the problem already exists. To determine this, check with state or county Extension agents, or search Land Grant university Websites for the existence of programs. If one is available, obtain the latest information for a particular crop/pest complex. If an IPM program does exist for a particular pest, formulate a program using the DT for the pest. If the DTs change with time or crop development stage, determine how this influences management practices, and make the appropriate adjustments to the program. If an IPM program does not exist for a particular crop/pest combination, follow the development steps below.

DEVELOPMENT STEPS

All steps are important in the process of developing an IPM program; however, there are issues from the beginning that need to be agreed upon in order to make the program run as efficiently as possible. Define who is responsible for the monitoring process, what is its purpose, what are the costs involved, what expertise is necessary, and are there any potential conflicts of interest. Deciding who will scout the field(s) is extremely important. As previously mentioned, monitoring is important to determine what pests are present, what their numbers and stage of development are, host development, what damage is being done (incidence, severity, and pattern in field), sampling methodology, soil type and moisture, dew periods, nutrient status, temperature, and relative humidity. Field books or software packages for the computer are places to store such information along with historic, but retrievable, information including field maps and history of pests, soil types, soil profiles, where problems have occurred, previous crops, and cropping systems.

Accurate pest identification is necessary, but determining who is available to help in the identification of the pests being monitored varies from state to state. University extension/research faculty and consultants traditionally have been employed to identify pests (Chapter 39). Accuracy in the identification will influence the success of the IPM program being developed. Inaccuracy can contribute to financial and ecological disasters.

Other considerations such as variety selection, crop rotation, plant populations, cultivation/tillage practices, planting/harvesting dates, and fertility levels influence the success of implementation of an IPM program. For example, use of **resistant** or **tolerant** varieties (Chapter 34) is the most economical and most widely used method for managing pests. Unfortunately, in many cases there is little or no pest resistance available, but where available, resistant or tolerant cultivars should be used. An example of success in this area is the use of resistant varieties to reduce the negative impact on yield and quality of wheat due to leaf rust development (Chapter 18).

Cultural practices (Chapter 35), such as crop rotation and cultivation/tillage, have been widely used historically to reduce inoculum levels of pathogens. For example, crop rotation has been extensively used to reduce disease inoculum by growing non-hosts. Subsequent hosts then benefit from the short-term reduction in pest populations, thus reducing their overall impacts on the yield. An example of

this practice is the reduction of *Gaeumannomyces graminis* var. *tritici* (take-all of wheat) populations by rotation to a nonsusceptible host (Chapter 35). Additionally, cultivation has been used to decrease weed populations and their competition to crops. The weeds that are alternate hosts for many pathogens are destroyed, thus short-term reductions of pathogen populations is achieved. With the advent of herbicides and their widespread use, and the increase of conservation practices, cultivation has played a diminished role in crop production. However, as an IPM practice, it is still a viable tactic and can greatly reduce pathogen populations when deployed in sites with a low potential for soil erosion.

Temporal and **spatial** dimensions of crop management also can influence pathogen levels. Variation of planting or harvesting dates can reduce or eliminate certain pests as a factor in causing damage or reducing yield. These shifts in dates should be more conducive to host development and less so for the pest. Some examples where these factors have reduced the inoculum load are early planting and/or planting early hybrids of corn to reduce late season Southern rust development or late planting of ornamental bedding plants, allowing for warming of the soil and reducing several cool-season soilborne plant pathogens.

Appropriate plant/row spacing allows for adequate growth of the host, but also allows for air circulation and reduction of moisture on leaves of the canopy. This alters the microenvironment so it is less conducive to development of some diseases. Conversely, narrower row spacing or dense canopies frequently favor development of certain plant diseases that thrive under high relative humidity and long dew periods.

Planting healthy seeds or propagating material is the backbone of successfully growing any type of crop or plant. If seeds or propagating material are damaged or infected by a pathogen, the germination rate will be reduced, increasing the likelihood that replanting will be necessary and the cost of production increased. Planting of disease-free seed and transplants, or using seed treated with fungicides will protect seed before and after germination (for a period) from seedling blight and many soilborne pathogens (Chapter 22).

The use of biocontrol as a tool in IPM has increased during the last decade as our understanding of its use and limitations has increased (Chapter 37). Introducing mycoparasites and antagonists as enemies of pathogens have reduced the inoculum level of pathogens in several cases. Biocontrol has the potential to reduce the use of the traditional pesticides, thus gaining support of many environmentally-concerned groups. Ecologically, biocontrol fits well into sensitive systems because nontarget organisms, such as beneficial insects, are not impacted. There are many within the nonagricultural community who feel that IPM should not include the use of pesticides to aid in the pest management process. In reality, however, the management of pests, which usually does include the judicious use of pesticides in conjunction with nonpesticide management tactics, would not be possible without the use of pesticides for many crop-pest combinations. Pesticides are a tool in a complete IPM system (Chapter 36).

EXAMPLE OF A PEST MANAGEMENT SYSTEM: WHEAT DISEASE MANAGEMENT IN THE SOUTHEASTERN UNITED STATES

Disease management is a key component of high-yielding wheat. Some diseases, such as take-all disease, must be managed proactively and cannot be affected once they are established. Other diseases, such as foliar diseases caused by fungi, can often be managed by the timely application of fungicides. Generally, wheat producers place too much emphasis on disease management using foliar fungicides only. Most diseases are best managed through the use of multiple tactics, both proactive (crop rotation, delayed planting, resistant varieties, proper fertility, and seed treatment fungicides) and reactive (application of foliar fungicides).

SCOUTING FOR DISEASES

Scouting for diseases is important for two reasons. First, yearly scouting helps to build an on-farm database that can be used to select appropriate disease management tactics for future crops. Second, scouting helps to determine if and when to apply fungicides. Once fields have been properly scouted, these data are helpful to determine disease management options. For help with this, contact your local Extension agricultural agent or a crop consultant for the latest recommendations. An appropriate course of action can begin only when up-to-date, accurate information is obtained.

HOW PREPLANT DECISIONS AFFECT DISEASES

Wheat producers have a significant portion of their total disease management program in place once the seed is in the ground. By that time, decisions have been made about crop and cultivar selection, method of tillage/seedbed preparation, variety selection, seed quality, seed treatment, planting date, seeding method and rate, and fall fertility. Individually and collectively, these decisions can play an important role in influencing which diseases develop, their severity and their effect on crop yield and test weight. Because preplant and planting decisions are so important in the management of wheat diseases, understanding how they affect disease is necessary.

VARIETY SELECTION

Decisions relating to variety selection are, perhaps, the most important decisions that can be made in managing diseases. Every commercially available wheat variety has a unique range of reactions to diseases common in the region. Which and how many varieties are planted determine the potential for certain diseases. Failure to consider the ramifications of variety selection in managing diseases can be a costly mistake. Selecting two or three varieties with the greatest amount of available resistance to the diseases most common on a farm or in the community is important. To do this, some idea about the disease history on the farm is necessary. If that information is not readily available, Extension agricultural agents or crop consultants will be of help. This information will not be as good as actual data from the grower's own farm, but it is far better than basing decisions on no information. It is important to plant more than one variety for this key reason; it is common for a single disease to severely damage a single variety (Chapter 34). When multiple varieties are planted, the risks of more significant losses are reduced. Planting more than one variety, especially when different maturities are represented, can help with the logistics of harvesting and soybean planting (double cropping), if this crop is used in rotation.

CROP ROTATION

Crop rotation helps to manage wheat pathogens that survive between wheat crops in wheat residue. When a crop other than wheat is grown in a field, levels of pathogens specific to wheat decline. This occurs simultaneously as the residue of previous wheat crops deteriorates. Lower levels of pathogens can translate into less disease pressure the next time wheat is produced. Crop rotation is helpful in the management of hidden diseases, such as Pythium root rot, and destructive diseases, such as take-all. In fact, rotating fields out of wheat is the only practical means of controlling take-all. Rotation also can reduce infections by the fungal species *Stagonospora* and *Septoria*. However, pathogen spores introduced from neighboring fields can negate the beneficial effects of rotation on these diseases.

By way of a specific example, in some areas of the soft red winter wheat-growing region of the United States, wheat is planted following corn harvest (i.e., corn matures before soybean). For many diseases, growing corn is a good non-host crop in non-wheat years. However, when wheat is planted after corn on a multi-county or regional basis, especially when conservation tillage planting methods are widely used, the chances that damaging levels of Fusarium head blight (FHB) might develop are increased. This is because the fungi that cause FHB also attack corn, causing stalk and ear rots; in addition, corn residue provides a substrate for the FHB fungi to survive between seasons, and often at high levels. The regional availability of fungal spores coming from corn residue on the soil surface, therefore, greatly increases the regional risk of

FHB if weather patterns favor disease development. On an individual field basis, planting wheat behind a non-corn crop, even in a conventionally-tilled environment, does not greatly reduce the potential for FHB as long as the regional situation just described exists. Research indicates that FHB inoculum is produced and blows around readily and, as long as conditions favor FHB, the disease will be a problem regardless of an individual farm's rotation scheme and tillage methods used. The importance of in-field inoculum production is overridden by regional inoculum production for FHB. In-field inoculum and residue management becomes more important where regional cropping and tillage practices do not favor FHB inoculum production and dispersion.

Tillage

Tilling wheat stubble hastens the breakdown of residue that harbors certain disease organisms. This can help reduce levels of take-all and foliar diseases, such as Septoria leaf blotch and tan spot. "Help" is the operative word here because it is unlikely that tillage will be of much good in the absence of other management methods. For fields in a wheat/double-crop soybean/corn rotation, tillage prior to planting corn should cause a significant decline in surviving wheat stubble. The year between wheat crops in this rotation also helps, except where high levels of the take-all fungus exist. In those cases, two or more years between wheat crops may be required to reduce inoculum.

Seed Quality, Seed Fungicides, Seeding Rate, and Planting Method

All of these can influence stand establishment and seedling development. To achieve the highest possible yields, sufficient stands are necessary. There must be excellent seed germination followed by emergence and development of seedlings to ensure the desired stands. Using high-quality (i.e., certified) seed treated with a broad-spectrum fungicide and good planting techniques foster good stand establishment. Excess stands, however, are undesirable because they encourage foliar and head diseases by reducing air circulation and light penetration into the canopy later in the season. Lodging of plants because of weakened stems is also a problem caused by excessively dense stands.

Planting Date

The trend in recent years has been to plant wheat earlier and earlier each year. Early-planted wheat, defined as wheat planted before the Hessian fly-free planting date, is at greater risk of damage caused by *Barley yellow dwarf virus* (BYDV), take-all disease, and Hessian fly than is later-planted wheat. If logistic considerations cause a grower to plant some wheat acres before the fly-free date

for that area, those acres should have been well rotated and planted to a variety that can tolerate some barley yellow dwarf virus. The use of seed treatment or fall-applied foliar insecticides may also be appropriate in these situations. These pesticides can help to reduce populations of the aphids, which transmit BYDV to wheat, barley, and oats. However, cost of treatment is a consideration and, thus, treatment is usually targeted at high risk, early-planted acres. Planting all wheat acreage before the fly-free date is extremely risky and is not recommended.

Nitrogen Fertility

Too much nitrogen in the fall can encourage excessive fall growth that can increase problems with barley yellow dwarf virus and most foliar diseases caused by fungi. Increased problems with barley yellow dwarf virus have to do with an extended period of activity by aphids that transmit the virus when stands are dense in the fall. The same situation encourages infection and over-wintering of pathogens causing foliar diseases, such as leaf rust, powdery mildew, and leaf blotch complex. Excessive spring nitrogen results in lush stands that promote disease in a manner similar to that associated with excessive seeding rates. Lodging may increase, too, resulting in higher moisture within the canopy of fallen plants and creating favorable conditions for fungal diseases.

Fungicide Seed Treatments

Obtaining and keeping a good stand of wheat is a key component of high yields. One management strategy many wheat producers use to attain excellent stands is to treat seed with a fungicide. Wheat seed treatment fungicides accomplish the following:

1. *Encourage good stand establishment.* Wheat is planted at a time of the year that can be hostile to germinating and emerging seedlings. Excessively cool, wet, or dry soils, seed planted too deep or too shallow, and no-till plantings where seed-to-soil contact is poor all slow the germination process and predispose developing seedlings to infection by seed- and soil-borne fungi. Species of *Pythium* are probably the main pathogens in excessively wet and warm soils. The other conditions mentioned favor infection of seedlings by species of the fungi including *Fusarium, Rhizoctonia, Septoria,* and *Stagonospora,* among others. Infection can result in fewer emerged seedlings and reduced vigor of the seedlings that do emerge.

   Most producers who plant wheat according to recommended guidelines have little difficulty achieving dense, vigorous stands of wheat seedlings. The fact that a good percentage of the wheat seed planted in many regions is treated with a fungicide probably has something to do with this situation. Routine use of high-quality, high-germ seed is probably a key contributing factor. In fact, historically, most small-plot seed treatment research done has shown that treating high-quality seed with a fungicide only rarely results in stand or yield increases. Nonetheless, it is still advisable to treat seed with a good general-use fungicide to protect seedlings from adverse soil conditions if they develop after planting. In this regard, seed treatment fungicides should be seen as a form of low-cost crop insurance. Environmentally, seed treatment fungicides are desirable because of their low toxicity, low use rates, rapid breakdown, and target application strategy.

2. *Enhance germination of marginal-quality seed lots.* If seed quality is marginal because of fungi such as *Fusarium*, seed treatment can be used to bring seed with less than optimal germination up to acceptable levels. In many cases, seed testing laboratories can provide tests that indicate whether or not fungicides will enhance germination of seed. Poor response of low-to-moderate germination seed lots to fungicides is indicative of a high percentage of dead seed, mechanical damage, or some factor apart from disease. Seed lots of low germination are not likely to be helped by any seed treatment fungicide and should not be used where high yield is a primary goal.

3. *Control loose smut.* Loose smut (Chapter 18) control is probably the main reason seed treatment fungicides are so widely accepted and used in the southeastern United States. Before seed certification, loose smut was a serious problem for both seed and grain producers. The use of seed treatment fungicides, such as carboxin, allowed seed lots to be "cleaned up" by eliminating loose smut from infected seed. Carboxin is still considered by many to be the standard for loose smut control. It is inexpensive and highly effective in most situations. However, sporadic, reduced activity of carboxin, caused by poor application procedures or soil conditions that caused the active ingredient to be washed off the seed, supported the development of a new generation of seed treatment materials that are highly effective against loose smut. All of these new-generation fungicides are extremely active at low use rates. In fact, because of the excessively low use rates of these products, application can often be done only by seed conditioners with the experience and equipment to do the job properly. This eliminates the option of on-farm seed treatment by producers and increases the cost in many cases.

4. *Control foliar diseases.* The new generation of systemic, sterol-inhibiting seed treatment fungicides (Chapter 36) can provide fall management of several fungal diseases, including *Septoria* and *Stagonospora* leaf blotches (leaf blotch complex), leaf rust, and powdery mildew (Chapter 33). Occasionally, management of these diseases extends into early spring as a result of reduced inoculum levels of the causal fungi in fields planted with treated seed. In some cases, this activity can be quite substantial, as is the case with triadimenol and powdery mildew. Management of powdery mildew through head emergence the following spring is not uncommon. Nonetheless, seed treatment fungicides should not be considered as a total replacement for spring-applied foliar fungicides as no seed treatment provides season-long control of foliar diseases.

In general, new-generation sterol-inhibiting fungicide seed treatments, because of their specific mode of action, are not highly effective against many common soilborne pathogens. For this reason, most are marketed as a mixture with either thiram or captan, both of which provide at least moderate activity against a wide range of soilborne fungi. Newer formulations may include metalaxyl to control soilborne *Pythium* species.

When contemplating the use of wheat seed treatment fungicides, consider these factors. Costs of materials and disease control vary widely, so it is critical to assess cost-benefit ratios of the various fungicides. The main consideration is to determine why a seed treatment fungicide should be used. Specifically, what diseases are to be managed and what is to be accomplished by seed treatment use? Fungicide labels (Chapter 36) should be read and followed completely. These labels refer to the specific diseases the products manage. Once these determinations have been made, selection of the most appropriate material (best disease control for the money) can be adopted. For example, triadimenol seed treatment is relatively expensive. If powdery mildew is an extensive problem on a farm, triadimenol may negate the need for an early spring foliar application of a fungicide on a mildew-susceptible variety. Taken in this context, the economics of triadimenol seed treatment become more favorable. If loose smut or general soil-borne pathogens are the main concern, and the risk of powdery mildew is minimal, triadimenol is not the most economical choice as less expensive materials are as good as triadimenol at managing these diseases.

Although fewer options are available to producers regarding on-farm treatment of seed, some hopper-box treatments are still available. If a hopper-box treatment is to be attempted, it is important to note that complete coverage of all seeds is essential. Poor coverage equals poor results and, perhaps, a waste of money. Even distribution can be accomplished on farm only with considerable effort and planning. Having seed treated by a professional eliminates potential problems of poor fungicide distribution and uneven rates on seeds.

FOLIAR FUNGICIDES

Deciding whether or not to apply foliar fungicides to wheat is one of the most difficult decisions a producer or crop consultant has to make because of the many variables that influence the need for and effectiveness of foliar fungicides. First, fungicides must be applied in the early stages of a disease epidemic to be very effective (Chapter 30). Applying fungicides too far in advance of an epidemic, or waiting too long to apply a fungicide, results in poor disease management and little or no economic benefit. Similarly, there is no economic gain from using foliar fungicides if yield-reducing levels of disease fail to develop or if crop yield potential is too low to cover costs. Finally, foliar fungicides manage only certain foliar and head diseases caused by fungi. They do not manage diseases caused by bacteria, viruses, or nematodes, and they have no effect on some fungal diseases, such as loose smut and take-all disease.

Below are listed the steps to take when making foliar fungicide use decisions:

Step 1. Commit to scouting fields. When considering the use of foliar fungicides, certain questions must be answered. Is there a commitment to scouting fields to determine the need to apply a fungicide? (If there is uncertainty about how to answer this question, refer to steps 3 through 6 below for specific field scouting requirements.) If there is no commitment to field scouting (whether it is done by the grower or by a consultant), there is a question as to the appropriateness of even considering fungicide use. Ultimately, a grower will need to decide how important wheat is to the total farming operation. If wheat is important to the profitability of the farm, it is advisable to make both time and monetary commitments to produce the best crop possible. If wheat is of only secondary importance relative to other farm operations, perhaps management of diseases using resistance and cultural practices should be considered.

Step 2. Determine the number of potential fungicide applications. Once the commitment has been made to scout wheat fields, the next significant determination is how many fungicide applications are you willing to make? Nearly every producer says "Only one!" and few indicate two. The answer to this question is important because it determines the approach to fungicide use. If a grower is going to make only one fungicide application, timing of the application is crucial. Research and experience show that a single application made during heading performs at least as well as, and usually better, than a single application made at flag leaf emergence, in most situations.

The problem with single applications at flag leaf emergence (regardless of the fungicide used) is that they frequently allow late-season disease pressure to build to excessive levels. As a result, the crop is damaged even though early diseases may have been kept in check. Heading applications, on the other hand, usually limit disease buildup on the flag leaf (F), the second leaf down (F-1) and the head, although disease is allowed to develop unchecked early in the season. Protection of the F and F-1 leaves and the head is much more important to yield and grain quality than is protection of lower leaves. The risk in making a single application is waiting too late to apply the heading treatment. Fungicides are of little or no value once the flag leaf and head are severely diseased. The best way to limit this risk is to start scouting operations in the stages just before flag leaf extension. It is unlikely that significant levels of most foliar diseases will have developed on the F and F-1 leaves by this time.

Crops receiving two fungicide applications (an early application followed by a late application) often yield significantly more than crops getting even the best single application. The question, however, is whether or not the economic benefit that results from the additional treatment is greater than its cost. As a general rule, the extra treatment at least pays for itself if early disease pressure is moderate to heavy, and crop prices are good. If early disease pressure is minimal or crop prices are low, it probably would not be an economically justified treatment.

Step 3. Know the disease reaction of the wheat variety planted. Typically, foliar fungicides are not necessary on wheat varieties rated as resistant or moderately resistant to a particular fungal disease. Careful scouting and observation are the keys. Leaf rust and powdery mildew can adapt to and attack a formerly resistant variety. This can happen in a single season, so growers and consultants need to be vigilant and still scout those crops.

Step 4. Estimate crop yield potential. Does the field have sufficient yield potential to justify a foliar fungicide application? Spraying with fungicides protects only yield already "built" into a crop; fungicides do not increase yield. Although various techniques can be used to estimate yield potential, most producers can look at a crop after greenup and know intuitively if the crop is worth protecting. In most cases, there will be a need to harvest an additional three to eight bushels per acre (depending on grain price and chemical/application cost) to offset the cost of a fungicide application. The higher the yield potential of a crop the more likely the economic benefit is to be realized from applying a foliar fungicide if disease becomes a problem.

Step 5. Know the disease(s). As indicated earlier, fungicides manage a relatively small number of fungal diseases. Fortunately, the diseases controlled are those that commonly reduce yields of soft red winter wheat crop in the southern United States, such as leaf rust, powdery mildew, leaf blotch complex and glume blotch, and tan spot. Other diseases are not managed with foliar fungicides. Thus, proper identification of the disease is critical in developing a control strategy.

Step 6. Scout fields. Scouting wheat fields to determine crop growth stage and current disease situation is critical to making good fungicide-use decisions. When scouting fields, observe the entire field. Decisions should not be based on what is found along the edges or what is seen from the seat of a moving or parked vehicle. The key is to make a decision based on the average disease situation in a field. This requires assessing disease levels in eight to ten randomly selected sites within the field.

Once in a field, it is important to determine the growth stage of the crop for the following two reasons:

1. All fungicides must be applied within specific growth-stage restrictions. Propiconazole products (e.g., Tilt), for example, cannot legally be applied once the crop is flowering. All fungicides have well-defined days-to-harvest restrictions.

2. Fungicides provide the greatest benefit when plants are protected from disease between flag leaf emergence and soft dough stage. In much of the southeastern United States, the most critical stage is typically from midhead emergence through flowering. This is the period in which fungicide applications are often most beneficial.

Step 7. Determine disease levels. To be effective, fungicides must be applied early in an epidemic. Too often, fungicides are applied too early, before any disease is visible. This approach results in no economic benefit if disease pressure remains low. Waiting too late to apply fungicides, although common, is equally ineffective. For leaf blotch complex, including

S. nodorum, if F-2 and lower leaves have symptoms and rain has been recent, there is a good chance that the flag leaf and head are already infected (with first symptoms appearing 7 to 12 d after). Therefore, examination of symptoms on lower leaves can help determine when to apply a fungicide. Herein lays the greatest obstacle to effective, economical use of foliar fungicides: How much disease is enough disease to justify a foliar application of a fungicide? There are no absolutes, but many states have developed various threshold guidelines to help producers make informed fungicide use decisions. Thresholds must be used along with some common sense. For example, if a specific threshold is reached for powdery mildew, application of a fungicide would not be recommended if an extended period of hot, dry weather is predicted. The threshold indicates that yield loss caused by one or more of the above diseases is likely; however, they do not mean losses will definitely occur. Weather can always intervene and impact the development and progression of a disease epidemic. There is no way to develop disease thresholds that are appropriate for all situations.

Step 8. Select a fungicide. Product labels provide detailed use instructions and product limitations. Apply all pesticides according to label specifications. You can make all the right fungicide use decisions, but if inferior application technique is used, all can be lost!

Step 9. Understand the risks. One problem is the inability to determine if disease-favorable conditions will persist after a fungicide is applied. Fungicides are valuable only if yields and test weights are threatened by disease. Similarly, fungicides are of limited value if other diseases develop that are not managed by those chemical treatments. Examples of organisms not responding to foliar fungicides are all virus and bacterial diseases and other fungal diseases including take-all, and loose smut. FHB is suppressed, but not controlled, by application of a foliar fungicide. Lastly, fungicides may be of limited value if yields and test weights are reduced by nondisease factors such as a spring freeze, lodging, delayed harvest, or poor grain fill period. Unfortunately, the above situations are always a risk to the fungicide user. Monitoring crop development throughout the season can reduce some risk by ascertaining yield-limiting factors that indicate fungicides would not be warranted. Of course, this is a moot point once a fungicide is applied. In all instances where fungicides are used, check the response of the crop to the treatment by leaving a nontreated strip in the field for comparison (Hollier et al., 2000).

The above example takes into consideration the discussions from the beginning sections of the chapter. It is important to review every aspect of how a crop is grown and look at a broad picture of the influences on the crop.

Integrated pest management is undergoing a tremendous transition as needs arise and the expectation of society demands change in the way the food supply is grown, handled, and stored. IPM offers a flexible tool for satisfying those demands. Conceptually, IPM is valuable, but in practice that value increases. IPM takes a closer look at the way in which agricultural products are grown and offers the producer the underlying reasons of why and how.

LITERATURE CITED

Anonymous. 1994. Integrated pest management practices on fruit and nuts, *RTD Updates: Pest Management*, USDA-ERS, 8 p.

Dufor, R. and J. Bachmann. 2001. Integrated pest management. *Appropriate Technology Transfer for Rural Areas (ATTRA)*. 41 p.

Hollier, C.A., D.E. Hershman, C. Overstreet and B.M. Cunfer. 2000. *Management of Wheat Diseases in the Southeastern United States: An Integrated Pest Management Approach*. Louisiana State University AgCenter. 44 p.

Jacobson, B. 2001. USDA integrated pest management initiative. In Radcliffe, E.B. and W.D. Hutchinson (Eds.). *Radcliffe's IPM World Textbook*, URL: http://ipmworld.umn.edu, University of Minnesota, St. Paul, MN.

McCarty, L.B., M.L. Elliott, D.E. Short, R.A. Dunn, G.W. Simone and T.E. Freeman. 1994. Integrated pest management strategies for golf courses, pp. 93–102. In McCarty, L.B. and M.L. Elliott (Eds.). *Best Management Practices for Florida Golf Courses*. University of Florida, Gainesville, FL, 174 p.

McMullen, M. 2001. Integrated pest management (IPM). URL: http://www.ag.ndsu.nodak.edu, North Dakota State University, Fargo.

Zadoks, J.C. and R.D. Schein. 1979. *Epidemiology and Plant Disease Management*. Oxford University Press. 427 p.

39 Plant Disease Diagnosis

Jackie M. Mullen

CHAPTER 39 CONCEPTS

- Disease diagnosis is the process of disease identification.

- Plants should be initially examined in the field, landscape, or garden setting, and the site, plant, and problem history should be determined and recorded.

- If the plant disease or other problem cannot be diagnosed in the field setting, a plant sample should be collected, packaged, and mailed so that the sample arrives at a diagnostic laboratory or clinic in a fresh condition with an adequate quantity of symptomatic tissue for examination and testing. The sample must be representative of the problem.

- Plant disease clinics or laboratories may be private, state supported, or land grant university supported.

- Clinics use a variety of techniques to diagnose a plant problem. Many clinics charge for services which consist of the diagnosis and disease control recommendations.

- Initial diagnostic techniques may involve visual study, use of references, soil pH analysis, total soluble salt analysis, and microscopy. The use of additional specialized procedures (including culture work, serology, and molecular testing) depends upon the disease suspected, the value of the crop, and the client. A diagnosis is typically based upon more than one procedure.

Disease **diagnosis** is the act or process of biotic or abiotic disease identification. Some plant disease agents cause visible symptoms distinct enough to allow for disease identification to be made relatively quickly on the basis of a visual study of only plant appearance. For example, crown gall caused by the bacterium *Agrobacterium tumefaciens* induces irregular galls to form on the lower portions of stems and upper sections of roots of infected plants. These galls are often distinctive and allow immediate diagnosis. However in many situations visual appearance of the diseased plant is not distinct enough to allow for an exact diagnosis of the problem. In these cases, diagnosis depends upon one or more assays or tests in addition to a visual inspection of the plant. These added studies may involve soil pH testing, soil analysis (total soluble salt and fertility and nematode assays), light and electron microscopic study of the damaged plant tissues, cultural and physiological studies of the isolated pathogen, molecular studies of the pathogen (**serology, gel electrophoresis, gas chromotography**, PCR [polymerized chain reaction] and DNA probe identification), plant tissue analysis, or soil analysis, and/or pathogenicity studies. The exact procedures used to diagnosis a plant disease depend upon the suspected disease and the plant or crop situation. The crop situation or the crop owner often dictates the level of specificity needed in the diagnosis. For a backyard garden, the identity of a corn (*Zea may*) leaf spot as a *Cercospora* species leaf spot may be sufficient, but for a plant breeder concerned with the leaf spot damage in a field situation, the leaf spot would be identified by genus and species, such as a *Cercospora zeae-maydis* leaf spot.

The process of disease diagnosis has existed, in some form, since diseases were first recognized. The Romans were aware of rust diseases of small grains as early as 310 B.C. The actual cause of rust diseases and other plant diseases was not known or objectively studied until after the development of the compound microscope in 1675. Pasteur and other early scientists in the mid-1800s dispelled the earlier belief of spontaneous generation and proved that microorganisms were present in our environment and were responsible for many diseases of plants and animals. De Bary in 1861 was the first to prove scientifically that the fungus *Phytophthora infestans* caused the disease of Irish potato called late blight, which had caused a widespread famine in Ireland in 1845–1846 (Agrios, 2005) (Chapters 1 and 33). Since that time many biotic plant diseases have been identified. Most diseases are caused by fungi (various chapters), bacteria (Chapter 6), viruses (Chapter 4), and nematodes (Chapter 8). The abiotic diseases (Chapter 25) are caused by such factors as temperature and moisture extremes, low or high soil **pH**, fertilizer excesses or deficiencies, pesticide damage, pollution effects, or weather/soil problems.

The hundred years following de Bary's first plant disease identification were very active in the area of new disease diagnosis. The procedure used to confirm the existence of a new plant disease was developed by Anton Koch in 1876 and is known as Koch's Postulates or Proof of Pathogenicity. This procedure is still used today for the diagnosis and identification of a new (previously unreported) plant disease. The procedure involved the following four steps:

1. The pathogen must found with all symptomatic plants.
2. The pathogen must be isolated and grown in pure culture, and its characteristics described. If the pathogen is a biotroph, it should be grown on another host plant and have the symptoms and signs described.
3. The pathogen from pure culture or from the test plant must be inoculated on the same species or variety as it was originally described on, and it must produce the same symptoms that were seen on the originally diseased plants.
4. The pathogen must be isolated in pure culture again and its characteristics described exactly like those observed in step 2.

The original Koch procedure (#1–3) procedure was amended by Robert Smith to include reisolation of the pathogen from the inoculated, symptomatic plant. If both cultures contain the same microorganism, then this organism was considered to be the pathogen or cause of the disease. Today, Koch's Postulates or Proof of Pathogenicity procedures include the last step added by R. Smith (Agrios, 2005). In diagnostic labs or clinics today, Koch's Postulates are performed infrequently, except when the disease agent is suspected to be new and previously unreported. Most of the plant disease diagnoses that are done today involve identification of plant diseases that have been described previously and named. There are several techniques that may be performed to determine the identity or causal agent (pathogen) of a disease. These diagnostic procedures are completed by or under the direction of plant pathologist diagnosticians. These procedures or analyses may include many of the ones mentioned earlier in this chapter. Recent advances in molecular techniques—such as the use of commercially available kits for DNA extraction from plant tissue and the development of real time PCR, which is a system that automatically controls the reaction, analysis, and computerized results production—have made PCR use in clinical settings much more practical. PCR methodology will allow for pathogen identification from very small samples. Visual studies of symptoms and signs, microscopy, culture media studies, and serology techniques are the most frequently used techniques in diagnostic clinics. Disease diagnosis may involve one or more procedures, depending upon the disease, the client, and the planting situation.

This chapter will describe the basic steps involved in plant disease diagnosis from the time the abnormality is noticed in the field or landscape until the diagnosis is accomplished. The chapter sections will be as follows: (1) field diagnosis and observations, (2) procedures for sample collection, packaging and mailing, (3) the overall operation of a diagnostic clinic, and (4) diagnostic methods used in clinics—a somewhat chronological listing of methods with a brief description and discussion of each procedure. The laboratory exercises in Chapter 40 emphasize practical experience with disease diagnosis.

FIELD DIAGNOSIS

Diagnosis begins in the field or backyard. In many cases, the grower or homeowner is able to diagnose his or her own problem by a field inspection and knowledge of the history of the field and plant problem. Knowledge of recent extreme weather conditions allow the grower to identify wilted plants as abiotic (drought conditions) or biotic diseases that are likely to develop under these conditions. Some diseases may be diagnosed by the presence of visible symptoms and/or signs and a previous experience may allow the grower/homeowner to recognize the malady by a field inspection only. For example, corn smut (Chapter 18), caused by *Ustilago maydis*, is a disease that can be easily recognized and diagnosed in the field by its large fleshy galls on the ears and stalks and by the black, powdery, sooty masses of spores contained inside the galls, which will crack open when mature.

Field diagnosis requires that the grower/homeowner be a good observer. Knowledge of the normal appearance of the plant and requirements for growth is paramount in the diagnosis process. Foliage may be inspected and found to have an abnormal appearance, but the inspection should proceed to include more than the foliage. Stems, fruit, and roots should be observed, if possible. In many cases, a dieback of twigs and branches may relate to a stem, trunk, or root problem. Soil also should be considered as a factor that may be contributing to a dieback or other foliage problem. Many plant problems involve more than one causal agent, so it is important to consider the condition of the whole plant and soil when trying to make a diagnosis. Also, consideration should be given to the site history, fertilization, lime and pesticide applications, and recent weather conditions. Not all plant problems are attributable to micro-pathogens (fungi, bacteria, and viruses). Think about the possibility of an insect or nematode infestation. Also, pay attention to the distribution of the disease. Biotic diseases often exhibit a scattered occurrence or incidence, whereas abiotic problems may occur uniformly in an area or in patterns of rows or edges of a field.

CASE STUDY 39.1

WHEN TEST RESULTS DO NOT AGREE

- A container juniper from a nursery showed dieback of branches and dying roots.
- ELISA testing of the dying roots for the presence of *Phytophthora* produced positive results.
- Culture isolations of the dying root tissues on selective media for *Phytophthora* produced no *Phytophthora* isolates or other fungi.
- Inquiries were made to determine what recent treatments were made on the plants at the nursery. The nursery records showed that a metalaxyl fungicide had been applied as a drench three days before the plants were brought to the clinic.
- Studies have shown that a metalaxyl treatment applied as a drench to roots shortly before roots are placed into a selective medium for *Phytophthora* growth may suppress or delay *Phyophthora* growth.
- The diagnostic reply was made as a "suspect Phytophthora root rot," as the ELISA and culture results did not agree.

COLLECTING, PACKAGING, AND MAILING PLANT SAMPLES

If diagnosis is not possible by inspection in the field, then it will be necessary to seek assistance. County extension agents are a good source for such assistance. They may be able to visit the planting or crop area and help arrive at a field diagnosis. If it is not possible for them to visit the site, it will be necessary to collect a plant sample. A few considerations should be given to collecting samples. First, the sample collected should be representative of the problem. If a leaf spot and fruit rot of apple (*Malus sylvestris*) are observed as the only type of abnormality present, then leaves and fruit should be collected showing early, middle, and late stages of the damage. An accurate diagnosis depends upon a sample that is representative of the situation. Second, the sample should be fresh. Ideally, the specimens collected should be recently infected. Samples should be collected for diagnosis when disease development is first noticed and during initial stages of disease spread. These types of recently infected plant tissues are most easily diagnosed because the pathogen is actively developing. For example, a disease that develops in rainy weather may be dependent upon those wet conditions for disease activity and spread. Collecting the sample during a later dry period may result in a sample where disease is no longer active. The symptoms may be dramatically altered, signs may not be present, and secondary decay bacteria and fungi may develop and prevent an accurate diagnosis. Collection of fresh, recently infected tissues is usually required for an accurate diagnosis. Third, the sample should be adequate in size. How large is an "adequate" sample? This answer varies with the sample and problem. For a leaf spot problem, usually about 20 leaves providing about 20 leaf spots is adequate for the diagnosis procedures, which may involve microscopy, culture work, and possibly other testing and study by more than one specialist. Of course, the exact size of "adequate"

will vary with the plant and disease agent. Large leaves with many spots will require fewer leaves for diagnosis than will small leaves with fewer spots. For galls and cankers, usually 4 or 5 specimens are sufficient to complete the diagnostic process. Whole plants are generally needed for all problems that are not leaf spots, galls, or cankers. In general, turfgrass samples should be 7 to 8 sq in. and 3 in. deep, and taken from the edge of the damaged area. About half of the sample should contain damaged plants; the other half should contain healthy plants. The pathogen is usually located at the margin of the damage area, and diagnosis is not possible if all of the turfgrass in the sample is dead. For seedlings, generally 25 plants or more are needed. With larger plants, fewer plants are needed in the sample; for example, one mature tomato plant is sufficient to diagnosis most problems. Determination of tree problems not involving leaf spots, cankers, or galls can be difficult as collecting the whole plant is obviously not possible. In these cases, sections of the plant must be sampled.

Never collect dead plant samples. When plant tissues die, plant pathogens may become overrun with saprophytic micro-organisms that make diagnosis impossible. Collect soil samples around damaged and healthy plants as soil pH or nutrients may play a key role in plant problems. When it is not possible to collect a whole plant, it is sometimes not possible to give an exact diagnosis. It is not unusual for samples from tree limbs showing leaf scorch to be diagnosed as "suspect the problem/disease involves the roots or lower trunk."

Once the sample is collected, it must be packaged to keep it as fresh as possible until it is examined by a specialist. Most samples should be placed into a plastic bag to prevent dessication. Damp paper towels are not needed to preserve a specimen; however, a dry paper towel can be placed in the bag to absorb excessive moisture and prevent rot in transit to the laboratory. If moisture is added, additional decay and increased populations of bacteria and fungi could make diagnosis difficult. Fruits and vegetables should be packaged in several layers of newspaper instead of plastic. The newspaper allows the vegetables and fruit to have some air exchange and, at the same time, samples will not dry out. Soil should be packaged separately in a plastic bag. Samples should be refrigerated as soon as possible after collections are made to help prevent secondary decay bacteria and fungi from developing on the sample.

Samples to be mailed should be placed into a padded envelope or box with ample packing material (styrofoam chips or other similar materials) to prevent damage during transit. Packages should be addressed clearly. Incorrectly addressed packages that stay in transit for long periods of time are usually not fresh enough for accurate diagnosis. It is always best to mail a sample during the first part of the week, rather than late in the week when there is a risk that the sample will stay in transit for the weekend. If you have any question about mailing procedures for your sample, call the clinic so you will be sure the sample will arrive promptly and will be processed in a timely manner.

OVERVIEW OF THE OPERATION OF A DIAGNOSTIC CLINIC

ORGANIZATION/SUPPORTING AGENCY

Diagnostic clinics may be privately owned and entirely supported by their fee structure, or they may be supported or partially supported by the state land grant university system. There are also some diagnostic laboratories that are associated with the department of agriculture in some states.

Private diagnostic clinics usually receive samples from homeowners and commercial operations. They are typically well-equipped for a variety of testing services including soil minerals and pH analysis, disease diagnosis, and nematode analysis. Fees for services vary, but generally charges are higher than at a land grant university diagnostic laboratory.

Most state land grant university systems support one or more plant diagnostic laboratories/clinics and are usually listed on the Web page of the university. Every state has a land grant university that offers research, teaching, and extension activities for agriculture and human sciences. Depending upon the specific state, the diagnostic clinics may be listed under the Extension component of the college or school of agriculture or it may be listed under a specific department such as plant pathology, botany, or possibly entomology. A table listing all university-related diagnostic clinics was prepared recently by Ruhl and coworkers (2001).

State department of agriculture diagnostic laboratories usually accept samples only from state inspectors who examine commercial samples to certify plants free of disease (or free of specific diseases) before they are shipped out of the state. Not all states have a department of agriculture diagnostic lab. In some states, agriculture inspectors cooperate with university clinics when laboratory diagnostic procedures are needed.

In 2002, as a consequence of the terrorist attack on New York City in 2001, the U.S. Congress created the Department of Homeland Security. Shortly thereafter, the National Plant Diagnostic Network (NPDN) was created by the United States Department of Agriculture (USDA) and the Animal and Plant Health Inspection Service (APHIS) in conjunction with the Department of Homeland Security. Land grant universities in each state have a designated plant diagnostic laboratory/clinic, and each of these clinics is a member of the NPDN and their regional network unit. There are five regional NPDN centers and coordinating laboratories located at Cornell University in New York (Northeast), Michigan State University (north-central), University of Florida (Southeast), Kansas State University (Great Plains), and University of California (West). The NPDN and its regional coordinating laboratories strive to assist individual state land grant university clinics by providing support for rapid and accurate diagnoses, a network communication system, a database for archiving diagnostic data on a national level (part of the National Agricultural Pest Information System), and a protocol for reporting to responders and decision makers. In addition, the NPDN and regional centers develop training programs for diagnostic workers and for agricultural workers who would be apt to encounter high-risk diseases ("First Detectors").

If a high risk disease is identified at a state clinic, the diagnosis can be transmitted and communicated to all other labs and agricultural specialists quickly. If the disease is quarantined in a particular state, the state department of agriculture will be contacted. Typically, a state or federally quarantined plant disease will be destroyed by the state department of agriculture inspectors or APHIS officers, respectively. If the disease is on the APHIS list of "Select" high risk plant diseases, the diagnosis must be communicated to APHIS quickly and steps will be taken by APHIS officials for the destruction of the diseased plants.

SUBMITTING AND RECEIPT OF PLANT SAMPLES

Land grant university diagnostic clinics usually receive plant samples from county Extension agents, homeowners, commercial growers, consultants, golf course superintendents, business operations, and researchers. When a plant sample arrives at a diagnostic clinic, it is recorded in a log book and/or in a computer data base and assigned a tracking number. Many clinics use their data bases for rapid information retrieval on samples. Also, annual reports are compiled with disease listings for each crop received.

CASE STUDY 39.2

WHEN MORE THAN ONE PROBLEM EXISTS, THERE MAY BE AN INITIAL PRIMARY PROBLEM CAUSING PLANTS TO BE STRESSED AND WEAKENED

- A nursery boxwood container plant was brought to the diagnostic lab. Foliage showed leaf blight and dieback. Root decay was also present.
- Visual observation of the leaf blight showed that tiny black specks were present on the leaf blight surfaces. Study of the leaf blight with a stereomicroscope showed that fungal pycnidia were present on the spot surfaces. Study with a compound microscope revealed that the spores and pycnidia were diagnostic for the fungus *Macrophoma* sp. *Macrophoma* is considered to be a weak pathogen, causing leaf blight on boxwoods that have been previously weakened or stressed.
- Visual observation of the trunk and branches showed that a waxy, orange growth was present on some branch and trunk areas. Study with the stereomicroscope showed that the orange areas were raised mounds (fungal sporodochia) present on sunken lesion areas on trunks and branches. Study with the compound microscope showed that the orange sporodochia consisted of masses of orange spores characteristic for the fungus *Volutella buxi*, which is reported to cause canker and dieback of boxwood.
- Root decay areas were examined visually and with the stereo- and compound microscopes. Evidence of fungal or bacterial structures was not present. ELISA testing for *Phytophthora* and *Pythium* was done. Results were positive for the presence of *Pythium*. Culture work on selective media for *Pythium* and *Phytophthora* isolated *Pythium spinosum*. The development of *Pythium* and/or *Phytophthora* requires the free water and usually a prolonged wet situation is required for disease to be a problem. Pythium root decay on woody ornamental plants in a landscape usually is associated with wet conditions and a stressed plant.
- Further examination of the plant in the container showed that the boxwood was planted 2–3 in. too deep. Deep planting of a woody ornamental will usually result in a stressed plant that will develop a variety of problems including root and foliage diseases.
- There are control recommendations for Macrophoma and Volutella blights and dieback and also for Pythium root decay, but the primary problem in this situation is the deep planting, which caused plants to be more susceptible than normal to a variety of root and foliage diseases.

PLANT DISEASE DIAGNOSES

Plant diagnostic clinics provide plant disease diagnosis of both biotic and abiotic diseases. Problems diagnosed include fungal, bacterial, virus, and nematode diseases, as well as damage caused by nutritional imbalances, abnormal soil pH, and total soluble salts, and pesticide, environmental, mechanical, and air pollution damage. Clinics may also provide soil nematode analysis, but in some states, nematode analysis is handled separately and may be administered by another department. Integrated clinics provide plant disease diagnosis, insect damage identification, and sometimes plant/weed identification and insect identification, and possibly soil nematode analysis. Integrated clinics usually are staffed by or cooperate with pathologists, entomologists, weed scientists, and possibly nematologists. Botanists, agronomists, and horticulturists may also be consulted on some samples.

In many diagnostic clinics, about 50% of the disease samples are diagnosed with biotic diseases, and the remaining samples are found to have problems caused by abiotic factors. These percentages vary from clinic to clinic and from year to year. In 2000 and 2006, when a severe drought

occurred during the summer in much of the Southeast, about 65% of the samples submitted to clinics in the region exhibited damage caused by drought or drought-related factors. In more typical years, about 80% of the biotic diseases seen in these clinics are caused by fungal plant pathogens. Bacteria and viruses together comprise about 15% or more. Plant parasitic nematode problems account for less than 5% of the plant problems submitted for diagnosis. Plants with insect damage are also submitted to many integrated clinics. Insect damage is considered to be an injury rather than a biotic or abiotic disease problem. Some clinics have an entomologist on their staff or refer their samples to cooperating entomologists.

RESPONDING TO THE CLIENT

The response from the clinic to the client consists of the diagnosis and the control recommendation. The name of the disease and the name of the pathogen are provided. The name of the disease may or may not include the genus name of the pathogen. Because the disease name is sometimes not specific for a particular pathogen, the scientific name of the pathogen is supplied to unequivocally identify the cause of the disease and avoid ambiguity. As an example, consider the disease name of "common leaf spot" of strawberry. In the United States, most people familiar with strawberry diseases will know that common leaf spot is caused by the fungus *Mycosphaerella fragariae*. However, the disease name "common leaf spot" may not provide adequate information to someone who is unfamiliar with strawberry diseases. The scientific name of the pathogen always identifies the disease and pathogen.

Control recommendations vary with the disease, the disease severity, and the cropping situation. Control recommendations given to a homeowner may not be practical or economical for a commercial grower. For example, removal of the infected plants and plant foliage in the fall may be the most practical and economical recommendation for control of common leaf spot of strawberry in a small garden where the leaf spotting is severe on only a few plants. For a commercial grower with a moderate amount of leaf spot throughout the planting, a regular fungicide spray program in the early spring of the following year may be the most practical and economical control recommendation.

SERVICE CHARGES

Clinics were typically begun as free services and were supported entirely by extension and/or university funding. This situation has changed in many states during the past 15 to 20 years. Today many of the university-associated diagnostic clinics charge for their diagnostic services and fees vary from state to state. Plant disease diagnostic service charges may vary from $5 to $25, but some highly technical and expensive analyses, such as DNA identification, may require additional charges. Soil nematode analysis fees also vary depending upon the specific state. Charges usually range between $5 and $15. When out-of-state samples are sent to clinics, higher charges are usually imposed.

DISTANCE DIAGNOSIS USING DIGITAL IMAGES

With the development of e-mail and capabilities to send images electronically, many clinics have adopted a program for receiving digital images of plant disease situations and samples from county Extension agents. These programs have been designed so that transmitted images of plant damage show the field situation, close-up views of the plant damage, and microscopic views of damaged tissues. Additionally, a digital diagnostic submission form may be transmitted from cooperating county offices or other university locations to the diagnostic clinic or pathology Extension specialists. All responses from the clinic are via e-mail, shortening the diagnostic response time from the 2–3 days taken for land mail. When the disease is one that may spread rapidly, a quick diagnosis

and control recommendation reply will facilitate a timely implementation of control practices, which may make the difference between a large or small amount of crop lost to disease. The digital image diagnosis program allows for the images, submission form, and the diagnosis to be kept in an electronic file for future educational or reference uses.

Not all samples may be diagnosed by images. Although fungal diseases that produce visible symptoms and macroscopic or microscopic signs are ideal for this system of sending digital images for diagnosis, this method is often not suitable for bacterial and viral diseases. When microscopic structures of fungal pathogens are not present, a plant sample must be sent to the clinic for culture work or other types of analyses. Digital diagnosis is also dependent on the photographic skills of the cooperator sending the images and the collection of vital information about the problem.

CLINIC DIAGNOSTIC PROCEDURES: CHRONOLOGY OF INITIAL TESTS AND SPECIALIZED ANALYSES, AND DISCUSSION OF INDIVIDUAL DIAGNOSTIC PROCEDURES

Initial diagnostic procedures followed in most laboratories or clinics are fairly standard and these procedures usually follow in roughly the same order as listed below.

1. Visual examination for symptoms and signs
2. Review of the information sent with the sample
3. Consultation of plant disease compendia and/or host indices
4. Microscopic examination with stereo-dissecting microscope
5. Microscopic examination with compound microscopes
6. Possible further consultation with references books
7. Soil pH determination

If visible symptoms and/or signs are distinct enough to be diagnostic of a particular disease, then the diagnosis may be completed after procedures 1, 2, and 3. If an experienced diagnostician is making the diagnosis and the disease is common, then consultation with compendia and/or host indices may not be necessary. If visible symptoms and/or signs are not diagnostic, then microscopy (procedures 4–6) will usually follow. Usually both the stereo-dissecting and compound microscopes are used and consultation with references will usually follow unless the microscopic structures are well-known to the diagnostic worker. The evidence may be sufficient for diagnosis of the disease if

---

CASE STUDY 39.3

MORE THAN ONE PRIMARY PROBLEM MAY BE PRESENT

- A Bermuda grass sample arrived showing small leaf spots, leaf blight, and poor growth.
- Microscopic study with stereo- and compound microscopes revealed that *Bipolaris cynodontis* was present on the leaf blight and leaf spot areas. This fungus is commonly seen as a leaf spot and blight on Bermuda grass.
- The soil pH was 4.4. This pH is much too low for good growth of Bermuda grass. It is low enough to cause poor growth of the grass and possibly some nutritional problems.
- A soil nematode analysis showed that the high number of ring nematodes (*Criconemoides* sp.) would also contribute to growth problems.
- Either of the above three problems—Bipolaris leaf blight, low pH at 4.4, and a very high population of ring nematodes—could cause a major problem with the growth of the Bermuda grass.
- Control of the foliage disease and nematode problem may be brought about by correcting the soil pH and checking the soil for mineral deficiencies. Bipolaris foliage disease on Bermuda has been shown to be related to low potassium levels in the soil in some cases. As there are no nematicide treatments that can be applied in a home setting, maintaining a stress free environment for the grass will help it develop in the presence of the high nematode populations.

---

microscopic fungal spores are observed or bacterial cells are observed oozing out of damaged tissues.

If soil is sent with the sample, the pH is usually measured. Some symptoms of damage, such as scorch, dieback, or poor growth, may directly relate to pH levels that are too alkaline or too acidic. An extremely acidic soil pH may cause some minor element toxicities. Inappropriate soil pH may also predispose a plant to certain diseases.

If visible symptoms are not diagnostic, microscopic structures are not observed and pH determination does not account for the damage, then other diagnostic testing procedures are used. The exact procedures used and the order followed in using each diagnostic procedure depends upon symptoms and the possible agents that could cause those symptoms. For example, if microscopy does not unequivocally confirm a disease and if the suspected pathogen is fungal or bacterial, then culture isolations usually follow. If an isolated plant pathogen has been previously associated with the symptoms observed on the specimen, the diagnosis may be considered complete. If culture results are not conclusive, then other analyses may follow, depending upon the suspected pathogen.

Diagnosis of a disease is usually based upon results of more than one type of study or analysis. Many diagnoses are based upon visual symptoms and possibly visual signs, accompanying information, microscopic evidence of signs, and literature descriptions of diseases. When trying to arrive at a disease diagnosis, review the following information: notes on symptoms and signs, the problem description and site/plant history, information from the host index, descriptions of possible diseases, microscopic evidence of pathogen structures, and results of other tests. Specialized analyses used to identify specific pathogens are the following:

1. Measurement of electrical conductivity as a criterion of total soluble salts in a soil sample. This is very useful in determining if excess fertilizer caused the damage.
2. Moist chamber incubations of plant materials to help determine the presence of a fungal pathogen.
3. Culture isolation techniques aid in the identification of a fungal or bacterial pathogen.
4. ELISA or other serology method. These techniques detect fungal, bacterial, and viral pathogens.
5. Specific physiological tests depending on bacteria suspected of causing the disease.
6. Gas chromatography of bacterial fatty acids.
7. Gel electrophoresis for identification of pathogen species or strain by specific enzyme or protein identification.
8. DNA identification by various molecular methods to identify specific pathogens; methods include PCR.
9. Proof of pathogenicity or Koch's Postulates protocol when fungal or bacterial agents are suspected to be the pathogen.

The diagnostic techniques used will vary according to the particular disease(s) suspected. However, there is an overall pattern and chronology that is often followed with the initial five studies listed above. The results of the initial studies will help determine which of several subsequent specialized analyses may be performed. Other factors that determine the diagnostic tests conducted are the value of the crop and the grower. For a home gardener, it may not be necessary to identify the exact species of the fungus

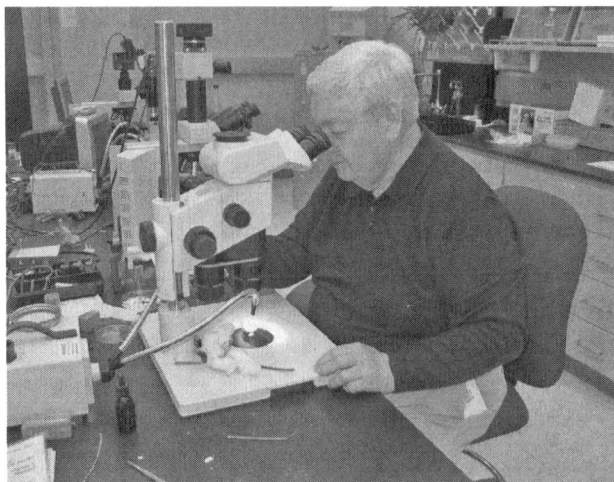

FIGURE 39.1   Examination of damaged plant tissue with a stereomicroscope.

(e.g., *Alternaria*) that is causing a leaf spot on a shade tree. However, a nursery grower or plant breeder may request an identification of the species of *Alternaria* on their containerized shade trees.

VISUAL SYMPTOMS AND SIGNS

Initially, the plant sample is examined for visual symptoms and/or signs of disease. For example, black spot (*Diplocarpon rosae*) on rose is often diagnosed on the basis of visible symptoms as the leaf spots are distinctively irregular with a feathery margin. Cedar-apple rust causes distinctive galls to form on juniper, and also unique spore-producing structures form on apple leaves. The presence of the symptoms and signs of this rust disease are sufficient for disease diagnosis. Most bacteria diseases are not identified by visual observations as bacteria do not typically produce visible diagnostic symptoms or diagnostic signs. Viruses also do not usually produce visible diagnostic symptoms and diagnostic signs are not produced. The only plant parasitic nematode that produces visible symptoms is the root knot nematode (*Meloidogyne* spp.), which causes irregular galls to form on roots. These galls may be diagnostic, but microscopy is usually used to confirm the presence of nematodes.

REVIEW OF THE INFORMATION SENT WITH THE SAMPLE

Every diagnostic clinic/lab has a diagnostic information sheet or questionnaire that should be completed and should accompany the plant sample. If the client does not have access to these questionnaires, then a letter should accompany the sample to describe the damage, the distribution of the damage, the history of the site, pesticides and/or fertilizers applied, and recent weather events. A description of the damage as it appeared when the sample was collected can be very helpful, especially if additional decay occurs in transit to the clinic. An account of the development of

the problems also is helpful to the diagnostician. Site history is important in diagnosis, especially if a herbicide carryover may be causing the damage. Pesticide or fertilizer applications may be directly or indirectly involved in problem development. Excess lime, fertilizer, or pesticides may cause damage or affect disease development.

REFER TO DISEASE HOST INDICES AND CROP COMPENDIA

At this point, references are often used to develop a list of possible problems or diseases, based upon visual symptoms and signs with supporting information. Disease compendia and disease host indices (Farr et al., 1989) are especially helpful. As indicated above, some diseases produce visual symptoms and signs that are diagnostic. However, most disease diagnosis requires further study before diagnosis can be completed. References provide clues of diseases or pathogens that may be involved. Additional assays or tests confirm the absence or presence of plant pathogens.

STUDY USING A STEREOMICROSCOPE

When diagnostic structures are not visible to the naked eye, damaged plant tissue is examined with a stereomicroscope (sometimes called a dissecting microscope) at a magnification of 0.5× to 60× (Figure 39.1). Low magnification is useful for viewing fungal fruiting bodies and large fungal spores. Stereo-dissecting microscopes are also useful for finding fruiting bodies or spore masses to transfer to a glass slide for examination at a higher magnification with a compound microscope.

If a fungal infection is suspected but fungal fruiting bodies are not found, it is possible that the fungus has not yet produced any fruiting bodies or doesn't produce fruiting bodies at all, but rather produces spores on individual conidiophores on the plant tissue surface (Chapter 16).

Damaged tissue devoid of fruiting bodies should still be examined for the presence of spores using a compound microscope.

Bacterial disease agents may produce ooze droplets that could be observed using a stereo-dissecting microscope. The presence of ooze indicates the possible presence of a bacterial disease, but the specific bacterial disease agent cannot be confirmed using microscopy. Viruses cannot be identified using conventional light microscopy. Root knot nematode may be identified using the stereo-dissecting microscope. Galls can be dissected and the mature root knot female can be seen as a pearly-white, moist, pear-shaped body.

STUDY USING A COMPOUND MICROSCOPE

The compound microscope is useful for examining fungal fruiting bodies and spores at 45× to 1000×. Many fungal diseases are identified based on spore morphology. This microscope is also used to view masses of bacteria associated with plant diseases as they ooze out of infected tissue. Occasionally, viruses may be identified with a compound microscope. Tissue may be stained with a protein stain or a nucleic acid stain, highlighting the protein or nucleic acid components of virus bodies. Virus inclusion bodies consist of aggregations of virus particles or virus products. The bodies are visible at 450× and are specific for individual virus groups, but not individual viruses (Christie et al., 1986).

Microscopic examination of plant parts and pathogen structures is an integral part of the diagnostic process at plant disease clinics. For example, *Septoria leaf spot*, a common disease on tomato, is usually identified by visual symptoms of small (approximately 2.6 mm diameter), gray, circular spots with dark brown or black borders; the presence of fruiting bodies called pycnidia, which are observed in the stereomicroscope as small (100–200 μm diameter), usually black, flask-shaped structures with a small hole or ostiole at the top or apex of the body; and the presence of very long (about 67 by 3.2 μm), thin, filiform spores (Jones et al., 1991) evident in the compound microscope. The diagnosis is based upon three pieces of information, two of which involve microscopic study. Sometimes microscopic study does not reveal any diagnostic fungal or bacterial structures. In these situations, additional studies/procedures are usually performed.

CONSULTATION WITH REFERENCE BOOKS

Even if observed structures are familiar to the diagnostician, references may be checked to confirm that structures viewed are exactly the size, shape, or arrangement diagnostic for a particular fungal disease. Small variations in spore morphology may indicate that a different or new fungal species is present and possibly the cause of the problem. Disease compendia are excellent references for identifications as they contain descriptions of visible symptoms and signs, microscopic structure descriptions, and comments on other tests that may be necessary for diagnosis of particular common pathogens of a specific crop or crop group. However, not all crops and landscape plants are covered by the compendia series. Many ornamentals are not included and other references must be consulted along with mycology texts and fungal genus keys and monographs. Disease host indices are often very helpful in disease diagnosis. They include listings of all disease agents reported on a particular plant up to the date of publication. Some disease and pathogen descriptions are only available in recent manuscripts in research journals where the disease was first described.

SOIL pH DETERMINATION

If soil is sent with the plant sample, it is always a good idea to check the soil pH (level of acidity or alkalinity of the soil). If the soil pH is inappropriate for a particular plant, stress could predispose the plant to a variety of weak disease agents. Also, the stressed plant might become more susceptible to some moderately pathogenic disease agents. Leaf spot diseases may be more severe on plants weakened by soil pH that is too acidic for normal plant growth. Some root diseases may be more prevalent and damaging when plants are weakened by a number of environmental stresses including soil that is too acidic or too alkaline. If the environmental stress problem is not corrected, the plant will continue to be highly susceptible to certain diseases and as soon as disease control treatments conclude, the plant may very quickly become reinfected and develop disease symptoms again. To determine soil pH, a small amount of soil is mixed with an equal volume of filtered water, allowed to stand for about 30 min, and the solution pH is measured.

TOTAL SOLUBLE SALT MEASUREMENTS BASED UPON SOIL ELECTRICAL CONDUCTIVITY

An electrical conductivity meter determines total soluble salts level in a sample of soil or soilless media. Measurements are given in units of electrical conductivity, mmhos/cm. The usual procedure is to mix 40 mL of soil with 80 mL of filtered water. After mixing is complete, the soil is allowed to settle and the water solution is poured into a graduated cylinder. The electrical probe is submersed into the solution and the total soluble salt reading is recorded. Charts are available for the specific ratio of soil to water (in most cases, 1:2) and for the type of soil (sandy soil mix, heavy soil mix, and soilless mix) to give an indication of whether the levels of salt are low, moderate, acceptable, or too high for most plants.

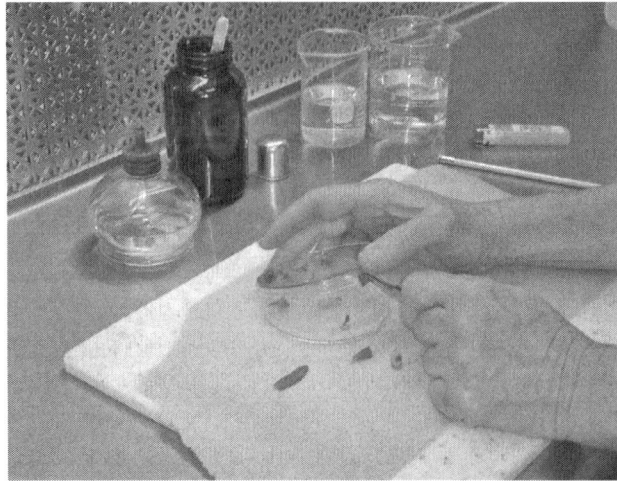

FIGURE 39.2 Surface sterilized tissue pieces cut from the edge of symptomatic leaf spot tissues are placed into a sterile culture dish containing a growth medium using aseptic technique.

Low readings may indicate nutrient deficiencies, whereas very high readings indicate that roots may have been injured by high concentrations of salts. Root injury from high salts often cause new foliage to show severe levels of leaf scorch. Sometimes new growth is entirely brown and dead as a result of root injury from salts. High salt levels kill root hairs and effectively simulate drought stress.

MOIST CHAMBER INCUBATIONS

Tissue incubation in a moist environment may stimulate fungi to produce spores within developing fruiting bodies or on specialized hyphae. Once spores and fruiting bodies are produced, diagnosis is possible.

Moist chambers should be set up only after visual symptoms and signs are documented as these characteristics are usually destroyed in moist chambers. Sometimes moist chambers are not helpful in the diagnostic process as secondary fungi completely overgrow the tissues, making it impossible to detect the pathogen spores and structures (Waller et al., 1998; Shurtleff and Averre, 1997). A moist chamber consists of a plastic bag or other closed container (not completely sealed from air exchange) that contains plant tissue and a moist paper towel. The goal is to provide an environment with high relative humidity without exposing the specimen to free water. High humidity favors the development of fungal and bacterial pathogens and saprophytes present on decaying tissues. Tissues in a moist chamber should be examined daily for a period of 1–5 d for the development of fungal structures. Microscopic examination of tissues after incubation may reveal several types of spores of which one may be the pathogen. Therefore, it is important to be familiar with common genera of plant pathogenic and saprophytic fungi.

ISOLATION OF THE PLANT PATHOGEN ON CULTURE MEDIA

When fungal disease is suspected and microscopic evidence is not present, moist chamber incubations are often used first. Culture work is used if microscopy methods and the moist chamber technique are not successful for fungal and bacterial detection and identification. Tissue used in isolations in culture media must be fresh, consisting of recently infected tissue areas bordering on healthy tissue. Tissue to be cultured is usually surface sterilized with a 10% (v/v) solution of household bleach. The duration of surface **sterilization** ranges from a few seconds to two min. Immediately after soaking the tissue pieces in the bleach solution, the tissue should be rinsed in sterile distilled or filtered water. Tissue should then be blotted dry (clean paper towels usually work fine for blotting). Tissue sections should be aseptically cut into small pieces (2–5 mm diameter) and placed into sterile culture medium in sterile petri dishes (Figures 39.2 and 39.3). Dishes are usually maintained at room temperature for 3–7 d and examined daily for fungal growth. Potato dextrose agar acidified with lactic acid to retard bacterial growth is often used as a general purpose medium for the culture of fungal foliage pathogens (Armentrout and Baudoin, 1988). When fresh, recently infected tissues are cultured from marginal areas of infection, cultures should produce a consistent type of fungal growth. Fungal isolates should be transferred to a sterile petri dish with sterile medium to produce a single fungal isolate in pure culture. The growth of certain fungal pathogens is very distinctive in culture aiding in identification of the pathogen. Conclusive evidence for fungal identification is the development of distinctive fungal spores and/or fruiting bodies. If tissue decay is advanced, several fungi may grow out in culture; each isolate should be examined to determine

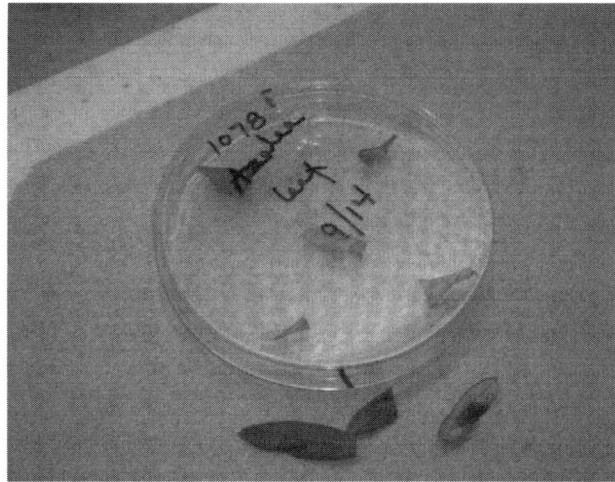

FIGURE 39.3 Usually four to five tissue pieces are placed equidistantly in a culture dish.

which ones are plant pathogens. In some cases, secondary pathogens or decay fungi may grow faster than the pathogen and the medium in the dish is overgrown by these undesirable fungi, preventing identification of the primary pathogen (Waller et al., 1998; Shurtleff and Averre, 1997; Fox, 1993).

If fungus fails to grow from the cultured plant tissues, surface sterilization may have been too severe. It is often a good idea to prepare 2–4 culture dishes using a range of surface sterilization times. The optimum sterilization time varies with the pathogen, the plant tissue type, and the secondary and decay organisms present.

If a bacterial disease is suspected, a different method is used for isolation in culture. After surface sterilization of the tissue, a piece of the damaged tissue is placed into a sterile plate and cut up or crushed in a drop of sterile filtered water. The macerated tissue remains in the water for 3–5 min and then a small amount of water is streaked across a culture dish using a sterile bacteriological transfer loop. A general purpose medium for bacterial growth is tryptic soy agar. Most diagnostic labs will streak the bacteria in four quadrants in a prescribed manner so that the bacterial concentration is diluted by the time the last quadrant streak is streaked. The objective of the streak technique is to dilute the bacteria enough so that single cells will be separated in the last quadrant (see Figure 7.1). The resulting colonies from single cells may be easily transferred to another culture dish, which is then a pure culture derived from a single bacterial cell. Bacteria are not identified easily as they do not produce microscopic structures that are diagnostic. Instead, bacteria are typically identified by specific physiological reactions or by specific molecular characteristics. Therefore, the isolation of bacteria in pure cultures is just the first step toward other procedures to identify the bacteria by genus and species (Waller et al., 1998; Shurtleff and Averre, 1997; Fox, 1993).

BAITING FOR FUNGAL PATHOGENS

Using a bait to isolate a fungal pathogen typically involves placing a piece of infected or damaged tissue into a healthy plant part where the specific pathogen will be stimulated to grow. Usually baits involve selective stimulation of growth for certain fungal pathogens. Carrot roots are baits for the fungal pathogen *Thielaviopsis basicola,* which causes black root rot of some plants. Green apples and green pears are used as baits for growth of the fungus *Phytophthora.* If culture methods are not available, successful baiting is an alternative strategy for isolating and identifying pathogens (Fox, 1993; Shurtleff and Averre, 1997; Waller et al., 1998).

USE OF SEROLOGY TECHNIQUES (OUCHTERLONY AND ELISA)

Serology, the study of immunological reactions, derives its name from serum, which is blood fluid after all the blood cells are removed. There are several types of tests that involve immune reactions; that is, several tests that involve the reaction of **antigens** (substances foreign to the body, usually proteins) and **antibodies** (specific molecules produced by mammals in response to the presence of the foreign protein or substance). One of the first serology tests to be conducted was known as the Ouchterlony test (Chapter 4) named for the scientist who designed the test. In this test, the antigen–antibody reaction takes place in water agar within a petri dish. The antigen or foreign protein is placed into a well in the center of the plate. This antigen is tested against several known antibodies, which are individually placed into wells around the edge of the agar dish. The antigen and antibodies diffuse out of the wells and move through the agar medium. As these substances come in contact with each other, about midway from the center and edge of the plate, an arc of white pre-

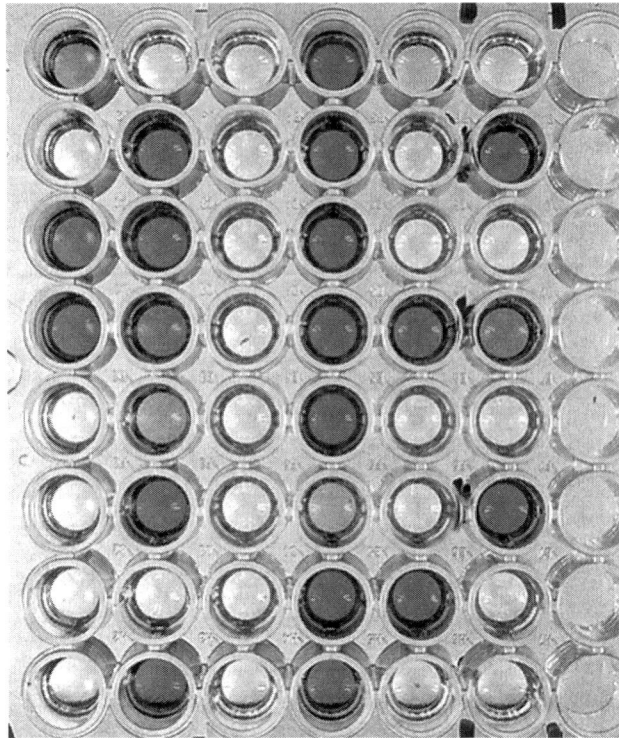

FIGURE 39.4 Multiwell plate shows positive reactions (dark) for samples tested for *Phytophthora* by the ELISA technique.

cipitate is formed between the antigen and its antibody. There is no precipitate formed between the antigen and the other nonreacting antibodies. An alternative arrangement for the test is for one antibody to be placed in the center well. Each well along the dish edge contains plant sap from a specific plant sample suspected to contain the antigen virus. Where the matching antibody and antigen virus contact, a white precipitate forms and identifies those samples as positive (Chapter 5; Figure 5.5).

Today the most common serology test used by most diagnostic clinics/labs is known by the acronym ELISA, enzyme-linked immunosorbent assay (Chapter 4; Figure 4.2). The immune-reacting molecules (antigen and antibody) are adsorbed onto the wells of a plastic multiwell, microtiter dish. One of the reacting antibodies is linked to an enzyme that allows for a color reaction to indicate a positive test reaction. The type of ELISA may be direct or indirect. The direct test is often called a double antibody sandwich, and it is the shorter procedure of the two. The indirect ELISA is a longer procedure, but the longer protocol allows for a more reactive antibody–enzyme component to combine with the antigen. The indirect test is often considered to be more sensitive than the direct method. For the purposes of this general plant pathology text, the shorter, simpler, direct ELISA will be described (Figure 39.4). The wells of a multiwell plate are coated with a specific antibody (AB). After this AB is allowed to dry and be adsorbed onto the wells, the unknown antigen (usually the sap expressed from the infected tissues) is added to the test wells. At the same time control wells are prepared by adding sap expressed from a healthy plant of the same age, variety and location as the infected test plant (a negative, healthy plant control); sap expressed from a plant known to contain the specific antigen or pathogen that is suspected to be present in the unknown test plant (a positive control); and buffer or water only (a negative control). The antibody (AB)–antigen (AG) reaction mixture is allowed to incubate for a period of time that varies depending upon the specific antigen test. After the incubation time is complete, the wells (test and controls) are washed several times (usually 6 times) and then allowed to drain. If the AG reacted with the AB, the bound mixture of AB-AG remains attached to the test and positive control wells. If the test plant sap did not contain the specific antigen reactive with the added AB, then no binding took place and only the AB remains attached to the wells. The next addition to the wells is the AB preparation attached to an enzyme, which is usually peroxidase or alkaline phosphatase. The AB-enzyme (AB-E) will attach to those wells that contain the AG bound to the originally added AB. This three-component mixture of AB-AG-AB-E is allowed to incubate for the recommended period of time. After the incubation time is completed, the wells are again washed several times (usually six times) to remove any unbound AB-E from wells. Next, the substrate (S) for the enzyme (hydrogen peroxide or p-nitrophenylphosphate, respectively, for the aforementioned enzymes) is added and allowed to react with the mixture. A color change indicates that the enzyme has bound to the substrate and is a positive test for the presence of the AB in question (Figure 39.4). If positive control wells and negative control wells react or not as they should, then the completed ELISA provides very valuable and specific information as to the identity of the pathogen causing the plant damage (Fox, 1993; Matthews, 1993; Schaad, 2001). ELISA results usually are ready in a matter of hours or after an overnight reaction so the client may obtain an indication of the cause of his or her problem in less than 24 h. This test procedure is very valuable for commercial growers where a rapid implementation of control measures could save the crop from widespread damage and economic loss.

SPECIFIC PHYSIOLOGICAL TESTING TO DETERMINE THE IDENTITY OF BACTERIAL PATHOGENS

Bacteria (Chapter 6) do not produce many characteristic microscopic structures that can be used as a tool for identification. Most plant pathogenic bacteria are short rods. However, bacteria do differ in their capability for specific physiological characteristics. An initial analysis is the determination of cell wall structure by the Gram stain or by bacterial reaction with 3% potassium hydroxide. Bacteria are generally divided into two groups on the basis of their cell wall structure. Either one of these two tests will identify the bacteria as having a tight cell wall structure (Gram positive or KOH viscosity negative) or having a loose cell wall structure (Gram negative or KOH viscosity positive). Many of the physiological characteristics involve the ability of bacteria to utilize specific carbohydrates or other carbon or protein molecules. Some tests determine the presence or absence of a particular enzyme capable of bringing about a particular chemical reaction. The identity of a bacterial genus may depend upon the presence (or absence) of growth and the appearance of the growth on a specific agar or broth medium. Some bacterial growth media are referred to as differential media because the bacteria may be differentiated from each other on the basis of the type of growth on the media. Yeast dextrose carbonate agar (YDC) and *Pseudomonas* agar F (PAF) are common bacterial differential media. Xanthomonads typically grow as yellow, mucoid colonies on YDC and can be distinguished from many other types of bacteria growth on the basis of appearance. However, not all yellow colonies on YDC are xanthomonads. Pseudomonas agar F is a differential medium for the group of pseudomonad bacteria that have the ability to produce a fluorescent pigment. Fluorescence of bacterial colonies under ultraviolet light on this medium is characteristic of a group of bacteria that cause plant diseases. There are many different types of differential media that help categorize and identify bacterial genus and species. Prior to the development of more rapid and accurate molecular testing, bacterial identification was based upon numerical taxonomy; that is, the genus groupings and species were identified on the basis of results of multiple tests. In order to identify a particular bacterial species, 10 to 20 different physiological tests might be necessary. Selective media are available as aids for identification of some bacteria. Depending upon on the degree of selectivity of the media, bacterial species may be identified with fewer tests (Schaad, 2001).

BIOLOG FOR IDENTIFICATION OF BACTERIAL PATHOGENS

BIOLOG is a combination of "numerical taxonomy," molecular analysis, and the use of a computer software library program. The analysis system consists of a multiwell dish where each well is coated with a different carbon or food source. A metabolically active bacterial preparation is added to each well. Positive and negative controls must be included with each test. After 1–2 d incubation, the mixture in the wells changes color if a degradation (oxidation) of the substrate took place. The optical densities of the wells are measured by a spectrophotometer and the readings are compared via a computer program. The pattern of positive reactions in the multiwell plate is compared to the known pattern for plant pathogenic bacteria that have been recorded in the software library program. Results are given as a percentage of similarity to a specific bacterium in the software library that is most similar to the test bacterium. This analysis system is considered to be fairly accurate for genus identification and sometimes accurate for species (Fox, 1993; Schaad, 2001).

GAS CHROMATOGRAPHY OF BACTERIAL FATTY ACIDS TO DETERMINE THE IDENTITY OF BACTERIAL PATHOGENS

A gas chromatographic analysis system is commercially available (MIDI, Inc.) for the identification of bacterial species and subspecies groups (pathovars or pathotypes). Studies have shown that fatty acid profiles are characteristic and specific for individual bacterial genera, species, or pathotypes (cultivars). A bacterial sample fatty acid profile consists of the identification of each fatty acid and the quantification of each fatty acid present in the bacterial sample.

Before the chromatographic analysis, the bacterial preparation in broth must be treated to disrupt or break the cell walls and release the fatty acids, which are then methylated so that they become volatile and move easily in the gas medium (Figure 39.5). The fatty acid methyl esters are separated on size and identified by their rate of movement as compared to the movement rates for known fatty acids (Schaad et al., 2001).

Results of the gas chromotography are transmitted to a specially designed computer program where the profile of the test bacterium is compared to a library of fatty acid profiles of other bacteria. Results are given as a similarity coefficient that expresses the degree or percentage of

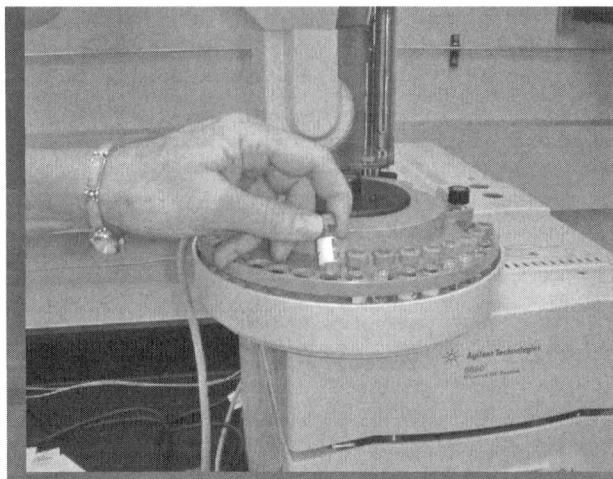

FIGURE 39.5 Vial contains methyl esters of fatty acids prepared from a lantana leaf spot sample bacterial culture.

agreement between the test organism and the most similar bacterium in the library system. A very high coefficient, such as 0.925, indicates very good agreement and the probability that the identification is accurate to the genus, species, and pathotype level. A lower coefficient, such as 0.456, usually indicates the identification is accurate at the genus and species level.

Some diagnostic clinics perform bacterial fatty acid gas chromatography, but, due to the expense of the equipment, many clinics cooperate with research labs that perform this identification procedure.

GEL ELECTROPHORESIS USED TO IDENTIFY SPECIFIC ENZYMES OR PROTEINS OF SPECIFIC PATHOGENS

Gel electrophoresis (Chapters 26, 27, and 30) is a method of identifying proteins or enzymes by the rate at which they move through an electric field. The rate of their movement through the gel depends upon their size and their electric charge. The rate of movement of a particular protein is compared to the rate of movement for known proteins. If production of a unique protein is specific for a specific pathogen or pathogen subspecies, the electrophoresis method can be used as a diagnostic method. This method was used to identify an enzyme specific to A2 mating types of metalaxyl-resistant *Phytophthora infestans* in the early 1990s. Gel electrophoresis assisted pathologists in advising growers whether they should use metalaxyl or one of the newer, more expensive fungicides for late blight control.

The analysis involved lysing the fungal cells, extracting the proteins by centrifuging and organic phase separations. Identification of the protein bands in the gel involved a specific staining technique. The known proteins (positive controls) and negative controls must always be included with the test protein sample (Goodwin et al., 1995). This method could be used with fungi or bacteria that produce unique enzymes or other proteins.

DNA MOLECULAR METHODS/PCR TO DETERMINE THE IDENTITY OF PATHOGENS

The most accurate method of organism identification is by DNA analysis (Chapters 26 and 27). Identification is always done by determining compatibility (or hybridization) of the test DNA single strand to the known DNA single strand. The "probe" refers to the marker attached to the known DNA single strand. The presence of the marker allows for visual identification of the double-stranded hybridized DNA. The attached probe might be radioactive P-32, biotin, a serology component of an ELISA reaction, or other marker.

Initially, the DNA must be extracted from the infected plant sample. There are several kits available for plant DNA extraction. The procedure is easy to follow and involves lysis of plant cells usually by grinding in liquid nitrogen (or other mechanical treatment at cold temperatures). The rest of the extraction procedure involves separation by column chromatography and centrifugation methods.

Once the DNA is extracted from the sample, the basic procedure involves the addition of the sample DNA and the known single-stranded DNA attached to a marker. The mixture must be heated to cause separation of the double strands into single strands. After heating for a specific time and temperature, the mixture is cooled to facilitate reassembly (annealing) of the compatible single-strands of DNA into double strands. If the unknown sample DNA was the same identity as the known probe-connected single-stranded DNA, then the probe would be present and detected in the double-stranded DNA. If the sample DNA is not compatible (does not hybridize) with the marker attached known single-stranded DNA, then double strands will not form with the probe attached single stranded DNA. The single stranded probe DNA will be removed from the test solution by separation procedures for double-stranded DNA. P-32 probes are detected

FIGURE 39.6 With RT-PCR, the DNA replication and the probe release detection both take place in the reaction mixture tube.

by x-ray film; biotin is detected by a visible color reaction as is the case for an ELISA probe system (Duncan et al., 1992; Matthews, 1993; Fox, 1993; Schaad, 2001).

PCR (polymerized chain reaction) (Chapters 26 and 27) is the method of DNA identification that is being used most often today in research and diagnostic laboratories. The method involves extracting the DNA from symptomatic plant tissues as was described above. The extracted DNA unknown sample is added to a reaction mixture containing single-stranded pieces of the known pathogen DNA called primers, appropriate DNA replicase enzymes, nucleotides, buffers, and a marker. The reaction mixture must be heated so that the sample double-stranded DNA is converted to single strands. The known DNA primers will cause the synthesis of new DNA single strands of the sample DNA if the primer DNA is the same pathogen identity as the sample. Multiple cycles of heating and cooling cause the synthesis of quantities double-stranded DNA that are detectable by gel electrophoresis methods. The markers are designed so that they are present with the double-stranded DNA product if newly synthesized DNA is present in the final double-strand product with the final cooling phase of the reaction. The PCR method will detect very small quantities of extracted pathogen DNA. This is one of the most sensitive and accurate methods available for pathogen identification. The method is laborious and requires a trained, skilled technician.

With the start of the 21st century, a new modification of the PCR method began to attract the attention of diagnostic laboratories and research laboratories. The method is called real time PCR. With this method, the polymerized chain reaction (synthesis of DNA) that takes place during heat and cool cycles described above occurs in a reaction mixture in a small plastic tube that is placed into a heating-cooling block. The markers used are designed to become detectable when the primers are synthesizing more DNA. With real time (RT) PCR, the DNA synthe-

sis and marker detection system occurs in the same reaction tube (Figure 39.6). As the DNA synthesis proceeds, marker is released in a form that is detectable. A computer software program is designed to report the release of the marker as the indication of a positive reaction. Once the reaction mixture is prepared, it is placed into the heat-cooling block and the computer program is turned on. Results are usually visualized on the computer screen within 30 min or less. This method is very sensitive, accurate, and compatible with other activities in a diagnostic laboratory. Training is minimal compared to the time and skill needed for the more traditional PCR method.

In the fall of 2004, Asian soybean rust first entered the continental United States and was initially detected in Louisiana (Case Study 18.2). The disease spreads easily by wind currents and has the potential to cause tremendous damage to soybean fields if fungicides are not applied at the appropriate time. RT-PCR has been used by several clinics to confirm the presence of the pathogen *Phakopsora pachyrhizi*. Microscopic study of the spore characteristics will identify the presence of *Phakopsora* at the genus level. Both *P. pachyrhizi* and a much less virulent species, *P. meiobomae*, produce spores that have the same microscopic characteristics. As more primers become available and equipment costs decline, this method will be used more extensively by research and diagnostic laboratories.

PROOF OF PATHOGENICITY

If the organism suspected to be the pathogen has not been reported to cause the symptoms and signs observed, then it is necessary to isolate the suspected pathogen in culture and inoculate a healthy plant of the same cultivar and age and nutritional status. If the original symptoms develop and the original organism can be isolated in culture again, then pathogenicity is considered to be proven (Baudoin, 1988) (Chapter 2).

TABLE 39.1

Commonly Used Diagnostic Procedures in Plant Diagnostic Clinics

Diagnostic Procedure	Plant Pathogen Group			
	Fungi	Bacteria	Viruses	Nematodes
Visual exam	+	+	+	+
Microscopy	+	+	(+)	+
Moist chamber	+	–	–	–
Baiting	+	–	–	–
Culture isolation	+	+	–	–
ELISA	+	+	+	–
Physiological tests	–	+	–	–
Gas chromatography	–	+	–	–
BIOLOG	–	+	–	–
RT-PCR (real time-polymerized chain reaction)	(+)	(+)	(+)	(+)

Note: + = used routinely, (+) = used occasionally with a specific technique/situation, – = not used.

SUMMARY

Plant disease diagnosis is the process or procedure of determining the cause of a plant disease. Initial steps in the diagnostic procedure for most diseases are very similar. Initially, the plant is examined visually for symptoms and signs, plant sample information is examined, and reference materials are consulted. These procedures may be done at the plant site situation or in the lab/clinic. If a diagnosis cannot be made with this information, microscopy often follows. In many cases, microscopic study will provide a confirmed specific diagnosis for fungal diseases. If initial diagnostic procedure results do not provide information that allows for a confirmed diagnosis, the diagnostician will try to compile a list of possible pathogens from previous test results. Methods used to prove or disprove a possible pathogen will vary, depending upon the pathogen suspect, the crop/plant situation, and the client needs. An exact species identification of a pathogen may not be needed in many situations. Moist chambers are often used if a foliage fungal pathogen is suspect. Culture work is often done when fungal or bacterial diseases are possible. A summary of procedures used for each major group of plant pathogens may be found in Table 39.1. When abiotic problems are suspected, some procedures used for biotic problems may be tried so as to help confirm abiotic problems through the negative results from such studies as microscopy and culture work.

COMMON DISEASE DIAGNOSIS SITUATIONS

A leaf spot disease such as early blight, caused by *Alternaria solani*, on tomato may be diagnosed using different methods, depending upon the sample. If a pathologist or an experienced grower sees the characteristic zonate oval leaf spots on the lower leaves and stems of the tomato, they might very well diagnose this disease in the garden on the basis of it being a commonly seen leaf spot of tomato with distinctive symptoms. However, if this leaf spot had progressed to involve the whole plant and dry conditions existed such that typical zonate patterns were not evident, the pathologist or grower might want the leaf spot examined using microscopic techniques. If the *A. solani* spores were observed, then diagnosis may be completed with symptoms and microscopic signs evidence. If spores were not evident with microscopic exam, the sample may be placed into a moist chamber. If spores developed after a few days, then diagnosis would be complete after visual symptoms and microscopy of moist chambered tissues. In the event that moist chamber incubation did not produce spores, culture work would be done. If spore production occurred, diagnosis would be completed after observation of symptoms and culture exam with microscopy.

Diagnosis of a bacterial leaf spot may be based upon visual symptoms and microscopic streaming. Streaming is usually considered to be positive evidence of bacterial leaf spot. Microscopic streaming does not confirm the identity of the bacteria, but control recommendations are very similar for all bacterial leaf spots. Literature reports and descriptions could help provide information for a suspect bacterial identity. If the client needs to know the bacterial identity, then culture isolations are done followed by physiological testing. When species and/or pathovar identity is needed, gas chromatography is the method considered to be most accurate.

Virus diagnosis is usually based upon visual assessment and ELISA results. With ELISA, you must have a good idea or suspicion of what viruses to test. Occasionally, other tests may be performed, but ELISA is the most popular method, if antibodies to the suspect viruses are available. Virus mechanical transmission to a healthy plant similar to the diseased plant will demonstrate that the problem is viral in nature. Transmission to a host range is still occasionally done, but results require 10–14 days, and results may not be reliable. A stain procedure is available to identify virus groups by the characteristic bodies formed by groupings of virus particles as viewed in the light compound microscope at 450×. This inclusion body technique produces easily recognized virus structures for some virus families. With some other viruses, the bodies are not as distinctive and might be confused with normal cell structures. Electron microscopy may be used if available. Electron microscopic studies are not specific for exact virus diagnosis. Virus particle shapes and size can be observed, but this character is not distinctive for specific virus identification. ELISA is the most specific, quickest, and easiest virus analysis available.

REFERENCES

Agrios, G.N. 2005. *Plant Pathology.* 5th ed. Academic Press, New York. 952 p.

Armentrout, V.N. and A.B.A.M. Baudoin. 1988. Isolation of fungi and bacteria. In Baudoin, A.B.A.M. (Ed.) *Laboratory Exercises in Plant Pathology: An Instructional Kit.* APS Press, St. Paul, MN. 213 p.

Baudoin, A.B.A.M. 1988. Diagnosis of disease and proof of pathogenicity (Koch's postulates). In Baudoin, A.B.A.M. (Ed.). *Laboratory Exercises in Plant Pathology: An Instructional Kit.* APS Press, St. Paul, MN. 213 p.

Christie, R.G. and J.R. Edwardson. 1986. Light microscopic techniques for detection of plant virus inclusions. *Plant Disease* 40: 273–279.

Duncan, J.M. and L. Torrance. 1992. *Techniques for the Rapid Detection of Plant Pathogens.* Blackwell Scientific Publications, Cambridge, MA. 235 p.

Farr, D.F., G.F. Bills, G.P. Chamuris, and A.Y. Rossman. 1989. *Fungi on Plants and Plant Products in the United States.* APS Press, St. Paul, MN. 1252 p.

Fox, R.T.V. 1993. *Principles of Diagnostic Techniques in Plant Pathology.* CAB International, Wallingford, U.K. 213 p.

Goodwin, S.B., R.E. Schneider, and W.E. Fry. 1995. Cellulose-acetate electrophoresis provides rapid identification of allozyme genotypes of *Phytophthora infestans. Plant Disease* 79: 1181–1185.

Jones, J.B., J.P. Jones, R.E. Stall, and T.A. Zitter (Eds.). 1991. *Compendium of Tomato Diseases.* APS Press, St. Paul, MN. 73 p.

Matthews, R.E.F. (Ed.). 1993. *Diagnosis of Plant Virus Diseases.* CRC Press, Boca Raton, FL. 374 p.

McCain, J.W. 1988. Use and care of the light microscope. In Baudoin, A.B.A.M. (Ed.). *Laboratory Exercises in Plant Pathology: An Instructional Kit.* APS Press, St. Paul, MN.

Ruhl, G., J. Mullen, and J. Woodward. 2001. Plant problem diagnosis and plant diagnostic clinics. In Jones, R.K. and D.M. Benson (Eds.). *Diseases of Woody Ornamentals in Nurseries.* APS Press, St. Paul, MN, pp. 442–450.

Schaad, N.W., J.B. Jones, and W. Chun (Ed.). 2001. *Laboratory Guide for Identification of Plant Pathogenic Bacteria,* 3rd ed. APS Press, St. Paul, MN. 373 p.

Shurtleff, M.C. and C.W. Averre III. 1997. *The Plant Disease Clinic and Field Diagnosis of Abiotic Diseases.* APS Press, St. Paul, MN. 245 p.

Waller, J.M., B.J. Ritchie, and M. Holderness. 1998. *Plant Clinic Handbook,* IMI Technical Handbooks No. 3. CAB International, Wallingford, U.K. 94 p.

40 Diagnostic Techniques and Media Preparation

Jackie M. Mullen

Plant disease diagnosis is a process that begins when a plant problem is first noticed at a field, garden, nursery, or greenhouse. A home gardener, farmer, or county agent observes the damaged plant(s) and the entire field and notes the obvious visible **symptoms** and **signs**. The distribution of the damage should be documented. Scattered or clumped distribution of the plant problem is often characteristic of a biotic disease whereas a row or obvious spray pattern is often indicative of pesticide injury or a cultural problem (Chapter 35). When entire fields or sections of fields are damaged and have symptoms of uniform age, the problem is often due to an abiotic factor related to weather, such as freeze, frost, or drought injury (Chapter 25). Information should be gathered on the history of the problem and site or planting area. All recent pesticide and fertilizer application dates and rates of application should be noted. Insect injury should be considered. The whole plant should be examined for evidence of damage.

If the problem can not be diagnosed at the site, then a sample must be collected, packaged, and sent for examination by a diagnostic specialist. The sample should consist of whole plants if possible as foliage damage may be related to root and/or soil problems. Care should be taken to select plants showing a variety of damage stages or development. The sample should be representative of the damage seen in the field. If the damage seen consists of leaf spots and cankers, then the sample collected should include leaves and stems with each symptom. A diagnosis is based upon the sample sent to the clinic. If good quality samples are not sent, then diagnosis may not be possible. If the sample is not representative of the problem, then the diagnosis will be inaccurate and the control recommendation will not be effective. If plants are small, then many plants should be collected. For example, when bedding plant plugs or seedlings are sent for diagnosis, a flat of the 1-inch tall seedlings is usually sufficient for diagnosis. If 5-inch tomato transplants are sent for diagnosis, usually five to ten plants are collected. When small shrubs are damaged, one or two whole plants are usually sufficient. Trees are not usually collected as whole plant specimens, but occasionally clinics do receive whole trees. In general, submit the portion of the tree exhibiting symptoms: leaves with leaf spots, anthracnose, rust, or mildew; branch segments with cankers or galls; etc. Roots are usually not requested unless the tree is small. Root disease is not easily diagnosed on large trees and root samples are often not requested until all other problems are eliminated. Also, send 1 to 2 pints of soil for analysis. Sending a soil sample from areas with healthy plants is helpful for comparison.

Plant disease diagnosis may involve several procedures, depending on the symptoms noted by the diagnostician. Initially, all samples are examined for symptoms and signs that usually give an indication of the causal agent. Depending upon whether the suspect pathogen is fungal, bacterial, or viral, the diagnosis will proceed to the next level or block of diagnostic tests. The following three exercises are described in this chapter: (1) diagnosis of a fungal leaf spot, (2) preparation of culture media and diagnosis of a root rot disease using media prepared, and (3) the complete process of disease diagnosis including field observations, collecting and packaging of a sample, and the performance of diagnostic techniques for diagnosis of the disease. The last exercise will require students to locate a plant problem in their home landscape. The plant and the area should be studied. The plant diagnostic questionnaire should be filled out, noting as much information available on the problem, its development, treatments made, recent weather, etc. A sample representative of the problem must be collected and packaged appropriately so as to keep the sample as fresh as possible. The sample will be diagnosed with the help of the instructor. At the end of the laboratory session, students should present the disease they collected and justify the diagnosis. Diagnostic reports should be turned in to the instructor.

Other exercises that appropriately could be part of this chapter include (1) the diagnosis of a bacterial leaf spot, (2) the diagnosis of a virus disease using mechanical transfer to a healthy plant (Chapter 5) and an **ELISA** method, and (3) the diagnosis of a nematode disease by extraction of the nematodes from a soil sample followed by the microscopic identification of a plant parasitic nematode by its stylet and other microscopically viewed body features (Chapter 9). These exercises are covered elsewhere in this text so it is not considered necessary to repeat these activities here.

EXERCISES

EXPERIMENT 1. DIAGNOSIS OF FUNGAL LEAF SPOTS

Fungal leaf spot identification is accomplished by associating a fungus with the symptoms on the foliage. Fungi are identified either by spore morphology and spore arrangement on conidiophores or by spore morphology and fruiting bodies observed. If the spores are present on the symptomatic tissues, then diagnosis may be completed with light microscopy. If spores are not present, then a moist chamber may be prepared to stimulate spore development and maturation. After 1 to 7 d, tissues are examined for signs of sporulation. If spores are not produced in a moist chamber, then diagnosis proceeds with preparation of tissue isolations in culture. After approximately 3 to 7 d, cultures are examined microscopically to observe fungal spore development. (If possible, culture dishes should be examined daily during the 3-to-7-d period.) In addition to or instead of cultures, ELISA methods may be used to detect the presence of the pathogen protein. If the previous methods are not successful, other methods may be used such as gel electrophoresis or DNA probe analysis.

A common fungal leaf spot of tomato is Septoria leaf spot. Leaf spots are small, circular, and gray-black in color. If moisture is sufficient, tiny black fruiting bodies (**pycnidia**) of the fungus (Chapter 16) will develop and be evident when the tissues are examined with a dissecting microscope. If the fruiting bodies are mature, spores or conidia that are long, thin, and multicelled will be present and can be seen with a compound microscope. Once the spores and spore-producing bodies are observed, identification of the fungus should be possible. If the fungus is reported to cause the symptoms on the plant, then diagnosis is complete with only a visual and microscopic study. If spore bodies are viewed in the tissue, but spores are not present, the diagnostician will probably place the tissues in a moist chamber for 1 to 7 d in hopes that the humidity will stimulate development of the spores. If spores do not develop, culture isolations will be prepared. Fungal growth from tissue in culture at room temperature usually requires about 3 to 4 d. Spore production in culture often requires 1 to 4 weeks.

If spores do not develop in culture, the diagnostician may provide the client with a tentative or suspect diagnosis on the basis of symptoms and fruiting bodies or just symptoms. With most diseases, the confidence level of a diagnosis based upon symptoms and fruiting bodies is higher than the confidence level when the diagnosis is based upon symptoms alone.

Another common leaf spot and foliage blight disease is called **anthracnose**. This disease is usually a leaf spot and/or blight caused by one of several fungi that produce their spores in a cup-shaped structure called an **acervulus**

(Chapter 16). Diagnostic procedures are much the same as was described above for Septoria leaf spot of tomato. If an anthracnose fungus is identified and has not been reported on the plant, then a proof of pathogenicity protocol must be followed to determine if the fungus is capable of causing the symptoms seen.

Materials

Students may work in pairs and the following is needed for each team of students.

- Plant sample showing leaf spots (Some common leaf spot diseases that might be used for this exercise are early blight on tomato, Septoria leaf spot on tomato, black spot or Cercospora leaf spot on rose, anthracnose on cucumber or other plants, Alternaria leaf spot on zinnia, or Entomosporium leaf spot of Photinia.)
- Dissecting and compound microscopes
- Microscope slides and coverslips
- Dropper bottle with water and cotton (aniline) blue stain (0.5%)
- Plastic bags and paper towels

Follow the protocols listed in Procedure 40.1 to complete this experiment.

Anticipated Results

Fungal fruiting bodies with spores or spores on conidiophores not within fruiting bodies should be observed at 100× and 450×. The fungus should be identified by spore morphology, fruiting bodies, or arrangement of spores on conidiophores. The moist chamber should stimulate sporulation of the causal fungus. After referring to a disease host index and a compendium or similar references, determine whether or not the symptoms and signs observed have been described previously on the host. Comments should be made on the diagnosis of the problem.

Questions/Activities

- Distinguish between recently developed leaf spots and old leaf spots. Describe.
- How many types of spores did you see when examining the tissues microscopically before moist chamber incubation and after incubation? Explain this difference? Draw the spores that you observed.
- When you see more than one spore type, how can you determine which one is the pathogen?
- When would a tape mount be more effective than a wet mount?
- Why should the entire specimen not be placed in a moist chamber?

Procedure 40.1

Diagnosis of a Fungal Leaf Spot Disease Using Visual Symptoms, Microscopic Signs, and a Moist Chamber Incubation

Step	Instructions and Comments
1	Select a sample (provided by instructor) for diagnosis. Leaf spots of varying size and color should be included in the sample.
2	Study and record observations on the symptoms present. Describe the leaf spots as to color, shape, size, texture, etc.
3	Select some leaf spots of varying appearance and observe them with the aid of a stereomicroscope. Look for evidence of small black or other colored bodies scattered over the leaf spots. Record observations.
4	If bodies are seen, cut out tissue sections with bodies and place one or two sections on top of a microscope slide and add drop of water. Place a cover slip on top of the tissue and water droplet (see Procedure 42.9 for more detail on how to make a wet mount). Observe slide mount (wet mount) for presence of bodies and spores.
5	Make a second wet mount after attempting to cut through some of the fruiting bodies. Again, examine slide wet mounts for presence of spores in fruiting bodies.
6	If bodies are not seen in the stereomicroscope, make a slide mount of two sections of leaf spots. Also prepare two tape mounts of the leaf spot tissue. Apply the sticky slide of cellophane tape to the leaf spot. Remove the tape and place it onto a microscope slide containing a drop of water. (The tape is used like a coverslip.) Observe the tape mount at 100× and 45× for the presence of spores.
7	Take a few leaves with leaf spots and place them into a plastic bag containing a damp paper towel. Incubate for 3–7 d at room temperature. Check the leaves for the presence of additional fruiting bodies and spores.
8	Refer to the *Illustrated Genera of Imperfect Fungi*, 4th edition (Barnett and Hunter, 1986) to identify the fungus. Consult host disease indices and compendia of crop diseases (see references in Chapter 38 for a partial listing) to determine whether or not the fungus has been associated with a previously reported leaf spot disease. Now make the diagnosis.

EXPERIMENT 2. CULTURE MEDIUM PREPARATION AND ITS USE IN DISEASE DIAGNOSIS

Isolation of fungi and bacteria in culture is an important component for diagnosis of many fungal and bacterial diseases. Diagnostic clinics often use several culture media, depending upon the pathogen to be isolated. Lists of culture media available for isolation of specific pathogens or pathogen groups are available in several references. A general-purpose medium is used for isolation when the exact fungal or bacterial pathogen is unknown. Potato dextrose agar (PDA) and nutrient agar (NA) or tryptic soy agar are general purpose media to isolate fungal and bacterial pathogens, respectively. These media may be made from either the individual ingredients as they were when initially formulated or purchased pre-mixed in a dry form. A commercially available medium is measured according to directions on the container and mixed with filtered or distilled water. The medium is then sterilized by heat at 121°C under pressure for 15–20 min so it does not boil. Usually medium is sterilized in an **autoclave**, but sterilization may be performed in a pressure cooker. The high temperature will kill all bacteria and fungi, and inactivate viruses. When medium is about 50°C, it can be poured into sterile petri dishes. **Aseptic** technique should be used

when pouring media into petri dishes to prevent contamination from airborne bacteria or fungi. Once the medium in the dishes is cool and solidified, the culture dishes may be used for isolation of pathogens from plant tissues.

Plant tissue should be cut from the edge or margin of the damage area. Four or five pieces of the tissue should be surface sterilized, rinsed with sterile water, blotted with clean paper towels, and placed equidistantly from each other on the medium in the dish. Again, aseptic technique must be used and care taken to sterilize the scalpel or other instrument used to move the small tissue pieces into the petri dishes. A 10% solution of household bleach is often used for surface sterilization. The time allotted for surface sterilization depends on the type of tissue. Thin, fragile tissue is surfaced sterilized for 0.5 to 1.0 min, whereas thicker tissue is surface sterilized for 1 to 2 min. In many situations, several surface sterilization times are used to see which time produces the best results.

When preparing bacterial isolations, whole leaf spots or tissue cut from the edge of the damaged area may be used for culture purposes. Surface sterilization of this tissue is similar to the technique used with fungal diseases. After surface sterilization and rinsing, the leaf spots or other damaged tissues with some marginal area are cut up or crushed with a sterile glass rod in a small drop of sterile water and allowed to sit for 5 min (or more) to allow the bacteria to ooze into the water. A sterilized bacterial loop is used to lift up a small quantity of water containing bacteria (a suspension) and streaked over the surface of a general purpose bacterial medium such as tryptic soy agar. Most clinics streak the bacteria in a four-quadrant dish to dilute the bacteria concentration so that single cells may be isolated in the last quadrant streak. Colonies from single cells may be transferred to sterile culture dishes for future studies.

In this experiment, students will prepare PDA, fill culture dishes with the acidified agar, and then prepare tissue isolations of infected roots using the prepared PDAa (acidified PDA) dishes in order to diagnose the cause of a root rot disease on garden bean. With many fungal root rot diseases, symptoms are not distinct enough to allow for a symptom-based diagnosis. In some situations, the causal agent may be observed microscopically with a compound microscope. When diagnostic fungal structures are not present, root segments must be placed onto a culture medium to allow for growth and identification of the pathogen. Some fungal root pathogens are isolated with specialized culture media that contain antibiotics (see Procedure 21.2 for details). Diseased root tissue may be placed on three or four types of media, each specific for different fungal pathogens.

Materials

The following materials are needed for each pair of students:

- Commercially prepared and dehydrated potato dextrose agar (PDA)
- 250-mL Erlenmeyer flask
- Balance with weighing papers or cups
- Magnetic stir plate
- Autoclave or pressure cooker
- 10-cm diameter petri dishes
- Alcohol burner and matches
- 10% household bleach solution
- 25% lactic acid solution in a dropper bottle
- Plants with root rot disease
- Dissecting and compound microscopes

Follow the protocols listed in Procedure 40.2 to complete this experiment.

Anticipated Results

After 3–4 days, fungal growth should be visible in culture dishes. After two or three more days, the fungal growth should be examined using the dissecting and compound microscopes. Identify the fungus isolated by using references provided (host indices, crop disease compendia,

etc.) Fungal pathogens should be consistently isolated from root segments.

Questions

- What would you conclude if fungal growth occurred all over the dishes rather than only from tissue placed into the dish?
- What would you conclude if no fungal or bacterial growth occurred in culture isolation dishes?
- What is your conclusion if several fungi grew out of isolations in your culture dishes?
- Most fungal root pathogens are identified by the characteristic spores that they produce. What is the fungal root pathogen that is identified by its characteristic hyphae? Does identification of this fungus in culture require more or less time than the other fungal root pathogens that are identified by spore characteristics? Explain your answer. Are there other root rot pathogens easily identified by their hypha-like characteristics?

Experiment 3. Plant Disease Diagnosis: The Complete Process of Field Observations, Sample Collection, Sample Packaging, and Laboratory Diagnosis

Successful diagnosis of plant problems often depends on observation of symptoms, history of the problem, and the site and knowledge of recent fertilizer, pesticide, or other chemical applications. The grower should record all of this information. Collect and package samples to provide representative, fresh samples of adequate quantity. If possible, collect whole plants with symptoms in early, middle, and late stages of disease development.

Materials

The following items are needed by each student:

- Plants with damage (Herbaceous plants such as bedding plants or vegetables are recommended as samples, because whole plants can usually be collected. Also, plants showing leaf spot, blight, or crown or root lesions are suggested. Some common diseases that might be used for this exercise are anthracnose of pansies; *Sclerotium rolfsii*—crown rot of pepper or other plants; *Fusarium* or *Rhizoctonia*—lower stem rot of bean; Alternaria leaf spot of zinnia, marigold, cabbage, or tomato; and Botrytis blight of rose or poinsettia.)
- Diagnostic questionnaire
- Plastic bags and twist ties
- Dissecting and compound microscopes

	Procedure 40.2
	Preparation of Acidified Potato Dextrose Agar (PDA) and Its Use in Root Disease Diagnosis
Step	Instructions and Comments
1	Prepare potato dextrose agar, following directions on the jar. Weigh out enough of the dehydrated medium to prepare 150 mL of medium. Place medium into a 250-mL Erlenmeyer flask and add 150 mL of distilled water.
2	Add a magnetic stir bar to the flask of medium and place on a stirrer unit for 3 min to disperse medium clumps and dissolve powdered medium in the water.
3	Place nonabsorbent cotton as a plug at the top of the flask and cover with aluminum foil. Autoclave (or use a pressure cooker) the medium at 121°C for 20 min.
4	Allow the unit to cool while slowly reducing the pressure. When pressure is reduced to normal and the internal temperature of the unit is below 100°C, remove the flask and allow it to cool to about 50°C, a temperature that is easy to handle, but still warm. Add 5 to 7 drops of 25% lactic acid and swirl the contents of the flask to mix.
5	Pour the medium into sterile 10-cm diameter petri dishes in a laminar flow hood. If a hood is not available, select a clean table surface away from traffic flow and air currents. Wipe the table with 70% ethanol or other disinfectant and allow it to dry. Begin by tilting the Erlenmeyer flask at an angle and remove the aluminum foil and cotton plug. Flame the top of the flask opening and allow it to cool. While keeping the flask tilted, pour about 15 mL of medium (about half the depth of the petri dish bottom) into each of five dishes. When pouring medium, hold the dish lid at an angle so that there is very little opportunity for microorganisms in the air to fall into the dish. Reflame the mouth of the flask and allow it to cool. Pour the remainder of the medium into the other five dishes. Allow the medium in the dishes to cool to room temperature.
6	Observe plants with root decay using a dissecting microscope for signs. Record symptoms of damage.
7	Cut roots into small (~1 cm) segments and place on a microscope slide. Prepare a wet mount. Observe tissues using the 100× and 450× magnification using the compound microscope. Make notes of observations.
8	Select some small root pieces that show slight discoloration and some that show a darker discoloration.
9	Surface sterilize cut segments of small roots by dipping them into a 10% solution of household bleach. Rinse with sterile water and blot dry with a clean paper towel.
10	Cut surface sterilized roots showing slightly discolored (SD) and moderately discolored (MD) regions into smaller lengths using sterile technique. Place three pieces of small roots in a clump onto one spot on the PDA dish by using an aseptic technique. When opening the petri dish, crack the cover of the dish so as to just allow space for placement of the root pieces. Do not take the lid completely off the bottom. Repeat this procedure placing three or four more root clumps onto the agar surface, spacing tissues equidistant from each other. Prepare two PDA isolation dishes and label each with your name and date. Label one dish SD and the other plate MD.
11	Incubate the tissue isolation dishes at room temperature for 7 days and observe for fungal growth. Record your observations each day.
12	Prepare wet mounts of the fungal growth and observe at 100× and 450× using a compound microscope. Record your observations.
13	Consult with your instructor if you need assistance with identification of the fungal isolate(s).
14	Refer to disease host indices and compendia of crop diseases to determine whether or not the fungi isolated have been associated with a previously reported root rot disease. Report your conclusions.

	Procedure 40.3
	Plant Disease Diagnosis: The Complete Process of Plant and Site Observations, Information Gathering, Sample Collection, Sample Packaging, and Laboratory Diagnosis
Step	Instructions and Comments
1	Inspect a garden, field, or landscape and locate damaged or disease plants. For purposes of this exercise, fungal leaf spot diseases are preferred, but some lower stem (crown) or root rot diseases may also be incorporated successfully into the experiment.
2	Thoroughly inspect the plants(s) for symptoms and signs. Make notes about the distribution of the disease—isolated, scattered, or the presence of a pattern.
3	Obtain information on the site and problem history. Determine dates and rates of fertilizers, lime, and pesticide applications, and note recent weather conditions in the area. Complete a plant diagnostic questionnaire.
4	Collect a sample of the problem. Collect whole plants if possible. If plants are small, collect several; if plants are 2-ft tall or more, one or two plants are sufficient. If it is not possible to collect a whole plant, collect several pieces of the plant showing varying stages of plant damage and symptom development. Do not collect dead plants. Ideally, damage should be recent. The sample should be representative of the problem, and the sample size must be large enough to subdivide for two to four different analyses and examinations.
5	Package the sample in plastic or several layers of newspaper as soon as possible after collection and place into a refrigerator or cool ice chest. Most samples may be placed into plastic bags. Fruits and vegetables keep best if wrapped in newspaper.
6	In the laboratory, make notes of the appearance and disease symptoms and signs of the sample.
7	Observe the damaged tissues for pathogen signs using the dissecting and compound microscopes. If diagnostic structures are not seen with microscopy, prepare a moist chamber and culture isolations (see Procedures 40.1 and 40.2)
8	After incubating the specimens in moist chambers and culture dishes for 7 d, examine the plant material/cultures microscopically for mycelia, spores, and fruiting bodies. Consult references to identify fungal structures and fungi and determine whether or not the disease can be confirmed by symptoms and microscopic structures.
9	Complete a diagnostic report including all notes on the plant problem, appearance in the field, pertinent site/plant information, symptoms and signs, and microscopic fungal structures observed in moist chambers and/or cultures. Conclude the report with a disease identification citing key diagnostic criteria and references.

- Microscope slides and coverslips
- Plastic bags and paper towels
- Two 10-cm diameter petri dishes of acidified potato dextrose agar

Follow the protocols listed in Procedure 40.3 to complete this experiment.

Anticipated Results

Most common plant diseases caused by fungi are identified by symptoms and signs observed with dissecting and compound microscopes. Many leaf spot diseases, powdery mildews, downy mildews, and rusts can be confirmed via light microscopy. Acidified PDA is very useful for isolating fungal pathogens from root and leaf tissue. If a fungal pathogen is the causal agent of a leaf spot or blight, it is usually consistently isolated from diseased tissue. Obligate parasites, such as powdery

mildew, downy mildews, and rusts will not grow on culture media used to isolate most fungal pathogens. Moist chambers are helpful for stimulating sporulation of most fungal foliage pathogens. However, saprophytic fungi, which often grow rapidly and sporulate profusely in moist chambers, may make identifying the causal agent difficult. Host indices and disease compendia are very helpful for information on potential pathogens, symptomatology, and conditions that favor disease development.

Questions

- How would you prove or disprove a diagnosis of pesticide spray damage?
- How could recent weather have affected the disease and diagnosis?
- What would you conclude if you consistently isolated a *Colletotrichum* sp. from a leaf spot on

cucumber, but the leaf spot symptoms did not resemble symptoms described for this disease?

LITERATURE CITED

Barnett, H.L. and B.B. Hunter. 1989. *Illustrated Genera of Imperfect Fungi*, 4th ed. Burgess, Minneapolis, MN, 218 p.

Baudoin, A.B.A.M. (Ed.). 1988. *Laboratory Exercises in Plant Pathology: An Instructional Kit*. APS Press, St. Paul, MN.

Farr, D.F., G.F. Bills, G.P. Chamuris, and A.Y. Rossman. 1989. *Fungi on Plants and Plant Products in the United States*. APS Press, St. Paul, MN.

Part 6

Special Topics

41 *In Vitro* Plant Pathology

Subramanian Jayasankar and Dennis J. Gray

CHAPTER 41 CONCEPTS

- Plants can be regenerated from single cells. Thus, a large number of single cells with potential to become plants can be grown in a container and subjected to intense screening with appropriate selection agents. Surviving cells are regenerated to produce resistant phenotypes. This process is called *in vitro* selection.

- To determine if resistance is actually induced immediately after selection process, a number of small tests using the surviving cell or callus culture and the selection agent [or the organism that produces the selection agent] can be performed under laboratory condition. These "bioassays" are very effective especially when dealing with plants with long regeneration cycles.

- When plant cells are subjected to such intensive *in vitro* selection, a number of defense genes are induced; however, only those cells whose defense mechanism is activated quickly and in a sustained manner remain viable and survive recurrent *in vitro* selection. Plants regenerated from such cells exhibit certain native defense genes in a constitutive manner and thus they are resistant than their parental material.

- This approach of activating plant's innate resistance using *in vitro* selection constitutes a viable, noncontroversial biotechnological approach of generating resistant plants.

In vitro culture and selection of plant cells and tissues has been used effectively as a tool for developing novel, disease resistant genotypes (Jayasankar and Gray, 2003). The first demonstration of this technique was to produce fire blight resistant tobacco plants, through *in vitro* selection using methionine sulfoximine, a structural analog of fireblight toxin. Since then plant cells have been successfully selected against an array of pathogenic microorganisms and regenerated into plants with enhanced disease resistance. In addition, plant cell and tissue culture has become an important tool in the study of plant–pathogen interactions at the cellular and molecular levels. Plant cells react to certain biotic and abiotic stresses in a manner similar to that of an intact plant. This makes such cell cultures ideal candidates for understanding the resistance responses and changes occurring at the cellular and subcellular levels when infected with pathogenic organisms (Jayasankar, 2000). Some of the common responses that are well-documented include changes in permeability of plasma membrane and triggering the synthesis of new biochemical compounds, especially defense-related enzymes (such as the pathogenesis-related proteins). Plant cell cultures provide an ideal population of homogeneous genetic material. A single flask of embryogenic cell suspension culture theoretically represents millions of plants that can be effectively screened. For instance, a suspension culture of grape proembryogenic masses (Figure 41.1) contain enough totipotent cells to regenerate plants for hundreds

FIGURE 41.1 Suspension culture of grapevine (*Vitis vinifera*). These suspension cultures contain highly embryogenic cell aggregates called proembryogenic masses, which are ideal for *in vitro* selection against plant pathogens.

of acres of vineyard. Furthermore, they are very useful for performing a number of genetic tests such as bioassays, and they can also be useful in the culture of biotrophic pathogens. This chapter will address how *in vitro* culture can be best utilized to study and understand plant pathogen interactions.

DEVELOPING DISEASE RESISTANT GENOTYPES OF CROP PLANTS

Though cell cultures of most species are homogeneous for a specific trait, subtle differences at the subcellular level do exist among the population. In order to isolate such cells with subtle genetic differences out of a huge population, we have to devise a method to select or screen the population with a selection agent. Sometimes such selection agents can also cause minor but specific mutations, thereby altering their genetic makeup. Plants regenerated from such mutant cells exhibit altered phenotypes. Among the most common selection agents that are used to generate disease resistant phenotypes are the metabolites such as **phytotoxin** produced by the pathogen, crude culture filtrate and, rarely, the pathogen itself (see Daub, 1984; 1986). A list of important work done in this field is summarized in Table 41.1.

The use of phytotoxins as selection agents has received the most attention mainly because of the ease in exposing a population of cells to controlled, sublethal doses of the isolated phytotoxin. In addition, toxin selection is easier to handle than pathogen selection, as it is a purified chemical compound that can be incorporated into the culture medium at measurable doses. In several instances, cells that were selected against the toxin retained significant levels of resistance to the toxin as well the pathogen when regenerated into plants. In a few studies, such resistance following selection against the toxin was transmitted to the progeny as well. However, this approach has a number of pitfalls as well. First, toxin selection has worked well only against host-specific toxins such as the HMT-toxin produced by *Cochliobolus heterostrophus* race T (syn. *Helminthosporium maydis*). Second, there are only a handful of characterized phytotoxins, whereas the number of pathogenic microbes is vast. Because selection is targeted against a particular pathogen, the regenerated plants often exhibit resistance only to the particular pathogen in question. When these plants are planted in the field, they are susceptible to other pathogens, thus making evaluation very difficult.

These pitfalls, perhaps, can be overcome by the use of crude culture filtrates, instead of purified phytotoxin, obtained by growing the pathogen in nutrition broth solutions. Culture filtrates of both pathogenic bacteria and fungi have been successfully used in establishing resistant cultures and regenerating resistant plants (Table 41.1). The use of culture filtrates, though not considered as the best approach by many (see Daub, 1986), has its own advantages. The culture filtrate approach more closely approximates aspects of the host–pathogen interaction than toxin selection, because culture filtrates have the entire spectrum of compounds (including the toxin) produced by the pathogen. In addition recent studies have shown that some compounds (e.g., proteins) produced by pathogens in culture that are not a critical factor in the disease process are often implicated in evoking a general disease resistance response in the host plant (Strobel et al., 1996). Similarly, culture filtrate selection has also evoked broad spectrum resistance in grapevine (Jayasankar et al., 2000).

Whether it is purified toxin or crude culture filtrate, the mode of selection is very important. In some studies, solid medium (medium that was gelled with agar or similar compounds) was used for effecting such selections, incorporating measured doses of toxin or culture filtrate in the medium. Plants regenerated from such selection schemes quite often showed epigenetic responses that faded away with time. The probable reason is that in a solid medium, the cultured explants are not in full contact with the medium, and often a gradient is established from the top to bottom of the explant (Litz and Gray, 1992). This results in **acclimatization** of those cells at the top of the gradient to the selection agent, and the plants regenerated from such cells are generally false positives, exhibiting **epigenetic** resistance. It is possible to circumvent this gradient factor by plating the cells as a thin layer, but several plant species require a mass of small cells rather than single cells for optimum regeneration. In a suspension culture system of selection, cell masses are completely and rapidly immersed in toxin containing medium. This not only helps to avoid the gradient factor, but the cell masses do not have sufficient time to elicit a epigenetic response and, hence, suspension cultures are better for such selections.

PROBLEMS ENCOUNTERED IN SELECTION

Selection against toxin or culture filtrate is not possible under certain circumstances and against certain diseases. To give a few examples, it is very difficult to select against fungal pathogens that cause powdery mildew since most of these fungi do not grow in culture. Some bacterial pathogens such as the xylem-limited bacteria (e.g., *Xylella fastidiosa* of grapevine that causes Pierce's disease) can kill the plant by physical means rather than chemical action. These bacteria are so restricted in their growth, that isolation and culture of these bacteria and addressing resistance through selection is often futile. To date, no successful selection has ever been carried out against viral diseases, because the only way possible to select against virus is to grow the pathogen in their hosts and look for resistance. To impart resistance under such conditions, one has to use gene or genome transfer techniques.

BIOASSAYS

In vitro bioassays are very useful tools in determining the level of resistance in a breeding program or for screening a population for sensitivity to a pathogen or pathogen-derived metabolites. Assays using intact plants (seedlings) or plant

TABLE 41.1

***In vitro* Selection for Disease Resistance**

Crop	Pathogen	Selection Agent	Results
Alfalfa			
Medicago sativa	*Colletotrichum gloeosporioides*	Culture filtrate	Enhanced disease resistance
Asparagus			
Asparagus officinalis	*Fusarium oxysporium* f. sp. *asparagi*	Pathogen inoculation	Increased disease resistance
Barley			
Avena sativa	*Helminthosporium sativum*	Partially purified culture filtrate	Plants with increased disease resistance and transmitted to progeny
Chinese cabbage			
Brassica campestris ssp. *pekinensis*	*Erwinia carotovora*	Culture filtrate combined with UV irradiation	Plants with increased disease resistance and transmitted to progeny
Coffee			
Coffea arabica L.	*Colletotrichum kahawae*	Partially purified culture filtrate	Plants with increased resistance
Eggplant			
Solanum melangena	*Verticillium dahliae*	Culture filtrate	Increased disease resistance
Grapevine			
Vitis vinifera	*Elsinoe ampelina*	Culture filtrate	Plants with increased resistance
Grapevine			
Vitis vinifera	*Colletotrichum gloeosporioides*	Culture filtrate	Plants with increased resistance
Maize			
Zea mays	*Helminthosporium maydis*	T-toxin	Phytotoxin resistant cells and plants
Mango			
Mangifera indica	*Colletotrichum gloeosporioides*	Colletotrichin and culture filtrate	Resistant embryogenic cultures
Peach			
Prunus persica	*Xanthomonas campestris pv. pruni*	Culture filtrate	Resistant clones to the disease
Potato			
Solanum tuberosum	*Phytophthora infestans*	Culture filtrate	Phytotoxin resistant plants
Rice			
Oryza sativa	*Xanthomonas oryzae*	Culture filtrate	Filtrate resistant plants
Strawberry			
Fragaria sp.	*Fusarium oxysporum* f. sp. *fragariae*	Fusaric acid	Resistant shoots
Sugarcane			
Saccharum officinarum	*Helminthosporium sacchari*	Culture filtrate	Disease resistant clones
Tobacco			
Nicotiana tabaccum	*Pseudomonas tabaci*	Methionine sulfoximine	Phytotoxin resistant plants
Tomato			
Lycopersicon esculentum	*Fusarium oxysporum*	Fusaric acid	Plants with elevated resistance

parts (leaves or shoots) against a pathogen or metabolites are being used routinely for screening purposes. More recently, callus or cell cultures are also used extensively for such screening. Barring a very few exceptions, *in vitro* response exhibits a direct correlation to *in planta* response. Hence, these assays, when combined with a selection program (as described previously in this chapter), are very effective, especially in perennial crops where the regeneration cycles are long. Another added advantage of using cell or callus cultures for screening against pathogen/pathogen-derived metabolites is that it facilitates studying the subcellular mechanisms underlying in those interactions (detailed in a later section of this chapter).

Germination of seeds in a toxin solution is a very easy and common method of testing phytotoxicity. Both host-specific and nonspecific toxins can be assayed in this manner. Typically, a wide range of concentration is used to determine the LD 50 levels and sublethal doses of toxin, based on germination inhibition. Additional parameters such as inhibition of root growth (especially for soil-borne pathogen), malformation of cotyledons, and chlorosis in the emerging leaves of intact plants can also be used to determine toxicity. The most common way to test live pathogen is to spray a measured quantity of spore or bacterial suspension on clean leaves and let it incubate under ideal conditions for 24–72 h. Alternatively, a drop of spore or bacterial suspension can be placed over the leaf lamina, which after sufficient soaking, are incubated and observed for disease development. In some cases, vacuum-infiltrate or pricking with a needle may be necessary to facilitate disease development. It is very important to include appropriate controls in these studies. The easiest method is by soaking water or buffer (in which the spores were suspended) preferably in the same leaf or on a similar leaf from the same plant that is used for pathogen infection. These tests are also equally effective when a pathogen-derived metabolite is tested (see Jayasankar et al., 1999 for details). For assaying "wilt toxins" (toxins that are involved in wilt diseases) or pathogens that cause wilt diseases, the best method is to place a live and clean cutting in a solution containing the metabolite or the pathogen for a predetermined time, and then incubating such treated cuttings in sterile water for disease symptom development. Parameters such as inhibition of shoot growth, chlorosis or necrosis of leaves, and time to taken for the plant organ to wilt as against the appropriate untreated control will serve to determine the resistance/susceptibility of the plant in question.

Callus cultures are routinely used for *in vitro* screening of multiple genotypes to assess their sensitivity to a pathogen. For instance, Nyange et al. (1995) screened nine genotypes of coffee against *Colletotrochium kahawae*, the fungal pathogen causing coffee berry disease. Early studies (before plant tissue culture was very common) involving plant tissue cultures aimed at only assessing the effect of phytotoxin. However, our current knowledge of plant tissue culture and plant–pathogen interactions at the molecular level has provided a powerful tool in understanding the disease process. A list of crop plants where tissue culture-based assays were used to study plant–pathogen interactions is furnished in Table 41.2. An effective variant among the tissue culture-based bioassays is the dual culture assay. Whereas other screening procedures are based on the growth of pathogen, dual culture assays are based on inhibition of the pathogen by plant cell culture. This assay is an effective tool where the objective is inducing disease resistance using *in vitro* selection, as shown in some perennial crops such as lemon, grapevine, and mango (Jayasankar et al., 2000).

MOLECULAR STUDIES

In vitro plant–pathogen interaction studies provide an ideally controlled environment to study events occurring at the molecular level. Synthesis and accumulation of defense-related proteins have been observed following a pathogen infection, and several species genes encoding these proteins have been identified, cloned, and characterized. Such studies lead to the identification of an important group of proteins, termed "pathogenesis-related proteins" (PR-proteins) (Linthorst, 1991). Examples of PR proteins include chitinases, glucanases, osmotin, and thaumatin-like proteins. To date, several PR-proteins have been identified and grouped into 14 "families," based on their serological affinity.

One of the earliest responses that plant cells exhibit following pathogen attack is the rapid increase in "reactive oxygen species" (ROS), which is also referred as the oxidative burst. This initial oxidative burst that can be detected within a few minutes after infection occurs regardless of the host's resistance or susceptibility. In a resistance reaction, a second oxidative burst develops a few hours later that sustains for longer periods. Such elevated levels of ROS can directly or indirectly inhibit the invading pathogen and also serve as intermediates in the activation of other defense responses (Baker and Orlandi, 1995).

Several days after these initial responses, other defense mechanisms are activated. Among these are the **hypersensitive responses (HR)** that result in the rapid and localized death of a small group of plant cells surrounding the site of pathogen infection in a programmed manner. As this programmed cell death proceeds around the infection site the invading pathogen is killed, thereby preventing any further spread of necrosis (Hammond-Kosack and Jones, 1996). As these cells prepare to die, they also set forth a series of other defense responses, which include the stimulation of genes encoding PR-proteins. These responses, culminating in the expression of PR-proteins, are also seen in other, non-infected tissues of the plant, which is termed as **systemic acquired resistance (SAR)** (Ward et al., 1991).

TABLE 41.2

***In vitro* Culture Based Bioassays Involving Plant–Pathogen Interaction**

Plant Species	Pathogen	Type of Bioassay	Purpose
Alfalfa			
Medicago sativa albo	*Verticillium albo-atrum*	Viability staining of cell cultures	Evaluation of phytotoxicity of culture filtrate
Alfalfa			
Medicago sativa	*Fusarium* spp.	Intact seedlings in hydroponic culture containing culture filtrate	Evaluation of phytotoxicity of culture filtrate
Coffee			
Coffea arabica	*Colletotrichum kahawae*	Growth of fungus on callus cultures	Screening of genotypes for resistance
Dogwood			
Cornus florida	*Discula destructiva*	Growth of callus on medium containing culture filtrate	Screening of genotypes for resistance
Grapevine			
Vitis vinifera	*Elsinoe ampelina*	Dual culture using proembryogenic mass	Assessing resistance after *in vitro* selection
Lemon			
Citrus limon	*Phoma tracheiphila*	Dual culture using callus	Assessing resistance after *in vitro* selection
Mango			
Mangifera indica	*Colletotrichum gloeosporioides*	Dual culture using proembryogenic mass	Assessing resistance after *in vitro* selection
Pine			
Pinus taeda	*Cronartium quercuum* f. sp. *fusiforme*	Inoculation of embryos with live fungus	Assessing *in vitro* resistance of embryos
Soy bean			
Glycine max	*Fusarium solani* f. sp. *glycines*	Viability staining of cell cultures; stem cutting assay	Evaluation of phytotoxicity of culture filtrate
Tomato			
Lycopersicon esculentum	*Alternaria alternata* f. sp. *lycopersici*	Detached leaf bioassay	Evaluation of phytotoxicity of toxin
Wheat			
Triticum aestivum	*Microdochium nivale*	Detached leaf and young seedlings	Screening of cultivars for resistance

In the past 10 years, SAR (Chapters 28 and 31) has become one of the most widely researched areas in plant–pathogen interaction. These studies lead to the finding of several interesting secondary signals such as ethylene, jasmonates, and salicylic acid (SA) that have crucial roles in mediating the induction of plant defenses. Although it has been over 25 years since White (1979) showed that application of SA or acetyl salicylic acid (aspirin) to tobacco leaves increased its resistance to TMV infection, molecular studies elucidating the role of SA in plant defense have been done only in the past 7–8 years. SA has been shown to be an endogenous plant signal that has a central role in plant defense against a variety of pathogens including viruses (see Dempsey et al., 1999). SA treatment also triggers the same set of PR-proteins that are expressed as a result of SAR. Plants treated with SA and those exhibiting SAR have heightened resistance to viruses as well, whereas none of the PR-proteins themselves have been shown to exhibit antiviral activity.

Conclusion

It is clear that *in vitro* plant pathology has come a long way over the past 25 years or so. What was once designed to study the plant–pathogen interaction in a controlled environment has slowly evolved into a decisive tool in generating disease-resistant crop plants and, in turn, greatly enhanced our understanding of these interactions at the molecular level. It is crucial that we understand the mechanisms by which plants defend pathogen attack so that it will be possible for us in future to activate their own defense, instead of using transgenes to confer resistance. Induced resistance conferred by "native genes" will be more stable and will also help us to generate "environmental friendly" genotypes of crop plants.

The following laboratory exercises will provide the students with some experience in performing experiments related to *in vitro* plant pathology. The experimental subjects provided in these exercises are primarily based on

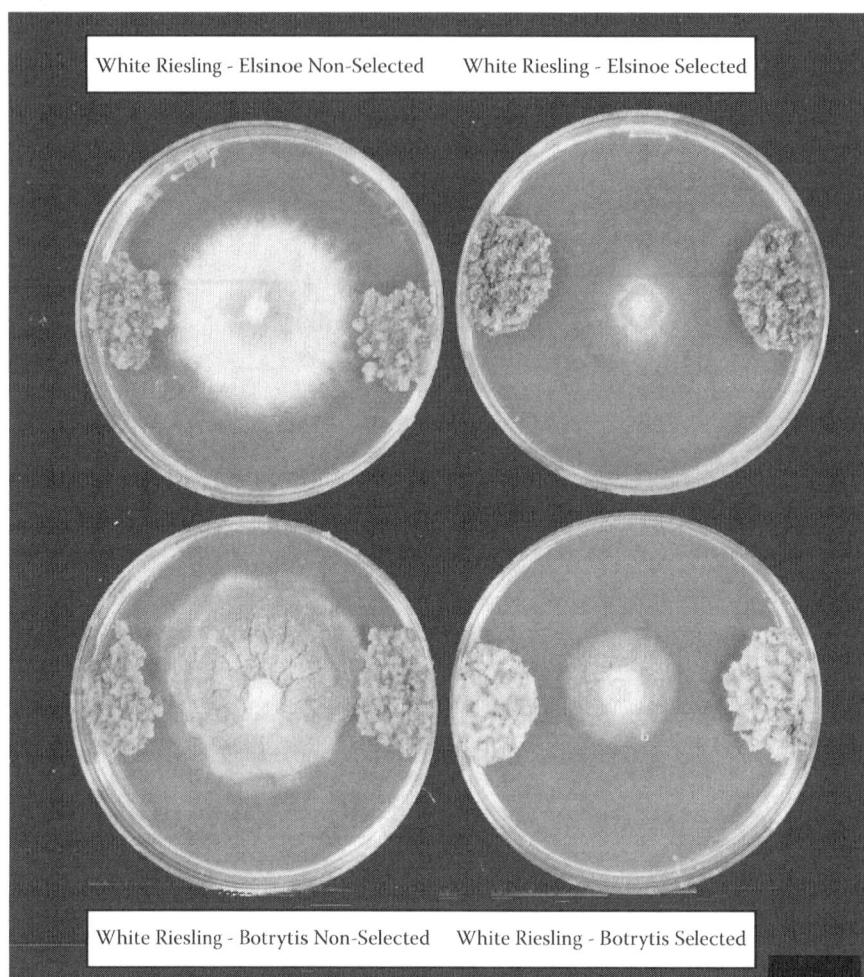

| White Riesling - Elsinoe Non-Selected | White Riesling - Elsinoe Selected |
| White Riesling - Botrytis Non-Selected | White Riesling - Botrytis Selected |

FIGURE 41.2 Dual culture assay: embryogenic cultures of grapevine "White Riesling" that were selected in vitro with culture filtrate of *Elsinoe ampelina* and tested for resistance against the *E. ampelina* [top] as well as *Botrytis cinerea* [bottom]. Fungal mycelium grew uninhibitedly in the plates containing nonselected control (left), while their growth was severely inhibited by the *in vitro* selected cultures (right).

ease of availability. However, these approaches are very broad based and can be adapted or modified to any other similar situations. As with any biological experiment, details often differ with plant and fungal species. Although the examples provided in these experiments are based on plant–fungal interactions, they can be easily adapted to address other plant–microbe interactions also.

GENERAL CONSIDERATIONS

PREPARATION OF MEDIA

The agar media for growing fungal pathogens are quite simple and are usually available as "ready-to-use" powders from many biochemical suppliers. Resuspend the required amount of powdered medium in distilled water in 500-mL Pyrex bottles and autoclave for 20 min at 121°C. After cooling to about 65°C (when they can be held in hand without too much discomfort), dispense the medium in petri dishes in a laminar flow hood. Avoid condensation in petri dishes by either stacking the dishes into columns of ten or so immediately after pouring or by keeping the lids of the dishes ajar inside the laminar flow hood until the medium solidified and cooled to room temperature. Medium in petri dishes can be sealed in polythene sleeves (reuse the sleeves in which the petri dishes are supplied) and kept in the refrigerator for 3 to 4 months.

HANDLING OF THE PATHOGEN AND PHYTOTOXINS

Most of the plant pathogenic fungi are harmless to human and generally does not require extra precaution while handling. In most instances, the same laminar flow hood may have to be used for both plant and pathogenic cultures. If the same laminar flow hood is used for both plant and fungal/bacterial cultures, at least 1 h should be allowed between working with fungal and plant cultures. If possible, the pathogen cultures should be done at the end of

	Procedure 41.1 Inhibition of Tobacco Seed Germination by *Alternaria* Toxin
Step	Instructions and Comments
1	Preparation of tobacco seeds for germination assay. Wash the seeds with 10% commercial bleach containing few drops of Tween 20 by vortexing in a microcentrifuge tube. Centrifuge briefly and remove the bleach solution using a pipette. Rinse the seeds three times with sterile distilled water. Transfer the seeds to a sterile filter paper and allow it to dry in the laminar flow hood. Now the seeds are ready for assay.
2	Germination. Place a sterile filter paper in each petri dish and label four dishes for each treatment. Wet the filter paper with 5 mL of appropriate toxin solution. Place 100 tobacco seeds in each petri dish. Close and gently seal the petri dishes with parafilm. Incubate at $28 \pm 2°C$ in a growth room with 16 h photoperiod.
3	Observations. After 1 week, count the number of seeds that have germinated in each plate. Repeat the count after two weeks. During the second week, count any abnormalities such as nonemergence of cotyledonary leaves and malformation of germinated seedlings in each dish separately.

the day, so that the laminar flow hood can be cleansed thoroughly and allowed to run overnight.

Some phytotoxins may be harmful to human beings even at very low concentrations. If commercially available phytotoxins are used for experiments, read the instructions very carefully before using it. Use appropriate precautions such as wearing gloves, mask, and fume hood as directed.

Materials that are used in experiment 3 [SDS-PAGE] contain chemicals such as acrylamide, mercaptoethanol, and TEMED that have been listed as carcinogenic chemicals. These chemicals should be handled with extra caution as instructed.

EXERCISES

EXPERIMENT 1. EFFECT OF PHYTOTOXIN ON TOBACCO SEED GERMINATION

Generally phytotoxins will have some adverse effect on seed germination. This is more conspicuous when nonspecific phytotoxins are used. Tobacco seeds are convenient models for assaying the efficacy of phytotoxins, because they are small, easy to handle, and germinate rather quickly. Although this experiment uses a commercially available phytotoxin as a model, any fungal [or bacterial] product, such as concentrated culture filtrate or partially purified extracts, can also be tested effectively. This experiment will require approximately two weeks.

Materials

The following material will be needed for each team of three students (the suggested number per team).

- Twenty disposable petri dishes
- Twenty 20 sterile Whatman No. 3 filter papers
- Five 100-mL beakers
- Ten 1-mL serological pipettes

General for the whole class

- Approximately 500 mg of tobacco seeds. Although cultivar is not a constraint, it will be good to use different cultivars with different teams, so that a comparison of the response can be made between the cultivars. These seeds are available with most seed vendors.
- Alternaria toxin stock solutions. Several toxins from *Alternaria* are commercially available from chemical companies. Prepare stock solutions altertoxin I of at least five strengths e.g., 0, 5, 10, 50, and 100 µM. Each team will need 20 mL of each stock and hence prepare stocks accordingly.

Follow the protocols in Procedure 41.1 to complete the experiment.

Anticipated Results

There will be a decline in germination percentage as the concentration of toxin increases. At higher concentrations, e.g., 50 µM or more, there may be noticeable abnormalities among the germinated seedlings. Some abnormalities that can be readily identified include "albino" types and variegated cotyledons or young leaves.

Question

- Why does the germination capacity of tobacco seeds decline when germinated in toxin solution?

EXPERIMENT 2. EVALUATION OF ANTHRACNOSE RESISTANCE IN COMMON BEANS USING A LEAF DISC BIOASSAY

Anthracnose is a very common disease of several crop plants, usually caused by the fungus *Colletotrichum*. In common beans (*Phaseolus vulgaris*), anthracnose is caused by *C. lindemuthianum*. The disease can be easily identified by spots, mostly on the underside of the leaf. Red-brown spots and streaks also develop on stems, petioles, and leaves. A characteristic symptom of the disease is veins turn brick-red to purple and eventually black on the underside of the leaf. In resistant varieties, small specks of infection can be noticed, and usually the disease spread is arrested. This experiment will provide the students an opportunity to evaluate the leaf disc bioassay, which is the most common assay when several cultivars have to be evaluated for disease resistance. Usually, the results can be seen within a week.

Materials

The following materials will be needed for each team of students.

- Four 50-mL beakers
- Six-to-eight week old bean plants, five of each variety
- 1-cm diameter cork borer
- 20 sterile disposable 10-cm diameter petri dishes
- 20 sterilized Whatman No. 3 filter papers
- 2 pairs of forceps
- 500 mL sterile, distilled water

General for the class

- Culture of *Colletotrichum lindemuthanium* (several accessions are available from ATCC, e.g., accession numbers 16341, 16342) on lima bean agar medium (Difco 0024) for 1 week or until orange-to-pinkish clusters of spores can be seen. Culture two dishes per team. Collect the spores in distilled water and adjust the spore density to 100,000 spores per mL. Each team needs 20 mL of spore suspension.

Follow the protocols outline in Procedure 41.2 to complete the experiment.

Anticipated Results

Most of the leaf discs will start developing lesions and in some cases will turn necrotic within one week. There will also be differences in sensitivity between the resistant and susceptible varieties. Resistant varieties may develop lesions much later than the susceptible varieties. Hypersensitive cell death around the sites of fungal infection may be present. Other saprophytic fungi and mold that may grow on the leaf discs.

Questions/Activities

- Define "resistance" and "susceptible," based on your test results.
- Describe "hypersensitive response."

Procedure 41.2

Evaluation of Anthracnose Resistance in Common Beans Using a Leaf Disc Bioassay

Step	Instructions and Comments
1	Preparation of leaf discs for bioassay. Wash young, fully expanded leaves bean leaves thoroughly in tap water and rinse well with deionized water. Gently blot dry the leaves using paper towels. Cut 30 leaf discs (1 cm diameter) using a clean cork borer. Keep the discs on a moist paper towel.
2	Inoculation with fungal spores. Soak 20 leaf disks in fungal spore suspension for at least 5 min. in a 5-mL beaker. Make sure that all the leaf discs are completely immersed in spore suspension by gently swirling the beaker. After 5 min gently decant the spore suspension into another beaker. As a control, soak ten leaf discs in sterile deionized water.
3	Incubation and observations. Place wet, sterile filter paper in a standard petri dish. Carefully place the leaf discs on the wet filter paper, five per petri dish, with adequate spacing. Gently wrap the petri dishes using parafilm and incubate in a growth room at $28 \pm 2°C$ in a growth room with 16 h photoperiod. Observe the leaf discs for any symptom development after 3 d. After 1 week, count the number of discs showing symptoms in each variety. Also try to grade the intensity of the symptom on a suitable scale (e.g., 1–5, with 1 representing no symptom to 5 representing very high symptoms).

Procedure 41.3

Induction and Analysis of PR Proteins Using SDS-PAGE

Step	Instructions and Comments
1	Induction. Spray 5 mL of spore suspension or salicylic acid solution using an atomizer to 2 sets of plants each and label them. As mock inoculation, spray sterile distilled water in one of the remaining 2 sets of plants. The sixth jar of plants serves as "absolute control." Seal the jars with parafilm and incubate at 16 h photoperiod.
2	Protein extraction. One week after spraying, collect young leaves from different sets of plants. Weigh 500 mg of leaves and wrap in aluminum foil. Immediately, plunge in liquid nitrogen and grind the frozen leaves to a fine powder with a pestle in a mortar. Add 1 mL of protein extraction buffer to the fine powder and homogenize further. Transfer the resulting slurry to a 1.5 mL microcentrifuge tube. Keep this on ice, until all the four samples are extracted. Centrifuge the slurry at 12,500 rpm for 10 min at 4°C and carefully pipette out 750 µL of the supernatent into a fresh tube. This constitutes crude protein extract. Add ammonium sulfate to this supernatent to 80% saturation and centrifuge again at 12,500 rpm for 10 min. The proteins will precipitate as a small, white pellet. Resuspend this pellet in 100 µL of sterile water. This is the protein sample to be analyzed
3	Protein estimation. Determine the protein concentration in the samples using a protein estimation kit (commercially available from several vendors). Readjust the protein sample volume to a final concentration of 1 or 0.5 µg per µl, depending on the reading. Keep this sample at –20°C, if it is not used immediately.
4	Loading and running the gel. After setting up the gel apparatus and the gel in the assembly (follow manufacturer's directions), load the protein samples carefully in appropriate wells using a gel loading pipette tip or Hamilton syringe. Care should be taken to avoid cross contamination of wells. Load a standard marker to one of the wells so that it will be convenient for reference. Connect the electrodes appropriately and run the gel at the recommended voltage.
5	Staining. A lot of commercial stains based on the principles of coomassie staining of proteins and peptides are available and are easy to use. Any of the stains (e.g., GelCode blue from Pierce Endogen) may be used for staining and visualizing the protein bands. These stains are fast, very sensitive, and only a small volume is required.

EXPERIMENT 3. INDUCTION OF PATHOGENESIS-RELATED (PR) PROTEINS IN TOBACCO AND *ARABIDOPSIS*

As discussed before (Chapters 28 and 31), plants respond to pathogen attack in a number of ways. Researchers have also discovered that certain chemicals such as salicylic acid can also evoke similar responses. Induction of PR proteins is one such common response. This experiment will enable the students to see if such response can be induced in tobacco or *Arabidopsis* using polyacrylamide electrophoresis (Laemmli, 1970). Nucleic acid and proteins are negatively charged and, hence, when subjected electric field they will migrate toward the cathode. All electrophoresis techniques are based on this principle. This experiment will suit graduate students studying plant pathogen interactions and will take approximately 10 days to conclude.

Materials

The following materials are needed for this experiment. Plants for this experiment should be germinated at least 6 weeks before the planned dates of experiment. If there are several groups, they can be split to test tobacco and *Ara-*

bidopsis. Two members are suggested for each team and both should be encouraged to collect and analyze their own samples.

- Six sets of axenic plants, consisting of three to four uniform plants grown in tall jars
- Sterile pestle and mortar
- 30 sterile 1.5-mL microcentrifuge tubes

General for the whole class:

- Micropipettes and sterile pipette tips
- Liquid nitrogen
- Spore suspension. Collect spores from a fresh plate of *Colletotrichum destructivum* (or *Alternaria alternata*) culture and suspend in 5 mL distilled water. After determining the density using a hemocytometer, adjust the spore suspension to 10,000 spores per mL and make 200 mL (or the desired quantity) of suspension.
- 200 mM salicylic acid solution. Dissolve 13.81 g of salicylic acid in 500 mL distilled water.

- Protein extraction buffer. Several buffers are used successfully to extract total proteins from leaves. A common buffer to extract proteins from leaves will be 100 mM Tris (pH 6.8), 100 mM sodium phosphate 2.5% sodium dodecyl sulfate (SDS), 5% glycerol, and 10 mM dithiotreitol (DTT). It is safe to add protease inhibitors such as phenylmethyl sulfonic acid (10 mM) or leupeptin (10 mM) to the buffer.
- Polyacrylamide gel electrophoresis units and required power supplies. Mini-gel units that are sold by several commercial vendors such as Bio-Rad or Hoefer scientific can be used. 10% or 12% SDS-PAGE gels. These gels can be prepared in the lab if gel casters are available or ready gels can be purchased from the makers. For casting gels in the lab please refer to Sambrook et al. (1989).
- Protein molecular weight standard. Available from several companies such as Bio-Rad and Sigma. Reconstitute as per manufacturer's directions.
- Protein sample loading buffer. Prepare the sample loading buffer as follows. Dissolve 2.4 mL of 0.5 M tris (pH 6.8), 2.0 mL glycerol, 4.0 mL of 10% (w/v) SDS, and 1.0 mL of 0.1% (w/v) bromophenol blue in 9.6 mL of sterile distilled water. This can be kept at room temperature for 3 months. Just before using add 25 µl ß-mercaptoethanol to 475 µl of the above buffer. This working buffer should be used within 48 h.

Follow the protocols described in Procedure 41.3 to complete the experiment.

Anticipated Results

Proteins and peptides can be seen as blue bands of varying intensities like a ladder in each lane. There should be additional bands and/or higher expression (more intensely stained) of certain bands in the lanes containing samples that were sprayed with spores or salicylic acid in comparison with untreated control and mock-inoculated control. It is likely (but not limited to) that these differences will be seen between 20–45 kDa region, which can be referenced using the marker standard.

Question

- What will happen if the protein samples are allowed to be at room temperature (not kept on ice, until denaturing) for extended period?

ACKNOWLEDGMENTS

We sincerely thank Dr. Richard Litz of the Tropical Research and Education Center, University of Florida, Homestead, for his encouragement and critical suggestions in the preparation of this chapter.

LITERATURE CITED/SUGGESTED READING

Baker, C.J. and E.W. Orlandi. 1995. Active oxygen in plant pathogenesis. *Annu. Rev. Phytopath.* 33: 299–321.

Daub, M.E. 1984. A cell culture approach for the development of disease resistance: Studies on the phytotoxin cercosporin. *HortScience.* 19: 382–387.

Daub, M.E. 1986. Tissue culture and the selection of resistance to pathogens. *Annu. Rev. Phytopathol.* 24: 159–186.

Dempsey, D.A., J. Shah, and D.F. Klessig. 1999. Salicylic acid and disease resistance in plants. *Crit. Rev. Plant Sci.* 18: 547–575.

Hammond-Kosack, K.E. and J.D.G. Jones. 1996. Resistance gene-dependent plant defense responses. *Plant Cell* 8: 1773–1791.

Jayasankar, S. Variation in tissue culture. 2000. In *Plant Tissue Culture Concepts and Laboratory Exercises (Second Edition)*. Trigiano R.N., and Gray, D.J., Eds. pp. 386–395.

Jayasankar, S.and D.J. Gray. 2003. In Vitro selection as an alternate technology for genetic improvement of disease resistance in crop plants. Mini-review. *AgBiotechNet* (May 2003), 1–5.

Jayasankar, S., Z. Li, and D.J. Gray. 2000. In vitro selection of *Vitis vinifera* "Chardonnay" with *Elsinoe ampelina* culture filtrate is accompanied by fungal resistance and enhanced secretion of chitinase. *Planta.* 211: 200–208.

Jayasankar, S., R.E. Litz, D.J. Gray, and P.A. Moon. 1999. Responses of embryogenic mango cultures and seedling bioassays to a partially purified phytotoxin produced by a mango leaf isolate of *Colletotrichum gloeosporioides* Penz. *In Vitro Cell. Dev. Biol. Plant.* 35: 475–479.

Laemmli, U.K. 1970. Cleavage of structural proteins during the assembly of the head of bacteriophage T4. *Nature.* 227: 680–685.

Linthorst, H.J.M. 1991. Pathogenesis-related proteins of plants. *Crit. Rev. Plant Sci.* 10: 123–150.

Litz, R.E. and D.J. Gray. 1992. Organogenesis and somatic embryogenesis. In *Biotechnology of Perennial Fruit Crops*. Hammerschalg, F.A. and Litz, R.E., Eds. CAB International, Wallingford, U.K., pp. 3–34.

Nyange, N.E., B. Williamson, R.J. McNicol, and C.A. Hackett. 1995. *In vitro* screening of coffee genotypes for resistance to coffee berry disease (*Colletotrichum kahawae*). *Ann. Appl. Biol.* 27: 251–261.

Sambrook, J., E.F. Fritsch and T. Maniatis. 1989. Molecular cloning. A laboratory manual. Cold Spring Harbor Laboratory Press. New York.

Strobel, N.E., C. Ji, S. Gopalan, J.A. Kuc, and S.Y. He. 1996. Induction of acquired systemic resistance in cucumber by *Pseudomonas syringae* pv. *syringae* 61HpZpss protein. *Plant J.* 9: 431–439.

Ward, E.R., S.J. Uknes, S.C. Williams, S.S. Dincher, D.L. Wiederhold, D. Alexander, P. Al-Goy, J.P. Metraux, and J.A. Ryals. 1991. Coordinate gene activity in response to agents that induce systemic acquired resistance. *Plant Cell* 3: 1085–1094.

White, R.F. 1979. Acetylsalicylic acid (aspirin) induces resistance to tobacco mosaic virus in tobacco. *Virology* 99: 410–412.

42 Proper Use of Compound and Stereo Microscopes

David T. Webb

This chapter is written as if the reader is a microscope novice. Experience has taught us that it is best to assume that most students will know next to nothing about using a microscope correctly, and it is best to start from scratch. In some cases you may have learned some bad practices that need to be corrected. The chapter also covers compound microscopes that have a field diaphragm, and a condenser that can be centered and focused to achieve Koehler illumination. Many student scopes do not have these features as their condensers and field diaphragms are fixed or of limited flexibility. In the course of your careers you will encounter microscopes that have the ability to achieve Koehler illumination. At that point this tutorial will be even more useful.

Although the compound microscope is the most commonly used biological instrument, it is often used improperly. This may not matter with very thin commercial slides at low to medium magnifications. However, proper alignment of the illumination system is essential for viewing thick sections, whole mounts, and for highly magnified samples of fungi and bacteria. It is also crucial for studying unstained specimens and for photomicroscopy.

You will be using microscopes throughout your career. If you learn the simple lessons contained within this chapter you will do much better work and see the exciting world of microscopy in a new light. The modified procedure we will present was developed by the German scientist, August Köhler (1866–1948), and bears his name (Köhler, 1893). Recently his ideas were used to make the EM 910 Electron Microscope by Zeiss. Thus, this procedure, which was introduced in 1893, has been of lasting value.

MONOCULAR, BINOCULAR, AND TRINOCULAR MICROSCOPES

Microscopes are partly categorized by the number of oculars they contain. The first microscopes had one ocular and therefore were monocular. Binocular scopes have two oculars, whereas trinocular scopes have three. The third ocular is modified typically for the use of a camera. This chapter is written for the use of binocular microscopes. The same principles apply to all of the preceding types.

However, stereo or dissecting microscopes differ significantly from the typical compound microscope.

THE COMPOUND MICROSCOPE

Because the optical systems in a microscope are composed of many lenses, the term compound microscope is used. This is applied specifically to microscopes used to study thin sections with high-power objectives (also known as objective lens) (Figure 42.1). Dissecting or stereo microscopes are used to examine larger, three-dimensional specimens at lower magnifications (Figure 42.2). These also have compound lenses, but they are not generally called compound microscopes. Both types of microscope use **transillumination** (illumination through), in which light passes from the microscope base through the specimen and travels to your eyes through oculars. This requires a special transillumination base for stereo microscopes (Figure 42.2). **Epiillumination** (illumination from above) is typically used with stereo scopes, but is not typically used with compound scopes. This chapter is illustrated with Zeiss microscopes and an American Optical stereo microscope. Your microscopes may be somewhat different, but you should be able to transfer the terminology and procedures described herein to your instrument. We will examine compound microscopes first and discuss stereo scopes later.

MICROSCOPE CARE AND HANDLING

Please treat these instruments with great care—they are expensive and somewhat fragile.

- Value what they can do, and handle them with respect.
- Always use two hands to carry microscopes. Place one hand on the arm, the curved area that connects the body to the stage and base, and the other hand under the base of the microscope (Figures 42.1 and 42.2).
- Do not carry scopes sideways or upside down, as the oculars and other parts will fall out.
- Use lens paper to clean all lenses on the compound scope before each lab and especially after using

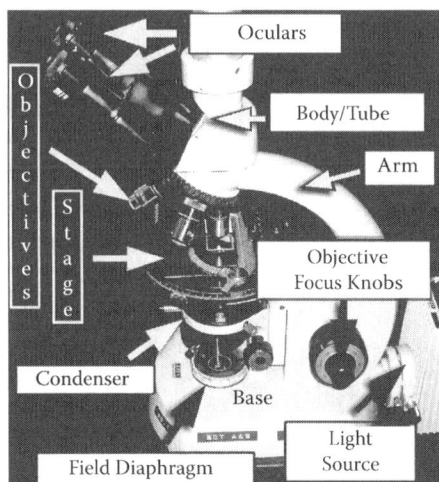

FIGURE 42.1 Zeiss Standard microscope showing major parts of a typical compound microscope.

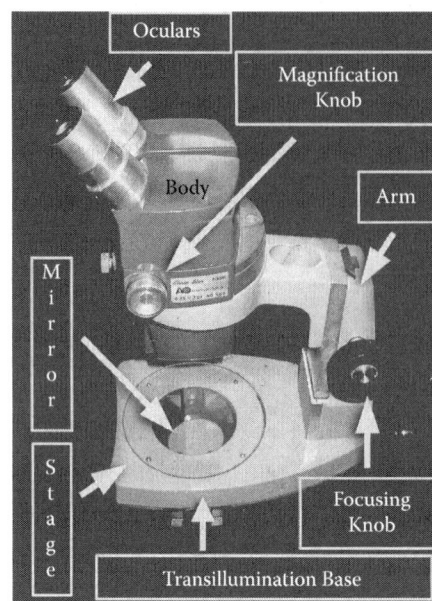

FIGURE 42.2 Typical stereo or dissecting microscope with transillumination base. This is an American Optical Stereo Zoom microscope.

immersion oil. Do not use any other kind of paper except lens paper to clean microscope lenses.

- Do not use liquids (except where specified) when cleaning the lenses (Duke, 2004).
- Always use the correct focusing technique to avoid contact between any objective and your slide.
- Turn off the light when not using microscopes for long time periods.
- Carefully place the power cord or any other cords out of the way at your workspace.
- Always replace the cover on the microscope when you put it away.
- Deal with any problems immediately. Do not use the microscope if you cannot see your specimens clearly.

THE PATHWAY OF LIGHT IN A TYPICAL MICROSCOPE

The light in a typical microscope traverses the following pathway:

Light Source → Mirror → Field Diaphragm → Condenser → Stage → Specimen → Objective → Body/Tube → Ocular → Eye or Camera

- Locate the major parts of your microscope by referring to Figure 42.1.

Light is provided by a bulb and is reflected through the field diaphragm, the condenser, the specimen, the objective, the tube, and the ocular. There are various control

knobs on the microscope that affect the light path. In addition, there are knobs for coarse and fine focus, as well as knobs to move the stage.

FOCUSING THE OBJECTIVES

The objectives are focused on the specimen by two sets of knobs that are located on both sides of your microscope (Figure 42.1). The large outer knob is for coarse focusing. This should be used at the lowest magnification when you first place a specimen on the stage. It should be used with caution at higher magnifications. The smaller, central knob is for fine focusing and used more at the higher magnifications, but is also used with low-magnification objectives.

- Locate the coarse and fine focusing knobs on each side of your scope.

MOVING THE MECHANICAL STAGE

Light passes through the stage opening so that it can illuminate the specimen. The knobs that control the mechanical stage (stage transport knobs) are usually on the left side of the microscope as it faces you. One of these moves the stage from side to side (X-axis), whereas the other moves the stage in and out (Y-axis). Most stages have X and Y scales. These will allow you to record the precise location of objects that you may want to relocate without searching the entire specimen. Slides are held in place by a mechanical slide holder.

- Locate the stage transport knobs on your microscope.
- Locate the *X* and *Y* scales on your stage.
- Locate the mechanical slide holder and explore its mode of action.

USING THE CONDENSER

The condenser aligns and focuses light on the specimen. It may be equipped with a high-power condenser lens (Figure 42.3). This is used with 10 to 100X objectives, but is removed from the light path with low-magnification objectives, which are typically 4× to 5×. The position of the high-power condenser lens (HPCL) may be controlled by a rotating knob. On some microscopes it is moved in and out of the light path by a push-pull plunger or a lever. Failure to use this lens properly is the most common mistake that people make. The lens is typically left out of the light path at low magnification because it limits the **field** of illumination. A fully illuminated field is achieved with the HPCL out of the light path with low-power objectives. There may not be a large penalty for examining commercial slides at higher magnification with the HPCL out, but there is a severe visual penalty at higher magnifications and with fresh mounts.

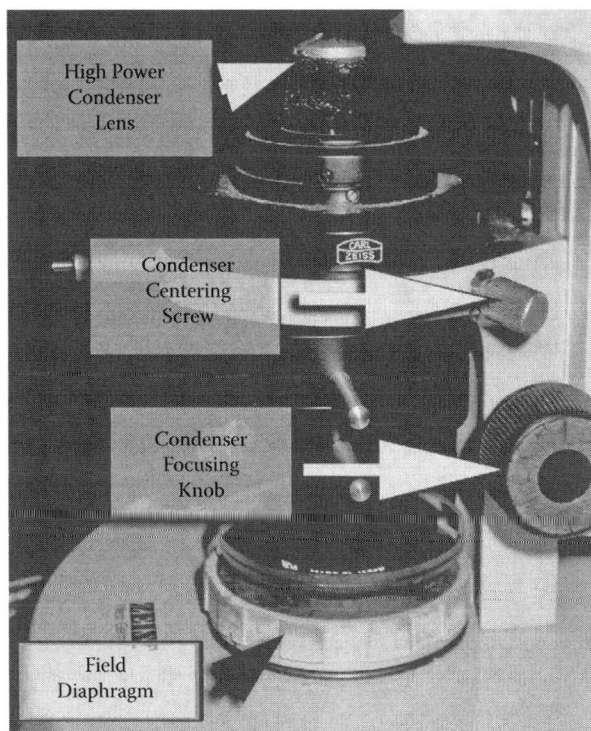

FIGURE 42.3 Side view of a Zeiss Standard microscope showing the high-power condenser lens, condenser-focusing screws, and the condenser-focusing knob as well as the field diaphragm.

- Locate the high-power condenser lens on your microscope, and determine how to move it into and out of the light path.

CONDENSER-CENTERING SCREWS

A pair of screws, set apart by a 45° angle, are used to center the condenser. These are located along the back right and left sides of the condenser on the Zeiss Standard (Figure 42.3). However, they may be found near the front of the condenser with other scopes.

- Locate the condenser-centering screws on your microscope.

The condenser is used to focus light onto the specimen from below. This is an extremely important, but poorly understood function of the condenser. It is obvious that you must focus the objectives onto the specimen to see it clearly, but it is less obvious that you need to focus the condenser on the specimen so that it is properly illuminated. Imagine that the condenser is a magnifying glass, and you want to start a fire. You need to move the magnifying glass to a position that produces the smallest focused beam of sunlight in order to start your fire. That is exactly the concept you need to keep in mind when you focus the condenser. This is done by rotating the condenser-focusing knob that is found on the right side of the condenser with the Zeiss Standard (Figure 42.3).

- Locate the condenser-focusing knob on your microscope.

CONDENSER (APERTURE) DIAPHRAGM

Finally, there is a lever that controls the aperture of the condenser or aperture diaphragm (Figure 42.4). I like to refer to this as the condenser diaphragm to prevent confusion with the field diaphragm. However, it has typically been called the aperture diaphragm. Partially closing this iris improves **contrast** (the difference between light and dark) especially at intermediate and high magnifications. It also increases the **depth of field**, which is very small at high magnifications.

Do NOT use this to increase or decrease brightness! This is the second most frequent mistake that people make! It is best to leave this completely open (rotated to the left on the Zeiss Standard) at the outset. Later, you will experiment with this to see its effects.

- Locate the condenser (aperture) diaphragm lever and manipulate it.
- Leave it in the open position for now.

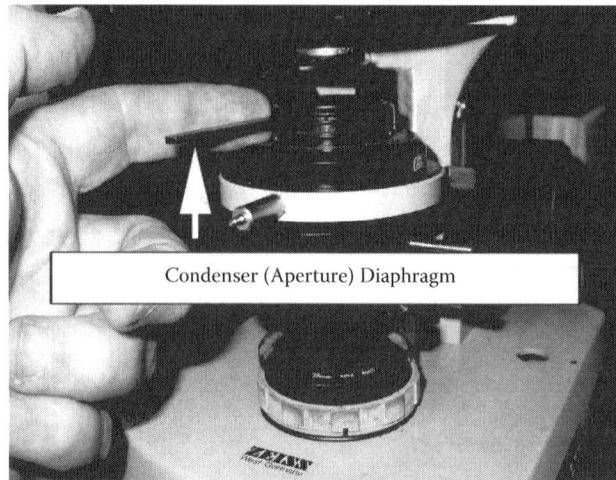

FIGURE 42.4 Lever that controls the condenser (aperture) diaphragm. It is completely open in this position. Rotating the lever to the right closes the iris diaphragm inside.

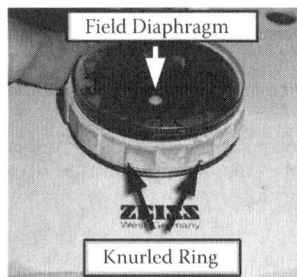

FIGURE 42.5 Field diaphragm that is almost completely closed. The knurled ring is used to open and close the iris inside.

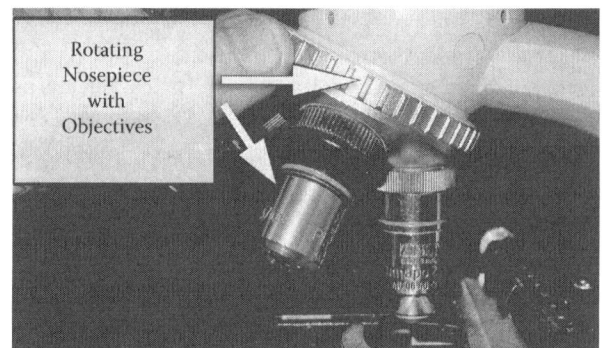

FIGURE 42.6 Rotating nosepiece with objectives. In this case the nosepiece is rotated clockwise (left to right) to change objectives to higher magnifications.

USING THE FIELD DIAPHRAGM

The light source is usually housed in the base of the microscope. Light passes through the field diaphragm, which also contains an iris (Figure 42.5). The size of its iris diaphragm is controlled by rotating a knurled ring, which is concentric with it. The field diaphragm controls the area of illumination.

- Locate the field diaphragm on your microscope.
- Manipulate the knurled ring to open and close the iris inside.
- Leave it in the fully open position for now.

USING THE OBJECTIVE LENS

The magnification of an image is regulated by the objectives, which are housed in a rotating nosepiece (Figure 42.6). To change objectives, you rotate the nosepiece.

Ensure that the low-power objective is in place before you start using your scope and when you are finished using it. This prevents damage to the objectives and your specimen.

- Always start viewing with the low-power objective (4× or 5×).
- Do not start by swinging in the 10× to 100× objectives into position.
- Be sure that you rotate the nosepiece in the right direction! You do not want to switch from 4× to 100×!
- Objectives may be damaged if they hit the specimen.
- You should focus on the sample with each objective before switching to the next. Focus on the sample using the low power objective and rotate to the next lens (10×), refocus and repeat this until you reach the magnification you plan to use. It is vital to focus on your specimen with the 40× objective before switching to 100×.

Really good microscopes have objectives that are **parfocal**, the ability to focus with one objective and switch to the next one and the one after that without refocusing. However, this rarely occurs, especially with student

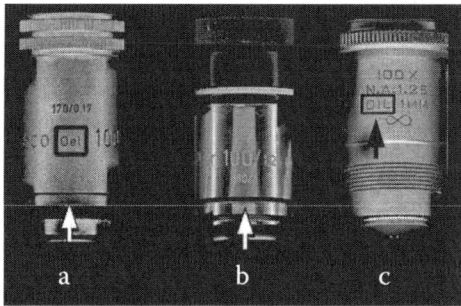

FIGURE 42.7 Oil immersion objectives from Leitz (A), Zeiss (B), and American Optical (C). Note the black lines (indicated by the white arrows) on A and B. This means that they are oil immersion objectives. The AO objective (C) lacks this line. However, it has the word Oil inscribed on it, as indicated by the black arrow and box. The word Oel (oil) is inscribed on the Leitz objective (A).

scopes. Special care must be used with fresh sections and whole mounts, which can be thick and irregular. Consequently, greater care must be taken when changing objectives. When in doubt, play it safe.

A typical microscope will have a series of objectives like the following: 5×, 10×, 20×, 40×, and 100×. The length of the objectives is a rough indication of their relative magnification. However, their magnifications are engraved on them.

• Check the magnification of your objectives.

Most oil immersion objectives have a black line near their tip (Figure 42.7). However, this is not true for all manufacturers. The words "Oil" or "Oel" indicate that it is an oil immersion objective. Oil improves the image because it unites the coverslip and the objective and replaces air with oil. Immersion oil has the same refractive index as glass. Thus, less light scattering and refraction occur. The oil also protects the objective lens from getting scratched.

The markings homogenous immersion (HI) or multiple immersion (Imm) indicate oil or multiple medium immersion objectives, respectively. Water immersion objectives are marked with the words "water," "waser," "water immersion," or with WI. Furthermore, an objective with a **numerical aperture** greater than 1.0 (Figure 42.7) is probably an immersion objective. Oil is the most common immersion medium, but water and glycerin also can be used with Imm objectives.

• Check your 100× objective to see its markings.

The instructions for properly using oil immersion objectives can be found in Procedure 42.1.

For optimum results, oil should also be placed between the uppermost condenser lens and the bottom of the slide. However, the condenser needs to have a numerical aperture greater than 1.0 for the oil to have a beneficial effect. This may be impractical for routine studies, but should be used for critical examinations and for the most detailed microphotography. However, most oil immersion objectives require you to partially close the condenser aperture. This alleviates the need to add oil to the condenser lens because it effectively lowers the

---

Procedure 42.1

Using an Oil Immersion Objective

Step	Instructions and Comments
1	Locate the area of interest in the center of the field. Focus on this with the 40× objective.
2	Raise the objective to its upper limit. Swing the oil immersion objective into viewing position.
3	Looking laterally with your eyes at the level of the stage, place a drop of oil on the 100× oil immersion objective and place a small drop of oil on the coverslip.
4	While observing from the side of the stage with your eyes, focus on the objective and the specimen. Your eyes must be at the same height as the stage. Lower the objective lens carefully (use the coarse-focusing knob) until it just touches the oil on the coverslip. A light flash may be observed when the oil on the objective meets the oil on the slide.
5	Now look through the oculars and use the coarse-focusing knob to bring the specimen into rough focus. Use the fine-focusing knob to complete focusing on the sample.
6	Hereafter, avoid focusing down on the specimen with an oil immersion lens. Change the focus so that the objective is traveling away from the slide. If the image does not come into focus readily, carefully reverse the direction until it does. When in doubt, stop and ask for help. The lens might be dirty, or there may be some other problem.
7	Important! Wipe the oil from the objective with lens paper when you are finished. Clean the objective until no more oil is visible on the lens paper. Wipe oil from the coverslip of the slide if it is to be saved.

Procedure 42.2	
Adjusting the Interpupillary Distance of the Oculars	
Step	Instructions and Comments
1	Position your head so that you can see through the oculars while focusing on a sample. Focus on a specimen using a 10×–20× objective.
2	Grasp the base of the ocular tubes or use the dial located between the oculars.
3	Move the oculars so that they are as close together as possible. Carefully move the oculars apart until you see only one image.

numerical aperture of the condenser, such that adding oil is no longer beneficial.

USING THE OCULARS

The oculars must be adjusted to suit both of your eyes. You should be able to adjust the interpupillary distance between the two objectives. This means that you can move them to match the distance between your eyes. Grasp the base of the ocular tubes or the plate and gently spread them apart or draw them together. In some cases there is a dial that you can use to move the oculars. Either one or both oculars may be movable. There should be a scale and a reference line or dot that allows you to record the best spacing for your eyes. Thus, you can readily readjust the oculars to your personal setting if they have been moved. Follow the steps in Procedure 42.2 to adjust the interpupillary distance of your oculars.

Head position is very important. You need to find a comfortable distance for your eyes from the oculars to see things properly. This depends on the type and quality of your oculars, and may require some experimentation. Most oculars can be used without eyeglasses that are corrected for near or far-sightedness. However, they are not compensated for astigmatism. If you have astigmatism you need to use your eyeglasses or contact lenses. Oculars with eyeglasses engraved on them are suitable for use with glasses (Figure 42.8), but you are not required to wear glasses to use these.

It is important that each ocular be in focus for your eyes when you examine samples. Oculars that are capable of independent focusing will have a scale, a reference line, and a knurled ring on them (Figure 42.8). These markings may be on the ocular tube rather than on the oculars themselves. Follow the instructions in Procedure 42.3 to focus your oculars.

KOEHLER ILLUMINATION

The best resolution occurs when all elements of the microscope are in perfect alignment, and the iris diaphragms are properly adjusted to the best apertures for the objectives you are using. On simple microscopes you may not

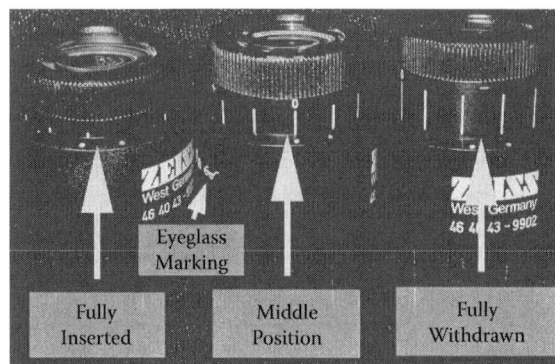

FIGURE 42.8 Zeiss oculars showing three focusing positions. These are focused by grasping the knurled ring at the top of the ocular and rotating it while holding onto the barrel below. The one on the left has been adjusted so that the rotating part of the ocular is fully inserted into the barrel. The one on the right shows the extreme opposite rotation. The one in the center is approximately in the middle. In this case the position of the ocular is indicated by the length of the white lines. In other cases there are numbers that can be used to designate the best focusing position for the ocular. Note the eyeglasses engraved on the middle objective. This indicates that these oculars were designed to be used with your glasses.

be able to focus and align the condenser, but on the Zeiss Standard and many other microscopes it is possible to do this and achieve "Koehler illumination." This makes a significant difference for viewing unstained and lightly stained samples, especially at high magnifications.

CENTERING THE LAMP FILAMENT

The first step in this process involves centering the lamp filament. This may not be possible with your microscope. Furthermore, it is best done by someone who is very familiar with this process. Check the illuminator housing on the back of your microscope to see if there are any adjustable screws. If not, you cannot do this. A generic description of this follows:

- Turn on the microscope illuminator.
- Pace a piece of paper over the field diaphragm.

Procedure 42.3	
Focusing Oculars for Your Eyes	
Step	Instructions and Comments
1	Before you make any adjustments, place a slide on the stage and focus on the central part of the specimen with a 10×–20× objective.
2	Block one of the oculars. Look through the other ocular with your matching eye (left eye ← → left ocular or right eye ← → right ocular), and focus on a fine detail in the center of the specimen with the objective-focusing knobs at the rear of the scope.
3	Switch to the other ocular and look through it with the matching eye. Do not look through the first ocular while you are doing this!
4	Rotate the knurled ring of the ocular to bring the fine detail into sharp focus. You will need to stabilize this with one hand while you turn it with the other.
5	Check the first ocular to see that the image is still in focus with your other eye.
6	Both oculars are now focused for your eyes.

- If the illumination is uneven, use the lamp-centering screws or rotate the lamp to get uniform illumination, or do both.

FOCUSING ON THE SPECIMEN FOR KOEHLER ILLUMINATION

The recommended procedure for focusing on a specimen as part of achieving Koehler illumination is given in Procedure 42.4. However you decide to proceed, it is very important to focus on a specimen before doing anything else. After you have focused on it, you may want to move the specimen out of the light path the first few times you do this.

FOCUSING AND CENTERING THE FIELD DIAPHRAGM

This is the heart of Koehler illumination. See Procedure 42.5 for the detailed steps in this process. Briefly, completely close the field diaphragm and use the condenser-focusing knob to focus the field diaphragm until you see that it is a small polygon of light (Figure 42.9A). Use the condenser-centering screws to center the image of the

field diaphragm (Figure 42.9B). Partially open the field diaphragm and center it again (Figure 42.9C). Open the field diaphragm until the field is completely illuminated and stop.

ADJUSTING THE CONDENSER (APERTURE) DIAPHRAGM

When working with 10× to 100× objectives, it is important to adjust the condenser diaphragm (Figure 42.4). This is especially true for translucent structures. Closing this iris increases contrast. Thus, something indistinct becomes sharp, and something faint becomes dark. It also improves the depth of field, which is critically small at high magnifications. It is usually possible to close the iris and judge its effects subjectively (Delly, 1988). However, there is a "tried and true" procedure (Procedure 42.6) that you should know.

In practice, you can experiment with this while viewing a specimen and adjust it without removing the ocular. This is what I do when I want to take photos. I slowly close this iris until I first see a perceptible change in the specimen and take a photograph. I close it some more and take another photo and repeat this for a third time. In reality,

Procedure 42.4	
Focusing on a Specimen as Part of Koehler Illumination	
Step	Instructions and Comments
1	Adjust the oculars so they have the correct interpupillary position and are in focus for your eyes.
2	Use a commercial slide with obvious well-stained contents, and move the specimen into the light path. Focus on the specimen with your low-power objective. Now, rotate the 10× objective into viewing position.
3	Watching from the side of the stage (not looking through the oculars), lower the 10× objective so that it comes close to the coverslip.
4	Look through the oculars, and rotate the objective-focusing knobs so the objective is retracted from the slide. Make a note on which direction retracts the objectives.
5	Stop when the sample comes into focus. Moving the mechanical stage slightly during this process may help.

	Procedure 42.5 Focusing and Centering the Condenser
Step	Instructions and Comments
1	Use the 10× objective to focus on the center of a specimen. Reduce the illumination to a moderate level so you do not hurt your eyes.
2	Check to see that the condenser (aperture) diaphragm is open. Also, ensure that the high-power condenser lens is in the light path.
3	Close the field diaphragm so that the circle of light becomes smaller. Observe the field diaphragm through the oculars when it is being closed (Figure 42.9). When the diaphragm is as small as possible, use the condenser-focusing knob (Figure 42.3) to make the "circle" of light as small as possible (Figure 42.9).
4	You should see that the field diaphragm is not circular in outline, but has a polygonal shape (Figure 42.9). You may see a red or blue fringe as you bring the field diaphragm into focus. The best position is the one in between the red and blue fringes.
5	Use the condenser-centering screws to center the field diaphragm. Open the field diaphragm by rotating its knurled ring. It may not be perfectly centered. Perform final centering of the field diaphragm when it fills most of the field. Expand the field diaphragm just beyond the field of view and stop!
6	Repeat this with the 20× and 40× objectives. For critical work, this should be done for each objective. This is especially important for taking photographs and for examining minute, translucent specimens such as fungi and bacteria. This is difficult to do this with the 100× objective. However, if you achieve proper alignment with the 40× objective, the 100× objective will be similar.

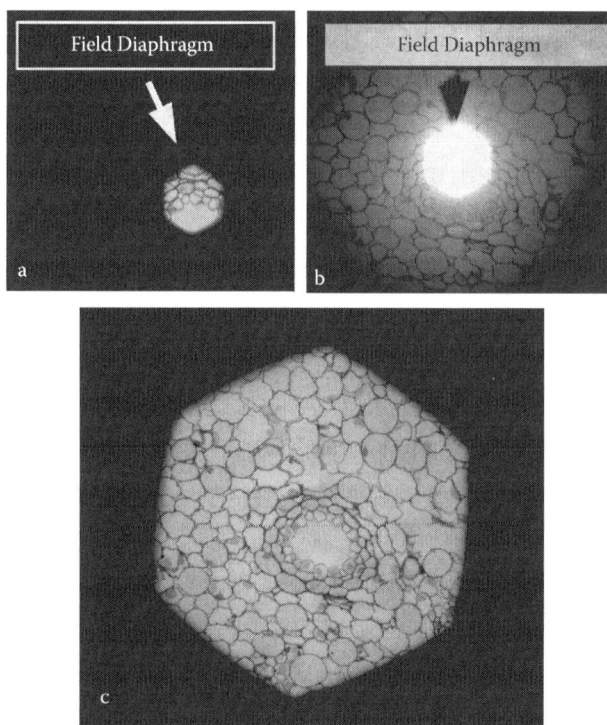

FIGURE 42.9 Focused and closed field diaphragm (A and B), uncentered (A) and centered (B). Partly open and centered field diaphragm (C).

Procedure 42.6	
Adjusting the Condenser (Aperture) Diaphragm	
Step	Instructions and Comments
1	Complete all of the preceding operations (Procedures 42.1–42.5).
2	Place a lightly stained specimen in the light path. Focus on the specimen using a 20×–40× objective lens.
3	Remove one of the oculars and look directly down the tube at the light field. Close the condenser diaphragm so that it occludes ¼ to ⅓ of the area. This should give the best contrast.
4	Examine a specimen before and after adjusting this iris.
5	This should be done for each objective for critical viewing.

each specimen is different, and strict rules such as those in Procedure 42.6 may not give the best results. Closing the condenser diaphragm also increases depth of field. Thus, more regions of a three-dimensional specimen will be in focus. However, if it is closed too much, a flat indistinct image results.

The examples in Figure 42.10 show how the condenser iris increases contrast and depth of field. It shows a diatom frustule that is very translucent. There is little detail when the condenser diaphragm is wide open (Figure 42.10a). When it is fully closed (Figure 42.10b), the contrast and depth of field are greatly increased. When the iris is closed 25–30%, there is improved contrast and depth of field with less theoretical potential for aberrations (Figure 42.10c). In this case Figure 42.10c should have been the best image, but Figure 42.10b appears to be the best.

- Experiment with the condenser diaphragm while viewing a lightly stained or unstained slide.
- Once you have achieved what you think gives the best image, remove one of the oculars and see how much of the field is occluded.

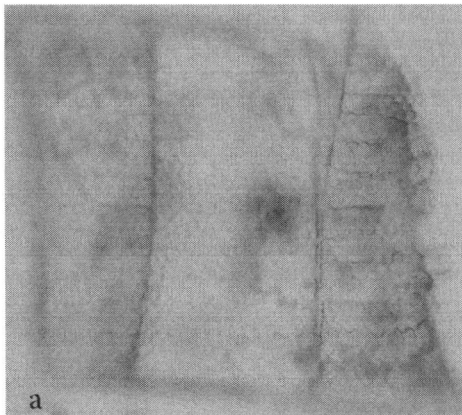

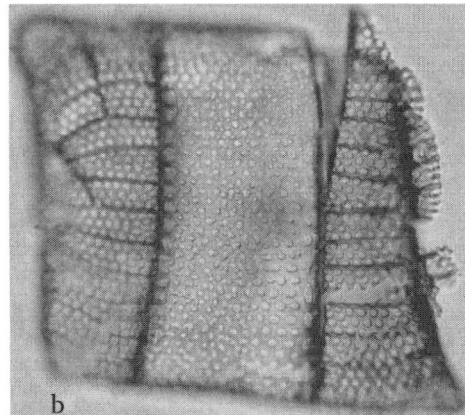

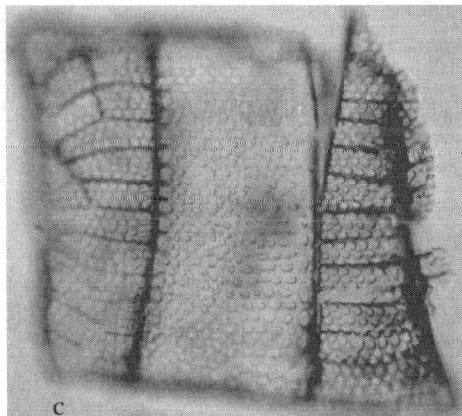

FIGURE 42.10 View of a diatom with condenser (aperture) diaphragm completely open (a), completely closed (b), and partially closed (c). It is hard to see any detail, and most of the subject is out of focus because of the shallow depth of field in Figure 42.10a. More details are visible because of increased contrast, and there is greater depth of field in Figure 42.10b compared to Figure 42.10a. More details are visible, and there is greater depth of field with Figure 42.10c compared to Figure 42.10a. In this case Figure 42.10c should have been the best. However, Figure 42.10b appears to be the best.

	Procedure 42.7
	Calibrating an Ocular Micrometer
Step	Instructions and Comments
1	Place the stage micrometer onto the stage of your microscope, and move it into the light path. Focus on its scale with the 4× objective. Switch to the 10× objective.
2	Move the stage micrometer so that some of the reference lines on it coincide with reference lines on the ocular micrometer (Figure 42.12).
3	Because the distances between the lines of the stage micrometers are known, you divide this known distance by the number of lines from the ocular micrometer. This gives you the distance measured by the intervals of the ocular micrometer at that magnification.
4	Repeat this for each objective lens on your microscope.

Throughout your career you will be using different stains to study their effects on fresh specimens. Experiment with the aperture diaphragm as you study these. The condenser diaphragm can be used to great effect with this type of material. Although some of these procedures may seem tedious, they will become routine as you progress in your work. Your results will be superior to others who do not know how to do this.

SIMPLE MEASUREMENTS WITH A COMPOUND MICROSCOPE

In most cases you cannot accurately determine the magnification of a compound microscope by multiplying the magnifications of the ocular and the objective lenses. This is a very common misconception. Microscope parts are not manufactured that precisely. Furthermore, the length of the microscope tube/body differs from one type of scope to another. This is especially true in photography because projection lenses of different magnifications are used in place of oculars, and the total distance of the light path is different from that used with the oculars. We will work through the procedure (Procedure 42.7) for calibrat-

ing an **ocular micrometer** that can be used to make direct measurements during observations.

Before we proceed, a quick review of the metric system will be helpful. A millimeter (mm) is 10^{-3} meters, whereas a micron (μm) is 10^{-6} meters. Consequently, 1 mm equals 1,000 μm, 0.1 mm equals 100 μm, and 0.01 mm equals 10 μm. A stage micrometer (Figure 42.11) is used to precisely determine magnification.

CALIBRATING AN OCULAR MICROMETER

The **stage micrometer** is the "known" in this process. It has finely etched distance calibrations on its surface. The largest dimensions from one end to the other are millimeters (Figures 42.11A and B). Each mm (1,000 μm) is divided into 0.1 mm (100 μm) segments (Figure 42.11B). Each 0.1 mm segment is divided into 0.01 mm (10 μm) segments.

Ocular micrometers have precisely etched lines engraved on them. However, because of differences in the optics of individual microscopes, they must be calibrated with a stage micrometer. Briefly, the stage micrometer and the ocular micrometer are brought together under the microscope at 10× so that the intervals of the optical microm-

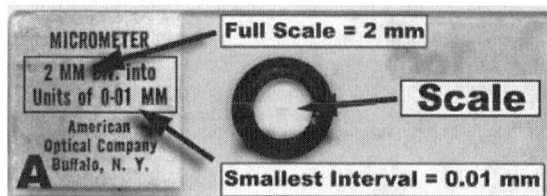

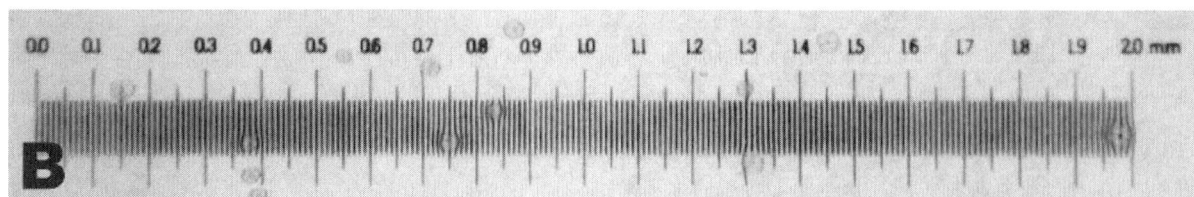

FIGURE 42.11 Stage micrometer (A) and magnified scale from the stage micrometer (B). The total length of the scale is 2 mm (2000 μm). It is divided into 20 intervals of 0.1 mm or 100 μm (B). Each of these is further divided into intervals of 0.01 mm or 10 μm.

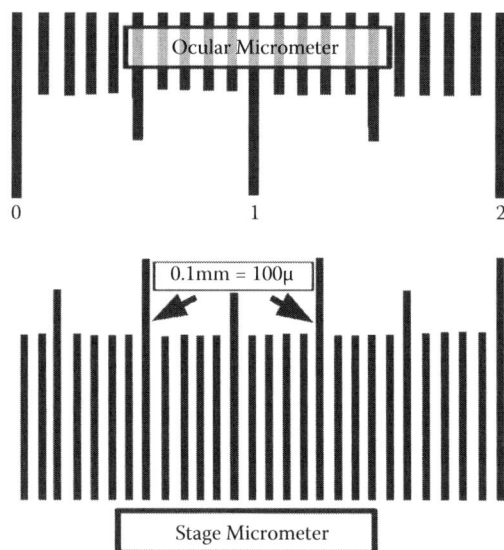

FIGURE 42.12 Diagram of a stage micrometer and ocular micrometer. In this case 2 large units on the ocular micrometer = 270 μm. One small unit on the ocular micrometer = 13.5 μm.

eter (unknown) are matched with intervals on the stage micrometer (known). The actual distance between units of the ocular micrometer is determined by dividing the known distance (stage micrometer) by the unknown units (ocular micrometer). This gives the actual distance for each unit on the ocular micrometer (Figure 42.12). In the example provided, two large units on the ocular micrometer equal 270 μm. Consequently, one large unit on the ocular micrometer equals 135 μm, and each small unit equals 13.5 μm. You need to record this value and repeat it for all of the objectives on your scope. If you transfer your ocular micrometer to another microscope, you must repeat this process for that scope. The ocular micrometer can now be used to precisely record the diameter of fungal hyphae in microns.

TYPICAL STAINS FOR COMMERCIAL SLIDES OF PLANTS

Commercial slides typically are stained with Safranin O and Fast Green. Safranin O appears brilliant red in chromosomes, nuclei, and in lignified, suberized or cutinized cell walls. Fast Green appears a brilliant green in cytoplasm and unlignified cellulosic cell walls. Fast Green turns blue in basic solutions and may appear blue to bluish green in the stems and leaves of aquatic plants and most gymnosperms (Johansen, 1940; Ruzin, 1999).

FRESH SECTIONS OF PLANT ORGANS

There are several ways to generate fresh sections of plant organs. These include hand-sectioning, use of a hand **microtome**, use of an inexpensive sliding microtome, or a traditional sliding microtome (O'Brien and McCully, 1981). Freezing microtome sections are nearly equivalent to fresh sections in many cases. This is a topic unto itself.

However, I will review the process of hand sectioning as this is a quick way to produce useful sections. Details are given in Procedure 42.8.

The most important element in this process is the blade that you use. Teflon-coated, stainless steel, safety razor blades or injector blades work fairly well. If you use a double-edged blade, be sure to put tape over one edge to prevent cutting your fingers. Use tape on your fingers as well. Single-edge utility blades give reasonable results, but they lose their edge quickly. An inexpensive sliding microtome is available from ScientificsOnline.com that surpasses most hand-sectioning attempts. Dip the blade in water before you use it. Also, wet the surface of the specimen. Water acts as a lubricant and helps the sections stay on the blade. You should slice quickly rather than try and make one perfect section by cutting slowly.

The ability to make fresh sections allows you to quickly analyze plant organs without resorting to laborious and hazardous procedures. A tremendous amount of information can be derived from these. They have natural colors because they have not been extracted with organic or caustic chemicals. This can be very important in plant pathology if you want to know the exact symptoms of a malady. Sections need not be extremely thin to be of use. In addition, they do not need to be complete or uniform. They provide three-dimensional information that is not available with extremely thin sections. Your initial attempts at hand sectioning will be frustrating; however, you will become proficient.

The following procedures work well for stems, petioles, and roots. Leaves are more difficult to section because they are so flexible. One way to overcome this is to make a sandwich of three (5 × 10 mm) leaf pieces containing the midrib and attached lamina. By squeezing these between your thumb and forefinger, reasonable cross sections can be obtained. Another way to do this is to use artificial cork as a support medium. This is far superior to real cork. Hold a 5 × 10 mm leaf piece between two layers of artificial cork to make cross sections. You may want to trim excess "cork" away from the end so you do not need to cut through too much of it. Real cork works, but is more difficult to use. You can purchase commercial pith, but it is expensive and delicate. Fresh carrot can be used, but you get a lot of debris from it, and it is rather slippery.

ADDING COVERSLIPS TO WET MOUNTS

It is best to use 22 × 40 mm rather than 22 × 22 mm coverslips with wet mounts. It is essential that air bubbles be avoided when adding coverslips to wet mounts. These will interfere with your observations. To avoid air bubbles, follow the steps outlined in Procedure 42.9. Basically, use forceps or a dissecting needle to support one end of the coverslip while it is slowly lowered onto the wet mount. This allows the specimen to be covered by any solution

Procedure 42.8

Making Hand Sections

Step	Instructions and Comments
1	Place a Band-Aid or adhesive tape on the thumb of your left hand. Position the cotton portion on the bottom of your thumb. The thumb is the backstop for this operation. Place another Band-Aid or adhesive tape on the end of your index finger. The index finger will control the height of the specimen and, thus, its thickness. Grasp the plant structure between your thumb and forefinger so that the top of the specimen extends above the level of your forefinger.
2	Take a single-edge razor blade in your right hand. Ensure that it is wet. Rest the blade on your forefinger, and use a slicing motion to cut off the top of the specimen. This is a thick section that is designed to produce a flat surface. Slice away from your thumb in order to avoid cutting your thumb with the blade.
3	Raise the specimen slightly by manipulating it with your fingers, and repeat the slicing motion. It is best to make a lot of quick slices rather than a few slow and careful sections.
4	Thin sections can often be obtained by pressing the blade down on your forefinger and then slicing through the specimen several times. After several sections have accumulated on the blade, wash them off in a petri dish of water. Keep on slicing until you have some thin sections. These will appear translucent when seen against a dark background. In most cases, the sections will have thin and thick regions. As long as part of the section is thin, you may be able to use it, and thick sections are frequently good for gross anatomy.
5	Sections can be picked up with forceps or a wet artist's brush. Place these in a drop of water or stain on a microscope slide. Sections will be released from the brush if you rotate it in the water on the slide. The brush is good for delicate specimens.
6	It is a good idea to view unstained sections prior to staining. Proper use of the condenser (aperture) diaphragm is important for viewing unstained samples.

Procedure 42.9

Proper Method for Adding a Coverslip to a Wet Mount

Step	Instructions and Comments
1	Place your samples in 2–3 drops of water or stain in the center of the slide.
2	Use a large (20 × 40 mm) coverslip. Place one end of the coverslip on the slide at a 45° angle without touching the solution containing the specimens. Steady this end with your thumb and index finger.
3	Grasp the other end of the coverslip with fine forceps. Alternatively, rest it on a dissecting needle. Slowly lower the forceps or needle until the coverslip touches the solution. Continue until the forceps or needle touches the slide. Release your grip on the forceps. Slowly remove the forceps or needle by sliding them along the slide.
4	Remove excess solution by touching the side of a Kimwipe or paper towel near one of the coverslip edges. Be careful to not sponge out your samples with the excess solution. Slowly remove the Kimwipe so that you do not drag the coverslip over the slide.
5	If you have been using a stain that must be removed, add water to one end of the coverslip. Withdraw the stain at the opposite end by blotting with a Kimwipe or paper towel.
6	Wipe excess fluid from the bottom of the slide, or it will stick on the stage and make slide transport difficult. Excess fluids may damage the stage or other microscope parts.
7	Carefully place the slide into the slide holder on your stage.

Procedure 42.10
Staining with Toluidine Blue O

Step	Instructions and Comments
1	Add several sections to a drop of water on a slide. Now add 2–3 drops of toluidine blue O to the water. Quickly add a coverslip as in Procedure 42.9. Remove the excess stain by blotting with a Kimwipe. Wipe excess fluid from the bottom of the slide (Procedure 42.9). View right away because this stain fades over the time span of the lab.
2	Caution: Toluidine blue O is hard to get out of clothing, so use it carefully, and clean up any spills with lots of water. In addition, it is mildly poisonous, so avoid getting it on your skin as much as possible. Use surgical gloves to protect your hands. Be sure to wash your hands well if they become stained. This applies to all of the stains you use.

without the formation of large air bubbles. Be sure to wipe excess fluid from the bottom of the slide. Otherwise, the slide will stick to the stage and prevent its transport. Furthermore, excess fluids can damage the stage.

STAINS FOR FRESH SECTIONS

It is important to examine unstained samples prior to staining. Plant parts often have natural pigmentation that may not be obvious. This may lead to a misinterpretation of staining reactions. Furthermore, staining often masks natural pigmentation. Consequently, you are losing data if you do not examine unstained sections.

SAFETY

All stains have some safety risk associated with them. Wear latex or Nitex gloves in the lab to prevent staining your hands. Nitex is preferable to latex. Never touch your eyes, and always wash your hands as soon as possible after using these stains, and after any lab work in which chemicals have been used.

STAINS

Phloroglucinol

Phloroglucinol dissolved in 20% HCl stains lignified, cutinized, and suberized walls a red to pink to an orange color (O'Brien and McCully, 1981; Berlyn and Miksche, 1976). It is good for staining lignified cells, such as sclereids, fibers, and xylem tracheary elements. It also stains the epidermal cuticle. The stain is very easy to use.

Place the sections in the stain, apply a coverslip, and wait a few minutes to observe. You do not need to remove the stain, but you can by applying water to one edge of the coverslip and drawing the stain out the other side with a Kimwipe or paper towel. Because it is dissolved in 20% HCl, it can damage the microscope or your clothes. Be careful to remove excess stain from your slides and work areas.

IKI

IKI stands for iodide (I) and potassium iodide (KI). It stains starch a blue-black to brown color (O'Brien and McCully, 1981; Berlyn and Miksche, 1976). It can be used exactly as described for phloroglucinol. It also imparts a golden color to cell walls and nuclei, although this incidental staining is not specific for any substance in them.

Toluidine Blue O

Toluidine blue O stains lignified walls a blue to blue-green color, and pectin-rich walls stain pink (Feder and O'Brien, 1968)—see Procedure 42.10 for details. Some walls, especially those of the phloem, may not stain at all. It is a fast-acting stain, and overstaining may destroy its specificity. You need to act quickly with this stain. It is good to compare results from toluidine blue O with those from phloroglucinol.

POLARIZING FILTERS

Polarizing filters cause light to vibrate in one plane and thus produce "plane-polarized light." Light traveling from a source vibrates in all possible planes. Imagine many radii emanating from a common center. These would represent the vibrational planes of the light beam. A polarizer cuts out all but one of these. One can think of polarizers as combs. A comb straightens tangled hair so that the strands are parallel to one another. Polarizing filters "comb" light so that only one plane passes through.

If two polarizers are parallel to each other, light will pass through because the plane-polarized light that passes through the first filter is parallel to the "teeth" in the second comb. However, if two polarizers are crossed at a 90° angle, no light passes through the second polarizer because the first polarizer eliminates all light that vibrates parallel to the "teeth" in the second polarizer. You can verify this by holding one polarizer while looking at a light source. Take a second polarizer in your other hand and superimpose it on the first. Turn either one until the light is completely blocked. You can do this with polarized sunglasses.

If a crystalline object is placed between crossed polarizers, it will depolarize the light that passes through the first polarizer (Berlyn and Miksche, 1976). This property is known as **birefringence**. The birefringent material will create light that vibrates in the same plane as the second polarizer, and it will be visible while all else will be dark. Cell walls, crystals, and starch grains are birefringent, and become apparent using crossed polarizers. This works with unstained and some stained sections. However, staining with IKI may destroy the birefringent properties of starch grains.

Inexpensive polarizing filters can be purchased from ScientificsOnline.com, and polarizing filters designed for cameras can be used. Use neutral gray polarizing filters. Circular polarizers work better than linear polarizers for microscopy.

- Place one polarizer over the field diaphragm.
- Place another over your ocular, or wear polarized sunglasses.
- Focus on your sample.
- Rotate the polarizer over the field diaphragm.

Thick cell walls, starch grains, or crystals will become bright while the background becomes dark. Intermediate effects are also possible, and can reveal subtle features that are not visible otherwise. If you want to photograph your results, you will need to place one polarizer between the specimens and your camera lens.

DISSECTING OR STEREO MICROSCOPES

Dissecting microscopes are also called stereo microscopes because they contain two separate light paths that travel to separate oculars. The specimen is seen from two different angles. This results in three-dimensional images. This is a vital feature for viewing and dissecting 3-D subjects. Early dissecting scopes consisted of two monocular scopes bound together. Compound scopes visualize only one light path that goes to both of the oculars. This produces a two-dimensional image. Dissecting scopes have many similarities to compound microscopes (Figures 42.1 and 42.2). The basic parts of a dissecting scope are as follows: base, stage, arm, focus knob, body, magnification knob, and oculars.

- Locate the major parts of the dissecting scope by referring to Figure 42.2.

Some dissecting scopes have a transillumination base. There is a mirror in the base that can direct light through the specimen. This is used to examine translucent specimens. It can provide an overview of a large translucent sample that

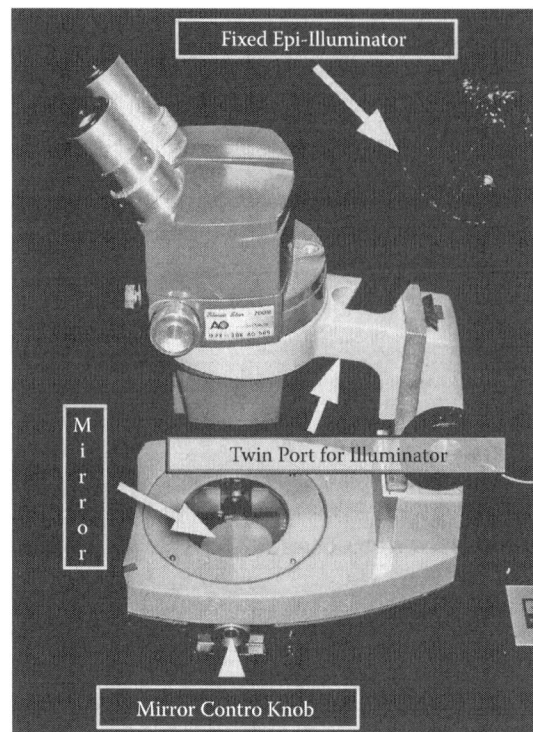

FIGURE 42.13 American Optical stereo zoom microscope with transillumination base showing the mirror and its control knob plus an Illuminator placed in one of the twin ports that are designed to provide two angles for epiillumination from behind the body of the microscope.

cannot be seen in its entirety with a compound scope. The mirror can be rotated using a knob in the base. This yields various angles of illumination. The mirror usually has a white opaque back that can provide diffuse reflected light.

In most cases epiillumination is used. There may be a "built-in" illuminator or a slot for placing an epiillumination light source above and behind the microscope body (Figure 42.13). This avoids creating shadows as the sample is manipulated on the stage. There may be a way to vary the illumination angle. The American Optical stereo zoom scope has two contiguous slots at different angles. In many cases you will use a separate epiilluminator or a pair of them that can be positioned around the scope on your lab bench. There is an adjustable arm that can be used to vary the direction of light. This provides considerable flexibility. Fiber-optic illuminators may have one or two flexible light guides that can be bent to provide many angles of illumination (Figure 42.14).

- Identify the types of illuminators available for your scopes.
- Explore their utility with various types of samples.

Uniform, shadow-free illumination can be obtained from ring illuminators mounted just below the objectives

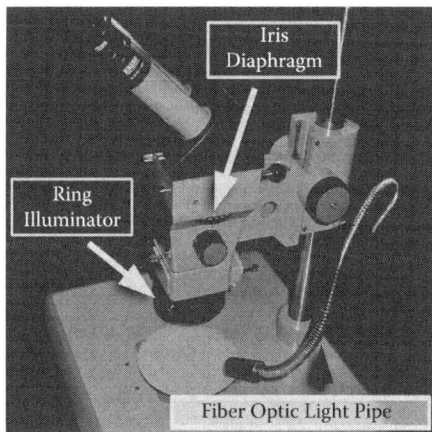

FIGURE 42.14 Zeiss stereo microscope with two types of illuminators. One is a ring illuminator, which provides shadow-free light. The other is a fiber-optic illuminator that has two flexible light pipes. These can be adjusted to provide light from various angles. This scope also has an iris diaphragm, which increases the depth of field. This is especially useful for photography.

(Figure 42.14). Fluorescent ring illuminators are inexpensive and do not produce damaging heat. Light emitting diode (LED) ring illuminators can produce more light than fluorescent illuminators and are relatively "heat-free." Fiber-optic ring illuminators are available, but are significantly more expensive than other light sources. They also produce "cool" light because the heat is dissipated by the transformer box. Shadow-free illuminators are extremely useful for most situations, but the ability to produce shadows in a controlled fashion can be helpful in obtaining 3-D relief from your specimen.

The stage either has a removable plate that is translucent or a plate that is white on one side and black on another. This allows you to vary the background. Stage clips can usually be attached to the base. These are used for samples mounted on glass slides, but can secure some other large samples. There are mechanical stages for stereo scopes. These work like the same devices on compound scopes and are useful when fine adjustments are necessary. There are also clever devices that allow you to tilt your samples. Modeling clay or plasticine can be used to stabilize small samples.

There is typically one focusing knob on each side of the arm, but there is generally no fine-focusing knob. Rotating the focusing knob moves the microscope body up and down. Objectives are located near the base of the body. Magnification is either controlled by a knob located on the side or the top of the body. Older and less expensive microscopes (Figure 42.14) contain a set of fixed objectives that are rotated into the light path by turning the knob. Most modern dissecting scopes have zoom objectives that can achieve continuous magnification over a range, which is typically 1×–3×. Auxiliary objective lenses can be fitted over the "built-in" objec-

tives much like camera filters. These can increase or decrease the magnification.

Some stereo scopes have an iris diaphragm inside the body (Figure 42.14). There is a dial or knurled ring that is used to control the opening of the iris. Closing this increases the depth of field and is useful for photography. It will diminish the amount of light that gets through, but can be useful during some dissections. This is usually not present on inexpensive microscopes.

The oculars are inserted into tubes that are attached to the body. These are similar to oculars used on a compound microscope. They must be adjusted for your eyes by following the steps in Procedures 42.2. and 42.3.

There may be an ocular micrometer in one of the oculars. To calibrate this, you may be able to use the same stage micrometer used to calibrate a compound scope. Otherwise, use an extremely accurate ruler, such as the one that can be obtained from Ted Pella Inc. (product # 54480). Similar calibration aids may be available, but when in doubt, use the best ruler you can obtain.

SUMMARY

In this chapter we document the proper way to use compound and stereo microscopes. We emphasize the compound microscope, which is more complex in design and more complicated to use. The process of achieving Koehler illumination is presented along with the proper use of the condenser and oculars. Instructions for the calibration of an ocular micrometer with a stage micrometer are provided. The production of fresh sections and the interpretation of various plant stains and polarizing filters are discussed. Various ways of illuminating specimens for stereo microscopes are reviewed.

LITERATURE CITED

Berlyn, G.P. and J.P. Miksche. 1976. *Botanical Microtechnique and Cytochemistry*. The Iowa State University Press, Ames, IA.

Delly, J.G., 1988, *Photography Through the Microscope*. The Eastman Kodak Company, Rochester, New York.

Feder, N. and T.P. O'Brien. 1968. Plant microtechnique: some principles and new methods. *Am. J. Bot.*, Vol. 55, No. 1, 123–142.

Johansen, D.A. 1940. *Plant Microtechnique*. McGraw-Hill, New York.

Köhler, A. 1893. A new system of illumination for photomicrographic purposes. Royal Microscopical Society, Oxford, United Kingdom, Koehler Illumination Centenary (1994), pp. 1–5.

O'Brien, T.P. and M.E. McCully. 1981. *The Study of Plant Structure Principles and Selected Methods*. Termarcarphi Pty. Ltd., Melbourne.

Ruzin, S.E. 1999. *Plant Microtechnique and Microscopy*. Oxford University Press, New York.

IMPORTANT WEBSITES

A-Z Microscope Glossary of MicroscopeTerms (http://www.az-microscope.on.ca/glossaryofterms.htm).

Duke, C. 2004. Lens Cleaning—Best Practice Review (http://www.microscopy-uk.org.uk/mag/indexmag.html? http: www.microscopy-uk.org.uk/mag/artfeb04/cdclean.html).

Microscopy Primer (http://www.microscopy-uk.org.uk/index.html?http://www.microscopy-uk.org.uk/primer/index.htm).

Molecular Expressions Optical Microscopy Primer (http://micro.magnet.fsu.edu/primer/index.html).

Nikon MicroscopyU (http://www.microscopyu.com/).

Part 7

Appendix

Appendix 1: Careers in Plant Pathology

Alan S. Windham and Mark T. Windham

Plant pathologists are the ultimate plant doctors. Just as your family doctor and medical researcher diagnose and study human diseases, respectively, and a veterinarian diagnoses and studies animal diseases, a plant pathologist studies diseases of plants. Just as there are specialists in animal and human medicine, there are specialists in plant pathology that study the causes of plant disease, research management strategies, diagnose plant problems, breed and select new cultivars of plants, and/or work in many related fields.

New and emerging diseases that threaten food, fiber, and ornamental crops are a constant danger to our agricultural economy. Professionals with training in plant pathology are needed to regulate the movement of these pathogens, educate the general public and agricultural professionals about the threat of these diseases, and conduct research on the nature and management of these diseases.

If you have decided upon a career in plant pathology, it is important to choose college courses that will increase your knowledge and be valuable in your career path. As an undergraduate, you may wish to take a diverse number of courses that cover disciplines such as genetics, mycology, botany, entomology, microbiology, virology, chemistry, botany, biochemistry, agronomy, soils, and horticulture. If you decide to specialize and take an advanced degree, you can tailor your curriculum to meet your specific goals. For example, if you are interested in nematology, you may wish to fashion an academic program that explores invertebrate biology, soils, ecology, and genetics and molecular biology. For research degrees such as M.S. and Ph.D., you will design an original research project on some aspect of nematology.

Professionals with undergraduate training in plant pathology may be employed as growers or pest management specialists at greenhouses or nurseries, as county Extension agents with the Cooperative Extension Service, research associates, or technicians with land grant universities, and in agrichemical companies or businesses specializing in biotechnology. Graduates with a master's degree in plant pathology may work as a sales or technical representative for seed or agrichemical companies. There are also opportunities to work as research technicians with land grant universities, federal agencies such as the United States Department of Agriculture's (USDA) Agricultural Research Service, agrichemical companies, or businesses specializing in biotechnology. There are also survey and regulatory positions available with USDA APHIS (Animal and Plant Health Inspection Service) and PPQ (Plant Protection and Quarantine). Each land grant university has a plant disease clinic that often hires graduates in plant pathology to diagnose plant problems. Diagnosticians serve as a valuable link to the general public. There are also teaching positions available at community and 4-year colleges in the biological sciences.

Graduates with a doctoral degree in plant pathology may work as Extension specialists for the Cooperative Extension Service, join the faculty of colleges and land grant universities, conduct research for the USDA Agricultural Research Service, the U.S. Forest Service, seed companies, agrichemical companies or companies specializing in biotechnology. There are also private practitioners that specialize in the diagnosis of plant diseases. These diagnosticians frequently work with farmers and growers of specialty crops such as greenhouse and nursery crops and turfgrass.

If you are interested in a career in plant pathology, contact a plant pathology department at a land grant university for educational and career opportunities. You may also contact the American Phytopathological Society. Find out more about plant pathology at: www.apsnet.org.

Part 8

Glossary and Index

Glossary

Abscission The loss of leaves, fruit, or other plant parts from the plant due to a breakdown of the specialized cells at the base of the structure; referred to as the abscission layer or zone.

Absorption The movement of water and solutes from the exterior environment into the various locations of the plant, including into cells.

Agarose A highly purified fraction of agar.

Aggressiveness Amount of disease caused by an isolate of a pathogen.

Agonomycetes See Mycelia Sterilia.

Aleuriospore Asexually produced fungal spore that originates by swelling of a terminal or lateral cell of a hypha.

Alkaloid A nitrogenous organic compound with alkaline properties, usually bitter in taste, poisonous to animals, and produced by certain plants.

Allele One of a pair of genes at a single locus.

Alloinfection Pathogen infection of a host plant by inoculum from another individual plant.

Aminopyrrolizidine An alkaloid known as a toxic principle in plants poisonous to animals and dangerous to man.

Amplicon A portion of DNA amplified or copied during the polymerase chain reaction.

Amplified fragment length polymorphism (AFLP) A method used for detecting polymorphisms in DNA by restriction enzyme digestion of DNA, amplification of restriction fragments, and visualization.

Amylase (amylolytic) Enzymes that degrade starch to maltose and D-glucose.

Anaerobic Does not require oxygen for respiration (fermentive).

Analysis The resolution of problems by representing component processes through mathematical equations.

Anamorph Conceptually, the asexual aspect of a fungus; contrast with holomorph and teleomorph.

Anastomosis Fusion of cell walls between two hyphae.

Anastomosis group (AG) Refers to Rhizoctonia isolates having three or more nuclei per hyphal cell (multinucleate).

Angiosperms Flowering plants.

Annealing temperature The temperature at which rejoining of single strands of DNA into the double helix occurs or temperature at which primers pair with DNA.

Annulations In nematodes, circular grooves around the outside of the body, providing flexibility to the nematode during movement. May be very shallow, appearing as just markings, or very deep, giving the appearance of distinct segmentations or rings.

Annulus The ring around the stipe that is a remnant of the inner veil in the developing mushroom that once enveloped only the gills or pores.

Antagonism An association between organisms in which one or more of the participants is harmed or has its activities limited by the other through antibiosis, competition, parasitism, or predation.

Antagonist An organism that harms another organism through its activities.

Antheridium (pl. antheridia) Specialized cell in which nuclei are formed that function as male gametes; therefore, a male gametangium; contrast with gynecium.

Antibiosis The inhibition or destruction of one organism by a metabolite produced by another organism.

Antibiotic Chemical produced by one organism that is harmful to another organism.

Antibody A new or modified protein produced by an animal's lymphatic system in response to a foreign substance (called an antigen). The antibody binds with the antigen, and this binding causes the antigen to become inactive.

Antigen Usually a foreign protein (occasionally a lipid, polysaccharide, or nucleic acid) that is introduced into animal tissue and reacts by producing antibodies that specifically bind and inactivate the protein.

Antisense RNA A RNA strand that is complementary or antisense to a positive sense RNA strand.

Antiserum (pl. antisera) The serum fraction of the blood from an animal that has been immunized with an antigenic substance; the serum contains a number of different immunoglobin types and reactivities.

Apothecium (pl. apothecia) An ascomycete fruiting structure that is cup-shaped or based on a cup shape in which exposed asci form; the structure develops as a result of nuclear exchange, typically involving an ascogonium and antherdium.

Appressorium (pl. appressoria) Terminal or intercalary cell of a germ tube of a fungus used for attachment to the host.

Arbitrarily amplified DNA (AAD) Collective term that describes those nucleic acid amplification techniques that use arbitrary primers. AAD includes several well-established techniques that provide genetic information from a plurality of sites (loci) in a genome or nucleic acid molecule, such as random amplified polymorphic DNA (RAPD), arbitrarily primed PCR (AP-PCR), DNA amplification fingerprinting (DAF) and AFLP.

Arbitrary primers Generally, short (7-to-12 base pairs) segments of DNA that do not correspond to known genomic sequences and thus may pair many sites (arbitrary). They are used in PCR methods such as RAPD and DAF.

Arbuscule Special, dichotomously branched haustorium that develops intracellularly and serves as the site of nutrient and carbon exchange in glomalean endomycorrhizae.

Area under curve The region of a graph bounded by axes and a line, where the line depicts disease levels in time.

Arthropods Any member of the Phylum Arthropoda; arthropods have distinct body segments, exoskeleton, jointed appendages, and bilateral symmetry.

Arthrospore An asexual spore formed when hyphal fragments break into unicellular sections.

Ascogenous hyphae Dikaryotic hyphae that will give rise to asci.

Ascogonium (pl. ascogonia) Specialized cell that functions to accept nuclei from an antheridium and is therefore a female gametangium.

Ascoma (pl. ascomata) Fungal tissue in or on which asci are formed; a fruiting body, syn. ascocarp.

Ascomycota (Ascomycetes) A diverse group (phylum) of fungi that are characterized by production of asci usually containing eight ascospores.

Ascospores Fungal spores produced in a sac-like structure called an ascus, most often following sexual recombination.

Ascus (pl. asci) Specialized cell in which nuclei fuse, undergo meiosis, and are subsequently incorporated into ascospores.

Aseptic Free of all living organisms.

Asexual reproduction Vegetative reproduction of an organism accompanied by mitotic divisions of nuclei in which the chromosome number is not reduced by meiosis.

AUDPC Area Under the Disease Progress Curve; see also area under curve.

Autoclave A chamber that sterilizes materials using pressurized steam; a temperature of 121°C for 15 min is required to kill most microbes.

Autoinfection Pathogen infection of a host plant by inoculum arising from that same individual plant.

Auxins Plant growth regulators (hormones) affecting shoot elongation and other functions.

Avirulence Inability of a pathogen to cause a compatible (susceptible) reaction on a host cultivar with genetic resistance.

Axenic Free from all microbial contamination (describing a pure culture).

Bactericide A chemical compound that kills or inhibits the growth of bacteria.

Bacteriocin Antibiotics that inhibit specific groups of microorganism that are closely related to the microorganism that produced the chemical.

Baiting The use of a living organism, instead of artificial culture media, to promote the growth of a particular microorganism.

Bark splitting Localized death of the cambium layer of tree trunk which leads to bark pulling away from the sapwood.

Base pairs A pair of complementary nitrogenous bases in a DNA molecule that form a "rung of the DNA ladder." Also, the unit of measurement for DNA sequences.

Basidiocarp Fruiting structure in the Basidiomycota.

Basidiomycota (Basidiomycetes) A diverse group (phylum) of fungi that are characterized by the production of basidia and basidiospores.

Basidiospore Ephemeral, thin-walled, haploid (N) spore of basidiomycetes following nuclear fusion and meiosis in basidia. In rusts and smuts formed upon germination of teliospores.

Basidium (pl. basidia) Specialized cell in the hymenium where basidiospores are formed following karyogamy and meiosis.

Bioassay A technique using a plant cell, tissue, or organ as a source to test the response against another organism or chemicals under controlled conditions. May also be used to detect chemical agents.

Biofilms Bacterial cells embedded in extracellular polysaccharides. Zoogloea are examples of biofilms occurring among plant cells.

Biological control The use of natural or modified organisms, genes, or gene products to reduce the effects of undesirable organisms such as plant pathogens, and to favor desirable organisms such as crop plants.

Biotic disease A disease caused by living pathogens including viruses.

Biotroph (biotrophic) An organism that can live and multiply only on or within another living organism (i.e., obligate parasite).

Birefringence (double refraction) The division of a light ray into two rays when it passes through certain types of material, such as crystals, starch grains, and thick cell walls.

Bitunicate ascus Having a double wall.

Blastospores Spores produced by budding.

Bordeaux mixture Copper sulfate and lime; one of the first fungicides.

Capsid The protein that encapsulates or encloses a viral genome.

Callus (pl. calli) Tissue mass (usually not uniform) arising from proliferation of cells and tissues.

Canker Localized infection of the bark on the stems or trunk.

Capillary water Soil water which is held cohesively as a continuous layer around soil particles and in pores between soil particles.

Cellulase (cellulolytic) A collection of hydrolytic enzymes capable of degrading native and modified celluloses to D-glucose.

Cellulose β-(1,4) linkages of D-glucose; major component of the cell wall of the Oomycota and plants.

Central vacuole The large membrane-bound fluid-filled body found in plant cells. It functions in storage of chemicals and enzymes and maintenance of osmotic potential.

Ceratobasidium anastomosis group (CAG) Rhizoctonia isolates having two nuclei per hyphal cell (binucleate).

Character An observable feature of an organism that can distinguish it from another.

Chelating agent A chemical with the ability to bind divalent ions such as calcium or magnesium.

Chemotaxis Swimming orientation of a zoospore in response to an external chemical stimulus.

Chitin β-(1,4) linkages of n-acetylglucosamine; chief structural component of the cell wall of higher fungi (kingdom Fungi).

Chlamydospore A thick wall resting spore characteristic of some fungi.

Chloroplast Specialized membrane bound cellular organelle containing the chlorophyll pigments that are involved in the photosynthetic processes of the plant.

Chlorosis Abnormal yellowing or fading of the normal green color of leaves due to loss of chlorophyll production or increase in chlorophyll degradation.

Chromosome One of the threadlike "packages" that contains genetic information such as genes and other DNA in the nucleus of a cell. Chromosomes generally occur in pairs: one obtained from the mother; the other from the father.

Chytridiomycota Fungal phylum where a true mycelium is lacking, uniflagellate zoospores are present, and where asexual reproduction occurs when the mature vegetative body transforms into thick-walled resting spores; in the kingdom Fungi.

Cibarial pump The muscularized chamber in Homopterans that serves to intake fluid during feeding.

Cirrus (pl. cirri) A mass of spores, usually in a gelatinous matrix, that form a droplet or ribbon as it is forced from the fungal fruiting body.

Clade A monophyletic (single lineage or origin inferred through cladistic analysis) that are each others' closest relatives.

Clamp connection A structure connecting two hyphal cells and involved in continuation of the dikaryotic condition for some basidiomycetous fungi.

Clavate Club-shaped.

Cleistothecium (pl. cleistothecia) Sexual fruiting body of an ascomycete that lacks a natural opening for spore discharge. Spores are usually dispersed following breakage by freezing, thawing, or wear and tear.

Codon Set of three adjoined nucleotides, or triplet, that specifies a particular amino acid or signal sequence for a polypeptide chain during protein synthesis.

Coelomycetes An informal class of mitosporic (Deuteromycota) fungi where spores are produced in conidiomata.

Coenocytic Lacking, or mostly lacking, cross walls or aseptate.

Colony (fungal) Discrete or diffuse, circular collections of hyphae and/or spores that arise from one cell or one grouping of cells.

Columella Swollen tip of a conidiophore that bears asexual sporangiospores (or mitospores); in the Zygomycota.

Commensalism Two organisms living in close association and neither having an obvious advantage.

Competition A process where two or more organisms try to utilize the same food or mineral source or occupy the same niche or infection site.

Competitive saprophytic ability An intrinsic characteristic of a fungal species that results in successful competitive colonization of organic substrates.

Competitors Chemical compounds that are utilized to slow a reaction by competing with the substrate of interest for enzymatic binding sites or for reactive chemicals.

Conidiomata Asexual fruiting structure (e.g., acervulus, pycnidium, synnema).

Conidiophore Simple or branched hyphae where mitosis occurs and conidia are produced.

Conidium (pl. conidia) Asexually produced spore borne on a conidiophore.

Contact fungicides　Chemicals applied as foliar sprays to protect above ground plant parts from infection; they may also be used as seed treatments.

Contrast　The ratio of light and dark. To produce a good image, you must have good contrast.

Crop rotation　Growing various plants that differ in disease susceptibility in a given area.

Cross protection　A form of competition in which an avirulent or weakly virulent strain of a pathogen (e.g., virus) is used to protect against infection from a more virulent strain of the same or closely related pathogen (e.g., virus).

Cultural practices　Methods used in planting, maintaining crops, and ultimately suppressing plant pathogens.

Culture　To grow a microorganism or other live cells or tissue on an artificially prepared nutritional medium.

Culture media　An artificially prepared solid or liquid media on which microorganisms or other cells are grown.

Cytokinins　Plant growth regulators (hormones) affecting cell division and other functions.

Damage threshold　The amount of crop damage that is greater than the cost of management measures.

Damping-off　Rotting of seeds in the soil or the death of seedlings soon after germination.

Denaturing temperature　Temperature at which there is loss of three dimensional structure of proteins and DNA; in PCR reactions, effectively the strand separation of DNA.

Depth of field　A distance along the optical path throughout the specimen that can be seen with clarity.

Diagnosis　The process or procedure for determining the cause of a plant disease or other problem; the process of disease identification.

Dicots　Plants having two cotyledons (as beans).

Dikaryotic　Having two haploid nuclei per cell (N + N: a condition found in most basidiomycete cells and the ascogenous hyphae of ascomycetes).

Diploid　Having a single nucleus containing two sets of chromosomes (2N), one from each parent.

Disease complexes　Diseases caused by a combination of pathogens such as fungi and nematodes.

Disease cycle　The series of events, processes, and structures involving both host and pathogen in development of a disease, repeated occurrence of the disease in a population, and injurious effects of the disease on the host.

Disease diagram　A pictorial representation of a disease level on a leaf or plant.

Disease-free (pathogen-free)　A loosely used phrase to denote that specific plant pathogens are not present on or in selected plants. Other pathogens may or may not be present.

Disease incidence　The number or proportion of plants in a population that are affected by disease.

Disease progress　The development of disease, usually in time.

Disease suppressive soils　Soils in which pathogens cannot establish, they establish but fail to produce disease, or they establish and cause disease at first, but disease becomes less important with continued culture of the crop.

Disinfectant　A chemical agent that is used to eliminate plant pathogens from the surface of plants, seed, or inanimate objects such as greenhouse benches, tools, pots, and flats.

Dispersal　To separate or move in different directions.

DNA amplification fingerprinting (DAF)　An arbitrarily amplified DNA technique that uses one or more synthetic oligonucleotide primers of typically 5 to 8 nucleotides in length and high primer-to-template ratios. Amplification products are generally visualized by polyacrylamide gel electrophoresis and silver staining and fingerprints are highly complex (syn. DNA profile).

DNA microarray　A collection of microscopic DNA spots attached to a solid surface, such as glass, plastic, or a silicon chip forming an array.

DNA polymerase　Enzyme that makes the DNA polymer.

DNA polymorphism　A variation in a DNA sequence detected through DNA sequencing or DNA analysis (fingerprinting, profiling, etc.).

DNAses　Enzyme that degrades or destroys DNA.

Dolipore septum　A complex septum found in hyphae of basidiomycetes that prevents movement of organelles between cells.

Downy mildews　Diseases caused by obligate fungal-like parasites classified in the phylum oomycota and the Peronsporaceae.

Durable resistance　Resistance that remains effective during a prolonged and widespread use in an environment favorable for disease.

Ebb and flow irrigation (flood and drain)　Potted plants are placed in a tray or nonporous floor and water is used to fill the tray or floor area. After the potting medium reaches the desired moisture level, the excess water is drained away. Often, the excess water is returned to a reservoir and used again at the next watering.

Ectomycorrhizae　Symbiotic, thought to be mutualistic, association between a plant root and fungus. The fungus forms a layer or mantle around the root and grows inward between the outer cortical root cells. The complex of hyphae around and in the root is referred to as a "Hartig Net." Fungus receives carbohydrates from the plant and the plant gains greater access to water and mineral nutrients and some protection from root pathogens.

Ectoparasite In nematology, parasitic nematodes that feed on roots from outside.

Ectotrophic hyphae Root-infecting hyphae that develop mainly on the root surface.

Egestion The expulsion of food materials through the aphid stylet; aphids alternate between ingestion (the intake of food materials) and egestion during feeding. It may serve to ensure that the stylet is not blocked during feeding.

Electrophoresis (gel) Movement of charged molecules, such as DNA and proteins, through a supporting medium (a gel) that is usually composed of either starch, agarose, or acrylamide. Molecules are usually separated by weight.

ELISA Enzyme linked immunosorbent assay; a serological technique that employs binding antibodies or antigens to a polystyrene microtiter plate and utilizes antibodies linked to an enzyme by gluteraldehyde for detection.

Endemic Prevalent in or peculiar to a particular locality or region.

Endobiotic Living entirely in the interior of a host; in the Plasmodiophoromycota.

Endoconidium (pl. endoconidia) A conidium produced in the interior of a conidiogenous cell.

Endonuclease Enzymes that cleave the phosphodiester bonds within a polynucleotide chain at a specific sequence.

Endoparasites In nematology, nematodes that penetrate the root and feed on tissue inside the root.

Endophyte Plant or fungus developing and living inside another plant.

Endophytic (prokaryotes) Endophytes live within plants; some are pathogens and others apparently are nonpathogenic inhabitants. The degree of endophytic intimacy varies. Some endophytes live within cells (liberobacteria, phytoplasmas, and spiroplasmas); other endophytes colonize the surface of living cells within the plant (pseudomonads and xanthomonads).

Environment The conditions, influences, or forces that influence living forms.

Enzyme A protein that is a catalyst for chemical reactions without being altered or degraded during the process; in fungi and bacteria, exoenzymes.

Epidemic Any increase in the occurrence of disease in a population.

Epidemiology The study of factors that lead to a change in disease.

Epidermis The outer layer of cells on a plant.

Epigenetic Environmentally induced changes in appearance or response of an organism that are not permanent.

Epiillumination Light is projected onto the specimen from above. This is typically used with opaque samples.

Epinasty Downward bending of the leaf from its normal position.

Epiphytic Growing upon a plant without harming it.

Eradicant fungicides Chemical compounds that halt fungal growth if applied shortly after infection.

Eradicate To completely eliminate a pathogen from a plant or field.

Ergot A sclerotium produced by *Claviceps purpurea* and related species in the floret of infected grains.

Etiolation Excessive elongation of internodes in plants usually in response to low light or disease.

Eukaroytic Organisms that have a nucleus as well as other double membrane organelles.

Exogenous Not produced within the organism.

Exponential model A mathematical equation used to represent disease development in time; this model allows infinite increase in disease.

Expressed sequence tag (EST) A short strand of DNA (approximately 200 base pairs long) that is part of a cDNA. Because an EST is usually unique to a particular cDNA, and because cDNAs correspond to a particular gene in the genome, ESTs can be used to help identify unknown genes and to map their position in the genome.

Extension temperature Temperature in PCR cycle that is optimum for the activity of DNA polymerase.

Extracellular enzyme A protein (enzyme) produced within the cytoplasm of fungi and bacteria that is transported across plasmalemma to act on a substrate in the environment, e.g., cellulases. May be involved in pathogenesis and nutrition.

Facultative fermentive An organism that can derive energy using oxygen (aerobic) or fermentive (anaerobic) processes.

Facultative parasite An organism that normally lives as a saprophyte but under certain conditions can become a parasite.

Facultative saprophyte An organism that normally lives as a parasite but under certain conditions can become a saprophtye.

Field The diameter of the viewing area. As magnification is increased, the field of view is decreased.

Flagellum (pl. flagella) A motile appendage projecting from a zoospore. Some flagella have hairs (e.g., straminipilous flagellum), whereas others are hairless.

Fluorometer A single wavelength spectrophotometer used to measure DNA concentration.

Forecast To estimate in advance; predict.

Forma(ae) specialis(es) Special forms of a pathogenic fungal species, separation based on morphology or host.

Formulation Pesticides consist of an active ingredient and inert ingredients in a dry or liquid form such as granules or emulsifiable concentrates.

Free-living nematodes Generally soil-inhabiting nematodes that are not plant-parasitic. Important links in the soil food web and nutrient recycling.

Fungicide A chemical compound that kills fungi.

Fungus (pl. fungi) Eukaryotic, heterotrophic, absorptive organism that develops a microscopic, diffuse, branched, tubular thread called a hypha.

Fusiform In nematodes, cigar-shaped body, relatively short and wide, descriptive of ring nematodes.

Gall An abnormal, somewhat spherical growth produced by the plant in response to some insects, nematodes, fungi, or bacteria.

Gametangial copulation Sexual reproduction in the Oomycota involving the fusion of gametangia and resulting in the formation of an oospore.

Gametangium (pl. gametangia) Sex organ that contains gametes.

Gametes Sexual cells (e.g., sperm and eggs or their functional equivalents).

Gas chromatography The movement of molecules through a gas medium so that molecules of different molecular weight will be separated; used to identify bacteria by profiles of bacterial fatty acids.

Gene The fundamental physical and functional unit of heredity, or an ordered sequence of nucleotides located in a particular position on a particular chromosome that encodes a specific functional product (i.e., a protein or RNA molecule).

General resistance Resistance effective against all biotypes of a pathogen.

Genetic engineering The addition of genetic material to the genome of the organism.

Genetic variation The variation in DNA from one individual to the next within a species which allows for selection, natural or guided, to have a foothold for implementing change.

Genomic DNA The entire genome of an organism that includes nuclear, mitochondrial, and sometimes chloroplast DNA.

Genus (pl. genera) Taxonomic classification below that of family that includes related species.

Giant cells Enlarged, specialized feeding cells formed in plant roots by root-knot nematodes.

Gill Plate on the undersurface of the pileus that supports the production of basidia and basidiospores.

Glucans Long chains of glucosyl residues; with chitin, a structural component of the cell wall of higher fungi (kingdom Fungi).

Glycocalyx Cell wall; composed chiefly of polysaccharide that is slimy as in slime molds or firm as in most other fungi.

Glycogen A storage polysaccharide used as a major carbon reserve compound in fungal cells.

Glycolysis Enzymatic breakdown of glucose and other carbohydrates to form lactic acid or pyruvic acid and the production of energy in the form of adenosinetriphosphate (ATP).

Gradient A slope, or the change in disease or inoculum concentration with distance from a source.

Gravitational potential Gravitational forces acting on soil water.

Hartig net Intercellular network of fungal hyphae in the root cortex that does not penetrate cortical cells and that serves as an exchange site for nutrients in certain groups of mycorrhizae; connected to the mantle.

Haustorium (pl. haustoria) Specialized branch that extends from an intercellular hypha into a living host cell and serves as a nutrient absorptive organ for the fungus; absorptive organ of parasitic plants.

Hemacytometer A type of counting chamber, originally designed for counting blood cells, that can be used to determine the concentration of fungal spores in a liquid suspension.

Hemibiotroph (hemibiotrophic) A plant pathogenic fungus that maintains its host cells alive for a few hours or days and then kills them for nutrition.

Hemiendophytic Living partly within plant tissue.

Hemolymph The circulatory fluid produced by arthropods.

Herbicide injury Plant injury caused by the misapplication, drift, or residue of a herbicide on non-target plants.

Heterokaryon Hyphal cells have two or more genetically different nuclei.

Heterothallic Self-sterile; a complementary mating type is needed for sexual reproduction.

Heterotroph (heterotrophic) An organism living only on organic food substances as primary sources of energy.

Holocarpic Entire thallus matures to form thick-walled resting spores (Chytridiomycota).

Holomorph Collective concept that refers to both the sexual and asexual aspects of a fungus; contrast with anamorph and teleomorph.

Homothallic Self-fertile; only one mating type required for sexual reproduction.

Hormone Organic compound produced by organisms at minute levels and are generally transported to other parts of the organism to control various growth and physiological processes of the organism; in plants also referred to as plant growth regulators.

Host growth stage Particular physical or physiological point in plant host development.

Host range The variety of plant genera and species that a pathogen can infect.

Hygroscopic water The component of soil water that is held so tightly on the surface of soil particles that it is not available to plant roots.

Hymenium (pl. hymenia) Layer of hyphae where sexual reproductive cells, such as asci and basidia, and other sterile cells occur.

Hyperauxiny Increased levels of indole acetic acid in host tissues that result in tumorous tissue development and hypocotyl elongation.

Hyperparasitism Parasitism of parasites by other organisms.

Hyperplasia Increased or abnormal frequency in division of cells.

Hypersensitive response or reaction (HR) Active defense mechanism that limits the progression of the infection by the rapid death of a few host cells. The HR is a manifestation of host recognition of the pathogen avirulence gene product. Typically, an HR includes signal transduction, programmed cell death, increased activation of defense-related genes, and a distant induction of general defense mechanisms that serve to protect the plant.

Hypertrophy Abnormal enlargement of plant cells.

Hypha (pl. hyphae) Thread-like filaments that form the mycelium or vegetative body of a fungus that may or may not have crosswalls.

Hyphomycetes An informal class of mitosporic fungi (Deuteromycota) where spores are produced on separate conidiophores.

Hyphopodium (pl. hyphopodia) Stalked, thick-walled, lobed cells that stick to plant surfaces; term used also to describe the infection structures produced by ectotrophic hyphae of certain root-infecting fungi such as Gaeumannomyces (take-all pathogen).

Hypovirulence Reduced level of virulence in a strain of pathogen resulting from genetic changes in the pathogen or from the effects of an infectious agent on the pathogen.

Immunobinding The technique of binding proteins to membrane sheets and utilizing antibodies for their detection.

Immunoglobulins Glycoproteins with the ability to antigenic regions or epitopes of proteins or other substances.

Incidence The proportion of a population affected by a disease.

Indexing The examination and testing of plants in order to detect the presence of certain specific pathogens (see Disease-free).

Indolediterpenes Mycotoxins of a diverse range of chemical structures that share a common biosynthetic origin.

Induced systemic resistance (ISR) An active defense mechanism in which a root colonizing bacterium causes distant induction of general defense mechanisms that serve to protect the plant. In ISR, jasmonic acid and ethylene signal distal portions of the plant and antimicrobial proteins are not produced.

Infection court Site of entry and establishment of a pathogen in or on a host plant.

Infectivity The ability of a pathogen (virus) to establish infection.

Infectivity curve The comparison of diluted virus inoculum and the number of infections caused by the dilution.

Inflorescences Flowering parts of plants.

Inhibitors Compounds with the ability to slow or stop a specified reaction.

Initial inoculum Parts or forms of a pathogen that are able to infect after survival through time periods when a host is absent.

Inoculation Placement of the propagule of a pathogen on or near a host cell or at a site where it can it can infect the host.

Inoculum Sexual or asexual spores, vegetative mycelium or other infectious bodies that serve to initiate infection in a host organism (being restricted to only specific spore stages in rusts and smuts).

Inoculum density Number of pathogen propagules per unit volume of soil or other medium.

Integrated pest management (IPM) A system for controlling disease, mite, and insect problems using multiple strategies (cultural, biological, and chemical).

Intensity The level or degree of occurrence; in plant pathology, this is often used when disease severity and incidence are combined.

Intercalary A structure formed as an interruption of a continuous strand, as with certain fungal spores (chlamydospores).

In vitro **selection** A process by which cells are selected by screening with an appropriate selection agent in a container or under laboratory conditions.

Isolates Any propagated culture of a virus (or other organism) with a unique origin or history.

Juveniles In nematodes, the immature life stages before mature adults are formed. Delineated by molts, but not referred to as larvae since there is no true metamorphosis.

Karyogamy Fusion of two haploid nuclei to form a diploid nucleus; final step in fertilization.

Koehler illumination A highly effective illumination designed by August Köhler that involves focusing and centering and all the elements in the light path.

Larva (pl. larvae) Juvenile organisms that undergo several growth stages prior to reaching adult development.

Latent period The duration of time between arrival of inoculum onto a suitable host until formation of the next generation of dispersal units (e.g., spores, etc.).

Lesion (local) Abnormal appearance in a localized area due to a wound or disease.

Life cycle All the successive stages, morphological and cytological, in an organism that occur between the first appearance of a stage and the next appearance of that same stage, and including all the intervening stages.

Lignin Complex organic material derived from phenylpropane that imparts rigidity and strength to woody tissues.

Lipase (lipolytic) Enzyme that hydrolyzes ester bonds between fatty acids and glycerol (or sorbitol in Tween) backbone found in fats.

Locule A small cavity.

Lodge A falling over of a plant generally at the soil line.

Logistic model A mathematical equation used to represent disease development in time; this model is symmetrical, with slower start and end, and a limit to disease increase.

Loline Fungal secondary metabolite present in grasses symbiotic with endophytes in the genera Epichlöe and Neotyphodium.

Lolitrems Indolediterpenes that are potent neurotoxins; cause tremors in mammals and deter insect feeding.

Maceration The blending, grinding, or pulverizing of tissue until a smooth homogenous mixture results; may also refer to sot rots caused by enzymatic degradation of host cell walls.

Mantle Fungal sheath enveloping the outside of infected feeder roots; in ectomycorrhizae and some ericaceous mycorrhizae.

Marginal Along the edge (e.g., leaf).

Material safety data sheet (MSDS) Contains important information about pesticides, such as adverse effects of overexposure, first aid measures, fire fighting measures, how to handle a spill or leak, proper storage conditions, and information about worker protection standards including required personal protection equipment.

Mating types "Sexes"; complementary mating types are needed for sexual reproduction in heterothallic fungi.

Matric potential Forces independent of gravity that act on soil water and are due to the adsorptive force by which soil particles attract water and capillary forces, resulting from the surface tension of water and its contact angles with soil particles.

Mechanical inoculation The ability to inoculate a virus into a host without the natural vector. For example, abrasives cause wounding that enable the mechanical inoculation of sap extracts.

Mechanical transmission The transmission of a virus into a host without the natural vector.

Meiosis A sequence of cellular events in sexual cell division, leading to the division of the nucleus into four nuclei that are genetically dissimilar due to chromosome recombination.

Melanin Dark brown or black pigment found in fungal cell walls.

Mesophile (mesophilic) An organism that grows well between 10° and 40°C.

Microsclerotium (pl. microsclerotia) A small group of dark-colored cells with thickened walls that can form a germ tube or hypha. Microsclerotia are characteristic of Verticillium.

Microtome A mechanical device used to cut biological specimens into very thin sections for microscopic examination.

Migratory Nematodes that move from one host plant to another or an individual nematode that feeds from many sites along a root.

Millardet Professor of botany at Bordeaux University, who studied and promoted the use of a mixture of copper sulfate and lime (Bordeaux mixture) to manage downy mildew of grape and other diseases.

Mitochondrion (pl. mitochondria) Cytoplasmic organelle that generates energy in the form of ATP.

Mitosis A sequence of cellular events in asexual cell division, leading to the division of the nucleus into two genetically similar nuclei.

Mitosporic fungi The imperfect or fungi belonging to the form-phylum, Deuteromycota.

Model Any physical, mathematical, or theoretical representation of some process.

Molecular marker A distinguishing molecular feature that can be used to identify a segment or region of a genome, chromosome, or genetic linkage group.

Mollicutes Cell wall-less prokaryotes thriving in high osmoticum habitats protecting the integrity of their plasma membranes from lysis. Mollicutes include the polymorphic phytoplasmas and helical spiroplasmas. These organisms are nutritionally fastidious, growing only in phloem sap in their plant and hemolymph in their insect vectors.

Monocots Plants having a single cotyledon (as grasses).

Monocyclic Relative to plant disease, a process of pathogen infection, growth, production of symptoms, and reproduction without repetition.

Monogenic resistance Resistance conditioned by a single gene.

Monokaryotic A cell having genetically identical haploid nuclei.

Monomolecular model A mathematical equation used to represent disease development in time; this model has a limit to disease increase.

Monophyletic Descendants from a single common ancestor.

mRNA A single-stranded molecule of "messenger" ribonucleic acid that directs protein production (translation).

Multigene resistance Resistance encoded by several to many genes; may be similar to horizontal resistance.

Mutualism A form of symbiosis in which two or more organisms of different species are living together for the benefit of both or all.

Mycelial fan Fan-shaped aggregation of hyphae found under the bark of trees that invades and kills the cambial tissues of the host; common in Armillaria species.

Mycelia Sterilia A group of fungi in the Deuteromycota that are characterized by not producing spores (see Agonomycetes).

Mycelium (pl. mycelia) Mass of highly branched thread-like filaments (hyphae) that constitute the vegetative body of a fungus.

Mycoparasitism One fungus being parasitic on another fungus.

Mycorrhizae "Fungus-root"; a mutualistic relationship between certain fungi and the roots of plants, also mycorrhizal fungi.

Mycotoxin Toxin produced by a fungus in infected grain or legumes that is poisonous to humans or animals when consumed.

Naked ascus Ascus not produced in an ascocarp.

Necrosis Death of plant tissue in a localized area resulting in brown or black lesion.

Necrotroph (necrotrophic) A plant pathogenic fungus that kills its host cells for nutrition.

Nematicides Chemicals used to kill or interrupt the life cycle of plant-parasitic nematodes.

Nematology The study of nematodes.

Nodulation Nitrogen-fixing swellings on the roots of legumes. Not to be confused with galls caused by root-knot nematodes.

Nucleic acid A single or double stranded polynucleotide containing either deoxyribonucleotides (e.g., DNA) or ribonucleotides (e.g., RNA) linked by 3′-5′-phosphodiester bonds.

Numerical aperture (NA) NA indicates the resolving power of a lens. A lens with a larger numerical aperture will be able to visualize finer details than a lens with a smaller numerical aperture. Lenses with larger numerical apertures also collect more light and will generally provide a brighter image.

Obligate aerobes (anaerobes) Require oxygen for respiration and metabolism. Obligate anaerobes are posioned by oxygen; they use compounds other than oxygen as electron acceptors for respiration and metabolism.

Obligate parasites Requires a living host to feed. Cannot survive on dead or decaying plant material.

Ocular micrometer A precisely etched circular glass element that is placed inside an ocular. It contains a scale that enables the observer to take accurate measurements. It is calibrated with a stage micrometer.

Odontostyle A stylet with or without flanges and found in dagger and needle nematodes.

Oidium (pl. oidia) Asexual spore that splits off in succession from the tip of a short branch or oidiophore; e.g., powdery mildews.

Oligogenic resistance Resistance conditioned by a few genes.

Oligonucleotide A DNA molecular consisting of a few to many nucleotides.

Onchiostyle A tooth-like stylet found in stubby-root nematodes.

Onset The beginning or early stages; point in time when disease is initially seen in a population of plants.

Oogonium Female gametangium of the fungi-like Oomycota species.

Oomycota "Water molds"; a fungus-like phylum where sexual reproduction results in production of oospores; in the kingdom Stramenopila.

Oospore A thick-walled resting spore that develops from the fertilization of an oogonium(ia) by an one or more antheridium(a).

Osmotic potential Forces that result from ions and molecules dissolved in soil solution. These forces predominate when water is moved across a semipermeable membrane.

Ostiole An opening or pore in a pycnidium or perithecium.

Oxidative phosphorylation Metabolic processes where energy in the form of adenosinetri-phosphate (ATP) is stored during the transfer of electrons through the electron transport pathway.

Papilla (pl. papillae) Deposit of callose and other material on the inside of a plant cell wall that are produced in response to pathogen ingress.

Parasite (parasitic or parasitism) An organism, virus, or viroid living with, in, or on another living organism and obtaining food from it.

Parfocal A microscope designed so that specimens that are in focus at one magnification will be focused when the objective is changed.

Partial resistance Resistant reactions in which some symptoms develop or pathogen reproduction occurs but the amount is less than on a fully susceptible reaction. Partial resistance implies nothing about inheritance, gene expression, or race specificity of the resistance.

Passive defense Defense mechanisms that are present before pathogen recognition.

Pathogen An organism or agent capable of causing a disease on a host plant.

Pathogenesis-related proteins (PR) Synthesis and accumulation of defense-related proteins following pathogen infection of plants.

Pathogen fitness The ability of the pathogen to survive or compete in the environment where the host is grown.

Pathogenicity The ability of a pathogen to cause disease on a specific host species.

Pectinase (pectinolytic) A complex battery of enzymes capable of degrading pectin to a number of products including glacturonic acid.

Peloton An intra- or inter-cellular hyphal coil in the roots of some endomycorrhizae.

Penetration peg Specialized, slender fungal hypha that penetrates the cell wall of a host plant through a combination of enzymatic and mechanical force. Usually produced from the undersurface of an appressorium.

Pericarp Outermost wall of a ripened ovary or fruit; the "seed" coat, i.e., the outer layer of the karyopsis in cereals and other grasses.

Perithecium (pl. perithecia) A flask-shaped structure with a hole (ostiole) at the top of the flask in which asci form in a single layer (hymenium); the structure develops as a result of nuclear exchange, typically involving an ascogonium and antheridium.

Permanent wilt Wilt due to dysfunction of vascular tissues.

Permeability The ability of a membrane to control the passage of materials through it.

Pest An organism that injures or has a harmful effect on a plant.

pH The measure of acidity or alkalinity within a range of 0 to 14 where pH 7 is neutral. The pH scale is logarithmic so a change of one unit is equal to a 10-fold change of hydrogen ion concentration.

Phenol (phenolic) Compound having one or more hydroxyl groups substituted in single or multiple benzene rings.

Phialide A bottle-shaped type of conidiogenous cell that produces conidia in basipetal succession. In basipetal succession, conidia mature from the apex toward the base, i.e., the conidium at the base is the youngest and most recently formed.

Phialoconidium (pl. phialoconidia) Asexual reproductive fungal spore formed by abstriction from the top of a phialide.

Phialospore Synonymous with phialoconidium (pl. phialoconidia).

Photosynthesis Process conducted by plants that transfers light energy with carbon dioxide and water into a chemical form of energy, carbohydrates, releasing oxygen in the process.

Phyllosphere The aerial surfaces of plants (leaves, stems, fruit, and flowers), which serve as habitats for phytopathogenic prokaryotes during the resident phase or pathogenesis.

Phylogenetic analysis The use of hypothesis of character transformation to group taxa hierarchically into nested sets and then interpreting these relationships as a phylogenetic tree. Phylogenetic relationships are "reconstructed" using distance, parsimony, and maximum likelihood methods.

Phylogenetic tree An hypothesis of genealogical or evolutionary relationship among a group of taxa.

Phylum (pl. phyla) A taxonomic rank between the kingdom and class (formerly referred to as division).

Physiology Study of basic activities conducted within the cells and tissues of a living organism as they are related to the chemical and physical processes of the organism.

Phytoalexins Active defense antimicrobial small molecular weight compounds that are produced after infection or elicitation by abiotic agents.

Phytoanticipins Passive defense antimicrobial small molecular weight compounds that are stored in the plant cell or are released from a glucoside.

Phytotoxic Causing injury to plants, as a pesticide or a phytotoxin.

Phytotoxin A chemical produced by a plant pathogenic microbe in the process of disease development.

Pileus The cap of a mushroom basidiocarp.

Plasmalemma Cell membrane that bounds the protoplast; found inside the cell wall in fungi and plants.

Plasmid A self-replicating piece of circular DNA that is present in some bacteria and is not a part of the chromosomal DNA. Its replication is not associated with cell division.

Plasmodiophoromycota A phylum within the kingdom Protists characterized by organisms that have cruciform nuclear division and produce plasmodia.

Plasmodium (pl. plasmodia) A cell without a wall (naked protoplast), usually consisting of a mass of protoplasm with many nuclei.

Plasmogamy Fusion of protoplasts to form a dikaryotic cell; first step in fertilization.

Polarizing filter A filter that produces polarized light in which the light waves are aligned in one plane.

Pollution Contamination of air, water, or soil by particulates or chemical substances.

Polycyclic Relative to plant disease, the repeating process of pathogen infection, growth, production of symptoms, and reproduction.

Polyetic Taking place over more than 1 year.

Polygenic resistance Resistance conditioned by several genes.

Polymerase chain reaction (PCR) An in vitro nucleic acid amplification technique capable of increasing the mass (number) of a specific DNA region (fragment). The method uses two synthetic oligonucleotide primers (15–30 nucleotide long) that specifically hybridize to opposite DNA strands flanking the region to be amplified. A series of temperature cycles involving DNA denaturation, primer annealing, and the extension of annealed primers by the activity of a thermostable DNA polymerase amplify the target region many millionfold.

Polymorphism More than one form. In DNA fingerprinting, the presence or absence of a specific band or PCR product at a specific locus.

Polyphenol oxidase An enzyme that oxidizes adjacent hydroxyl groups on phenolic compounds.

Powdery mildew A plant disease caused by a fungus in the order Erysiphales, generally causing white, powdery patches on the surface of leaves, stems, or flower parts.

Predispose To make more susceptible (to disease).

Programmed cell death (PCD) An active defense mechanism commonly associated with reproductive and xylem tissue development. In PCD the attacked cell and several plant cells around it die in response to chemical signals. The sacrifice of these cells isolates the pathogen.

Prokaryotes Bacteria and mollicutes that differ from eukaryotes (higher plants, fungi, and animals) in that they lack organelles (membrane-bound nuclei, mitochondria, and chloroplasts) and their proteins are translated from mRNA exclusively on 70S ribosomes.

Proprioceptor Sensory receptor located deep in the tissues such as skeletal muscles, tendons, etc.

Protein subunits The individual proteins composing the viral capsid.

Proteomics The study of the structure and function of proteins, including the way they work and interact with each other inside cells.

Pseudothecium (pl. pseudothecia) A flask-shaped structure with a hole (ostiole) at the top of the flask, in which asci form in a single layer (hymenium); the structure develops before any nuclear exchange involving an ascogonium and antheridium.

Psychrophile (psychrophilic) An organism that grows well at colder temperatures (<10°C).

Puccinia path The path, some 600–900 miles (1000–1300 km) wide, extending from northern Mexico and southern Texas to the Prairie Provinces of Canada, over which urediniospores of stem and leaf rust fungi annually blow northward and successively reproduce, extending those diseases from Mexico and Texas to Canada.

Pycnidium (pl. pycnidia) Conidiomata that are flask-shaped and contain exposed conidia and conidiophores.

Pycniospore Thin-walled, monokaryotic (N) gamete produced in a pycnium (spermatium), the nuclei of which pair, but not fuse, with compatible nuclei in receptive, flexuous hyphae to establish the dikaryotic (N+N) condition of the rust fungus life cycle.

Pyriform In nematodes, largely swollen or flask-shaped adult female body, found in root-knot, cyst, and several other nematode species.

Qualitative resistance Resistance that can be placed into distinct categories that form the basis of Mendelian ratios.

Quantitative resistance Resistant host reactions that are not differentiated into distinct classes because there is continuous variation from resistant to susceptible phenotypes.

Quarantine The exclusion of pathogens that are not established in an area; quarantines are a form of legal control for plant diseases.

Race Population of pathogen isolates that have the same virulence.

Randomly amplified polymorphic DNA (RAPD) Amplified DNA polymorphism uncovered by the use of arbitrarily amplified DNA.

Rating scale A progressive classification, as of size, amount, importance, or disease grades.

Reaction mixture or cocktail (slang) The components of a PCR reaction. Usually includes buffer, primer, nucleotides, template DNA, magnesium ion, DNA polymerase, and water.

Reducing agent A chemical that donates electrons to reduce another chemical reaction.

Reducing sugar (compound) Any sugar (or sugar-acid) that has an anomeric carbon; one in which the hydroxyl group around a carbonyl carbon can exist in different isomeric forms (e.g., D-glucose).

Regurgitant The fluid mixture produced by beetles during feeding; it contains high levels of ribonucleases, deoxyribonucleases, and proteases.

Repeating spore A spore stage that gives rise to the same type of mycelium and subsequent type of spore production as that from which it developed.

Resident phase Epiphytic colonization of plants by phytopathogenic prokaryotes without apparent disease production. Among phytopathogenic prokaryotes lacking resistant spores, resident phases are important for survival of the pathogen until conditions favor pathogenesis.

Resistance Ability of the host to reduce the growth, reproduction, and/or disease-producing activities of the pathogen. In pathogens, lack of sensitivity to a pesticide, acquired through a genetic change.

Resistant Describes a condition where plants are not hosts for a specific pathogen, and the pathogen cannot cause a disease on that plant. Often used

to describe varieties of a plant species on which the pathogen cannot cause disease.

Respiration Metabolic processes where carbohydrates and lipids are converted into energy with the uptake of oxygen from the environment.

Resting spore (or cyst) A spore that functions as a survival structure under conditions of extreme temperature and moisture.

Restriction fragment length polymorphisms (RFLP) The variation in the length of a DNA fragment produced by a specific restriction endonuclease and generally detected by Southern hybridization with a nucleic acid probe.

Rhizoid A branched fungal hypha that resembles a root and penetrates a substrate; in the Zygomycota.

Rhizomorph Root-like structure composed of thick strands of somatic hyphae; facilitates the dispersal of some fungi to new substrates.

Rhizoplane The plant surface/soil interface on roots.

Rhizosphere The volume of soil that is physically, chemically, or biologically affected by the presence of living roots.

Ribozyme Self-cleaving RNA strand that can specifically bind and cleave other RNA strands.

RNAses Enzymes that destroy or degrade RNA.

Rogue (rouging) Removal and destruction of undesired plants.

Root cap A mass of cells that covers and protects the growing root tip.

Root exudates A mixture of compounds, mainly sugars and amino acids, that leak from growing and expanding sections of roots, and from broken cells at exit points of lateral roots, and diffuse into soil.

Saline Containing a salt or salts.

Salivary glands The organs in aphids that produce the salivary excretions.

Sanitation Control methods used to prevent the introduction or spread of plant pathogens.

Saprophyte (saprophytic) An organism that obtains its food from a nonliving (dead) host.

Sclerotium (pl. sclerotia) Spherical resting structure 1 mm to 1 cm in diameter with a thick-walled rind and a central core of thin-walled cells with abundant lipid and glycogen reserves; individual hyphae have lost identity.

Scouting Comprehensive, systematic check of fields at regular intervals to gather information on the crop's progress and pest level.

Secondary metabolite A compound not directly associated with the processes that support growth but that presumably plays some other role (e.g., defense compounds).

Seed certification The practice of providing disease-free seed by verifying that source plants used in seed production or the seed themselves are free of infection or contaminating pathogens.

Selection The process of isolating and preserving individuals or characters from a group of individuals or characters.

Sensillum(a) An epithelial sense organ composed of one or a few cells with a nerve connection and taking the form of a spine, plate, rod, cone, or peg.

Serology The study of antigen and antibody reactions; a method used to identify antigens and the organisms that produce them.

Seta (pl. setae) A sterile, bristle-like hair.

Severity The proportion of an individual (leaf, plant, single field) affected by disease.

Sexual dimorphism When the male and female of a nematode species exhibit distinct morphological forms.

Sexual reproduction Reproduction of an organism requiring the union of compatible nuclei, forming the diploid chromosomal number, followed by a meiotic division that reduces the chromosomal number to the haploid state, followed by the subsequent formation of gametic cells that unite, reestablishing the diploid condition.

Siderophore Compounds released by an antagonist that inhibit the growth of other organism by sequestering iron.

Signal recognition/transduction Active defense mechanism by which the plant recognizes the pathogen and begins the production of induced defenses. Ion fluxes, oxidative bursts, protein phosphorylation, and signal molecules are involved in signal recognition and transduction.

Signal words Words such as "Caution," "Warning," or "Danger" appear on pesticide labels depending on the toxicity of the product.

Signs The presence of the pathogen or its parts on a host plant.

Simple sequence repeats (SSR) One of many DNA sequences dispersed throughout fungal, plant, and animal genomes, composed of short (2–10 base pairs) sequences that are repeated in tandem and are usually highly variable. Also known as microsatellites.

Single gene resistance See Monogenic resistance.

Smut A basidiomycete fungus (Ustilaginales) producing usually brown teliospores in sori.

Soilborne plant pathogen Diverse group of plant pathogens that can survive in soil for an extended period of time in the absence of host plants. Group includes fungi, bacteria, viruses, nematodes, and parasitic higher plants.

Soil fungistasis Phenomenon in natural soils whereby fungal growth or spore germination is inhibited. The effect of fungistasis is often overcome by root or seed exudates.

Soil inhabitant Organism that survives in soil as propagules and through saprophytic colonization of dead organic matter. Soil inhabitants can maintain their population density or increase in number in the absence of a host plant.

Soil invader Organism that survives in soil as propagules, but does not have an active phase in soil, such as saprophytic colonization. Over time, in the absence of a host plant, the population of a soil invader declines.

Soil water potential The difference in potential energy per unit quantity of water between soil water and pure, free water (reference state set at 0 water potential). The components of soil water potential include gravitational, matric, osmotic, and pressure potential.

Soilless media Material used in pots, flats or other containers in which plants are grown. Components of soilless media may include peat moss, sand, vermiculite, perlite, rockwool, styrafoam, compost, bark, or other organic materials. Soil test laboratories usually consider a potting medium "soilless" if it is composed of less than 25% field soil.

Solarization Use of sunlight to heat soil to temperatures lethal to plant pathogens and other biological contaminants.

Sorus (pl. sori) A group of fruiting bodies that contain resting spores and/or sporangia.

Spatial Of, or pertaining to, existing in space.

Species A group of individuals that are genetically or morphologically distinct from other organisms; specific epithet.

Specific resistance (syn. race-specific resistance) Resistance effective against certain biotypes or races of a pathogen but ineffective against other biotypes or races.

Spectrophotometer An instrument used to measure the intensity of light of a specific wavelength transmitted by a substance or a solution. This measurement indicates the amount of material in the solution absorbing the light.

Spermatia Spores that function as male gametes in the Ascomycota and Basidiomycota.

Spermosphere The volume of soil that is physically, chemically, or biologically affected by the presence of living seeds.

Sporangiophore The hyphal stalk that bears a sporangium.

Sporangiospore Motile or nonmotile asexual spore produced in a sporangium.

Sporangium (pl. sporangia) A sac-like structure that is internally converted into spores or a single spore (downy mildews).

Spore The cellular reproductive unit of many organisms (similar in function to seeds of plants).

Sporocarp Fruiting structure that bears spores.

Sporodochium (pl. sporodochia) Conidiomata with short conidiophores which are clustered into a rosette and produced on a superficial mycelial layer.

Sporophore Stalk that supports a spore (see conidiophore; sporangiophore).

Stage micrometer A glass slide that contains a precisely etched scale. The largest unit is typically a millimeter (1000 μm) and the smallest unit is usually 0.01 mm (10 μm). It is used to calibrate an ocular micrometer. It is also used to determine the magnification of images recorded with a microscope.

Sterilization The process for killing all forms of life.

Stipe The stalk of a mushroom basidiocarp that supports the pileus.

Stolon Branch of hyphae that skips over the substrate—a runner; in the Zygomycota.

Strain A virus isolate that differs from the type isolate of the species in a definable character, but does not differ enough to be a new species.

Stylet A needle-like structure found in the anterior end of nematodes and used to penetrate cell walls for feeding and movement inside roots. Also, a type of piercing-sucking mouthpart found in Homoptera.

Subcuticular Beneath the waxy cuticle.

Sun scald Injury that occurs to fruit or leaves when exposed to a sudden increase in light intensity.

Susceptibility (susceptible) Inability of the host to reduce the growth, reproduction, and/or disease-producing activities of the pathogen when environmental conditions are favorable (also susceptible).

Symptom The response of the plant to the presence of a pathogen or adverse conditions.

Syncytium A specialized, multinucleate feeding cell induced by cyst nematodes inside plant roots.

Synnema (pl. synnemata) Conidiomata with compact or aggregated clusters of erect conidiophores with conidia formed at or near the apex.

Systemic acquired resistance (SAR) An active defense mechanism in which a necrotizing pathogen causes distant induction of general defense mechanisms that serve to protect the plant. Salicylic acid or methyl salicylate are produced as primary and secondary responses, and antimicrobial proteins are produced.

Systemic fungicides Fungicides that have one or more of the following characteristics: the ability to enter a plant through roots or leaves, water solubility to enhance movement in the plants vascular system, and stability within the plant. Systemic fungicides are often more vulnerable to the development of resistance in the target fungus population.

Tannins Polyphenolic compounds occurring in plants that have the ability to bind and precipitate proteins.

Teleomorph Conceptually, the sexual aspect of a fungus; contrast with holomorph and anamorph.

Teliospore Thick-walled, overwintering or resting spore of rusts and smuts, in which karyogamy (nuclear fusion) occurs and from which the basidium arises.

Temporal Of, or limited by, time.

Thallus The vegetative (nonreproductive) body of a fungus.

Thermalcycler The machine that is programmed for automatic temperature changes (cycles) required for the PCR reaction. Typically has set points for denaturing, annealing, and extension temperatures.

Thermophile (thermophilic) An organism that grows well at warmer temperatures (>40°C).

Thigmotropism Contact response to a solid or rigid surface that results in orientation of an organism or one of its parts.

Tolerance Ability of a cultivar to perform well under adverse conditions. "True tolerance" occurs if one cultivar sustains less damage than another cultivar when the amount of infection is the same for both cultivars. Tolerance also is used synonymously with partial or general resistance in which case the improved performance under adverse conditions is the due to a lesser amount of disease on the "tolerant" cultivar (i.e., partial resistance).

Transcription The transfer of genetic information associated with deoxyribonucleic acid (DNA) to messenger ribonucleic acid (mRNA).

Transillumination The inspection of a specimen by passing a light through it. This is typically used with compound microscopes and can be used with stereoscopes that have transillumination bases.

Translation The transfer of genetic information associated with the messenger ribonucleic acid (mRNA) into a sequence of amino acids forming a polypeptide chain ultimately forming a protein.

Translocation Movement of materials, including water, minerals, and organic materials, through the vascular system of the plant.

Transpiration Loss of water in the vapor phase from leaves and other above-ground plant parts.

Trehalose Disaccharide of glucose used as a carbon reserve compound in fungal cells.

Trichomes Plant hairs.

Tylosis(tyloses) Occurs in xylem tissue; outgrowth of a parenchyma cell that partially or completely blocks the lumen of the vessel. Often associated with wilts.

Unitunicate ascus An ascus having a single wall.

Urediniospore Dikaryotic (N+N), repeating, "summer" spore stage of rust fungi.

Vector A nematode, insect, or other organisms that can transmit a virus or other agent into or onto a plant.

Vegetative compatibility Vegetative hyphae that are able to fuse and maintain the heterokaryotic genetic state.

Veraison Period of grape ripening during which berry growth slows down and coloration develops.

Vermiform In nematodes, worm-shaped body, much longer than it is wide. Threadlike. Common shape of most plant-parasitic nematodes.

Vesicle Intra- or intercellular, ovate to spherical structure that contains storage lipids and may also serve as a propagule in some endomycorrhizae. Also, the thin walled, balloon-like structure in which zoospores of Pythium species (Oomycota) are differentiated.

Viroid Small particles (250–400 nucleotides) of circular, single-stranded RNA whose genome is too small to code for proteins and thus has no protein coat.

Virulence Ability of a pathogen to cause a compatible (susceptible) reaction on a host cultivar with genetic resistance.

Virulence formula Identification of isolates of pathogens based on effective and ineffective host genes.

Viruliferous A vector that has acquired virus and is capable of transmitting it.

Virus A pathogen that is comprised of either RNA or DNA and is enclosed in a protective protein coat.

Viscin A sticky mucilaginous pulp that coats the seed of dwarf mistletoes.

Volva At the base of the stipe, a remnant of the universal veil that once enveloped the entire developing mushroom.

White rusts Diseases caused by Albugo species (Oomycota, Albuginaceae).

White smut A basidiomycete fungus producing spores in white sori.

Witches' brooms Distortions of normal shoot growth in which a loss of apical dominance creates a bunched growth of shoots.

Wood decay Enzymatic and chemical process by which microorganisms, such as basidiomycetes, degrade plant cell walls to obtain nutrients for growth and reproduction. Generally white and brown rots are recognized.

Zone of root elongation The region of a root tip, just behind the root apical meristem, where cells undergo elongation. Cellular elongation pushes the root cap and apical meristem through the soil.

Zoospore A spore produced during reproduction capable of moving in water (i.e., "swimming spore").

Zygospore A diploid, thick-walled spore that results from sexual reproduction; in the Zygomycota.

Zygomycota Fungal phylum where sexual reproduction results in the production of zygospores; includes common bread mold and decay fungi; in the kingdom Fungi.

Index